Proceedings of the Eighth GAMM-Conference on Numerical Methods in Fluid Mechanics

Edited by Pieter Wesseling

Notes on Numerical Fluid Mechanics (NNFM) Volume 29

Proceedings of the Eighth GAMM-Conference on Numerical Methods in Fluid Mechanics

Edited by Pieter Wesseling

Manuscripts should have well over 100 pages. As they will be reproduced photomechanically they should be typed with utmost care on special stationary which will be supplied on request. In print, the size will be reduced linearly to approximately 75 per cent. Figures and diagramms should be lettered accordingly so as to produce letters not smaller than 2 mm in print. The same is valid for handwritten formulae. Manuscripts (in English) or proposals should be sent to the general editor Prof. Dr. E. H. Hirschel, Herzog-Heinrich-Weg 6, D-8011 Zorneding.

Vieweg is a subsidiary company of the Bertelsmann Publishing Group International.

Produced by W. Langelüddecke, Braunschweig
Printed in the Federal Republic of Germany

ISSN 0179-9614

ISBN 3-528-07629-1

PREFACE

This volume contains the Proceedings of the Eighth GAMM Conference on Numerical Methods in Fluid Mechanics, held in Delft, The Netherlands, September 27–29, 1989, at Delft University of Technology. This international conference is organized bi–annually at various locations in Europe by the Committee on Numerical Methods in Fluid Dynamics of the Gesellschaft für Angewandte Mathematik und Mechanik, chaired at present by W. Kordulla. At the moment, the organizing committee draws its members from ten European countries.

The conference attracted a worldwide attendance of about 140 registered participants. The book contains the written versions of the 61 papers that were presented and a report on the GAMM–Workshop "Numerical Simulation of Oscillatory Convection in Low–Pr Fluids" held in Marseille, October 12–14, 1988. The organizing committee had a difficult task in selecting the papers from 145 submitted abstracts. In view of the high quality of the abstracts it was decided to allow more papers than in previous conferences in this series.

The contents of this volume reflect current trends in computational fluid dynamics. The renaissance of hypersonic aerodynamics manifests itself clearly, strongly stimulated in Europe by the Hermes project and plans for a later future. Notable advances are reported in adaptive and multigrid computing methods, visualization and computing in three space dimensions.

The local organization was taken care of by the Congress Office of Delft University of Technology, under the able directorship of Mrs. M.H.P. Komen–Zimmerman and Mrs. A. de Bruin.

The conference was supported financially by the following sponsors: Royal Academy of Sciences, Shell International Petroleum Company Ltd., N.V. Nederlandse Gasunie, Philips Research Laboratories, IBM Nederland N.V., AKZO, Gesellschaft für Angewandte Mathematik und Mechanik, Office of Naval Research Branch Office London, Delft University of Technology. This support is gratefully acknowledged.

P. Wesseling, conference chairman and editor — Delft, October 30, 1989
W.J. Bannink, conference secretary

CONTENTS

Extension of Roe's approximate Riemann solver to equilibrium and nonequilibrium flows

Rémi Abgrall *

Abstract

In this paper, we first develop a generalization of Roe's Riemann solver to the case of a frozen mixture of perfect gases which equation of state contains vibrational terms. We assume that each species are at thermal equilibrium with the others. We then show how to obtain a class of Roe average in the thermal and chemistry equilibrium limits.

These Riemann solvers are tested on a two dimensional test case. The scheme is semi implicit for the source terms because of stiffness. The results show the capacity of these methods.

Introduction

The development of trans-atmospheric vehicle projects has considerably revived research in hypersonic. The success of upwind schemes in transonic and supersonic applications gives the legitimate desire of extending these methods to hypersonic designs. A fundamental tool in an upwind scheme is the Riemann solver. One of the most popular is Roe's. Various authors have attempted to generalize it to chemically equilibrium flow (see [2,5,4] for example) or to nonequilibrium flow (see [3]).

In this paper, we describe a generalization of Roe's Riemann solver for a mixture of perfect gases with chemical reactions. We assume that each species are at thermal thermal equilibrium with the others. Both equilibrium and nonequilibrium chemistry cases are considered.

Our methodology is the following. We first describe an extension of Roe's Riemann solver to *frozen* mixtures. Some of it properties are mentioned. We then show how to build a Riemann for equilibrium flow from that previous solver.

These extensions are very different from those given by previous authors for both kind of flows. In each case, even if physical justifications can be argued, the expressions of these Roe's averages mainly rely on algebraic tools and not that much on the expression of the state equation.

We illustrate how the method works on a two dimensional test case with Park's model [1] of air. The results are presented for both equilibrium and nonequilibrium chemistry.

*INRIA, 2004, route des Lucioles, Sophia Antipolis, 06560 Valbonne, France

1 Equation of state, Jacobian matrix and physical model

1.1 Equation of state

We will consider a mixture of ns perfect gases which interact with each other through a chemical mechanism.

The mixture obeys the Dalton's law : $p = \sum_{i=1}^{ns} \rho_i c_{v_i} \kappa_i T = \rho c_v \kappa T$, where $c_v = \sum_{i=1}^{ns} Y_i c_{v_i}$ and $\kappa = \dfrac{\sum_{i=1}^{ns} Y_i c_{v_i} \kappa_i}{\sum_{i=1}^{ns} Y_i c_{v_i}}$. Here, c_{v_i} is the specific heat of species i ; if γ_i is the ratio of specific heats for species i, $\kappa_i = \gamma_i - 1$.

The relation between the temperature, the mass fractions Y_i and the total energy e is given by $e = \rho\frac{u^2}{2} + \sum_{i=1}^{ns} \rho_i c_{v_i} T + \sum_{i=1}^{ns} \rho_i {h_i}^0 + \sum_{i=1}^{ns} \rho {e_{vib}}^i(T)$ where ${h_i}^0$ is the enthalpy of formation of species i, ${e_{vib}}^i$ is the specific vibrational energy for species i.

The state variable $W = (\rho_1, \cdots, \rho_{ns}, \rho u, e)^T$, satisfy the system of Euler equations for reactive flows which is (in 1 D) :

$$\frac{\partial W}{\partial t} + \frac{\partial F(W)}{\partial x} = \Omega.$$

The flux term is $F(W) = (\rho_1 u, \cdots, \rho_{ns} u, \rho u^2 + p, u(e+p))^T$, and the source term $\Omega = (\Omega_1, \cdots, \Omega_{ns}, 0, 0)^T$ will be given by Park's model [1].

When the characteristic times of the chemistry and the thermal relaxation are much smaller than that of the convective part of the equations, the mixture is considered to be at chemical and vibrational equilibrium. Then, it is possible to include all chemistry effects directly in the state equation. The pressure p is the a highly nonlinear function of two independent thermodynamical variables, for instance the density ρ and the internal energy ϵ. The flow is governed by the Euler system of equations where the state variable is $W = (\rho, \rho u, e)^T$ and a flux term $F(W) = (\rho u, \rho u^2 + p, u(e+p))^T$.

1.2 Jacobian matrices

1.2.1 Frozen case

In this paragraph, we give the expression of the Jacobian matrix of the convective part of the Euler equation in the frozen case where no chemical reaction occur. We have (H is the total enthalpy and Y_i is the mass fraction of species i and p_{z_i} is $\frac{\partial p}{\partial z_i}$ for any of the independent variables of W) :

$$\frac{\partial F}{\partial W} = \begin{pmatrix} (1-Y_1)u & -Y_1 u & \cdots & -Y_{ns} u & Y_1 & 0 \\ -Y_2 u & (1-Y_2)u & \cdots & -Y_2 u & Y_2 & 0 \\ \vdots & \vdots & \vdots & \vdots & \vdots & 0 \\ -Y_{ns} u & -Y_{ns} u & \cdots & (1-Y_{ns})u & Y_{ns} & 0 \\ -u^2 + p_{\rho_1} & -u^2 + p_{\rho_2} & \cdots & -u^2 + p_{\rho_{ns}} & 2u + p_{\rho u} & p_e \\ (-H + p_{\rho_1})u & (-H + p_{\rho_2})u & \cdots & (-H + p_{\rho_{ns}})u & H + p_{\rho u} u & (1+p_e)u \end{pmatrix}. \quad (1)$$

The derivatives of the pressure are given by :

$$p_{\rho_i} = {\chi^{fr}}_i + \frac{1}{2}\kappa_{fr} u^2 \ , p_m = -\kappa_{fr} u \ , p_e = -\kappa_{fr} \ , \quad (2)$$

with

$$\kappa_{fr} = \kappa \left[1 - \frac{\sum_{i=1}^{ns} Y_i \frac{de_{vib}^i}{dT}}{c_v + \sum_{i=1}^{ns} Y_i \frac{de_{vib}^i}{dT}} \right] \quad \text{and} \quad {\chi^{fr}}_i = c_{v_i}\left(\kappa_i - \kappa_{fr}\right) T - \kappa_{fr} {h_i}^0 - \kappa_{fr} {e_{vib}}^i. \quad (3)$$

1.2.2 Equilibrium case

We immediately get

$$\frac{\partial F(W)}{\partial W} = \begin{pmatrix} 0 & 1 & 0 \\ -u^2 + \chi_{eq} + 1/2\kappa_{eq} u^2 & (2 - \kappa_{eq})u & \kappa_{eq} \\ (-H + \chi_{eq} + 1/2\kappa_{eq} u^2)u & H - \kappa_{eq} u^2 & (1 + \kappa_{eq})u \end{pmatrix} \quad (4)$$

where $\chi_{eq} = \left(\frac{\partial p}{\partial \rho}\right)_\epsilon$ and $\kappa_{eq} = \left(\frac{\partial p}{\partial \epsilon}\right)_\rho$.

To end with this paragraph, we should notice that, since the partial densities are functions of the total density and the internal energy, we have the following expression for χ_{eq} and κ_{eq} :

$$\chi_{eq} = \sum_{i=1}^{ns} {\chi^{fr}}_i \left(\frac{\partial \rho_i}{\partial \rho}\right)_\epsilon \quad , \quad \kappa_{eq} = \sum_{i=1}^{ns} {\chi^{fr}}_i \left(\frac{\partial \rho_i}{\partial \epsilon}\right)_\rho + \kappa_{fr}. \quad (5)$$

1.3 Chemical model

We will assume that the composition of the mixture is uniform and that standard air contains 21% of oxygen and 79% of nitrogen :

$$\frac{[O] \; + \; [NO] \; + \; 2[O_2]}{21} = \frac{[N] \; + \; [NO] \; + \; 2[N_2]}{79} \quad (6)$$

In the equilibrium limit, it will reduce to a set of three equations defining, with equation (6), the composition of the mixture. A form of it may be :

$$\frac{[O]^2}{[O_2]} = K_1(T), \quad \frac{[N]^2}{[N_2]} = K_2(T), \quad \frac{[O]\,[N]}{[NO]} = K_3(T). \quad (7)$$

K_1, K_2 and K_3 are given by Park [1]. They only depend on the temperature.

We will assume the harmonic oscillator model for the vibrational energies of NO, O_2 and N_2.

2 A Roe average for a frozen mixture

2.1 Introduction

Let us recall what the Roe average is for an hyperbolic system of conservation laws with flux function F.

Two states W_L and W_R being given, we seek for a matrix $\overline{A}(W_L, W_R)$ which satisfies the following properties :

1. $F(W_L) - F(W_R) = \overline{A}(W_L, W_R)(W_L - W_R)$;

2. $\overline{A}(W_L, W_R)$ have real eigenvalues and a complete set of eigenvectors ;

3. $\overline{A}(W, W) = A(W)$.

In all what follows, we will adopt the classical convention : for any function $\mathcal{D}$ of the state W, $\Delta\mathcal{D} = \mathcal{D}_L - \mathcal{D}_R$.

A natural wish is to get an expression with reduces to the one given by Roe in his original paper [6] for a calorically perfect gas. The simplest way to perform this is to seek for an "average" state $\overline{W}$ such that $\overline{A}$ can be written in the form $\overline{A}(W_L, W_R) = A(\overline{W})$. Unfortunately, for complex equations of state, this is probably impossible. Nevertheless, the expression of A shows that its W dependency is expressed through the total enthalpy H, the mass fraction Y_i, the velocity u, the temperature T, the vibrational energy $e_{vib}{}^i$ and their derivatives. So we will assume the following form for $\overline{A}$:

$$\overline{A} = \mathcal{H}(\overline{Y}_i, \overline{u}, \overline{H}, \overline{e_{vib}{}^i}, \overline{\frac{de^i_{vib}}{dT}}) \,. \tag{8}$$

Hence, it is sufficient to find "averaged" values of H, T, u, Y_i and $e_{vib}{}^i$ and $\frac{de^i_{vib}}{dT}$ to define an "averaged" Jacobian matrix.

As we noticed it before, the expression of A is *formally* identical to the one of a non-reactive flow. In [7], the determination of $\overline{u}$, $\overline{Y_i}$ and $\overline{H}$ makes no use of the state equation provided the pressure jump between the two states W_L and W_R obeys the following equation :

$$\Delta p = p_L - p_R = \sum_{i=1}^{ns} \overline{\chi}_i \Delta\rho_i + \overline{\tilde{\kappa}} \Delta\epsilon. \tag{9}$$

(In this section, χ_i will stands for $\chi^{fr}{}_i$ and $\tilde{\kappa}$ for κ_{fr}.)

If (9) is satisfied, the "averaged" quantities are uniquely determined if we want to find back the classical expression of Roe :

function	u	Y_i	T	H	$e_{vib}{}^i$	$\frac{de^i_{vib}}{dT}$
"average"	$\mathcal{R}(u)$	$\mathcal{R}(Y_i)$	$\mathcal{R}(T)$	$\mathcal{R}(H)$	$\mathcal{R}(e_{vib}{}^i)$	$\frac{de^i_{vib}}{dT}$ if $\Delta T = 0$ $\frac{\Delta e_{vib}{}^i}{\Delta T}$ else

In this table, we have set $\mathcal{R}(a) = \dfrac{\sqrt{\rho_R} a_R + \sqrt{\rho_L} a_L}{\sqrt{\rho_R} + \sqrt{\rho_L}}$. The details of that derivation can be found in [8].

One can prove the the eigenvalues of $\overline{A}$ are *real* [8]. Moreover the Riemann solver which is defined with $\overline{A}$ respects the local proportion of atoms of any species. More precisely, if a linear condition is imposed on the distribution of mass fraction such as (6) for any points of the domain of interest, the first order scheme with Roe's Riemann solver as numerical flux function does respect this condition for any time step.

3 A Roe average for equilibrium flows

As in the frozen case, the major difficulty for finding a Roe average for the Jacobian matrix of the Euler equations in the equilibrium case is to find "averaged" values of χ_{eq}

and κ_{eq} which satisfy the following relation on the pressure jump Δp , the variation of density $\Delta\rho$ and that of internal energy $\Delta\epsilon$ between two states W_L and W_R :

$$\Delta p = \overline{\chi}\Delta\rho + \overline{\kappa}\Delta\epsilon \ .$$

The basic remark for building our version of a Roe average for the jacobian matrix of the Euler equations in the equilibrium case is the following : since χ_{eq} and κ_{eq} are defined by equations (5), we could find "averaged" derivative values for the partial densities ρ_i, $\overline{\frac{\partial\rho_i}{\partial\rho}}$ and $\overline{\frac{\partial\rho_i}{\partial\epsilon}}$ and set :

$$\overline{\chi_{eq}} = \sum_{i=1}^{ns} \overline{\chi}^{fr}{}_i \left(\widetilde{\frac{\partial\rho_i}{\partial\rho}}\right)_\epsilon \quad , \quad \overline{\kappa_{eq}} = \sum_{i=1}^{ns} \overline{\chi}^{fr}{}_i \left(\widetilde{\frac{\partial\rho_i}{\partial\epsilon}}\right)_\rho + \overline{\kappa}_{fr} \tag{10}$$

where $\overline{\chi}^{fr}{}_i$ and $\overline{\kappa}_{fr}$ appear in the expression of the frozen Roe average (8).

The only condition to impose on the averaged derivative of the partial density is to fulfill :

$$\Delta\rho_i = \left(\widetilde{\frac{\partial\rho_i}{\partial\rho}}\right)_\epsilon \Delta\rho + \left(\widetilde{\frac{\partial\rho_i}{\partial\epsilon}}\right)_\rho \Delta\epsilon \ .$$

This can be easily verified with equation (9). In the next paragraph, we will give the principles of algorithm to compute these "averaged" derivatives.

How is it possible to define them ? This is not a trivial question since there is no more guide in finding the right answer as the Roe average for perfect gases was a guide in section 2.

To do this, we will rely on a simple principle which is verified by the exact derivatives : first, their expressions do not depend on any algebraic combinations of the chemical reactions which are used to define a set of equations verified by the equilibrium mixture ; second, when one changes the set of variables used to write these equations (say mass fractions instead of molar masses), the expressions of the derivatives are only modified by the Jacobian matrices of the change of variables.

The discretized system should preserve this "independence principle" as much as possible. One can easily show this is impossible [8]. One has to make a choice on the set of variables and the form of the system which gives the equilibrium mass fractions.

In this paper, we have chosen to rewrite system (7) as :

$$\begin{array}{ll} [O] = \sqrt{K'_1(T)}\sqrt{[O_2]} & K'_1(T) \equiv \sqrt{K_1(T)} \\ [N] = K'_2(T)\sqrt{[N_2]} & K'_2(T) \equiv \sqrt{K_2(T)} \\ [NO] = K'_3(T)\sqrt{[O_2]}\sqrt{[N_2]} & K'_3(T) \equiv \dfrac{K_3(T)}{\sqrt{K_1(T)}\sqrt{K_2(T)}} \end{array} \tag{11}$$

because the equilibrium constants K'_1, K'_2, K'_3 are smooth functions which limits, for small temperature, are zero. Equation (6) must be added.

To simplify the writing, let us denote $x \equiv \sqrt{[O_2]}$ and $y \equiv \sqrt{[N_2]}$. The system (6)-(11) is then equivalent to :

$$\begin{array}{l} x^2 + \left(K'_3\, y + K'_1\right) x - \alpha\rho = 0 \\ y^2 + \left(K'_3\, x + K'_2\right) y - \beta\rho = 0 \ . \end{array} \tag{12}$$

The values of α and β translate equation (6) and that the sum of the densities of each species is the total density ρ. If we apply the Δ operator to (12), we easily get a linear system like :

$$\begin{cases} A\,\Delta x + B\Delta y = \alpha\,\Delta\rho + \mathcal{A}\Delta T \\ A'\,\Delta x + B'\Delta y = \beta\,\Delta\rho + \mathcal{B}\Delta T \end{cases} \tag{13}$$

where A, B, A', B', $\mathcal{A}$ and $\mathcal{B}$ depends on the arithmetic averages of the left and right state through x, y, ΔT and K'_i $(i = 1,2,3)$. It is always possible to solve the linear system (13) : the determinant remains positive.

Now, from (11), we can recover the variation of the molar masses. For example, $\Delta[O_2] = 2\overline{x}\Delta x$. It is possible to express ΔT in term of $\Delta\rho$ and $\Delta\epsilon$ since :

$$\Delta\epsilon = \left(\sum_{i=1}^{ns}\left[c_{v_i}\overline{T} + h_i^{\,0} + \overline{e_{vib}{}^i}\right]\Delta\rho_i\right) + \left(\sum_{i=nv+1}^{ns}\overline{Y_i}\left[c_{v_i} + \frac{\Delta e_{vib}{}^i}{\Delta T}\right]\right)\Delta T\,.$$

4 A semi-implicit numerical scheme for the Euler equation with source terms

We will assume the systems we are interested in are those where the Jacobian matrix of Ω with respect to W have eigenvalues with a negative real part. In the case of air, this seems to be true though there are no theoretical evidence of this, as far as we know.

For the different test cases, we have used a finite volume formulation on either structured (equilibrium calculation) or non structured meshes (non equilibrium ones). Only first order accurate calculation have been performed. The first order scheme we have used in the equilibrium calculation is uses Roe's solver and is very classical. In the F.E.M. calculation, we have used the method developped by Fezoui [9].

In the following, we will only give the main features of the extension to non equilibrium flows of the F.E.M. in a 1 D context for the sake of clarity. It is explicit for the convective terms and semi-implicit for the source terms.

$$\left(I - \Delta t\frac{\partial\Omega}{\partial W}(W_j{}^n)\right)\delta W_j = \lambda\left(\mathcal{F}(W_{j+1}{}^n, W_j{}^n) - \mathcal{F}(W_j{}^n, W_{j-1}{}^n)\right) + \Delta t\Omega(W_j{}^n) \tag{14}$$

where $\mathcal{F}$ is the numerical flux which will be the (frozen) extension of Roe's Riemann solver defined in the previous subsection. In (14), λ is the ratio of the time step over the local mesh size.

This is a $(ns+2)\times(ns+2)$ system. In fact, (14) is much simpler to solve due to the conservation relations (6). After some algebraic manipulations, the system (14) can be rewritten :

$$(I - \Delta tA)\,(\delta\rho_1, \delta\rho_2, \delta\rho_3)^T = \left(RHS'_1, \ldots, RHS'_3\right)^T \tag{15}$$

$$\frac{\delta\rho_1}{m_1} + \frac{\delta\rho_3}{m_3} + 2\,\frac{\delta\rho_4}{m_4} = \frac{RHS_1}{m_1} + \frac{RHS_3}{m_3} + 2\,\frac{RHS_4}{m_4} \equiv \mathcal{B} \tag{16}$$

$$\frac{\delta\rho_2}{m_2} + \frac{\delta\rho_3}{m_3} + 2\,\frac{\delta\rho_5}{m_5} = \frac{RHS_2}{m_2} + \frac{RHS_3}{m_3} + 2\,\frac{RHS_5}{m_5} \equiv \mathcal{C} \tag{17}$$

where the matrix A in (15) is :

$$A' = \begin{pmatrix} \frac{\partial\Omega_1}{\partial\rho_1} - \frac{\partial\Omega_1}{\partial\rho_4} & \frac{\partial\Omega_1}{\partial\rho_2} - \frac{\partial\Omega_1}{\partial\rho_5} & \frac{\partial\Omega_1}{\partial\rho_3} - \frac{m_1}{m_3}\frac{\partial\Omega_1}{\partial\rho_4} - \frac{m_2}{m_3}\frac{\partial\Omega_1}{\partial\rho_5} \\ \frac{\partial\Omega_2}{\partial\rho_1} - \frac{\partial\Omega_2}{\partial\rho_4} & \frac{\partial\Omega_2}{\partial\rho_2} - \frac{\partial\Omega_2}{\partial\rho_5} & \frac{\partial\Omega_2}{\partial\rho_3} - \frac{m_1}{m_3}\frac{\partial\Omega_2}{\partial\rho_4} - \frac{m_2}{m_3}\frac{\partial\Omega_2}{\partial\rho_5} \\ \frac{\partial\Omega_3}{\partial\rho_1} - \frac{\partial\Omega_3}{\partial\rho_4} & \frac{\partial\Omega_3}{\partial\rho_2} - \frac{\partial\Omega_3}{\partial\rho_5} & \frac{\partial\Omega_3}{\partial\rho_3} - \frac{m_1}{m_3}\frac{\partial\Omega_3}{\partial\rho_4} - \frac{m_2}{m_3}\frac{\partial\Omega_3}{\partial\rho_5} \end{pmatrix}$$

and

$$\begin{aligned} (RHS_i) &= \lambda\,[\mathcal{F}(W_{j+1}{}^n, W_{j1}{}^n) - \mathcal{F}(W_j{}^n, W_{j-1}{}^n)] + \Delta t\ \Omega(W_j{}^n) \quad \text{for } i = 1,7 \\ RHS' &= (RHS)_{i=1,3} \ + \ \Delta t \left(\tfrac{\partial\Omega_i}{\partial m}\delta m + \tfrac{\partial\Omega_i}{\partial e}\delta e + m_1 \tfrac{\partial\Omega_i}{\partial\rho_4}\mathcal{B} + m_2 \tfrac{\partial\Omega_i}{\partial\rho_5}\mathcal{C}\right)_{i=1,3}. \end{aligned} \tag{18}$$

Since no atom of oxygen or nitrogen can be created (remark 2 on the fluxes, section 2.1), we notice that (16) and (17) make no use of the source terms despite the presence of Ω terms in (18). Thus, the resolution of (14) has been reduced to the solution of a 3×3 system of linear equations.

5 Numerical results

We have chosen to show the capacities of the method on two dimensional tests cases : a double ellipse shuttle which length is 6 meters. The flight conditions are summarized in the following table :

Mach	Temperature	Pressure	Density	Incidence
17.9	204 K	2. Pa	.42 10^{-4} Kg/m^3	0.º

These condition have been used for both equilibrium and non equilibrium computations. We have represented the temperature and c_p lines and also the O_2 and N_2 lines.

5.1 Comparisons

Some results concerning the equilibrium simulation are displayed in figure 1 while the non equilibrium results are given in figure 2.

The main differences between the two calculations are the following : first, the position of the shock is closer to the double ellipse in the equilibrium simulation than in the non equilibrium one. On the contrary, the canopy shock is more detached in the non equilibrium simulation than in the equilibrium one.

The temperature levels are displayed in the following way : it reach a peak just behind the shock at about 8500 K in figure 2 while in figure 1 the maximum is only about 5000 K and is reached at the stagnation point.

An other difference lies in the mass fraction levels : the show the chemistry is more intense in the equilibrium simulation than in the non equilibrium one.

All these results have to be confirmed and improved by second order calculations.

6 Conclusion

In this paper, we studied the Euler system of equation for the flow of a gas which is a mixture of several perfect gases in which chemical reactions occur. For that purpose, we developed two Roe average formulas for the convective part of the Euler equations, one adapted to chemically equilibrium simulations, the other one to non equilibrium ones. In that case, we have developed a semi-implicit scheme for solving the whole system of equations. We believe that our solution for the Roe average is more consistent with the physic involved by the Euler system of Equations than that of previous authors.

In the non equilibrium situation, the solution of the linear system of equations resulting from the semi-implicit scheme is simplified by the use of conservation relations. Hence we got a much simpler system to solve. We can also show that the eigenvalues of the Roe average remains real.

The numerical tests we performed show the capacity of these methods to handle very large discontinuities. It is very easy to extend these Riemann solvers to thermal non equilibrium situations. Second order accurate schemes are under development.

Acknowledgements : I wish to acknowledge my colleagues L. Fezoui and J. Talandier who help me to obtain the non equilibrium results presented here.

References

[1] C. Park. On the Convergence of Chemically Reacting Flows. AIAA Paper no 85-0247, 23rd Aerospace Sciences Meeting, Reno, Nevada, January, 14-17 1985.

[2] M. Vinokur *Flux Jacobian Matrices and Generalized Roe Averaging for Real Gas* NASA Contractor Report 177512, December 1988.

[3] M.S. Liou J.S. Shuen, B. Van Leer. A Detailed Analysis of Inviscid Flux Splitting Algorithms for Real Gases with Equilibrium of Finite-Rate Chemistry. International Conference in Numerical Methods, Williamsburg, June 1988.

[4] J.S. Shuen M.S. Liou, B. Van Leer. *Splitting of Inviscid Fluxes for Real Gases.* Technical Report NASA TM 100856, NASA, April 1988.

[5] P. Glaister. An Approximate Linearised Riemann Solver for the Euler Equation for Real Gases *Journal of Computational Physics*, vol. 74, 1988. pages 382-408

[6] P.L. Roe. Approximate Riemann solvers, parameter vectors and difference schemes. *Journal of Computational Physics*, Vol 43, 1981, pages 357,372 .

[7] R. Abgrall. Généralisation du solveur de Roe pour le calcul d'écoulements de mélanges de gaz parfaits à concentrations variables. *La Recherche Aérospatiale*, no 1988-6, 1988, pages 31-43.

[8] R. Abgrall. Extension of Roe's Riemann solver to equilibrium and non equilibrium flows *submitted to Computer and Fluids*

[9] L. Fezoui. *Résolution des équations d'Euler par un schéma de Van Leer en éleémants finis.* INRIA report *n°* 358 January 1985.

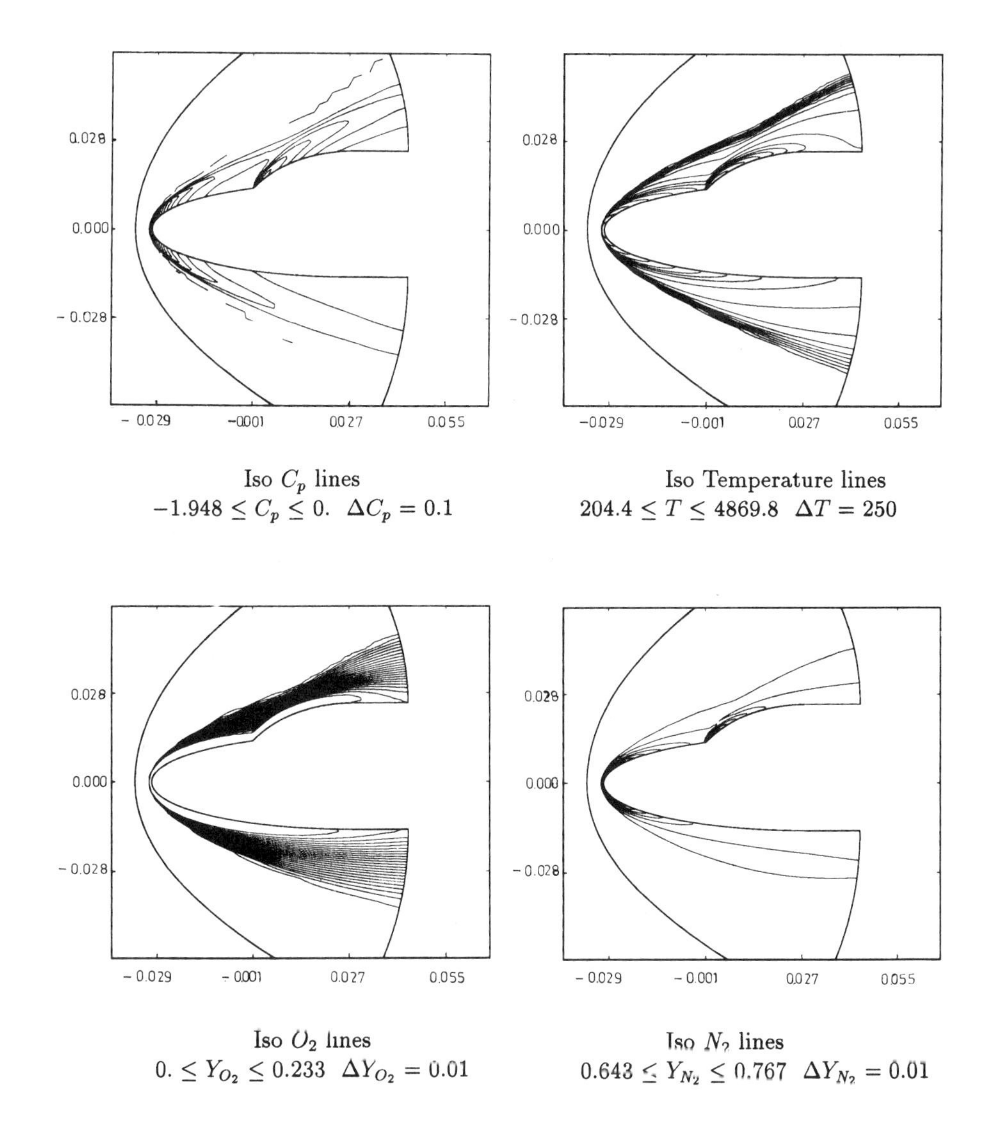

Iso C_p lines
$-1.948 \leq C_p \leq 0. \quad \Delta C_p = 0.1$

Iso Temperature lines
$204.4 \leq T \leq 4869.8 \quad \Delta T = 250$

Iso O_2 lines
$0. \leq Y_{O_2} \leq 0.233 \quad \Delta Y_{O_2} = 0.01$

Iso N_2 lines
$0.643 \leq Y_{N_2} \leq 0.767 \quad \Delta Y_{N_2} = 0.01$

Figure 1: Equilibrium Chemistry

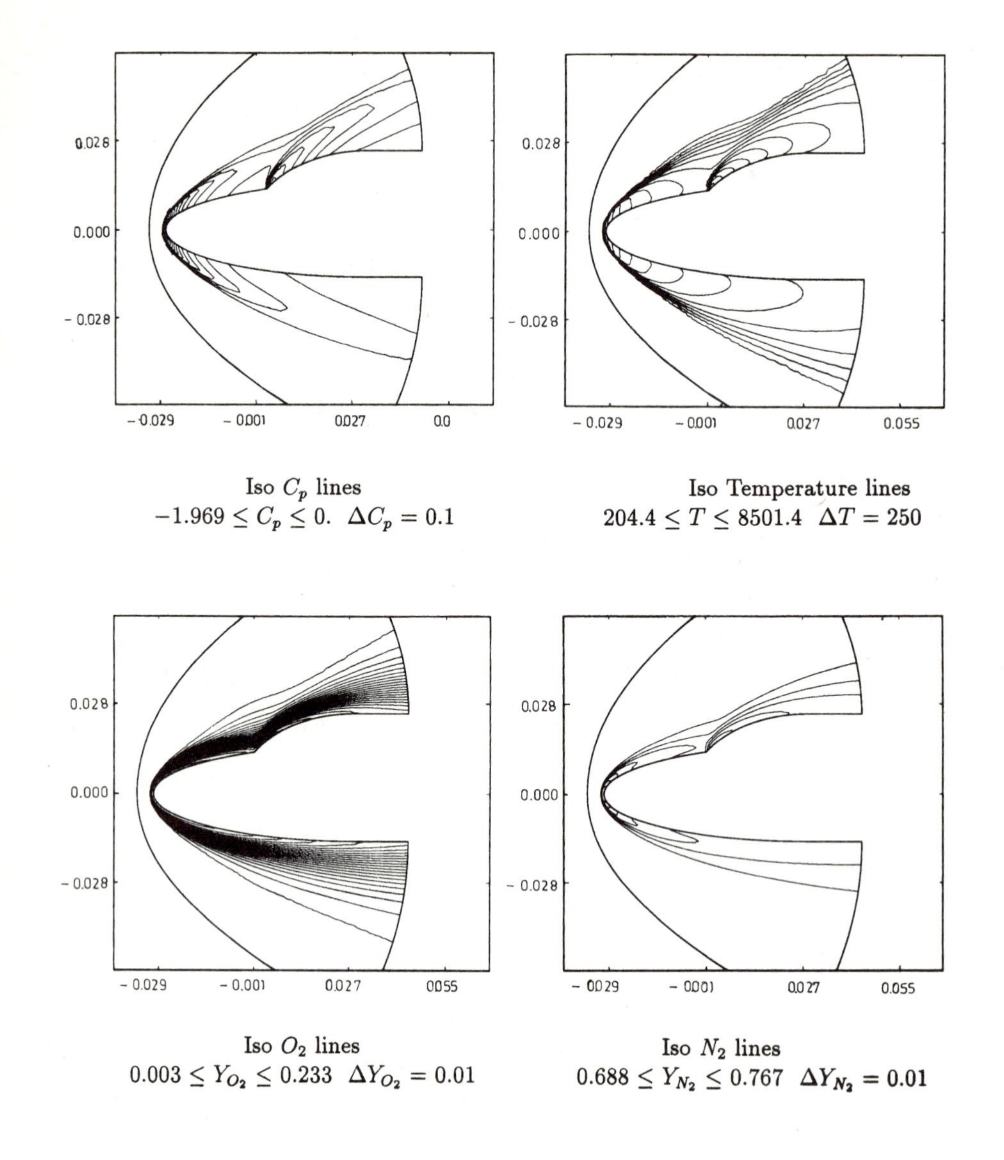

Iso C_p lines
$-1.969 \leq C_p \leq 0.\ \ \Delta C_p = 0.1$

Iso Temperature lines
$204.4 \leq T \leq 8501.4\ \ \Delta T = 250$

Iso O_2 lines
$0.003 \leq Y_{O_2} \leq 0.233\ \ \Delta Y_{O_2} = 0.01$

Iso N_2 lines
$0.688 \leq Y_{N_2} \leq 0.767\ \ \Delta Y_{N_2} = 0.01$

Figure 2: Non Equilibrium Chemistry

DEVELOPMENT AND VALIDATION OF CHARACTERISTIC BOUNDARY CONDITIONS FOR CELL-CENTERED EULER FLOW CALCULATIONS *)

J.I. van den Berg, J.W. Boerstoel,
National Aerospace Laboratory NLR,
P.O. Box 90502,
1006 BM Amsterdam,
Telephone 31-(0)20-5113113, Telex 11 11 8 (nlraa nl),
Fax (CCITT code 2/3) : 31 (0)20-178024

SUMMARY

An overview of the development, analysis, and numerical validation of a solid-wall boundary condition for cell-centered Euler-flow calculations is presented. This solid-wall boundary condition is provided by the theory of characteristics, and is based on a central-difference scheme. The boundary condition was developed to clearify the question what the effect is of various boundary-condition algorithms on the accuracy of calculation results for 3D Euler flows around delta wings.

A mathematical analysis of the boundary condition was performed. It consists of a theoretical study to the consistency, well-posedness, and stability. The numerical validation consists of a comparison of calculation results with various boundary conditions with each other. Also discretization and convergence errors were investigated.
As a test case, the NLR 7301 profile under supercritical, shock-free, flow conditions of $M = 0.721$, $\alpha = -0.194$ deg, was chosen.

1.INTRODUCTION

The subject of this paper is a study of the effect of solid-wall boundary conditions on the numerical accuracy of calculation results of 3D Euler flows around aerodynamic configurations, like delta wings.
The conventional boundary condition (normal velocity component zero and second-order accurate extrapolation of the pressure towards the solid wall) leads to a false or numerical entropy layer along the surface. This entropy layer can affect the accuracy of the inviscid-flow solution and could render future extensions to viscous-flow computations useless. Also the false entropy layers can have a large effect on flow separation from sharp edges, which is of prime importance for the simulation of flow around configurations with leading-edge vortices.
At the start of this study, the effects of the solid-wall boundary condition on the accuracy were not well understood. Questions to be answered were:
- what is the effect of the boundary-condition algorithm,
- what is the effect of the pseudo-time integration to a steady-state solution, and
- what is the effect of grid resolution and grid quality.

*) This investigation has been carried out under contract with the Netherlands Agency for Aerospace Programs (NIVR) for the Netherlands Ministry of Defence.

The conclusions of this study are presented in this paper, together with a numerical example. It is sufficient to restrict this example to that of an aerofoil flow.

The Euler-flow equations are discretized by a fully-conservative cell-centered difference scheme [1]. The solid-wall boundary condition introduced in the present paper, is called characteristic solid-wall boundary condition. It is provided by the theory of characteristics, and based on a cell-centered central-difference scheme [2].
Prior to the numerical validation of the boundary conditions, a theoretical analysis of its consistency, well-posedness and stability is performed. In the numerical validation, we compare to each other results of calculations with the new solid-wall boundary condition and with the standard solid-wall boundary condition. Also numerical error sources, such as the influence of the grid-quality and the number of iterations in the convergence history are investigated.
As a test case, the NLR7301 profile under so-called supercritical shock-free flow conditions of $M = 0.721$, $\alpha = -0.194$ deg., is chosen.

For this study, the NLR information system for 3D Euler-flow simulation has been used [3,4,5,6]. This system consists of (c.f. fig. 1):
- subsystems for geometry processing, grid generation, flow calculation, and flow visualization and data-processing,
- subsystems for method management and data management.

2. DISCRETIZATION OF THE EULER CONSERVATION EQUATIONS

The Euler conservation equations form a set of five coupled non-linear first-order hyperbolic partial differential equations describing the conservation of mass, of momentum, and of energy for the inviscid flow of an ideal gas,

$$U_t + F_x + G_y + H_z = 0. \qquad (2.1)$$

Here U is the state vector,

$$U = [\ \rho,\ \rho u,\ \rho v,\ \rho w,\ \rho E]^T\ , \qquad (2.2)$$

F,G,H are the flux-vectors,

$$[F,G,H] = \begin{bmatrix} \rho u & \rho v & \rho w \\ \rho u^2 + p & \rho uv & \rho uw \\ \rho uv & \rho v^2 + p & \rho vw \\ \rho uw & \rho vw & \rho w^2 + p \\ uH & vH & wH \end{bmatrix} , \qquad (2.3)$$

and

$(u,v,w)^T$ = the velocity vector,
ρ = the density,
p = the pressure,
E = the total energy,
H = the total enthalpy.

The equations are solved numerically with a cell-centered central-difference scheme. Artificial dissipative terms have to be added to the discretized conservation equations. They consist of a second-order term that should be active near shocks, and a fourth-order term active outside shocks which eliminates odd-even decoupling. A switch factor dependent on the second-order derivative of the pressure is used as shock sensor, to switch the dissipative terms on or off. See [2] for a detailed description of this artificial dissipation model.
The time integration of the discrete Euler equations is performed with a 4-stage Runge-Kutta scheme. For each grid cell, the discrete conservation equations can be written as

$$U_{ijk}^{(m+1)} = U_{ijk}^{n} - dt \left(Q_{ijk}^{(m)} + D_{ijk}^{n} \right) \tag{2.4}$$

where

U_{ijk}^{n} = the state vector at cell-centre ijk at time level n ,

(m) = stage in Runge-Kutta scheme, m=0..3, $U_{ijk}^{(0)} = U_{ijk}^{n}$, $U_{ijk}^{(4)} = U_{ijk}^{n+1}$,

n = time level,

$Q_{ijk}^{(m)}$ = convective divergence at cell-centre ijk at stage (m),

D_{ijk}^{n} = dissipative divergence at cell-centre ijk at time level n ,

dt = time-step [2].

3. DEVELOPMENT OF CHARACTERISTIC BOUNDARY CONDITIONS

The discretization of the Euler equations with a cell-centered scheme requires the flow state at each cell-centre of an extra layer of cells on the flow boundary, but outside the flow domain. The state vector in these so-called halo cells has to be determined from the boundary conditions.

Along a solid-wall boundary, a layer of auxiliary cells of half a mesh thickness above the surface is introduced (fig. 2). For each auxiliary cell first a set of five semi-discrete conservation equations is defined,

$$(U_t)_{ij\frac{1}{2}} + Q_{ij\frac{3}{4}}(U^n) + D_{ij\frac{3}{4}}(U^n) = 0. \tag{3.1}$$

where

$(U_t)_{ij\frac{1}{2}}$ = the time derivative of the state vector at the boundary point $ij\frac{1}{2}$,

$Q_{ij\frac{3}{4}}(U^n)$ = the divergence of the convective fluxes, at the cell centre $ij\frac{3}{4}$ of the auxiliary cell, at time level n,

$D_{ij\frac{3}{4}}(U^n)$ = the divergence of the dissipative fluxes at the cell centre $ij\frac{3}{4}$ of the auxiliary cell, at time level n.

These equations are transformed into characteristic form. This transformation is performed by premultiplying the equations by the 5*5 matrix

$$T_{ij\frac{1}{2}} = R^{-1}_{ij\frac{1}{2}} C_{ij\frac{1}{2}} S_{ij\frac{1}{2}} . \tag{3.2}$$

S introduces a transformation from the (x,y,z) coordinate system to a local coordinate system $(\underline{n},\underline{s},\underline{t})$, with $\underline{n}$ the unit normal to the cell face $ij\frac{1}{2}$ on the boundary, and $(\underline{s},\underline{t})$ tangential to the surface along the boundary. C is a scaling matrix, scaling eigenvalues (velocities) to dimensionless quantities (Mach number). R^{-1} is the transformation of flow-state changes in primitive variables to flow-state changes in Riemann variables. This leads to

$$(\phi_t)_{ij\frac{1}{2}} + T_{ij\frac{1}{2}} [Q_{ij\frac{3}{4}}(U^n) + D_{ij\frac{3}{4}}(U^n)] = 0, \tag{3.3}$$

a set of five conservation equations. Here ϕ_t is the vector of time-derivatives of the Riemann-variables,

$$\phi_t = [r_t^+ , r_t^S , r_t^{t1}, r_t^{t2}, r_t^-], \tag{3.4}$$

where

$$\begin{aligned} r_t^+ &= \rho\tilde{u}_t + \frac{1}{c} p_t , \\ r_t^S &= \frac{p}{cS} S_t , \\ r_t^{t1} &= \rho\tilde{v}_t , \\ r_t^{t2} &= \rho\tilde{w}_t , \\ r_t^- &= \rho\tilde{u}_t - \frac{1}{c} p_t , \end{aligned}$$

These time derivatives correspond to the eigenvalues $\lambda_1 = \tilde{u}+c$, $\lambda_2 = \lambda_3 = \lambda_4 = \tilde{u}$, and $\lambda_5 = \tilde{u}-c$, respectively. Furthermore,

$\tilde{u}$ = velocity component in $\underline{n}$ direction (normal to the boundary) (positive if u is directed into the flow),
$\tilde{v},\tilde{w}$ = velocity components tangential to the boundary,
c = speed of sound,
S = $p\rho^{-\gamma}$ = total entropy.

Time derivatives of the Riemann variables representing information entering the flow domain through the boundary are replaced by boundary conditions. They are indicated by a negative sign of the corresponding eigenvalue.
At the solid wall, a boundary condition has to be prescribed for r_t^-. The boundary condition is : the normal component of the velocity vector must tend to zero, when time goes to infinity.
For the Riemann variables with the zero eigenvalues ($\lambda_2 = \lambda_3 = \lambda_4 = 0$), it is permitted to prescribe boundary conditions but it is not neccesary. Physically the three corresponding conservation equations for r_t^S, r_t^{t1}, r_t^{t2}, describe the transport of entropy and vorticity in a direction tangential to the solid wall. From numerical experiments with transonic shock-free flows around aerofoils, it was concluded that it is useful to apply also a boundary condition for the Riemann variable r_t^S. The boundary condition chosen in the present work is that the entropy gradient in the direction normal to the solid wall tends to zero on the solid wall, when time goes to infinity (steady state).
In formulae these two boundary conditions are

$$(r_t^-)_B = - d_{uB}\,\rho_B \;(\tilde{u}_B - \tilde{u}_{oB}) / \Delta t_B - (r_t^+)_B \tag{3.5.a}$$

$$(r_t^S)_B = - d_S \left(\frac{p}{cS}\right)_B (S_B - S_{oB}) / \Delta t_B \tag{3.5.b}$$

where subscript B denotes the multi-index on the solid-wall boundary,

$$.B = .ij\tfrac{1}{2} .$$

Using $r_t^+ + r_t^- = 2\rho(\tilde{u}_t)_B$, and giving prescriptions that drive $(\tilde{u}_t)_B$ and $(S_t)_B$ to zero, gives the desired boundary conditions,

$\tilde{u}_B$ = actual value of normal velocity at cell-face centre B,

$\tilde{u}_{oB}$ = desired value of normal velocity at cell-face centre B,

S_B = actual value of entropy at cell-face centre B,

S_{oB} = desired value of entropy at cell-face centre B, $S_{oB} = S_{ijl}$,

d_{uB} , d_S scaling parameters (> 0), specifying how fast $\tilde{u}_{oB}$, and S_{oB} are driven to their desired values $\tilde{u}_{oB}$, S_{oB} when $t \to \infty$.

These boundary conditions may be written in the general form,

$$(\phi_t)_B + f_B(U^n) = 0. \tag{3.6}$$

This is a set of five equations, two of them are the boundary conditions (3.5), the remaining three can be arbitrarily defined, because they are eliminated, see below.
Of the five conservation equations (3.3), two equations have been replaced by boundary conditions. The remaining three equations are also required to obtain a total of five difference equations for the flow-state vector at each halo cell centre. They are called auxiliary equations. These auxiliary equations are thus semi-discretized conservation equations.
The boundary-condition equations (3.6) and the remaining auxiliary equations are combined to a set of five new equations, by a so-called incidence matrix I_B. It is defined as a diagonal matrix whose elements are 0 for Riemann variables which corresponding conservation equation has been replaced by a boundary condition, and 1 for the other Riemann variables. In our case $I_B = \mathrm{diag}(0,1,0,0,1)$.
Now the boundary conditions and conservation equations can be combined using the incidence matrix,

$$(\phi_t)_B + I_B T_B [Q_{ij\frac{3}{4}}(U^n) + D_{ij\frac{3}{4}}(U^n)] + (I - I_B) f_B(U^n) = 0. \tag{3.7}$$

This set of equations is mapped back into equations for the primitive state variable U by premultiplying with the matrix T_B^{-1}.

4. ANALYSIS OF THE CHARACTERISTIC BOUNDARY CONDITIONS

4.1 CONSISTENCY

The concept of consistency requires that the finite-difference equations including the boundary conditions should approach the differential equations, for time step and mesh size tending to zero. The consistency follows immediately from the discretized equations, by applying Taylor expansions. It then follows that the discrete conservation equations and discrete boundary conditions are of second order in space, provided the numerical dissipative terms and entropy gradient normal to the solid wall are also of second order. Further the semi-discretized boundary conditions are first-order accurate in time because, in equation (3.7), ϕ_t is discretized on the boundary, while the remaining terms are discretized at the centres of the auxiliary cells.

4.2 WELL-POSEDNESS

Well-posedness of the boundary conditions of the discrete set of Euler conservation equations is analyzed following the theory of Higdon [7]. His analysis is based on a linearized set of hyperbolic equations and boundary conditions.
An initial-boundary-value problem for these hyperbolic equations is then said to have well-posed boundary conditions if disturbing solutions can not enter the flow domain through the boundary. Solutions that grow in time and decay in space away from the boundary are classified as disturbing. This can be understood as follows. Consider the space domain [0,X], and let a disturbing solution enter the flow field through the boundary at x=0. When this solution reaches the boundary at x=X, it is damped out, hence it will not be 'seen' by the boundary condition at x=X. The boundary condition at x=0 must supress this disturbing solution.
The solution of an initial-boundary-value problem can be expressed as a sum or integral of elementary waves, also called 'normal modes'. An expression for these normal modes may be derived using Fourier transforms in the directions tangential to the boundary, Laplace transforms in time, and an amplitude function in normal direction. Substitution of these normal modes in the initial-boundary-value problem leads to a system of ordinary differential equations for the amplitude function. Using hyperbolicity, it can be shown that the general solution of this system of ordinary differential equations can be written as a sum of solutions either growing or decaying in normal direction. According to the above mentioned criteria of Higdon, the decaying solutions must be eliminated by the boundary conditions in order to have well-posed boundary conditions.
It can be shown that the solid-wall boundary conditions (3.7) for the Euler conservation equations are well-posed. In [2] the complete analysis is described.

4.3 STABILITY

Because we want to analyse not only the discrete Euler conservation equations but also the boundary conditions, we use the concept of practical stability. It is introduced by Richtmyer and Morton [8], and is based on the principle that the growth of any Fourier component of the discrete approximation should not be allowed to be larger than the most rapid growth of the continuum solution of the initial-boundary-value problem. This gives a bound to the amplification factor. We are interested in solutions that do not grow unlimited in space and time, so a time-independent bound of unity can be imposed.
With the aid of discrete Fourier analysis, and using linearized conservation equations and linearized boundary conditions, an expression for the amplification factor can be derived. In our case a 4-stage Runge-Kutta scheme is used for the numerical time integration.

This criterium leads to a maximum CFL number, or a the maximum allowable time step.

5. NUMERICAL VALIDATION OF THE BOUNDARY CONDITIONS

As a test case the NLR 7301 profile is chosen, with as free-stream conditions M =.721 , α = -.194 deg. This leads to a so-called shock-free super-critical flow i.e. the flow has subsonic and supersonic regions but no shock. It is a widely used and essential test case for numerical validation of Euler codes. All calculations have been performed on the NEC SX-2 supercomputer at NLR.

In table 1 an overview of the grids used for the calculation around the NLR 7301 profile is given.
The grid A is an O-type grid with 256*160 cells on the finest grid level (256 cells around the profile, 160 in direction normal to the surface), and depicted in figure 3. The grid has a cell-aspect-ratio of 1 at the profile, and 2 at the outer boundary. The outer boundary is located 100 chords from the profile.
On the grids B and C, it was not possible to derive a converged shock-free solution. The difference between grid A and B is the location of the outer boundary, not the grid quality. The difference between grid A and C is the grid resolution in normal direction. From this we can conclude that not only the location of the outer boundary is of importance, also the grid quality and grid resolution.

Calculation results obtained with the characteristic solid-wall boundary condition were compared to results derived with the solid-wall boundary condition of a zero normal velocity component and extrapolation of the pressure.
Figure 4 depicts the rate of convergence for both calculations. The calculations are performed on a coarser grid until a converged solution is computed. This solution is used as an initial solution on the finer grid. The choice of boundary condition does not influence the rate of convergence. The convergence history is of importance for the computed solution. For example, the large difference between the Mach-number distributions along the profile of figure 6 (residue norm 10^{-5}) and of figure 9 (residue norm 10^{-2}), shows that the residu norm must be reduced in this case by 5 decades.

Figure 5 compares the pressure distribution along the profile obtained with the characteristic boundary conditions with the pressure distribution from hodograph theory [9]. They coincide well.
In figure 6 the Mach-number distributions along the profile of both calculations are compared to each other. The Mach-number distribution around the profile obtained with the characteristic boundary conditions is given in figure 7.
The entropy production on the surface of the profile of both calculations are compared to each other in figure 8.a. Also the entropy production in the flow field around the profile of both calculations, with enlargements around the leading and trailing edges,is given in figure 8.b and 8.c. The calculation with the characteristic solid-wall boundary condition gives a more accurate solution at the profile. The entropy in the numerical entropy-layer has been reduced by a factor of two. Only near the stagnation point, a production of entropy can be observed. A study of the origin of this entropy 'bubble' leads to the conclusion that it is due to discretization errors.

6. CONCLUSIONS

The characteristic solid-wall boundary condition leads to a more accurate solution of the Euler equations than the standard solid-wall boundary condition with extrapolation of the pressure, because the numerical entropy layer appeared to be reduced.
The two boundary conditions have comparable convergence rate.

Further study of the grid-dependence of the solution, and of the influence of the location of the outer-boundary, and thus the boundary conditions at the outer boundary, is required.

7. REFERENCES

[1] JAMESON,A. ,SCHMIDT,W. :"Recent developments in numerical methods for transonic flows.", Computer methods in Applied Mechanics and Engineering 51, (1985), pp. 407-463.

[2] VAN DEN BERG,J.I. ,BOERSTOEL,J.W. :"Theoretical and numerical investigation of characteristic boundary conditions for cell-centered Euler flow calculations.", NLR TR 88124 L.

[3] BOERSTOEL,J.W. :"Progress report of the development of a system for the numerical simulation of Euler flows, with results of preliminary 3D propellor-slipstream/exhaust-jet calculations.", NLR TR 88008 L.

[4] BOERSTOFL,J.W. et al.:"Design and testing of a multiblock grid-generation procedure for aircraft design and research.", NLR TP 89146 L.

[5] BUYSEN,F.A. :"Flow visualisation at NLR : VISU3D ", NLR TP 89317.

[6] SCHUURMAN,J.J. ,KASSIES,A. ,MEELKER,J.H. :"Method Management for the benefit of large software packages for analyses in CAE.", NLR TP 89027.

[7] HIGDON,R.L. :"Initial boundary value problems for linear hyperbolic systems.", SIAM Review, Vol. 28, No. 2, June 1986, pp. 177-217.

[8] RICHTMYER,R.D. , MORTON,K.W.:"Difference methods for initial-value problems.", Interscience Publishers, second edition, (1967).

[9] YOSHIHARA,H. , SACHER,P. (ed.) :"Test cases for inviscid flow field methods.", Advisitory Report 211, AGARD-AR-211, 1985.

Table 1. Overview of the grids

GRIDS	A	B	C
# cells along profile	256	256	256
# cells in normal direction	160	128	64
cell-aspect ratio at solid-wall	1.	1.	.5
cell-aspect ratio at outer boundary	2.	2.	7.
location of outer boundary (chords from profile)	100	40	100

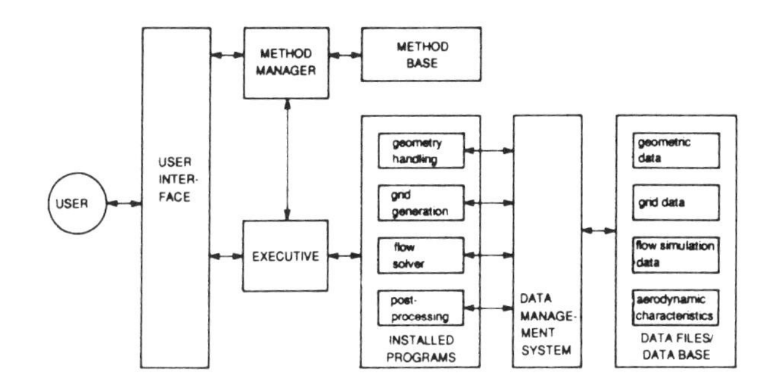

Fig. 1 Conceptual design of an information system for flow simulation.

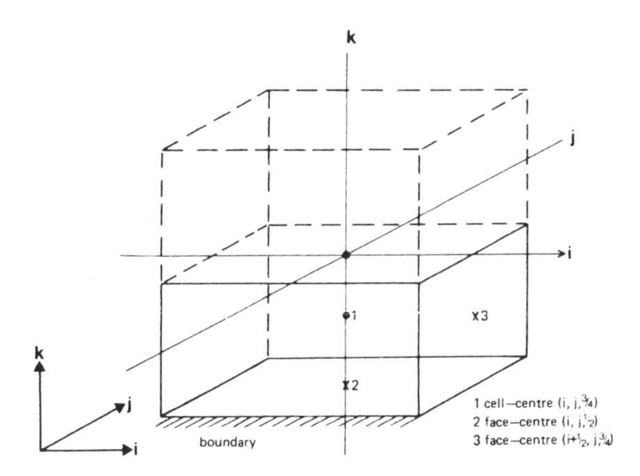

Fig. 2 Auxiliary cell along the boundary of the computational domain

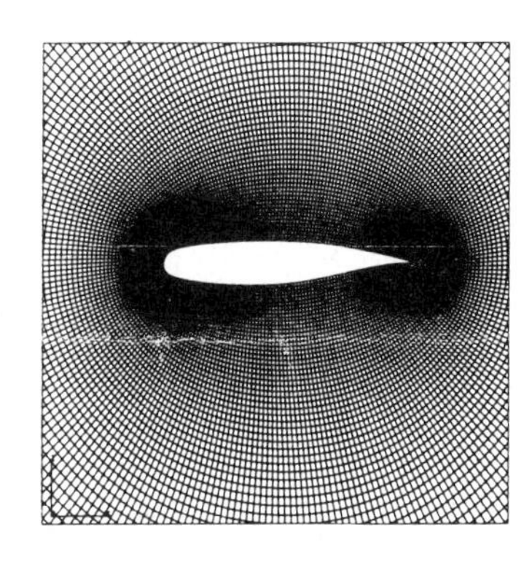

Fig. 3 Grid around the NLR 7301 profile

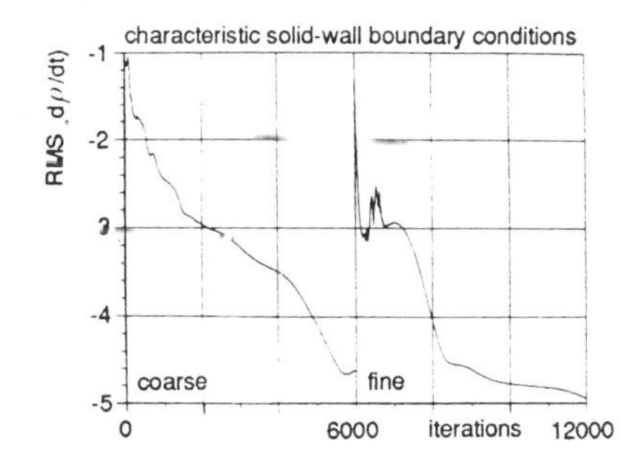

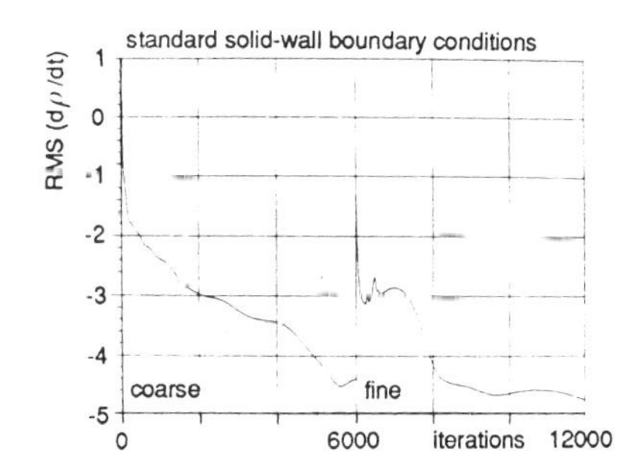

Fig. 4 Rate of convergence for the calculation around the NLR7301 profile on two grid levels, with the characteristic solid-wall boundary conditions (left) and the standard solid-wall boundary conditions (right)

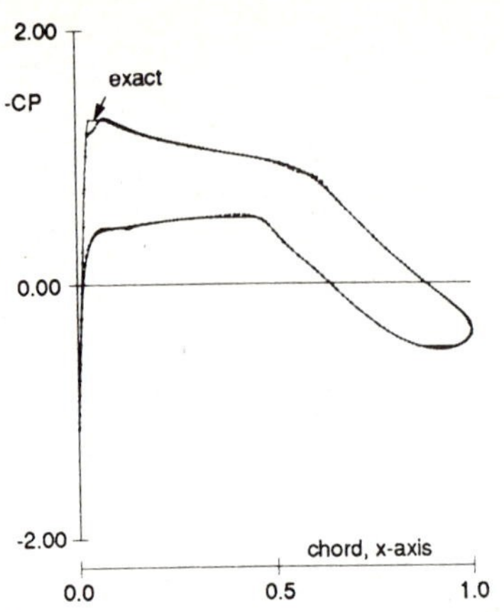

Fig. 5 Comparison of exact pressure coefficient distribution and calculated pressure coefficient distribution along the profile.

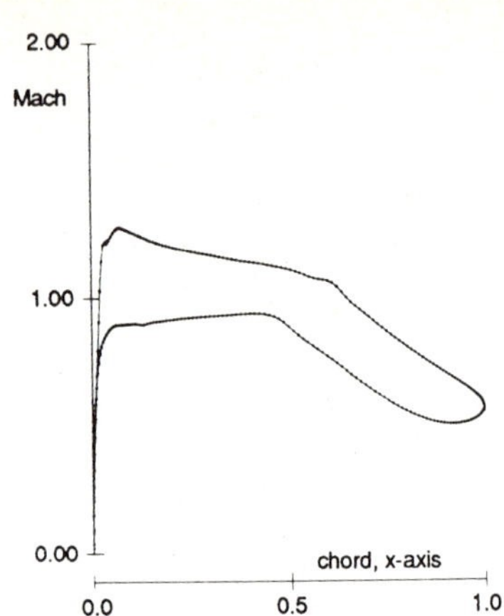

Fig. 6 Comparison of calculated Mach number distribution along the profile with characteristic and standard solid-wall boundar conditions.

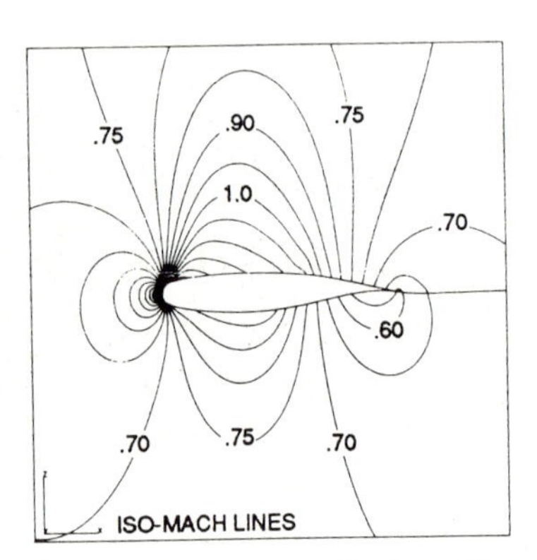

Fig. 7 Isoplot of the Mach number around the NLR7301 profile.

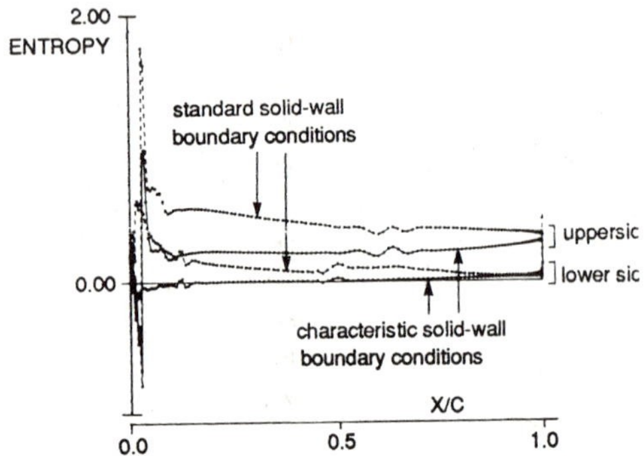

Fig. 8a Comparison of entropy on the profile, calculated with characteristic and standard solid-wall boundary conditions.

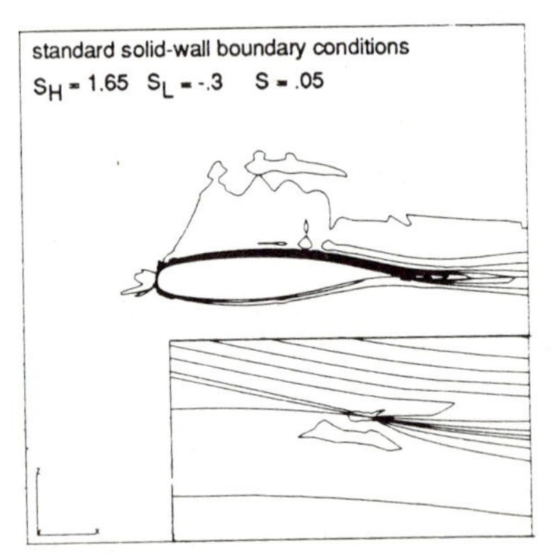

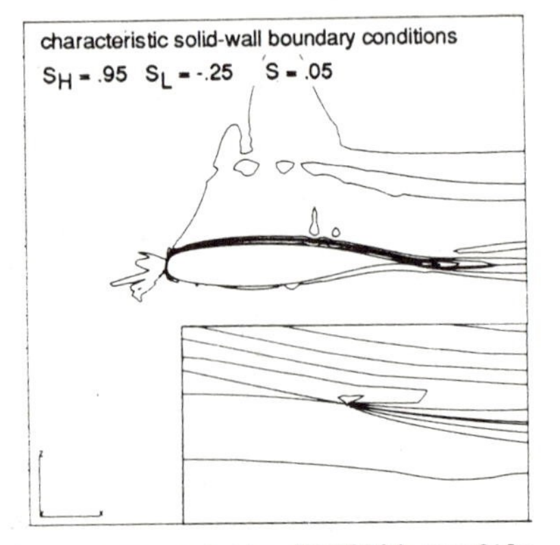

Fig. 8b Isoplot of the numerical entropy layer around the NLR7301 profile, with enlargement around the trailing edge.

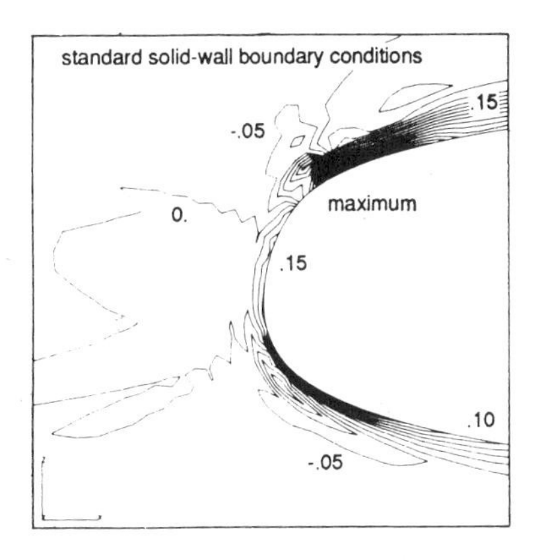

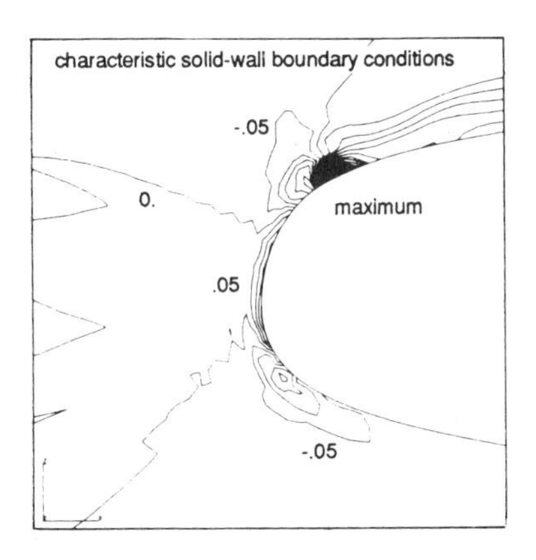

Fig. 8c Enlargement around the leading edge of the numerical entropy layer around the NLR7301 profile.

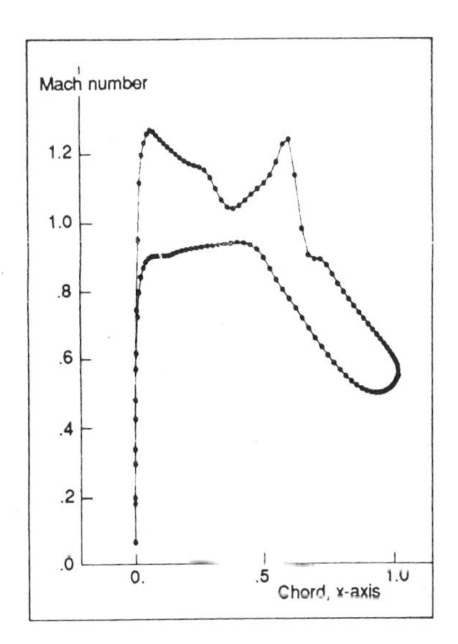

Fig. 9 Mach number distribution along the profile, convergence until residu = 10^{-2}

SOLUTION ADAPTIVE LOCAL RECTANGULAR GRID REFINEMENT FOR TRANSONIC AERODYNAMIC FLOW PROBLEMS *

Michael B. Bieterman, John E. Bussoletti, Craig L. Hilmes,
Forrester T. Johnson, Robin G. Melvin, Satish S. Samant, and David P. Young

The Boeing Company
P.O. Box 3707, MS 7K-06
Seattle, Washington 98124, USA

SUMMARY

We describe the use of solution adaptive local grid refinement in a numerical method for solving transonic flow problems about complex three dimensional aircraft configurations. The method is implemented in the TRANAIR code, which has been applied to help solve many practical engineering problems. Attention is focused here on the principal components of the solution adaptive grid algorithms currently being developed and on two applications that demonstrate the capabilities of the algorithms.

1 INTRODUCTION

Aerodynamic analyses of real aircraft designs often must include effects of complex multiple component geometries. Commercial airplanes have nacelles, struts, slats, flaps, ailerons and stabilizers, in addition to wings and fuselages. Military aircraft configurations can be even more complex. The aerospace engineering community needs analysis tools that can be quickly applied to such general geometries. Over the past few years, we have developed and integrated numerical methods to create one such tool, the TRANAIR code, which runs on Cray computers. This paper describes numerical method developments directed toward improving this tool.

The mathematical problem to be numerically solved is summarized in the next section along with some of the features of the underlying (nonadaptive grid) version of the present method. Section 3 describes the solution adaptive grid algorithms that are being developed. Section 4 gives an account of results of two applications of the adaptive grid method. This is followed by conclusions.

2 PROBLEM AND UNDERLYING METHOD

The governing mathematical problem is to solve the full potential equation

$$\nabla \cdot (\rho \nabla \Phi) = 0, \tag{1}$$

*Partially supported by NASA contract NAS2-12513

describing steady irrotational compressible flow in three dimensional space, supplemented by boundary conditions including zero normal mass flux $\rho \nabla \Phi \cdot \hat{n}$ through impermeable surfaces, specified mass flux on engine fan faces, Φ coinciding with the freestream total potential in the far field, and appropriate jump conditions on wake surfaces. The density ρ depends nonlinearly on velocity magnitude $|v| = |\nabla \Phi|$ according to

$$\rho = \rho_\infty \left[1 + \frac{\gamma - 1}{2} M_\infty^2 (1 - \frac{|v|^2}{|v_\infty|^2}) \right]^{\frac{1}{\gamma - 1}}, \tag{2}$$

where M denotes Mach number, γ is the ratio of specific heats, and subscripts indicate specified freestream values. Equation (2) is modified (cf., e.g., [1],[2],[5]) in regions such as engine exhaust plumes when power effects are modeled by assuming total temperature and pressure there are constants, but different from their freestream values.

The underlying (nonadaptive) numerical method and applications of the TRANAIR code implementing it have been described in many recent articles [1–8]. Here we briefly summarize some of the features of the method and its operation.

A finite element method based on an extension [1] of the Bateman variational principle [10] for equation (1) is used to discretize the problem. Continuous piecewise trilinear approximations of the potential Φ are employed on locally refined Cartesian grids. The density ρ is approximated to be a constant on each box element in the grid and is upwind biased [1] in supersonic flow regions in a first order manner to produce needed artificial dissipation. The surface of a configuration is embedded in the grid, thus circumventing the need to generate a body conforming grid. Zero normal mass flux conditions on the intersection of a surface with a rectangular box finite element are treated "naturally" in the finite element formulation via volume integration of element basis functions on the region in the box element exterior to the surface. Other boundary conditions are treated similarly, with local surface integration supplementing the aforementioned volume integration.

In an application of the method, local resolution of the finite element grid is determined near surfaces by their curvature and away from surfaces by specified refinement levels in given zones (hexahedra). Near-body grid tends to be smoothly graded into the field while off-body grid takes on the appearance of blocks having the shapes of the zones. Using an oct-tree data structure, the construction of the entire grid is carried out automatically through hierarchical local refinement of a specified uniform Cartesian grid.

The nonlinear system of equations resulting from this discretization is approximately solved with an inexact Newton method [2] utilizing a preconditioned GMRES iterative algorithm for each linearized problem [1],[9]. A recently implemented technique [2] that reliably accelerates convergence of Newton's method and that provided a gateway to the presently described adaptive grid work is the technique of grid sequencing. It consists of solving a coarse grid problem, transferring the solution Φ to a finer grid (via piecewise trilinear interpolation), using the solution as an initial guess on the finer grid, and repeating these steps on subsequent grids. The sequence of grids is created at the beginning of a problem's solution by recursively derefining (coarsening) a given (finest) grid. The numbers of elements in the nested grids used during the solution process increase fairly rapidly (typically by a factor of three to five per grid), which contributes significantly to the efficiency of the technique.

3 SOLUTION ADAPTIVE GRID METHOD

The solution adaptive grid method operates like the grid sequenced method once the latter has begun work on the coarsest grid, with two exceptions. In the adaptive method, local grid resolution is determined by *a posteriori* computed local error estimates, rather than being *a priori* determined. The second difference is that single-level local grid refinement *and* derefinement are used. In refining, a rectangular box element is replaced by eight smaller similar elements. In derefining, eight sibling elements are coalesced to form a larger similar element.

The goal in adapting grids to solve a problem is to obtain at moderate cost a final grid whose number of elements approximately equals a specified target number N and to obtain a numerical solution on the grid that is nearly as accurate as the best solution one could obtain using N elements in a grid. Details of the present solution adaptive grid method are given in descriptions of the five sequential steps that are carried out in creating a new grid.

STEP 1. COMPUTING LOCAL ERROR ESTIMATES

For results presented in this paper, local differences of velocity component values were used as error indicators. The indicator for an element outside of all surfaces is defined as

$$errind \equiv \max_j \{(\Delta v_1^j)^2 + (\Delta v_2^j)^2 + (\Delta v_3^j)^2\}^{\frac{1}{2}}, \tag{3}$$

where Δv_i^j denotes the difference across the element's jth face of the element centroid values of the ith velocity component. The maximum in (3) is taken over all faces not connected to a larger element. For an element near or intersecting a surface, the error indicator is defined similarly, but where element faces lying completely inside of a configuration are excluded in computing the maximum in (3) and where the velocity component values are those at the centroids of the regions in the elements that are exterior to surfaces. Figure 1 illustrates in the case of a two dimensional airfoil the directions in which velocity components are differenced to compute error indicators for five elements, labeled A–E.

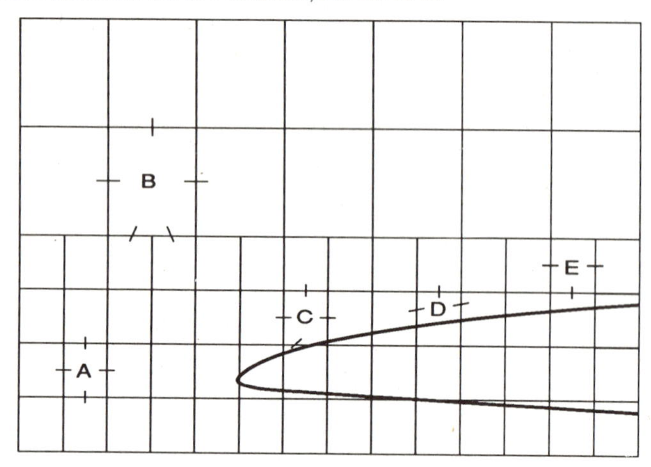

Figure 1: Directions for Velocity Component Differences in Error Indicators for Elements A–E

Each local error indicator computed via (3) is an estimate of the local maximum norm error in velocity. This error measure was chosen because it provides a natural way to detect flow features having different length scales near different configuration components and it does not tend to lead to excessive grid in the far field. Other local error indicators we are examining include: local L_1 and L_2 norm velocity error indicators (obtained from the one used here by weighting according to local element volume); one using ρv instead of v; and an indicator in which the sum in (3) is replaced by the element face-normal component, and for elements intersecting a surface (e.g., C and D in figure 1), $|v \cdot \hat{n}|$ is included in the maximum in (3), where $\hat{n}$ is the surface normal vector. This indicator is equivalent to (3). in the limit of decreasing local grid size.

STEP 2. COMPUTING LOCAL ERROR PREDICTORS

Local error predictors are formed from the error indicators. For results presented here, a local smoothing algorithm was used. In it, nodal values are first set equal to the largest of the error indicators for adjacent elements and then interpolated at element centroids to form the predictors. This algorithm spreads the effect of large estimated errors by one or two elements and tends to prevent "holes" in resulting grids. We also are examining ways of extending this smoothing process in a more grid independent manner and are investigating the use of local history information in the prediction process.

STEP 3. APPLYING A PRIORI GRID REFINEMENT CONTROLS

For the most accurate and efficient analyses of complex configurations, it is important that one is permitted to communicate regions of greatest interest to a solution adaptive grid code. The mechanism for this in the present method is an extension of the specified hexahedral zone concept used in the nonadaptive method (cf. Section 2). With each hexahedral shaped zone, one specifies minimum and maximum levels of refinement allowed in the zone and a weight used to scale the corresponding local error predictors. The results of applying these zonal restrictions are an element refinement/derefinement eligibility list and a list of scaled error predictors ordered by size.

STEP 4. APPLYING GRID REFINEMENT STRATEGY

This step consists of marking some of the refinement-eligible elements with the largest scaled error predictors for subdivision and marking some of the derefinement-eligible elements with the smallest scaled predictors for coarsening. In examining strategies for deciding how many elements of each type to mark, two principles have proven useful. First, direct control should be exercised over the rates at which numbers of elements in successive grids increase. This excludes, for example, a prevalent strategy that refines/derefines solely on the basis of cut-off values which are proportional to the current mean local error indicator. Direct control is important because problem size increases *very* rapidly with grid refinement in three dimensions. Second, for early and intermediate grids, grid refinement should be limited in regions where dominant flow features have been detected and should be forced to occur elsewhere. Failure to adhere to this principle can allow some flow features (e.g., leading edge expansions) to attract all available grid before other important features (e.g., shocks) develop.

A simple strategy following these principles was used in the applications here. It consists of (i) refining and derefining fixed percentages of elements for most grids, (ii) attempting to more equally distribute local errors without significantly changing the number of elements in an intermediate grid, and (iii) only refining on the last grid. More specifically, with given intermediate and final target numbers of elements N_{Itarg} and N_{Ftarg} satisfying $N_{Itarg} < N_{Ftarg}$ and $N_{Itarg} > .25N_{Ftarg}$, and $N < N_{Itarg}$ denoting the number of elements in the current grid; if $N < .4N_{Itarg}$, 20% of the elements are marked for refinement and up to 50% for derefinement; if $.4N_{Itarg} < N < .9N_{Itarg}$, only refinement is used to increase the number of elements to about N_{Itarg}; and if N is approximately equal to N_{Itarg}, 2% of the elements are refined and up to 20% are derefined, and the next (final) grid adaptation consists of (only) refining enough elements so that the final grid has about N_{Ftarg} elements.

STEP 5. CONSTRUCTING A GRID

Using a list of marked elements and a grid regularization constraint [1] that prohibits face-neighbor and edge-neighbor elements from differing by more than one refinement level, a grid is constructed by building a new oct-tree data structure describing it. Since relatively few grids typically are used in an application, it has not been found necessary to incorporate a more dynamic data structure that is retained from grid to grid.

4 APPLICATIONS

Described here are applications of the solution adaptive grid method to two cases, the ONERA M6 wing with freestream Mach number $M_\infty = .84$ and angle of attack $\alpha = 3.06°$ and a Boeing 747-200 wing/body/struts/nacelles transport configuration at $M_\infty = .80$ and $\alpha = 2.70°$. The first case was chosen because of the strong leading edge expansion and normal shock on the wing and the presence of an oblique shock. The second case involves less pronounced flow features, but a much more complex geometry.

ONERA M6 WING

The initial grid had about $15K$ box elements. Intermediate and final target numbers of elements were specified as $N_{Itarg} = 300K$ and $N_{Ftarg} = 440K$, respectively. No minimum-level refinement restriction was employed. The maximum level of refinement throughout the flow field was specified to be 4 levels finer than the given uniform grid except for the final grid to be constructed, for which 5 levels of refinement were permitted. One specified zone was used to guide the solution adaptive gridding process. This was located about the wing tip to prevent refinement finer than one level below the uniform grid there. No scaling of local error predictors was done.

In the resulting adaptive grid run, 6 grids were created. These contained approximately $39K$, $86K$, $184K$, $286K$, $312K$ and $423K$ elements. Figure 2 shows 70% span station cuts through the initial grid and final adaptive grid. Cuts through the final adaptive grid at 0% and 70% span are shown in figure 3. Figure 4 displays graphs of the computed pressure coefficients versus percentage wing chord at 0% and 70% span for the initial grid, the second adaptive grid and the final adaptive grid. These results are generally quite accurate. One notices, however, that the oblique shock present on the wing at about 25% chord on the 70% span station is somewhat smeared. The oblique shock is a relatively weak phenomenon in this problem that can only be detected once very fine grid is present.

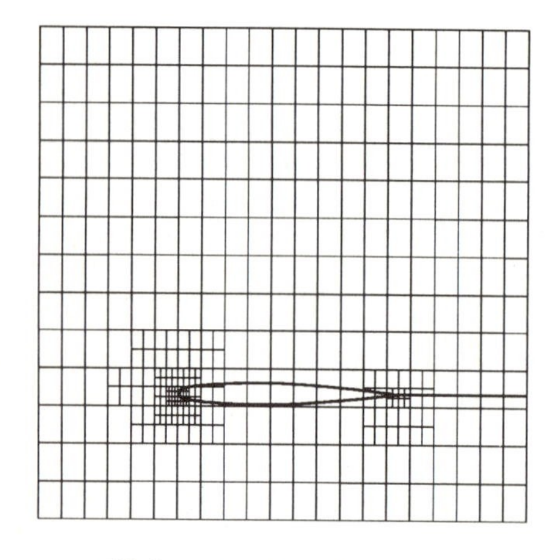

70% Span, Initial Grid

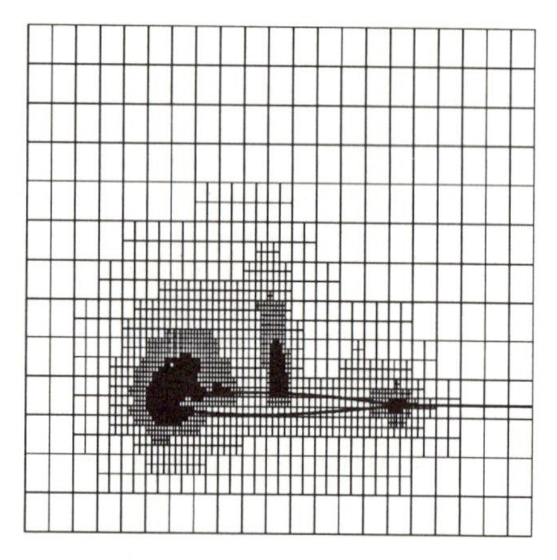

70% Span, Final Adaptive Grid

Figure 2: Grid Cuts at 70% Span for the ONERA M6 Wing, $M_\infty = .84, \alpha = 3.06°$.

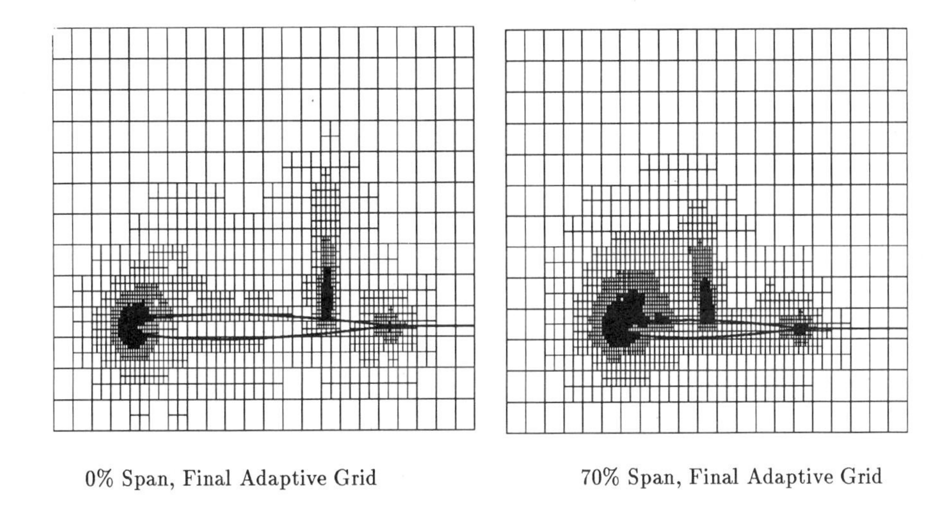

0% Span, Final Adaptive Grid

70% Span, Final Adaptive Grid

Figure 3: Grid Cuts at 0% and 70% Span for the ONERA M6 Wing, $M_\infty = .84, \alpha = 3.06^o$.

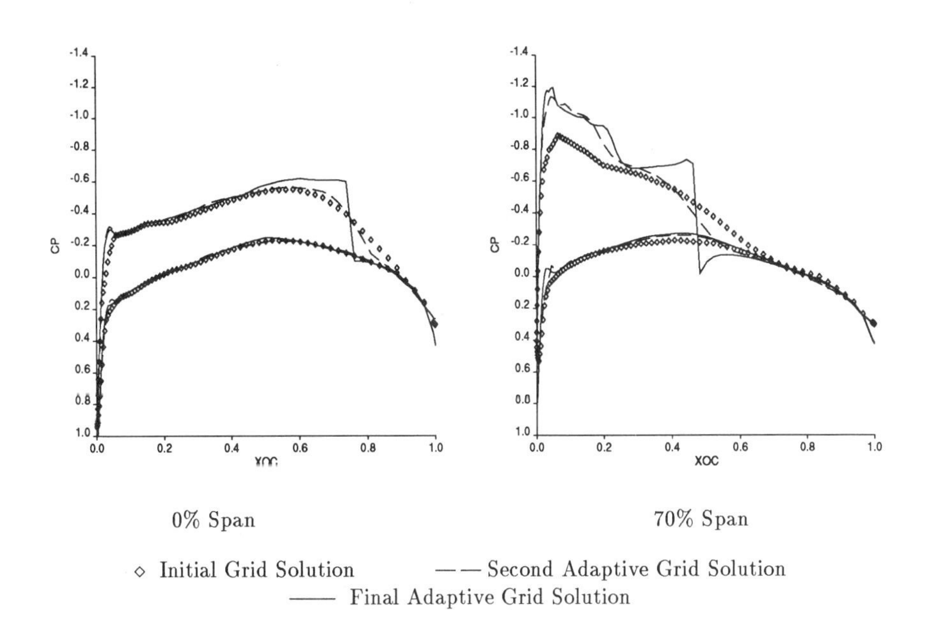

0% Span

70% Span

◇ Initial Grid Solution — — Second Adaptive Grid Solution
—— Final Adaptive Grid Solution

Figure 4: Pressure Coefficients for ONERA M6 Wing, $M_\infty = .84, \alpha = 3.06^o$.

BOEING 747-200 CONFIGURATION

This is a geometrically complex configuration (cf. figure 5). In applying the solution adaptive grid method to a half-model symmetric about the centerbody, an initial grid containing about $26K$ elements was used. The specified intermediate and target numbers of box elements were $125K$ and $250K$, respectively. Three zones were specified to guide the adaptive grid method. These included one about the wing tip as in the ONERA wing case, one under the wing enclosing the nacelles, and one about and above the wing. No minimum-level refinement restriction was employed. Four levels of refinement were permitted for all elements except in the zone about and above the wing. There, 4 to 6 levels were permitted from the body to the wing tip, respectively, in all grids except the final grid, for which 5 to 7 levels of refinement were permitted. Since wind tunnel test data for comparison were only available on the wing, the importance of this region was (further) emphasized by specifying a scaling factor of 8 for the local error predictors in the zone containing the wing. A scaling factor of 2 was used in the zone containing the nacelles. Predictors in other regions were not scaled.

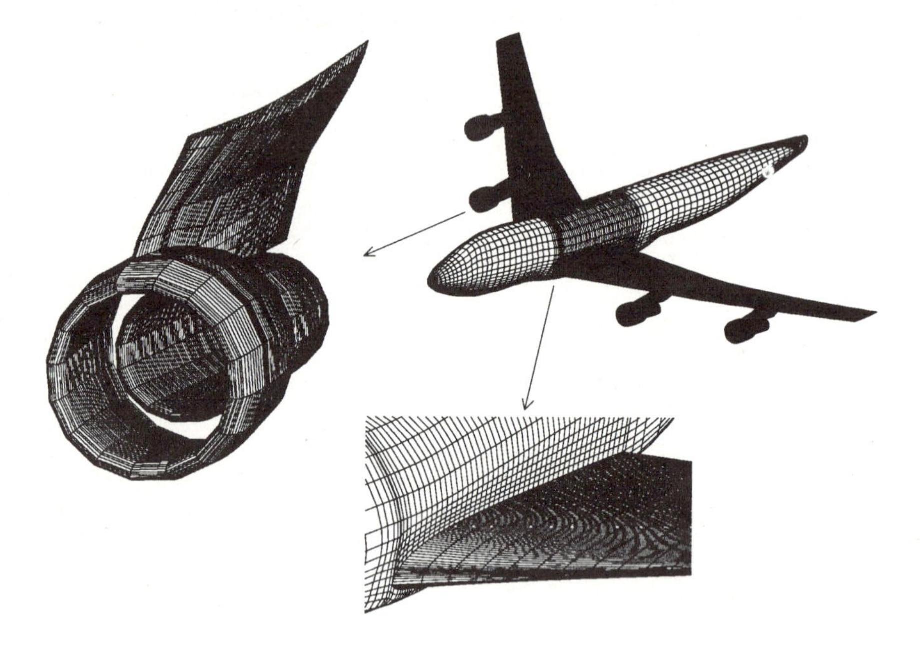

Figure 5: 747-200 Transport Configuration.

Four grids were created in a run with the solution adaptive grid method. These contained approximately $58K$, $123K$, $133K$ and $243K$ elements. Figure 6 shows 69% and 96% wing span station cuts through the initial grid and final adaptive grid. Figure 7 compares computed wing pressures with wind tunnel test data at four wing span stations. The method yielded results that compare favorably with the wind tunnel experimental data. Most of the differences are attributable to viscous boundary layer effects not modeled here. At the 69% span station, which is located above the outboard nacelle, one sees the effect of the nacelle on the lower wing surface pressure. The very high speed local flow near the leading edge on the upper surface at this span station is due to the presence of a strut cap connected to the outboard strut and wing leading edge.

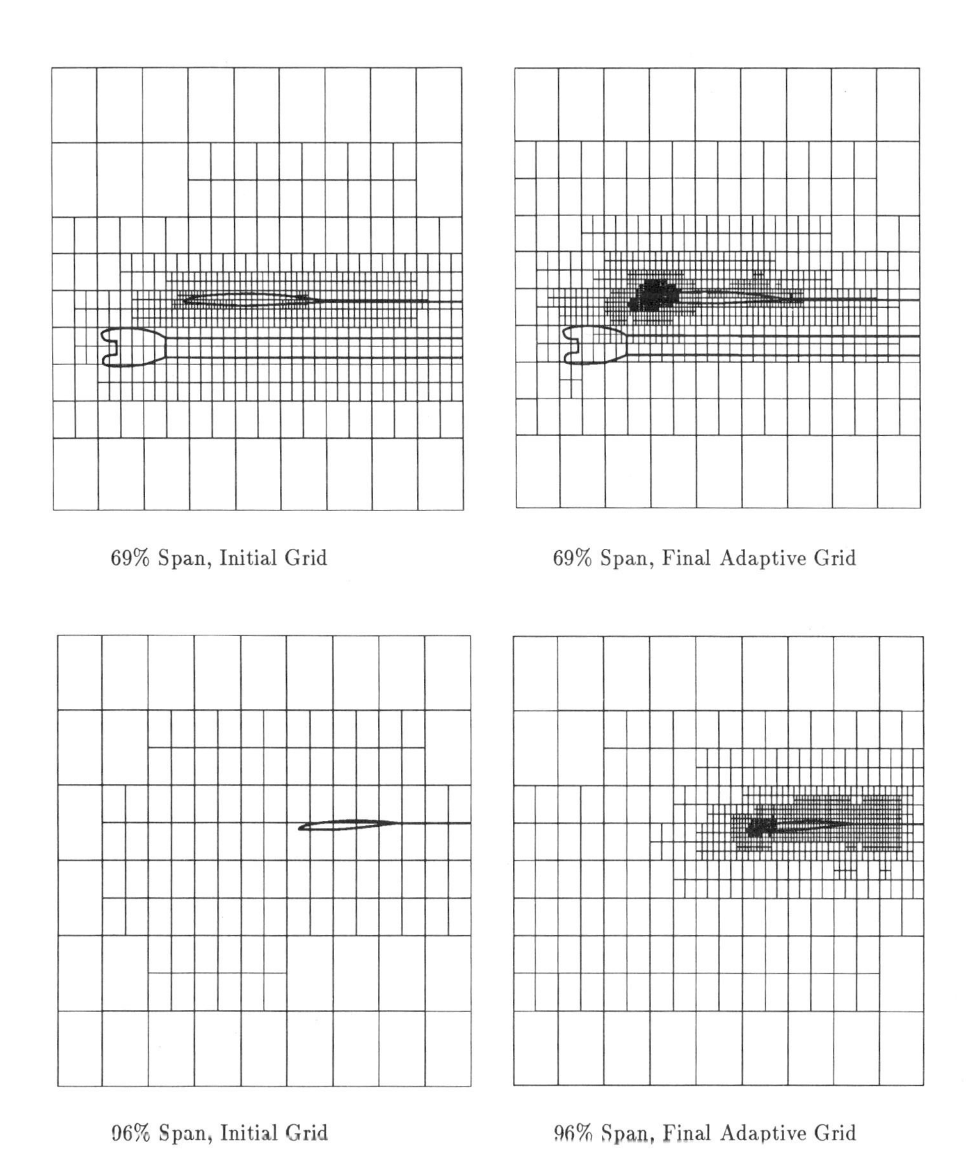

Figure 6: Grid Cuts at 69% and 96% Wing Span for 747-200, $M_\infty = .80, \alpha = 2.70°$.

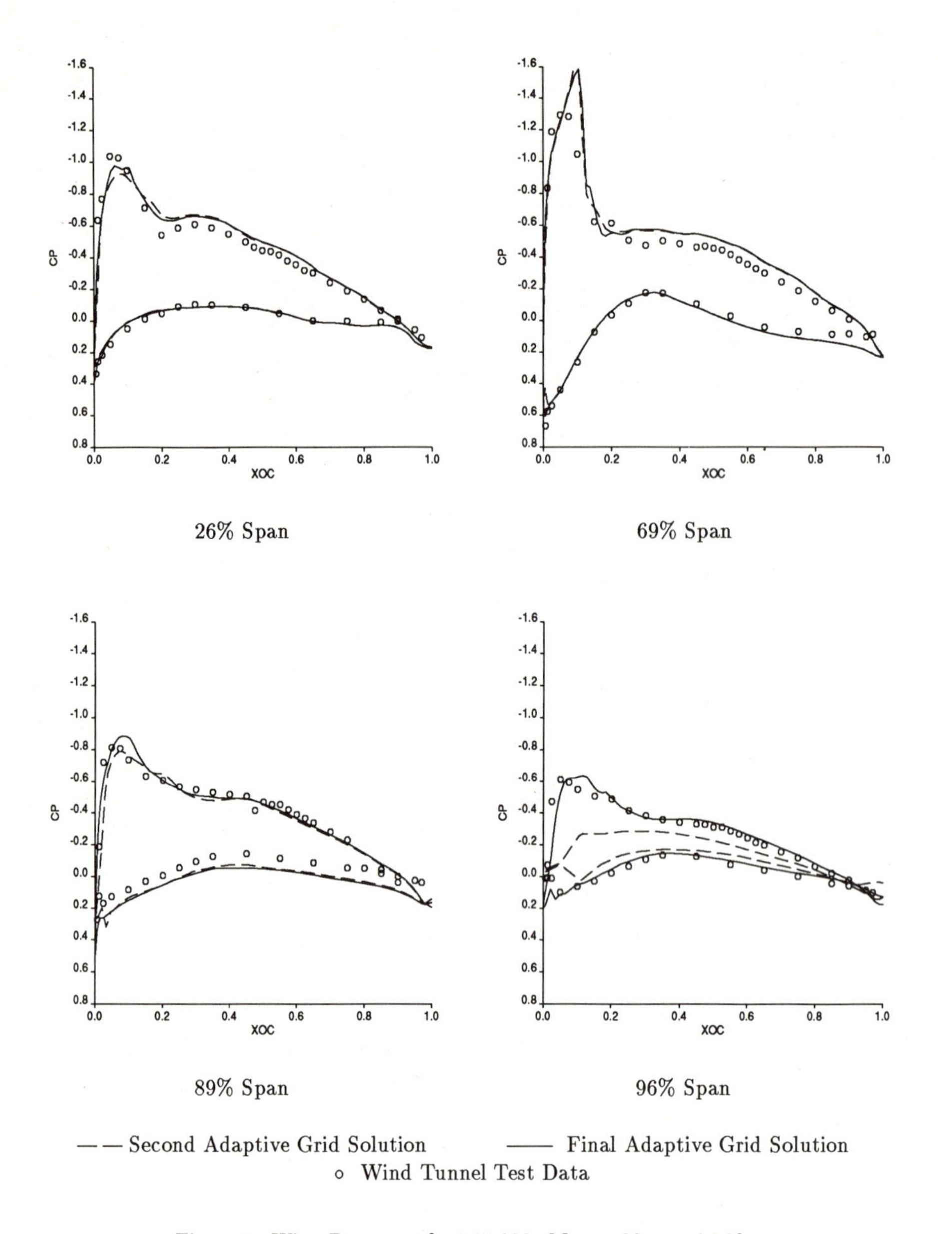

Figure 7: Wing Pressures for 747-200, $M_\infty = .80, \alpha = 2.70°$.

CONCLUSIONS

We have described some of the key components of a solution adaptive grid method for complex three dimensional transonic flow problems. Results presented for two applications demonstrate the method's ability to adapt grids automatically for difficult problems and to obtain accurate solutions.

References

[1] D.P. YOUNG, R.G. MELVIN, M.B. BIETERMAN, F.T. JOHNSON, S.S. SAMANT and J.E. BUSSOLETTI, "A Locally Refined Rectangular Grid Finite Element Method," Report SCA-TR-108-R1, Boeing Computer Services, Seattle, Washington, April, 1989 (to appear in *Journal of Computational Physics*).

[2] D.P. YOUNG, R.G. MELVIN, M.B. BIETERMAN, F.T. JOHNSON and S.S. SAMANT, "Global Convergence of Inexact Newton Methods for Transonic Flow," Report SCA-TR-124, Boeing Computer Services, Seattle, Washington, August, 1989.

[3] R.G. MELVIN, M.B. BIETERMAN, D.P. YOUNG, F.T. JOHNSON, S.S. SAMANT and J.E. BUSSOLETTI, "Local Grid Refinement for Transonic Flow Problems," pp. 939-950, Vol. 6, *Proceedings of the Sixth International Conference on Numerical Methods in Laminar and Turbulent Flow*, (C. Taylor, P. Gresho, R.L. Sani and J. Häuser, eds.), Pineridge Press, 1989.

[4] J.E. BUSSOLETTI, F.T. JOHNSON, D.P. YOUNG, R.G. MELVIN, R.H. BURKHART, M.B. BIETERMAN, S.S. SAMANT and G. SENGUPTA, "TRANAIR Technology: Solutions for Large PDE Problems," to appear in *Solution of Super Large Problems in Computational Mechanics* (J.H. Kane and A.D. Carlson, eds.), Plenum Publishing, 1989.

[5] F.T. JOHNSON, S.S. SAMANT, M.B. BIETERMAN, R.G. MELVIN, D.P. YOUNG, J.E. BUSSOLETTI and M.D. MADSON, "Application of TRANAIR Rectangular Grid Approach to Aerodynamic Analysis of Complex Configurations," to appear in *Proceedings of the 64th Meeting of the AGARD Fluid Dynamics Panel, Specialist's Meeting on Applications of Mesh Generation to Complex 3-D Configurations*, Loen, Norway, 1989.

[6] A.W. CHEN, M.M. CURTIN, R.B. CARLSON and E.N. TINOCO, "TRANAIR Applications to Engine/Airframe Integration," AIAA Paper 89-2165, 1989.

[7] A.M. GOODSELL, M.D. MADSON and J.E. MELTON, "TRANAIR and Euler Computations of a Generic Fighter Including Comparisons with Experimental Data," AIAA Paper 89-0203, 1989.

[8] W. TSENG and A. CENKO, "TRANAIR Applications to Fighter Configurations," AIAA Paper 89-2220, 1989.

[9] D.P. YOUNG, R.G. MELVIN, F.T. JOHNSON, J.E. BUSSOLETTI, L.B. WIGTON and S.S. SAMANT, "Application of Sparse Matrix Solvers as Effective Preconditioners," to appear in *SIAM Journal on Scientific and Statistical Computing*, pp. 1186-1199, Vol. 10, No. 6, 1989.

[10] H. BATEMAN, "Irrotational Motion of a Compressible Inviscid Fluid, pp. 816-825, Vol. 16, *Proceedings of the National Academy of Sciences*, 1930.

NUMERICAL SIMULATION OF LAMINAR HYPERSONIC FLOW PAST BLUNT BODIES INCLUDING HIGH TEMPERATURE EFFECTS

G. Brenner, S. Riedelbauch, B. Müller
DLR Institute for Theoretical Fluid Mechanics,
Bunsenstr. 10, D-3400 Göttingen, FR Germany

SUMMARY

Axisymmetric laminar hypersonic flows of perfect gas and equilibrium air past blunt bodies are simulated. The thin-layer Navier-Stokes equations are solved by a semi-implicit central finite-difference method. The thermodynamic and transport properties of equilibrium air are obtained from curve-fitting routines and from a Gibbs-energy minimization procedure. Results are obtained for a flow past a hemisphere and a hyperboloid.

INTRODUCTION

For the aerothermodynamic design of reentry vehicles, e.g. space shuttle, Hermes and hypersonic planes like NASP or Sänger, the prediction of the aerodynamic efficiency and the thermal load is of great importance. Thereby questions arise associated with the physical understanding of high-temperature effects of air e.g. the excitation of internal degrees of freedom of nitrogen and oxygen molecules and chemical reactions like dissociation and ionization. Some of these phenomena are studied by means of a numerical simulation of axisymmetric, laminar, hypersonic flow of air in thermal and chemical equilibrium past blunt bodies, namely a hemisphere and a hyperboloid.

The thin-layer Navier-Stokes equations are solved by a second-order central finite-difference method in conservation law form with added fourth-order numerical damping on a surface oriented grid system. A shock-fitting procedure is employed for the bow-shock. The thermodynamic and transport properties of equilibrium air are computed using two different approaches. The first approach employs the curve-fitting routines of Tannehill et al.. In the second approach the thermodynamic properties of five and nine component models of air are obtained by a procedure which is based on the minimization of the Gibbs-energy.

GOVERNING EQUATIONS

The viscous, three-dimensional flow of a compressible fluid is governed by the unsteady Navier-Stokes equations. To simplify the treatment of the boundaries and to cluster grid points in flow field regions where the dependent variables are expected to undergo rapid changes the arbitrary transformation [1]

$$
\begin{aligned}
\tau &= t && \text{time ,} \\
\xi &= \xi\,(x,y,z,t) && \text{streamwise coordinate ,} \\
\eta &= \eta\,(x,y,z,t) && \text{circumferential coordinate and} \\
\zeta &= \zeta\,(x,y,z,t) && \text{wall normal coordinate}
\end{aligned}
\tag{1}
$$

is chosen. The time-dependent physical domain to be discretized is transformed onto a time-independent rectangular hexahedron. The Cartesian velocity components are retained as dependent variables. The main interest of the current work is the simulation of high Reynolds number laminar hypersonic flows. Thus, all ξ- and η-derivatives of the viscous terms are neglected to obtain the thin-layer Navier-Stokes equations:

$$
\frac{\partial \hat{q}}{\partial \tau} + \frac{\partial \hat{E}}{\partial \xi} + \frac{\partial \hat{F}}{\partial \eta} + \frac{\partial \hat{G}}{\partial \zeta} = \frac{1}{Re_{\infty, L}} \; \frac{\partial \hat{S}}{\partial \zeta} \tag{2}
$$

where

$$
\hat{q} = J^{-1}\,[\rho,\ \rho u,\ \rho v,\ \rho w,\ e]^T\ ,
$$

$$
\left.\begin{matrix} \hat{E} \\ \hat{F} \\ \hat{G} \end{matrix}\right\} = J^{-1} \begin{bmatrix} \rho \;\; W^{(k)} \\ \rho u \;\; W^{(k)} + k_x p \\ \rho v \;\; W^{(k)} + k_y p \\ \rho w \;\; W^{(k)} + k_z p \\ (e+p)W^{(k)} - k_t p \end{bmatrix} \quad \text{for } k = \left\{\begin{matrix} \xi \\ \eta \\ \zeta \end{matrix}\right. ,
$$

$$
\hat{S} = J^{-1} \begin{bmatrix} 0 \\ \mu\, a\, u_\zeta + (\mu/3)\, b\, \zeta_x \\ \mu\, a\, v_\zeta + (\mu/3)\, b\, \zeta_y \\ \mu\, a\, w_\zeta + (\mu/3)\, b\, \zeta_z \\ a \left((\mu/2)\, \dfrac{\partial}{\partial \zeta}\, (u^2+v^2+w^2) + \dfrac{\bar{c}_{p\,\infty}\, \bar{T}_\infty}{Pr_\infty\, \bar{u}_\infty^2}\, \kappa\, \dfrac{\partial T}{\partial \zeta} \right) \\ \qquad + \dfrac{\mu}{3}\, (\zeta_x u + \zeta_y v + \zeta_z w)\, b \end{bmatrix}
$$

with

$$W^{(k)} = k_t + k_x u + k_y v + k_z w \ ,$$

$$a = \zeta_x^2 + \zeta_y^2 + \zeta_z^2 \ ,$$

$$b = \zeta_x u_\zeta + \zeta_y v_\zeta + \zeta_z w_\zeta \ .$$

All quantities are nondimensionalized with their respective freestream values (index "∞"), the pressure p and the total energy e with twice the freestream dynamic head $\overline{\rho}_\infty \, \overline{u}_\infty^2$. The bar indicates dimensional quantities. In the perfect gas case, the pressure and temperature are calculated from

$$p = (\gamma_\infty - 1)\, \rho\, \varepsilon \ ,$$
$$T = \gamma_\infty (\gamma_\infty - 1)\, M_\infty^2\, \varepsilon \ , \tag{3}$$

where γ_∞ is the ratio of specific heats ($\gamma_\infty = 1.4$ for two atomic-gases) and ε is the specific internal energy per unit mass,

$$\varepsilon = \frac{e}{\rho} - \frac{1}{2} (u^2 + v^2 + w^2) \ . \tag{4}$$

The viscosity μ is calculated from the Sutherland law with the constant $S = 110\, K/\overline{T}_\infty$ and the thermal conductivity κ is obtained assuming a constant Prandtl number.

In the equilibrium air case, the thermodynamic and transport properties are obtained by two different approaches. On the one hand, the curve-fitting routines of Tannehill and co-workers [2,3] are used. These curve fits are piecewise continuous approximating polynomials of the thermodynamic data of Bailey [4] and the transport properties of Peng and Pindroh [5]. These calculations are based on a 9 component reacting gas model including molecular, atomar and ionized oxygen, nitrogen and nitric-oxide and free electrons. The curve-fitting routines contain the following relations:

$$p = p(\rho, \varepsilon) \ , \qquad \mu = \mu(\rho, \varepsilon) \ , \qquad c = c(\rho, \varepsilon) \ ,$$
$$T = T(p, \rho) \ , \qquad \kappa = \kappa(\rho, \varepsilon) \ , \tag{5}$$

where c is the equilibrium speed of sound, which is defined by

$$c^2 = \frac{p}{\rho^2} \cdot \left.\frac{\partial p}{\partial \varepsilon}\right|_\rho + \left.\frac{\partial p}{\partial \rho}\right|_\varepsilon \ . \tag{6}$$

The partial derivatives are computed analytically from the curve-fits.

In the second approach, the equilibrium composition, temperature and pressure of a reacting mixture of five and nine components are obtained by a procedure which is based on the minimization of the Gibbs-energy. The five component model contains molecular and atomar nitrogen, oxygen and nitric-oxyde, the nine component model additionally includes the respective ionized particles and free electrons. Consider the following generalized system of m chemical reactions of s components

$$\sum_{i=1}^{s} v_{i,k}^{f} A_i \Leftrightarrow \sum_{i=1}^{s} v_{i,k}^{r} A_i \quad , \quad k = 1 , \ldots, m \ , \tag{7}$$

where $v_{i,k}^{f}$ and $v_{i,k}^{r}$ are the forward and backward stoichiometric coefficients and A_i is the chemical symbol of the i-th component. For a system of s components consisting of l elements, m = s-l independent reactions can be found.

From the law of mass-action and the theory of the irreversible processes the following relation holds at equilibrium [11].

$$\sum_{i=1}^{s} (v_{i,k}^{f} - v_{i,k}^{r}) \hat{\mu}_i = 0 \quad , \quad k = 1 , \ldots, m \ , \tag{8}$$

$$\hat{\mu}_i = \mu_i^o (T) + R \cdot T \ln p_i \ , \tag{9}$$

where $\hat{\mu}_i$ is the chemical potential and μ_i^o the standard chemical potential, which is only a function of the temperature. p_i is the partial pressure of component i.
From these relations the following equilibrium constants are derived [11].

$$K_{p,k}(T) = \exp \left\{ - \frac{\sum_{i=1}^{s} (v_{i,k}^{f} - v_{i,k}^{r}) \cdot \mu_i^o (T)}{R\,T} \right\} = \prod_i \cdot p_i^{(v_{i,k}^{f} - v_{i,k}^{r})} \ , \tag{10}$$

and

$$K_{m,k}(T) = K_{p,k}(T) \left(\frac{R \cdot T}{\rho} \right)^{\sum (v_{i,k}^{f} - v_{i,k}^{r})} = \prod_i \left(\frac{y_i}{m_i} \right)^{(v_{i,k}^{f} - v_{i,k}^{r})} \ , \tag{11}$$

$K_{p,k}$ and $K_{m,k}$ are the equilibrium constants referred to the partial pressure and the mass fractions y_i, respectively. The chemical potential can be expressed in terms of the Gibbs-energy

$$\mu_i^o = G_i^o - RT \ln p_o \ , \tag{12}$$

where p_o is the pressure at standard state. The Gibbs-energy of every component has to be computed from the statistical mechanics in terms of the partition functions. Equations (11) constitute a system of m nonlinear equations for s unknown mass fractions. The remaining l equations, which are necessary to solve this system, can be obtained from the constraint, that the mass of every element in the system is conserved,

$$\sum_{i=1}^{s} y_i \cdot a_{ij} \cdot m_j / m_i = z_j \quad ; \quad j = 1, l \ . \tag{13}$$

Here a_{ij} is the number of particles j in component i, m_i the molar mass of component i and z_j the mass fraction of element j in the mixture (for air: $z_N = 0.76$, $z_O = 0.34$, $z_{e^-} = 0$).

This system of equations is solved for the s mass fractions in terms of temperature and density by a Newton-Raphson method. If the independent variables in the iteration procedure are density and specific internal energy ε, then the equation

$$0 = \varepsilon - \sum_{i=1}^{s} y_i \, \varepsilon_i (T) \tag{14}$$

closes the system of nonlinear algebraic equations with s+1 unknowns y_i and T . The specific internal energy of every component has to be computed from the partition functions, where the particles are treated as rigid rotators and harmonic oszillators. The global pressure can be obtained from the relation

$$p = R \cdot \rho \cdot T \sum \frac{y_i}{m_i} \quad . \tag{15}$$

The derivatives of the pressure with respect to density and internal energy, needed to calculate the equilibrium speed of sound, are obtained after the Newton-iteration by means of the implicit function theorem.

The transport properties of the single species were calculated according to the rigorous theory [6]. The molecular interactions are described by the (6-12) Lennard-Jones potential. The viscosity and the thermal conductivity of the mixture are calculated using appropriate mixing rules, namely Wilke's rule.

NUMERICAL METHOD AND BOUNDARY CONDITIONS

The present conservative finite-difference scheme employs a hybrid evaluation of the time derivative to first order. The formulation is implicit in the wall normal direction to remove the stiffness caused by the small mesh increments in this direction and explicit in the streamwise directions [1]. Fourth-order damping terms are employed for stability reasons.

The physical domain considered here is bounded by the body, the bow-shock and the outflow surface. For axisymmetric flow simulations, only three planes in the circumferential direction are used which are normal to the body and separated by a constant angle. The calculations are performed in the central plane, while the two neighbouring planes enter for the axisymmetric flux evaluation in the 3-D code. At the wall, the no-slip condition and the boundary layer approximation of a zero wall-normal pressure gradient are employed. The wall is assumed to be isothermal. At the outflow, the variables are extrapolated because of the supersonic flow in this region. At the inflow boundary the shock-fitting technique is employed, see e.g. [7,8]. The shock location is determined after each time step from the Rankine-Hugoniot relations of a moving shock and a compatibility relation which describes the signal propagation from the post-shock region onto the shock. This compatibility relation is obtained by a transformation of the Euler equation onto the characteristic direction, defined by the eigenvector of $\partial\hat{G}/\partial\hat{q}$ corresponding to the only positive eigenvalue $W^{(\zeta)} + c\sqrt{a}$ of $\partial\hat{G}/\partial\hat{q}$ behind the bow shock. The grid in the present calculations has a singular axis, which coincides with the stagnation stream line. There, the conservative variables are interpolated [1].

RESULTS

The freestream conditions for the flow about the hemisphere were chosen according to the Navier-Stokes simulation of Vigneron [9], (Case I: $M_\infty = 11.26$, $Re_{\infty,R} = 4984$, $\bar{T}_{wall} = 1000\,K$, $\bar{T}_\infty = 182.3\,K$, $\bar{\rho}_\infty = 3.9 \cdot 10^{-5}\,kg/m^3$) and the experimental investigation of Miller III [10] (Case II: $M_\infty = 7.72$, $\bar{\rho}_\infty = 7.18 \cdot 10^{-3}\,kg/m^3$, $\bar{T}_\infty = 1255\,K$, $Re_\infty/m = 7.9\,10^5\,/m$, $D_{sphere} = 6.35$ cm , $\bar{T}_{wall} = 290\,K$). Figure 1 shows the grid and temperature isolines for the perfect gas and equilibrium air simulation using the curve-fitting routines on a grid with 51 x 43 (normal x streamwise) points. Figure 2 shows, for comparison with the simulation of Vigneron, the temperature and density along the stagnation streamline for a perfect gas and equilibrium air. Here the same grid dimensions (21 x 43) and grid clustering as in the Vigneron simulation was chosen. Due to the energy consumption of the dissociation reactions ahead of the bow shock, the temperatures behind the shock are lower than in the perfect gas simulation. Since the pressure is not much affected by the real gas effects, the density increases whereas the shock-stand of distance is reduced. Using this coarse grid, the gradients near the wall are not resolved sufficiently; the heat flux distribution shown in figure 3 was computed on the finer grid. Although the temperature in the entire flow field is lower in the equilibrium flow case, the heat flux has the same magnitude compared to the perfect gas. This effect is caused by an heat release near the "cold" wall, because there the recombination reactions predominate over the dissociation. A quantitative analysis of the composition along the stagnation streamline (figure 4), which is based on a computation using the Gibbs-energy minimization procedure, shows, that the degree of dissociation approaches zero near the wall. Figure 5 shows the comparison of the predicted and measured stand-off distance for case II for equilibrium air. In general, good agreement is obtained; the deviations in the streamwise direction are probably caused by nonequilibrium effects in the experiment.

The flow about the hyperboloid was simulated for the freestream conditions $M_\infty = 25.0$, $Re_{\infty,R} = 7654$, $R = 0.44\,m$, $\bar{T}_\infty = 192\,K$, $\bar{\rho}_\infty = 3.2 \cdot 10^{-5}\,kg/m^3$ and $\bar{T}_{wall} = 800\,K$. This flow corresponds to that one along the windward symmetry plane of a generic HERMES vehicles at an angle of attack of 30° and for an altitude of 77 km. Figure 6 shows the corresponding temperature distribution along the stagnation streamline for the perfect gas and equilibrium air using the Gibbs-minimization technique. No ionization was observed in the entire flow field. The same effects as in the previous flow cases can be observed, but due to the higher temperatures dissociation of nitrogen occurs (figure 7).

Figure 8 shows the heat-flux distribution along the body surface on the hyperboloid for the perfect gas and equilibrium air using the curve fits and the Gibbs-technique. Both equilibrium air procedures lead to the same results. Near the stagnation point numerical problems occur in the equilibrium flow case, leading to oscillations or kinks in the heat-flux distributions in this region. They are caused by stronger gradients in the streamwise direction which are not resolved sufficiently with the present grid.

CONCLUSIONS

The thin-layer Navier-Stokes equations are solved with a hybrid finite-difference method for axisymmetric laminar hypersonic flow past blunt bodies. Solutions employing perfect gas relations and equilibrium air chemistry are compared. The temperature and density distributions in the flowfield show significant differences. Note, however, that the surface heatflux has nearly the same magnitude for both models. Two different procedures for the calculation of equilibrium air properties (curve fit, Gibbs-energy minimization) lead to same results. Using the Gibbs approach further informations about the physics of the flow field, namely the reactions and the composition, can readily be obtained. Some applications has been made to the flow past the windward symmetry plane of a generic HERMES vehicle at an altitude of 77 km.

ACKNOWLEDGEMENT

Part of this work has been supported by a HERMES R & D contract of AMD-BA.

REFERENCES

[1] RIEDELBAUCH, S., MÜLLER, B.: "The Simulation of Three-Dimensional Viscous Supersonic Flow Past Blunt Bodies with a Hybrid Implicit Explicit Finite-Difference Method", DFVLR-FB 87-32 (1987).

[2] TANNEHILL, J.C., SRINIVASAN, S., WEILMUENSTER, K.J.: "Simplified Curve Fits for the Thermodynamic Properties of Equilibrium Air", NASA RP 1181 (1987).

[3] SRINIVASAN, S., TANNEHILL, J.C., WEILMUENSTER, K.J.: "Simplified Curve Fits for the Transport Properties of Equilibrium Air", Final Report, NASA CR 180422 (1987).

[4] BAILEY, H.E.: "Programs for computing equilibrium Thermodynamic properties of gases", NASA TN D-3921, April 1967.

[5] PENG, T., PINDROH, A.L.: "An Improved Calculation of Gas Properties at High Temperatures Air", D-11722, The Boeing Company, 1962.

[6] HIRSCHFELDER, J., CURTIS, C., BIRD, B.: "Molecular Theory of Gases", John Wiley & Sons, New York, 1954.

[7] PFITZNER, M., WEILAND, C.: "3-D Euler Solution for Hypersonic Mach Numbers", AGARD Conference Proceeding, AGARD-CP 428 (1987).

[8] BRENNER, G., RIEDELBAUCH, S., MÜLLER, B.: "The Computation of Axisymmetric Hypersonic Air Flow in Chemical Equilibrium. Part I: The Use of Curve-Fitting Routines", DLR-IB 221-89 A10, 1989.

[9] VIGNERON, Y.C.: "Hypersonic Flow of Equilibrium Air around Blunt Bodies". Thesis, Iowa State University, 1976.

[10] MILLER, III, C.G.: "Shock Shapes on Blunt Bodies in Hypersonic-Hypervelocity Helium, Air, and CO_2 Flows, and Calibration Results in Langley 6-inch Expansion Tube". NASA TN-D-7800, 1975.

[11] VINCENTI, W.G., KRUGER, C.H.: "Introduction to Physical Gas Dynamic". J. Wiley, New York, 1965.

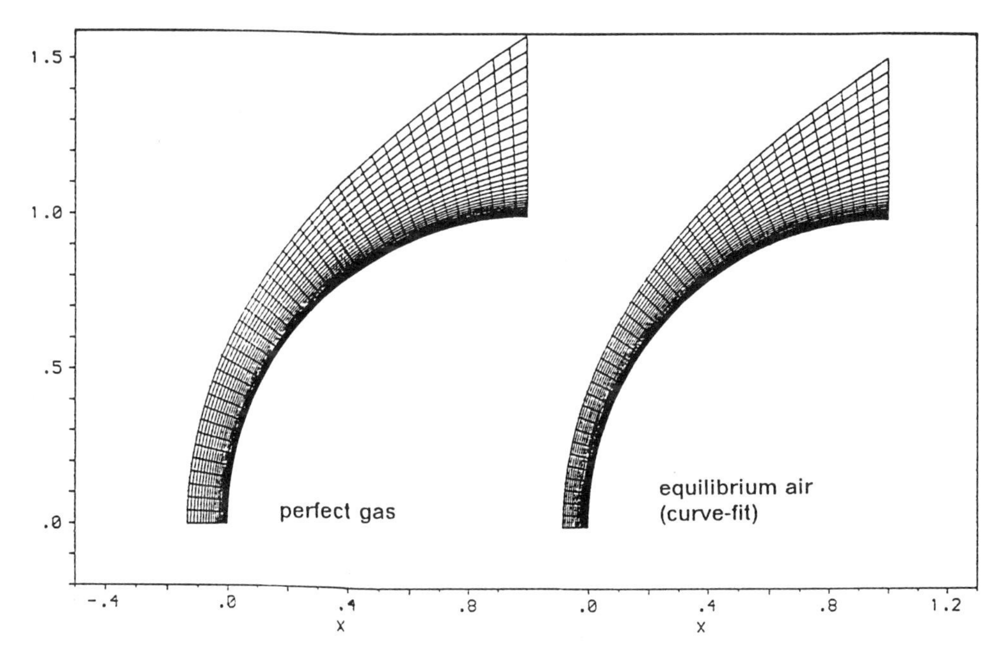

Fig. 1a Computational grid (converged solution) for hemisphere

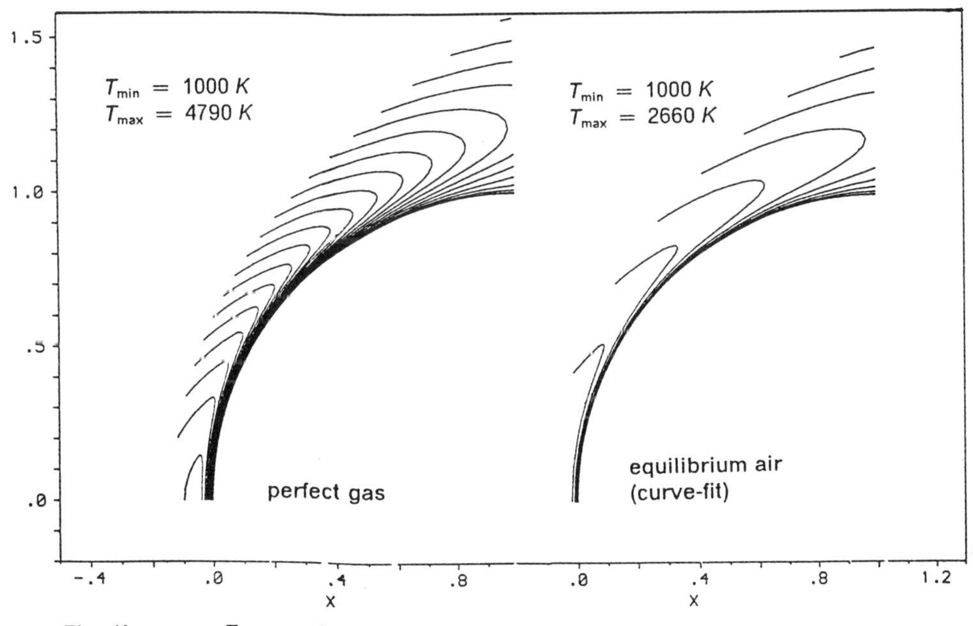

Fig. 1b Temperature contours

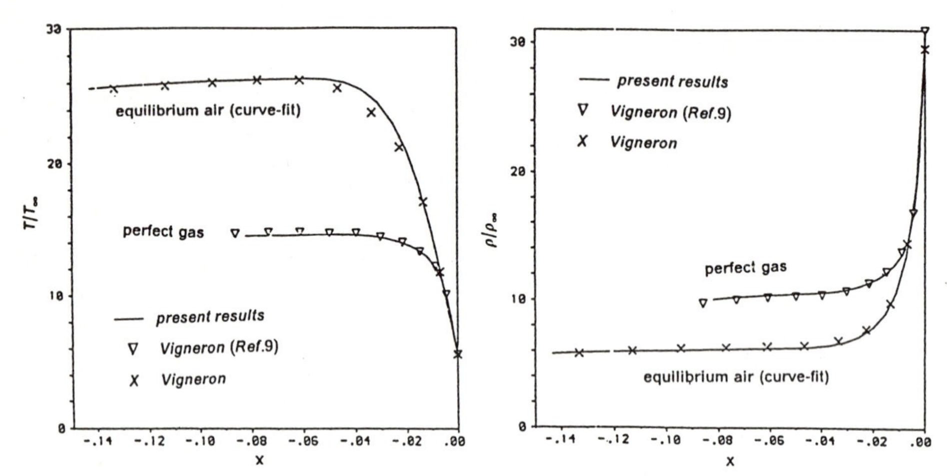

Fig. 2 Temperature and density along the stagnation streamline for hemisphere

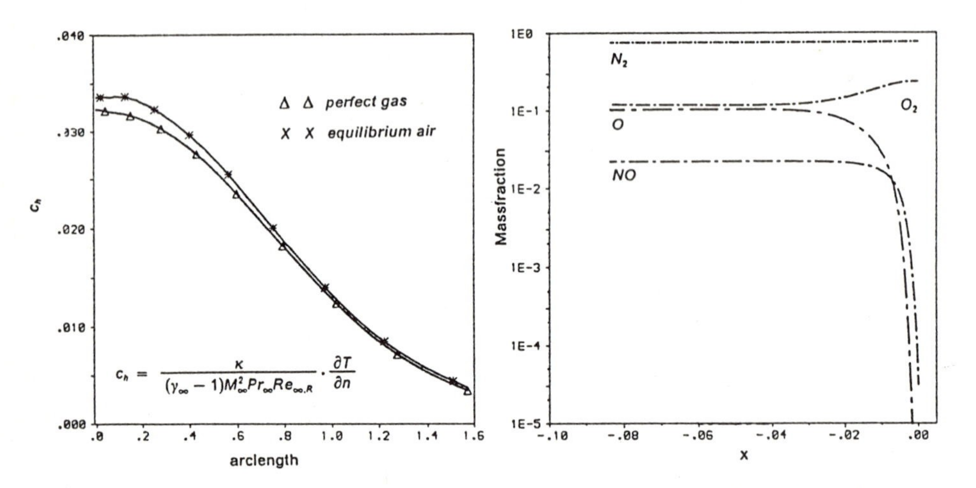

Fig. 3 Heatfluxdistribution along bodysurface

Fig. 4 Composition of air along the stagnation streamline

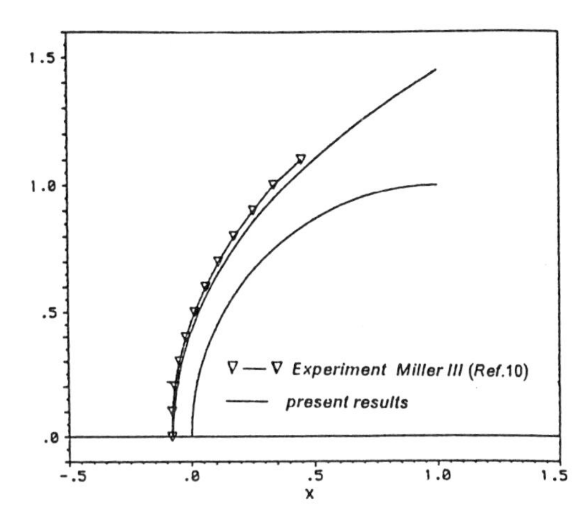

Fig. 5 Comparison of predicted and measured shock stand-off distance

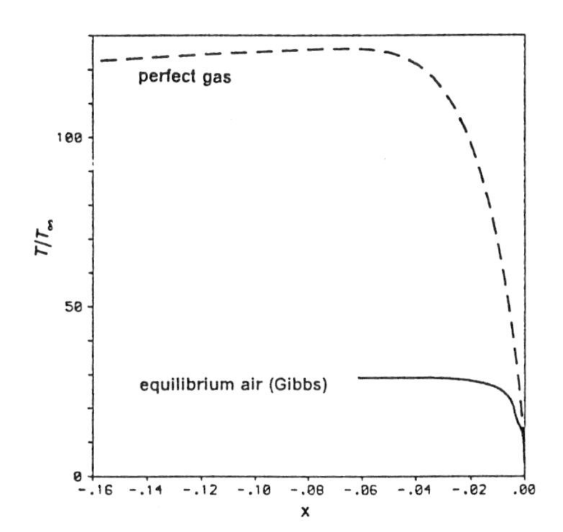

Fig. 6 Temperature along the stagnation streamline of the hyperboloid

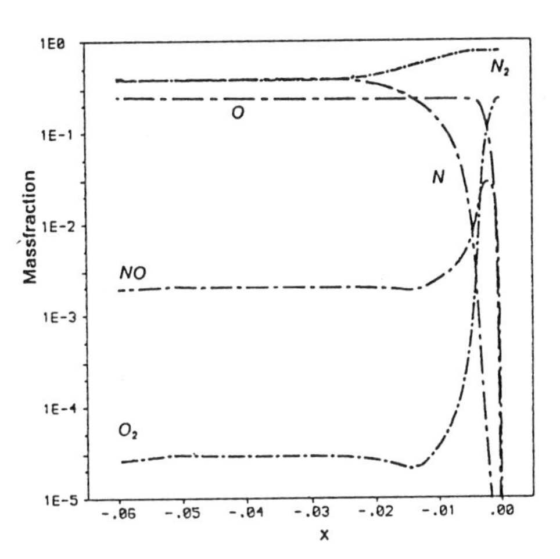

Fig. 7 Composition along the stagnation streamline of the hyperboloid

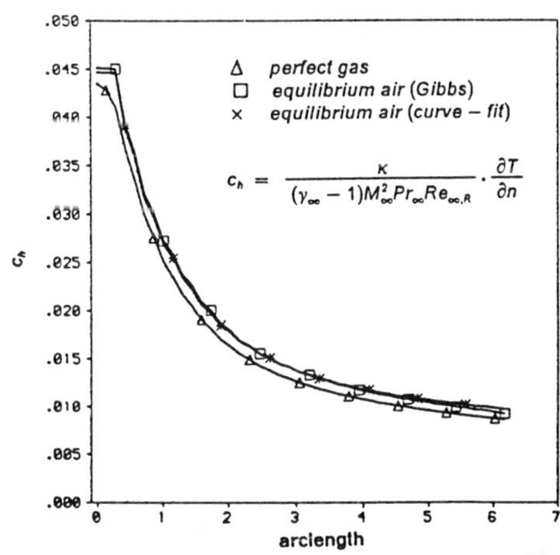

Fig. 8 Heatfluxdistribution along the body surface of the hyperboloid

SOLUTION OF THE 3-D, INCOMPRESSIBLE NAVIER-STOKES EQUATIONS FOR THE SIMULATION OF VORTEX BREAKDOWN

M. Breuer, D. Hänel
Aerodynamisches Institut, RWTH Aachen
Wüllnerstr. zw. 5 und 7, 5100 Aachen, W.-Germany

SUMMARY

A method of solution is presented for the threedimensional, unsteady Navier - Stokes equations of an incompressible fluid. The method is based on the principle of artificial compressibility, extended to time-dependent solutions by the concept of dual time stepping. Higher order upwinding (Quick interpolation, and Roe-type splitting) is used for the spatial discretization. The properties of the method were tested by means of different validated flow problems. Present computations concern with the unsteady, and three-dimensional problem of the breakdown of an isolated vortex. The temporal evolution and the internal structure of the breakdown bubble is investigated and compared with experimental observations.

INTRODUCTION

Typical problems, where vortex breakdown has to be considered are the flow around a delta wing, in a combustion chamber, and in a vortex tube. Vortex breakdown characterizes an abrupt change in the structure of the core of a swirling flow. Its occurrence is often marked by the presence of a free stagnation point on the axis of an axial vortex flow combined with flow reversal. Basically, three types of vortex breakdown are observed, mild (double helix) breakdown, spiral-type breakdown and nearly axisymmetric bubble-type breakdown.

The phenomenon of vortex breakdown was studied in many theoretical and experimental investigations, but no general agreement in the explanation of its fundamental nature could be achieved. Numerical investigations were made under the assumption of axisymmetric flow e.g. [1] , [2] , [3] , [4], which however seems to be to restrictive, and only applicable for bubble-type vortex breakdown. Threedimensional investigations are conducted by Menne [5] with a Fourier decomposition, by Spall & Gatski [6] for the bubble-type and by Leibovich & Ma [7] for the spiral-type of breakdown.

The present study is concerned with the fully threedimensional breakdown process, likewise. In order to concentrate to pure vortical flow, and to reduce the computational work an isolated vortex is considered. The starting conditions of the vortex were predicted by a rotational vortex core embedded in an irrotational stream. The temporal evolution and the internal structure of vortex breakdown flow field is investigated and compared with experimental observations.

The solution method for the unsteady incompressible Navier-Stokes equations in primitive variables is based on the principle of artificial compressibilty, originally proposed by

Chorin [8] for solving steady state problems. By adding an artificial time derivative to the conservation laws the method of artificial compressibility can be extended to unsteady fluid motions (dual time stepping). Using two different higher order upwind discretization schemes (Quick interpolation [9], [10] and flux difference splitting of Roe [11]) for the Euler terms and central differencing for the viscous parts combined with a matrix splitting of the flux Jacobians the resulting system of equations is solved by an implicit Line-Gauss-Seidel relaxation method in each physical time step. The method of solution was tested for several flow problems such as flat plate boundary layer, driven cavity, Couette flow, and flow around a circular cylinder.

GOVERNING EQUATIONS

The vortex motion is described by the Navier - Stokes equations. With the assumption of laminar and incompressible flow the 3-D unsteady equations in a Cartesian coordinate system (x,y,z,t) read :

$$\bar{R} * Q_t + F_x + G_y + H_z = Re^{-1} * (R_x + S_y + T_z) \tag{1}$$

with

$$Q = \begin{pmatrix} p \\ u \\ v \\ w \end{pmatrix} \quad F = \begin{pmatrix} u \\ u^2 + p \\ uv \\ uw \end{pmatrix} \quad G = \begin{pmatrix} v \\ vu \\ v^2 + p \\ vw \end{pmatrix} \quad H = \begin{pmatrix} w \\ wu \\ wv \\ w^2 + p \end{pmatrix}$$

$$\bar{R} = \begin{pmatrix} 0 & 0 & 0 & 0 \\ 0 & 1 & 0 & 0 \\ 0 & 0 & 1 & 0 \\ 0 & 0 & 0 & 1 \end{pmatrix} \quad R = \begin{pmatrix} 0 \\ u_x \\ v_x \\ w_x \end{pmatrix} \quad S = \begin{pmatrix} 0 \\ u_y \\ v_y \\ w_y \end{pmatrix} \quad T = \begin{pmatrix} 0 \\ u_z \\ v_z \\ w_z \end{pmatrix}.$$

Herein Q is the vector of the primitive variables , F,G and H are the Euler fluxes and R,S and T are the viscous terms. $\bar{R}$ corresponds to the matrix of the time derivatives.

METHOD OF SOLUTION

The properties of the solutions of the conservation equations for strictly incompressible flow are different to that of compressible flow. Numerically, major differences are caused by the lack of the time derivative of the density in the continuity equation which results in a singular coefficient matrix $\bar{R}$ for the time derivatives. The methods of solution differ mainly in the way to overcome this difficulty. For threedimensional flows the methods of solution can generally be grouped into three types. These include methods based on pressure iteration to satisfy the continuity equation, methods using the vorticity-velocity formulation, and methods based upon artificial compressibility.
The concept of artificial compressibility, successfully used for steady-state computations, removes the matrix singularity by adding an artificial time derivative to the continuity

equation. This renders the incompressible Euler equations totally hyperbolic which allows to utilize the highly developed algorithms of compressible flow to incompressible problems. An extension of this concept to unsteady flows can be carried out by dual time stepping. Based on the dual time stepping an algorithm was developed, where in the pseudo-time the artificial compressibility and approximate Rieman solvers are used to solve the incompressible Navier-Stokes equations for each level of the physical time. The solution for each time step is carried out by an implicit relaxation method. Particular advantages are the flexibility of implementing different accurate and stable schemes for incompressible flow at high Reynolds numbers, and the straightforward extension to 3-D curvilinear coordinates.
This concept is the basis of the present solution method. In the following the original principle of artificial compressibilty, as suggested by Chorin, and the extension of this method to the solutions of the unsteady Navier- Stokes equations will explained briefly.
The basic idea of the artificial compressibility method was the evolution of a solution method for the steady incompressible equations. The introduction of an artificial compressibility term into the original continuity equation

$$\nabla \cdot \vec{v} = 0 \tag{2}$$

yields the following form of the altered continuity equation :

$$\frac{\partial p}{\partial t} + \beta^2 \ \nabla \cdot \vec{v} = 0 \tag{3}$$

where β is an artificial speed of sound which represents an ' artificial compressibility ' according to an equation of state $p = \beta^2 \cdot \rho$, similar to compressible flows. This altered continuity equation (3) has the advantage to form a fully hyperbolic system of partial differential equations together with the inviscid momentum equations. However the formulation has the disadvantage, that only at steady state the flow field is divergence free. The extension of the method to unsteady flows is possible by adding artificial time derivatives to all equations (dual time stepping). If applying the artificial compressibility only to the pseudo-time the physical time behaviour is not influenced by that.
Then the system of equations reads:

$$\widehat{R} * Q_\tau + \bar{R} * Q_t + F_x + G_y + H_z = Re^{-1} * (R_x + S_y + T_z) \tag{4}$$

$$\bar{R} = \begin{pmatrix} 0 & 0 & 0 & 0 \\ 0 & 1 & 0 & 0 \\ 0 & 0 & 1 & 0 \\ 0 & 0 & 0 & 1 \end{pmatrix} \qquad \widehat{R} = \begin{pmatrix} \frac{1}{\beta^2} & 0 & 0 & 0 \\ 0 & 1 & 0 & 0 \\ 0 & 0 & 1 & 0 \\ 0 & 0 & 0 & 1 \end{pmatrix} .$$

The difference between eq.(1) and eq.(4) is the additional term $\widehat{R} * Q_\tau$. In contrast to the matrix $\bar{R}$ the matrix $\widehat{R}$ is not singular. If the term Q_τ converges to zero, the original equation (1) remains.
For the discretization in the artificial time (index ν) a formulation of first order accuracy can be used, whereas for the physical time (index n) a second order accurate formulation in time is chosen.

$$Q_\tau = \frac{Q^{n+1,\nu+1} - Q^{n+1,\nu}}{\Delta\tau} = \frac{\Delta Q^\nu}{\Delta\tau} \tag{5}$$

$$Q_t = \frac{1.5 * Q^{n+1,\nu+1} - 2. * Q^n + 0.5 * Q^{n-1}}{\Delta t} \quad . \tag{6}$$

The Euler and the viscous fluxes are developed in Taylor series with respect to the pseudo-time:

$$F^{n+1,\nu+1} = F^{n+1,\nu} + A^{n+1,\nu} * \Delta Q^\nu \tag{7}$$

$$A = \frac{\partial F}{\partial Q} \quad .$$

Herein A is the Jacobian matrix of the flux F. If these approximations are inserted into equation (4) the resulting quasi-discrete delta-form (for inviscid flow) reads :

$$\left(\frac{\hat{R}}{\Delta\tau} + \frac{1.5 * \bar{R}}{\Delta t} + \left(\frac{\partial}{\partial x} A + \frac{\partial}{\partial y} B + \frac{\partial}{\partial z} C \right)^{n+1,\nu} \right) * \Delta Q^\nu =$$

$$- \left(\frac{1.5 * Q^{n+1,\nu} - 2. * Q^n + 0.5 * Q^{n-1}}{\Delta t} + (F_x + G_y * H_z)^{n+1,\nu} \right) . \tag{8}$$

The left handside of this equation, called the numerical part of the equation, consists mainly of the Jacobian matrices. On the right handside one can recognize the basic equation (1), which describes the physical part of the equation.
The spatial discretization is different for the left and the right handside of eq.(8). The right handside is discretized by either the ' Quadratic Upstream Interpolation for Convective Kinematics ' (QUICK) as proposed by Leonard [9],[10], or by a simplified flux-difference splitting of Roe. The Quick scheme is an upwind discretization of second order accuracy. For the Roe splitting we achieve a second order formulation by using the MUSCL - approach. The simplification of the splitting is given by the averaging of the variables at the cell interface with constant density in the flow field. For the viscous terms central differences of second order accuracy are used.
The goal for the discretization of the left handside is to achieve a diagonal-dominant solution matrix, which guarantees the stability of the relaxation method. An upwind formulation of first order combined with a matrix splitting technique is used. For that reason the Jacobian matrices are diagonalized by an appropiate similarity transformation:

$$A = M \Lambda M^{-1} \tag{9}$$

where M is the transformation matrix and Λ the diagonal matrix :

$$\Lambda = \begin{pmatrix} \lambda_1 & 0 & 0 & 0 \\ 0 & \lambda_2 & 0 & 0 \\ 0 & 0 & \lambda_3 & 0 \\ 0 & 0 & 0 & \lambda_4 \end{pmatrix} .$$

The diagonal matrix Λ can be split into parts with positive and negative eigenvalues, Λ^+ and Λ^-:

$$\Lambda = \Lambda^+ + \Lambda^- \tag{10}$$

The eigenvalues of $\Lambda^\pm$ are defined as :

$$\lambda^\pm = \frac{\lambda \pm |\lambda|}{2} \tag{11}$$

$$\lambda_{1,2} = u \qquad \lambda_3 = u + c \qquad \lambda_4 = u - c \qquad c = \sqrt{u^2 + \beta^2}\,.$$

Back transformation allows to split the matrix A in the matrices A^+ and A^- :

$$A = M\Lambda^+ M^{-1} + M\Lambda^- M^{-1} = A^+ + A^- \,. \tag{12}$$

The matrices $A^\pm$ are updated with values extrapolated from the left and right, respect..

Now the solution method can be summarized as follows. The resulting system of equations takes the form :

$$LHS * \Delta Q^\nu = RHS \,. \tag{13}$$

The left handside of this equation describes the numerical part. Here a first order upwind discretization with a matrix splitting technique is applied. The right handside of eq.(13) consists of the physical part, the discrete unsteady incompressible Navier-Stokes equations. Its discretization is of second order accuracy in the physical time and in the space.
The solution of this system of equations is achieved by an implicit Line-Gauss-Seidel relaxation method in each physical time step until the condition is reached :

$$\max(RHS) \leq \varepsilon \qquad with \qquad \varepsilon = O(0.01 \cdot \Delta x^2)\,. \tag{14}$$

Altough it is an implicit method, the algorithm is highly vectorized.

The properties of the method, in particular convergence and accuracy, were tested by means of validated flow problems. For example the spatial accuracy was checked by means of flat plate boundary layer calculations. Both the Quick interpolation as well the Roe splitting have compared well with the exact solution of Blasius. The flux-difference splitting has shown a slightly stronger dissipation. Other test cases have confirmed the same tendency. The temporal accuracy was verified for time-dependent problems like the temporal evolution of the separation zone behind an impulsively started circular cylinder. These results (not shown here) for both discretizations were found to be in close agreement with experimental investigations.

INITIAL- AND BOUNDARY CONDITIONS

To study the development of an isolated vortex a straight vortex embedded in an inviscid irrotational flow was prescribed as the initial condition. The circumferential velocity distribution, assumed to be independent of the axial coordinate z, is given by a polynomial in the radial coordinate r :

$$v_\theta(r) = \widehat{\beta} * g(r) \qquad g(r) = \begin{cases} r(2-r^2) & for \quad r \leq 1 \\ \frac{1}{r} & for \quad r > 1 \end{cases} . \tag{15}$$

The swirl parameter $\widehat{\beta}$ is set to 0.68 in all calculations. This distribution corresponds to solid body rotation in the core and to a potential vortex outside.

The initial field of the axial velocity is given by :

$$w(z) = \begin{cases} w_0 + bz + cz^2 + dz^3 & for \quad z \leq z_1 \\ 1 + e * (z - z_1) & for \quad z > z_1 \end{cases} \tag{16}$$

where $w(z)$ is assumed to be constant in the radial direction. The field of the initial pressure can be computed from the Poisson equation for the pressure.
The choice of the axial velocity distribution, eq.(16) , is motivated by the fact that the breakdown bubble tends to move upstream to the inflow boundary, as observed e.g. in [2], [6], [12]. The polynomial in eq.(16) represents a slight acceleration to stabilize the flow near the inflow boundary. In the further downstream part, $z > z_1 = 4$, the flow is deccelerated with a constant velocity gradient e in order to provoke breakdown by an imposed positive pressure gradient in the axial direction.
To control the evolution in time the initial axial distribution of $w(z)$ (eq.(16)), and the corresponding pressure is hold fixed at the lateral boundaries. Furtheron v_θ is taken to be constant, and the radial velocity is computed from continuity equation.
At the inflow boundary the axial and circumferential velocities are prescribed according to eq.(15) and (16), the axial gradient of the radial velocity $\frac{\partial v_r}{\partial z}$ is set to zero, and the pressure is calculated from the axial momentum equation. At the outflow boundary the velocity components are extrapolated from the interior, the pressure is determined from axial momentum equation.

RESULTS

The method of solution for the unsteady incompressible Navier - Stokes equations is used to study the 3-D problem of vortex breakdown. Two different Reynolds numbers are investigated, where the Reynolds number is based on the outer axial velocity at the plane $z = z_1$ and the radius of the vortex core. A Cartesian grid with $41 \times 41 \times 60$ grid points is used, where the computational domain is bounded by $-5 \leq x \leq 5$, $-5 \leq y \leq 5$ and $0 \leq z \leq 18$.
First the temporal evolution of the vortex breakdown process is examined. Figure 1 shows the streaklines of four different time levels for a Reynolds number of 200. In the first time period $T < 60$ (not shown here) the streaklines, starting at the inflow section, have a slightly diverging structure because of the deccelerated outer flow field. For the time level $T = 60$ first changes in the structure of the vortex filaments can be observed. In the middle of the computational domain a swelling of the vortex core takes place on the axis of the vortex, combined with the development of a free stagnation point. The breakdown bubble evolves from the swelling of the vortex core. For $T = 72$ a first small bubble can

be observed, which grows with the time ($T = 84$), moves upstream and reaches the inflow section ($T = 99$). In several experimental investigations the breakdown process was studied in a vortex tube. Sarpkaya [13] found, that the axisymmetric breakdown evolves either from a double helix, or from a spiral, or directly from an axisymmetric swelling of the vortex core. The mode of evolution depends on the Reynolds and circulation number, whereas for sufficiently high values only the last form is observed. Escudier [14] documented the temporal evolution of the breakdown process by a series of photographs, showing the swelling of the core, the formation of a first small breakdown bubble, which dimensions increase in time and the movement of the bubble towards the inflow section. Altough the conditions of the numerical and the experimental investigations are not identical, the results are qualitatively in close agreement but quantitative comparisons are not meaningful.

Of particular interest is the structure of the internal breakdown region, e.g. the vorticity distribution, which is virtually unobtainable by experimental measurements. In Figure 2 the distribution of the vorticity component ω_x normal to the y-z plane ($x = 0$) is sketched for the same four time levels as in Fig.1. For $T = 60$ the breakdown regions consists of a single vortex ring. With increasing time the ring moves upstream while its strength grows. For $T = 99$ the formation of a second counter rotating vortex ring, which lies inside the first ring, is indicated by the lines of constant vorticity ω_x. How the internal structure with two counter rotating rings is comparable to the experimental observations of Leibovich [15], who found a double ring structure, is difficult to say, since the bubble has already reached the inflow section at this time and the results must be interpreted with caution.

Fig. 3 and 4 shows the streamlines and the vorticity distribution of the calculated vortex breakdown for a higher Reynolds number of 2000 at the time $T = 84$. The evolution of the bubble is similar to the lower Reynolds number case. Two counter rotating vortex rings appear. It is very conspicuous, that small asymmetries can be observed, which increases in time. The breakdown bubble however reaches the inflow section and it makes no sense to continue the simulation.

This problem of upstream movement is documented in numerous works [2], [4], [12]. By the reduction of the imposed pressure gradient on the outer flow and by special lateral boundary conditions, which will be discussed in more details in a later paper, it is possible to stabilize the location of the breakdown region. Fig. 5 shows lines of constant vorticity ω_x in the y-z plane for four different time levels (Re = 200) at larger times. The solution is highly asymmetric. The flow field consists of a number of vortex rings, which are located staggered. The vortex rings develop in the front section alternating in the upper and the lower part. Then they move with the flow to the outflow section and leave the integration domain. The flow field remains unsteady and asymmetric all the time. In Fig. 6 a typical picture of the streaklines is sketched. The breakdown bubble is vanished and a structure appears, which looks like a spiral-type of breakdown. The sense of the spiral is opposite to that of the basic flow, whereas the spiral itsself precesses with the flow.

CONCLUSIONS

A method of solution for the threedimensional, unsteady Navier - Stokes equations of an incompressible, laminar flow has been developed. The algorithm is based on the principle

of artificial compressibility but extended to unsteady flows by a dual time stepping procedure. Second order accuracy in time and space is achieved by employing two different upwinding schemes and a three point backward time derivative. Particular advantages are the use of primitive variables which allows the straightforward extension to 3-D curvilinear coordinates and the flexibility of implementing different higher order stable and accurate Euler schemes. Although the numerical method is implicit in the pseudo-time, the code is highly vectorized. The method was tested at several validated flow problems and the accuracy was found to be quite well.
In the second part the breakdown of an isolated vortex is investigated. The present computations concern with the fully threedimensional breakdown process. The temporal evolution of the breakdown process and the internal structure of the flow field are examined and the results are found to be in qualitative agreement with experimental studies.
Nearly axisymmetric bubble-type solutions are achieved as well as highly asymmetric structures, which are similar to the experimentally observed spiral-type breakdown. The mode of breakdown seems to depend on the imposed pressure gradient.

References

[1] Grabowski, W.J. , Berger, S.A. : *Solutions of the Navier-Stokes equations for vortex breakdown,* J. Fluid Mech., vol.75, part 3, pp.525-544, (1976).

[2] Menne, S. : *Rotationssymmetrische Wirbel in achsparalleler Strömung,* Diss. RWTH Aachen, (1986).

[3] Hafez, M. , Ahmad, J. ,Kuruvila, G. , Salas, M.D. : *Vortex breakdown simulation,* Part I, AIAA Paper 87-1343., (1987).

[4] Menne, S. : *Vortex breakdown in an axisymmetric flow,* AIAA 26th Aerospace Sciences Meeting, Reno, Nevada, Jan.11-14, (1988).

[5] Menne, S. : *Simulation of vortex breakdown in tubes,* AIAA-Paper 88-3575, 1st National Fluid Dynamics Congress, Cincinatti, Ohio, July 25-28, (1988).

[6] Spall, R.E. , Gatski, T.B. : *A Numerical Simulation of Vortex breakdown,* Forum on Unst. Flow Sep., ASME Fluids Eng. Conf., June 1987.

[7] Leibovich, S. , Ma, H.Y. : Phys. Fluids, vol. 26,3173 (1983).

[8] Chorin, A.J. : J. Comput. Phys. 2 , 12-26, (1967).

[9] Leonard, B.P. : Proceedings of the 2nd National Sym. on Numerical Properties and Methologies in Heat Transfer, Univ. of Maryland, 211, (1983).

[10] Leonard, B.P. : Proceedings of the 1983 Int. Conf. on Computational Techniques and Applications, Univ. of Sydney, Australia, 106 , (1984).

[11] Roe, P.L. : *Approximate Rieman solvers, parameter vectors and difference schemes,* J. Comp. Phys., vol 22, pp 357, (1981).

[12] Spall, R.E. , Gatski,T.B. , Grosch C.E. : *A Criterion for vortex breakdown,* Phys. Fluids 30, (11), (Nov. 1987).

[13] Sarpkaya, T. : *On stationary and travelling vortex breakdowns,* J. Fluid Mech. , vol.45, part 3,pp.545-559, (1971).

[14] Escudier, M. : *Vortex breakdown: Observation and Explanations,* Progress in Aerospace Sciences, vol.25, No.2, pp.189-229, (1988).

[15] Leibovich, S. : *The structure of vortex breakdown* , Ann. Rev. of Fluid Mechanics, vol.10, pp.221-246, (1978).

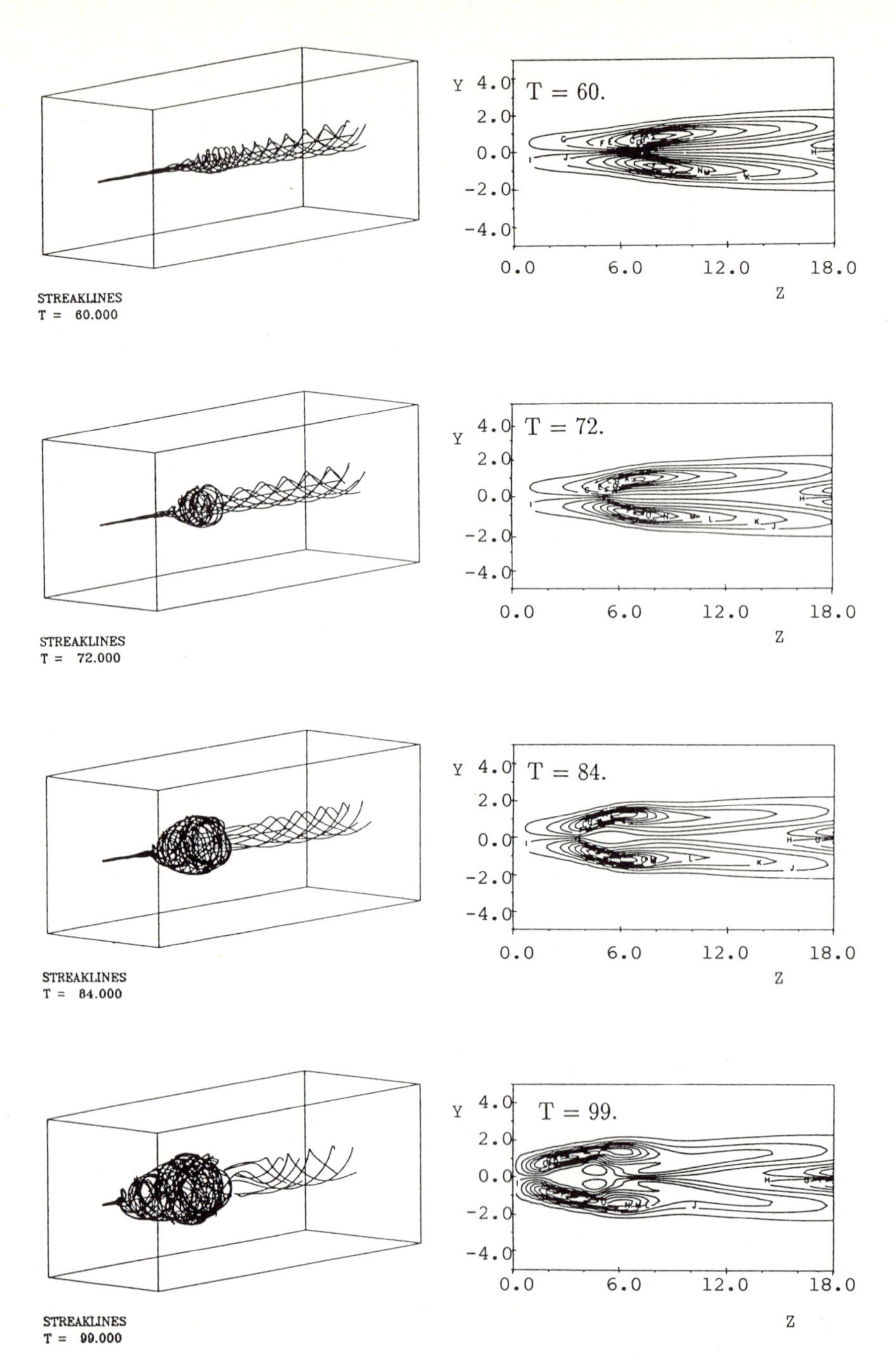

Fig. 1 Streaklines of four different time levels, Re=200., $\hat{\beta} = 0.68$

Fig. 2 Lines of constant vorticity ω_x in the y-z plane for x=0., Re=200., $\hat{\beta} = 0.68$

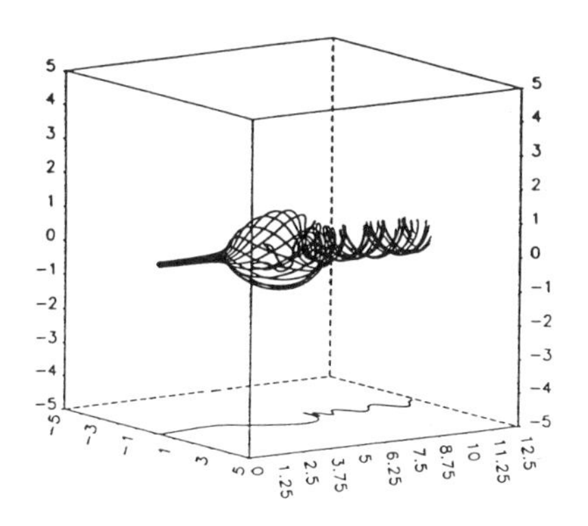

Fig. 3 Streamlines
T=84., Re=2000., $\hat{\beta} = 0.68$

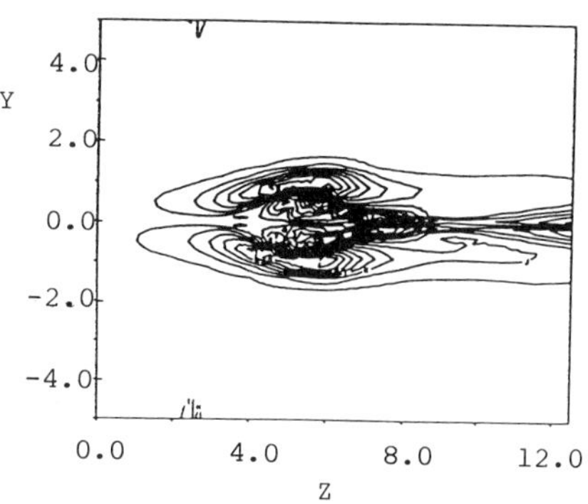

Fig. 4 Lines of constant vorticity ω_x in the y-z plane for x=0., T=84., Re=2000., $\hat{\beta} = 0.68$

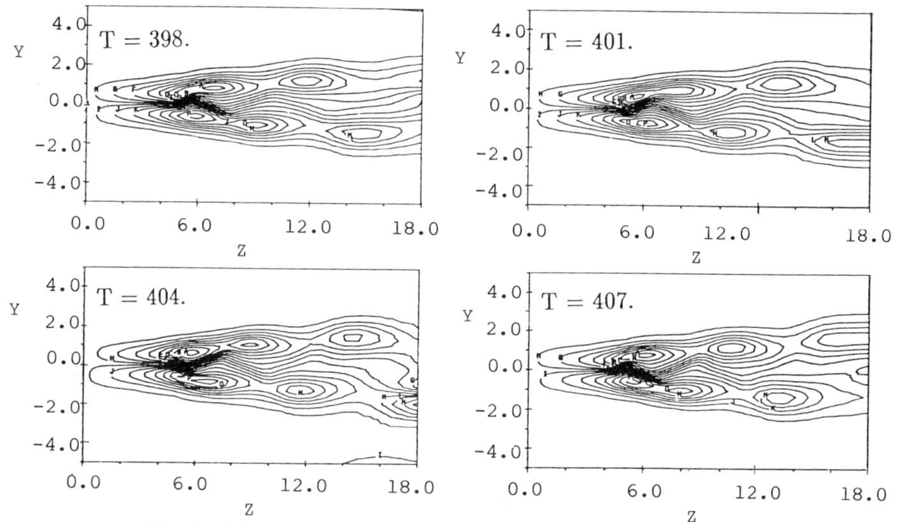

Fig. 5 Lines of constant vorticity ω_x in the y-z plane for x=0., four different time levels, Re=200., $\hat{\beta} = 0.68$

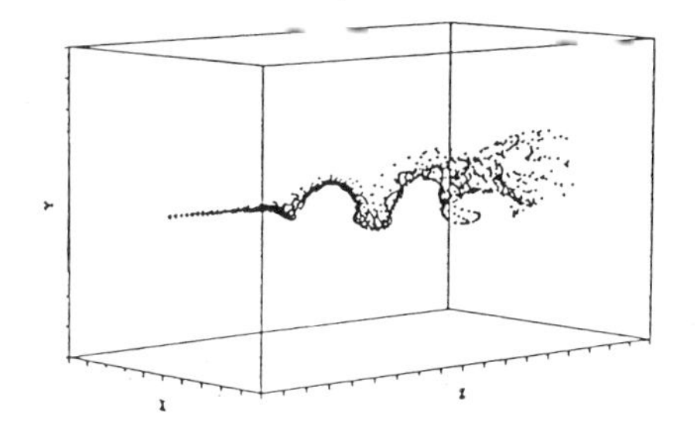

Fig. 6 Streaklines of an asymmetric solution, Re=200., $\hat{\beta} = 0.68$

COMPUTATION OF VISCOUS FREE-SURFACE FLOW AROUND A SINGLE BUBBLE

D. CLAES and W. LEINER
Institut für Thermo- und Fluiddynamik, Ruhr-Universität Bochum
Postfach 10 21 48, D-4630 Bochum 1, FRG

SUMMARY

A numerical method for two-dimensional unsteady free-surface flows including effects of surface-tension was developed to investigate the convection generated by vapour bubbles in nucleate boiling systems. This paper presents the incorporation of the free-surface conditions into the implicit segregated solution approach for the incompressible Navier-Stokes-equations. The finite-volume discretization is based on the physical contravariant velocity components in a staggered boundary-fitted grid. The results for non-steadily departing and rising bubbles compare well with experimental data from the high-speed cinematography.

INTRODUCTION

The heat transfer from a heating wall to a boiling fluid is highly enhanced by the occurance of vapour bubbles. Not only the transport of latent heat with the vapour, but mainly the convection of superheated liquid is responsible for this increase of the heat transfer coefficient. Early approaches to compute the flow field around growing bubbles [11] [2] utilized the MAC-Method [7] with a movement of the free surface through the fixed cartesian grid. In the case of departing and rising vapour bubbles the influence of surface-tension is very strong. The low resolution of the free surface in the cartesian grid resulted either in a failure to compute the departure and rise of the bubbles [11] or the bubble surface had to be geometrically prescribed as part of a sphere [2].

Different methods based on a boundary-fitted coordinate system have been proposed to improve the free-surface approximation. RYSKIN & LEAL [15] solved the vorticity/stream-function equations with finite-differences in an orthogonal boundary-fitted grid for steadily rising bubbles. RIEGER [14] applied a finite-volume method with contravariant velocity components in non-orthogonal staggered grids to melting problems, while SAITO & SCRIVEN [16] and KISTLER & SCRIVEN [10] developed a finite-element method to compute steady coating flows.

The advantages of the primitive variables in defining the free-surface conditions and the relative small implementation expense of FVM's compared to FEM's suggested the application of the finite-volume formulation with contravariant velocity components to the problem of bubble separation and rise. Details of the method can be found in [3].

MATHEMATICAL MODEL

Geometry and vector basis

The extremly complicated flow situation of multiple interacting bubbles in nucleate boiling was reduced to the incompressible laminar flow around a single axisymmetric bubble of constant volume and constant internal pressure. Fig. 1 provides a sketch of the axisymmetric boundary-fitted coordinate system. The time-dependent transformation from the cylindrical system to the axisymmetric boundary-fitted system and its inverse are

$$x^i = (\xi,\eta,\phi) \;=\; [\xi(r,y,t),\eta(r,y,t),\phi] \qquad \tau = t, \tag{1}$$

$$x^{i'} = (r,y,\phi) \;=\; [r(\xi,\eta,\tau),y(\xi,\eta,\tau),\phi] \qquad t = \tau. \tag{2}$$

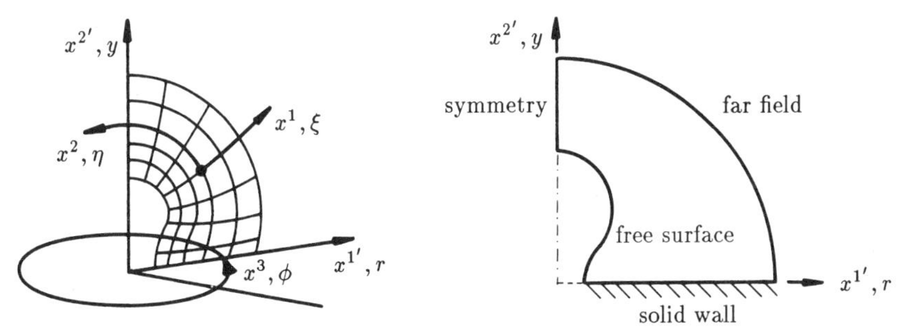

Figure 1: *The axisymmetric boundary-fitted coordinate system.*

The Jacobian $\sqrt{g}$ of the transformation $\beta_j^{i''} = \partial x^{i''}/\partial x^j$ from the cartesian system $x^{i''} = (x, y, z) = (r\cos\phi, y, r\sin\phi)$ to the system x^i is derived as

$$\sqrt{g} = r(r_\xi y_\eta - r_\eta y_\xi) = r\sqrt{g'}. \tag{3}$$

The grid generation technique described below provides orthogonal grid lines at the boundaries, thus $g^{12} = g^{21} = 0$ at the free surface. The time derivative of a variable f at a fixed point of the moving coordinate system x^i is denoted as $\partial f/\partial\tau = f_\tau$. The fluid velocity $\underline{v}$ and the grid motion $\underline{w} = x_t^j \underline{g}_j$ yield the relative velocity $\underline{u}$ between a fluid particle and the grid as $\underline{u} = \underline{v} + \underline{w}$. Following DEMIRDŽIĆ et. al. [4] [5] the base vectors are normalized to reduce the numerical sensitivity of the discretized equations to non-smooth grids:

$$\underline{v} = v^{(i)}\underline{g}_{(i)} \quad \text{with} \quad v^{(i)} = v^i\sqrt{g_{ii}} \quad \text{and} \quad \underline{g}_{(i)} = \underline{g}_i/\sqrt{g_{ii}}. \tag{4}$$

Introducing the Christoffel symbols Γ_{ij}^k and the terms Π_j^k, which describe the variation of base vectors in time [4],

$$\Gamma_{ij}^k = \beta_{m''}^k \left(\beta_i^{m''}\right)_{,j} \qquad \text{and} \qquad \Pi_j^k = \beta_{i'}^k \left(\beta_j^{i'}\right)_{\tau}, \tag{5}$$

the space and time derivatives of the normalized base vectors can be written as:

$$\underline{g}_{(i),(j)} = \frac{\sqrt{g_{kk}}}{\sqrt{g_{ii}g_{jj}}}\left[\Gamma_{ij}^k - \frac{g_{il}}{g_{ii}}\Gamma_{ij}^l\delta_i^k\right]\underline{g}_{(k)} = \Gamma_{(ij)}^{(k)}\underline{g}_{(k)}, \tag{6}$$

$$\left(\underline{g}_{(i)}\right)_\tau = \frac{\sqrt{g_{kk}}}{\sqrt{g_{ii}}}\left[\Pi_i^k - \frac{g_{il}}{g_{ii}}\Pi_i^l\delta_i^k\right]\underline{g}_{(k)} = \Pi_{(i)}^{(k)}\underline{g}_{(k)}. \tag{7}$$

Governing equations

The continuity equation for an arbitrary moving control volume V with surface F is

$$\oint_F \underline{n}\cdot\underline{v}\,dF = 0. \tag{8}$$

The differential form of the Navier-Stokes equation with Christoffel symbols is integrated over the control volume,

$$\frac{\partial}{\partial\tau}\int_V v^{(j)}\,dV + \oint_F \underline{n}\cdot\left[(u^{(i)}v^{(j)} - \mu\,D^{(ij)}/\rho)\underline{g}_{(i)}\right]dF = \tag{9}$$

$$-\int_V\left[\left(u^{(i)}v^{(k)} - \mu\,D^{(ik)}/\rho\right)\Gamma_{(ki)}^{(j)} + \frac{g^{(ij)}}{\rho\sqrt{g_{ii}}}P_{,i} + v^{(k)}\Pi_{(k)}^{(j)}\right]dV,$$

where $\underline{\underline{D}}$ is the deformation tensor and P is the modified pressure $\nabla P = \nabla p - \rho\underline{f}$.

Boundary conditions

Let the free surface be represented by a coordinate line $x^1 = const.$ At the free surface the normal component of the relative velocity between grid and fluid vanishes by the kinematic boundary condition (KBC):

$$\underline{n} \cdot \underline{u} = \underline{n} \cdot (\underline{v} + \underline{w}) = 0 \qquad \longrightarrow \qquad v^{(1)} + w^{(1)} = 0. \tag{10}$$

The tangential stress exerted by the vapour onto the liquid is negligible. This tangential stress boundary condition (TSTBC) is specified in the boundary orthogonal grid as

$$\underline{t} \cdot (\underline{n} \cdot \underline{\underline{D}}) = 0 \qquad \longrightarrow \qquad v^{(2)}|_{(1)} + v^{(1)}|_{(2)} = 0, \tag{11}$$

where $v^{(i)}|_{(j)} = v^{(i)}{}_{,(j)} + v^{(k)}\Gamma^{(i)}_{(lj)}$ denotes the physical covariant derivative. By the normal stress boundary condition (NSTBC) the pressure p and the normal viscous stress $\mu\, \underline{n} \cdot (\underline{n} \cdot \underline{\underline{D}})$ in the liquid together with the capillary pressure $\sigma(1/R_1 + 1/R_2)$ are balanced to yield the vapour pressure p_V inside the bubble,

$$p_V - p + \mu\, \underline{n} \cdot (\underline{n} \cdot \underline{\underline{D}}) - \sigma(1/R_1 + 1/R_2) = 0, \tag{12}$$

where R_1 and R_2 are the principle radii of curvature. Introducing the modified pressure and setting $p_V = 0$ we get for the orthogonal grid at the free surface

$$- P + \rho g\, x^{2'} + 2\mu\, v^{(1)}|_{(1)} - \sigma(1/R_1 + 1/R_2) = 0. \tag{13}$$

Dimensional analysis

Density ρ, viscosity μ and surface-tension σ of the liquid as well as the gravitational acceleration g are parameters of the flow. The diameter d of the volume-equivalent sphere for the bubble volume V ($d = \sqrt[3]{6\,V/\pi}$) serves as reference length. In the case of steadily rising bubbles with $U = |\underline{U}_\infty|$, the following nondimensional groups enter the mathematical model:

$$Re = \frac{\rho\, d\, U}{\mu}, \quad We = \frac{\rho\, d\, U^2}{\sigma} \quad \text{and} \quad Fr = \frac{U^2}{g\, d}. \tag{14}$$

In the numerical solution with given values of Reynolds and Weber number, the Froude number is iteratively corrected until the steady state is reached. Replacing U by $\sqrt{gd}$ in the case of unsteady flow, the governing non-dimensional groups are the Galilei and the Eötvös number:

$$Ga = \left[\frac{\rho\, d \sqrt{gd}}{\mu}\right]^2 = \frac{\rho^2\, g\, d^3}{\mu^2} \qquad \text{and} \qquad Eo = \frac{\rho\, g\, d^2}{\sigma}. \tag{15}$$

DISCRETIZATION

Grid generation

The elliptic grid generation is used to construct the two-dimensional grid in one half of the (r, y)-plane. Grid orthogonality and grid spacing at the boundaries are specified by an iterative correction of the source terms introduced by HILGENSTOCK [8].

A robust method for the distribution and adjustment of grid points on the boundaries is crucial for the computation of moving free surfaces. Here a parametric formulation of the boundaries is used, which describes the coordinates x_{FS} and y_{FS} respectively by a cubic spline fit. In order to prevent an influence of the spline boundary conditions, the free surface shape is mirrored at the symmetry axis to yield a closed curve with periodic boundary conditions in case of the rising bubble. The periodic tridiagonal systems of spline curvature are solved by the approach given by AHLBERG et. al. [1]. In case of the departing bubble, zero curvature is used as a spline boundary condition at the contact line. The grid points at the free surface are redistributed during the flow calculation on the analytical spline curve by a simple iterative algorithm.

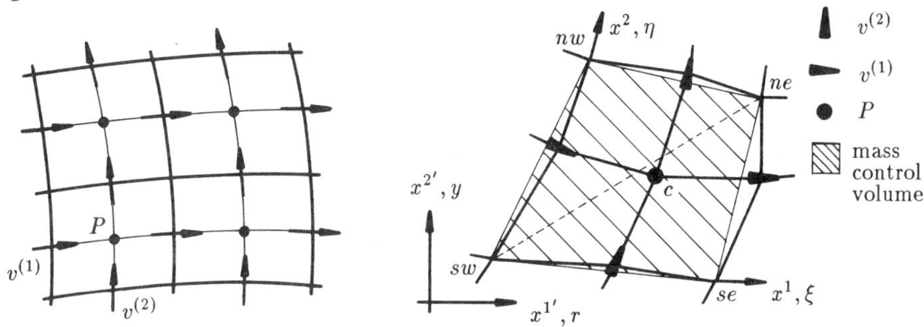

Figure 2: *Staggered arrangement of variables and definition of control volumes.*

Finite-volume approach

The variable locations in the staggered grid arrangement are defined by a double-fine grid, Fig. 2. To reduce the effort in computing surface integrals, the surfaces of the control volumes are given by the straight connections of the corner points nw, ne, sw, se. The determinants $\sqrt{g'}$ and $\sqrt{g}$ represent the area and the rotational volume of the control volumes, which are easily computed from the triangles (sw, se, ne) and (sw, nw, ne) [17]. With the usual interpretation of surface integrals by the mean value theorem, the discrete continuity equation (8) can be written as

$$\left(\sqrt{g}\,v^1\right)_e - \left(\sqrt{g}\,v^1\right)_w + \left(\sqrt{g}\,v^2\right)_n - \left(\sqrt{g}\,v^2\right)_s = 0. \tag{16}$$

To satisfy the geometric conservation law, the grid velocities are computed from the discret volume changes $\delta V_n, \delta V_e \ldots$ produced by the respective moving cell surfaces [6]. The surface integral of the Navier-Stokes equation is discretized for component 1 at surface e:

$$\int_{F_e} \underline{n}_e \cdot [\ldots]\, dF_e \approx \left[\sqrt{g}\left(v^{(1)}v^{(1)} - \mu\, D^{(11)}/\rho\right) + \frac{\delta V_o}{\Delta\tau}\, v^{(1)}\right]_e . \tag{17}$$

The volume integral for component 1 is written as a mean value at point c:

$$\int_V [\ldots] dV \approx \sqrt{g}\left[\left(u^{(i)}v^{(k)} - \mu\, D^{(ik)}/\rho\right)\Gamma^{(1)}_{(ki)} + \frac{g^{(i1)}}{\rho\sqrt{g_{ii}}} P_{,i} + v^{(k)}\Pi^{(1)}_{(k)}\right]_c . \tag{18}$$

The computation of variables at the cell surfaces is performed by linear interpolation from neighbouring grid points, thus the approximation is second order accurate in space. Upwind interpolation or adding artificial viscosity proved not to be necessary even for flow cases with $\sqrt{Ga} = 4000$. This might be due to the existance of the free surface, where disturbances can leave the computational domain.

Temporal discretization

To achieve a stable coupling of the equations of momentum and continuity with the free-surface conditions an implicit time marching algorithm is required. Here the implicit backward Euler time step is employed to obtain maximum stability, though the temporal accuracy is only of first order.

NUMERICAL SOLUTION

Segregated solution method

The segregated solution method approximates the pressure field by a Poisson type equation separate from the momentum equations. Often the solution of the continuity equation is divided into two steps [12] [13]: First the pressure correction equation is solved for a given divergence field. Secondly, the velocities are updated according to the pressure change to yield a divergence-free flow field. Another possibility was originally proposed as a simplification of the MAC-method: The SOLA-algorithm [9] corrects pressure and velocities successively, this causes a slight increase of the computational time, however, the Neumann pressure boundary condition is simplified.

The momentum equation is either implicitly treated by an iterative solver (e.g. in SIMPLE and deduced methods [12]) or explicitly updated by a single Jacobi step (e.g. in PRIME [13]). For an implicit time step the solution of the momentum and the continuity equation alternate until convergence is reached up to the desired level.

Incorporation of free-surface conditions

Different iterative algorithms for free-surface flows have been developed in the finite-element context. First a direct boundary condition for either the fluid velocity or the free-surface position was derived from the normal stress condition [16]. Convergence for all ranges of physical parameters could not be achieved with either of this schemes. Later a coupled algorithm was introduced, which solves all variables including the free-surface position in a "full" Newton iteration [10].

In contrast to this, the free-surface conditions can be selectively incorporated into the different iteration cycles of the finite-volume or finite-difference techniques. RYSKIN & LEAL [15] suggest a heuristic correction of the free-surface position depending on the actual imbalance of normal stress. Since the free-surface movement determines the normal velocity at the free surface via the KBC, the free-surface position and the continuity equation are strongly coupled. This leads to the requirement of solving the normal stress condition together with the continuity equation. To estimate the actual imbalance in normal stress at the free surface during the solution of the continuity equation, the velocities are successively updated with the pressure-velocity correction scheme of the SOLA-algorithm [9].

The boundary condition for the tangential velocity at the free surface is derived from the TSTBC using the normal gradient of the tangential velocity $(v^{(2)}|_{(1)})$ to compute the value of $v^{(2)}$ in the imaginary boundary cells. This boundary condition is applied once at each outer iteration step together with the no-slip, the far field and the symmetry conditions.

The definition of the contravariant velocity components as dependent variables and the orthogonality of the grid lines at the boundaries simplify the formulation and the iterative solution of the free-surface conditions considerably.

Outer Iteration

The continuity equation and the normal stress condition at the free surface are the dominating parameters of the current problem. Thus, a single explicit Jacobi update of the momentum equation is sufficient to obtain good convergence rates in the outer iteration loop of the implicit time step. The grid is re-generated at each outer iteration step by the four-colour SOR using the grid of the previous step as an initial grid. After updating the geometrical variables as metric coefficients and Christoffel symbols, the boundary conditions for the tangential velocity are applied. In the present scheme the Jacobi step and the tangential velocity boundary conditions are fully vectorized for the CYBER-205.

Inner Iteration

The inner iteration loop solves the continuity equation by a Red-Black SOR step for the pressure-velocity correction alternating with a Jacobi step for the normal stress condition. Again both steps work fully vectorized. After the pressure-velocity correction step KI, the new position of the free surface is computed from the KBC. The curvature of the new free surface can then be determined to give the actual imbalance Θ^{KI} of the normal stress:

$$\Theta^{KI} = \left[P - \rho g\, x^{2'} - 2\mu\,(v^{(1)}|_{(1)}) + \sigma(1/R_1 + 1/R_2)\right]^{KI} . \tag{19}$$

Introducing an artificial inertia for the reaction of the normal flux $(\sqrt{g}\, v^1)_c$ at a point c of the free surface to the imbalance Θ^{KI}, the heuristic correction of the normal velocity at the free surface can be derived in analogy to the pressure gradient in the momentum equation,

$$(\sqrt{g}\, v^1)_c^{KI+1} = (\sqrt{g}\, v^1)_c^{KI} - \omega_{FS}\,\Delta\tau \left[\sqrt{g}\,\frac{g^{(11)}}{\rho\, g_{11}}\right]_c \Theta_c^{KI} , \tag{20}$$

where ω_{FS} is a relaxation parameter. After the pressure-velocity iteration has converged below the specified divergence level, the normal stress balance is reached up to the same level of accuracy.

RESULTS

Steadily rising bubbles

To validate the numerical scheme steadily rising bubbles were simulated and compared in terms of drag coefficient and aspect ratio to the results of RYSKIN & LEAL [15]. In the case $We = 0$ and $Re = 5$ (sperical bubble) we obtain a drag coefficient of 4.26 on a grid with 79 * 79 control volumes. The difference with the value of RYSKIN & LEAL is about 0.5 %. In Fig. 3 a part of the grid (39 * 39 control volumes, single-fine grid shown) and the flow field of a bubble with $We = 15$ and $Re = 20$ are displayed. Here the differences increase to 2.7 % in the aspect ratio and 7 % in the drag coefficient. A good agreement is achieved concerning the geometry of the backflow region.

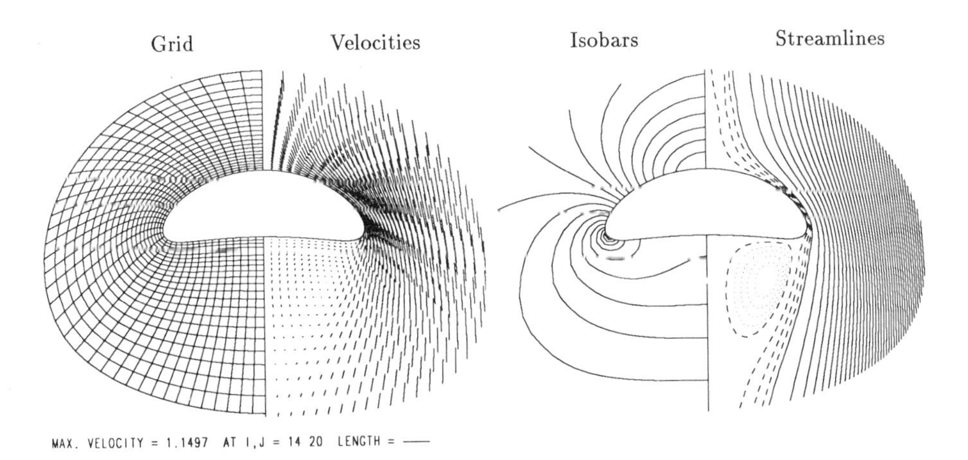

Figure 3: *Flow field near a steadily rising bubble with* $We = 15$ *and* $Re = 20$.

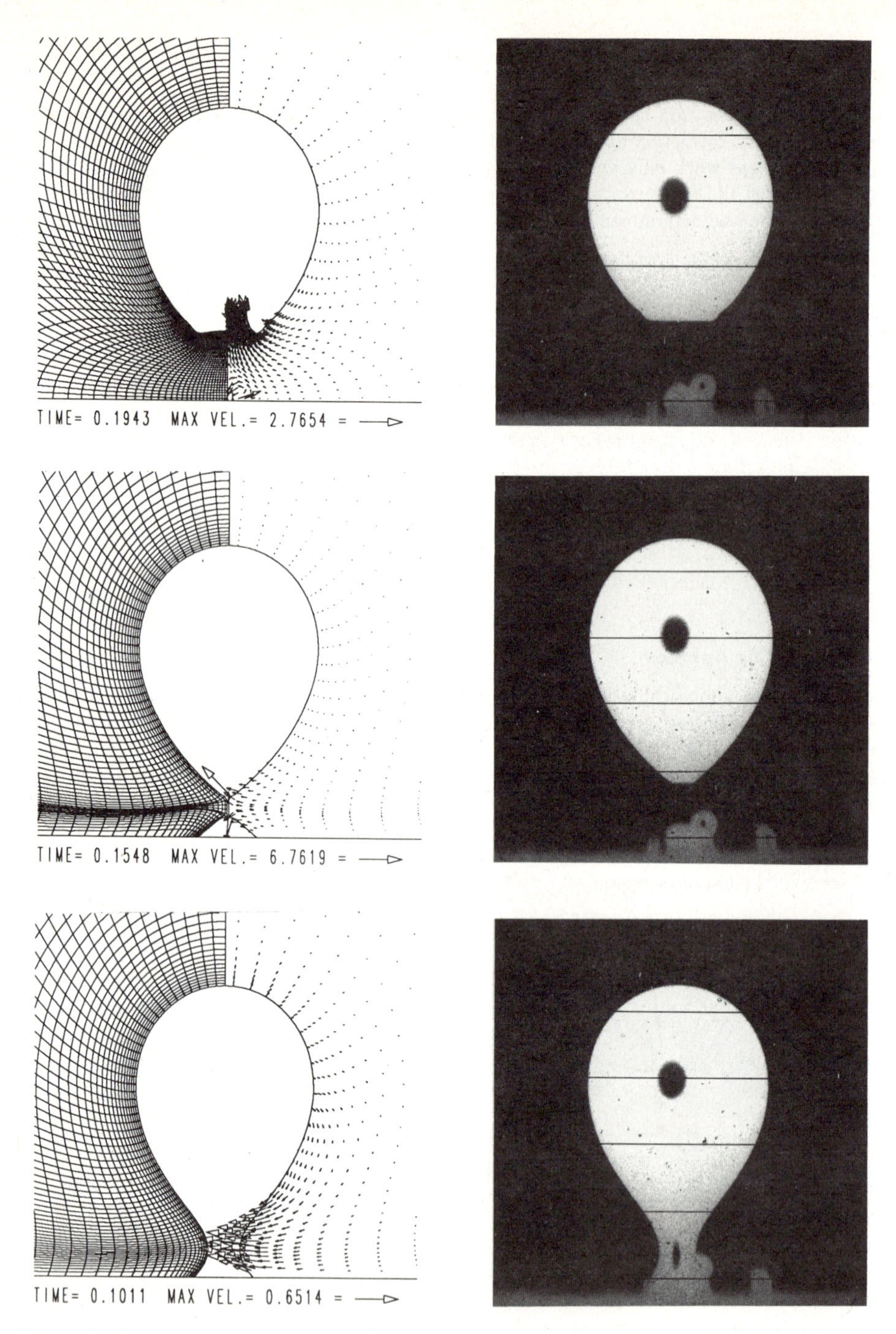

Figure 4: *Computed and experimentally observed results for the departure of an air bubble in water* ($Eo = 2.14$, $Ga = 6.9 \cdot 10^5$).

Non-steadily departing and rising bubble

To provide experimental results for comparison of the unsteady movement of the free surface, an air bubble of $33\,mm^3$ volume was released into isothermal water ($22.8°C$) and filmed by a high-speed camera ($3940 fps$). A maximum bubble size was obtained by covering the solid wall with poorly-wetted PVC-tape, which leads to an almost fixed contact line. Fig. 4 gives a comparison of flow fields (single-fine grid shown) and bubble contours for selected stages of the bubble separation.

The bubble necks at a certain distance above the wall and finally breaks at this indentation. The smaller part of the bubble remains attached to the solid wall. The acceleration by capillary pressure produces large velocities near the bubble neck. In order to retain a reasonable resolution of the bubble during the further rise, the grid topology was switched from C-type to O-type after the bubble separation. Fig. 5 illustrates the free-surface movement in form of the aspect ratio development for the numerical and the experimental bubble. After departure, the part of the bubble remaining on the wall is neglegted and replaced by liquid. The aspect ratio is then measured for the rising part of the bubble only. This creates the discontinuity of the aspect ratio shown in diagram 5. The results illustrate correctly the evolution of the bubble oscillation, represented by the local extrema of the curves in Fig. 5.

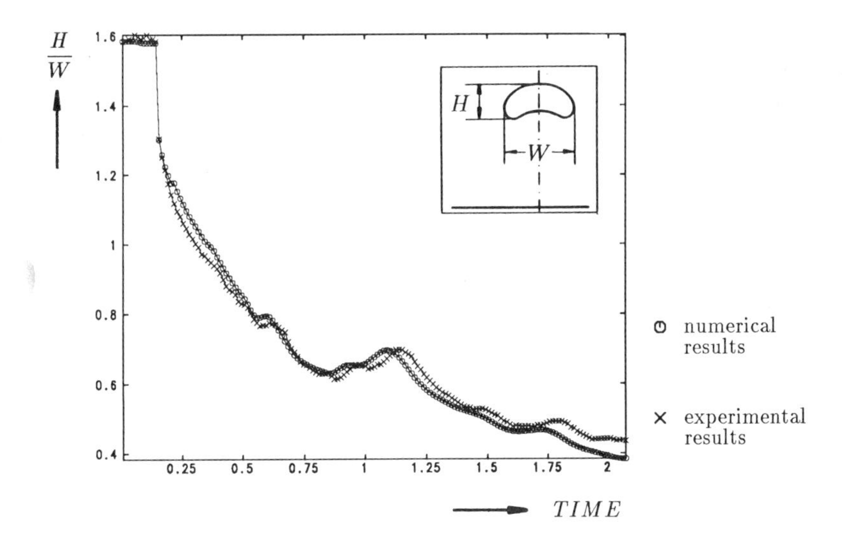

Figure 5: *Development of aspect ratio of the computed and filmed air bubble.*

CONCLUSIONS

The simulation of axisymmetric flow fields around separating and freely rising bubbles has been proved to be achievable with the current numerical approach. A straightforward formulation of the free-surface conditions is obtained by defining the physical contravariant velocity components as dependent variables. For an efficient solution of these conditions, the dominating NSTBC is incorporated into the pressure-velocity correction of the implicit segregated solution method. The computed results are in good agreement with the literature as well as the experimental observations and demonstrate the advanced capabilities of the presented method to predict free-surface phenomena compared to previous cartesian approaches [11] [2].

ACKNOWLEDGEMENTS

The current research was financially supported by the *Deutsche Forschungsgemeinschaft* (DFG) under grant Le444/3. The computations were performed on the CYBER-205 of the Ruhr-University Bochum.

REFERENCES

[1] AHLBERG, H.H., NILSON, E.N., WALSH, J.L. *The Theory of Splines and their Applications.* New York: Academic Pr., 1967.

[2] BUROW, P. *Die Berechnung des Blasenwachstums beim Sieden von Flüssigkeiten an Heizflächen als numerische Lösung der Erhaltungsgleichungen.* Dissertation, TH Darmstadt, 1979.

[3] CLAES, D. *Numerische Simulation von instationären Strömungen mit freien Oberflächen am Beispiel ablösender und aufsteigender Blasen.* Dissertation, Ruhr-Univ. Bochum, in preparation.

[4] DEMIRDŽIĆ, I.A. *A Finite Volume Method for Computation of Fuid Flow in Complex Geometries.* Ph.D. Thesis, Imperial College, London, 1982.

[5] DEMIRDŽIĆ, I., GOSMAN, A.D., ISSA, R.I., PERIĆ, M. "A calculation procedure for turbulent flow in complex geometries" *Computers & Fluids.* **15**, 3 (1987): 251–273.

[6] DEMIRDZIC, I., PERIC, M. "Space conservation law in finite volume calculations of fluid flow" *Int. J. Num. Meth. Fluids.* **8** (1988): 1037–1050.

[7] HARLOW, F.H., WELCH, J.E. "Numerical calculation of time-dependent viscous incompressible flow of fluid with free surface" *Physics of Fluids.* **8**, 12 (1965): 2182–2189.

[8] HILGENSTOCK, A. "A fast method for the elliptic generation of three-dimensional grids with full boundary control" *Proc. 2. Int. Conf. on Numerical Grid Generation in Computational Fluid Dynamics.* Swansea: Pineridge Pr., 1988.

[9] HIRT, C.W., NICHOLS, B.D., ROMERO, N.C. *SOLA — A numerical solution algorithm for transient fluid flows.* Los Alamos Scientific Lab., Rep. LA-5852, 1975.

[10] KISTLER, S.F., SCRIVEN, L.E. "Coating flow theory by finite element and asymptotic analysis of the Navier-Stokes system" *Int. J. Num. Meth. Fluids.* **4** (1984): 207–229.

[11] MADHAVAN, S. *An Experimental and Mathematical Study of the Shapes of Bubbles Growing on Surfaces in an Isothermal Superheated Fluid.* Ph.D. Thesis, Univ. of Kansas, 1970.

[12] RAITHBY, G.D., SCHNEIDER, G.E., "Elliptic Systems: Finite-Difference Method II" *Handbook of Numerical Heat Transfer.* Eds. W.J. MINKOWYCZ et. al., New York: Wiley, 1988, 949–999.

[13] PERNG, Ch.-Y., STREET, R.L. "Three-dimensional unsteady flow simulations: Alternative strategies for a volume-averaged calculation" *Int. J. Num. Meth. Fluids.* **9** (1989): 341–362.

[14] RIEGER, H. *Numerische Berechnung von Wärmetransportvorgängen bei laminaren freien Konvektionsströmungen in beliebigen festen sowie zeitabhängigen ebenen Geometrien.* Dissertation, TH Darmstadt, 1984.

[15] RYSKIN, G., LEAL, L.G. "Numerical solution of free-boundary problems in fluid mechanics — Part 1. The finite-difference technique — Part 2. Buoyancy driven motion of a gas bubble through a quiescent liquid." *J. Fluid Mech.* **148** (1984): 1–35.

[16] SAITO, H., SCRIVEN, L.E. "Study of coating flow by the finite element method" *J. Comput. Physics.* **42** (1981): 53–76.

[17] VINOKUR, M. *An Analysis of Finite-Difference and Finite-Volume Formulations of Conservation Laws.* NASA Contractor Report 177416, 1986.

A Convenient Entropy Satisfying T.V.D. Scheme for Computational Aerodynamics

by

N Clarke, R Saunders and D M Causon
Department of Mathematics and Physics
Manchester Polytechnic
Chester Street
Manchester M1 5GD
United Kingdom

ABSTRACT

Conservative shock capturing methods for the unsteady Euler equations are reviewed and it is shown that the concepts of entropy satisfaction and total variation diminution can be applied to the classical Lax-Wendroff scheme. For an associated scheme to be efficient in applications, it is necessary that it be able to capture strong shock waves with high resolution. We describe a scheme which is efficient in both respects.

INTRODUCTION

The last fifteen years has seen considerable advancement in the design of shock capturing finite difference schemes for the solution of practical high speed aerodynamic flow problems. Much recent research has led to the development of total variation diminishing (T.V.D.) schemes and the closely related essentially non-oscillatory (E.N.O.) schemes. The former, pioneered by Harten [1], will not allow the creation of new extrema in the numerical solution so that no non-physical spurious oscillations can occur around discontinuities. This paper is concerned with the *convenient* T.V.D. correction of popular MacCormack codes for the solution of the Euler equations. A practical aerodynamic application of this approach can be found in [2]. The work of Sweby [3] describes how a non-linear term can be appended to the Lax-Wendroff scheme [4] to satisfy Harten's T.V.D. constraints. Davis [5] removed the necessity to upwind-weight this appended term and so developed a fully symmetric scheme which can be applied very easily to the Euler equations. Unfortunately, solutions obtained with this scheme are rather more dissipative than the best of the modern shock capturing schemes, particularly on contact discontinuities. The use of artificial compression techniques [6] can remove this deficiency. Figures 1 and 2 demonstrate this for the simple one-dimensional Riemann problem with initial condition

$$\underset{\sim}{u}_L = (0.445,\ 0.311,\ 8.928), \qquad \underset{\sim}{u}_R = (0.5,\ 0.0,\ 1.4275).$$

However, numerical experiments have shown that the addition of artificial compression *explicitly* can lead to trouble. It can cause compression of profiles that it should not, producing entropy-violating "expansion shocks". Fortunately, it is possible to design schemes which employ more compressive T.V.D. flux limiters so that no artificial compression term needs to be added explicitly. Our experiments show that this technique is more robust, though problems of entropy-violation may occur in certain circumstances. In this paper, we discuss these problems and present some new results enabling the design of entropy-satisfying, T.V.D. versions of the MacCormack scheme which are second order accurate except at points of extrema, where they may drop to first order.

PRELIMINARIES

We begin by considering the solution of the scalar convective wave equation

$$\partial_t u + c\partial_x u = 0 \ , \tag{1}$$

This is a non-trivial problem. "Good" schemes for its solution can be extended successfully to the Euler system by the use of an approximate Riemann solver [9], [10] and to multi-dimensions by means of operator-splitting. We use a classical oscillatory second-order scheme with corrective wind-weighted T.V.D. term:

$$u^j = L_2 u_j + G_{j+1/2}\Delta_{j+1/2} - G_{j-1/2}\Delta u_{j-1/2} \ . \tag{2}$$

Here, $u^j = u_j^{t=n+1}$, $\Delta u_{j+1/2} = u_{j+1} - u_j$ and L_2 is the Lax-Wendroff or MacCormack-variant operator. The function G can be constructed so that scheme (2) is T.V.D.. If we solve (1) with $c>0$, the function

$$G_{j+1/2} = (v/2)(1-v)(1-\Phi_j), \tag{3}$$

in which $v=c\Delta t/\Delta x$, $\Phi_j = \Phi(r_j) = \Phi(\Delta u_{j-1/2}/\Delta u_{j+1/2})$,

will guarantee that the T.V.D. conditions are satisfied by (2) provided that

$$\left.\begin{array}{l} \max(\Phi) - \min(\Phi/r) \leqslant 2/(1-v) \\ \text{and } \max(\Phi/r) - \min(\Phi) \leqslant 2/v \end{array}\right\} . \tag{4}$$

We define the T.V.D. region by taking bounds on Φ and Φ/r:

$$-k \leqslant \Phi \leqslant \epsilon \ ; \ -\ell \leqslant \Phi/r \leqslant \theta ; \ k, \ \ell, \ \epsilon, \ \theta \geqslant 0$$

and illustrate the region in Figure 3.

The T.V.D. conditions (4) can then be written

$$\epsilon + \ell \leqslant 2/(1-v), \qquad k + \theta \leqslant 2/v. \tag{5}$$

For $c<0$ we take

$$G_{j+1/2} = (|v|/2)(1-|v|)(1 - \Phi_{j+1}/r_{j+1}) \tag{6}$$

and require

$$\epsilon + \ell \leqslant 2/|v|, \ k + \theta \leqslant 2/(1-|v|) \ . \tag{7}$$

In addition, we need to design the limiter Φ such that

$$\Phi(1+\delta) = 1 + 0(\delta), \ \delta \text{ small} \tag{8}$$

for second order accuracy in smooth regions. Taking k and ℓ to be zero for the moment and choosing the limiter shown in Figure 4:

$$\Phi = \max\,[\min(\theta r, 1), \ \min(r, \epsilon), \ 0] \tag{9}$$

$$\theta, \ \epsilon \geqslant 1,$$

experiments show that θ and ϵ are parameters which control the amount of artificial compression applied by the scheme *implicitly*. Figures 5a-c show a convected square wave. Figure 5a utilises the Davis [5] type limiter for which $\epsilon=1$, $\theta=2$.

More compression is apparent in Figure 5b where maximum values of θ and ϵ are used. The best result for this problem can be seen in Fig. 5c where the Davis limiter and *explicit* artificial compression have been used. Comparing these results for the convected sine wave in Figures 6a-d we can see that artificial compression can lead to a flattening of the sine wave. We comment on this in more detail in the next section.

We also note a loss of peak amplitude of the sine wave as time evolves. This may be caused by the drop to first order accuracy of the T.V.D. scheme at points of extrema. The limiter

$$\left.\begin{aligned} \Phi &= \max\,[0,\ \min\,(\theta r,1),\ \min\,(r,\epsilon)] && r>0 \\ &= r && -1 \leqslant r \leqslant 0 \\ &= 1 && r < -1 \end{aligned}\right\} \qquad (10)$$

(where $k = \ell = 1$ in (4) so that θ, ϵ must be adjusted accordingly)

does help to reduce marginally the loss of peak amplitude. We observe that extrema are characterised by $r<0$ so that, for $c>0$ in (1), should $-1 \leqslant r_j, r_{j-1} \leqslant 0$ or $r_j, r_{j-1} <-1$ then second order accuracy will be maintained in these regions. The limiter is shown in Figure 7.

ENTROPY SATISFACTION

Taking values of ϵ and θ in (5) to be smaller than their allowable maximum will reduce the amount of artificial compression applied by the scheme implicitly. There will then be less chance of encountering entropy-violation. We remark that utilising a negative T.V.D. reigon will require the definition of non-zero values of $-\min(\Phi)$, $-\min(\Phi/r)$; i.e. non-zero values of k and ℓ. It can be seen from (5) that the values of θ and ϵ must be reduced from their allowable maxima to allow non-zero k and ℓ. Limiters designed to be non-zero for $r<0$ may help to convect smooth data accurately, with little non-physical "squaring", but only because of the inevitable reduction of θ and ϵ values. We can, of course, reduce the values of θ and ϵ without introducing non-zero values of k and/or ℓ.
The hyperbolic system of conservation laws, of which the Euler equations are a subset, can be written

$$\partial_t \underset{\sim}{u} + A\partial_x \underset{\sim}{u} = 0 \qquad (11)$$

where $A = \partial f(\underset{\sim}{u})/\partial \underset{\sim}{u}$ has distinct eigenvalues. If A is considered as a matrix of locally frozen constant coefficients then the system of equations can be uncoupled into a set of locally-linear conservation laws. The technique we use to solve the wave equation is therefore of importance even though the concept of entropy satisfaction may seem abstract when applied to scalar schemes.

We would expect physically relevant weak solutions of the Euler equations to be limit solutions of the system

$$\partial_t u^j + \partial_x f^j = \xi\, \partial_{xx}^2 u^j, \quad j = 1,2,3 \qquad (12)$$

as the viscosity term on the right-hand side becomes negligible. If a solution of (12) is $u_1\,(x,t;\xi)$ and a weak solution of the Euler system is $u_2\,(x,t)$, then u_2 is physically admissable if

$$\operatorname*{Lt}_{\xi\to 0} u_1 = u_2 \,.$$

Lax [7] has shown that this will be guaranteed provided that, for the scalar non-linear conservation law

$$\partial_t u + \partial_x f = 0 \tag{13}$$

a generalised entropy function E(u) and associated flux F(E) are constructed such that

$$\partial_u E \partial_u f = \partial_u F \tag{14}$$

$$\partial^2_{uu} E > 0 \tag{15}$$

$$\partial_t E + \partial_x F \leqslant 0 \,. \tag{16}$$

The latter inequality is known as the *entropy condition*. Equations (14) and (15) can be satisfied by $E=u^2$, $F=cu^2$ for the linear wave equation with $c>0$, so that (16) becomes

$$\partial_t u^2 + c\,\partial_x u^2 \leqslant 0. \tag{17}$$

The Lax-Wendroff or MacCormack T.V.D. scheme (2) with (3) can be written

$$u^j = u_j - \upsilon[h_{j+1/2}(u) - h_{j-1/2}(u)] \tag{18}$$

where $h_{j+1/2}(u) = u^j + \frac{1}{2}(1-\upsilon)\Phi_j \Delta u_{j+1/2}$.

Applying this to (17) we find the inequality

$$(u^j)^2 - u_j^2 + \upsilon[h_{j+1/2}(u^2) - h_{j-1/2}(u^2)] \leqslant 0. \tag{19}$$

This inequality is difficult to manipulate, but we find that the quantities

$h_{j+1/2}(u^2)$ and $[h_{j+1/2}(u)]^2$ have the same upper and lower bounds.

Hence, a sufficient entropy condition is

$$(u^j)^2 - u_j^2 + \upsilon\left([h_{j+1/2}(u)]^2 - [h_{j-1/2}(u)]^2\right) \leqslant 0 \,. \tag{20}$$

Using (18) in (20) gives the explicit entropy inequality

$$[h_{j+1/2}(u) - h_{j-1/2}(u)][(1+\upsilon)h_{j+1/2}(u) + (1-\upsilon)h_{j-1/2}(u) - 2u_j] \leqslant 0. \tag{21}$$

Solution of (21) then gives the rather restrictive condition for entropy satisfaction:

$$\frac{1+\upsilon}{1-\upsilon}\,\theta + \epsilon \leqslant \frac{2}{1-\upsilon} \,. \tag{22}$$

For entropy satisfaction and second order accuracy we require

$$\theta = \epsilon = 1 \,. \tag{23}$$

Accordingly, we are able to construct an entropy-satisfying T.V.D. version of the Lax-Wendroff or MacCormack scheme which is second order accurate except at points of extrema. Artificial compression is applied to the scheme implicitly by increasing the values of θ and ϵ in the limiter (9) beyond unity (which is the lower bound for second order accuracy) up to the maximum allowable bound for satisfaction of the T.V.D. constraints (5). The entropy condition (22) will be violated by the implicit introduction of compression which explains why non-physical results can occur when using this technique.

Clearly the scheme is more robust if we choose to satisfy the entropy condition (22).

SYSTEMS OF CONSERVATION LAWS

The scalar scheme can be extended to systems of conservation laws by the approximate Riemann solver approach of P.L. Roe [9]. We simply add to the corrector step of the MacCormack scheme, the uncoupled T.V.D. term with k th component:

$$\frac{1}{2}\sum_{\ell}\Big\{ |\nu^{\ell}_{i+1/2}|(1-|\nu^{\ell}_{i+1/2}|)e^{k\ell}_{i+1/2}G^{\ell}_{i+1/2}\,\alpha^{\ell}_{i+1/2} - |\nu^{\ell}_{i-1/2}|(1-|\nu^{\ell}_{i-1/2}|)e^{k\ell}_{i-1/2}G^{\ell}_{i-1/2}\,\alpha^{\ell}_{i-1/2}\Big\}, \quad (24)$$

where, for the Euler equations k, ℓ = 1,2,3. The subscripts $i + 1/2$ require the associated quantity to be evaluated at averages of grid point values i, i+1 with a similar meaning for $i-1/2$. The right eigenvectors of the Jacobian matrix A are represented by e and expressions for α^{ℓ} can be found in [10]. The function G has the same structure as in (3) or (6), but Φ now becomes a function of α. The Courant numbers ν^{ℓ} are defined in terms of the eigenvalues of the Jacobian i.e. $\nu^{\ell} = \lambda^{\ell}\Delta t/\Delta x$ where $\lambda^1 = u - c$, $\lambda^2 = u$ and $\lambda^3 = u + c$.

Figures (8) and (9) show the entropy–satisfying T.V.D. scheme aplied to convect a square wave and sine wave respectively whilst figures (10) and (11) show the sacrifice made by using an entropy–satisfying scheme on the Riemann problem. We are currently investigating methods of improving the resolution of entropy satisfying T.V.D. schemes.

CONCLUSIONS

The use of compressive limiters in a T.V.D scheme without further constraints being imposed to guarantee entropy–satisfaction can lead to non–physical results. Non–physical squaring of smooth data and staircasing phenomena can occur. The analysis presented shows that an entropy condition is violated by the use of compressive limiters, but that it is possible to design a second order accurate T.V.D. version of the Lax–Wendroff or MacCormack scheme which does satisfy this condition.

REFERENCES

[1] A. Harten, "High resolution schemes for hyperbolic conservation laws", J. Comp. Phys, 49 pp357–393, 1983.

[2] D.M. Causon, "Numerical computation of external transonic flows", in Proc. of the 7th Gamm Conf. on Num. Meth. in Fl. Mech., Vieweg 1988.

[3] P.K. Sweby, "High resolution schemes using flux limiters for hyperbolic conservation laws", Siam J. Num. Anal. 21 No 5 pp995–1011, 1984.

[4] Lax P. and Wendroff B., "Systems of conservation laws", Comm. on Pure and App. Math., 13 pp217–237, 1960.

[5] Davis S.F. "T.V.D. finite difference schemes and artificial viscosity", ICASE report

84-20, 1984.

[6] Harten A., "The artificial compression method for computation of shocks and contact discontinuities: III. Self-adjusting hybrid schemes", Math. of Comp. 32 No 142, pp363-389, 1978.

[7] Lax P. "Shock waves and entropy", Proc. Symp at the Univ. of Wisconsin. E.H. Zarantonello, 110, ed. pp603-634, 1971.

[8] Merriam M.L. "Smoothing and the second law", Comp. Meth. in App. Mech. and Eng. 64 pp177-193, 1987.

[9] Roe P.L. "Approximate Riemann solvers, parameter vectors and difference schemes", J. Comp. Phys., 43 pp357-372, 1981.

[10] Roe P.L and Pike J., "Efficient construction and utilisation of approximate Riemann solutions", Comp Meth. in App. Sc. and Eng. 6. Ed. R. Glowinski and J.L. Lions. pp.499-518, 1984.

FIGURES

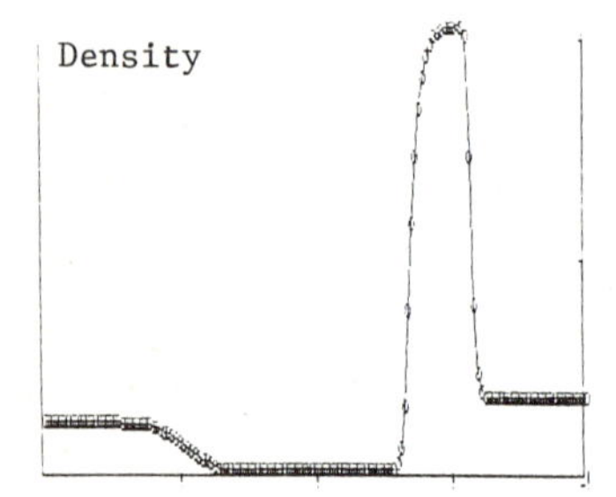

Fig. 1
Davis [5] limiter $\theta=2$, $\varepsilon=1$

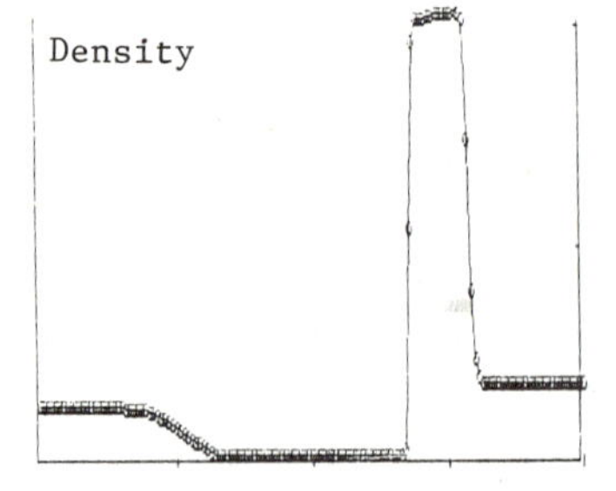

Fig. 2
Davis scheme with explicit artificial compression.

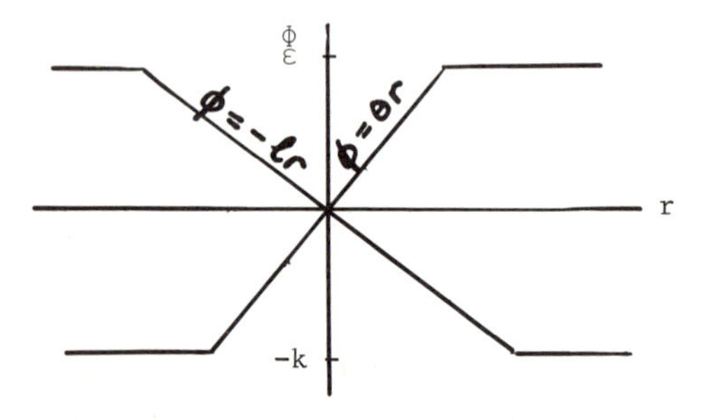

Fig. 3
General T.V.D. region.

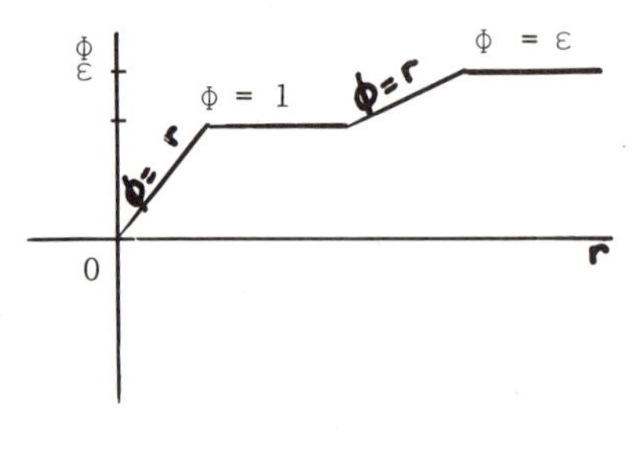

Fig. 4
Second order accurate limiter.

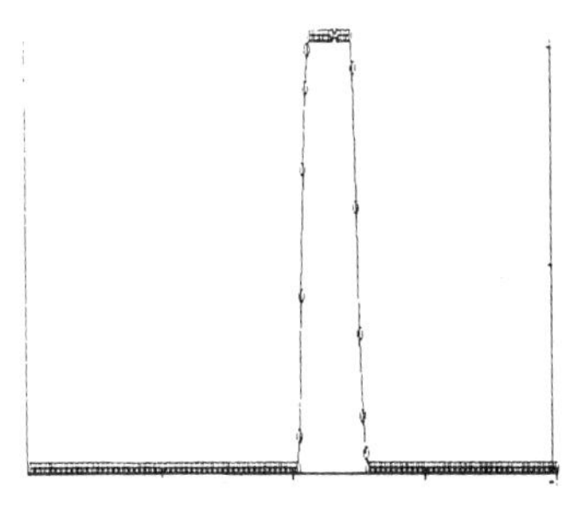

Fig. 5(a)
$\theta=2$, $\varepsilon=1$ in (9).

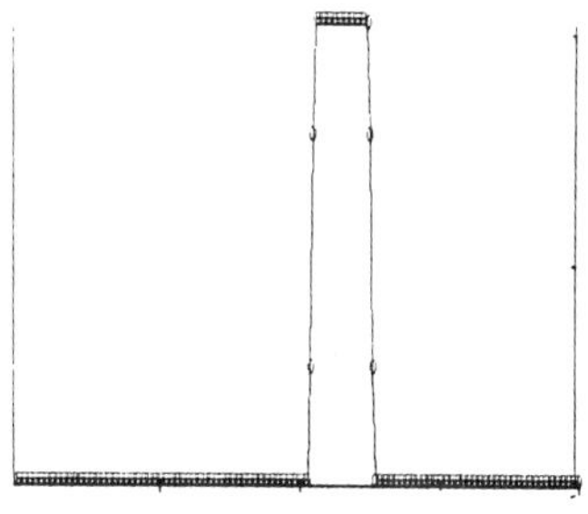

Fig. 5(b)
$\theta=2/\nu$, $\varepsilon=2/(1-\nu)$ in (9).

Fig. 5(c)
$\theta=2$, $\varepsilon=1$, with explicit artificial compression.

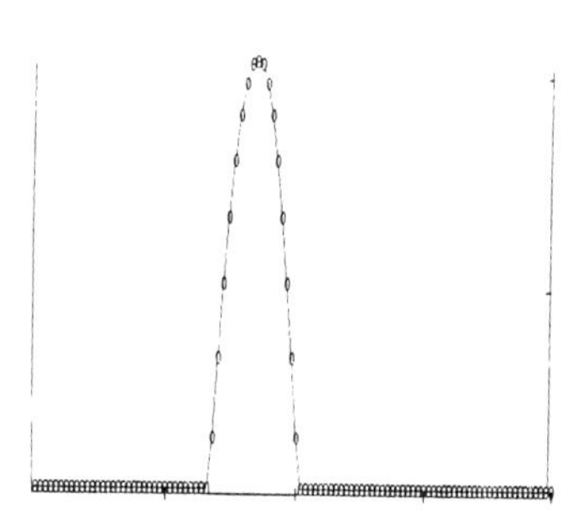

Fig. 6(a)
Sine wave initial conditions.

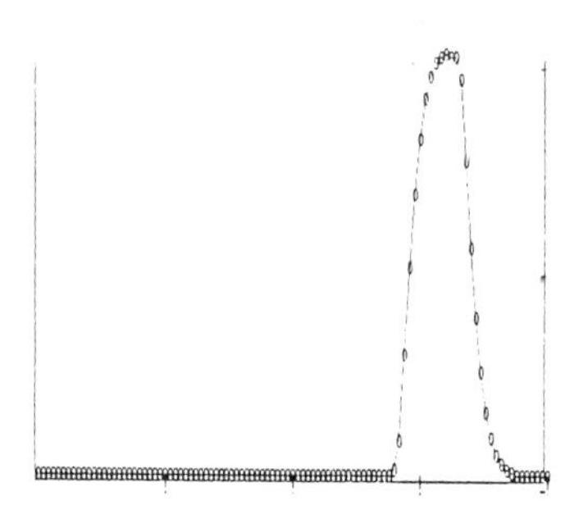

Fig. 6(b)
Sine wave with
$\theta=2$, $\varepsilon=1$ in (9).

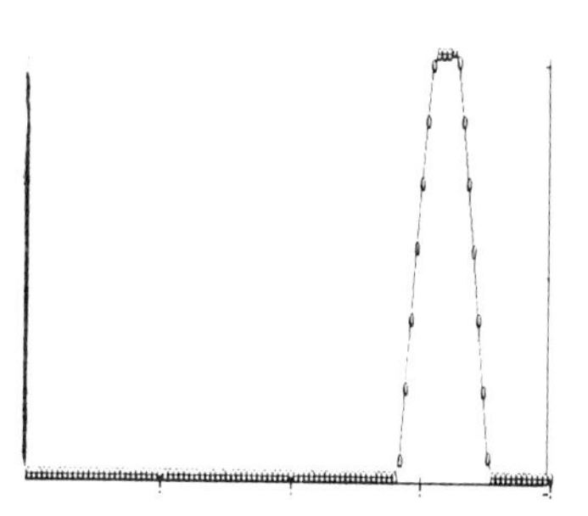

Fig. 6(c)
Sine wave with
$\theta=2/\nu$, $\varepsilon=2/(1-\nu)$ in (9).

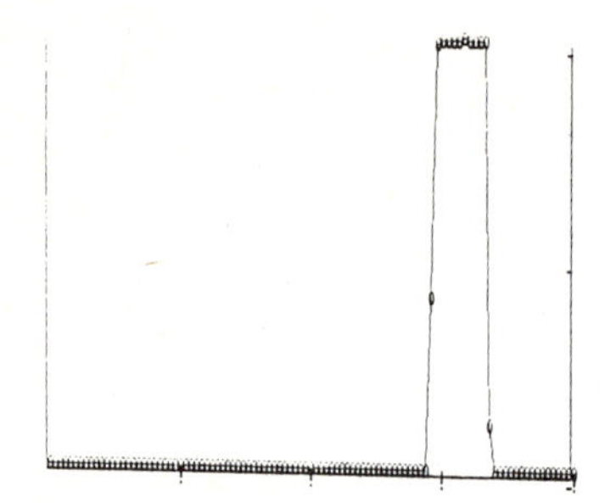

Fig. 6(d)
Sine wave with $\theta=2, \varepsilon=1$
in (a) and explicit artifial compression.

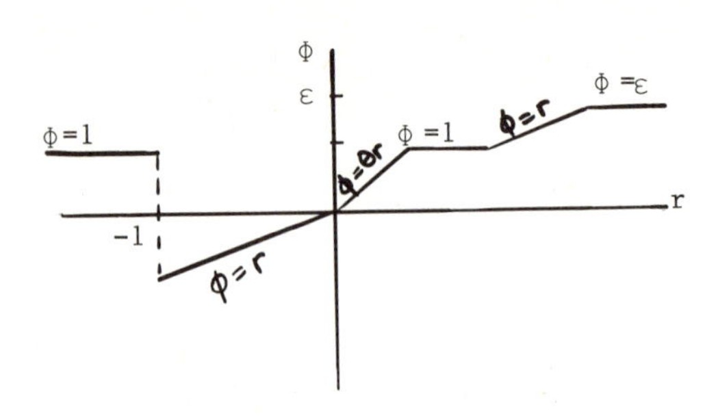

Fig. 7
Discontinuous limiter.

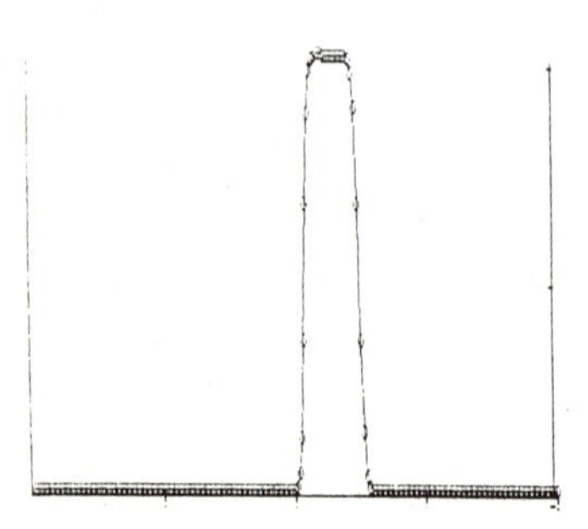

Fig. 8
Entropy satisfying scheme
$\theta=\varepsilon=1$. Square wave.

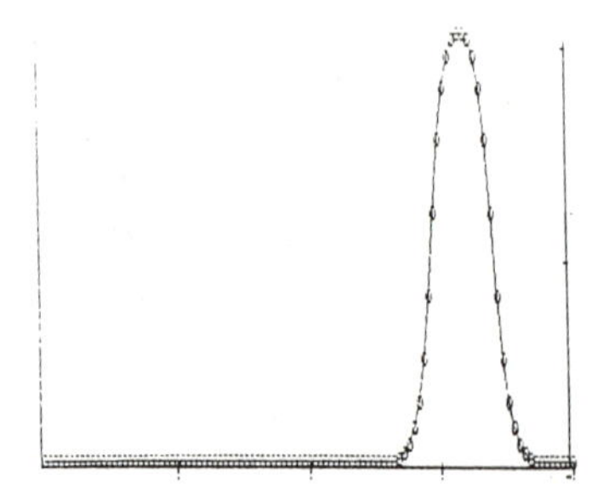

Fig. 9
Entropy satisfying scheme
$\theta=\varepsilon=1$. Sine wave.

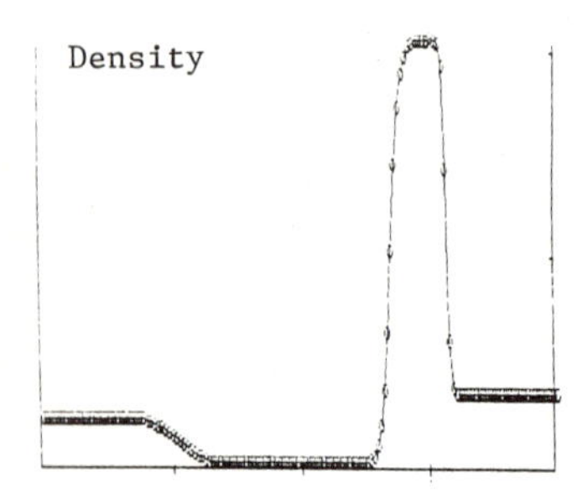

Fig. 10
Entropy satisfying scheme
$\theta=\varepsilon=1$.

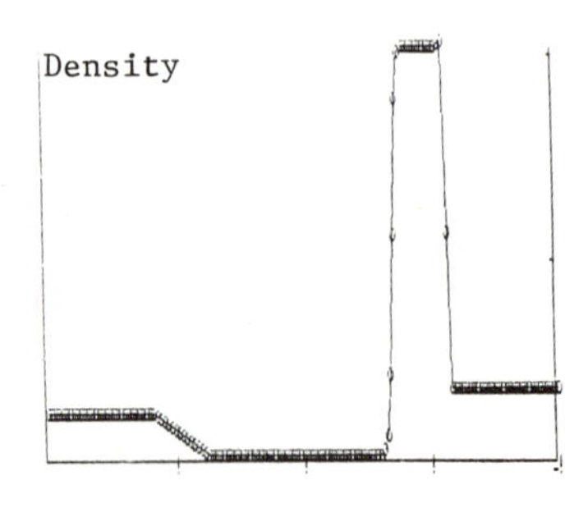

Fig. 11
Ultra-compressive limiter dependent on local C.F.L. number.

MULTIGRID METHOD FOR SOLVING EULER AND NAVIER-STOKES EQUATIONS IN TWO AND THREE DIMENSIONS

V. Couaillier

Office National d'Etudes et de Recherches Aérospatiales (ONERA)
BP 72 - 92322 CHATILLON (France)

SUMMARY

In this paper we describe the multigrid method we use for solving Euler and Navier-Stokes equations and which is based upon the method proposed by Ni [1] [2]. Because the multigrid acceleration technique used for the Navier-Stokes equations is straightforwardly derived from that used for the Euler equations, the main part of the presentation is dedicated to the Euler multigrid solver. Comparisons between calculations performed on a 3D complex inviscid flow allow us to optimize the convergence rate of the multigrid process by modifying the transfer operator. We indicate how the multigrid method can be easily extended to the solution of the Navier-Stokes equations.

I - INTRODUCTION

In 1981, Ni introduced a multigrid method for solving the Euler equations using a scheme based on an interpretation of the well-known Lax-Wendroff scheme [1]. Since then, several methods have been derived from Ni's scheme and several codes have been developed in 2-D, and some in 3-D. This type of multigrid method is also useful for solving the steady Navier-Stokes equations [3] [4].

II - MULTIGRID METHOD FOR THE 3D EULER EQUATIONS

The Euler code we use is based on the following numerical techniques : an explicit centered scheme of Lax-Wendroff type, a multidomain approach with or without overlapping, a numerical boundary condition and matching condition treatment based on the use of the characteristic relations [5]. The implementation of these techniques in 3-D is described in [6].
In this section, we describe the implementation of the multigrid method in 3-D which is analogous in the principle to the one described in 2-D [7].
The Euler equations can be written in the following conservative and compact form :

$$\frac{\partial f}{\partial t} + div\,(F) = 0\,, \tag{1}$$

where f denotes the conservative variables $(\rho, \rho\overline{V}, \rho E)^T$ and F denotes the fluxes $(\rho\overline{V}, \rho\overline{V}\otimes\overline{V} + p\overline{\overline{I}}, (\rho E + p)\,\overline{V})^T$, ρ, $\overline{V}$, p et E being respectively the density, the flow velocity, the static pressure and the total energy.

II-1 Basic numerical scheme

a) Time discretization

We use the two following time discretization schemes, where the solution f^{n+1}, at time $t^{n+1} = t^n + \Delta t$ is obtained with second order Taylor expansion :

- The one-step Lax-Wendroff scheme, denoted $S1$, which can be written as follows :

$$\delta f^n = -\Delta t \; div(F^n) \tag{2.a}$$

$$\delta f^{n+½} = \delta f^n - \frac{\Delta t}{2} div(A^n . \delta f^n), \tag{2.b}$$

$$f^{n+1} = f^n + \delta f^{n+½}, \tag{2.c}$$

where $\delta f = \Delta t \; \partial f / \partial t$. The relation (2.b) can be considered as a time integration of the following equation for the residual δf :

$$\frac{\partial \delta f}{\partial t} + div(A.\delta f) = 0, \tag{3}$$

- The two-step Lax-Wendroff scheme, denoted $S2$, :

$$f^{n+½} = f^n - \frac{\Delta t}{2} div(F^n), \tag{4.a}$$

$$f^{n+1} = f^n - \Delta t \; div(F^{n+½}), \tag{4.b}$$

Because the space discretization of the scheme $S2$ can be easily deduced from the space discretization of the scheme $S1$, and because the multigrid method relies upon an interpretation of the scheme $S1$, we will only describe the numerical implementation of the latter scheme.

b) Space discretization

At each time step t^n, the solution f^n is known at each point of the mesh , and the space discretization of the one-step scheme (2) is made of two stages.

i) the calculation of the first order term (2.a),
ii) the calculation of the second order term (2.b).

In the following we denote by A,B,C,D,E,F,G,H the eight mesh points defining the cell Ω_a the center of which is the point a, and by a,b,c,d,e,f,g,h the eight centers of the cells surrounding the mesh point G and defining the staggered cell ω_G (see figure 1). The first order term δf (2.a) is evaluated by the formula of the mean

$$\delta f_a = \frac{-\Delta t}{v(\Omega_a)} \int_{\Omega_a} div(F) \, dv = \frac{-\Delta t}{v(\Omega_a)} \int_{\partial\Omega_a} (F).\overline{n} \, dS, \tag{5}$$

by assigning the mean value δf_a of δf to the barycenter a of the cell Ω_a.

In a first version of the code the calculation of δf was performed by assuming that each face of the control volume Ω was a piece of hyperboloid which contained the four vertices of the face. Thus, for the face $\partial\Omega_{ABCD}$ of Ω_a the vertices of which are A,B,C and D, each physical coordinate x,y and z is expressed as a bilinear function of the coordinates X and Y of the unit square $\overline{\partial\Omega}$ (figure 2) with the following relation :

$$\Phi(X,Y) = (1-X)(1-Y).\Phi_A + X(1-Y).\Phi_B + XY.\Phi_C + (1-X)Y.\Phi_D \tag{6}$$

where Φ is replaced by each coordinate x,y,z.

In a same way the fluxes being known at the vertices A,B,C and D of $\partial\Omega_{ABCD}$, the fluxes at each point of the face were expressed with the relation (6).

In the last version of the code, we have chosen another way to compute δf_a. We approximate the flux of F through the face $\partial\Omega_{ABCD}$ by using the relation :

$$\int_{\partial\Omega_{ABCD}} (F).\overline{n} \, dS = \frac{F_A + F_B + F_C + F_D}{4} \int_{\partial\Omega_{ABCD}} \overline{n} \, dS, \tag{7}$$

where $\overline{ndS_a} = \int_{\partial\Omega_{ABCD}} \overline{n} \, dS$ is evaluated as the half cross product of the two diagonal line segments of the face.

We can remark here that this approximation is equivalent to assuming that $\overline{n}$ at each point of the face is given by the relation (6) with $\overline{n}$ known at A, B,C and D. The two approximations are identical for a plane cell, then for sufficiently regular mesh they are very closed. The approximation (F2) is more simple than the approximation (F1), and one can compute and store the quantities $\overline{ndS}$ at the beginning of the computation; then it is clearly evident that (F2) is less expensive than (F1) but requires in this case more memory storage.

Notice that the volume $v(\Omega_a)$ is defined as the flux through Ω of the vector $\overline{x}$.

At last, the first order term of (2.b) at the mesh point G is obtained by taking the arithmetic mean of the values of δf computed at the barycenters a, b, c, d, e, f, g et h of the cells surrounding G.

The first order term values δf being known at the barycenters of the cells, the quantity $div(A\,\delta f\,)$ is directly determined at the mesh point G with a formula of the mean on the control volume ω_G.

The second order term is evaluated as follows :

$$[\,div(A\,\delta f\,)\,]_G = \frac{1}{v(\omega_G)} \int_{\partial\omega_G} A\,\delta f\,.\,\overline{n}\,dS\,. \tag{8}$$

The contour integral is calculated by assuming that the operator A (jacobian matrix) and the first order term δf are constant on each cell Ω. The operator A is defined, for example on the cell Ω_a, by $A_a = A\,(\,f_a\,)$, where f_a is the arithmetic mean of the unknowns at the nodes of Ω_a :

$$f_a = (\,f_A + f_B + f_C + f_D + f_E + f_F + f_G + f_H\,)\,/\,8$$

The expression (8) can then be written :

$$[div(A\,\delta f\,)]_G = \sum_a (\,A\,\delta f\,)_a\,.\,\overline{N}_a \quad with \quad \overline{N}_a = \frac{1}{v(\omega_G)} \int_{\partial\Omega_a} \overline{n}\,dS\,, \tag{9}$$

where the summation Σ is done on the eight cells Ω_a, Ω_b, Ω_c, Ω_d, Ω_e, Ω_f, Ω_g and Ω_h surrounding the point G, and where the surface $\partial\Omega_a$ in the definition of the vector $\overline{N}_a$ is the intersection of the boundary $\partial\omega_G$ of the volume ω_G with the volume Ω_a.

In the computation code, we used the "distribution formulas" introduced by Ni [1], which can be obtained by defining the vector $\overline{N}_a$ by an expression which at least for sufficiently regular meshes, is similar to the one defined above and which brings in only the geometry of the cell Ω_a; we then get the following relations (fig. 1) :

$$\overline{N}_a^{\,G} = \frac{1}{v(\Omega_a)} [\,ci_a^{\,G}\,\overline{ndS}_I + cj_a^{\,G}\,\overline{ndS}_J + ck_a^{\,G}\,\overline{ndS}_K\,]\,, \tag{10}$$

where

$$\overline{ndS}_I = [\,\overline{ndS}_{ADHE} + \overline{ndS}_{BCGF}\,]\,/\,8\,, \tag{11.a}$$

$$\overline{ndS}_J = [\,\overline{ndS}_{ABFE} + \overline{ndS}_{DCGH}\,]\,/\,8\,, \tag{11.b}$$

$$\overline{ndS}_K = [\,\overline{ndS}_{ABCD} + \overline{ndS}_{EFGH}\,]\,/\,8\,. \tag{11.c}$$

The three "distribution coefficients" $ci_a^{\,G}$, $cj_a^{\,G}$ and $ck_a^{\,G}$ are the distribution coefficients of the cell Ω_a relative to the corner point G of Ω_a. For each point of Ω_a the geometrical coefficients defined by (11) are the same; only the distribution coefficients are different and, the vectors $\overline{ndS_l}$ $(l=i,j,k)$ being oriented in the direction of the increasing indices, these coefficients are defined as follows :

$$ci_a^{\,A} = +1\,,\quad cj_a^{\,A} = +1\,,\quad ck_a^{\,A} = +1\quad;\quad ci_a^{\,B} = -1\,,\quad cj_a^{\,B} = +1\,,\quad ck_a^{\,B} = +1$$

$$ci_a^{\,C} = -1\,,\quad cj_a^{\,C} = -1\,,\quad ck_a^{\,C} = +1\quad;\quad ci_a^{\,D} = +1\,,\quad cj_a^{\,D} = -1\,,\quad ck_a^{\,D} = +1$$

$$ci_a^{\,E} = +1\,,\quad ci_a^{\,E} = +1\,,\quad ck_a^{\,E} = -1\quad;\quad ci_a^{\,F} = -1\,,\quad cj_a^{\,F} = +1\,,\quad ck_a^{\,F} = -1$$

$$ci_a^{\,G} = -1\,,\quad cj_a^{\,G} = -1\,,\quad ck_a^{\,G} = -1\quad;\quad ci_a^{\,H} = +1\,,\quad cj_a^{\,H} = -1\,,\quad ck_a^{\,H} = -1$$

The space discretization of the divergence operators in (2) being performed with these approximations, the scheme (2) can then be written :

$$f_G^{\,n+1} = f_G^{\,n} + \sum_a \left[\frac{\delta f_a}{8} - \frac{\Delta t}{2} (\,A\,\delta f\,)_a^n\,.\,\overline{N}_a^{\,G} \right]. \tag{12}$$

The time step Δt used to integrate the solution in time with the above scheme is a local time step which is assumed to verify :

$$\Delta t = \eta \frac{h}{V + C},$$

where h is a characteristic length of the mesh size, representing an evaluation of the dimension of the numerical dependence domain, C the local sound speed, and where η is a numerical coefficient of the order of unity introduced because of the approximate character of the criterion.

c) Numerical smoothing

To ensure the stability of the scheme and capture the flow discontinuities correctly an additional artificial damping is needed. The numerical smoothing we use is that proposed by Jameson and Schmidt [8] with a boundary numerical treatment described by Eriksson [9] and Rizzi [10]. This additional term, which is a combination of a second order derivative and of a fourth order derivative, is introduced in the process as follows :

$$f^{n+1} = f_{sch.}^{n+1} + \frac{\Delta t}{v(\omega_G)} [D_2(f^n) + D_4(f^n)], \tag{13}$$

where "$sch.$" denotes the values obtained after application of the basic numerical scheme.

At a mesh point G of coordinates i , j , k we have:

$D_l(f) = D_l^I + D_l^J + D_l^K$, for $l = 2$ and $l = 4$.

The second order term is defined with the following relations :

$$D_2^I = d_{i+1/2,j,k}^{(1)} - d_{i-1/2,j,k}^{(1)} \tag{14.a}$$

$$\text{with } d_{i+1/2,j,k}^{(1)} = \epsilon_{i+1/2,j,k}^{(2)} \cdot \lambda_{i+1/2,j,k} \cdot (f_{i+1,j,k} - f_{i,j,k}). \tag{14.b}$$

In this expression λ is a scaling factor which represents the spectral radius of the characteristic system in the direction normal to the face of ω_G which is crossed by the segment [(i,j,k) ; (i+1,j,k)] (If the mesh is cartesian this segment and the normal of the face are aligned).

Let $\partial\omega_a$ be this face, thus :

$$\lambda_{i+1/2,j,k} = | \overline{V.ndS_a} | + C \, || \overline{ndS_a} || . \tag{15}$$

The coefficient $\epsilon_{i+1/2,j,k}^{(2)}$ is defined as follows :

$$\epsilon_{i+1/2,j,k}^{(2)} = \min [1 , K^{(2)} \max (\nu_{i,j,k}^I , \nu_{i+1,j,k}^I)] \tag{16}$$

$\nu_{i,j,k}^I$, a sensor which has to detect the discontinuities in the flow is often taken as a second order derivative of the pressure. Starting from this choice and because we want to detect contact discontinuities, we combine a second order derivative of the pressure and a second order derivative of the velocity.

$$\nu_{i,j,k}^I = \frac{1}{2} \frac{P_{i+1,j,k} - 2P_{i,j,k} + P_{i-1,j,k}}{| P_{i+1,j,k} + 2P_{i,j,k} + P_{i-1,j,k} |} + \frac{1}{2} \frac{V_{i+1,j,k} - 2V_{i,j,k} + V_{i-1,j,k}}{| V_{i+1,j,k} + 2V_{i,j,k} + V_{i-1,j,k} |} . \tag{17}$$

The fourth order term has the following form :

$$D_4^I = d_{i+1,j,k}^{(2)} - 2\, d_{i,j,k}^{(2)} + d_{i-1,j,k}^{(2)} \tag{18.a}$$

$$\text{with } d_{i,j,k}^{(2)} = \epsilon_{i,j,k}^{(4)} \cdot \lambda_{i,j,k} \cdot (f_{i+1,j,k} - 2 f_{i,j,k} + f_{i-1,j,k}). \tag{18.b}$$

This fourth order term is discarded in the regions of the flow where the gradients are high by using the following relations :

$$\epsilon_{i,j,k}^{(4)} = \max [0 , K^{(4)} - \epsilon_{i,j,k}^{(2)}] . \tag{19}$$

$K^{(2)}$ and $K^{(4)}$ are two constant coefficients respectively equal to 1 and to $\frac{1}{64}$ in the code.

II-2 Multigrid acceleration

a) Principle of the method

The multigrid acceleration ([3],[4] and [1]) applied to system (1) consists in integrating on coarse grids the following residual system based on the time derivation of system (1) :

$$\frac{\partial \delta f}{\partial t} + div\ A(f)\ \delta f = 0. \tag{20}$$

The residual δf is defined by $\delta f_h = \Delta t\ \partial f / \partial t$ in the fine grid G_h, and initialized in the k-th coarse grid $G_{2^k h}$ by a transfer of the residual obtained in the previous grid $G_{2^{k-1}h}$.

If $T_{2^{k-1}h}^{2^k h}$ denotes a transfer operator from $G_{2^{k-1}h}$ to $G_{2^k h}$ and $I_{2^k h}^{h}$ denotes an interpolation operator from $G_{2^k h}$ to G_h, the solution u in the fine grid is defined after a multigrid cycle with use of p coarse grids by :

$$f_h^{n+1+\bar{\alpha}} = f_h^{n+1} + \sum_{k=1}^{p} \alpha_k\ I_{2^k h}^{h}\ \delta f_{2^k h}^{\ n+1/2+\bar{\beta}_k}, \tag{21}$$

with $\delta f_{2^k h}^{\ n+1/2+\bar{\beta}_k}$ calculated by integration of (20) on the time interval $\beta_k\ \Delta t$ (with $\bar{\beta}_k = \Sigma \beta_k$) starting from the initialization:

$$\delta f_{2^k h}^{\ n+1/2+\bar{\beta}_{k-1}} = T_{2^{k-1}h}^{2^k h}\ \delta f_{2^{k-1}h}^{\ n+1/2+\bar{\beta}_{k-1}} \tag{22}$$

where $\delta f_h = f_h^{n+1} - f_h^{n}$.

In equation (21), $\bar{\alpha} = \Sigma \alpha_k$ denotes the time-advance coefficient of the solution in the fine grid, whereas β_k denotes the time-advance coefficient used for the integration of (3) in the k-th coarse grid.
The efficiency of the method is due to the fact that the residual long waves relative to the fine grid are very slow to vanish whereas they become small waves on the coarse grids, and are then well damped on these grids.
Starting from this point of view , it is clear that it is necessary that the basic scheme applied in the fine grid has to damp rapidly the high frequency errors (or residuals). In fact the multigrid process does not take into account high frequency terms and reintroduces high frequency errors with the interpolation stage. Moreover, the converged solution retains the space accuracy of the basic scheme in the fine grid because the quantity δu_h vanishes at convergence.

b) Coarse grid operators

The notations used in this paragraph refer to the figure (2).

- Transfer operator $T_{2^{k-1}h}^{2^k h}$:

This operator initializes the quantity $\delta f_{2^K h}$ in a cell center of the mesh $G_{2^K h}$. We have used two types of transfer operator :
The first one is a simple arithmetic mean transfer operator which has been used for the 2-D codes previously mentioned [7] and [4]. For this operator a one-dimensional theoretical study, confirmed by two dimensional Euler and Navier-Stokes numerical tests, to optimize the parameters α_k and β_k, had been done. This operator, denoted by T_1, can be chosen as follows (see figure 2) :

$$(\delta f_{2^K h})_{11} = \xi (\delta f_{2^{K-1}h})_{11} + \frac{1-\xi}{8} \sum_{1,..,8} (\delta f_{2^{K-1}h}) \tag{23}$$

where ξ has to be taken between 0.125 and 0.25 to ensure the stability of the multigrid process (this is in agreement with 1-D stability analysis [11]).

The second operator, denoted by T_2, described by Ni [4] and Usab [12] and called Injection-Distribution method, combines a direct transfer and a transfer defined by distribution formulas, and is more into accord with the physics of the problem. This operator is defined as follows :

$$(\Delta f_{2^K h})_{11} = \mu_d (\delta f_{2^{K-1}h})_{11} + \mu_i \sum_{1,..,8} (Df_{2^K h}) \tag{24}$$

$$(Df_{2^K h})_1 = \left[\frac{(\delta f_{2^{K-1}h})_1}{8} - \frac{\Delta t}{2} (A\,\delta f_{2^{K-1}h})_1 \cdot \bar{N}_1^{\,11} \right] \tag{25}$$

where $(Df_{2^K h})_1$ denotes the contribution of the residual $(\delta f_{2^{K-1}h})_1$ defined at point N_1 to the point N_{11} considered as a corner point of a cell the center of which would be the point N_1. The geometrical coefficients of this contribution are those of the coarse mesh which is centered at N_{11}. The coefficients μ_d and μ_i are both equal to 1 in the transfer operator T_2 proposed by Ni [4]. Whereas for the transfer operator T_1 the time-advance coefficients α_k cannot be chosen greater than 1 if one wants to ensure a stable process in most of the computational cases, the transfer operator T_2 defines a stable process with coefficients α_k equal to 2^k.

- Integration of the residual equation :

The integration of the residual equation on the coarse grid is formally identical to the integration of the residual δf for the calculation of the second order term in the basic scheme (2.b). The residual being known at the centers of the coarse mesh by using one of the above transfer operator, the new residuals at the nodes of the coarse mesh are obtained by using the distribution formulas on the coarse grid; for instance the contribution from the cell with center N_{11} to the point N_1, has the following form :

$$(\delta f_{2^K h})_1^{11} = \left[\frac{(\Delta f_{2^{K-1}h})_{11}}{8} - \beta_k \frac{\Delta t}{2} (A \Delta f_{2^K h})_{11} \cdot \overline{N}_{11}^{1} \right]. \tag{26}$$

$\overline{N}_{11}^{1}$ is defined with the same geometrical coefficients than $\overline{N}_1^{11}$; on the other hand these two vectors have opposite distribution coefficients. The coefficient β_k is equal to 2^k, whereas Δt is the time step evaluated in the fine grid.
In the expressions (25) and (26) the vectors $\overline{N}$ are defined on the coarse grids in the same way than those defined on the fine grid with the relation (10). Moreover the vector $\overline{ndS}$ of a coarse grid face is defined as the summation of the vectors $\overline{ndS}$ of the finer grid cells defining this face, and the volumes are defined in the same way.

- Interpolation operator $I_{2^k h}^{2^{k-1}h}$:

This operator is defined for the four types of points of the mesh $G_{2^K h}$ with the following arithmetic mean relations :

$$(I_{2^K h}^{2^{K-1}h} \cdot \delta f_{2^{K-1}h})_{11} = \frac{1}{8} \sum_{1,..,8} (\delta f_{2^K h}) \tag{27.a}$$

$$(I_{2^K h}^{2^{K-1}h} \cdot \delta f_{2^{K-1}h})_{10} = \frac{1}{4} \sum_{1,..,4} (\delta f_{2^K h}) \tag{27.b}$$

$$(I_{2^K h}^{2^{K-1}h} \cdot \delta f_{2^{K-1}h})_{9} = \frac{1}{2} \sum_{1,2} (\delta f_{2^K h}) \tag{27.c}$$

$$(I_{2^K h}^{2^{K-1}h} \cdot \delta f_{2^{K-1}h})_{1} = (\delta f_{2^K h})_1 \cdot \tag{27.d}$$

II-3 Boundary conditions

The treatment of the boundary conditions is based on the use of the characteristic relations associated with the normal to the boundary. This technique and its numerical implementation is detailed in [5]. It consists in keeping only the characteristic relations that transport information outside the computational domain and in replacing the discarded relation, (i.e. the relations corresponding to a transport inside the computational domain), by some boundary conditions.
The numerical treatment of boundary conditions is applied after each coarse grid integration to ensure a stable process, even if it is not necessary in all the cases. This treatment at the coarse grid boundary points is applied with the normal vectors calculated from the fine grid geometry to retain the fine grid accuracy for the converged solution.

II-4 Flow around a swept wing at high incidence

We present here calculations performed with the multigrid method for a flow around a swept wing with

a thin profile. The actual leading edge, with a small curvature radius, is approximated by a sharp one. Thus, the computational domain discretization is realized with two sub-domains (one upper domain and one lower domain). The two sub-domains are symmetric with respect to the wing plane, and each of them contains 81 points in the flow direction with 33 points on the wing, 45 points in the span direction with 31 points on the wing and 21 points in the direction normal to the wing plane. On the figure (3) is plotted a perspective view of the mesh in a neighbouring of the wing in the upper domain. The applied boundary conditions are the following : slip condition on the wing, non-reflective condition on the external boundaries and matching condition between the two subdomains.

We present now comparisons of five calculations using different transfer operator coefficients for the case M=0.84 and θ=25 ° ; these calculations are the following :

- calculation $c\,1$: without multigrid stage
- calculation $c\,2$: 2 coarse grids and transfer operator T_1 , ξ = 0.125
- calculation $c\,3$: 1 coarse grid and transfer operator T_2 , μ_d = 1 , μ_i = 1
- calculation $c\,4$: 2 coarse grids and transfer operator T_2 , μ_d = 1 , μ_i = 1
- calculation $c\,5$: 2 coarse grids and transfer operator T_2 , μ_d = 0 , μ_i = 2 .

The figures (4),(5) and (6) present a comparison of the evolution of the quadratic mean residual of ρ and of the lift coefficient C_z versus the number of iterations respectively for the calculations $c\,1$, $c\,2$ and $c\,3$, for the calculations $c\,1$, $c\,3$ and $c\,4$, and for the calculations $c\,1$, $c\,3$ and $c\,5$. One can remark that $c\,2$ and $c\,3$ using an identical time-advance coefficient in the multigrid stage (α_1=1 and α_2=1 for $c\,2$ whereas α_1=2 for $c\,3$), the evolutions of the C_z are very close, but $c\,3$ converges better. The calculation $c\,4$ does not converge, and several reasons can be a priori retained : the time-advance coefficient in the second coarse grid is perhaps too great, the wing tip is not represented on this grid (it corresponds to a mesh line equal to 31 in the fine grid), or the coefficients μ_d and μ_i can be chosen different. Starting from the fact that the multigrid process is unstable with T_1 if the coefficient ξ is greater than 0.25, one tried to apply the same repartition with the transfer operator T_2, μ_d=0.5 and μ_i=1.5. In this case the process converged well, but the best result we obtained which is plotted on figure (6) is for μ_d=0 and μ_i=2. In fact, the transfer operator T_2 with this last choice of coefficient is corresponding to a time integration residual in the coarse grid with a time-step corresponding to this grid. For this case and considering that the number of iterations needed to reach the steady state with the calculation $c\,5$ is divided by a factor 3 with respect to the calculation $c\,1$, and that the added cost per iteration with multigrid is 15%, we have saved 62% of the computational cost by using the multigrid technique.

We present also another calculation on figure (7) performed with the same flow conditions but with a shifted mesh so that the wing tip is represented in the second coarse grid (it corresponds to a mesh line equal to 29 in the fine grid). One remarks that the calculation $c\,4$ does not diverge on this mesh and that the calculation $c\,5$ converges much better.

The figure (8) presents the isobaric lines on the upper surface of the wing, where we can observe the trace of the large vortex starting from the leading-edge of the wing.

III MULTIGRID METHOD FOR THE 3D NAVIER-STOKES EQUATIONS

We just recall here how one can extend the multigrid method for solving the Navier-Stokes equations. For more details we refer to the following papers [4],[13] which indicate the different physical and numerical aspects to consider.

The Navier-Stokes equations being written in the following compact form,

$$\frac{\partial u}{\partial t} + div\ (F - F_v) = 0\ , \tag{28}$$

F and F_v being respectively the inviscid and viscous fluxes.

The one-step Lax-Wendroff scheme $S1$ can be adapted to solve the Navier-Stokes equations as follows :

$$\delta f^{\,n} = -\Delta t\ div(\,F^n\ - F_v^{\ n}) \tag{29.a}$$

$$f^{\,n+1} = f^{\,n} + \delta f^{\,n} - \frac{\Delta t}{2}\ div(\,A^n\,.\,\delta f^{\,n}\ )\ , \tag{29.b}$$

where $\delta f = \Delta t \, \partial f / \partial t$, and A denotes the inviscid jacobian matrix. Thus the residual equation for the Navier-Stokes equations is strictly the same as that defined for the Euler equations, and the multigrid process developed in inviscid fluid can be used without modifying it. We present here the result of a Navier-Stokes calculation performed by Cambier and Escande [14] with the scheme $S2$ as basic scheme, and with the multigrid acceleration technique (transfer operator T_1). This calculation concerns a shock-boundary-layer interaction in a transonic channel, and we refer to [14] for the physical description of the computation and for the analysis of the results. We just indicate that the mesh contains 600,000 points and that plotting accuracy (decrease of the residuals by three orders of magnitude) is obtained after 5,000 iterations. The figure (9) represents the iso-Mach number lines in three planes parallel to the lateral walls.

IV - CONCLUSION

The multigrid method has proved its ability to accelerate the convergence rate in 3-D complex calculations for the Euler and for the Navier-Stokes equations. The numerical study done on the 3D Euler calculation presented has proved the importance of a good choice for the transfer operator used to optimize the convergence.

REFERENCES

[1] Ni R.H. , A Multiple-Grid Scheme for Solving the Euler Equations, AIAA J., vol. 20,n° 11, (1982).

[2] Ni R.H. and Bogoian J.C., Prediction of 3D Multi-Stage Turbine Flow Field Using a Multiple-Grid Euler Solver, AIAA Paper n° 89-0203, (1989).

[3] Davis R.L., Ni R.H. and Carter J.E., Cascade Viscous Flow Analysis Using the Navier-Stokes Equations, AIAA Paper n° 86-0033, (1986).

[4] Cambier L., Couaillier V. and Veuillot J.P., Numerical Solution of the Navier-Stokes equations by a Multigrid Method, La Recherche Aérospatiale n° 1988-2, pp. 23-42 (English Edition).

[5] Viviand H. and Veuillot J.P., Méthodes pseudo-instationnaires pour le calcul d'écoulements transsoniques, ONERA Publication n° 1978-4, (English translation, ESA TT 561).

[6] Vuillot A.M., A Multi-Domain 3D Euler Solver for Flows in Turbomachines, Proceedings of the 9th ISABE Symposium, Athens (Sept. 1989).

[7] Couaillier V., Solution of the Euler Equations : Explicit Scheme Acceleration by a multigrid Method, 2nd European Conference on Multigrid Methods, GAMM Cologne, Oct. 1985, GMD-Studien n° 110 and ONERA TP n° 1985-129.

[8] Jameson A. and Schmidt W., Some Recent Developments in Numerical Methods for Transonic Flows, Computer Methods in Applied Mechanics and Engineering 51, pp. 467-493, (1985) North-Holland.

[9] Eriksson L.E., Boundary Conditions for Artificial Dissipation Operators, FFA TN 1984-53.

[10] Rizzi A., Spurious Entropy Production and Very Accurate Solutions to the Euler Equations, AIAA paper n° 84-1644, (1984).

[11] Couaillier V. and Peyret R., Theoretical and Numerical Study of the Ni's Multigrid Method, La Recherche Aérospatiale, n° 1985-5, French and English Editions.

[12] Usab W.J., Embedded Mesh Solutions of the Euler Equations Using a Multiple-Grid Method, Ph.D. Thesis, Dept. of Aeronautics and Astronautics, MIT (1983).

[13] Veuillot J.P. and Cambier L., Computation of High Reynolds Number Flows around Airfoils by Numerical Solution of the Navier-Stokes equations, 11th International Conference on Numerical Methods in Fluid Dynamics, Williamsburg, Virginia, (June 1988).

[14] Cambier L. and Escande B., Navier-Stokes Simulation of a Shock Wave-Turbulent Boundary Layer Interaction in a Three-Dimensional Channel, AIAA 20th Fluid Dynamics and Lasers Conference, Buffalo (N.Y.), (June 1989).

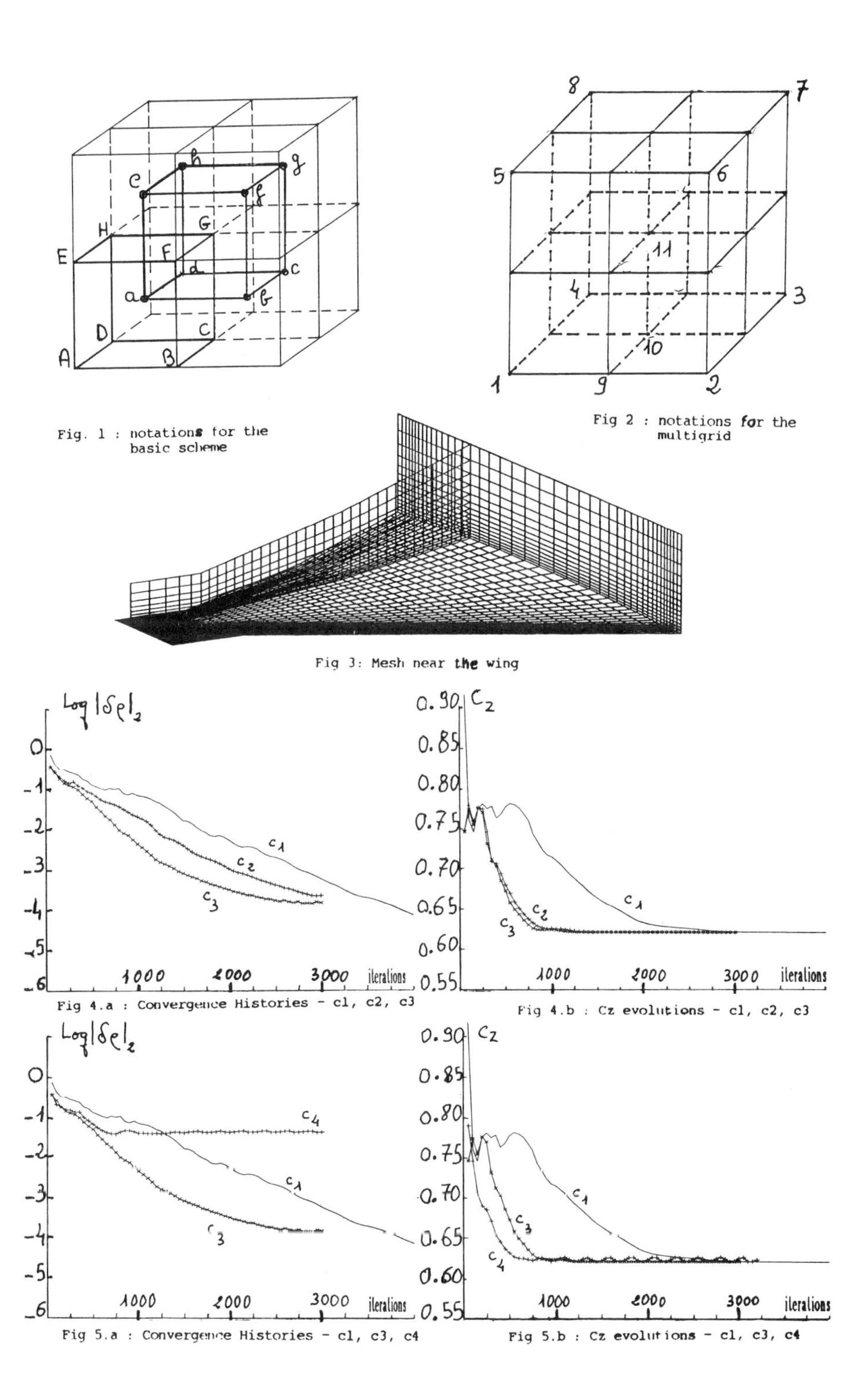

Fig. 1 : notations for the basic scheme

Fig 2 : notations for the multigrid

Fig 3: Mesh near the wing

Fig 4.a : Convergence Histories - c1, c2, c3

Fig 4.b : Cz evolutions - c1, c2, c3

Fig 5.a : Convergence Histories - c1, c3, c4

Fig 5.b : Cz evolutions - c1, c3, c4

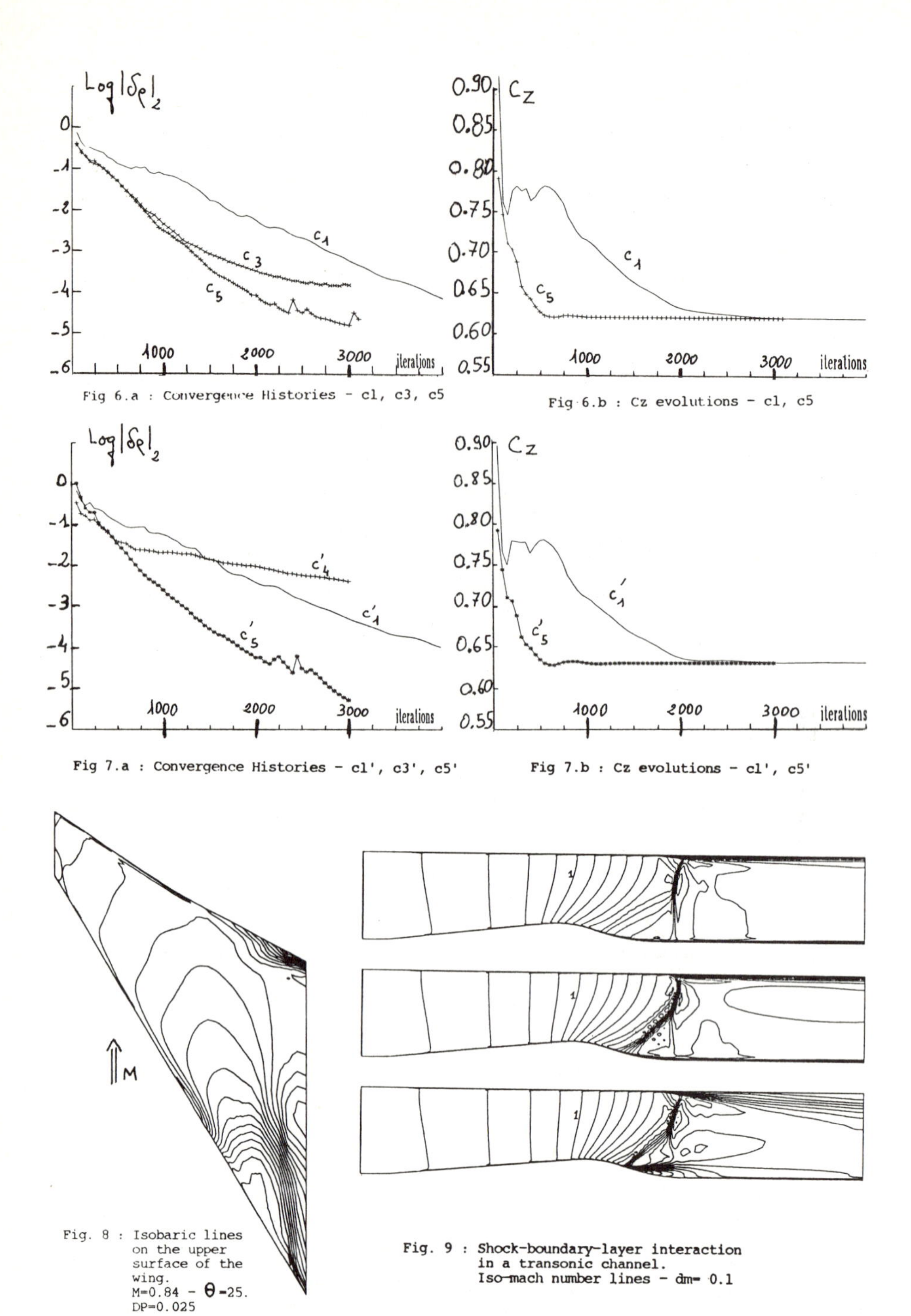

Fig 6.a : Convergence Histories - c1, c3, c5

Fig 6.b : Cz evolutions - c1, c5

Fig 7.a : Convergence Histories - c1', c3', c5'

Fig 7.b : Cz evolutions - c1', c5'

Fig. 8 : Isobaric lines on the upper surface of the wing. M=0.84 - θ=25. DP=0.025

Fig. 9 : Shock-boundary-layer interaction in a transonic channel. Iso-mach number lines - dm= 0.1

UNSTEADY VISCOUS FLOW CALCULATIONS INCLUDING SURFACE HEATING AND COOLING EFFECTS

C. P. van Dam, M. Hafez, D. Brucker, S. Elli

Division of Aeronautical Science and Engineering
Department of Mechanical Engineering
University of California
Davis, CA 95616, U.S.A.

ABSTRACT

A fully implicit solution technique is applied to study the unsteady flow characteristics including surface heating and cooling effects of an infinite circular cylinder at a Reynolds number of 100. The fluid flow problem is governed by the Boussinesq approximation of the Navier-Stokes equations. The equations in the finite-difference form are linearized using Newton's method and the resulting system of equations including the boundary conditions are solved simultaneously using a direct solution technique. In a previous paper, the solution technique was presented and results (coarse grid) were discussed. Here, results are presented for higher grid densities, different boundary conditions, and various angles between the freestream velocity vector and the gravitational acceleration vector.

INTRODUCTION

With the recent advances in the memory size and the speed of computers it has become practical to solve fluid flow problems with direct solution techniques. The main advantages of direct solution methods over iterative solution methods are their robustness and their grid-independent convergence characteristics [1]. However, the implementation of direct solvers, especially on smaller computers, is still hampered by the relatively large memory requirements. For two-dimensional incompressible viscous flow calculations the memory and the CPU time can be minimized by using the biharmonic equation to solve for the stream function [2]. This formulation has the advantage that only one fourth-order differential equation must be solved. The use of the biharmonic equation has not been very popular. This is mainly because of the slow rate of convergence obtained by iterative methods when solving this type of equation. However, the biharmonic formulation becomes quite attractive when a direct solution method is used. This approach to solve steady viscous flow problems has been applied by several researchers [2-7]. Recently, the unsteady flow problem was solved by Schütz and Thiele [8] and the authors [9] using the biharmonic equation and a direct solver at each time step.

The problem of the unsteady flow past an infinite circular cylinder is often used to benchmark two-dimensional time-accurate algorithms; see [10] for an overview. Extensive sets of experimental data for unsteady flow around circular cylinders across an extended Reynolds number range are available in the literature; e.g. [11]. For Reynolds numbers greater than about 60 alternating vortices are shed with a nondimensional frequency (Strouhal number) of about 0.21 for the Reynolds number range from about 200 to 10^5. The Strouhal number is lower below this Reynolds number range and higher for Reynolds numbers greater than 10^5. This cyclic flow field is called the Karman vortex street.

The introduction of a temperature difference between the cylinder surface and the freestream fluid can significantly affect the wake development. Noto et al. have shown some interesting flow developments for heated and cooled cylinders; see [12] for a summary. The temperature difference gives rise to variations in the fluid properties. By introducing the Boussinesq approximation, only the variation of the density in the buoyancy force is included [13]. Under the influence of the buoyancy force the Karman vortex street disappears. A steady flow field develops when the freestream velocity and the buoyancy force point in the same direction. An unsteady flow field appears to develop when the two have the opposite direction [9].

In the present paper the effects of temperature variations on the time-dependent viscous flow characteristics of a circular cylinder are studied using a fully implicit solution technique which is based on the combination of biharmonic equation, Boussinesq approximation, Newton's linearization, and direct solver. The calculations are conducted for a Reynolds number of 100 and a Prandtl number of 0.72 (air). The influence of different farfield boundary conditions on the solution are also studied.

GOVERNING EQUATIONS

Consider an infinite circular cylinder of radius a that has an isothermal surface temperature T_0 located in a fluid at rest of identical temperature. At time $t \geq 0$ the cylinder obtains the temperature T_1 and a uniform velocity U_∞ is applied to the fluid at infinity. The freestream velocity vector and the gravity vector are at an angle α (Fig. 1). The problem is governed by the strict Boussinesq equations; thus, the fluid is considered incompressible except for the density in the buoyancy force term [13]. The governing equations can be written in the form of the biharmonic equation and the energy equation. In strictly conservative form and transformed cylindrical coordinates these equations are:

$$(\nabla^2\psi)_t + (\tfrac{1}{G}\,\psi_\xi\nabla^2\psi)_\eta - (\tfrac{1}{G}\,\psi_\eta\nabla^2\psi)_\xi = \frac{2}{Re}\,\nabla^2(\tfrac{1}{G}\,\nabla^2\psi)$$
$$- \frac{1}{2}\frac{Gr}{Re^2}\left[\left\{\sqrt{G}\,\sigma\,\sin\pi(\eta-\eta^*)\right\}_\xi + \left\{\sqrt{G}\,\sigma\,\cos\pi(\eta-\eta^*)\right\}_\eta\right] \tag{1}$$

$$G\,\sigma_t + (\sigma\,\psi_\xi)_\eta - (\sigma\,\psi_\eta)_\xi = \frac{1}{Pr}\frac{2}{Re}\,\nabla^2\sigma \tag{2}$$

where t, ξ, η, ψ and σ represent the dimensionless time, radial coordinate, transverse coordinate, stream function, and temperature, respectively. The dimensionless variables in these equations are:

$$t = t'\,U_\infty/a \qquad \psi = \psi'/(U_\infty a)$$
$$\sigma = (T-T_0)/(T_1-T_0) \qquad \xi = \{\ln(r/a)\}/\pi$$
$$\eta = \theta/\pi \qquad \eta^* = \alpha/\pi \tag{3}$$
$$Re = 2aU_\infty/\nu_0 \qquad Gr = \alpha_0\,g_0\,(T_1-T_0)\,(2a)^3/\nu_0^2$$
$$Pr = \rho_0\,c_p\,\nu_0/k_0$$

where Re, Gr, and Pr denote the Reynolds number, the Grashof number and the Prandtl number, respectively. In addition, the parameter G is defined as $G = \pi^2 e^{2\pi\xi}$. These equations are solved in the computational domain $0 \leq \xi \leq \xi_\infty$ and $0 \leq \eta \leq 2$. The initial conditions are $\psi(\xi,\eta,t=0)=0$ and $\sigma(\xi,\eta,t=0)=0$. The boundary conditions on the cylinder surface ($\xi=0$) are $\psi=\psi_\xi=0$ and $\sigma=1$. The farfield boundary conditions are discussed in more detail in a following section.

SOLUTION TECHNIQUE

The biharmonic equation including the buoyancy term and the energy equation are solved numerically using a method similar to that described in [9]. The time discretization for both equations is conducted using the implicit trapezoidal or Crank-Nicolson scheme. Centered differences are used for all the spatial derivates in both equations. The difference equations are linearized using Newton's method. At each time step the system of linear algebraic equations including the boundary conditions is solved simultaneously using a Gaussian elimination scheme for banded matrices [14]. Depending on the Reynolds number, Grashof number, time level and time step, one-to-four Newton iterations are generally required to obtain convergence at each time step.

The drag coefficient $c_d = D/(q_\infty 2a)$ and the lift coefficient $c_\ell = L/(q_\infty 2a)$ of the cylinder are calculated through surface integration of the pressure coefficient c_p:

$$\left.\frac{\partial c_p}{\partial \eta}\right|_{wall} = \frac{4}{Re} \frac{\partial \omega_{wall}}{\partial \xi} - \sqrt{G}\, \frac{Gr}{Re^2}\, \sigma_{wall} \sin\pi(\eta-\eta^*) \tag{4}$$

and the skin friction coefficient c_f:

$$c_f = \frac{4}{Re}\, \omega_{wall} \tag{5}$$

where the vorticity is obtained from:

$$\omega = \nabla^2\psi/G \,. \tag{6}$$

FARFIELD BOUNDARY CONDITIONS

For the biharmonic equation (without the buoyancy force) the conventional boundary conditions at r_∞ for the flow around a cylinder are based on the assumption that the flow field is potential. Therefore, the vorticity is zero and the stream function $\psi' = U_\infty r_\infty (1-a^2/r_\infty^2)\sin\theta$ or $\psi = 2 \sin(\pi\eta)\sinh(\pi\xi)$. For $r_\infty \gg a$ or $\xi_\infty \gg 0$ the condition for the stream function simplifies to the value for a uniform potential flow. These potential flow conditions work often quite well. However, they tend to affect the solution accuracy near the outflow boundary of the domain.

A physically more correct condition for the outflow segment of the farfield boundary is the thin-layer approximation of eq. (1) which is obtained by setting $\omega_{\xi\xi}=0$ and by neglecting the contribution of the vorticity to $\psi_{\xi\xi}$. Thus, $\psi_{\xi\xi}$ is determined from the potential solution at the farfield boundary; $\psi_{\xi\xi} = 2\pi^2 \sin(\pi\eta)\sinh(\pi\xi)$. Generally, the wake cuts the farfield boundary at $\eta \simeq 1$ and thus, $\psi_{\xi\xi} \simeq 0$. In addition to these two conditions it is necessary to apply backward differencing instead of centered differencing for the radial convective term in eq. (1). These conditions follow naturally from the governing equation. Note that the thin-layer approximation for the Navier-Stokes equations is not valid for low Reynolds numbers. Thus, the Reynolds numbers based on the wake length must be moderate-to-high to allow implementation of this condition at the outflow segment of the farfield boundary.

For the remainder of the farfield boundary, the zero-vorticity condition is retained. In addition, a Neumann or Dirichlet condition for ψ based on the potential solution can be implemented. In Fig. 2 the time histories of the lift coefficient are presented for these two farfield boundary conditions. In order to excite the Karman street the cylinder is given an oscillatory motion for $10<t<24$. After applying this disturbance the calculations are continued until a fully-developed Karman street, with a repeatable limit cycle, exists. The lift coefficient oscillation has a Strouhal number St=0.172 and an amplitude of about $c_\ell=\pm0.40$ for the Neumann condition, whereas St=0.159 and $c_\ell=\pm0.30$ for the Dirichlet condition. These values of the Strouhal number fall within the range of the experimentally measured values for the Strouhal number at Re=100. A reduction in the grid density from 91x91 to 61x61 results in an increase in the amplitude to about $c_\ell=\pm0.45$ for the Neumann method, whereas the amplitude is virtually unchanged for the Dirichlet method. To further analyze this discrepancy in the results for the two farfield boundary conditions, the steady flow around a cylinder at Re = 40 is studied using the same solution method. In Fig. 3 the effects of the farfield boundary condition for ψ are shown for an outer boundary location $r_\infty/a=100$ and a grid density variation ranging from 61x61 to 121x121. The results are presented for both a local quantity (maximum wall vorticity) and a global quantity (drag coefficient). A change in grid density produces a very slight linear variation in both quantities for the Dirichlet condition and a much larger nonlinear variation for the Neumann condition. Very high grid densities are required to obtain accurate results with the Neumann condition for ψ at the farfield boundary. The performance of the Dirichlet condition is obviously superior. This observation is in contrast to Fornberg's [15] who concluded that the Neumann condition for ψ is the preferred boundary condition for Reynolds up to about 40.

However, the previously listed Dirichlet condition for the stream function must be modified when the flow field contains vorticity. In that case the farfield value is:

$$\psi' = \psi'_{pot} + \frac{1}{2\pi} \iint \omega' \, \ell n \, \frac{r^*}{a} \, dA \tag{7}$$

where ψ'_{pot} represents the potential solution and r^* is the distance between the location of the vorticity and the farfield boundary. If this distance is very large, eq. (7) can be approximated as follows:

$$\psi' = \psi'_{pot} + \frac{1}{2\pi} \ell n \frac{r}{a} \iint \omega' dA \,. \tag{8}$$

Thus the contribution of the vorticity to the farfield value of ψ' is not negligible even for very large values of $r=r_\infty$. It is time consuming to calculate the surface integral on the right-hand-side of eq. (8) at each time step. The circulation $c' = \iint \omega' dA$ in eq. (8) can be approximated by averaging the local values from the previous time step at the circumferential grid line $\xi_{NJ-1} = \xi_\infty - \Delta\xi$ near the farfield boundary. Thus, the nondimensional circulation is:

$$c = \frac{2}{\xi_{NJ-1}} \frac{1}{NI} \sum_{i=1}^{NI} (\psi_{pot} - \psi)_{i,NJ-1} \,. \tag{9}$$

Now, a modified Dirichlet condition can be obtained:

$$\psi_{i,NJ} = (\psi_{pot})_{i,NJ} + 0.5\ c\ \xi_{NJ} \,. \tag{10}$$

This boundary condition is implemented and the effect on the lift coefficient can be observed in Fig. 4. For the cyclic flow of a cylinder the effect of the additional term in eq. (10) on both the Strouhal number and the amplitude is small. Therefore, the original Dirichlet boundary condition with the thin-layer approximation for the wake region is used in the following calculations.

The farfield boundary condition for the temperature is treated similarly. For the entire farfield boundary except the wake region the freestream temperature $\sigma=0$ is specified. In the outflow region of the farfield boundary the thin-layer approximation for the energy equation is assumed to be valid and, thus, $\sigma_{\xi\xi}=0$. In addition backward differencing is used to approximate partial derivatives with respect to ξ.

RESULTS

The following results are calculated with the method for $\Delta t=1.0$, $r_\infty/a=100$, and Re=100. The forced convection problem (Gr=0) is analyzed with the biharmonic equation without the buoyancy term and without the energy equation for a 91x91 grid. The mixed convection problem is solved with the complete equations for a 61x61 grid.

In Figs. 2 and 4 the lift history of the cylinder is shown for the flow without the buoyancy force (Gr=0). The resulting forced convection flow field in terms of the instantaneous stream lines and the equivorticity lines is depicted at $t\simeq150$ in Fig. 5. The Karman vortex street is fully developed and clearly visible in the streamline pattern. Also, the effects of vorticity diffusion and dispersion are noticable within two cylinder diameters downstream of its trailing edge.

The buoyancy term in eq. (1) dominates the development of the combined convection flow field for sufficiently large values of the

Richardson number $Ri=Gr/Re^2$. Previously, we examined the resulting flows for $Ri<1$ [9]. Here, $Ri=1.0$ and the angle α between the freestream velocity vector and the gravitational acceleration vector is varied from 0° to 180°. In Fig. 6 the time trace for the lift coefficient is shown for the parallel flow case ($\alpha=180°$). The oscillations between $10<t<24$ are introduced to excite the instabilities in the flow field. The force data demonstrate that the flow field is stable for these conditions and that the Karman vortex street does not develop. As a result of these conditions the flow over the cylinder is nearly fully attached. In Fig. 6, the flow field in terms of the instantaneous streamlines and the isotherms is depicted at $t\simeq150$. A small separation region is still visible in the streamline pattern near the trailing edge of the cylinder. A slight rise in the Grashof number will suppress separation completely and cause the flow to become fully attached.

In Fig. 7 lift coefficient is plotted as function of time for the contra flow case ($\alpha=0°$). Also here, the oscillations between $t=10$ and $t=24$ are intentionally introduced. The time evolution of c_ℓ shows that the flow field is nearly steady at these conditions. However, an instability appears to be growing very slowly at these conditions. The instantaneous streamlines depict two very large counterrotating flow structures downstream of the cylinder. A very slight asymmetry is visible in the flow field. The Grashof number is raised to $Gr=15000$ ($Ri=1.5$) to accelerate the growth of the instabilities in the flow field. Now a very strong asymmetry develops as can be interpreted from the c_ℓ-history. The stream function contours in Fig. 8b are generated by the code at $t\simeq190$ and depict the collapse of these flow structures. In Fig. 9 c_ℓ versus time and contour plots are shown for the mixed convection problem of a heated cylinder in a horizontal flow ($Ri=1.0$, $\alpha=90°$). At these conditions the vortex shedding characteristics are very similar to those for the forced convection problem ($Ri=0$).

CONCLUSIONS

The unsteady viscous flow around an infinite circular cylinder has been studied with a fully implicit time-dependent solution technique. Several boundary conditions for the farfield are studied. At the outflow segment of the farfield boundary the thin-layer Navier-Stokes approximation is applied to improve the local accuracy of the solution. The Dirichlet condition for the stream function is superior to the Neumann condition at the conditions analyzed. A modified Dirichlet condition that accounts for the vorticity in the flow field is proposed.

The method is used to study the mixed convection flow field for a heated/cooled cylinder. The results obtained for higher grid densities are very similar to the previously published coarse grid results. The calculations for $Ri=Gr/Re^2$ of $O(1)$ demonstrate a steady parallel flow for $\alpha=0°$, and an unsteady flow (quite different from the Karman street) for $\alpha=180°$. For $\alpha=90°$ and Ri of $O(1)$ the vortex shedding characteristics are still very similar to those of $Ri=0$.

ACKNOWLEDGEMENTS

The authors would like to thank the NASA Ames Research Center and the College of Engineering of the University of California, Davis, for their support in the form of computer time.

REFERENCES

[1] WIGTON, L. B., "Application of MACSYMA and Sparse Matrix Technology to Multielement Airfoil Calculations," AIAA Paper 87-1142, June 1987.

[2] VAN DAM, C. P., HAFEZ, M. and AHMAD, J., "Calculation of Viscous Flows with Separation Using Newton's Method and Direct Solver," AIAA Paper 88-0412, Jan. 1988; also AIAA Journal, in print.

[3] ROACHE, P. J., and ELLIS, M. A., "The BID Method for the Steady-State Navier-Stokes Equations," Computers and Fluids, Vol. 3, 1975, pp. 305-320.

[4] TUANN, S. Y., and OLSON, M. D., "Numerical Studies of the Flow Around a Circular Cylinder by a Finite Element Method," Computers and Fluids, Vol. 6, 1978, pp. 219-240.

[5] CEBECI, T., HIRSH, R. S., KELLER, H. B., and WILLIAMS, P. G., "Studies of Numerical Methods for the Plane Navier-Stokes Equations," Computer Methods in Applied Mechanics and Engineering, Vol. 27, 1981, pp. 13-44.

[6] WALTER, K. T., and LARSEN, P. S., "The FON Method for the Steady Two-Dimensional Navier-Stokes Equations," Computers and Fluids, Vol. 9, 1981, pp. 365-376.

[7] SCHREIBER, R., and KELLER, H. B., "Driven Cavity Flows by Efficient Numerical Techniques," Journal of Computational Physics, Vol. 49, 1983, pp. 310-333.

[8] SCHÜTZ, H., and THIELE, F., "An Implicit Method for the Computation of Unsteady Incompressible Viscous Flows," GAMM Workshop on Numerical Methods in Fluids, Vieweg, 1987.

[9] VAN DAM, C. P., HAFEZ, M., and BRUCKER, D., "Unsteady Navier-Stokes Calculations Using Biharmonic Formulation and Direct Solver," AIAA Paper 89-0465, Jan. 1989.

[10] LECOINTE, Y., and PIQUET, J., "On the Use of Several Compact Methods for the Study of Unsteady Incompressible Viscous Flow Round a Circular Cylinder," Computers and Fluids, Vol. 12, 1984, pp. 255-280.

[11] SCHLICHTING, H., Boundary-Layer Theory, 7th ed., McGraw-Hill, 1979.

[12] NOTO, K., "Computation on Disappearance of the Karman Vortex Street past Heated Cylinder Submerged in Horizontal Main Flow," Proceedings of ISCFD Nagoya, Aug. 1989, pp. 605-610.

[13] GRAY, D. D., and GIORGINI, A., "The Validity of the Boussinesq Approximation for Liquids and Gases," International Journal of Heat and Mass Transfer, Vol. 19, 1976, pp. 545-551.

[14] DONGARRA, J. J., MOLER, C. B., BUNCH, J. R., and STEWART, G. W., "LINPACK User's Guide," SIAM, Philadelphia, 1979.

[15] FORNBERG, B., "A Numerical Study of Steady Viscous Flow past a Circular Cylinder," Journal of Fluid Mechanics, Vol. 98, 1980, pp. 819-855.

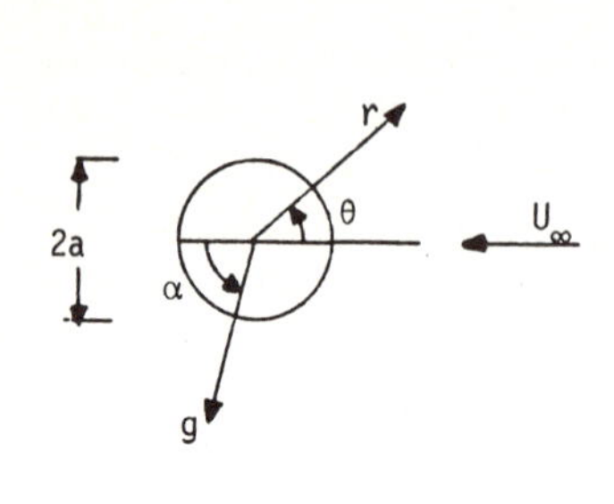

Fig. 1 Physical geometry for the flow problem.

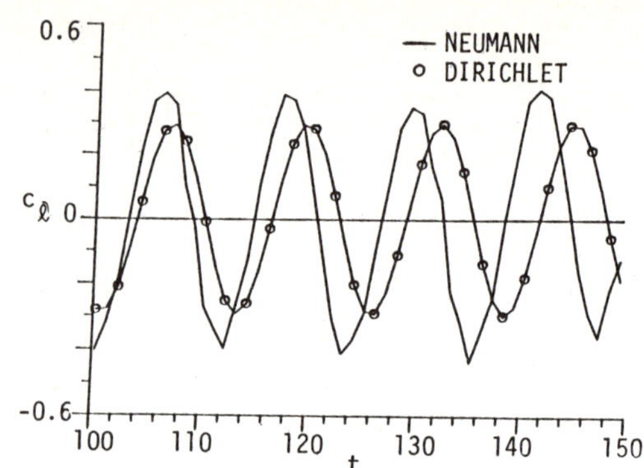

Fig. 2 Effect of farfield boundary condition on the lift coefficient; 91x91 grid, $r_\infty/a=100$, $\Delta t=1.0$, Re=100.

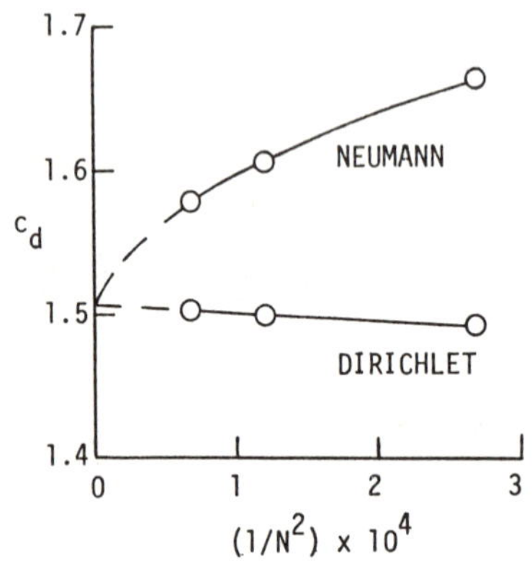

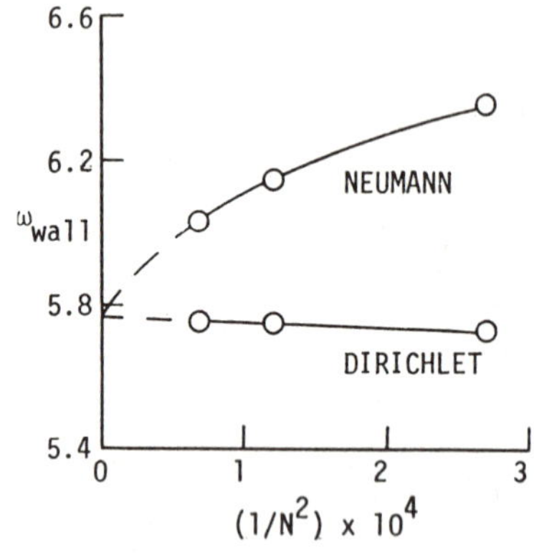

Fig. 3 Effect of grid density and farfield boundary condition on the drag coefficient and the maximum wall vorticity of a cylinder at Re=40; $r_\infty/a=100$ (N = number of grid points in each direction).

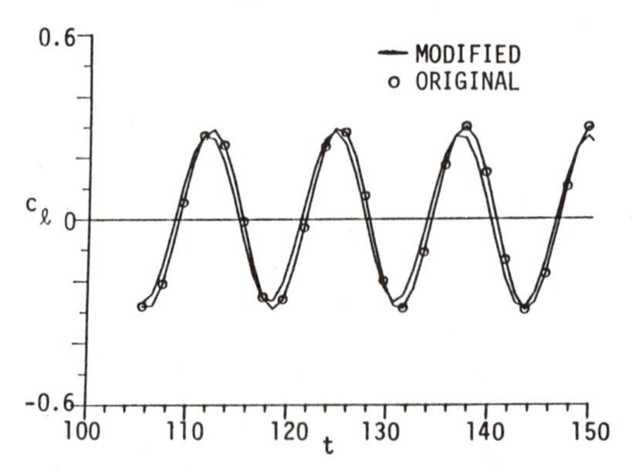

Fig. 4 Effect of modification of Dirichlet farfield boundary condition on the lift coefficient; 91x91 grid, $r_\infty/a=100$, $\Delta t=1.0$, Re=100.

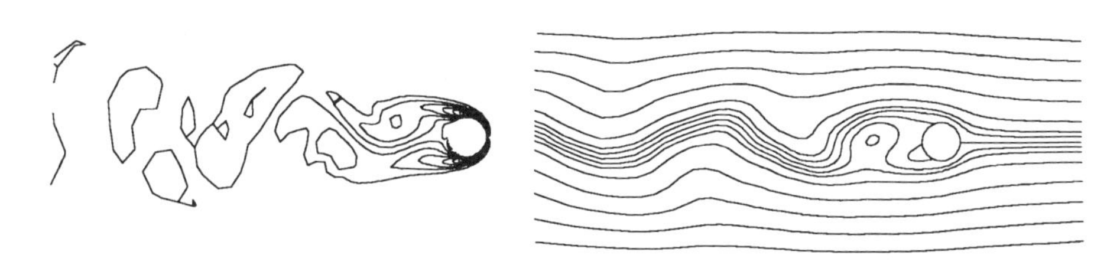

Fig. 5 Vorticity and stream function contours for forced convection flow (Gr=0).

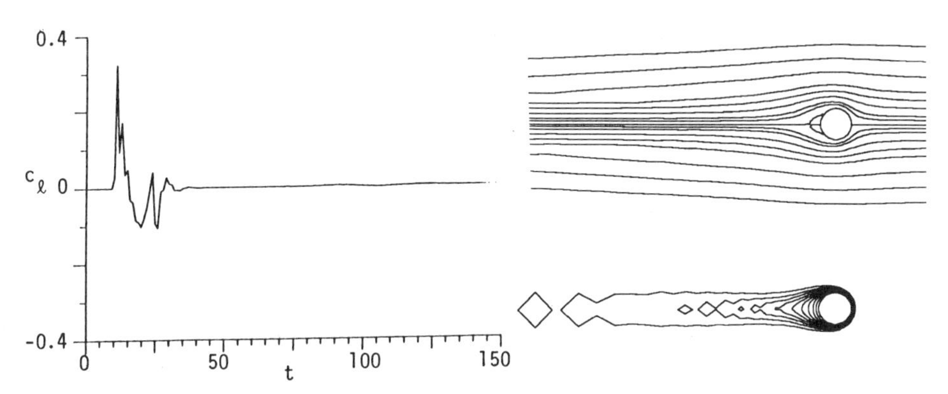

Fig. 6 Time history of lift coefficient, and stream function and temperature contours for combined free and forced convection flow (α=180°, Gr=10000).

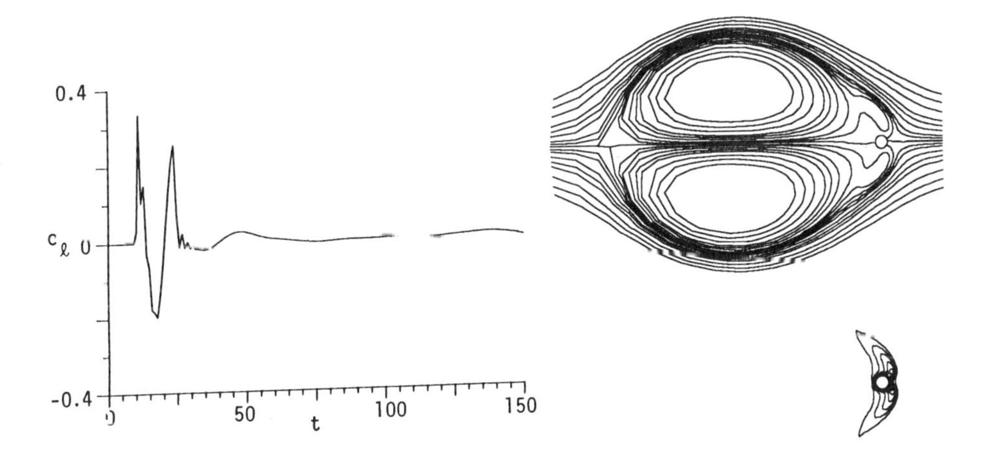

Fig. 7 Time history of lift coefficient, and stream function and temperature contours for combined free and forced convection flow (α=0°, Gr=10000).

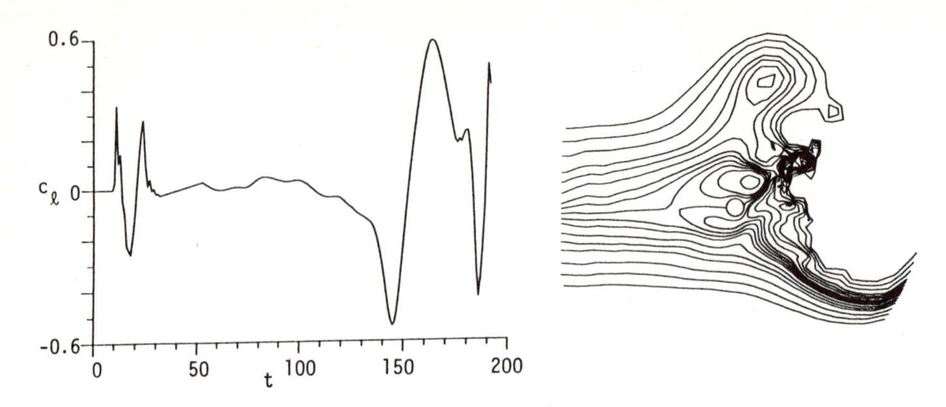

Fig. 8 Time history of lift coefficient, and instantaneous streamlines at t≃190 for combined free and forced convection flow (α=0°, Gr=15000).

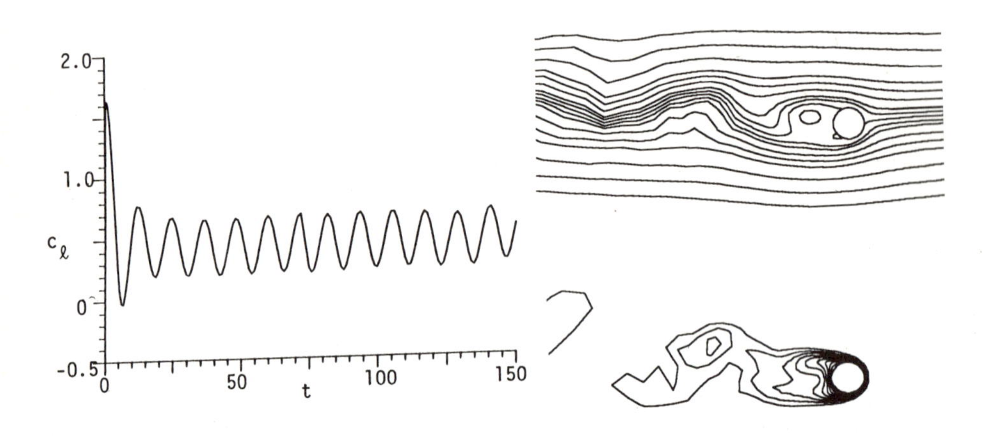

Fig. 9 Time history of lift coefficient, and stream function and temperature contours for combined free and forced convection flow (α=90°, Gr=10000).

SOLUTION METHOD FOR VISCOUS FLOWS AT ALL SPEEDS IN COMPLEX DOMAINS

by

I. Demirdžić[1], R.I. Issa[2] and Ž. Lilek[3]

[1]Mašinski fakultet Sarajevo
Omladinsko šetalište bb, 71000 Sarajevo, Yugoslavia

[2]Department of Mineral Resources Engineering
Imperial College, London SW7 2BP

[3]Unis-Institut
Tršćanska 7, 71000 Sarajevo, Yugoslavia

SUMMARY

An existing numerical method for the solution of laminar and turbulent incompressible flows in complex geometries is extended to the calculation of steady compressible flows. As a result a method is obtained which is equally valid for both incompressible and compressible flows, including transonic and supersonic regimes. The method is verified on a number of test cases, including inviscid and viscous both internal and external flows. The overall performance of the method may be considered good, except for the smearing of the shocks, which is the consequence of the curently emoployed first order differencing scheme. As an illustration of the capabilities of the method it is applied to the prediction of the flow around a projectile with and without base bleed.

INTRODUCTION

There are a number of methods for calculation of incompressible flows [1,2] and even more ones for the solution of compressible Euler and Navier-Stokes equations [3,4,5]. However, there are only few methods that are equally valid for all flow speeds which can be used to solve the forementioned flows [6,7].

The methods for compressible flows mainly have their origins in external aerodynamics and are characterized by the fact that they consider density as a main dependent variable extractable from the continuity equation and the pressure is obtained from an equation of state. While being able to predict high Mach number flows quite well, with reasonably good shock capturing abilities, the accuracy and efficency of these methods at low Mach number becomes questionable, because in these situations the density changes are very small and the pressure-density coupling becomes very week.

The methods for incompressible flows are mainly of the pressure-correction type, i.e. they combine momentum and continuity equations in order to get a predictor-corrector algorithm for establishing the pressure field and any changes in density are then calculated via an equation of state. These methods are in widespread use for calculation of incompressible (and low Mach number compressible) flows. However, as the Mach number increases, the governing equations change their character from an elliptic one to hyperbolic, with the cross-over occuring at Mach number of unity. This change is not reflected in the incompressible-flow pressure (or pressure-correction) equation which retaines an elliptic character and is hence not suitable for the prediction of high Mach number flows.

Methods for the calculation of flows at all speeds are based on the use of pressure rather then density as the main dependent variable. It is this which enables the application of the methods to both subsonic and supersonic regimes (including the totally incompressible case) without modification. Such schemes were first developed by Hirt et al [6] using a semi-implicit time marching approach and by Issa and Lockwood [8] for steady state flows using an iterative technique. Later, Hah [9], Rhie [10] and Karki and Patankar [7] adopted similar approaches to that of [8] but using different practices for the solution algorithm.

The present method is an extention of the scheme of Demirdžić et al [11,12] for incompressible flow in complex geometries to the compressible flow regime along similar lines as those of [8] and [7]. The original method uses primitive variables (contravariant velocity components and pressure) as the computing variables and a staggered mesh arrangement. For the velocity-pressure coupling the SIMPLE algorithm was modified to accomodate general non-orthogonal mesh. All these features are retained in the present method. However, in order to take into account the hyperbolic nature of the equations in the case of supersonic flows and to retain their elliptic nature in the case of subsonic flows, the treatment of pressure-velocity coupling is modified in the manner that was first suggested by Issa and Lockwood [8] and more recently by Karki and Patankar [7]. The new method is tested on a number of inviscid and viscous, both laminar and turbulent flow cases. Finally, as an illustration of the capabilities of the method, it is applied to the prediction of the flow around a projectile with and without base bleed.

SOLUTION METHOD

Finite volume discretization

Demirdžić at al [11] showed that the governing flow equations can be defined as an instance of the following generic equation

$$\frac{\Delta}{\Delta x^{(j)}}\left(\rho v^{(j)}\varphi - g^{(jm)}\Gamma_{\varphi}\frac{\partial\varphi}{\partial x^{(j)}}\right) = S_{\varphi}, \tag{1}$$

where φ represents contravariant velocity components $v^{(i)}$, any transported scalar (temperature T, kinetic energy of turbulence k, its dissipation rate ε etc) or 1 in the case of continuity equation; ρ is density, $g^{(i,j)}$ is contravariant metric tensor physical component, Γ_{φ} is the diffusion coefficient, S_{φ} represents any remaining terms (e.g. pressure gradients in momentum equations) and $\Delta/\Delta x^{(j)}$ and $\partial/\partial x^{(j)}$ are differential operators defined in [11]. By deviding the solution domain into a finite number of contiguous control volumes or cells (fig. 1) and by applying the finite volume discretization procedure described in detail in [11], the following set of non-linear algebraic equations can be obtained

$$a_P\varphi_P = \sum_M a_M\varphi_M + b \quad (M=E,W,N,S), \tag{2}$$

where coefficients are defined as follows

$$\begin{aligned} a_E &= D_e - (1-\omega_e)C_e, \\ a_W &= D_w + \omega_w C_w, \qquad & a_P &= \sum_M a_M + S_{2\varphi} \quad (M=E,W,N,S;\ S_{2\varphi} > 0), \\ a_N &= D_n - (1-\omega_n)C_n, \qquad & b &= S_{1\varphi}, \\ a_S &= D_s + \omega_s C_s, \end{aligned} \tag{3}$$

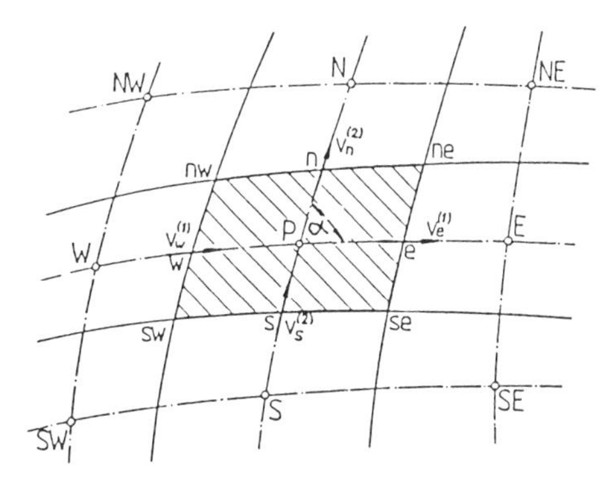

Fig. 1 Control volume and labelling scheme

where the spatial weighting factors (ω , i=e,w,n,s) are calculated according to the upwind differencing scheme, e.g.

$$\omega_e = \begin{cases} 1, & V_e^{(1)} \geqslant 0, \\ 0, & V_e^{(1)} < 0. \end{cases} \tag{4}$$

The convection (C_i) and diffusion (D_i) contributions to the coefficients are defined for the east cell face, for example

$$C_e = (\rho V^{(1)} \sin\alpha\, A)_e, \quad D_e = (g^{11} \Gamma_\varphi \frac{\sin\alpha\, A}{\delta x^{(1)}})_e \tag{5}$$

and the components of the linearized source term $s_{1\varphi}$ and $s_{2\varphi}$ contain, appart from the physical sources, the cross-derivative diffusion fluxes which do not feature in the a_M coefficients.

Pressure-velocity coupling

In the previous section, the difference equations for transported variables are presented. In what follows, the method for calculating the pressure field is outlined.

The algorithm employed is SIMPLE, in which a pressure-correction equation is derived from a combination of the continuity and momentum equations; this equation serves to yield the pressure increment in a predictor-corrector sequence necessary to drive the velocity field to satisfy continuity. This scheme is well documented [11] and needs not be described here. What is presented, are the necessary modifications to extend the algorithm to high Mach number flows.

The crucial step made in the extention of the scheme is in upwinding the density in the calculation of the mass flux crossing a cell face. Thus, for example, the mass flux crossing an east cell face is teken as

$$C_e = \rho_P V_e^{(1)} (\sin\alpha\, A)_e \tag{6}$$

for $V_e^{(1)} > 0$, with a corresponding expression for $V_e^{(1)} < 0$. If the predictor stage values of $V^{(1)}$ and ρ (which do not satisfy continuity) are denoted by * superscript, and the corrections to those values are denoted by ' superscript, then

$$C_e = (\rho_P^* + \rho_P')(V_e^{(1)*} + V_e^{(1)\prime})(\sin\alpha\, A)_e \doteq (\rho_P^* V_e^{(1)*} + V_e^{(1)*}\rho_P' + \rho_P^* V_e^{(1)\prime})(\sin\alpha\, A)_e. \quad (7)$$

Now, $V_e^{(1)\prime}$ is related to $(p_P' - p_E')$ by an expression derivable from the momentum equation as in the standard practice in the SIMPLE algorithm. The additional term $V_e^{(1)*}\rho_P'$ appearing in equ. (7) relates to compressibility and is related to p' by

$$\rho_P' = Q p_P', \quad (8)$$

where $Q = (\partial\rho/\partial p)^*$ and is determined from the equation of state.

Similar expressions to equ. (7) for the mass fluxes C_w, C_n and C_s are obtained, all of which are substituted into the continuity equation

$$C_e - C_w + C_n - C_s = 0 \quad (9)$$

to yield an equation for p' in the form

$$a_P p'_P = \sum_M a_M p'_M + b \quad (M=E,W,N,S), \quad (10)$$

where the summation on M is only over four principal nodes and coefficients a_M are always positive

$$a_E = (\rho^* d \sin\alpha\, A)_e - (1 - \omega_e)\, Q_E\, (v^{*(1)} \sin\alpha\, A)_e,$$

$$a_W = (\rho^* d \sin\alpha\, A)_w + \omega_w\, Q_W\, (v^{*(1)} \sin\alpha\, A)_w,$$

$$a_N = (\rho^* d \sin\alpha\, A)_n - (1 - \omega_n)\, Q_N\, (v^{*(2)} \sin\alpha\, A)_n,$$

$$a_S = (\rho^* d \sin\alpha\, A)_s + \omega_s\, Q_S\, (v^{*(2)} \sin\alpha\, A)_s,$$

$$\begin{aligned} a_P = {} & (\rho^* d \sin\alpha\, A)_e + \omega_e\, Q_P\, (v^{*(1)} \sin\alpha\, A)_e + \\ & + (\rho^* d \sin\alpha\, A)_w - (1 - \omega_w)\, Q_P\, (v^{*(1)} \sin\alpha\, A)_w + \\ & + (\rho^* d \sin\alpha\, A)_n + \omega_n\, Q_P\, (v^{*(2)} \sin\alpha\, A)_n + \\ & + (\rho^* d \sin\alpha\, A)_s - (1 - \omega_s)\, Q_P\, (v^{*(2)} \sin\alpha\, A)_s, \end{aligned} \quad (11)$$

$$b = -(C_e^* - C_w^* + C_n^* - C_s^*).$$

It should be noted that the coefficients of the pressure-correction equation (11) are similar to those of the transport equations (3) in the sense that they have diffusion-like first and convection-like second part, whose relative importance depends on the local Mach number. In the limit of incompressible flow (M=0 i.e. Q=0) the second part of the coefficients (11) wanish and they reduce to those of Demirdžić et al [11] leading to the fully elliptic pressure-correction equation. As the Mach number increases the relative importance of the convection-like terms increases and the pressure-correction equation gradually and continually becomes upstream weighted, a behaviour corresponding to the physical situation when the flow becomes supersonic and reflected by the hyperbolicity of the parent differential equations.

It should be noted that the upwinding practice on density is carried out also in the evaluation of the convective coefficients of all transport equations.

Boundary conditions

The treatment of boundary conditions in the case of subsonic flow appropriate for the present choice of dependent variables is given in Dermirdžić et al [11]. In the case of compressible flows walls and axes of symmetry are treated the same way as in incompressible case. At inflows and outflows the number of variables that have to be specified is equal to the number of incoming signals (characteristics). At free stream boundaries non-reflective boundary conditions, described by Issa and Lockwood [8], are used.

Overall solution algorithm

In this section the main steps in solving the discretized equations are summerized as follows:

1. Calculate the density field from the current pressure and temperature fields via an equation of state,
2. Solve the momentum equations to obtain an intermediate velocity field,
3. Solve the pressure-correction equation and update the velocity, pressure and density fields,
4. Solve the energy equation and obtain temperature field,
5. Repeat the steps 1 to 4 until convergence is reached.

RESULTS OF CALCULATIONS

Validation tests

Since it is shown that in the case of incompressible flow the present method reduces to the method of Demirdžić et al [11], only tests related to compressible flow calculations are reported herein. The capabilities of the solution procedure to handle compressible subsonic, transonic and supersonic flows are demonstrated by way of several test cases including inviscid and viscous flows in domains which require general boundary-fitted mesh.

Inviscid flow in a channel with a bump

Flow in a channel with a circular-arc bump on one of the walls suggested by Moretti [13] has been used as a test case by many authors [14,15,7]. In this paper three inlet Mach numbers corresponding to subsonic (M_{in} = 0.5), transonic (M_{in} = 0.675) and supersonic (M_{in} = 1.65) regimes have been considered. Computations were made on 54x20 CV non-uniform mesh and the temperature is calculated from a constant stagnation enthalpy instead of the energy transport equation.

Results of calculations are shown in fig. 2. In fig. 2a isomach lines are presented, showing almost symmetric solution in the case of subsonic flow, which indicates good accuracy in this subsonic application, and severely smeared shock waves in the case of transonic and supersonic flows, which is a consequence of the first order difference scheme employed. In fig. 2b the calculated Mach number distributions on the walls are compared with results of Eidelman et al [15] which are obtained by the Godunov method (M_{in} = 0.675 and 1.65) and its second order extention (M_{in} = 0.5). In the case of transonic and supersonic flows the agreement with the Godunov scheme is good except for the underprediction of the shock strength of about 10%. This agrees very closely with results of Karki and Patankar [7], who also use upwind differencing for convection and an upwind biasing of density in the continuity equation.

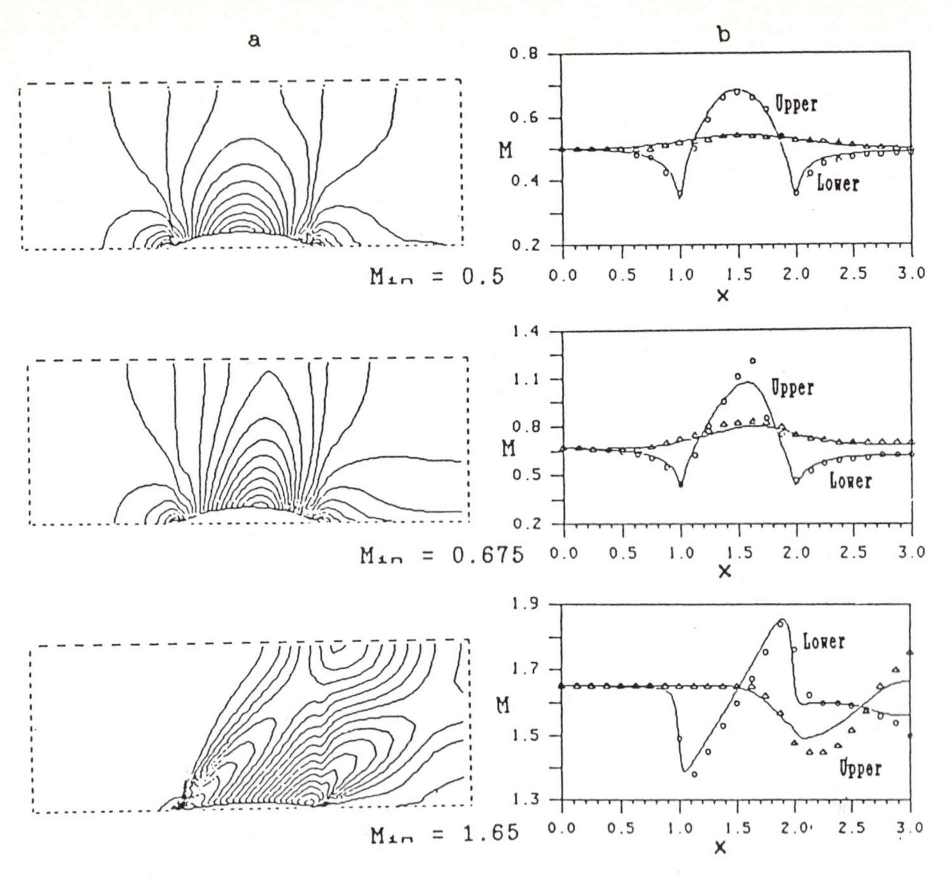

Fig. 2 Flow in a channel with bump. a) Isomach contours, b) Surface Mach number distribution (—— Present pred.; △ o - Predictions [15])

Viscous flow in a convergent-divergent nozzle

In this section experiments of Mason et al [16] are simulated by assuming laminar flow in a convergent-divergent nozzle shown in fig. 3. Calculations were performed on a 53x18 CV non-uniform mesh with more grid nodes near the wall and near the throat. The agreement of the calculated pressure distribution at the wall and the centerline with measurements for the ratio of the upstream stagnation pressure to the exit static pressure of 0.1135

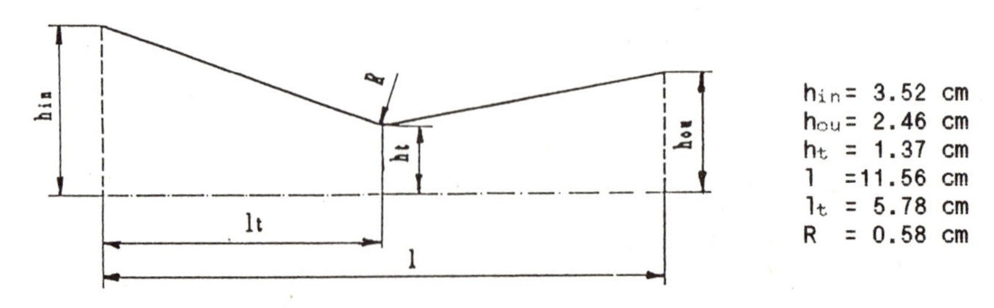

Fig. 3 Geometry of the nozzle

(fig. 4) may be considerd good. However, when that pressure ratio was increased to the value of 0.47, which resulted in a shock wave in the nozzle, the computed shock was at the correct location but considerably smeared due to excessive numerical dissipation.

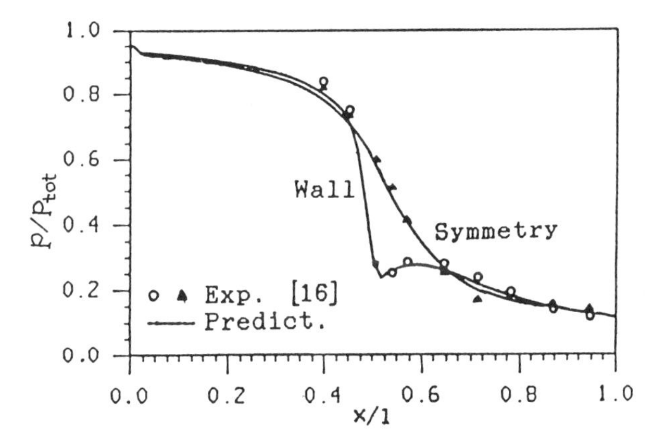

Fig. 4. Pressure distribution at the wall and the centerline of a convergent-divergent nozzle

Turbulent flow around a pipe with a bump

In order to test the remaining feature of the present method, namely the k-ε model of turbulence, the transonic turbulent flow around a pipe with circular bump (fig. 5), for which experimental results of Johnson et al [17] exist, is calculated. The agreement of calculated and measured pressure coefficient (fig. 6) is in accord with the fact that the method in the case of transonic flow produces an excessive dissipation which results in smeared shocks.

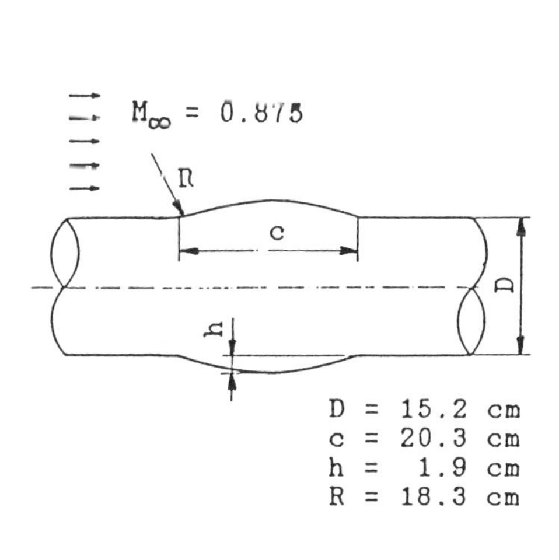

Fig. 5 Pipe with a bump

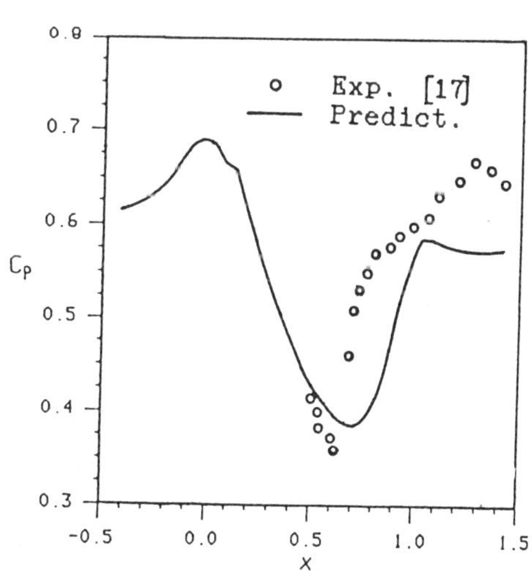

Fig. 6 Surface pressure coefficient

Example of application

As an illustration of the capabilities of the method, the predictions of turbulent flow around a projectile (fig. 7) with and without base bleed are presented. Calculations are made for two free stream Mach numbers ($M_\infty = 0.9$ and 1.5) and for a number of mass-injection parameters ($I = 4\dot{m}/\pi\rho u_\infty D^2$). As

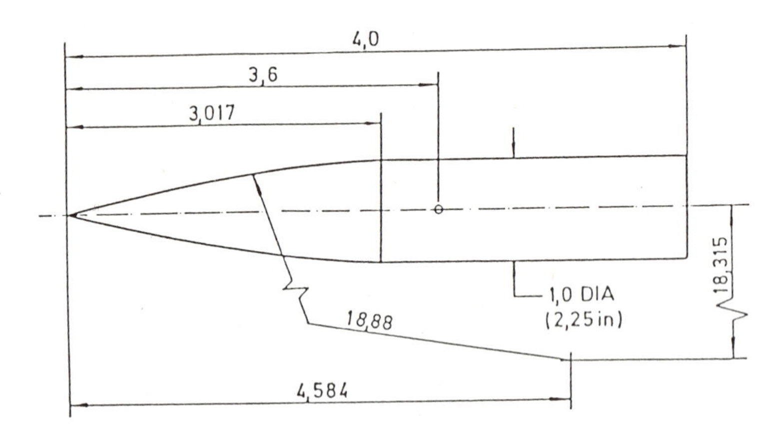

Fig. 7 Projectile geometry

an illustration the isomachs calculated on a 44x23 CV mesh are given in fig. 8. The isomach contours for $M_\infty = 0.9$ and $I = 0$ agree qualitatively with those of Lai and Khosla [18], while in the case of $M_\infty = 1.5$ the (smeared) shock vaves eminating from the tip and the base of the projectile are clearly seen. In order to show the effect of the mass injection at the projectile base the flow field in the base region for 6 different values of I is shown in fig. 9. As expected, increase in injected mass flow rate has a significant influence on the flow field, resulting first in moving further downstream and then completely removing the recirculation region, a behaviour also reproduced by Sahu et al [19].

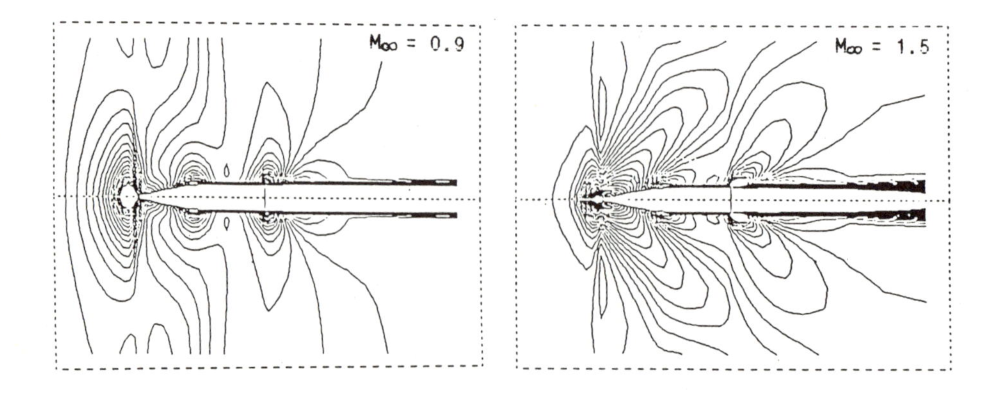

Fig. 8 Isomach contours for the flow around projectile

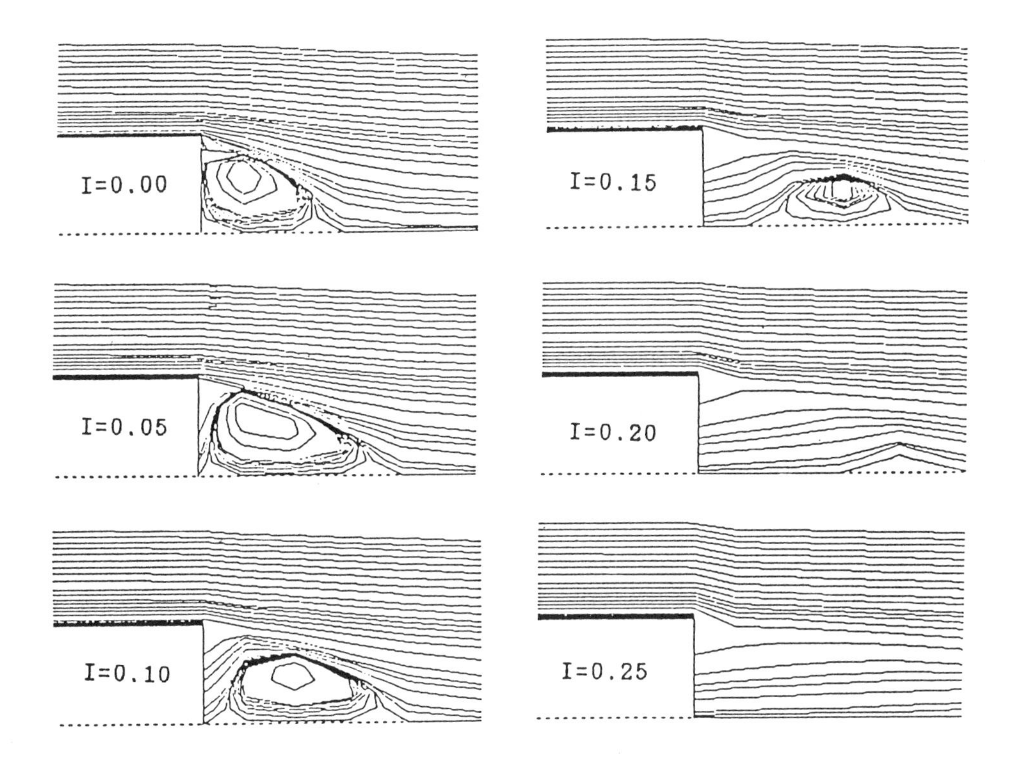

Fig. 9 Influence of the mass-injection parameter (I) on the flow field in the base region (M_∞ = 0.9)

CONCLUSIONS

A finite volume method that can be used to calculate fluid flow at all speeds is presented. Comparisons of the calculations with available numerical and experimental data show that: (i) the method is stable for all flow speeds, (ii) it handles the transition from the subsonic to supersonic regime smoothly, (iii) it captures shocks automatically and at the correct location.

It is observed that shocks are smeared to a large extent, and this is the consequence of the first order spatial difference scheme for the convective terms (upwinding). This, however, can be remedied either by the intoduction of a higher order difference scheme or by mesh refinement.

In order to improve the efficiency of the method one more corrector may be added to the pressure-correction procedure which would lead to the PISO method of Issa [20].

REFERENCES

[1] HARLOW, F.H., WELCH, J.E. "Numerical Calculation of Time-Dependent Viscous Incompressible Flow of Fluid with Free Surface", Physics of Fluids, 8 (1965) pp. 2182-2189.

[2] PATANKAR, S.V., SPALDING, D.B. "A Calculation Procedure for Heat, Mass and Momentum Transfer in Three-Dimensional Parabolic Flows", Int. J. Heat and Mass Transfer, 15 (1972) pp. 1787-1806.

[3] McDONALD, H, BRILEY, W.R.: "Three-Dimensional Supersonic Flow of a Viscous or Inviscid Gas", J. Comp. Physics, 19 (1975) pp.150-178.

[4] BEAM, R.M., WARMING, R.F.: "An Implicit Finite Difference Algorithm for Hyperbolic Systems in Conservation Law Form", J. Comp. Physics, 22 (1976) pp. 87-110.

[5] MacCORMACK, R.W.: "A Numerical Method for Solving the Equations of Compressible Viscous Flows", AIAA Journal, 20 (1982) pp. 1275-1281.

[6] HIRT, C.W., AMSDEN, A.A., COOK, J.L.: "An Arbitrary Lagrangean-Eulerian Computing Method for All Flow Speeds", J. Comp. Physics, 14 (1974) pp. 227-253.

[7] KARKI, K.C., PATANKAR, S.V.: "A Pressure Based Calculation Procedure for Viscous Flows at All Speeds in Arbitrary Configurations", AIAA Paper 88-0058 (1988).

[8] ISSA, R.I., LOCKWOOD, F.C.: "On the Prediction of Two-Dimensional Supersonic Viscous Interactions Near Walls", AIAA Journal, 15 (1977) pp. 182-188.

[9] HAH, C.: "Navier-Stokes Calculation of Three-Dimensional Compressible Flow Across a Cascade of Airfoils with an Implicit Relaxation Method", AIAA Paper 86-0555 (1986).

[10] RHIE, C.M.: "A Pressure Based Navier-Stokes Solver Using the Multigrid Method", AIAA Paper 86-0207 (1986).

[11] DEMIRDŽIĆ, I., GOSMAN, A.D., ISSA, R.I.: "A Finite Volume Method for the Prediction of Turbulent Flow in Arbitrary Geometries", Lecture Notes in Physics, Vol. 141, Springer-Verlag, 1980, pp. 144-150.

[12] DEMIRDŽIĆ, I., GOSMAN, A.D., ISSA, R.I., PERIĆ, M.: "A Calculation Procedure for Turbulent Flow in Complex Geometries", Computers and Fluids, 15 (1987) pp.251-273.

[13] MORETTI, G.: "Experiments on Initial and Boundary Conditions", Numerical and Physical Aspects of Aerodynamics Flows (Ed: R.Cebeci), Springer-Verlag, 1981.

[14] NI,R.-H.: "A Multiple Grid Scheme for Solving the Euler Equations", AIAA Journal, 20 (1982) pp.1565-1571.

[15] EIDELMAN, S., COLELLA, A.D., SHREEVE, R.P.: "Application of the Godunov Metod and Its Second-Order Extention to Cascade Flow Modelling", AIAA Journal, 22 (1984) pp. 1609-1615.

[16] MASON, M.L., PUTNAM, L.E., RE, R.J.: "The Effect of Throat Contouring on Two-Dimensional Converging-Diverging Nozzle at Static Conditions", NASA Technical Paper 1704 (1980).

[17] JOHNSON, D.A., HORSTMAN, C.C, BACHALO, W.D.: "Comparisson Between Experiment and Prediction for a Transonic Turbulent Separated Flow", AIAA Journal (1982) pp. 737-744.

[18] LAI, H.T., KHOSLA, P.K.: "Global Pressure Relaxation Procedure for Compresible Turbulent Strong Interaction Flows", Computers and Fluids, 16 (1988) pp. 217-228.

[19] SAHU, J., NIETUBITCZ, C.J, STEGER, J.L.: "Navier-Stokes Computations of Projectile Base Flow with and without Base Injection, AIAA Journal, 23 (1985) pp. 1348-1355.

[20] ISSA, R.I.: "Solution of the Implicitly Discretized Fluid Flow Equations by Operator-Splitting", J. Comp. Physics, 62 (1986) pp. 40-65.

A MULTIGRID FLUX-DIFFERENCE SPLITTING METHOD FOR STEADY INCOMPRESSIBLE NAVIER-STOKES EQUATIONS

E. DICK
Department of Machinery, State University of Ghent
Sint Pietersnieuwstraat 41, B-9000 Gent, Belgium

J. LINDEN
Gesellschaft für Mathematik und Datenverarbeitung, F1
D-5205 St. Augustin-Birlinghoven, W. Germany

SUMMARY

The steady Navier-Stokes equations in primitive variables are discretized in conservative form by a vertex-centered finite volume method. Flux-difference splitting is applied to the convective part.

In its first order formulation flux-difference splitting leads to a discretization of so-called vector positive type. This allows the use of classical relaxation methods in collective form. An alternating line-Gauss-Seidel relaxation method is chosen here. This relaxation method is used as a smoother in a multigrid method. The components of this multigrid method are : full-approximation scheme with F-cycles, bilinear prolongation, full-weighting for residual restriction and injection of grid functions.

Higher order accuracy is achieved by the Chakravarthy-Osher method. In this approach the first order convective fluxes are modified by adding second order corrections involving flux-limiting. Here, the simple minmod-limiter is chosen. In the multigrid formulation, the second order discrete system is solved by defect correction.

Computational results are shown for the well-known GAMM-backward facing step problem. The relaxation is performed on two blocks.

INTRODUCTION

Modern upwind discretization methods in compressible flow make use of flux-splitting concepts in flux-matrix, flux-vector or flux-difference form. Currently very popular are the flux-vector splitting method of Van Leer [1], the flux-difference method of Roe [2] and the flux-difference method of Osher and Chakravarthy [3]. It is not generally recognized that these concepts can also be applied to incompressible flow. Some recent applications of flux-difference splitting methods in incompressible flow are due to Hartwich and Hsu [4] and to Gorski [5]. These methods, however, resort to the concept of artificial compressibility in order to construct, through time integration, a solution of the steady incompressible Navier-Stokes equations. The use of artificial compressibility is, however, not necessary to apply flux-splitting concepts to incompressible flows. This was demonstrated by the first author [6,7] using non-conservative flux-matrix splitting. In this paper the flux-difference splitting concept is used in an analogous way in order to come to a conservative discretization. The flux-difference splitting is of polynomial type, very similar to the one developed in [8] for compressible Euler equations.

The steady Navier-Stokes equations for an incompressible fluid are, in conservative form :

$$\frac{\partial}{\partial x} u^2 + \frac{\partial}{\partial y} uv + \frac{\partial}{\partial x} p = \nu\left(\frac{\partial^2 u}{\partial x^2} + \frac{\partial^2 u}{\partial y^2}\right) , \qquad (1)$$

$$\frac{\partial}{\partial x} uv + \frac{\partial}{\partial y} v^2 + \frac{\partial}{\partial y} p = \nu\left(\frac{\partial^2 v}{\partial x^2} + \frac{\partial^2 v}{\partial y^2}\right) , \qquad (2)$$

$$c^2\left(\frac{\partial u}{\partial x} + \frac{\partial v}{\partial y}\right) = 0 , \qquad (3)$$

where u and v are the Cartesian components of velocity, c is a reference velocity introduced to homogenize the eigenvalues of the system matrices, defined in the sequel, p is pressure divided by density and ν is the kinematic viscosity coefficient.

The set of equations (1-3) can be written in system form as

$$\frac{\partial f}{\partial x} + \frac{\partial g}{\partial y} = \frac{\partial f_v}{\partial x} + \frac{\partial g_v}{\partial y} , \qquad (4)$$

where f and g are the convective fluxes, f_v and g_v are the viscous fluxes. These are :

$$f = \begin{pmatrix} u^2+p \\ uv \\ c^2 u \end{pmatrix} , \quad g = \begin{pmatrix} uv \\ v^2+p \\ c^2 v \end{pmatrix} , \quad f_v = \begin{pmatrix} \nu \frac{\partial u}{\partial x} \\ \nu \frac{\partial v}{\partial x} \\ 0 \end{pmatrix} , \quad g_v = \begin{pmatrix} \nu \frac{\partial u}{\partial y} \\ \nu \frac{\partial v}{\partial y} \\ 0 \end{pmatrix} . \qquad (5)$$

Differences of the convective fluxes can be written in algebraically exact form as follows :

$$\Delta f = \begin{pmatrix} 2\bar{u} & 0 & 1 \\ \bar{v} & \bar{u} & 0 \\ c^2 & 0 & 0 \end{pmatrix} \Delta \begin{pmatrix} u \\ v \\ p \end{pmatrix} , \qquad \Delta g = \begin{pmatrix} \bar{v} & \bar{u} & 0 \\ 0 & 2\bar{v} & 1 \\ 0 & c^2 & 0 \end{pmatrix} \Delta \begin{pmatrix} u \\ v \\ p \end{pmatrix} , \qquad (6)$$

where the bar denotes the algebraic mean of the differenced variables.

The matrices defined by (6) are discrete Jacobians. In the sequel, these are denoted by A_1 and A_2. Any linear combination of these Jacobians has the form

$$A = n_x A_1 + n_y A_2 = \begin{pmatrix} n_x\bar{u}+\bar{w} & n_y\bar{u} & n_x \\ n_x\bar{v} & n_y\bar{v}+\bar{w} & n_y \\ c^2 n_x & c^2 n_y & 0 \end{pmatrix} , \qquad (7)$$

where $\bar{w} = n_x\bar{u} + n_y\bar{v}$.

For $n_x^2 + n_y^2 = 1$, the eigenvalues of the Jacobian A are :

$$\lambda_1 = \bar{w} , \qquad \lambda_{2,3} = \bar{w} \pm a ,$$

with $a = \sqrt{\bar{w}^2+c^2}$.

The corresponding left and right eigenvector matrices are given by :

$$L = \begin{pmatrix} \frac{\bar{v}\,\bar{w} + n_y c^2}{a^2} & -\frac{\bar{u}\,\bar{w} + n_x c^2}{a^2} & \frac{n_x\bar{v} - n_y\bar{u}}{a^2} \\ \frac{n_x}{2}(\frac{\bar{w}}{a} + 1) & \frac{n_y}{2}(\frac{\bar{w}}{a} + 1) & \frac{1}{2a} \\ \frac{n_x}{2}(\frac{\bar{w}}{a} - 1) & \frac{n_y}{2}(\frac{\bar{w}}{a} - 1) & \frac{1}{2a} \end{pmatrix} , \tag{8}$$

$$R = \begin{pmatrix} n_y & \frac{\bar{u}}{a} - n_x(\frac{\bar{w}}{a} - 1) & \frac{\bar{u}}{a} - n_x(\frac{\bar{w}}{a} + 1) \\ -n_x & \frac{\bar{v}}{a} - n_y(\frac{\bar{w}}{a} - 1) & \frac{\bar{v}}{a} - n_y(\frac{\bar{w}}{a} + 1) \\ 0 & a - \bar{w} & a + \bar{w} \end{pmatrix} , \tag{9}$$

where $R = L^{-1}$.

The matrix A can be split into positive and negative parts by :

$$A^+ = R\,\Lambda^+\,L\ , \qquad A^- = R\,\Lambda^-\,L\ , \qquad A = A^+ + A^-\ ,$$

where $\Lambda^+ = \mathrm{diag}(\lambda_1^+, \lambda_2^+, \lambda_3^+)\ , \qquad \Lambda^- = \mathrm{diag}(\lambda_1^-, \lambda_2^-, \lambda_3^-)\ ,$

with $\lambda_i^+ = \max(\lambda_i, 0)\ , \qquad \lambda_i^- = \min(\lambda_i, 0)\ .$

By positive and negative matrices, matrices with respectively non-negative and non-positive eigenvalues are meant. This allows a splitting of any linear combination of flux-differences by

$$\Delta\phi := n_x\Delta f + n_y\Delta g = A^+\Delta\xi + A^-\Delta\xi\ ,$$

where ξ is the vector of dependent variables, $\xi^T = \{u,v,p\}$.

VERTEX-CENTERED FINITE VOLUME FORMULATION

Treatment of inviscid fluxes

Figure 1 shows the control volume centered around the node (i,j).

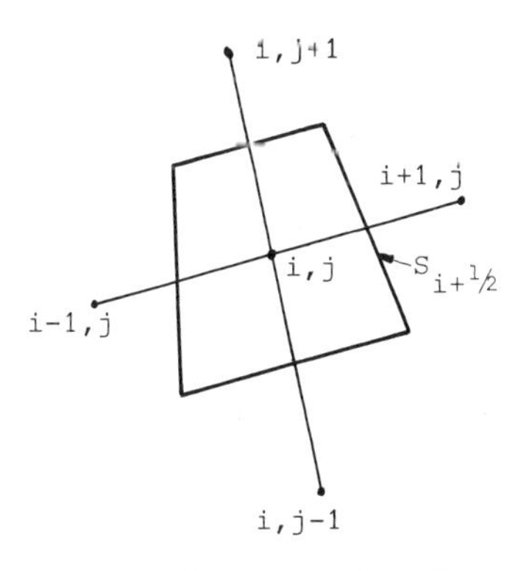

Fig. 1. Interior control volume

The inviscid part of (4) is

$$\frac{\partial f}{\partial x} + \frac{\partial g}{\partial y} = 0\ . \tag{10}$$

With piecewise constant interpolation of variables, the flux-difference over the surface $S_{i+\frac{1}{2}}$ of the control volume can be written as

$$\Delta F_{i,i+1} :=$$

$$\Delta s_{i+\frac{1}{2}}\,(n_x\,\Delta f_{i,i+1} + n_y\,\Delta g_{i,i+1}), \tag{11}$$

where $\Delta s_{i+\frac{1}{2}}$ denotes the length of the surface $S_{i+\frac{1}{2}}$ and n_x and n_y the components of the unit outward normal.

Using the notation of the previous section, the flux-difference is

$$\Delta F_{i,i+1} = F_{i+1} - F_i = \Delta s_{i+\frac{1}{2}} A_{i,i+1} \Delta\xi_{i,i+1} . \tag{12}$$

Furthermore, the matrix $A_{i,i+1}$ can be split into positive and negative parts. This allows the definition of the absolute value of the flux-difference by

$$|\Delta F_{i,i+1}| = \Delta s_{i+\frac{1}{2}} (A^+_{i,i+1} - A^-_{i,i+1}) \Delta\xi_{i,i+1} . \tag{13}$$

Based on (13) an upwind definition of the flux is

$$F_{i+\frac{1}{2}} = \frac{1}{2}[F_i + F_{i+1} - |\Delta F_{i,i+1}|] . \tag{14}$$

That this represents an upwind flux can be verified by writing (14) in either of the two following ways, which are completely equivalent :

$$\begin{aligned} F_{i+\frac{1}{2}} &= F_i + \frac{1}{2} \Delta F_{i,i+1} - \frac{1}{2} |\Delta F_{i,i+1}| \\ &= F_i + \Delta s_{i+\frac{1}{2}} A^-_{i,i+1} \Delta\xi_{i,i+1} , \end{aligned} \tag{15}$$

$$\begin{aligned} F_{i+\frac{1}{2}} &= F_{i+1} - \frac{1}{2} \Delta F_{i,i+1} - \frac{1}{2} |\Delta F_{i,i+1}| \\ &= F_{i+1} - \Delta s_{i+\frac{1}{2}} A^+_{i,i+1} \Delta\xi_{i,i+1} . \end{aligned} \tag{16}$$

Indeed, when $A_{i,i+1}$ has only positive eigenvalues, the flux $F_{i+\frac{1}{2}}$ is taken to be F_i and when $A_{i,i+1}$ has only negative eigenvalues, the flux $F_{i+\frac{1}{2}}$ is taken to be F_{i+1}.

The fluxes on the other surfaces of the control volume are treated in a similar way. Using (15) and (16), the flux balance on the control volume of figure 1 can be brought into the form

$$\begin{aligned} &\Delta s_{i+\frac{1}{2}} A^-_{i,i+1}[\xi_{i+1} - \xi_i] + \Delta s_{i-\frac{1}{2}} A^+_{i,i-1}[\xi_i - \xi_{i-1}] \\ &+ \Delta s_{j+\frac{1}{2}} A^-_{j,j+1}[\xi_{j+1} - \xi_j] + \Delta s_{j-\frac{1}{2}} A^+_{j,j-1}[\xi_j - \xi_{j-1}] = 0 . \end{aligned} \tag{17}$$

The set formed by the equations (17) for all nodes is both conservative and positive. The positivity can be seen by putting it into the form

$$\begin{aligned} C\xi_{i,j} &= \Delta s_{i-\frac{1}{2}} A^+_{i,i-1} \xi_{i-1,j} + \Delta s_{i+\frac{1}{2}} (-A^-_{i,i+1}) \xi_{i+1,j} \\ &+ \Delta s_{j-\frac{1}{2}} A^+_{j,j-1} \xi_{i,j-1} + \Delta s_{j+\frac{1}{2}} (-A^-_{j,j+1}) \xi_{i,j+1} , \end{aligned} \tag{18}$$

where C is the sum of the matrices in the right hand side, all of which have non-negative eigenvalues.

As a consequence of the positivity, a solution can be obtained by a collective variant of any scalar relaxation method. By a collective variant, it is meant that at each node all components of the vector of dependent variables ξ are relaxed simultaneously.

In practice, the flux-balance (17) is formed by summing expressions of type (15) over all surfaces, using the appropriate components of the unit outward normal n_x and n_y in the definition of the Jacobian (7).

Treatment of viscous fluxes

In order to define the viscous fluxes, f_V and g_V, in a piecewise constant way on the surfaces of the control volume, approximations to derivatives of u and v are to be calculated at midpoints of the surfaces. This is done here by the well-known Peyret-control volume technique. Figure 2 shows the Peyret-control volume around the point $(i+\frac{1}{2},j)$.

Integration over the control volume gives :

$$\left(\frac{\partial u}{\partial x}\right)_{i+\frac{1}{2}} \Omega_{i+\frac{1}{2}} \approx \int_{\Omega_{i+\frac{1}{2}}} \frac{\partial u}{\partial x}\, d\Omega = \int_{\Gamma_{i+\frac{1}{2}}} u\, n_x\, d\Gamma ,$$

where $\Gamma_{i+\frac{1}{2}}$ denotes the surface of the control volume.

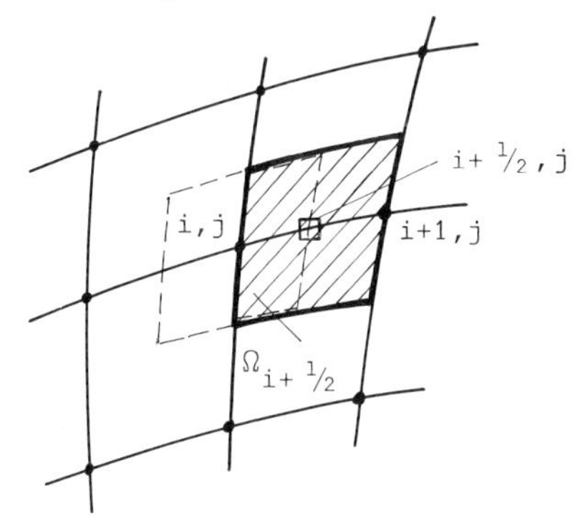

Fig. 2. Peyret-control volume

On each side of $\Omega_{i+\frac{1}{2}}$, midpoint-rules are used to evaluate the surface integral. Vertex values (at gridpoints (i,j) and (i+1,j)) or mean values over the four surrounding nodes are used.

By the Peyret-control volume technique, a central discretization of the viscous flux-balance is obtained.
The structure of the discrete set, when transferred to the left hand side in (4), is positive in the sense of the previous section. Adding the viscous flux-balance to the inviscid flux-balance enforces the positivity.

BOUNDARY CONDITIONS

Solid boundaries

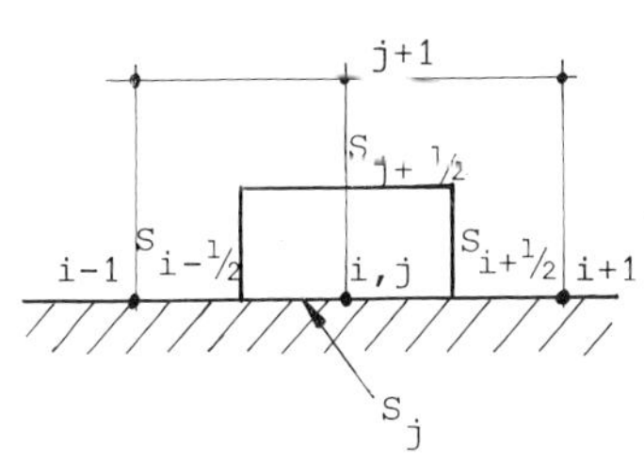

Fig. 3. Control volume on a solid boundary

Figure 3 shows the half-volume centered around a node on a solid boundary.

Fluxes on the surfaces $S_{i+\frac{1}{2}}$, $S_{i-\frac{1}{2}}$ and $S_{j+\frac{1}{2}}$ are defined as described above. On the surface S_j the convective flux is calculated with the values of the variables at the node (i,j). Thus, the convective flux balance takes the form (17), without contribution from a node (j 1).

The mass equation is taken from the convective flux balance, supplemented with the boundary conditions u = 0 and v = 0.

The mass equation takes the form of a Poisson equation for pressure. As a consequence, the system of equations at a solid boundary can be relaxed in the same way as the system of equations at an interior node.

In- and outflow

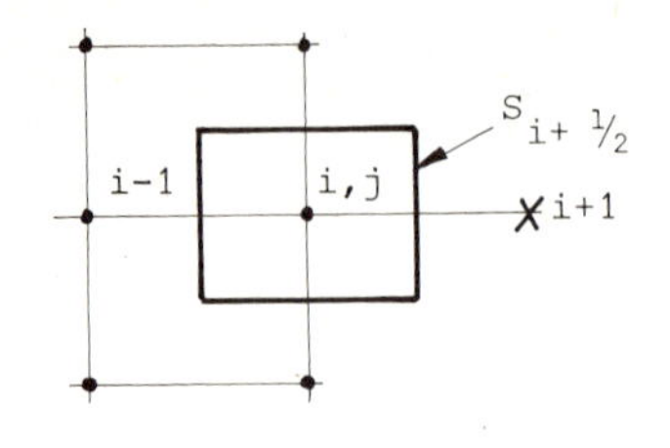

Fig. 4. Control volume at outflow

Figure 4 shows the control volume at an outflow boundary.

The grid is extended by a row of external nodes, so that complete control volumes can be defined. The convective flux balance at the outflow boundary is constructed as for an interior node. A contribution from an external node comes in. However, this contribution can be eliminated.

Using the left and right eigenvector matrices (8-9), the general expression for A^- is found to be

$$A^- = \begin{bmatrix} n_y\bar{w}^-\beta + n_x\lambda^2\hat{u}/2 & -n_y\bar{w}^-\alpha + n_y\lambda^2\hat{u}/2 & (n_y\bar{w}^-\gamma + \lambda\hat{u}/2)/a \\ -n_x\bar{w}^-\beta + n_x\lambda^2\hat{v}/2 & n_x\bar{w}^-\alpha + n_y\lambda^2\hat{v}/2 & (-n_x\bar{w}^-\gamma + \lambda\hat{v}/2)/a \\ -n_x c^2\lambda/2 & -n_y c^2\lambda/2 & -c^2/2a \end{bmatrix}, \quad (19)$$

where $\hat{u} = \bar{u} - n_x(\bar{w} + a)$, $\hat{v} = \bar{v} - n_y(\bar{w} + a)$, $\lambda = (\bar{w} - a)/a$,

$\alpha = (\bar{u}\,\bar{w} + n_x c^2)/a^2$, $\beta = (\bar{v}\,\bar{w} + n_y c^2)/a^2$, $\gamma = (n_x\bar{v} - n_y\bar{u})/a$.

At outflow $\bar{w}^- = 0$ and (19) simplifies to

$$A^- = \begin{bmatrix} n_x\lambda^2\hat{u}/2 & n_y\lambda^2\hat{u}/2 & \lambda\hat{u}/2a \\ n_x\lambda^2\hat{v}/2 & n_y\lambda^2\hat{v}/2 & \lambda\hat{v}/2a \\ -n_x c^2\lambda/2 & -n_y c^2\lambda/2 & -c^2/2a \end{bmatrix}. \quad (20)$$

Because rank $(A^-) = 1$, two independent combinations of the discrete set of equations corresponding to the convective flux balance exist, eliminating the outside node.

Useful combinations are given by

$$d_{1,2}^T\, A^- = 0 \,, \quad (21)$$

for $$d_1^T = [1,\ 0,\ \lambda\hat{u}/c^2]\ ,\ d_2^T = [0,\ 1,\ \lambda\hat{v}/c^2]\ . \quad (22)$$

These combinations are momentum equations corrected with a contribution from the mass equation. For the momentum equations the viscous flux balance is required. With an auxiliary assumption of fully developed outflow, the normal derivatives of u and v on the surface $S_{i+\frac{1}{2}}$ can be considered to be zero. For a grid that is Cartesian at outlet, this allows the formulation of the viscous flux balance without contributions from external nodes.

The resulting modified momentum equations are used as equations for u and v. The boundary condition added is the prescription of pressure ($p = 0$).

At inflow, a similar reasoning applies. In this case, the matrix A^- given by (19) cannot be simplified. Now rank $(A^-) = 2$, so that there exists one combination of the equations, eliminating the external node in the convective flux balance. This combination is

$$d_3^T A^- = 0 \; , \tag{23}$$

for

$$d_3^T = [- \frac{n_x c^2}{\lambda a} \; , \; - \frac{n_y c^2}{\lambda a} \; , \; 1] \; . \tag{24}$$

This combination is a mass equation corrected with contributions from the momentum equations. Again, in the viscous flux balance for the momentum equations, an auxiliary assumption of fully developed inflow can be used, so that there is no contribution from external nodes.

The resulting equation is used as an equation for pressure. The boundary conditions added are prescription of u and v.

COMPUTATIONAL EXAMPLE FOR FIRST ORDER DISCRETIZATION

Results are presented for the GAMM-backward facing step problem. The grid shown in figure 5 is the second coarsest in a series of four. The grids are almost rectangular. The finest grid has 2834 nodes. At inflow, velocity is prescribed (fully developed profile). At outflow, pressure is given. The Reynolds number is 300 based on maximum inlet velocity and inlet height.

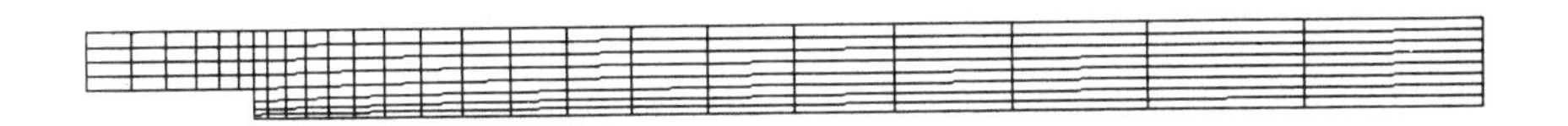

Fig. 5. Second coarsest grid for the backward facing step

A multigrid iteration with F-cycles is used for solving the discrete non-linear system of equations. As usual in multigrid methods, the non-linearity is treated implicitly by the full approximation scheme (FAS). Alternating line-Gauss-Seidel relaxation in lexicographic order is used for error smoothing where the coefficients of the equations (17) are always formed with the latest available information. The relaxation is organized blockwise. A first block is formed by the nodes upstream of the step (25 x 17 nodes in the finest grid). A second block is formed by the nodes at the step and downstream of the step (73 x 33 nodes in the finest grid). The first block is relaxed first. One alternating sweep is performed before and one after the coarse grid correction step. Within the flow field, restriction of residuals is done by full weighting, while injection is used at the boundaries. Coarse grid corrections are transferred back to finer grids by bilinear interpolation. The FAS-restriction of function values is injection. The calculation starts from zero initial values for all of the unknowns on the finest grid.

Figure 6 shows the convergence behaviour. The residuals are the maximum residuals over the discrete equations in the interior flow field and the boundary. The average convergence factor per cycle is approximately 0.24.

Figure 7 shows the streamlines obtained after postprocessing. The ratio of the reattachment length to the step height is about 4.77. That this ratio is too short is due to the first order accuracy.

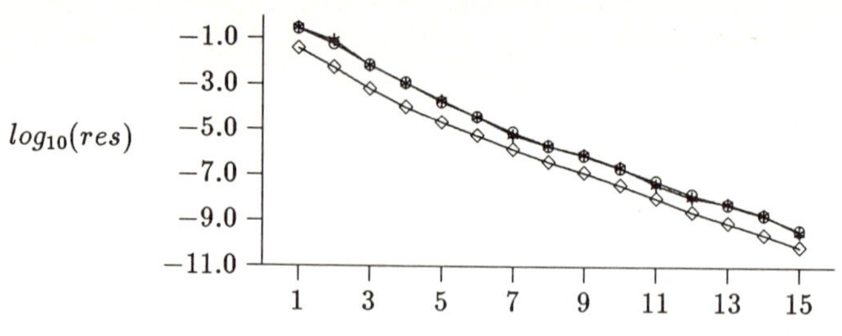

Fig. 6. Convergence history

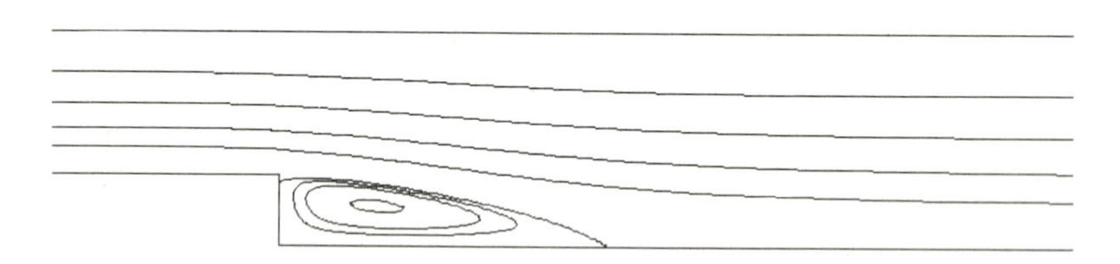

Fig. 7. Streamlines for the first order solution

SECOND ORDER FORMULATION

In order to obtain second order accuracy, the definition of the flux (14) is to be modified.

First we remark that the flux-difference (12) can be written as

$$\Delta F_{i,i+1} = \sum_n \Delta F^n_{i,i+1} := \Delta s_{i+\frac{1}{2}} \sum_n \lambda^n_{i+\frac{1}{2}} < l^n_{i+\frac{1}{2}}, \Delta\xi_{i,i+1} > r^n_{i+\frac{1}{2}} , \tag{25}$$

where the superscript n refers to the nth eigenvalue λ^n, and r^n and l^n are the corresponding right and left eigenvectors.

The first order flux (14) can now be rewritten as

$$F_{i+\frac{1}{2}} = \frac{1}{2}(F_i + F_{i+1}) - \frac{1}{2}\sum_n \Delta F^{n+}_{i,i+1} + \frac{1}{2}\sum_n \Delta F^{n-}_{i,i+1} , \tag{26}$$

where the + and - superscripts denote the positive and negative parts of the flux-difference, i.e. those parts obtained by replacing the eigenvalue by its positive or negative part in (25).

According to Chakravarthy and Osher [9], a second order flux corresponding to (26) can be defined by

$$\begin{aligned} F^u_{i+\frac{1}{2}} = \frac{1}{2}(F_i + F_{i+1}) - \frac{1}{2}\sum_n \Delta F^{n+}_{i,i+1} + \frac{1}{2}\sum_n \Delta F^{n-}_{i,i+1} \\ + \frac{1}{2}\sum_n \Delta\tilde{F}^{n+}_{i-1,i} - \frac{1}{2}\sum_n \Delta\tilde{F}^{n-}_{i+1,i+2} , \end{aligned} \tag{27}$$

where :

$$\Delta\tilde{F}^{n+}_{i-1,i} = \Delta s_{i+\frac{1}{2}} \lambda^{n+}_{i+\frac{1}{2}} < l^n_{i+\frac{1}{2}}, \Delta\xi_{i-1,i} > r^n_{i+\frac{1}{2}} , \tag{28}$$

with a similar definition for $\Delta\tilde{F}^{n-}_{i+1,i+2}$.

Clearly, (28) is constructed by considering a flux-difference over the surface $S_{i+\frac{1}{2}}$, i.e. using the geometry of the surface, with data shifted in the negative i-direction.

The definition (27) corresponds to a second order upwind flux. This can easily be seen by considering the case where all eigenvalues have the same sign.

Second order accuracy can also be achieved by taking a central definition of the flux vector

$$F^{c}_{i+\frac{1}{2}} = \frac{1}{2}\,(F_i + F_{i+1}) \ . \tag{29}$$

As is well known, using either (27) or (29) leads to a scheme which is not monotonicity preserving, so that wiggles in the solution become possible. Following the theory of flux limiters, a combination of (27) and (29) is to be taken. This has the form

$$\begin{aligned} F_{i+\frac{1}{2}} = \frac{1}{2}\,(F_i + F_{i+1}) &- \frac{1}{2}\sum_n \Delta F^{n+}_{i,i+1} + \frac{1}{2}\sum_n \Delta F^{n-}_{i,i+1} \\ &+ \frac{1}{2}\sum_n \Delta\tilde{\tilde{F}}^{n+}_{i-1,i} - \frac{1}{2}\sum_n \Delta\tilde{\tilde{F}}^{n-}_{i+1,i+2} \ , \end{aligned} \tag{30}$$

with

$$\Delta\tilde{\tilde{F}}^{n+}_{i-1,i} = \mathrm{Lim}(\Delta\tilde{F}^{n+}_{i-1,i},\ \Delta F^{n+}_{i,i+1}) \ , \tag{31}$$

$$\Delta\tilde{\tilde{F}}^{n-}_{i+1,i+2} = \mathrm{Lim}(\Delta\tilde{F}^{n-}_{i+1,i+2},\ \Delta F^{n-}_{i,i+1}) \ , \tag{32}$$

where Lim denotes some limited combination of both arguments. We choose here the simplest possible form of a limiter, i.e. Lim = MinMod, where the function MinMod returns the argument with minimum absolute value if both arguments have the same sign and returns zero otherwise. By the use of the limiter to the vectors (31) and (32) it is meant that the limiter is used component-wise.

At boundaries and in the vicinity of boundaries, some of the flux-differences in (31) or (32) do not exist. In this case a zero is entered.

Since, for the discretization obtained by the second order formulation, the positivity is not guaranteed, direct multigrid with collective relaxation as smoother is questionable. Therefore, defect-correction was used here as the solution procedure. Only one multigrid cycle (for the first order discretization) was used per defect-correction step.

The convergence of this combined iteration is dominated by the defect-correction part. The residual reduction per iteration can be as high as 0.9 A better coupling of the defect-correction and multigrid concepts should lead to improved efficiency. The bad convergence result obtained here is very similar to the result obtained by Koren and Spekreijse [10] with the same defect-correction approach applied to steady Euler equations. The result is similar although they use a different type of splitter and a different type of second order formulation.

Figure 8 shows the result obtained with the second order formulation, using the same grid as in the figures 6 and 7. Here, the reattachment length is 6.44.

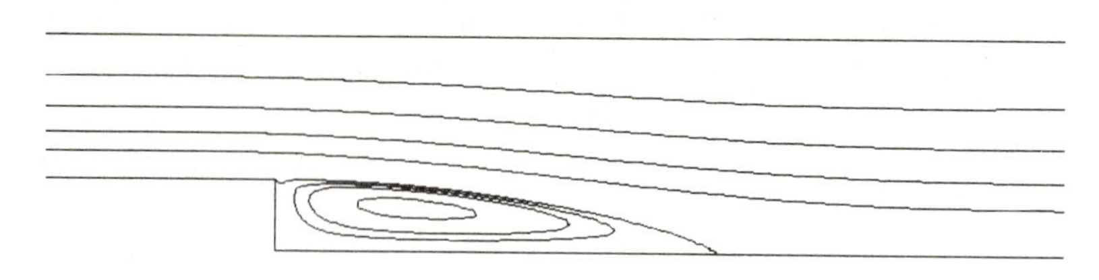

Fig. 8. Streamlines for the second order solution

CONCLUSION

It has been shown that for the case of incompressible Navier-Stokes equations, flux-difference splitting can also be applied, leading to a set of discrete equations which can be solved by multigrid methods.

REFERENCES

1. VAN LEER B. : "Flux-vector splitting for the Euler equations", Lecture Notes in Physics, 170 (1982) 507-512.

2. ROE P.L. : "Approximate Riemann solvers, parameter vectors and difference schemes", J. Comp. Phys., 43 (1981) 357-372.

3. OSHER S., CHAKRAVARTHY S.R. : "Upwind schemes and boundary conditions with applications to Euler equations in general geometries", J. Comp. Phys., 50 (1983) 447-481.

4. HARTWICH P.M., HSU C.H. : "High resolution upwind schemes for the three-dimensional, incompressible Navier-Stokes equations", AIAA-87-0547.

5. GORSKI J.J. : "Solutions of the incompressible Navier-Stokes equations using an upwind-differenced TVD scheme", Lecture notes in Physics, 323 (1989) 278-282.

6. DICK E. : "A flux-vector splitting method for steady Navier-Stokes equations", Int. J. Num. Meth. Fluids, 8 (1988) 317-326.

7. DICK E. : "A multigrid method for steady incompressible Navier-Stokes equations based on partial flux splitting", Int. J. Num. Meth. Fluids, 9 (1989) 113-120.

8. DICK E. : "A flux-difference splitting method for steady Euler equations". J. Comp. Phys., 76 (1988) 19-32.

9. CHAKRAVARTHY S.R., OSHER S. : "A new class of high accuracy TVD schemes for hyperbolic conservation laws", AIAA-85-0363.

10. KOREN B., SPEKREIJSE S. : "Solution of the steady Euler equations by a multigrid method", Lecture Notes in Pure and Applied Mathematics, 110 (1988) 323-336.

Hypersonic Leeside Flow Computations Using Centered Schemes for Euler Equations

by

Peter Eliasson and Arthur Rizzi
FFA The Aeronautical Research Institute of Sweden
S-161 11 Bromma, Sweden

1 SUMMARY

A centered finite-volume scheme using artificial viscosity is applied to hypersonic flow past a sphere. The windside of the flowfield is characterized by the strong bow shock, and the leeside by separation and vortical flow. In order to obtain the correct bow shock standoff distance, we have improved our scheme in terms of conservation and viscosity switches. A number of computed examples of hypersonic flows past a sphere demonstrate these improvements and illustrate the features of the leeside flowfield.

2 INTRODUCTION

Renewed interest in hypersonic vehicles has motivated the development of methods to numerically simulate such flowfields. The flow about these vehicles has two distinct characters. On the windside the bow shock is intense and the flow is compressed and attached to the body. On the leeside the bowshock is weak, the flow is thin and generally separated, forming vortex sheets and vortices.

In the past we have developed and used a centered finite volume scheme with artificial viscosity to study transonic flow around delta wings, and we have obtained good results for the leeside flow phenomena [1,2]. As part of the Hermes project we have been applying this method to the problem of hypersonic flow around a delta wing with blunt leading edges. Here spurious results were obtained for the windside flow, namely an incorrect shock standoff distance.

We decided to study this problem in the context of hypersonic flow past a sphere. The sphere was chosen as the suitable test case because it possesses the same type of windside and leeside flow phenomena, and because a lot of experimental and numerical data are available for comparison. We found that by handling conservation correctly in the vicinity of the mesh singular line and by refining the switches in the artificial viscosity model we do obtain the correct shock location even for very strong shocks. The purpose of the paper is to explain why the original scheme had difficulty with the windside flow and what modifications were necessary to correct it. The paper then discusses how well an inviscid model can represent the flow on the leeside where cross-flow shocks induce separation and produce vortices.

3 NUMERICAL APPROACH

The standard approach to inviscid hypersonic flow computations in the 1970s was the MacCormack scheme together with a bow shock fitting procedure. Recently we have been investigating two finite volume alternatives to this approach: 1) centered flux differences with artificial viscosity and bow shock fitting, and 2) centered flux differences with artificial viscosity and bow shock capturing. The latter approach has been well validated for transonic flow problems, and we briefly review it here.

3.1 Original Scheme

The scheme we have used for transonic flow is an explicit three-stage Runge-Kutta finite volume method with centered flux evaluations and artificial viscosity[1]. We solve the Euler equations in three dimensions, but to simplify the notation we discuss the semidiscrete form of the centered finite volume scheme in one space dimension

$$\frac{d}{dt} q_j = \delta F_j = F_{j+1/2} - F_{j-1/2} \tag{1}$$

where

$$F_{j+1/2} = 1/2(F_j + F_{j+1}) \ . \tag{2}$$

In the original approach we scale the convective difference by the local time step and add artificial viscosity terms

$$\Delta q_j = \Delta t_j \delta F_j + \varepsilon_2 \delta(s_j \delta q_j) + \varepsilon_4 \delta^4 q_j \tag{3}$$

where

$$s_j = \frac{1}{2}(dp_j + dp_{j+1})$$

and

$$dp_j = (p_{j+1} - 2p_j + p_{j-1})/(p_{j+1} + 2p_j + p_{j-1}) \, .$$

This scheme is semiconservative because the local time step varies in space. This usually has no effect if the flow contains only weak shocks and if the variation in Δt_j is smooth. This scaling has been found to be useful or computing transonic flows.

Figure 1 demonstrates the loss of conservation when Δt_j varies rapidly, as across a strong shock. Around the polar singular line the gradient of the local time step is large. When the singular line passes through the normal bow shock, a poor result is obatined. When the singular line passes through the normal bow shock, a poor results is obtained. When we rotate the grid 90 deg so that the grid is smooth near the normal part of the bow shock, a better result is obtained.

3.2 Conservation at Steady State

For cases with strong shocks it is better to make the scheme conservative (at least at steady state) by e.g.

$$\Delta \mathbf{q}_j = \Delta t_j \, \delta(\mathbf{F}_j + \alpha_j \, \delta \mathbf{q}_j + \beta_j \, \delta^3 \, \mathbf{q}_j) \tag{4}$$

where

$$\alpha_{j+1/2} = \frac{1}{2} \varepsilon_2 \left(\frac{1}{\Delta t_j} + \frac{1}{\Delta t_{j+1}}\right) \max (dp_j, dp_{j+1})$$

$$\beta_{j+1/2} = \frac{1}{2} \varepsilon_4 \left(\frac{1}{\Delta t_j} + \frac{1}{\Delta t_{j+1}}\right)$$

and ε_2 and ε_4 are constants.

If the steady state is reached, $\Delta \mathbf{q}_j = 0$ and full conservation is achieved. But as long as transients exist, it is semiconservative with local time stepping.

3.3 Flow Variable Averaging

Scheme (4) is in the form (1) where the numerical flux is

$$\mathbf{F}^{num}_{j+1/2} = \frac{1}{2}(\mathbf{F}_j + \mathbf{F}_{j+1}) + \alpha_{j+1/2} \delta \mathbf{q}_{j+1/2} + \beta_{j+1/2} \, \delta^3 \, \mathbf{q}_{j+1/2} \, . \tag{5}$$

An alternative is to average the flow variables

$$F^{num}_{j+1/2} = F\left(\frac{q_j + q_{j+1}}{2}\right) + \alpha_{j+1/2}\delta q_{j+1/2} + \beta_{j+1/2}\delta^3 q_{j+1/2}. \quad (6)$$

Our experience finds form (6) to be somewhat more robust than (5), but we have no explanation for this.

4 ARTIFICIAL VISCOSITY MODEL

It has been generally believed that a centered scheme could not capture a strong bow shock satisfactorily with artificial viscosity. Here we present an argument based on the analysis of Kreiss (private communication) that shows that the strength of the shock scales with the equations.

4.1 Kreiss Analysis

The analysis is based on the Burgers equation $u_t + uu_x = \varepsilon u_{xx}$, $u(\pm\infty) = \pm a$.

4.1.1 *Differential Model*

Integrate the steady equation $\frac{1}{2}(u^2)_x = \varepsilon u_{xx}$ from a to u to obtain $\varepsilon u_x = \frac{1}{2}(u^2 - a^2)$. The maximum derivative $u_x = -(1/2)(a^2/\varepsilon)$ occurs at $x = 0$. This shows that the shock width is $0(\varepsilon/a)$. To resolve the shock over, say, 3 cells with $\Delta x = h = 0(1)$ we need $\varepsilon/a = 3h$ or $h \le 0.3\ \varepsilon/a$. Hence, for stronger shocks there are two possibilities,

1) reduce h
2) increase ε: $\varepsilon \approx 3ha$

where, **note**, a is the shock jump and not the solution u. Next question is what the correct model for an artificial ε is. Away from the shock linearize the solution $u = -a+\eta$, thus $\varepsilon\eta_x = -a\eta$ and hence $\eta = \exp(-a/\varepsilon\ x)$. The shock strength scales with the viscosity.

4.1.2 *Difference Approximation*

The difference solution should mimic the differential solution. Centered differences give $\varepsilon\delta^2 u_j = \frac{1}{2}\delta u_j^2$. Sum this from 1 to j to obtain

$$\varepsilon(u_{j+1} - u_j) = \frac{1}{2}\left(\frac{1}{2}(u_j^2 + u_{j+1}^2) - a^2\right). \quad (7)$$

Then linearize with $u_j = -a + \eta_j$ and we find $\eta_{j+1} = ((1-ah/2\varepsilon)/(1+ah/2\varepsilon))\eta_j \approx \exp(-ah/\varepsilon)\eta_j$ for small h (cf. exact solution)

We then have three cases: If $ah/2\varepsilon < 1$, the solution is monotone. If $ah/2\varepsilon > 1$, but O(1), there are decaying oscillations. If finally, $ah/2\varepsilon >> 1$ there are persisting oscillations. Hence as shock strength grows, there are two alternatives: either reduce h, or increase ε so that $ah/2\varepsilon \approx 1$. The usual trick is to make ε proportional to some measure of solution variation, as evidenced by the pressure switch dp_j.

4.2 Blended Nonlinear Viscosity

Kreiss's analysis applies to the variable ε case as well. He finds that regardless of the model used for ε, sawtooth waves persist away from the shock, and deduces that one needs fourth order dissipation to smooth them. However, the fourth difference across the shock generates oscillations proportional to the shock jump, so this difference should be switched off across the shock.

If we simply implement the conclusions of the Kreiss analysis in the conservative scheme (4) for the Euler equations, we must construct the switch

$$\beta_{j+1} = \max\left(\frac{1}{2}\varepsilon_4\left(\frac{1}{4\Delta t_j} + \frac{1}{\Delta t_{j+1}}\right) - \alpha_{j+1/2},\ 0\right) \tag{8}$$

and we arrive at a scheme very similar to the one proposed by Jameson[3]. The switched fourth order dissipation must also be treated correctly near boundaries in order to maintain good convergence and to keep the variations in entropy small[4].

5 COMPUTED RESULTS

We demonstrate the scheme (1), (4) and (8) by computing flows past the sphere at Mach numbers 2, 4, 8, and 20. These have been computed in the so-called "block marching" mode (Fig. 2). Since the flow at the flanks of the sphere is supersonic, we compute first only in the windside block (1). When that is converged, the solution at the flanks are boundary conditions to solve the flow in the leeside block (2). Two mesh sizes are used 55x21 and 109x41. Only one layer of cells are needed for the third dimension because of axisymmetry.

5.1 Windside Flow

The agreement between numerical and experimental results are very good at the windside of the sphere. The standoff distance for the shock and the total pressure loss over the shock are calculated almost exactly compared to theoretical and shock-fitting results (Fig. 3 and table 1).

Figure 4 presents the isoMach lines from our computation for $M_\infty = 20$. The bow shock is reasonably captured. On the windside the convergence is rather strong, on the leeside the residuals are reduced by only two decades. On the leeside the isoMach contour indicate two cross-flow shocks.

5.2 Leeside Flow

Cross-flow shocks induce separation (Fig. 4b) and produce structure in the wake. We want to discuss this further. Table 2 presents the comparison of computed and measured separation angles for two Mach numbers. They do not compare very well. This has to be expected with an Euler simulation. The Mach number just ahead of the cross-flow shock appears to reach a limit as the freestream Mach number increases.

Figure 5 presents the distribution of Mach number along the line ξ for four cases. First it passes through the bow shock, then the stagnation point, the cross-flow shock, and the wake. Figure 6 shows the distribution of total pressure loss 1 - $p_t/p_{t\infty}$ along ξ starting at the flank of the sphere. The loss is greatest at the sphere, varies through the entropy layer, and jumps across the shock.

6 FINAL REMARKS

It has been thought that a centered scheme with artificial viscosity could not capture a strong bow shock satisfactorily. However the paper shows both theoretically and by actual computations that accurate results are obtained for very strong shocks if the scheme observes conservation and if the artificial viscosity is blended. The theoretical argument is based on analysis that shows that the strength of the shock scales with the equations. Results are presented for the Mach numbers, Mach = 2, 4, 8, 20. The agreement between numerical and experimental results are very good at the windside of the sphered. The standoff distance for the bow-shock and the total pressure loss across the shock are computed almost exactly compared to theoretical values. The results on the leeside of the sphered compares only approximately to experiments. We have discussed the position of the cross-flow shock and the structure of the wake.

7 REFERENCES

1 Rizzi, A. and Eriksson, L.E.: Computation of Flow Around Wings Based on the Euler Equations. J. Fluid Mech. 148, 1984, pp. 45-71.

2 Rizzi, A., and Purcell, C.J.: On the Computation of Transonic Leading-Edge Vortices Using the Euler Equations, J.Fluid Mech. 181, 1987, pp. 163-195.

3 Jameson, A.: Numerical Solution of the Euler Equations for Compressible Inviscid Fluids, in Numerical Methods for the Euler Equations, eds F. Angrand et at., SIAM, 1985.

4 Olsson, P.: Flow Calculations Using Explicit Methods on a Data Parallel Computer, Rep. No.117, Uppsala Univ. 1989.

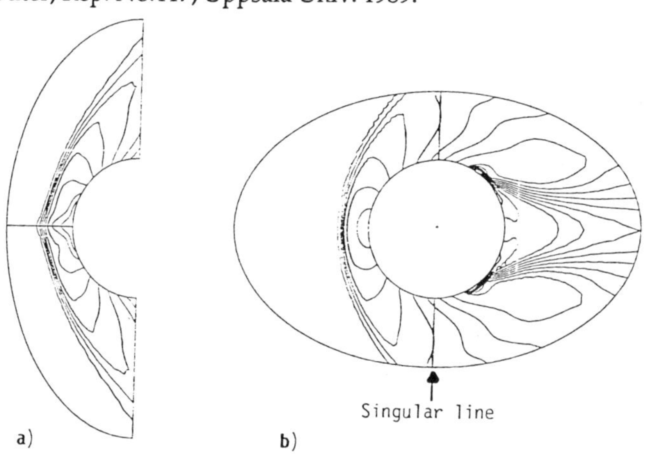

Fig. 1 Conservation is lost a) where the singular meets the shock, but not b) when the line is rotated 90 deg. $M_\infty = 2$.

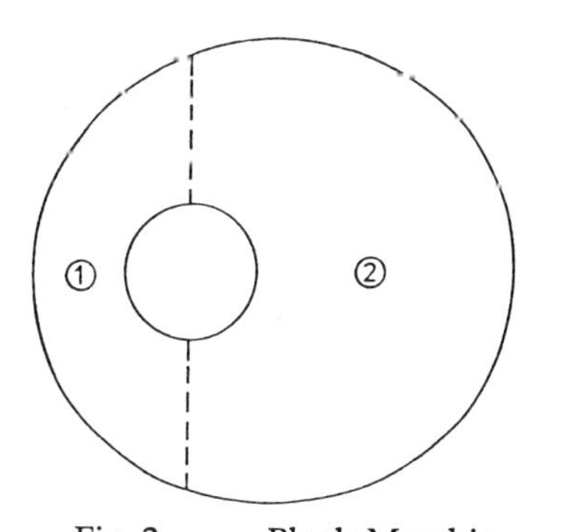

Fig. 2 Block-Marching mode.

Compute first in block 1, then block 2.

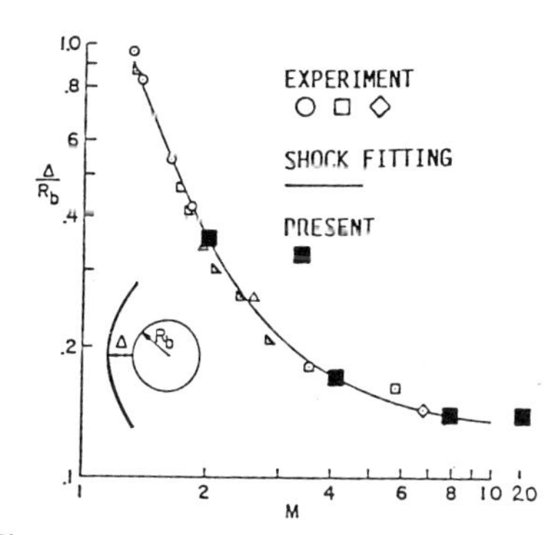

Fig. 3

Comparison of shock standoff distance.

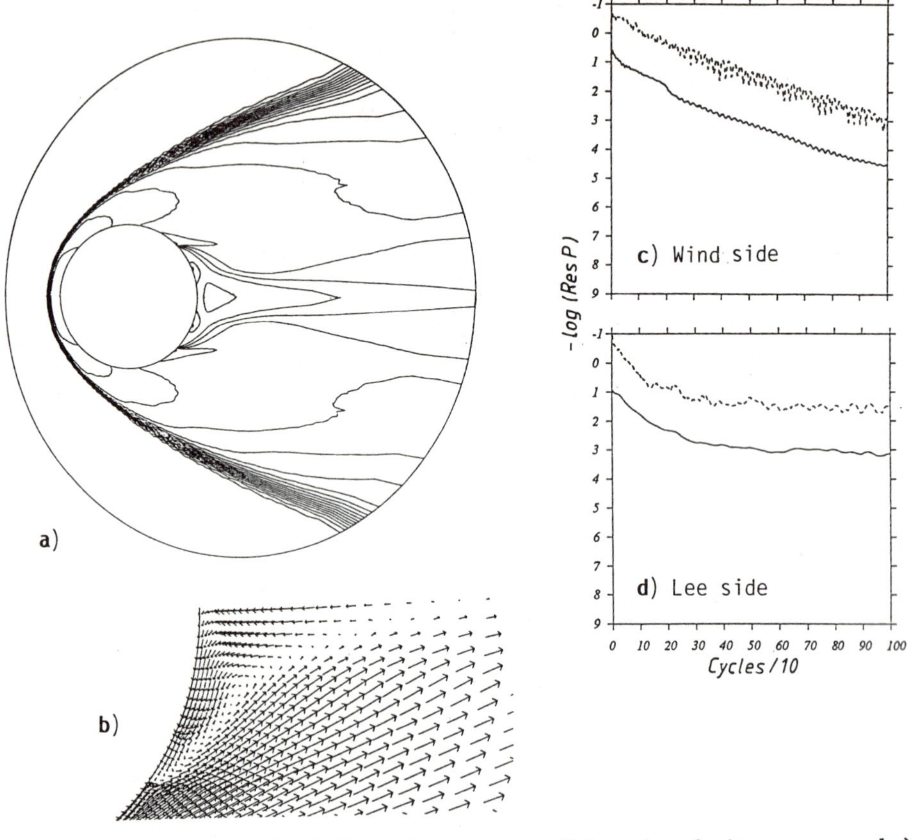

Fig. 4 Computed a) Mach number contours, b) leeside velocity arrows and c) and d) convergence history for $M_\infty = 20$.

M_∞	Δ/R present	Δ/R fitted	$1-p_t/p_{t\infty}$ across shock present	$1-p_t/p_{t\infty}$ across shock exact
2	.34	.342	.289	.28
4	.19	.191	.867	.86
8	.15	.156	.991	.9915
20	.15	.146	1.000	.999

Table 1

Comparison of computed and measured windside data

Table 2 Comparison of computed and measured leeside data.

M_∞	α deg. present	α deg. experiment	Mach before cross flow shock
2	55.2	77	3.5
4	52.0	73	4.4
8	47.3	-	5.4
20	47.3	-	5.7

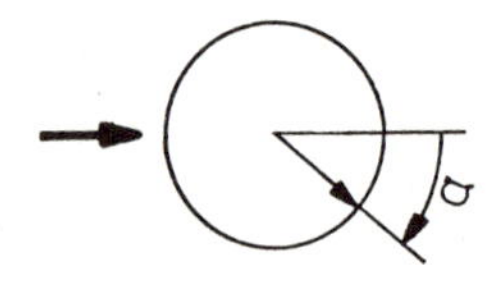

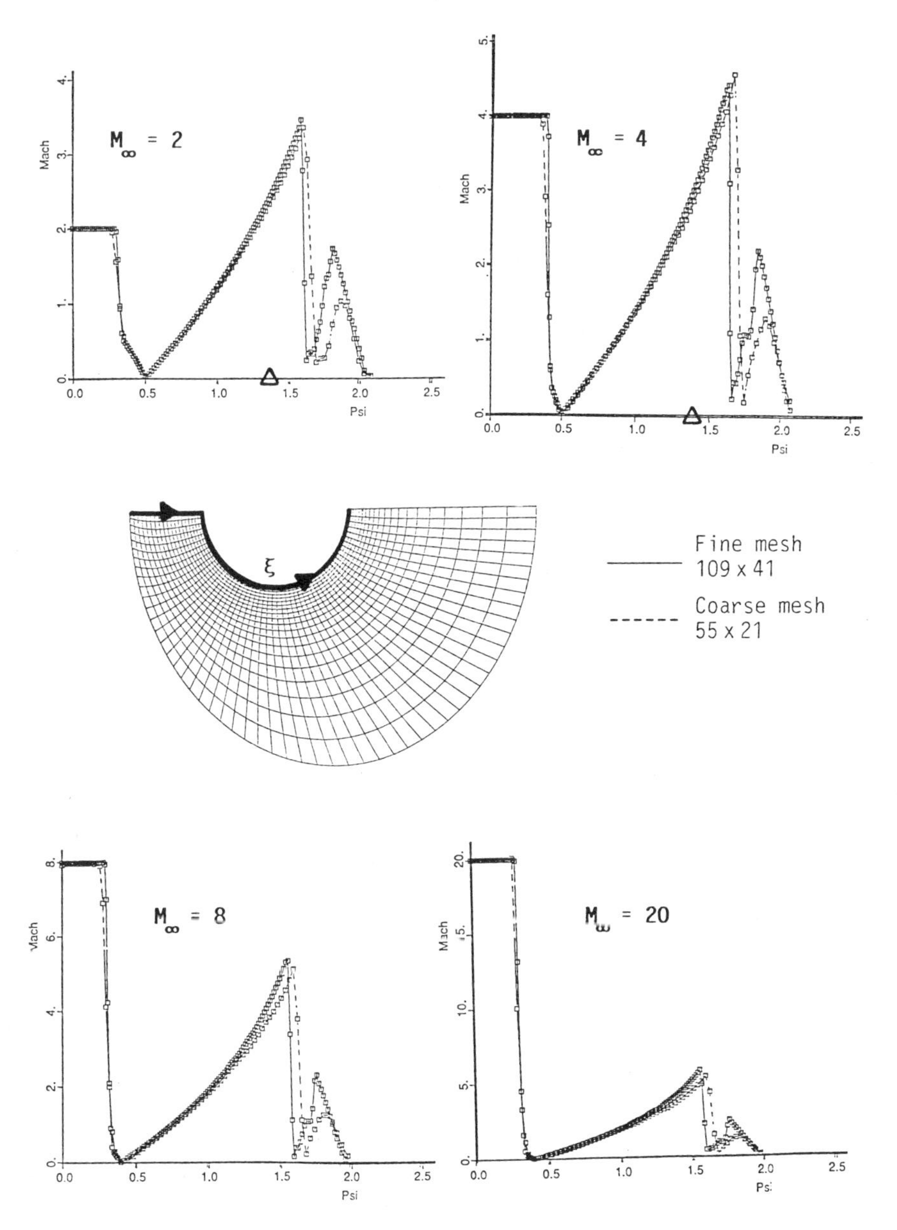

Fig. 5 Variation of computed Mach numbr along line ξ .

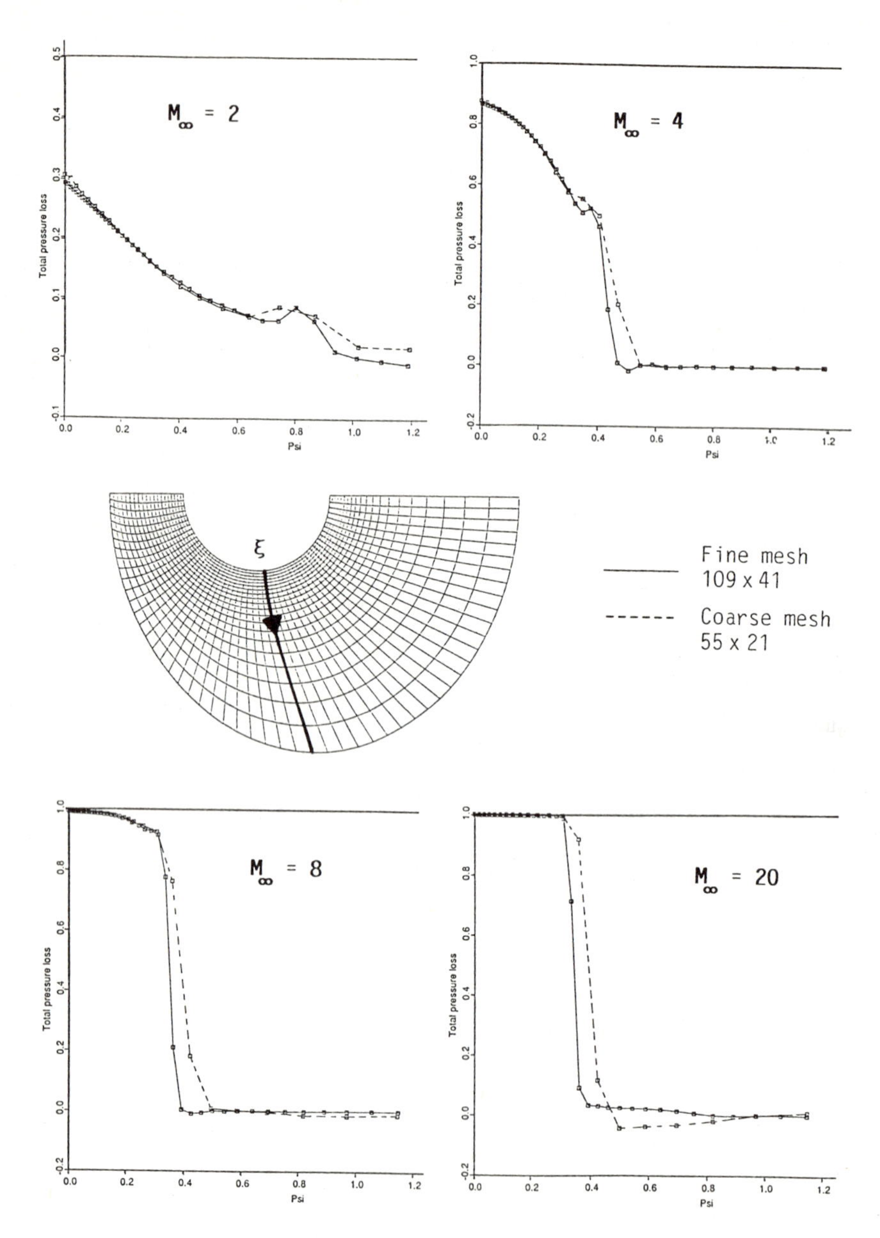

Fig. 6 Distribution of total pressure loss along line ξ .

Assessment and Improvement of Fast Euler Solver

Bernardo Favini

Università di Roma, "La Sapienza"
Dipartimento di Meccanica e Aeronautica

SUMMARY

A generalization of the Fast Solver of G. Moretti [1] for the two-dimensional Euler equations written for generalized curvilinear coordinates is presented. The normalized contravariant base is introduced. The Euler equations are recast in diagonalized form and discretized in time in implicit Δ-form. A system of four pseudo compatibility equations is obtained by means of local one-dimensional analysis. The integration of this system requires the inversion of four bidiagonal matrices instead of 3×3 block-tridiagonal matrices. Preliminary results for subsonic flows are presented.

INTRODUCTION

Implicit integration of Euler equations in quasilinear form for one-dimensional problems can be efficiently performed if the equations are reformulated as compatibility equations and the three invariants, defined by the simple wave solutions, are chosen as dependent variables [1]: the 3×3 block–tridiagonal matrix inversion required for the solution of the original system reduces to three bidiagonal inversion.

However the straightforward extension of this procedure to two-dimensional problems produces a redundant system of equations [2,3], with one equation in excess. As pointed out in [4], the redundancy is not a difficulty *per se*, but it causes an ambiguous definition of the speed of sound.

In order to avoid this awkward outcome, in Ref.s [4, 5] it has been proposed first to recast the equations in lambda-form [6] and then to formulate a sort of modified compatibility equations by means of a linear combination of the equations of motion.

In the present paper, a different approach is followed: first the Euler equations, rewritten in diagonalized form [7] are discretized implicitly in time; only after that, in order to improve the computational efficiency, a suitable form of the system is obtained by means of local one–dimensional analysis. This step consists of premultipling for each coordinate line, the system of equations by the left eigenvector matrix related to the jacobian matrix in the same direction. The outcoming equations resemble the compatibility equations but any ambiguity in variable's definition is avoided. A crucial step in the above outlined analysis is the adoption of the normalized contravariant base which allows the introduction of the appropriate set of dependent variables, a sort of invariants as for the one-dimensional case. For an orthogonal coordinate system the proposed scheme is almost equivalent to that presented in Ref. [4]. The adoption of the Δ-form [8] into the implicit technique allows an easier definition of the covariant partial derivative of velocity components. The covariant derivatives are expressed in terms of partial derivatives of the cartesian components avoiding the introduction of metric source terms, i.e. Christoffel symbol.

MATHEMATICAL MODEL

The Euler equations for a curvilinear coordinate system with a scaled speed of sound $E = 2a/(\gamma - 1)$, the fluid velocity $\underline{v}$ and the entropy as dependent variables are written in standard intrinsic form [9]:

$$\begin{aligned} & E_{,t} + v^i E_{,\xi_i} + a\, v^i\big|_{\xi^i} = 0 && (a) \\ & v^i{}_{,t} + v^j\, v^i\big|_{\xi^j} + g^{ik}\left(aE_{,\xi^k} - \theta s_{,\xi^k}\right) = 0 && (b) \\ & s_{,t} + v^i s_{,\xi^i} = 0 \qquad i = 1,2\,. && (c) \end{aligned} \tag{1}$$

The curvilinear coordinates are defined by the transformation,

$$\xi^k = \sigma^k(x^i) \qquad i = 1,2$$

where x^i are the cartesian coordinates.

If the contravariant base is normalized the equations (1) assume the form:

$$\begin{aligned} & E_{,t} + h^i\left[u^i E_{,\xi^i} + a\, u^i\big|_{\xi^i}\right] = 0 && (a) \\ & u^i{}_{,t} + h^j\left[u^j\, u^i\big|_{\xi^j} + \frac{g^{ij}}{h^i h^j}\left(aE_{,\xi^j} - \theta s_{,\xi^j}\right)\right] = 0 && (b) \\ & s_{,t} + h^i u^i s_{,\xi^i} = 0 && (c) \end{aligned} \tag{2}$$

where h^i is the scaling factor defined by the expression

$$h^i = \sqrt{g^{ii}}$$

and u^i is the scaled contravariant component of the velocity

$$u^i = v^i / h^i\,.$$

The system (2) can be diagonalized following the wave front model [7]: in vector form it results

$$\underline{U}_{,t} + R^1\Lambda^1 L^1\, \underline{U}|_{\xi^1} + R^2\Lambda^2 L^2\, \underline{U}|_{\xi^2} = 0 \tag{3}$$

where the vector of the dependent variable is

$$\underline{U} = \left\{E, u^1, u^2, s\right\}^T$$

the eigenvalue matrices are

$$\Lambda^1 = h^1 \begin{bmatrix} u^1 - a & 0 & 0 & 0 \\ 0 & u^1 & 0 & 0 \\ 0 & 0 & u^1 & 0 \\ 0 & 0 & 0 & u^1 + a \end{bmatrix} \qquad \Lambda^2 = h^2 \begin{bmatrix} u^2 - a & 0 & 0 & 0 \\ 0 & u^2 & 0 & 0 \\ 0 & 0 & u^2 & 0 \\ 0 & 0 & 0 & u^2 + a \end{bmatrix}$$

the right eigenvector matrices are defined by the expressions

$$R^1 = (L^1)^{-1} \qquad , \qquad R^2 = (L^2)^{-1}$$

and the left eigenvectors matrices are

$$L^1 = \begin{bmatrix} 1 & -1 & 0 & -\theta/a \\ 0 & 0 & 0 & 1 \\ 0 & -\alpha & 1 & 0 \\ 1 & 1 & 0 & -\theta/a \end{bmatrix} \qquad L^2 = \begin{bmatrix} 1 & 0 & -1 & -\theta/a \\ 0 & 0 & 0 & 1 \\ 0 & 1 & -\alpha & 0 \\ 1 & 0 & 1 & -\theta/a \end{bmatrix}$$

with

$$\alpha = g^{12}/h^1/h^2.$$

It is worthwhile to point out that the effect of normalizing the contravariant base shows up in the structure of the left eigenvectors with the first and fourth eigenvectors, which is related to a pressure waves, defined by constant elements except for the coefficient of the entropy as in the cartesian case. Unfortunately the third eigenvector, which is related to a shear wave, still contains a non linear term depending on the skewness of the grid.

IMPLICIT SCHEME

The system (3), implicitly discretized in time and recast in Δ-form [8], results

$$\frac{\Delta \underline{U}}{\Delta t} + \left(R^1 \Lambda^1 L^1\right)^n \Delta \underline{U}_{,\xi^1} + \left(R^2 \Lambda^2 L^2\right)^n \Delta \underline{U}_{,\xi^2} = \underline{B} \tag{4}$$

$$\underline{B} \;=\; = -\left[\left(R^1 \Lambda^1 L^1\right)^n \underline{U}^n|_{\xi^1} + \left(R^2 \Lambda^2 L^2\right)^n \underline{U}^n|_{\xi^2}\right]$$

with $\Delta \underline{U} = \underline{U}^{n+1} - \underline{U}^n$.

The covariant derivatives of the variation $\Delta \underline{U}$ reduce to partial derivatives, since the metric source terms, hidden behind the covariant derivative, are evaluated at the previous time level.

The equation of the entropy, as it appears in formula (2.c), is already in a suitable form for implicit scheme and can be integrated independently from the other equations [1,4,5]. At each time step we can suppose to perform a first step to integrate the entropy all over the field: this procedure enables us to include in the right-hand side the partial derivative of the entropy appearing in the mass and momentum equations.

Therefore the system (4) can be reformulated in the reduced form, where the dependence on the time step is omitted for brevity:

$$\frac{\Delta \tilde{\underline{U}}}{\Delta t} + \tilde{R}^1 \tilde{\Lambda}^1 \tilde{L}^1 \Delta \tilde{\underline{U}}_{,\xi^1} + \tilde{R}^2 \tilde{\Lambda}^2 \tilde{L}^2 \Delta \tilde{\underline{U}}_{,\xi^2} = \tilde{\underline{B}} \tag{5}$$

with

$$\Delta \tilde{\underline{U}} = \{\Delta E, \Delta u^1, \Delta u^2\}^T$$

$$\tilde{\Lambda}^1 = \begin{bmatrix} u^1 - a & 0 & 0 \\ 0 & u^1 & 0 \\ 0 & 0 & u^1 + a \end{bmatrix} \qquad \tilde{\Lambda}^2 = \begin{bmatrix} u^2 - a & 0 & 0 \\ 0 & u^2 & 0 \\ 0 & 0 & u^2 + a \end{bmatrix}$$

$$\tilde{L}^1 = \begin{bmatrix} 1 & -1 & 0 \\ 0 & -\alpha & 1 \\ 1 & 1 & 0 \end{bmatrix} \qquad \tilde{L}^2 = \begin{bmatrix} 1 & 0 & -1 \\ 0 & 1 & -\alpha \\ 1 & 0 & 1 \end{bmatrix}$$

where $\tilde{\underline{B}}$ contains also the terms proportional to the entropy derivatives other than the equation evaluated at the previous time level.

Premultiplying the system (5) by the $\tilde{L}^1$ matrix of left eigenvectors, related to $\xi^2 = const.$ coordinate lines, we obtain:

$$\begin{aligned} \frac{\Delta E - \Delta u^1}{\Delta t} &+ h^1(u^1 - a)\left(\Delta E,_{\xi^1} - \Delta u^1,_{\xi^1}\right) \\ &+ \frac{1+\alpha}{2} h^2(u^2 - a)\left(\Delta E,_{\xi^2} - \Delta u^2,_{\xi^2}\right) \\ &- h^2 u^2 \left(\Delta u^1,_{\xi^2} - \alpha \Delta u^2,_{\xi^2}\right) \\ &+ \frac{1-\alpha}{2} h^2(u^2 + a)\left(\Delta E,_{\xi^2} + \Delta u^2,_{\xi^2}\right) = b_1 - b_2 \qquad (a) \end{aligned}$$

$$\begin{aligned} \frac{\Delta u^2 - \alpha \Delta u^1}{\Delta t} &+ h^1 v^1 \left(\Delta u^2,_{\xi^1} - \alpha \Delta u^1,_{\xi^1}\right) \\ &- \frac{1-\alpha^2}{2} h^2(u^2 - a)\left(\Delta E,_{\xi^2} - \Delta u^2,_{\xi^2}\right) \\ &- \alpha h^2 u^2 \left(\Delta u^1,_{\xi^2} - \alpha \Delta u^2,_{\xi^2}\right) \\ &+ \frac{1-\alpha^2}{2} h^2(u^2 + a)\left(\Delta E,_{\xi^2} + \Delta u^2,_{\xi^2}\right) = b_3 - \alpha b_2 \qquad (b) \quad (6) \end{aligned}$$

$$\begin{aligned} \frac{\Delta E + \Delta u^1}{\Delta t} &+ h^1(u^1 + a)\left(\Delta E + \Delta u^1,_{\xi^1}\right) \\ &+ \frac{1-\alpha}{2} h^2(u^2 - a)\left(\Delta E,_{\xi^2} - \Delta u^2,_{\xi^2}\right) \\ &+ h^2 u^2 \left(\Delta u^1,_{\xi^2} - \alpha \Delta u^2,_{\xi^2}\right) \\ &+ \frac{1+\alpha}{2} h^2(u^2 + a)\left(\Delta E,_{\xi^2} + \Delta u^2,_{\xi^2}\right) = b_1 + b_2 \ . \qquad (c) \end{aligned}$$

When instead the system (5) is premultiplied by the $\tilde{L}^2$ matrix of the left eigenvectors, related to $\xi^1 = const.$ coordinate lines,

$$
\begin{aligned}
\frac{\Delta E - \Delta u^2}{\Delta t} \quad &+ \quad h^2(u^2 - a)\left(\Delta E,_{\xi^2} - \Delta u^2,_{\xi^2}\right) \\
&+ \quad \frac{1+\alpha}{2} h^1(u^1 - a)\left(\Delta E,_{\xi^1} - \Delta u^1,_{\xi^1}\right) \\
&- \quad h^1 u^1 \left(\Delta u^2,_{\xi^1} - \alpha \Delta u^1,_{\xi^1}\right) \\
&+ \quad \frac{1-\alpha}{2} h^1(u^1 + a)\left(\Delta E + \Delta u^1,_{\xi^1}\right) = b_1 - b_3 \qquad (a)
\end{aligned}
$$

$$
\begin{aligned}
\frac{\Delta u^1 - \alpha \Delta u^2}{\Delta t} \quad &+ \quad h^2 u^2 \left(\Delta u^1,_{\xi^2} - \alpha \Delta u^2,_{\xi^2}\right) \\
&- \quad \frac{1-\alpha^2}{2} h^1(u^1 - a)\left(\Delta E,_{\xi^1} - \Delta u^1,_{\xi^1}\right) \\
&- \quad \alpha h^1 u^1 \left(\Delta u^2,_{\xi^1} - \alpha \Delta u^1,_{\xi^1}\right) \\
&+ \quad \frac{1-\alpha^2}{2} h^1(u^1 + a)\left(\Delta E,_{\xi^1} + \Delta u^1,_{\xi^1}\right) = b_2 - \alpha b_3 \qquad (b) \quad (7)
\end{aligned}
$$

$$
\begin{aligned}
\frac{\Delta E + \Delta u^2}{\Delta t} \quad &+ \quad h^2(u^2 + a)\left(\Delta E,_{\xi^2} + \Delta u^2,_{\xi^2}\right) \\
&+ \quad \frac{1-\alpha}{2} h^1(u^1 - a)\left(\Delta E,_{\xi^1} - \Delta u^1,_{\xi^1}\right) \\
&+ \quad h^1 u^1 \left(\Delta u^2,_{\xi^1} - \alpha \Delta u^1,_{\xi^1}\right) \\
&+ \quad \frac{1+\alpha}{2} h^1(u^1 + a)\left(\Delta E,_{\xi^1} + \Delta u^1,_{\xi^1}\right) = b_1 + b_3 \qquad (c)
\end{aligned}
$$

where b_i are the elements of the right hand side vector $\underline{\tilde{B}}$.

The systems (6) and (7) are a sort of pseudo-compatibility equations related to two sets of waves travelling independently along each coordinate lines: equations (6.a) and (6.c) represents two pressure waves moving with velocity $h^1(u^1 - a)$ and $h^1(u^1 + a)$ respectively, while equation (6.b) is associated with a shear wave travelling with velocity h^1u^1. The main difference of the system (6) with respect to the system of compatibility equations related to the same set of waves is the presence of the cross terms, representing the effect of the waves propagating along the other coordinate lines. A similar analysis applied to the system (7). This peculiar structure ensures that both system (6) and (7) are perfectly equivalent to the original Euler equations (3), and any redundancy or ambiguity are avoided. However the terms proportional to α, which depends on the skewness of the grid, creates a coupling between the three equatirons of system (6), or equivalently, (7).

When an orthogonal coordinate system is considered and the cross terms are evaluated at the previous time level the coupling disappears and each equation can be integrated independently, strongly reducing the computational work necessary for the matrix inversion step.

For non-orthogonal coordinates a possible way to overcome this difficulty is to drop the equations related to the shear waves and maintains only the four pressure wave equations (6.a,c) and (7.a,c). In effect this procedure spoils the consistency of the system (6), or (7), with the original system (3), but, it can be easily proven that by linear combination of the four remaining equations the original system (3) is recovered exactly.

The mass equation can be obtained both by the sum of (6.a) and (6.c) or (7.a) and (7.c); the ξ^1-component of momentum equation subtracting (6.a) to (6.c), while the ξ^2-component subtracting (7.a) to (7.c).

FAST SOLVER

The perfomed analysis suggests as implicit time integration technique for the Euler equations in quasilinear form the following scheme: at each time step $n+1$,

- the entropy equation is integrated first
- the right hand side vector $\underline{\tilde{B}}$ is evaluated at the previous time level n except for the entropy gradient terms defined at the new time level
- the four equations (6.a,c), (7.a,c) are integrated independenly each other, and, therefore, the cross terms can be set equal zero, while the coupling effects are ensured by the right hand side.

  The equations can be recast in the more compact form:

$$\begin{aligned}
&\frac{R^1}{\Delta t} + \lambda^1 R^1{}_{,\xi^1} + \lambda^6 R^1{}_{,\xi^2} = b_1 + b_2 && (a)\\
&\frac{R^2}{\Delta t} + \lambda^2 R^2{}_{,\xi^1} + \lambda^6 R^2{}_{,\xi^2} = b_1 - b_2 && (b)\\
&\frac{R^3}{\Delta t} + \lambda^3 R^3{}_{,\xi^2} + \lambda^5 R^3{}_{,\xi^1} = b_1 + b_3 && (c)\\
&\frac{R^4}{\Delta t} + \lambda^4 R^4{}_{,\xi^2} + \lambda^5 R^4{}_{,\xi^2} = b_1 - b_3 && (d)
\end{aligned} \qquad (8)$$

  where

$$\begin{aligned}
R^1 &= \Delta E + \Delta u^1 & \lambda^1 &= h^1(u^1 + a) & \lambda^5 &= h^1 u^1/2\\
R^2 &= \Delta E - \Delta u^1 & \lambda^2 &= h^1(u^1 - a) & \lambda^6 &= h^2 u^2/2\\
R^3 &= \Delta E + \Delta u^2 & \lambda^3 &= h^2(u^2 + a) & &\\
R^4 &= \Delta E - \Delta u^2 & \lambda^4 &= h^2(u^2 - a)\,. & &
\end{aligned}$$

  The equations are integrated sweeping in the direction specified by the sign of the characteristic speed.

- The values of the dependent variables at the new time level are obtained by the relations:

$$\begin{aligned}
E^{n+1} &= E^n + \Delta E = E^n + \frac{R^1 + R^2}{2} = E^n + \frac{R^3 + R^4}{2}\\
(u^1)^{n+1} &= (u^1)^n + \Delta u^1 = (u^1)^n + \frac{R^1 + R^2}{2}\\
(u^2)^{n+1} &= (u^2)^n + \Delta u^2 = (u^2)^n + \frac{R^3 - R^4}{2}\quad .
\end{aligned}$$

- The right hand side is evaluated by second order accurate one-sided finite differences, each term biased by the sign of the related characteristic speed. For the left hand side first order accuracy in space is sufficient since only steady state solutions are considered.

TEST CASE

The subsonic flow into a duct with a bump has been considered as a test case. Set equal to 1 the throat of the duct, perfectly symmetric, the widths at the inlet and at the outlet of the duct are equal to 1.8: Fig. 1 show the geometry of the problem together with the grid. As boundary conditions at the inlet it is held constant the total temperature, and at the exit the pressure. The test case appears strongly attractive since is both simple and severe: simple because the flow at steady state should result completely symmetric and the analysis of the numerical results are easily performed; severe because the strong variation of the geometry about the throat region stress the use of non-orthogonal coordinate lines, and the evaluation of covariant derivatives by means of partial derivatives of cartesian components of the fluid velocity. The numerical solution show some inaccuracies near the throat with a maximum error of the total temperature equal to .002. The plot of the pressure field appears almost symmetric.

The efficiency of the fast solver has been evaluated by comparison with a two-step explicit scheme of the λ-type [9] which runs with a CFL limit number less than 1. The implicit scheme is theoretically stable for CFL numbers greater than one, but in order to obtain stable solutions, values less than .5 have been taken. A cause of this limit relies on the peculiarity of the problem: indeed, during an accelerated transient shocks may appear but the scheme is not able to handle them. In spite of this the enhancement of the convergence rate is satisfying: for instance, with the residual just below of 10^{-5} the number of iteration is reduced by a factor 4-5. It should be pointed out that the increase in computational work is minimal since it consists of the integration of the four equations (8).

REFERENCES

1. G. Moretti, "Fast Euler Solver for Steady One-Dimensional Flows", Computers & Fluids, vol. 13, 1985, pp. 61-81.
2. G. Moretti, "A Fast Euler Solver for Steady Flows", Proc. AIAA 6th Computational Fluid Dynamics Conference, Danvers, MA, 1983, pp. 357-362.
3. M. Onofri, D. Lentini, "Fast Numerical Solver for Transonic Flows", Computers & Fluids, 17, 1, 1989, pp.
4. A. Dadone, G. Moretti, "Fast Euler Solver for Transonic Airfoils. Part I: Theory", AIAA Journal, 26, 4, 1988, pp. 409-416.
5. G. Moretti, M. Onofri, "Fast Euler Solver for Blunt Body Flows, Poly M/AE Report no. 86-13.

6. G. Moretti, "A Technique for Integrating Two-Dimensional Equations, Computers & Fluids, 15, 1986, pp. 59-75.

7. L. Zannetti, B. Favini, "About the Numerical Modeling of Multidimensional Unsteady Compressible Flows", Computers & Fluids, 17, 1, 1989, pp. 289-299.

8. R.M. Beam, R.F. Warming, "An Implicit Factored Scheme for the Compressible Navier-Stokes Equations", AIAA Journal, 16, 1978, pp. 393-402.

9. B. Favini, R. Marsilio, L. Zannetti, "Numerical Approximation of Boundary Conditions for Initial-Boundary-Value Problem", proc. ISCFD-Nagoya, Nagoya, 1989.

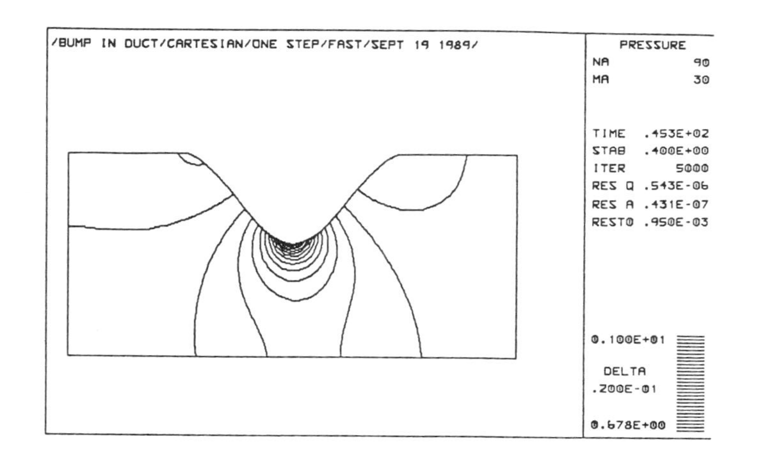

/BUMP IN DUCT/

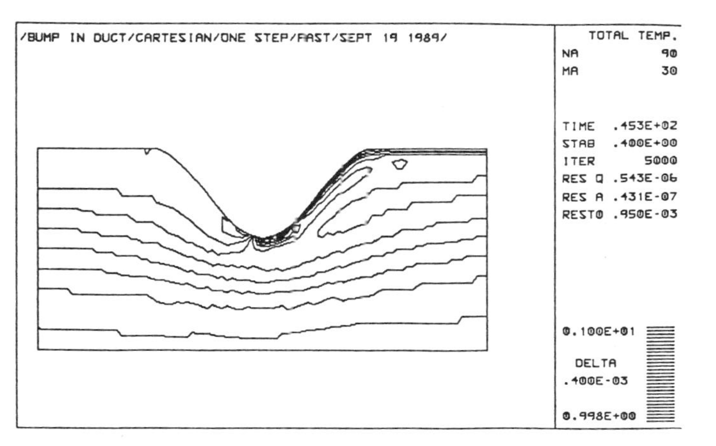

/BUMP IN DUCT/CARTESIAN/ONE STEP/FAST/SEPT 19 1989/
PRESSURE
NA 90
MA 30
TIME .453E+02
STAB .400E+00
ITER 5000
RES Q .543E-06
RES A .431E-07
RESTO .950E-03
0.100E+01
DELTA
.200E-01
0.678E+00

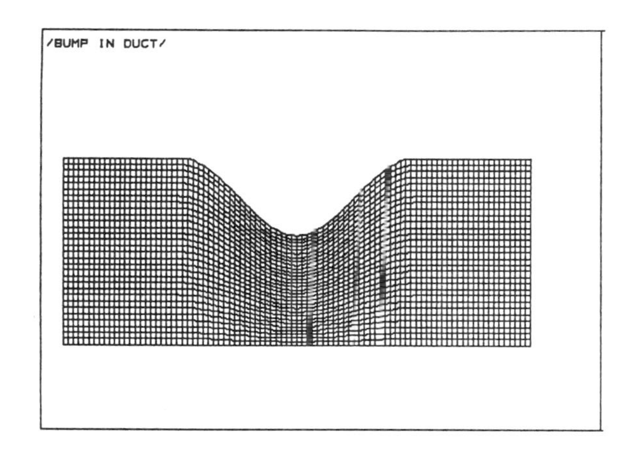

/BUMP IN DUCT/CARTESIAN/ONE STEP/FAST/SEPT 19 1989/
TOTAL TEMP.
NA 90
MA 30
TIME .453E+02
STAB .400E+00
ITER 5000
RES Q .543E-06
RES A .431E-07
RESTO .950E-03
0.100E+01
DELTA
.400E-03
0.998E+00

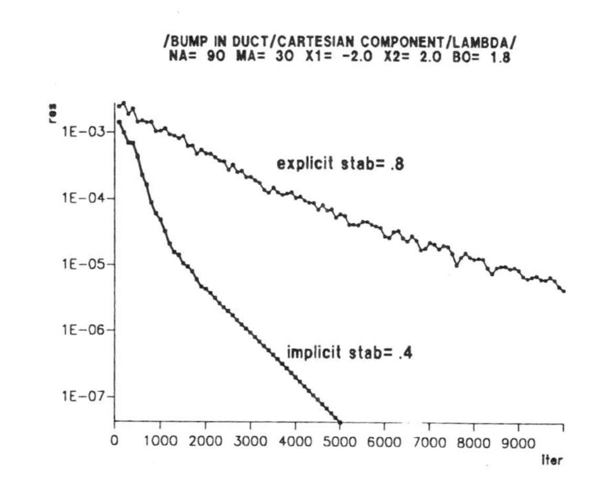

/BUMP IN DUCT/CARTESIAN COMPONENT/LAMBDA/
NA= 90 MA= 30 X1= -2.0 X2= 2.0 BO= 1.8
res
1E-03
1E-04
1E-05
1E-06
1E-07
explicit stab= .8
implicit stab= .4
0 1000 2000 3000 4000 5000 6000 7000 8000 9000
Iter

Chemical and Vibrational Non-Equilibrium Nozzle Flow Calculation by an Implicit Upwind Method

by C. FLAMENT

ONERA, 29 av. Division Leclerc, 92320 Chatillon, France.

Summary

The equations governing chemical and vibrational non-equilibrium nozzle flows are first presented. The resulting differential system is then discretized using a fully implicit non centered finite volume approach. The method is finally applied to two different hypersonic nozzle geometries. Results are compared with equilibrium as well as with previous space marching calculations.

Introduction

Hypersonic (or hyperenthalpic) flows over re-entry bodies develop non-equilibrium effects (dissociation of air, vibrational energy relaxation and ionization) which affect the aerothermal loading of vehicles and condition their design. These chemical and vibrational non equilibrium phenomena can be studied through detached shock waves or hyperenthalpic nozzles. The first case directly interests hypersonic flights and has been intensively studied over these past years (see for example [1]). However, hyperenthalpic nozzle flows reveal physical features not present in open flows and can give insight in the mechanisms of the vibrational energy relaxation and the vibration-dissociation coupling [4,9-10].

Hyperenthalpic nozzle flows generally comprise three different regions, according to their thermodynamic behaviour: an equilibrium flow region at the entrance, a non-equilibrium flow region beginning near the throat followed by a frozen flow region until the exit [2-5].

In the computation of nozzle flows, numerical stiffness problems generally arise near the equilibrium where the characteristic times of chemical kinetics and vibrational relaxation are very short. One way to overcome these difficulties is to treat separately the chemical and aerodynamic parts of the equations in order to directly compute the equilibrium as long as the conditions of the flow are not too far from it. The numerical schemes can then be either of space marching or time marching explicit type [3].

However, a more complete approach consists in computing the near equilibrium flow region as well as the other parts of the flowfield with a single method, the more appropriate one being a time-dependent fully coupled implicit one. This method reduces the need of a short time step, and enables the near equilibrium calculations [12-14].

In this paper, we describe an implicit finite-volume method of this type, which solves in a fully coupled manner the governing equations for non-equilibrium air flows. This method is then applied to two different hyperenthalpic nozzles, and results are compared with those given by an other method [3].

1 Governing Equations

Hypotheses of the study

Neglecting ionization, air can be modelized as a five species mixture (N_2,O_2,NO,N,O) each of them being treated as a perfect gas. Furthermore, radiation and external forces are assumed to be negligible. In a quasi-1D approach, diffusion and viscosity loose somewhat of their physical meaning, and are not treated here.

Translational, rotational and electronic modes are supposed to be in full equilibrium along the nozzle, giving a single temperature T, while vibrational nonequilibrium gives rise to as many vibrational temperatures as diatomic species in the model of air.

Differential system

The system of governing equations reduces to:

$$\frac{\partial U}{\partial t}+\frac{\partial F}{\partial x}+H=S \tag{1a}$$

$$U=W\begin{pmatrix}[\rho_\alpha]\\ \rho u\\ \rho E\\ [\rho_\beta e_{v;\beta}]\end{pmatrix},\quad F=W\begin{pmatrix}[\rho_\alpha u]\\ \rho u^2+p\\ \rho E u+pu\\ [\rho_\beta e_{v;\beta} u]\end{pmatrix},\quad H=\begin{pmatrix}[0_\alpha]\\ -p\,\partial W/\partial x\\ 0\\ [0_\beta]\end{pmatrix},\quad S=W\begin{pmatrix}[\hat{m}_\alpha]\\ 0\\ 0\\ [S_{t-v;\beta}+e_{v;\beta}\hat{m}_\beta]\end{pmatrix} \tag{1b}$$

where $W=W(x)$ is the section of the nozzle, ρ_α the density of species α ($\alpha=1,..,5$), ρ the mixture density, u the mixture velocity, E the total energy per unit mass, $e_{v;\beta}$ ($\beta=1,..3$) the vibrational energy of species β per unit mass, p the mixture pressure, $\hat{m}_\alpha$ the chemical volumic mass production of species α and $S_{t-v;\beta}$ is the volumic vibrational energy production of species β resulting from translational-vibrational interactions. Brackets [ ] represent column vectors with $\alpha=1,2,3,4,5$ and $\beta=1,2,3$.

Thermodynamic model

The expressions of mixture temperature and pressure are:

$$T=\frac{1}{\sum\rho_\alpha c_{v;t,r;\alpha}}\left(\rho E-\frac{1}{2}\rho u^2-\sum\rho_\alpha e_{el;\alpha}-\sum\rho_\beta e_{v;\beta}-\sum\rho_\alpha h^o_\alpha\right)\quad,\quad p=\sum\frac{\rho_\alpha}{M_\alpha}RT \tag{2}$$

where $e_{el;\alpha}$ is the electronic energy of species α per unit mass, h^o_α is the heat of formation of species α per unit mass, $c_{v;t,r;\alpha}$ is the specific heat of species α per unit mass taking into account translational and rotational modes only (5/2 for atoms and 7/2 for molecules), R is the universal gas constant and M_α is the atomic mass of species α.

Chemical model

The volumic mass production of species α writes:

$$\hat{m}_\alpha=\sum a_{\alpha,r}M_\alpha\left(k_{f,r}\prod\left(\frac{\rho_\beta}{M_\beta}\right)^{\nu'_{\beta,r}}-k_{b,r}\prod\left(\frac{\rho_\beta}{M_\beta}\right)^{\nu''_{\beta,r}}\right) \tag{3}$$

where $\nu'_{\alpha;r}$ and $\nu''_{\alpha;r}$ are stoechiometric coefficients and $k_{f;r}$, $k_{b;r}$ are the forward and backward reaction rate respectively of the reaction r. These last coefficients obey Arrhenius' law [8]:

$$k_{f,r}=B_{f;r}\,T^{C_{f;r}}\exp(-D_{f;r}/kT)\quad,\quad k_{b,r}=B_{b;r}\,T^{C_{b;r}}\exp(-D_{b;r}/kT) \tag{4}$$

where k is the Boltzmann's constant and where the other coefficients are different for each chemical model. In this paper, we use Gardiner and Moss models [7,8] which involve the following reactions:

$$N_2+M == 2N+M \qquad M=N_2,O_2,NO,N,O \tag{5a}$$
$$O_2+M == 2O+M \tag{5b}$$
$$NO+M == N+O+M \tag{5c}$$
$$N_2+O == NO+N \tag{5d}$$
$$NO+O == O_2+N \tag{5e}$$

Vibrational model

In the present approach, the vibrational characteristic time takes into account the contributions of all the different species [6]. The translational-vibrational interactions give the following contribution:

$$S_{t-v;\beta} = \rho_\beta \frac{e_{v;\beta}(T) - e_{v;\beta}}{\tau_{vib;\beta}} \quad , \quad \frac{1}{\tau_{vib;\beta}} = \sum \frac{X_\gamma}{\tau_{v;\beta;\gamma}} \tag{6}$$

where X_γ is the molar concentration of species γ ($\gamma = 1,..,5$) and where the characteristic times $\tau_{v;\beta;\gamma}$ are given by Millikan and White [9].

The vibrational temperature $T_{v;\beta}$ of the diatomic species β is given by the transcendental equation:

$$e_{v;\beta}(T_{v;\beta}) = \frac{\rho_\beta e_{v;\beta}}{\rho_\beta} \tag{7}$$

where $e_{v;\beta}(T_{v;\beta})$ is an analytical function of $T_{v;\beta}$ taking into account anharmonicity and rotation-vibration coupling [5].

Vibration-Dissociation coupling model

We use here the model of Park [10]. The coupling between vibration and chemistry only exists in the forward reaction rate of dissociations:

$$k_{f,r} = B_{f;r}\, T_{a;\beta(r)}^{C_{f;r}} \exp(-D_{f;r}/kT_{a;\beta(r)}) \quad , \quad T_{a;\beta(r)} = (TT_{v;\beta(r)})^{1/2} \tag{8}$$

where $T_{a;\beta(r)}$ is an average temperature, taking into account the mixture temperature T and the vibrational temperature $T_{v;\beta(r)}$ of the diatomic species β being dissociated in reaction r.

The influence of chemistry on vibrational relaxation is assumed to be negligible in this model.

2 Numerical Analysis

Numerical scheme

The ten equations of system (1) are solved in conservative form by an extension of a fully implicit non-centered finite-volume method [12] derived from Harten-Yee scheme [13]. Modified Harten-Yee flux limiters are generalized to the solution of the ten differential equations as in [14].

In this approach, aerodynamics, chemistry and vibration are strongly coupled. The scheme is fully implicit (except for the term H) i.e. the flux term F and source term S are treated implicitly at each time-step.

- The discretization starts from the following evaluation of the system (1):

$$\frac{U_j^{n+1} - U_j^n}{\Delta t} + \frac{\tilde{F}_{j+1/2}^{n+1} - \tilde{F}_{j-1/2}^{n+1}}{\Delta x} + H_j^n = S_j^{n+1} \tag{9}$$

where $U_j^n = U\,(t_n = n\Delta t\,, x_j = j\Delta x)$, ($\Delta t$ is the time step and Δx is the space step) and where $\tilde{F}$ is an approximation of the inviscid flux F including a flux correction of Yee type.

For this modified flux, we consider the diagonalization of the Jacobian matrix A of F [5]:

$$A = R\, diag(a^l)\, R^{-1} \quad , \quad (a^l) = (u, u, u, u, u, u-c, u+c, u, u, u)\,, \ (l = 1,10) \tag{10}$$

where R is the matrix with the right eigenvectors as columns, R^{-1} is the matrix with the left eigenvectors as rows, a^l is the lth eigenvalue of A and c is the frozen speed of sound.

The approximated inviscid flux writes:

$$\tilde{F}_{j+1/2} = \frac{1}{2}\,[\, F_j + F_{j+1} + W_{j+1/2} R_{j+1/2} \cdot \Phi_{j+1/2}\,] \tag{11a}$$

where Φ is the flux limiting function:

$$\Phi^l_{j+1/2} = \frac{1}{2}\psi(a^l_{j+1/2})(g^l_j + g^l_{j+1}) - \psi(a^l_{j+1/2} + \gamma^l_{j+1/2})\,\alpha^l_{j+1/2} \quad . \tag{11b}$$

The $\alpha^l_{j+1/2}$ are the differences of the characteristic variables:

$$\alpha^l_{j+1/2} = [R^{-1}_{j+1/2}(U'_{j+1} - U'_j)]^l \quad , \quad U' = U/W \tag{12a}$$

and the g^l_j are limiter functions operating on the $\alpha^l_{j+1/2}$:

$$g^l_j = minmod(\alpha^l_{j-1/2}\,;\alpha^l_{j+1/2}) = sign(\alpha^l_{j+1/2})\max\left[0\,,\min[|\alpha^l_{j+1/2}|, sign(\alpha^l_{j+1/2})\,\alpha^l_{j-1/2}]\right] . \tag{12b}$$

The function ψ is the entropy correction function:

$$\psi(z) = \begin{cases} |z| & si\ |z| > \delta_1 \\ (z^2 + \delta_1^2)/2\delta_1 & si\ |z| < \delta_1 \end{cases} \quad , \quad (\delta_1)_{j+1/2} = \tilde{\delta}\,(u + c)_{j+1/2} \tag{12c}$$

where δ_1 is a parameter which prohibits the annulation of the numerical dissipation in regions where the eigenvalues are equal to zero and $\tilde{\delta}$ is a coefficient equal to 10^{-8} in this paper.

The $\gamma^l_{j+1/2}$, introduced in the last term of (11b):

$$\gamma^l_{j+1/2} = \frac{\psi(a^l_{j+1/2})}{2}\begin{cases} (g^l_{j+1} - g^l_j)/\alpha^l_{j+1/2} & \text{if } \alpha^l_{j+1/2} \neq 0 \\ 0 & \text{if } \alpha^l_{j+1/2} = 0 \end{cases} \tag{12d}$$

modify the caracteristic speeds a^l in the entropy function in order to take into account the presence of flux limiters in (11b) and to restore the spatial second order accuracy in regions of smoothness.

The treatment of inviscid flux and source term in (9) at time level $n+1$ requires a subsequent linearization of the approximated flux $\tilde{F}$ with respect to the unknown U^{n+1}. This is done by using classical Taylor expansion of the flux with respect to the time. In the case of the correction term $[R_{j+1/2}\Phi_{j+1/2}]^{n+1}$ of equation (11a) we only take in the derivation with respect to time the dominating part of Φ, i.e. the last term of equation (11b).

The modified inviscid flux $\tilde{F}^{n+1}_{j+1/2}$ (Eq.9) and the source term S^{n+1} finally write:

$$\tilde{F}^{n+1}_{j+1/2} = \tilde{F}^n_{j+1/2} + \frac{1}{2}\left[A^n_{j+1/2} + \Omega^n_{j+1/2}\right](U^{n+1}_{j+1} - U^n_{j+1}) + \frac{1}{2}\left[A^n_{j+1/2} - \Omega^n_{j+1/2}\right](U^{n+1}_j - U^n_j) \tag{13a}$$

$$S^{n+1}_j = S^n_j + D^n_j(U^{n+1}_j - U^n_j) \tag{13b}$$

where $\Omega = [R\ diag(-\psi(a^l + \gamma^l))\,R^{-1}]$ and D is the Jacobian matrix of S [5].

- The solution is computed at each time step in two stages:

- an explicit stage:

$$d_j\hat{U}' = \frac{1}{U_j}\left[-\lambda(\tilde{F}^n_{j+1/2} - \tilde{F}^n_{j-1/2}) + \Delta t.S^n_j - \Delta t.H^n_j\right] \quad , \quad \lambda = \frac{\Delta t}{\Delta x} \tag{14}$$

- an implicit stage:

$$(e_1 d_{j-1} + e_2 d_j + e_3 d_{j+1})U' = d_j\hat{U}' \quad , \quad d_jU' = U'^{n+1}_j - U'^n_j \tag{15}$$

where e_1, e_2, e_3 are given by:

$$e_1 = -\frac{\lambda}{2}(A_{j-1/2} - \Omega_{j-1/2})^n \tag{16a}$$

$$e_2 = I - \frac{\lambda}{2}(-A_{j+1/2} + A_{j-1/2} + \Omega_{j-1/2} + \Omega_{j+1/2})^n - \Delta t.D_j^n \tag{16b}$$

$$e_3 = \frac{\lambda}{2}(A_{j+1/2} + \Omega_{j+1/2})^n. \tag{16c}$$

Initial flowfield and boundary conditions

The initial flowfield is taken to be in equilibrium , first computed with a Mollier diagram [11] for the mixture velocity, pressure and temperature. The mass concentrations are then computed in two steps consisting in the minimization of the total free energy followed by the annulment of the species chemical productions, adapted to the chosen chemical model. The vibrational energies are calculated using the mixture temperature.

The hypersonic downstream boundary conditions are simple extrapolations of the conservative quantities:

$$d_N U^n = 2d_{N-1}U^n + d_{N-2}U^n \tag{17}$$

The subsonic upstream boundary conditions are more complicated. We assume the fluid to be in full equilibrium at reservoir temperature and pressure, and extrapolate the flow rate from the inner meshes values. These conditions assume that the fluid velocity in the first mesh is negligible against the reservoir total enthalpy.

3 Application

Configurations

The method has been applied to two hypersonic nozzles. For both applications the fluid is assumed to be in equilibrium state at the entrance of the nozzle, with a pressure and a temperature equal to the reservoir ones. The behaviour of the flowfield is shown in Figure 1. As the flow is subsonic before the throat, the mass flow rate is extrapolated at this boundary. These upstream boundary conditions only hold if the kinetic energy of the fluid is negligible against the reservoir enthalpy.

The method has been first applied to a Mach 6 nozzle, R6 [3], with reservoir conditions $Pi = 50bars$ and $Ti = 4800K$. Figure 2 gives the geometry of the nozzle via distribution of sections. The throat is located at $34mm$ from the entrance. The calculations are performed until $x = 225mm$.

The method has been then applied to a hyperenthalpic nozzle , F4 [4], with a 11.3^o cone and reservoir conditions $Pi = 2000bars$ and $Ti = 8508K$. The throat is located at $94mm$ from the entrance. The calculations are performed until $x = 200cm$, corresponding to about 2/3 of the nozzle divergent length.

Results for R6 nozzle

The study has been conducted in four steps. The first step is the computation at full equilibrium. The second step concerns the chemical non-equilibrium only. The third step takes into account the chemical and thermal non-equilibrium without coupling between chemistry and vibration. The final step treats the complete problem, including the vibration-dissociation coupling.

The four steps have been compared, with calculations performed in a 225 uniform mesh. It has been shown that thermo-chemical non-equilibrium flows are very different from the chemical non-equilibrium flows, which in turn are very different from the full equilibrium flows, while vibration-dissociation coupling has only small effects.

Figure 3 shows the temperature distributions in the four cases, where the results from non vibration-dissociation and vibration-dissociation coupling are identical at the chosen scale. The vibrational equilibrium holds untill the throat and the vibrationally frozen flow begins at $100mm$. The frozen Mach number distributions have been reported in Fig. 4, confirming the strong influence of thermal non-equilibrium and the poor effect of vibration-dissociation.

Figure 5 shows the mass concentration distributions for NO in the four cases. Here, thermal non-equilibrium has only slight effects. Chemical mass productions for each species have been reported in Fig. 6. It shows in detail the different zones of chemical equilibrium, chemical non-equilibrium and chemically frozen flow.

The comparaison between full equilibrium and full non-equilibrium in Figures 3 and 5 shows that the implicit method computes the near equilibrium with very good accuracy, despite the stiffness of the problem.

The chemical models of Gardiner [7] and Moss [8] have been also compared, and differences in the results with the two models appear only when the distributions of concentrations are concerned. Figure 7 shows the temperature curves using the two chemical models, where only slight differences occur.

The results have been tested against earlier explicit space marching calculations [3], where a full equilibrium state was supposed upstream of the throat. Figure 8 compares the translational temperature obtained by the different approaches. The results fit fairly well, the differences in the upstream part coming from slightly different reservoir conditions ($4730K$ instead of $4800K$).

Results for F4 nozzle

For this configuration, we compute the chemical and thermal non-equilibrium flow field without coupling between chemistry and vibration, using the chemical model of Moss [8] and the vibrational model of Millikan-White [9]. A non-uniform mesh made of 214 points following geometric distributions on both sides of the throat is used in the present calculations.

Figure 9 and 10 show respectively temperatures and N_2 mass concentration distributions for nonequilibrium flow calculations and compare them with distributions at equilibrium. For both chemistry and vibration, equilibrium extends far after the throat. However the implicit method allows to compute near equilibrium in spite of the stiffness difficulty.

The results have been compared with more simple calculations [4] given by using the same algorithm as in [3]. Figure 9 compares the translational temperature obtained by the different approaches. Explicit and implicit results show a rather good agreement as in the previous configuration.

Conclusion

An implicit finite-volume method has been developed for non-equilibrium flow calculation and applied to quasi-one dimensional flow problems in two different hypersonic nozzle geometries.

The fully coupled implicit upwind scheme has shown its ability to overcome the stiffness difficulty due to the near-equilibrium region. The calculations have shown the importance of the vibrational energy relaxation, and the small effects of both vibration-dissociation coupling and differences between chemical models.

Although the present calculations show good results, nozzle flow calculations may be very sensitive to many physical problems as the relaxation times, the anharmonicity of the vibrational model and the vibration-dissociation coupling [15]. Theoretical approach and experimental results are therefore needed to validate the future calculations.

References

[1] **G. Candler and R. MacCormack**: "The computation of hypersonic ionized flows in chemical and thermal nonequilibrium", AIAA-88-0511, 1988.

[2] **J. G. Hall and C. E. Treanor**: "Nonequilibrium effects in supersonic-nozzle flows", CAL RT No. CAL-163, 1968.
[3] **Ph. Sagnier and L. Marraffa**: "Parametric Study of Thermal and Chemical Non Equilibrium Nozzle Flow", AIAA-89-1856.
[4] **C. Flament, L. Le Toullec, L. Marraffa, Ph. Sagnier**: "Inviscid Nonequilibrium Flow in ONERA F4 Wind Tunnel", Int. Conf. on Hyp. Aerodynamics, Manchester, 4-6 Sept. 1989.
[5] **C. Flament**: "Cinétique chimique et relaxation vibrationnelle dans une tuyère hypersonique", ONERA Report No 10/3637 AN, 1989.
[6] **J-H. Lee**: "Basic governing equations for the flight regimes of Aeroassisted Orbital Transfer Vehicles", AIAA-84-1729.
[7] **W. C. Gardiner Jr.**: "*Combustion chemistry*", Springer Verlag, 1984.
[8] **J. L. Shinn, J. N. Moss and A. L. Simmonds**: "Viscous shock layer heating analysis for the shuttle winward plane with surface finite recombination rates", AIAA-82-0842.
[9] **R. C. Millikan and D. R. White**: "Systematics of vibrational relaxation", J.Chem.Phys., Vol.36, N.12, pp 3209-3213, 1963.
[10] **C. Park**: "Assessment of two-temperature kinetic model for ionizing air", AIAA-87-1574.
[11] **J. Hilsenrath, M. Klein**: "Tables of Thermodynamic Properties of Air in Chemical Equilibrium including Second Virial Correction from 1500 to 15000K",AEDC TR65-58,1965.
[12] **H. Hollanders and C. Marmignon**: "Navier-Stokes high speed flow calculations by an implicit non-centered method", AIAA-89-0282.
[13] **H. C. Yee and A. Harten**: "Implicit TVD Schemes for Hyperbolic Conservation Laws in Curvilinear Coordinates", AIAA-85-1513.
[14] **H. C. Yee and J. L. Shinn**: "Semi-implicit and fully implicit shock-capturing methods for hyperbolic conservation laws with stiff source terms", NASA-TM-89415, 1986.
[15] **C.E. Treanor, J.W. Rich, R.G. Rehm**: "Vibrational relaxation of anharmonic oscillators with exchange-dominated collisions", J. of Ch. Phys., Vol.48, pp 1798-1807, 1967.

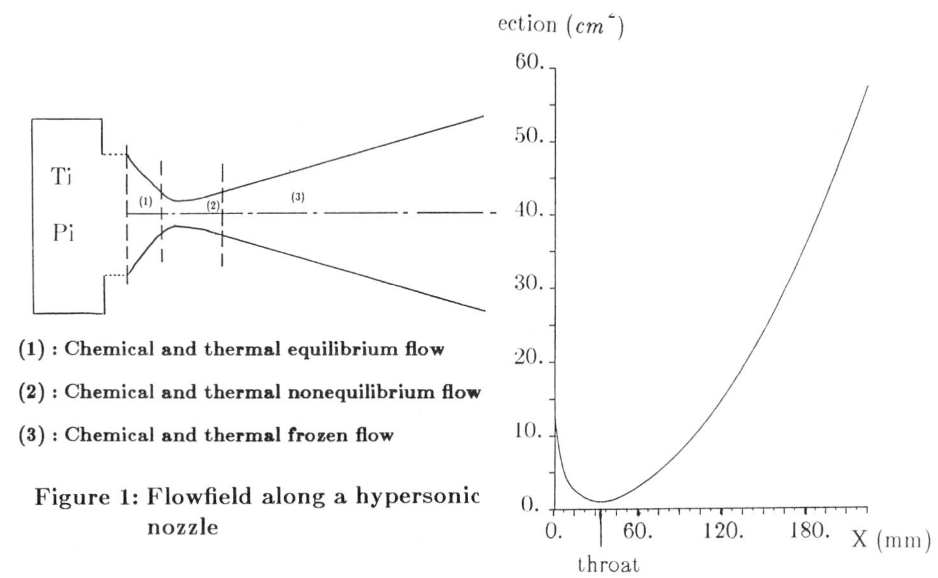

Figure 1: Flowfield along a hypersonic nozzle

Figure 2: Geometry of the nozzle R6

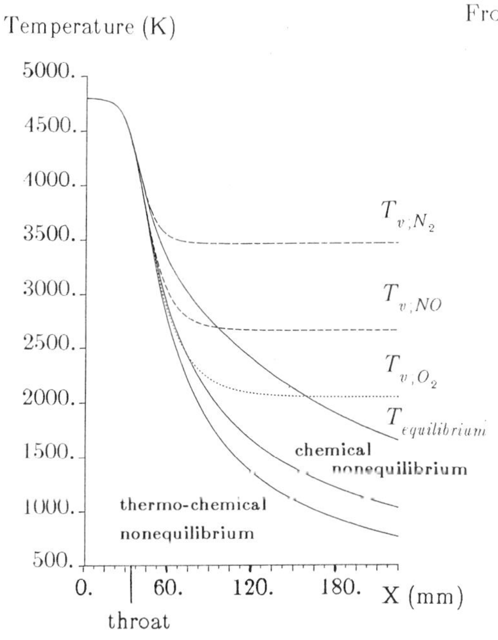

Figure 3: Mixture temperatures and vibrational temperatures distributions

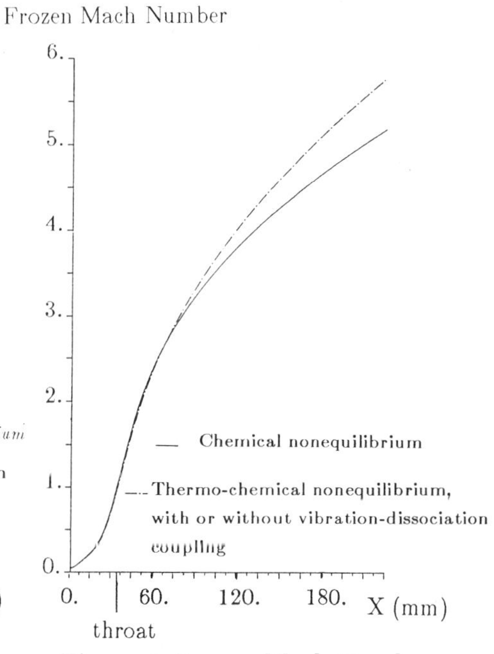

Figure 4: Frozen Mach Number distributions

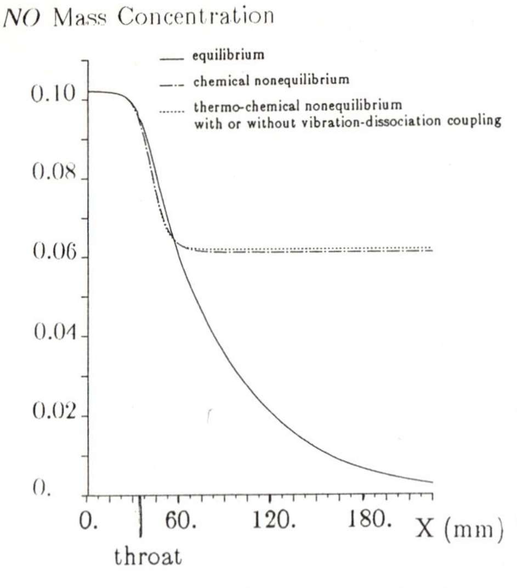

Figure 5: NO Mass Concentration Distributions

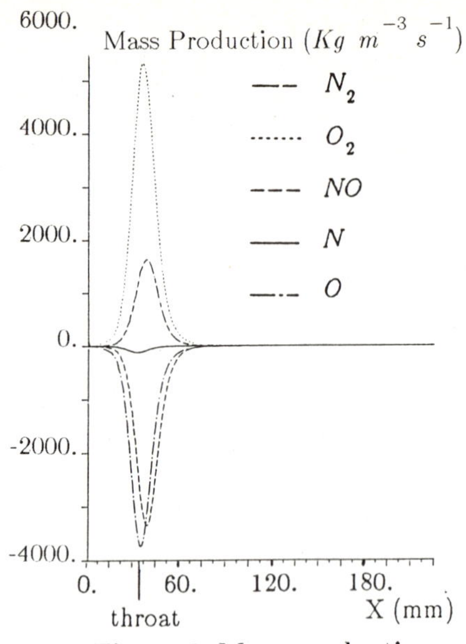

Figure 6: Mass production distributions

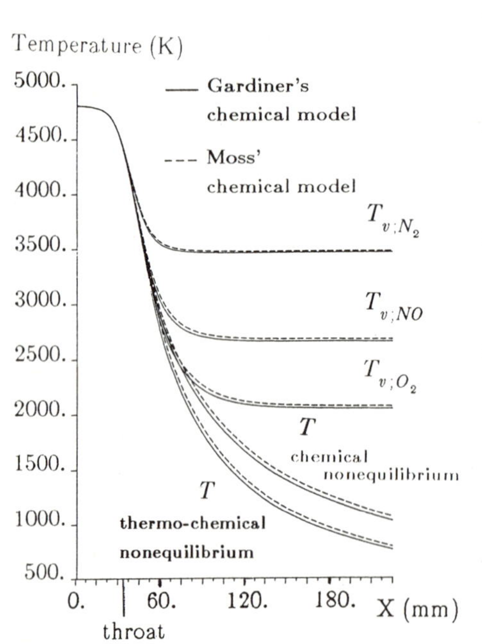

Figure 7: Comparaison between Moss and Gardiner chemical models

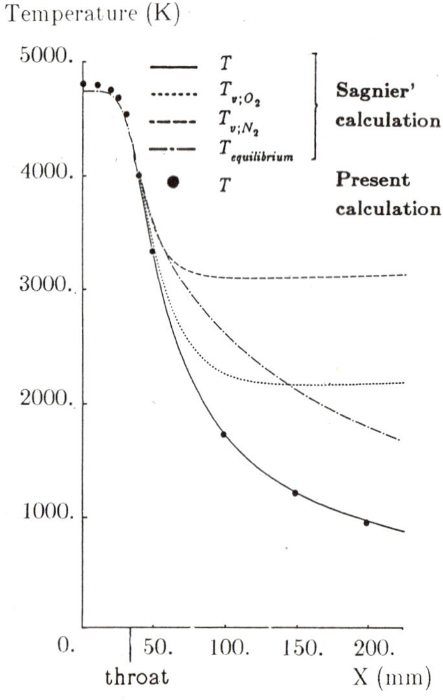

Figure 8: Comparaison with Sagnier's calculations

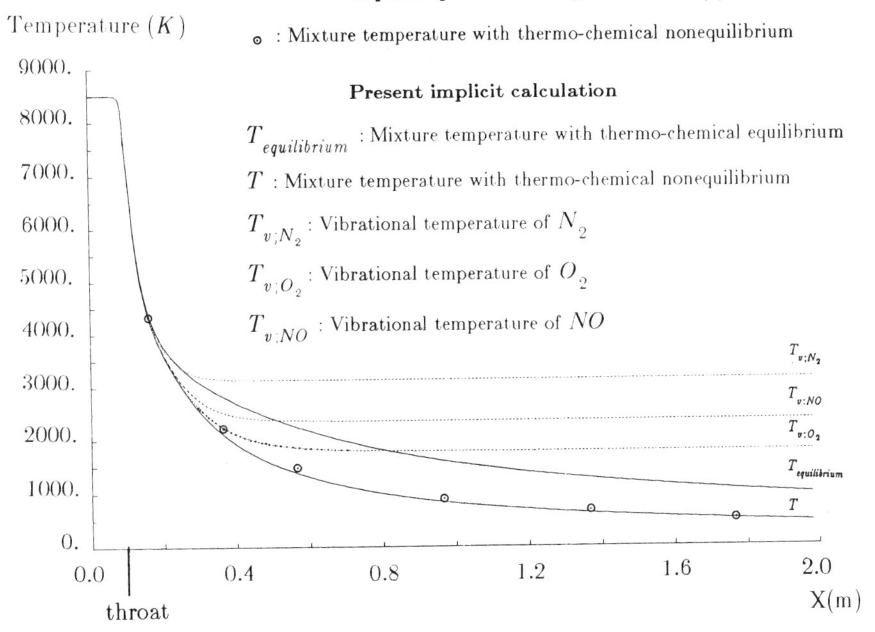

Figure 9: Temperatures distributions with F4 nozzle

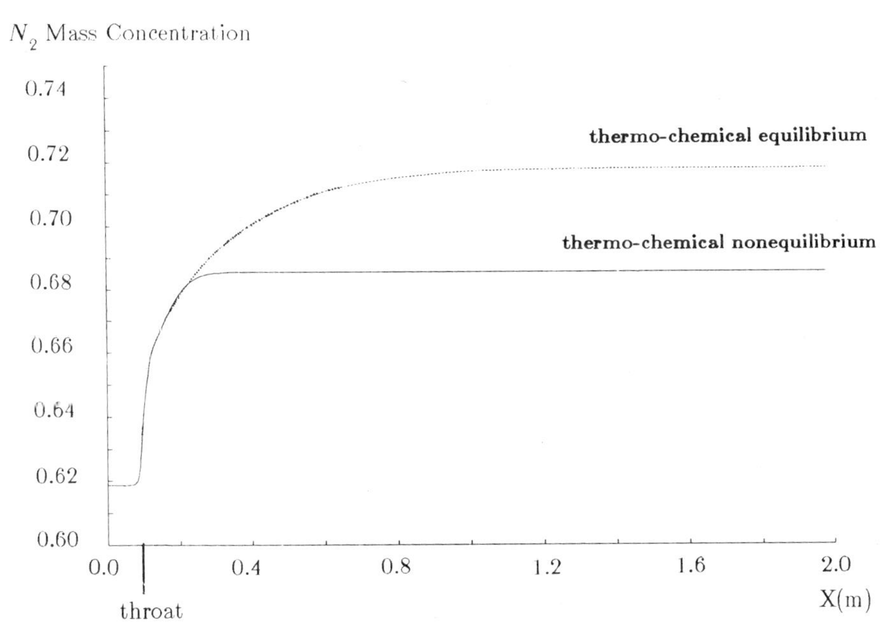

Figure 10: N_2 Mass Concentration Distributions with F4 nozzle

CALCULATION OF FLOW FIELDS USING OVERLAPPING GRIDS

Laszlo Fuchs
Department of Gasdynamics,
The Royal Institute of Technology,
S-100 44 Stockholm, Sweden.
and
Scientific and Technical Computing group, ACIS
IBM Svenska AB,
S-163 92 Stockholm, Sweden.

SUMMARY

A numerical scheme that is based upon a combination of Multi-Grid methods and systems of overlapping grids is presented. The main advantages of the the overlapping grid system are as follows: it allows easy generation of grids for complex geometries; it makes it possible to solve problems with variable geometries, and that all this is gained without the penalty of slow convergence. Some new features have been added recently to the scheme. One of these is the possibility to treat colliding objects. The flexibility of the method is demonstrated in calculating some steady and time dependent flows.

INTRODUCTION

The flow of a in a bounded domain is considered. The domain may contain several solid bodies and it may be composed of several different branches. Flows in such geometries is rather common in engineering. The common numerical approach for such problems is to generate a body-fitted mesh so that solid surfaces become coordinate lines. Such typical grid generations techniques are described in [1]. An application of the method for the computation of the creeping flow past a cylinder placed in a channel is given in [2]. An alternative grid generation method is using a zonal (domain decomposition) technique. Local meshes are generated in each zone, independently of each other. The meshes in the different zones may overlap one or more meshes that belong to other zones. The problem can be solved by computing alternatingly a smaller problem in each zone and then updating the 'internal boundary' values. This scheme is not new and it is known as Schwarz algorithm (see e.g. [3,4]). A special variant of zonal- grid methods is the usage of local mesh refinements. This approach is a natural extension of (global) Multi-Grid (MG) methods and they has been applied to transonic flows in [6,7], and to incompressible flows in [5] and recently in [15]. An extension of the zonal-grid algorithm is to include local mesh refinement into the overlapping grid system so that local grids are defined only when required on the appropriate sub-zones. The application of local grids, super-imposed on a global grid, for hyperbolic equations is described by Berger [8]. The basic approach in [7] is to use a global body fitted mesh and to refine it locally, by halving the mesh spacing at some sub-regions. In [8], only cartesian grids are used. Curved shocks are embedded in several small rectangular boxes. In each of these boxes a local cartesian mesh is used.

With all the variants of zonal grid methods one has to resolve several difficulties:

A. Defining a proper data structure so the data can be accessed easily and with little 'overhead'. In addition, the data management should allow addition of zones (i.e. extending the computational domain) and local mesh refinements. The former requirement is of importance if the boundary conditions are not known exactly at a given location of the computational boundary.

B. Information exchange procedure across the zonal interfaces. This procedure may require that certain properties (i.e. compatibility conditions such as total mass balance) has to be satisfied, for the existence of a solution. In addition, the procedure must be stable, for the convergence of the numerical algorithm.

C. The solution algorithm must be such that the convergence rate is not affected by the usage of zonal grids. The slow convergence rate of the basic Schwarz algorithm seems to be the underlying reason for the limited usage of zonal-grid techniques. One of the reasons for the renewed interest in the zonal technique is because it lends itself naturally for parallel computing.

The efficiency of the zonal MG algorithm has been demonstrated in [5,9 and 14]. Here we demonstrate the extension of the zonal MG method to allow treatment of special situations as arbitrary movement of some parts of the solid geometry including collision of objects.

Further, we consider the application of the extended Multi-Grid method applied to a zonal grid system. The basic grid system is composed of a uniform cartesian mesh and several local body-fitted meshes. The basic zonal-grid system may be extended by adding locally refined meshes that 'cover' only parts of some of the zones (i.e. local mesh refinements). Each of these local meshes is derived from its 'parent' grid that contains it. The data management is flexible enough so that new zones and new locally refined sub-grids can always be added. The 'inter-zonal' information exchange is described shortly, with further details given in [14]. The presence of solid bodies inside the computational domain requires special care to maintain the efficiency of the algorithm.

In the following we describe the application of the zonal Multi-Grid algorithm for the computation of two-dimensional viscous incompressible flows.

GOVERNING EQUATIONS AND THEIR APPROXIMATION

We solve the incompressible Navier-Stokes equations in either primitive variables or streamfunction and the vorticity formulations. Here, in all cases except the one specifically stated we use the later formulation. In both cases the system of PDE's require two conditions on the whole boundary. On solid boundaries the velocity vector vanishes, providing the two required conditions. No slip boundary conditions mean that the surface of solid bodies describe a streamline and that the derivative of the streamfunction normal to the body surface vanishes. The reference value of the streamfunction is arbitrary. If, however, there are solid bodies in the flow field the (constant) value of the streamfunction on such surfaces must be determined as part of the solution. To determine this value uniquely, an additional condition must be satisfied. This condition is derived from the momentum equations and is found by using the fact that the value of the pressure at any point must be independent of the path of integration [9].

At inflow and outflow boundaries one may specify the and the vorticity themselves (instead of the velocity vector). On a partitioned (zonal) domain, one must add 'inter-zonal-boundary conditions'. Such conditions must be defined so that the problem is well posed in each zone. In our case this would imply that two types of data must be given. There is, however, some freedom in choosing such data. One may choose the components of the velocity vector or the pair streamfunction and vorticity or the combinations of these. Using the velocity vector implies a compatibility condition on each sub-grid; i.e. that the total mass flux vanishes. Since this condition is not satisfied explicitly and since it is satisfied only at convergence, this choice of inter-zonal exchange results in a substantial slow-down of the convergence of the iterative solver. Therefore, the streamfunction and the vorticity are used to exchange information across the overlapping grids.

A zonal grid system is constructed by generating a local mesh in each zone. Uniform and cartesian grids are used as much as possible with the possible exception near solid bodies, where a local (body-fitted) grid is generated. The governing equations are approximated by finite-differences on each grid, using central differences for all the terms and upwind differences for approximating the first derivatives of the vorticity. The boundary vorticity (on surfaces of solid bodies) is computed by using the no-slip condition.

DATA ORGANIZATION AND MANAGEMENT

The current zonal scheme is an extension of the local mesh refinement technique that is described in [5]. The present scheme includes the treatment of locally and independently defined (overlapping) grids and multiply-connected domains. The scheme maintains the capability of accommodating locally refined subgrids. These refinements cannot, however, be defined arbitrarily. If an arbitrarily shaped region is to be refined, a completely new zone has to be added (during the solution procedure).

The data management must be sophisticated enough to allow the flexibility that is described above. Some more details are given in [5, 14]. Recently, we have included new features that allow the code to treat cases where object collide or move partially out of the computational domain. The new extensions may also be used to treat collision of objects.

In all, the grid management system is more complicated than that in simpler MG solvers. Nevertheless the total overhead is small compared to the total memory needed for storing the dependent variables. This is why the overlapping grid approach is a practical method for calculating flows past complex geometries.

NUMERICAL PROCEDURE

The discrete approximations to the governing equations are solved by a straightforward MG method: Both equations are 'smoothed' by using a Successive Point Relaxation (SPR) procedure. The non-linear problem is transferred to coarser grids (using the FAS scheme, see e.g. [10]) and the coarse grid corrections are interpolated linearly to the fine grid. Details of the basic scheme are given in [11]. The extensions that are added to the basic MG scheme include the treatment of zonal grids, possibilities for local mesh refinements and a procedure for updating the streamfunction value on the surface of inner bodies.

The basic Schwarz algorithm implies the solution of the discrete equations, in each zone separately and then allowing inter-zonal exchange. The procedure is repeated until convergence is achieved. Such updating procedure results in non-smooth residuals and results in slow convergence. The scheme that we employ transfers data among the zones on the finest grid <u>and</u> on the coarse grids. After an inter-zonal exchange, the approximation is smoothed out by some relaxation sweeps. On the coarsest grid only the corrections are interpolated among the zones. In contrast to the current correction scheme that is used on coarse grids, one may use a FAS variant (similar to the one described below for updating the boundary value of the streamfunction). The FAS variant has not yet been implemented in our code.

As noted above, by interpolating the streamfunction and the vorticity instead of the velocity vector, no additional conditions are needed for the existence of a solution. Therefore, we may use simple interpolations for information exchange among the zones. In our code, it is possible to chose either a bi-cubic or a bi-quintic interpolation scheme. In fact, the lower order scheme is adequate on the coarse grids (where the corrections are interpolated), even when the bi-quintic interpolation is used on the finest mesh. When locally refined grids are present, the interpolation of data from the coarser grid to the boundaries of the refined mesh is simpler. In such cases, the node points on the boundaries of refined regions lie on a coordinate line in common to the next coarser grid. (This is so by the way we introduce locally refined meshes). Therefore, a one-dimensional interpolation scheme can be used on these boundaries.

The value of the streamfunction at the surfaces of different objects is computed, when it cannot be set to a given value, only on the coarsest grids. Details of the scheme are described elsewhere [9].

COMPUTED EXAMPLES

In this section we shall demonstrate some of the basic features of the inter-zonal scheme. The special case of local mesh refinements alone is important to achieving desired accuracies (with or without automatic grid adaptation). An application in which grid refinements are important from resolution point of view is the flow in a room. ventilation situations imply often very small inlet and outlet openings. These inlets and outlets can be resolved only by clustering grid points at these locations. standard (structured) grids result in cells with very small and large aspect (height to length) ratio. This grid non-uniformity may have implication on the global accuracy and certainly causes slower convergence rate (even in the MG solver). The approach of local grid refinement results arbitrary good resolution while the cell aspect ratio can be made equal to one. Figures 1 show the flow field in a simulated ventilation problem. Local grids are imposed at the inlet and the outlet boundaries.

A typical problem with multiple objects is the flow around several cylinders. A global cartesian grid is used to discretize the computational domain. Around each cylinder, a local polar grid is placed. Information exchange can take place between any of the grids. This recent extension allows also situation of colliding objects, since the previous restriction of communication with a main grid has been removed. Figure 2 show a case of flow around several cylinders. For low Reynolds numbers, and a given configuration (distance among the cylinders), the flow is steady. As the Reynolds number increases beyond a certain (critical value) or when the distance between the two upstream cylinders (for sub-critical Re, but large enough), the flow becomes unsteady. This non-steadiness is detected and the solver switches over from a steady to a time-dependent procedure. The flow on the global and a locally refined region are shown in the figure.

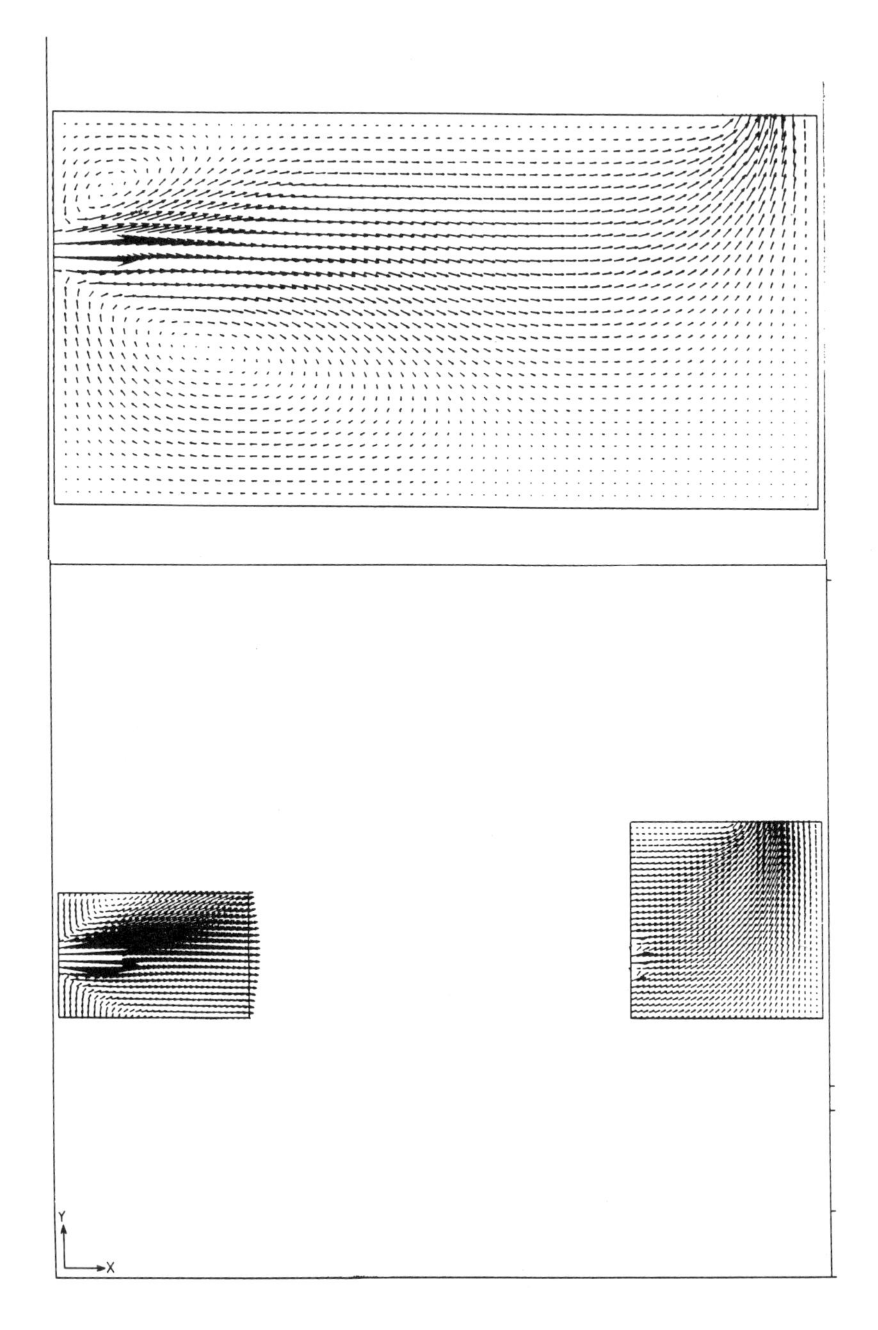

Figures 1: A typical ventilation problem. Fluid enters the room through a small entrance and leaves trough another small outlet in the opposite wall. Figure 1.a (above) shows the global field, and Fig. 1.b (below) shows the velocity vectors in the locally refined sub-regions at the in- and the outlets.

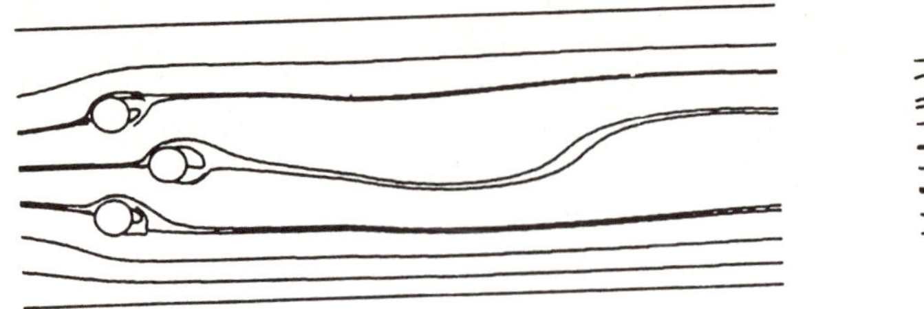

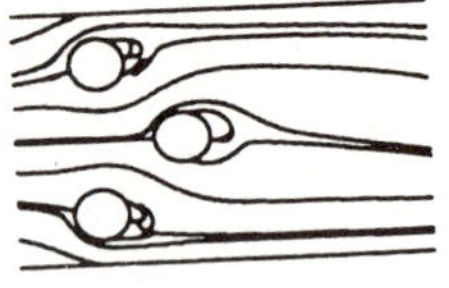

Figures 2: Unsteady flow past three cylinders. Global grid and local grid streamlines.

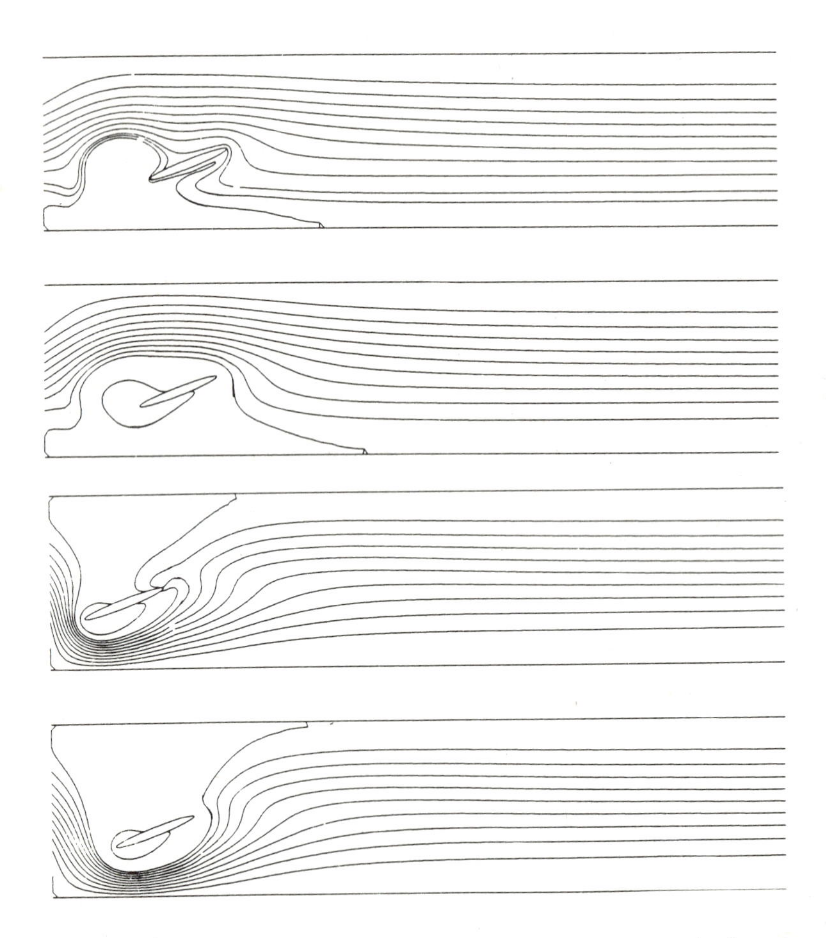

Figure 3.a: the instantaneous streamline at certain phases during the period. The occluder moves periodically in the right-left directions.

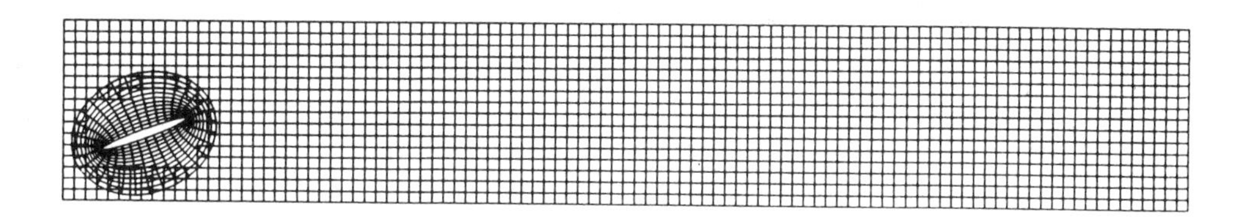

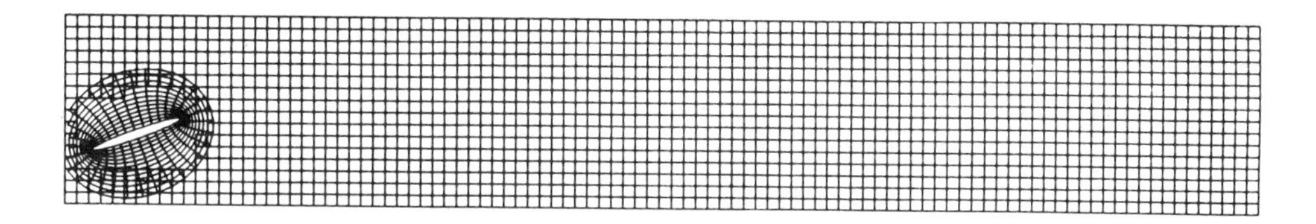

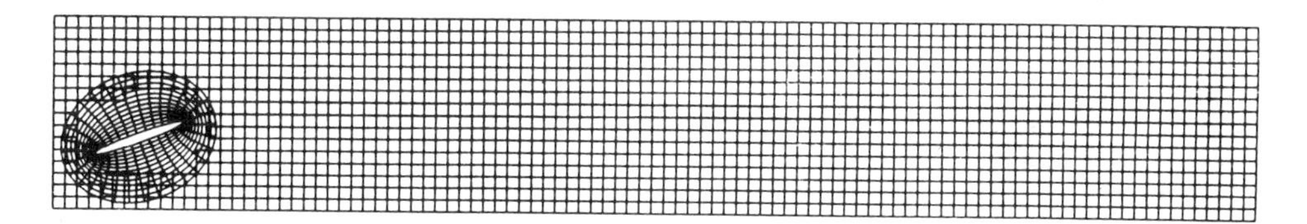

Figure 3.b: The time-dependent zonal grid system ("snapshots") as used to compute the periodical flow of Fig. 3.a.

Another problem that has been computed by the present scheme is the flow through a valve with an occlude (ellipsoid). The channel at the inflow plane is partially blocked. The occlude may be placed at a variable distance from the entrance plane and at an arbitrary angle of inclination. Since we are also interested in unsteady pulsating flows, with freely moving occlude, the use of body fitted grids (that must be define separately for each time step) is excluded for practical reasons. The present method, on the other hand, can be implemented in a straightforward manner. Here, we consider periodical motion of the occlude (along the axis of the channel) with a periodical mass flow through the channel. The mass flow is minimal when the occlude is placed closest to the inlet. Fig. 3.a show some "snapshot" pictures taken at different phases of the periodic motion. The grid in some of these time-steps is shown in Fig. 3.b. As seen, the same basic grids are used for all time-steps with the exception that some computational cells may be "shut" and thus allowing movement of the occlude inside the channel and even its movement out of the computational domain. An example for this case is shown in Fig. 4. where the ellipsoid is placed half way outside the channel. In this case the ellipsoid is allowed to move periodically toward and from the entrance. The time-dependent streamlines and the grids are shown in the figure (Fig. 4). The grid around the ellipsoid is logically the same as in the previous case with the exception that part of the body-fitted grid is logically deactivated.

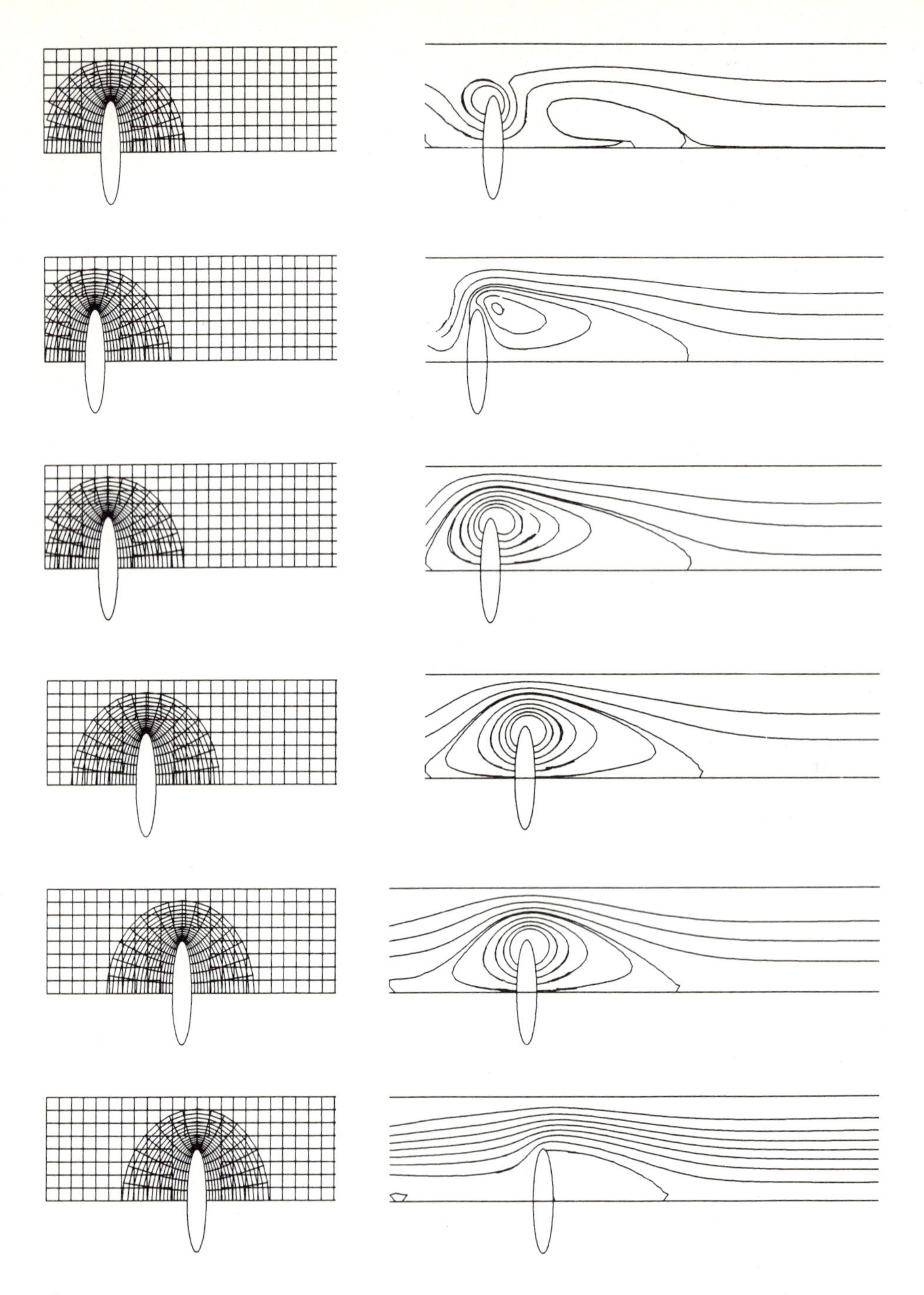

Figure 4: The streamlines and the grids at some time-steps in a pump model.

CONCLUDING REMARKS

The main features of the zonal-MG scheme that has been developed are: Flexibility in treating complex and possibly non-fixed geometries, as well as having the possibility to adding local mesh refinements. The scheme has good numerical efficiency due to the use of local body fitted grids, the use of uniform grids (as much as possible) and the MG procedure that yields good convergence rate. Recent improvements allow the treatment of more general cases as collision and partial or complete object elimination or addition.

The extension of the scheme to 3-D cases has been started by including local mesh refinements. Arbitrary grid overlapping is being currently developed.

REFERENCES

1. J. F. Thompson, Z.U.A. Warsi, and C.W. Mastin; "Numerical Grid Generation Foundations and Applications". North-Holland (1985).
2. A.S. Dvinsky and A.S. Popel; J. Comp. Phys. 67, 73 (1986).
3. L.W. Ehrlich; SIAM J. Sci. Stat. Comp. 7, 989 (1986).
4. Q.V. Dinh, R. Glowinski, B. Mantel, J. Periaux and P. Perrier; in "Computing Methods in Applied Sciences and Engineering", V. Edited by R. Glowinski and J.L. Lions. (North-Holland, 1982), p. 123.
5. L. Fuchs; Computers & Fluids, 15, 69 (1986).
6. C-Y. Gu and L. Fuchs; in "Numerical Methods Laminar and Turbulent Flow-IV", Edited by C. Taylor et. al., Pineridge Press, p. 1501 (1985).
7. W.J. Usab and E.M. Murman; AIAA P-83-1946 (1983).
8. M.J. Berger; SIAM J. Sci. Stat. Comp. 7, 905 (1986).
9. L. Fuchs; "Proc. 7th GAMM conf. on Numerical Meth. in Fluid Dynamics", Vieweg Verlag, 20, p. 96 (1988).
10. A. Brandt; Math. Comp. 31, 331 (1977).
11. T. Thunell and L. Fuchs; in "Numerical Methods Laminar and Turbulent Flow-II". Edited by C. Taylor and B.A. Schrefler. Pineridge Press, p. 141 (1981).
12. L. Fuchs; in "Numerical Methods for Fluid Dynamics-II", edited by K.W. Morton and M.J. Baines, Clarendon Press, Oxford, p. 569, (1986).
13. W. Cherdron, F. Durst and J.H. Whitelaw; J. Fluid Mech. 84, 13, (1978).
14. L. Fuchs; Proc. of the fifth GAMM-Seminar. Eds. R. Rannacher and W. Hackbusch, Notes on Numerical Fluid Mechanics, Vieweg Verlag (1989).
15. M.C. Thompson and J. H. Ferziger; J. Comp. Phys. 82, 94-121, (1989).

Multicolored Poisson Solver for Fluid Flow Problems

S. FUJINO
Institute of Computational Fluid Dynamics,
1-22-3 Haramachi, Meguro-ku Tokyo 152, Japan

T. TAMURA
ORI of Shimizu Corporation,
2-2-2 Uchisaiwai-cho, Chiyoda-ku, Tokyo 100, Japan

K. KUWAHARA
Institute of Space and Aeronautical Science,
3-1-1 Yoshinodai, Sagamihara, Kanagawa 229, Japan

Abstract

We propose the multicolored Poisson solver suited to vector computing. This solver is designed for solving the Poisson's equation which has a large sparse 19-points FDM (Finite Difference Method) coefficient matrix in the 3-D generalized coordinate system. Moreover this multicolored technique can be extended to the solvers for sparse 13 and 25-points FDM coefficient matrices which arise from the Navier-Stokes equations, in the 2-D and 3-D generalized coordinate system. We discuss the applicability of the present solver to the unsteady flow problem around an obstacle at high Reynolds number.

1. Introduction

In general, vector computers require long vectors and non-recursive operations in order to gain a highly efficient performance. Concerning the original SOR and line SOR method with which we often use for solving various type of equations, the vectorization is difficult. Therefore several techniques have been developed, for example, the pseudo SOR method for long vectors and the checker-board and the hyperplane SOR method for non-recursive operations.

The pseudo SOR method means that the vectorization is made possible due to Jacobi method applied to one directional array. In compensation for obtaining long length of vectors, the convergence rate of the pseudo SOR method is reduced as compared with that of the original SOR method. The convergence rates of the checker-board and the hyperplane SOR method are equal to that of the original SOR method. However the vector length of these methods is short of $O(cN_x)$ (N_x is the number of grid in one direction and c is a constant). It is very important to develop a method which makes possible to operate with not only sufficient vector length and but also fast convergence rate.

In the Cartesian coordinate system, the Red-Black SOR method is utilized. In both cases of 5 points in the 2-D and 7 points in the 3-D, the vector operations are of length

$O(N/2)$, where N is the number of grid points. In the generalized coordinate system, owing to the cross derivative terms, the points of difference approximation increase extremely as compared with the Cartesian coordinate system. For example, the 5-points difference scheme for the 2-D Poisson's equation turns into the 9-points difference scheme and the 7-points difference scheme for the 3-D Poisson's equation turns into the 19-points difference scheme. It is few cases for us to analyze the fluid flow problem in the Cartesian coordinate system. Accordingly we must develop a efficient solver in the generalized coordinate system.

In the analysis of fluid flow problem, the MAC method is often used[1] and we must solve the Poisson's equation for pressure and it needs much computational time to solve it. The 2-D Poisson's equation discretized with the 9-points difference scheme can be solved efficiently by the 4-colored ordering method by J.M. Ortega [2]. The 3-D Poisson's equation discretized with the 19-points difference scheme can be solved efficiently by the 7-colored ordering method which we proposed [3].

Moreover the Navier-Stokes equations are often integrated directly without any turbulent models in analyzing the fluid flow problem at high Reynolds number. The instability on computation with using the central finite-difference occurs due to the non-linearity of convection terms of them. So we often use some upwind schemes, such as the third-order accuracy upwind scheme [4]. In this case, the points of difference scheme increase, that is, 5-points in one direction is often used for approximation of the derivatives. The equation is discretized with the the 13-points difference scheme in the 2-D generalized coordinate system and also the equation in the 3-D, is discretized with the 25-points difference scheme.

In this study, first, we discuss on efficiency of the 7-colored ordering method to the matrices with not only constant coefficients and but also space-varying coefficients. Secondly we discuss on efficiency of a solver for the higher-order accuracy difference schemes.

2. Seven color ordering for 3-D Poisson's equation

In the 3-D generalized coordinate system, the Poisson's equation is discretized with the 19-points difference scheme by using the central finite-difference as illustrated Fig.1 (a). Since it needs many points for approximating it, as concerns the SOR method, the efficient computation based on the vectorization is not easy. Because the sequential use of update data results in recursive operations and the use of old data reduces the convergence rate.

The present technique is based on non-recursive, one-dimensional representation of multi-dimensional array. All the grid points are divided into seven colored lines as each point of the grid has a color different from that of all other points to which it is connected as illustrated Fig.1 (b). Moreover for using one dimensional array and simplifying to calculate the address of array, with regards to the grid points in three axes we make the assumption as follows:

$$N_x = 7m + P,\ N_y = 7n - P, N_z : free \quad (m, n = 1, 2, 3, \ldots).$$

Here P is 2,3,4,5 and N_x, N_y and N_z are the number of grids in X, Y and Z direction respectively. We can make non-recursive SOR algorithm by using 7-colored lines. Therefore this 7-colored SOR method can be called 'RAINBOW SOR'.

3. Efficiency of multicolored SOR method for Poisson's equation

For comparison purpose on efficiency of this technique, three problems are tested. First is the problem of 2-D Laplace's equation with Dirichlet boundary conditions in the generalized coordinate system. The equation is discretized by using the 9-points difference scheme and thier coefficients are variable in space.

The second problem is governed the 3-D second order derivative equation with cross terms in the Cartesian coordinate system, imposed by the Dirichlet boundary conditions. The equation is discretized with the 19-points difference scheme and the number of grid points is the same as the fluid flow case in the generalized coordinate system. However the test case has constant coefficients.

The third problem is the 3-D Poisson's equation with Dirichlet boundary conditions in the generalized coordinate system. As the previous problem, we discretize the equation using the 19-points difference scheme and thier coefficients are variable in space.

(3-1) 2-D problem(9-points) in the generalized coordinate system

First we apply the multicolored technique to the Laplace's equation in the 2-D generalized coordinate system.

$$\Delta U = 0 \;\; in \, \Omega, \qquad U = U(x, y) \;\; on \, \partial\Omega \, .$$

Here Ω is the analytic domain and the grid system is shown as illustrated in Fig.2. The value:U on $\partial\Omega$ is a function of x, y. The discretized equation is solved by the hyperplane SOR, the checker-board SOR and the 4-colors SOR method. The criterion of convergence is 10^{-6} in the L_2-norm of absolute residual and the initial values of iterations are all equal to 0.

The computational rates of various solvers are exhibited in Fig.3 ((a) the result of VP-400E, (b) the result of S-820/80 and (c) result of SX-2). The maximum speed presented in the catalogue is 1700 Mflops for VP-400E, 2000 Mflops for S-820/80 and 1300 Mflops for SX-2. We introduce the idea of $Mflops$ for estimation of efficiency. That is, if the number of grid points is N, $22N$ operations($11N$ multiplications and $11N$ additions) are required per one iteration of the 9-points difference scheme. The amount of operations is same for all SOR methods in this study. Therefore we can estimate nearly the computational rates(in $Mflops$, say) due to the following equation:

$$Mflops = 22N \times \; Iter/(CPU \; time(sec) \times 10^6). \tag{1}$$

Here $Iter$ is the number of iterations. Figure 4 shows the comparison of CPU time about various solvers. Even in the problem that the coefficients are changing in space, the multicolored technique has a great efficiency as compared with the other SOR techniques.

(3-2) 3-D problem(19-points) in the Cartesian coordinate system

The second problem is the second order derivative equation with cross terms and it is solved by the pseudo SOR, the checker-board SOR and RAINBOW SOR method.

$$U_{xx} + U_{yy} + U_{zz} + U_{xy} + U_{yz} + U_{zx} = f \tag{2}$$

where $(X, Y, Z) \in \Omega = [0,1] \times [0,1] \times [0,1]$ and U is given on the boundary of the domain Ω and f is a given function of X, Y, and Z as follows:

$$f(X,Y,Z) = e^{XYZ}((XY)^2 + (YZ)^2 + (ZX)^2 + (X+Y+Z)(1+XYZ)). \tag{3}$$

The equation has an analytical solution of $U = e^{XYZ}$. In order to obtain numerical solution the central finite-difference is employed. The criterion of convergence is 4×10^{-4} in the L_2-norm of relative residual and the initial values of iterations are set to 0. The computational rates of various solvers are presented in Fig. 5. In the 3-D problem, $42N$ operations($21N$ multiplications and $21N$ additions) are carried out per one iteration. The computational rate can be estimated similarly as follows:

$$Mflops = 42N \times \ Iter/(CPU\ time(sec) \times 10^6). \tag{4}$$

The comparison of CPU time about various solvers are shown in Fig.6. The ratio of the CPU time on various solvers to that of the 7-colored SOR method is given. It is discernible that the 7-colored SOR method is extremely efficient in the 3-D problem.

(3-3)3-D problem(19-points) in the generalized coordinate system

Thirdly we apply the multicolored technique to the problem of the 19-points difference scheme for the 3-D Poisson's equation:

$$\Delta\, U = -f \qquad in\ \Omega \tag{5}$$

with Dirichlet boundary conditions and f chosen so that the exact solutions are all 1.0 . Here Ω is the analytic domain and the grid system is shown as illustrated in Fig.7. The discretization grid is $30 \times 68 \times 20$ yielding 40800 equations. The criterion of convergence is 10^{-6} in the L_2 of absolute residual and initial values of iterations are all equal to 0.

The results of computation on VP-400E are exhibited in Table. 1 ((a)the concentrated grid and (b) the non-concentrated grid). The discretized equation is solved by the checker-board SOR method and the 7 colored SOR method. The results show that in both cases the latter is more efficient than the former.

4.Multicolored ordering for higher-order accuracy difference scheme

On occasion of using higher-order accuracy difference schemes for discretizing the equation, many points of approximation is necessary for the improvement of precision. For example, the following the 4th-order accuracy difference schemes are well known.

$$U_{xx} + U_{yy} = ((-20U_{i,j} + 4(U_{i,j-1} + U_{i-1,j} + U_{i+1,j} + U_{i,j+1}) +$$
$$(U_{i-1,j-1} + U_{i+1,j-1} + U_{i-1,j+1} + U_{i+1,j+1}))/(6h^2) + O(h^4)$$
$$U_x = (U_{i-2,j} - 8U_{i-1,j} + 8U_{i+1,j} - U_{i+2,j})/(12h) + O(h^4)$$
$$U_y = (U_{i,j-2} - 8U_{i,j-1} + 8U_{i,j+1} - U_{i,j+2})/(12h) + O(h^4)$$

where h is the grid spacing. Moreover for analyzing the fluid flow problem at high Reynolds number, the following the third-order accuracy upwind scheme for the convection terms is used: (Kawamura et al.[4])

$$(U\frac{\partial U}{\partial x})_i = U_i\frac{-U_{i+2} + 8(U_{i+1} - U_{i-1}) + U_{i-2}}{12h} + |U_i|\frac{U_{i+2} - 4U_{i+1} + 6U_i - 4U_{i-1} + U_{i-2}}{4h.} \tag{6}$$

The second term in the right hand side of Eq.(6) represents the numerical diffusion by a fourth-order derivative. For comparison purpose on efficiency of the multicolored technique, two problems are examined. The problems are governed the 2-D and 3-D second order derivative equation with cross terms in the Cartesian coordinate system, imposed by the Dirichlet boundary conditions.

(4-1)2-D problem(13-points) in the Cartesian coordinate system

In this section, the following equation is solved by the 5-colored SOR method.

$$U_{xx} + U_{yy} + U_x + U_y = -f \tag{7}$$

where $(X,Y) \in \Omega = [0,1] \times [0,1]$ and U is given on the boundary of the domain Ω and f is chosen that the exact solution is $U = Xe^{XY}sin(\pi X)sin(\pi Y)$. All the spatial derivatives are represented by using of the 4th-order accuracy central finite-difference and the equation is discretized with the 13-points difference scheme as illustrated in Fig. 8. The results of computation are exhibited in the Table. 2.

(4-2)3-D problem(25-points) in the Cartesian coordinate system

As the previous section, the next equation is solved by the 9-colored SOR method.

$$U_{xx} + U_{yy} + U_{zz} + U_{xy} + U_{yz} + U_{zx} = f \tag{8}$$

where $(X,Y,Z) \in \Omega = [0,1] \times [0,1] \times [0,1]$ and U is given on the boundary of the domain Ω and f is chosen that the exact solution is $U = e^{XYZ}sin(\pi X)sin(\pi Y)$. The spatial derivatives are represented by using the 2th-order and the 4th-order accuracy central finite-differences and the equation is discretized with the 25-points difference scheme as illustrated in Fig. 9. The results of computation are shown in the Table. 3.

5. Application to Fluid Flow problem

Next we apply the present SOR technique to the unsteady flow around a hemisphere at high Reynolds number. The Navier-Stokes equations in the generalized coordinate system are directly integrated, using third order upwind scheme without any turbulence models. The governing equations, i.e., the continuity equation and the Navier-Stokes equations, may be expressed as, in a properly non dimensionalized form,

$$divV = 0 \tag{9}$$

$$\frac{\partial V}{\partial t} + VgradV = -gradP + \frac{\Delta V}{Re} \tag{10}$$

where V, P, t and Re denote the velocity vector, pressure, time and the Reynolds number,respectively. All the spatial derivatives,with the exception of the convection terms, are represented by use of the second order accuracy central finite-differences. The third-order accuracy upwind scheme for the convection terms is adopted. The grid points of velocity and pressure are defined at identical place on the MAC method. The Poisson's equation for pressure is solved by the RAINBOW SOR method. The coefficient of the derivative terms is not constant. Fig.10 shows the computational model. The number of grid points is $51 \times 33 \times 100 = 168300$ points and Reynolds number is 1000. Fig.11 shows instantaneous pressure contours around a hemisphere.

6. Conclusion

We proposed the multicolored, that is, the 7-colored SOR method for the 19-points FDM scheme of the Poisson's equation in the 3-D generalized coordinate system. With this SOR method, we solve the equation with constant coefficients and with space-varying coefficients. Consequently it is discernible that this multicolored technique has a extreme efficiency. Moreover with the 5-colored and the 9-colored SOR method, we solved the equation with constant coefficients and we showed that these method are very efficient. It is comfirmed that the proposed multicolored SOR method is possible to apply to the fluid flow problem.

Acknowledgement

The authors would like to thank Dr. S., Shirayama of ICFD for their use of graphic systems. The authors are also grateful to T., Takeuchi of KAO corporation for several valuable suggestions, and in particular for pointing out of 5-colored and 9-colored SOR method presented (4-1), (4-2).

References

[1]Harlow,F.H. and Welch,J.E.,"Numerical Calculation of Time-Dependent Viscous Incompressible Flow of Fluid with Free Surface", Phys. Fluids, Vol.8, 1965

[2]Ortega,J.M.,"Introduction to Parallel and Vector Solution of Linear Systems", Plenum Press ,New York and London, 1988

[3]Fujino,S., Tamura,T. and Kuwahara,K.,"Application of the RAINBOW SOR technique to Fluid Flow Analysis in the 3-D Generalized Curvilinear Coordinate System",proceedings of 6th International Conference on 'Numerical methods in Laminar and Turbulent Flow, Swansea, U.K.,1989

[4]Kawamura,T. and Kuwahara,K.,"Computation of high Reynolds Number around a Circular Cylinder with Surface Roughness",AIAA paper,84-0340,1984

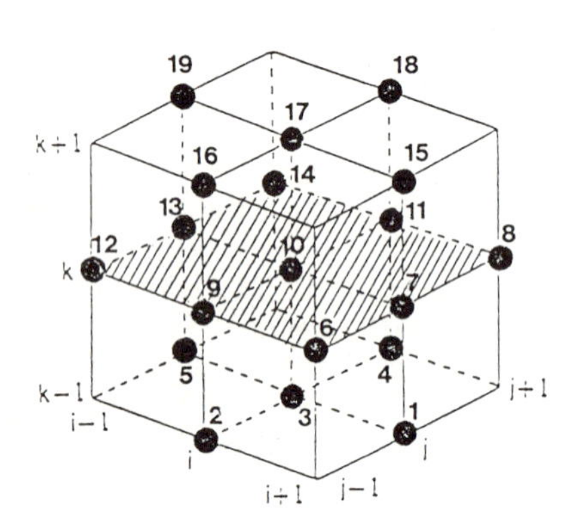

Fig.1(a) distribution of 19–points difference scheme

Fig.1(b) Seven color ordering for 19–points difference scheme (V:Violet,I:Indigo,B:Blue, G:Green,Y:Yellow,O:Orange,R:Red)

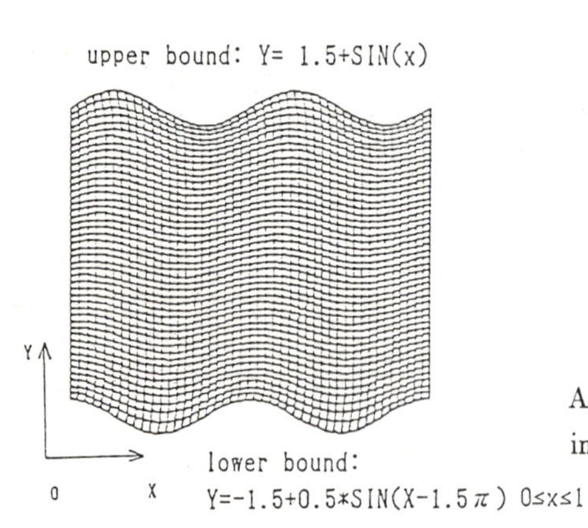

Fig.2

A grid system of 2-D Laplace's problem in the generalized coordinate system.

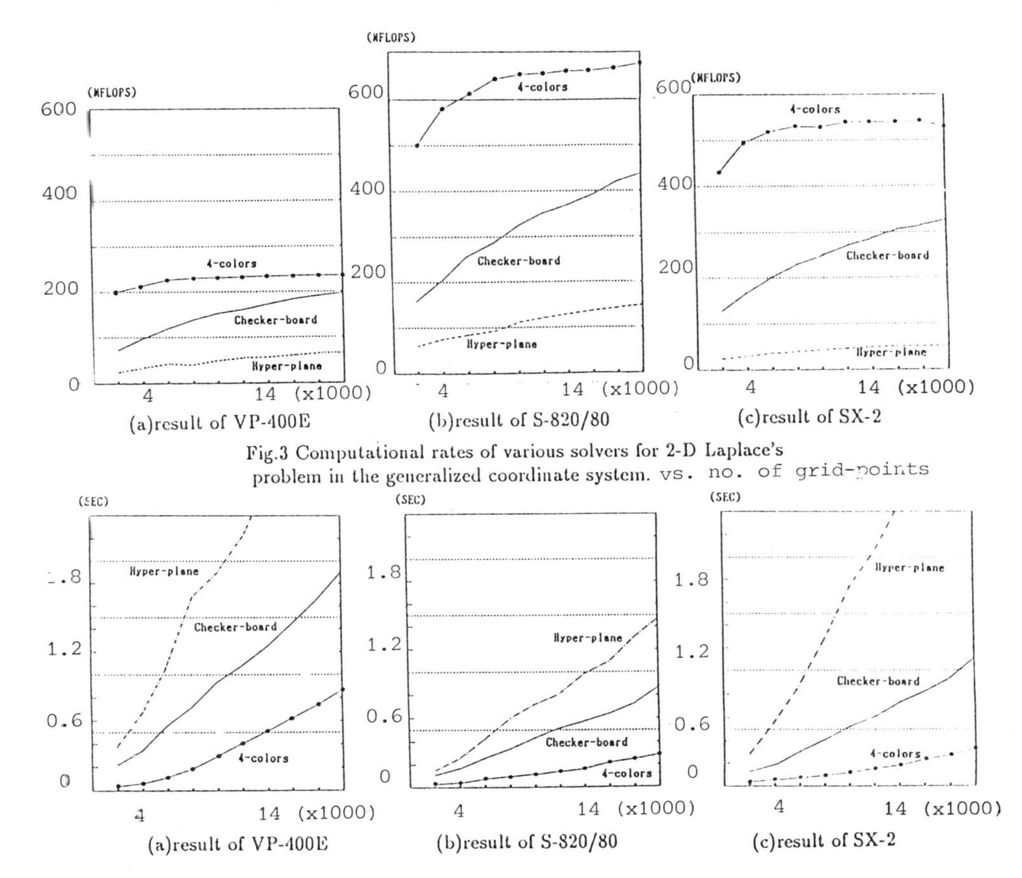

Fig.3 Computational rates of various solvers for 2-D Laplace's problem in the generalized coordinate system. vs. no. of grid-points

Fig.4 Comparison of CPU time about various solvers for 2-D Laplace's problem in the generalized coordinate system. vs. no. of grid-points

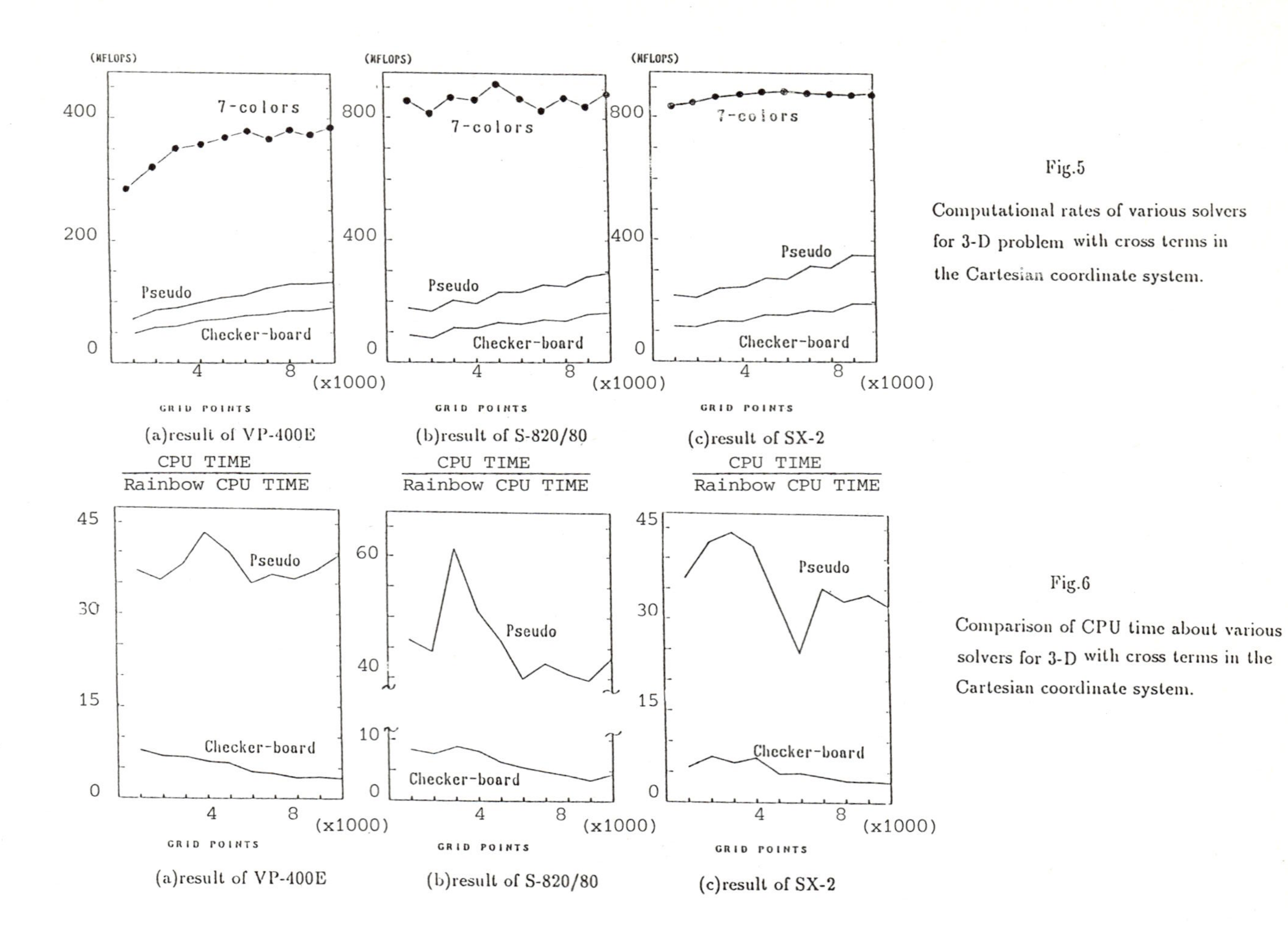

Fig.5

Computational rates of various solvers for 3-D problem with cross terms in the Cartesian coordinate system.

Fig.6

Comparison of CPU time about various solvers for 3-D with cross terms in the Cartesian coordinate system.

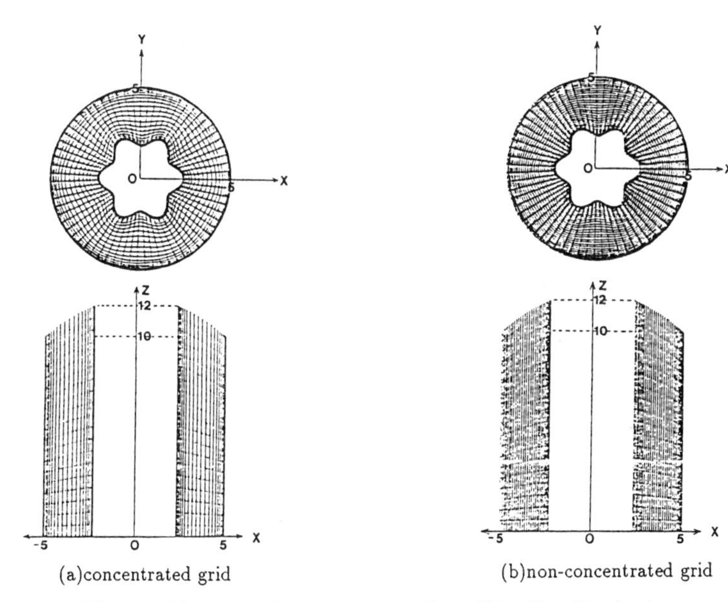

(a)concentrated grid

(b)non-concentrated grid

Fig.7 A grid system of 3-D Poisson's problem (30 × 68 × 20 points)

Table.1 The result of 3-D Poisson's equation on VP-400E

(a)concentrated grid

	RAINBOW (7colors)	CHECKER-BOARD
ACCELATION PARAMETER	ω =1.70	ω =1.60
ITERATION NUMBER	39	95
CPU TIME	0.185(1.0)	2.05(11.1)
MFLOPS	378 (4.6)	83 (1.0)
VECTOR LENGTH	5830	34

(b) non-concentrated grid

	RAINBOW (7colors)	CHECKER-BOARD
ACCELATION PARAMETER	ω =1.80	ω =1.75
ITERATION NUMBER	51	106
CPU TIME	0.240(1.0)	2.30(9.6)
MFLOPS	382 (4.6)	83 (1.0)
VECTOR LENGTH	5830	34

Table.2

Computational rate of 5-colored SOR method

MESH SIZE	VP-400E	S-820/80	SX-2
(1) 10[illegible] X 10[illegible]	287	731	804
(2) 142 X 142	299	770	837
(3) 172 X 172	299	776	834
(4) 202 X 202	300	779	847
(5) [illegible] X [illegible]	300	781	846
(6) 247 X 247	300	782	848
(7) 267 X 267	300	784	850
(8) 282 X 282	301	779	847
(9) 302 X 302	301	777	849
(10) 322 X 322	301	783	847
Average:	299	774	841

Table.3

Computational rate of 9-colored SOR method

MESH SIZE	VP-400E	S-820/80	SX-2
(1) 22x21x22	286	588	835
(2) 31x21x31	289	627	875
(3) 31x30x3[illegible]	[illegible]91	640	886
(4) 40x30x34	291	644	886
(5) 40x30x42	292	648	891
(6) 40x30x50	292	639	879
(7) 49x39x37	292	642	891
(8) 49x39x42	292	639	893
(9) 58x39x40	293	645	890
(10) 58x39x45	293	644	889
Average:	291	636	882

(unit: Mflops)

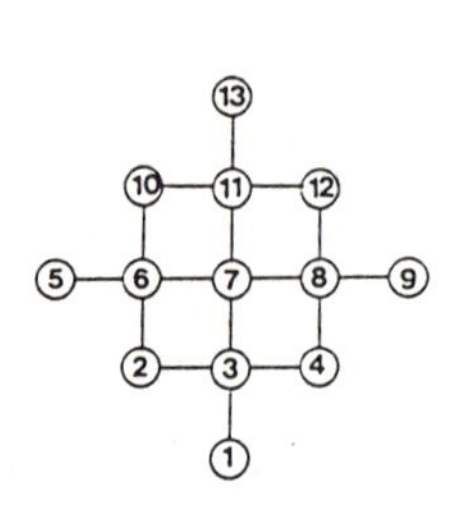

Fig.8

13-points difference scheme in the 2-D

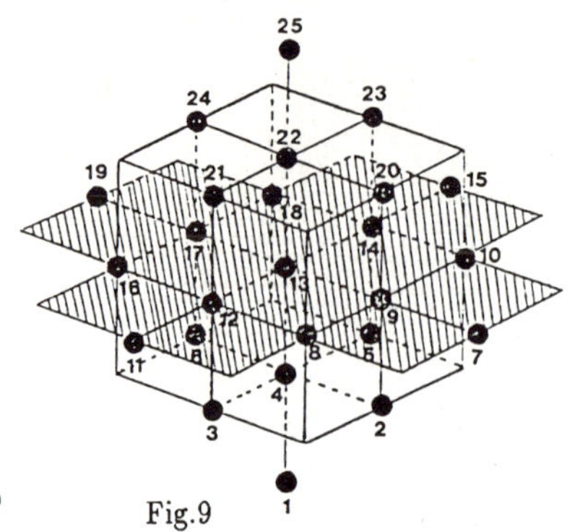

Fig.9

25-points difference scheme in the 3-D

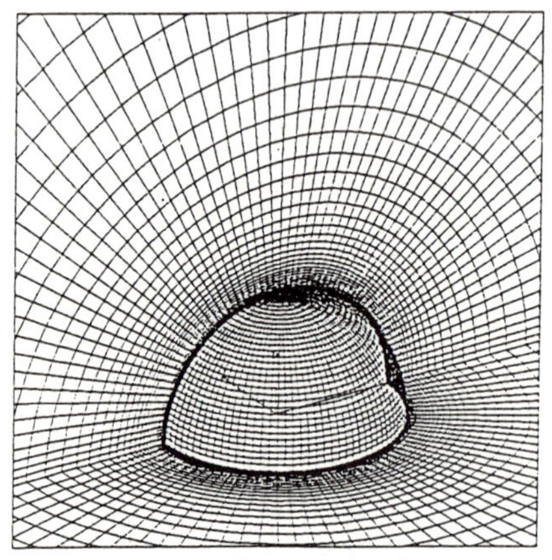

a hemisphere on the plane

Fig.10

A grid system for a hemisphere on the plane
(51 × 33 × 100 points)

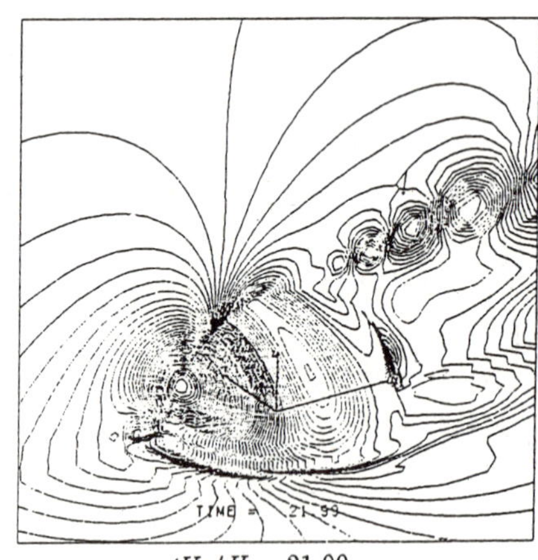

$tU_0/H = 21.99$

Fig.11

Instantaneous pressure contours ($R_e = 1000$)
(U_0 : the velocity of uniform flow,
H : Hemisphere height)

NON-EQUILIBRIUM HYPERSONIC FLOW COMPUTATIONS BY IMPLICIT SECOND-ORDER UPWIND FINITE-ELEMENTS

N. Glinsky, L. Fezoui, M. C. Ciccoli, J-A. Désidéri *

September 20, 1989

Abstract

Second-order implicit schemes are constructed for the solution of steady hypersonic chemically-reacting inviscid flows. In a standard non-equilibrium model accounting for 5 species (N_2, O_2, NO, N and O), the Euler equations are coupled with species convection equations. In a fractional-step approach, the set of fluid motion equations and the set of chemical kinetics equations are alternatively time-marched, both sets implicitly. The basic approximation method employs a finite-volume formulation applicable to arbitrary Finite-Element-type unstructured triangulations. The accuracy is enhanced by the M.U.S.C.L. extrapolation, and quasi-second-order solutions are obtained by slope-limitation or TVD-averaging. The merits of the various proposed approximations are evaluated by numerical experiments, computing the hypersonic flows over a cylinder and around a model geometry for the HERMES space shuttle.

1 Introduction

This article follows the previous work of the authors and their co-authors in the development of efficient numerical methods, mainly of Finite-Element-type, for the computation of steady chemically-reacting hypersonic flows at equilibrium or not, [1], [2].

The basic first-order explicit method has been developed in [1]. It is based on a hybrid Finite-Element/Finite-Volume formulation applicable to unstructured grids. For non-equilibrium flow, the operator is split into the fluid-motion equations and the species-convection equations (also referred to by others as a "weak-coupling" approach). Originally, only the source term of the chemistry equations (production rates) was treated implicitly. The approximation is of upwind type, and in the fluid-motion step it accounts for the spatially-variable composition by modifying the van Leer flux-vector splitting according to an "equivalent-γ". These techniques are reviewed in Section 3.

*Institut National de Recherche en Informatique et en Automatique, Centre de Sophia Antipolis, 2004, Route des Lucioles, 06560 Valbonne (FRANCE).

In [2] three-dimensional applications of interest for industry (space-shuttle like geometries) have been presented, demonstrating the importance of accounting for chemistry in the description of hypersonic flow, particularly to position the shocks accurately.

In this article, we present an enhanced version of our method for non-equilibrium flow, where the improvements consist of implicit time solutions (Section 4) and quasi-second-order approximation schemes including limitation devices (Section 5). The new method is tested by computing the hypersonic flows around a cylinder and a two-dimensional model geometry for the HERMES space shuttle (Section 6).

2 Governing Equations for Non-Equilibrium Inviscid Flow

We consider the case of dissociated air, assuming 5 species (N_2, O_2, NO, N and O) are subject to 17 chemical reactions (dissociation and exchange reactions; [3]):

$$\left\{\begin{array}{lcl} O_2 + X & \rightleftarrows & O + O + X \\ N_2 + X & \rightleftarrows & N + N + X \\ NO + X & \rightleftarrows & N + O + X \\ NO + O & \rightleftarrows & O_2 + N \\ N_2 + O & \rightleftarrows & NO + N \end{array}\right. \tag{1}$$

in which X is a collision partner.

Conservative steady solutions to the Euler equations are captured by pseudo-time integration of the usual conservation laws in Cartesian (untransformed) coordinates, here in two dimensions, coupled with species convection equations:

$$\left\{\begin{array}{l} W_t + F(W)_x + G(W)_y = 0 \\ \chi_t + (u\chi)_x + (v\chi)_y = \Omega \end{array}\right. \tag{2}$$

in which W is the vector of conservative variables,

$$W = (\rho, \rho u, \rho v, E)^T \tag{3}$$

where ρ is the density, u and v the x and y velocity components, and E the total energy, and χ the vector of partial densities,

$$\chi_s = \rho_s = \rho Y_s \qquad (s = 1, \ldots, 5) \tag{4}$$

where Y_s is the mass fraction of species s, and the vector $\Omega = \Omega(W, \chi)$ contains the production rates; for these, as in [1], [2], we are employing the curvefits proposed by Park [4].

As usual,

$$\left\{\begin{array}{l} F(W) = (\rho u, \rho u^2 + p, \rho uv, u(E+p))^T \\ G(W) = (\rho v, \rho uv, \rho v^2 + p, v(E+p))^T . \end{array}\right. \tag{5}$$

A mixture of perfect gases is assumed, and the state equation writes

$$p = \rho \mathcal{R} T \sum_{s=1}^{5} \frac{Y_s}{M_s} \tag{6}$$

where M_s is the molar mass of species s.

Our model assumes only one temperature, which is calculated from the total specific energy which is the sum of kinetic energy, formation enthalpy, translation-rotation energy, and vibrational energy (here assumed at equilibrium at the temperature T):

$$E = E_K + H_0 + E_{TR} + E_{VIB} \tag{7}$$

where

$$\begin{cases} E_K = \frac{1}{2}\rho(u^2+v^2) \\ H_0 = \sum_{s=1}^{3} \rho Y_s h_s^0 \\ E_{TR} = \sum_{s=1}^{5} \rho Y_s C_{v_s} T \\ E_{VIB} = \sum_{s=3}^{5} \rho Y_s R_s \Theta_{vs} / (\exp(\Theta_{vs}/T) - 1) \end{cases} \tag{8}$$

$(R_s = \mathcal{R}/M_s)$ where $s = 1, 2, 3$ corresponds to the products O, N and NO, and $s = 3, 4, 5$ to the diatomic species.

Note that the definition of the mass fractions and the global conservation of oxygen and nitrogen atoms allow us to write two independent algebraic equations relating the Y_s's:

$$\sum_{s=1}^{5} Y_s = 1\,; \quad \frac{\frac{Y_{N_2}}{M_N} + \frac{Y_N}{M_N} + \frac{Y_{NO}}{M_{NO}}}{\frac{Y_{O_2}}{M_O} + \frac{Y_O}{M_O} + \frac{Y_{NO}}{M_{NO}}} = \frac{79}{21} \tag{9}$$

Consequently, the mass fractions of N_2 and O_2 are eliminated from the system, and only three species convection equations $(s = 1, 2, 3)$ are retained along with the global mass conservation, momentum and energy equations.

As a result, for a two-dimensional problem, we are led to solve a hyperbolic system of 7 coupled (nonlinear) PDE's, along with a state equation in which the mass fractions appear; in addition, we note that temperature enters nonlinearly the definition of energy.

3 Basic (Explicit, First-Order) Finite-Element Solver

3.1 Fluid-Motion Step

Starting with a (possibly unstructured) Finite-Element triangulation of the domain, dual Finite-Volume cells are constructed around nodes whose edges are portions of the medians of the triangles.

The conservation laws are written in integral form

$$\frac{d}{dt}\iint_{\mathcal{C}_i} W\,dxdy + \int_{\partial\mathcal{C}_i} (F(W)\,n_x + G(W)\,n_y)\,ds = 0 \tag{10}$$

in which $\mathcal{C}_i$ is an arbitrary cell. A first-order upwind approximation scheme is then expressed:

$$\mathcal{A}_i \frac{W_i^{n+1} - W_i^n}{\Delta t} + \sum_{j \text{ neighbor of } i} \Phi\left(W_i^n, W_j^n, \vec{\eta}_{ij}\right) = 0 \tag{11}$$

in which W_i is the cell-average of W over $\mathcal{C}_i$, $\mathcal{A}_i$ the area of the cell, $\vec{\eta}_{ij} = \int_{\text{AIB}} \vec{n}\,ds$ is the integrated normal to the interface between i and j, and $\Phi\left(W_i, W_j, \vec{\eta}_{ij}\right)$ is the numerical flux through the interface. An appropriate rotation $\mathcal{R}_\theta$ is introduced to evaluate this flux as in one dimension,

$$\Phi\left(W_i, W_j, \vec{\eta}_{ij}\right) \approx F(W_I)\,\vec{\eta}_{ij,x} + G(W_I)\,\vec{\eta}_{ij,y} = \|\vec{\eta}_{ij}\|\,\mathcal{R}_\theta^{-1} F(\mathcal{R}_\theta W_I) \tag{12}$$

and (flux-vector splitting):

$$F(\mathcal{R}_\theta W_I) = F^+(\mathcal{R}_\theta W_I) + F^-(\mathcal{R}_\theta W_I)\,. \tag{13}$$

Finally, the van Leer [5] flux-vector splitting is employed:

$$\begin{aligned} F^\pm(W) &= \begin{pmatrix} \pm D \\ DU/\gamma \\ \frac{1}{2} DU^2/(\gamma^2-1) \end{pmatrix} \quad \text{if } M < 1\,, \\ F^+(W) &= F(W)\,,\ F^-(W) = 0 \quad \text{otherwise,} \end{aligned} \tag{14}$$

in which u denotes the velocity, $D = \rho(u \pm c)^2/(4c)$, $U = 2c \pm (\gamma - 1)u$ where c is the soundspeed and M the local Mach number. In these expressions, for chemically reacting gas, we employ an "equivalent-γ" defined by

$$\gamma = h/\epsilon \tag{15}$$

(h, ϵ: specific enthalpy and internal energy) and "false" soundspeed and Mach number:

$$c = \sqrt{\gamma p/\rho}\,,\ M = u/c\,. \tag{16}$$

This approximation affects the way subsonic regions are identified and discriminated in the upwinding scheme; however, the resulting approximation is consistent and converges to the proper weak solution.

3.2 Chemistry Step

At each timestep, after the fluid-motion step has been completed at frozen (but not uniform) composition, a chemistry step is made to update the mass fractions. For this, a similar Finite-Volume scheme is constructed in which the source term Ω is treated implicitly (to remove possible severe timestep restrictions) and time-linearized:

$$\mathcal{A}_i \frac{\chi_i^{n+1} - \chi_i^n}{\Delta t} + \sum_{j \text{ neighbor of } i} \Psi\left(\chi_i^n, \chi_j^n, \vec{\eta}_{ij}\right) = \mathcal{A}_i\left(\Omega_i^n + \Omega_i'^n(\chi_i^{n+1} - \chi_i^n)\right) \tag{17}$$

in which $\Omega' = \partial\Omega/\partial\chi$, and now the numerical flux associated with species convection is modeled on the numerical flux of the global mass conservation equation, as proposed in [6]; that is, if as a result of (13-14)

$$\Phi\left(W_i, W_j, \vec{\eta}_{ij}\right) = \Phi_{ij}^+\left(W_i, W_j, \vec{\eta}_{ij}\right) + \Phi_{ij}^-\left(W_i, W_j, \vec{\eta}_{ij}\right) \tag{18}$$

we let

$$\Psi\left(\chi_i, \chi_j, \vec{\eta}_{ij}\right) = \left[\Phi_{ij}^+\left(W_i, W_j, \vec{\eta}_{ij}\right)\right]_1 \left(\frac{\chi}{\rho}\right)_i + \left[\Phi_{ij}^-\left(W_i, W_j, \vec{\eta}_{ij}\right)\right]_1 \left(\frac{\chi}{\rho}\right)_j \tag{19}$$

in which the subscript 1 refers to the first component, and χ/ρ is the vector of mass fractions. In this way, the flux-vector splitting of the species convection equations reinforces the coupling with the fluid motion equations, and accounts for all three eigenvalues that appear when the Jacobian of the Euler equations is diagonalized. Also, solutions of the form Y_s = const. can be captured exactly at the steady state.

4 Implicit Formulation

The basic explicit scheme of the previous section is subject to the CFL condition. In order to remove this timestep limitation, an implicit formulation was sought.

The fluid-motion step was made implicit (Euler Implicit Method in Δ-form), linearized and solved by Gauss-Seidel iteration, a technique that has been described for unstructured grids in many contributions, e.g. [7], [8].

Hence, in this section we only concentrate on defining an implicit chemistry step. The following implicit analog of (17) is constructed [12]:

$$\mathcal{A}_i \frac{\chi_i^{n+1} - \chi_i^n}{\Delta t} + \sum_{j \text{ neighbor of } i} \Psi\left(\chi_i^{n+1}, \chi_j^{n+1}, \vec{\eta}_{ij}\right) = \mathcal{A}_i \left(\Omega_i^n + \Omega_i'^n (\chi_i^{n+1} - \chi_i^n)\right). \quad (20)$$

One could solve this nonlinear implicit equation by some iterative process, e.g. Newton's method. Instead we prefer to employ the following approximate linearization of the flux term, which is inspired from (19):

$$\begin{aligned} \Psi\left(\chi_i^{n+1}, \chi_j^{n+1}, \vec{\eta}_{ij}\right) &= \Psi\left(\chi_i^n, \chi_j^n, \vec{\eta}_{ij}\right) \\ &+ \frac{\left[\Phi_{ij}^-\left(W_i^n, W_j^n, \vec{\eta}_{ij}\right)\right]_1}{\rho_i^n} \delta\chi_i^{n+1} + \frac{\left[\Phi_{ij}^-\left(W_i^n, W_j^n, \vec{\eta}_{ij}\right)\right]_1}{\rho_j^n} \delta\chi_j^{n+1} \end{aligned} \quad (21)$$

where $\delta\chi^{n+1} = \chi^{n+1} - \chi^n$. These terms are substituted in (20) and grouped by nodes (summation on j); if N_s is the number of nodes, this implicit formulation results in a linear system for the N_s unknown 3-component vectors $\{\delta\chi_i^{n+1}\}$ $(i = 1, \ldots, N_s)$, where the right-hand side is the usual explicit update. The system is typically solved by some 10 Gauss-Seidel sweeps.

The implicit solver can operate with much larger timesteps, usually controlled by the norm of the residual (or right-hand side), and the iteration number n. Typically, the CFL number is gradually increased from 0.4 (usual limit of the explicit method) to 20 or 50, and this results in a reduction of the number of the necessary iterations by 40, and of the computation time by 15.

5 Second-Order Approximation Schemes

Second-order approximation schemes are constructed by the van Leer M.U.S.C.L. approach [9], in which the flow variables at nodes are replaced in the expression of the numerical fluxes, by linear extrapolates at the cell interface:

$$\Phi_{ij} = \Phi\left(W_{ij}, W_{ji}, \vec{\eta}_{ij}\right) \, , \quad \Psi_{ij} = \Psi\left(\chi_{ij}, \chi_{ji}, \vec{\eta}_{ij}\right) \, . \quad (22)$$

Two extrapolation procedures are proposed. However, in both cases, the working variables are not the conservative variables (the vectors W and χ at the interface being formed after the extrapolation), but for inert gas [8], the primitive variables ρ, u, v and p. For chemically-reacting gas, the necessary control of the temperature leads to replace p by T which is efficient but costly, or by the internal energy (here per unit mass) ϵ, as suggested by Montagné et al. [10], which is more economical. This alternative is adopted here. The extrapolation of the chemistry variables is performed on the mass fractions. In what follows, the symbol w refers to any of these 7 variables.

5.1 Extrapolation with slope limitation

In the P1-Lagrange approximation, gradients are constant by triangles. Thus, the most natural way to extrapolate nodal values, is to first compute at each node i the average gradient of the variable ∇w_i, as the weighted average of the gradients evaluated in the triangles surrounding the node, the weights being the areas of the respective triangles; and secondly, to perform the extrapolation:

$$w_{ij} = w_i + \nabla w_i . \frac{\vec{ij}}{2} \tag{23}$$

(and similarly, $w_{ji} = w_j - \nabla w_j . \frac{\vec{ij}}{2}$). In doing this, in the implicit formulation, while the preconditioner is a first-order upwind operator, the explicit update (or right-hand side) is a "half-fully-upwind" second-order operator, since the corrections to the first-order upwind nodal values are based on centered gradients. This combination is recommended in [8] for best iterative convergence.

In fact, the above second-order approximation is not robust enough for being practical. In an attempt to construct a monotonic scheme, limitation is applied to the gradient, or "slope", prior to the extrapolation, by the Min-Mod function: the variations $\nabla w_T . \vec{ij}$ are calculated for all triangles T surrounding the node; if all of these numbers are of the same sign, $\nabla w_i . \vec{ij}$ is set equal to the one of smallest modulus; otherwise it is set equal to zero, and the approximation is locally only first-order accurate.

In conclusion, this construction results in a quasi-second-order half-fully upwind scheme.

5.2 Quasi-TVD scheme based on the upwind triangle

Another route to construct a half-fully upwind scheme, is to average the prediction of a fully upwind scheme with that of a centered scheme. For ϵ sufficiently small, the point $i - \epsilon\vec{ij}$ belongs to the same triangle, defined as the "upwind triangle" T_{ij}. A fully upwind extrapolated value w_{ij} can then be computed by letting $\nabla w_i = \nabla w_{T_{ij}}$ in (23). Hence the corrections brought to the nodal value w_i by the fully upwind scheme and the centered scheme are respectively:

$$a = \frac{1}{2} \nabla w_{T_{ij}} . \vec{ij} , \; b = \frac{1}{2}(w_j - w_i) \; . \tag{24}$$

The half-fully upwind scheme is then obtained by letting

$$w_{ij} = w_i + Ave(a, b) \tag{25}$$

in which the symbol Ave stands for some averaging function.

Again, if the averaging function is the arithmetic mean, the approximation lacks robustness. Instead, in an attempt to construct a TVD-like scheme, the van-Albada average [11] is employed:

$$\begin{aligned} Ave(a, b) &= \frac{(a^2 + \epsilon)b + (b^2 + \epsilon)a}{a^2 + b^2 + 2\epsilon} && \text{si } ab > 0 \\ &= 0 && \text{otherwise} \end{aligned} \tag{26}$$

where ϵ is some small number to avoid zero divide.

6 Numerical Experimentation and Flow over a Double-Ellipse

All the flow calculations presented in this section were made assuming a hypersonic freestream Mach number $M_\infty = 17.9$ to facilitate the comparison with some other data available in [1].

In a first series of experiment, the hypersonic flow over a half-cylinder is computed using a medium-size mesh of 1540 points whose upper part is depicted on Fig. 1.

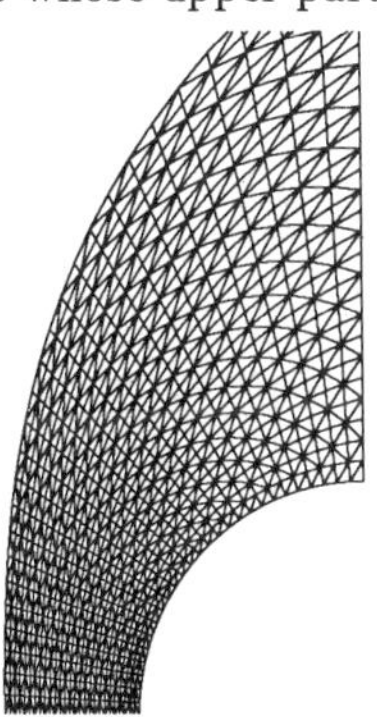

FIG. 1: MESH AROUND HALF-CYLINDER

Three approximation schemes are compared:

(a) the first-order scheme,

(b) the quasi-second-order method with slope limiters, and

(c) the quasi-second-order method based on the upwind triangle and TVD-averaging.

The plots of Mach number and temperature (Figs. 2-3) demonstrate in particular that the predicted stand-off distances by the second-order methods are visibly smaller.

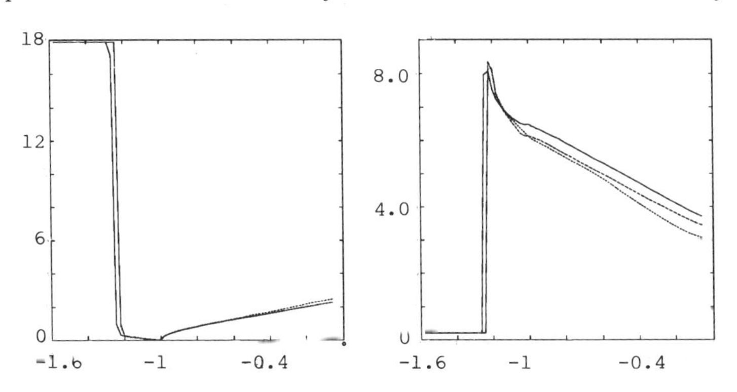

FIG. 2: MACH FIG. 3: TEMPERATURE

Concerning accuracy and numerical dissipation, it appears that the results of Method (b) (slope limiters) are essentially intermediate between those of Methods (a) (first-order) and (c) (upwind triangle). This is confirmed by Fig. 4 on which the temperature field is shown in the three cases.

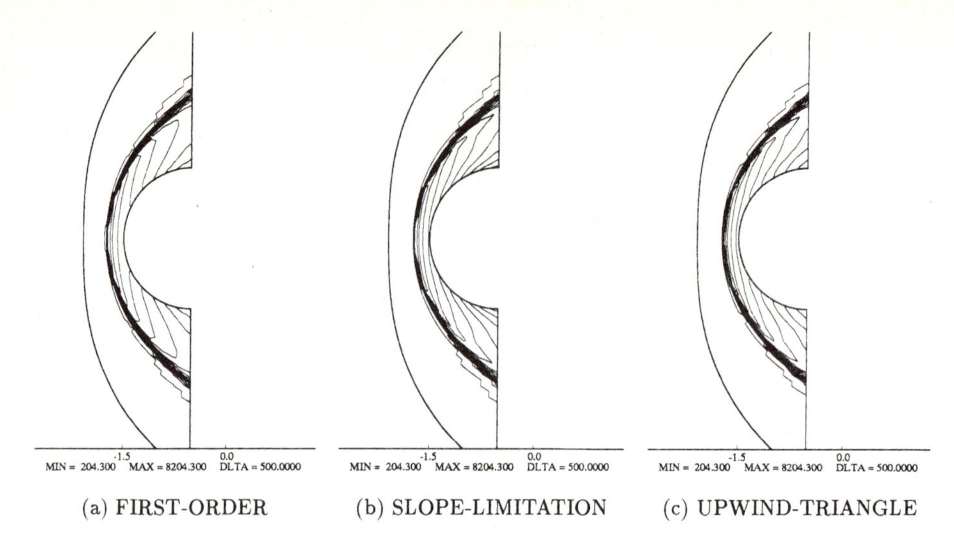

(a) FIRST-ORDER (b) SLOPE-LIMITATION (c) UPWIND-TRIANGLE

FIGURE 4: ISO-TEMPERATURE CONTOURS BY THREE METHODS

In addition, the quasi-second-order method with slope limiters is more robust than the quasi-second-order method based on the upwind triangle, but the latter is more accurate.

In a second series of experiment the hypersonic flow over a double ellipse was calculated using the first-order approximation and the quasi-second-order approximation with limitation by triangle.

The efficiency of the implicit formulation is demonstrated on Fig. 5 in the case of a first-order approximation, where a gain in computation time close to 15 is achieved.

The remaining computations were made with a triangular mesh of 3275 points shown on Fig. 6.

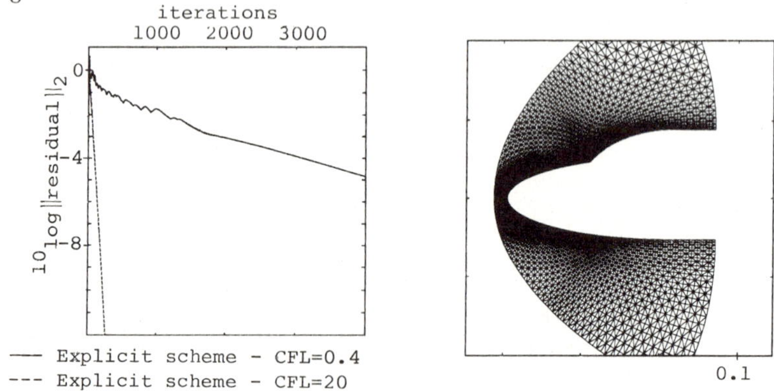

FIG. 5: CONVERGENCE HISTORY FIG. 6: MESH FOR DOUBLE-ELLIPSE

The quality of the results is observed on Figs. 7-8-9 and particularly on Fig. 10 where it appears that the excessive amount of artificial dissipation inherent to the first-order approximation alters considerably the solution, since only the second-order method predicts the complete dissociation of O_2.

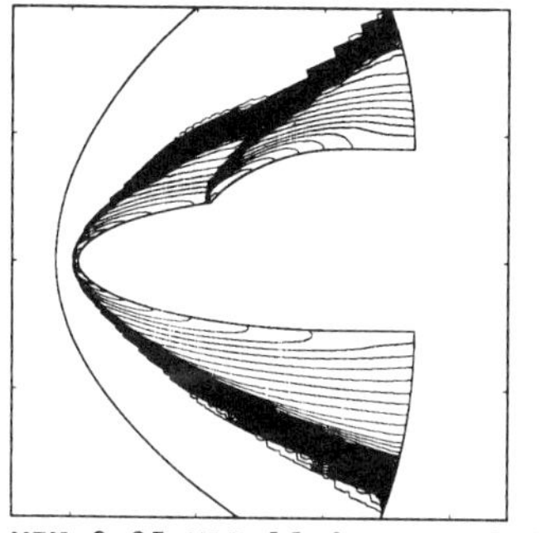

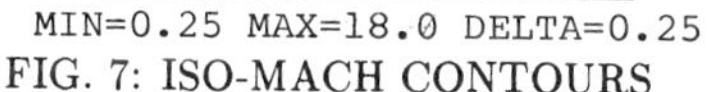

FIG. 7: ISO-MACH CONTOURS

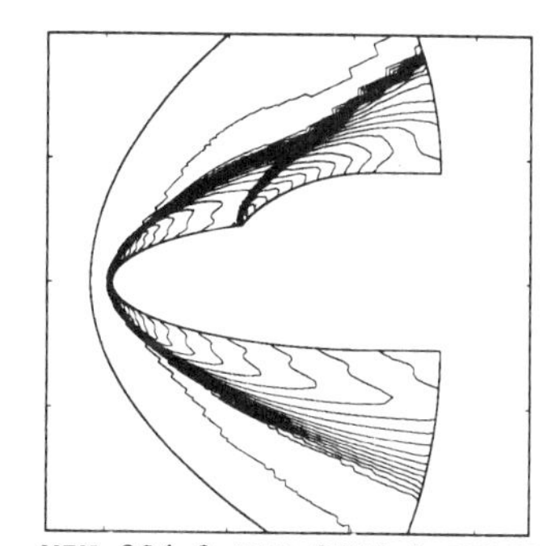

FIG. 8: ISO-TEMPERATURE CONTOURS

FIG. 9: MACH

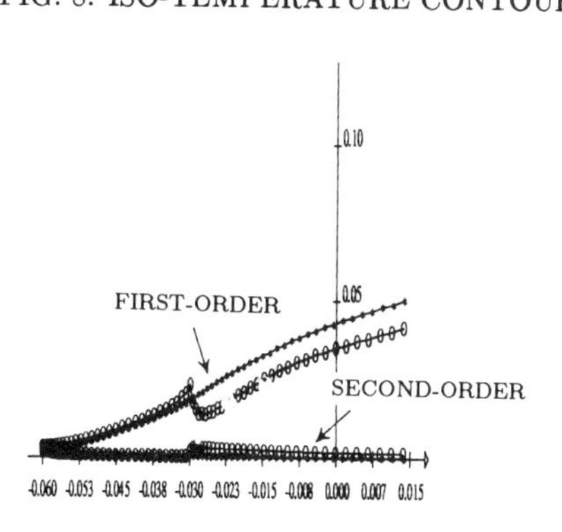

FIG. 10: O_2 MASS FRACTION

7 Conclusions

This article has presented a robust and accurate method for the computation of hypersonic non-equilibrium flow.

The approximation is of mixed Finite-Element/Finite-Volume type which permits the usage of unstructured grids. Quasi-second order half-fully upwind schemes have been constructed by the van Leer M.U.S.C.L. approach in two different ways, and limitation devices appropriate to each have been experimented. Efficient implicit solvers are employed in an operator-split algorithm, and typical reductions in computing costs by factors of 15 have been observed.

Finally, the accuracy of the method has been demonstrated by the computation of the hypersonic reactive flow over a double-ellipse, a difficult test-case of a forthcoming workshop (Workshop on Hypersonic Flows for Reentry Problems, Jan. 22-25, Antibes, France).

References

[1] J-A. DESIDERI, N. GLINSKY, E. HETTENA, "Hypersonic Reactive Flow Computations", *Computers and Fluids, 1989*, to appear.

[2] J-A. DESIDERI, L. FEZOUI, N. GLINSKY, E. HETTENA, J. PERIAUX, B. STOUFFLET, "Hypersonic Reactive Flow Computations around Space-Shuttle-Like Geometries by 3-D Upwind Finite Elements", *AIAA Paper No. 89-0657, AIAA 27th Aerospace Sciences Meeting, Jan. 9-12, 1989/Reno, Nevada.*

[3] Y. P. RAIZER, Y. B. ZELDOVICH, *Physics of Shock Waves and High-Temperature Hydrodynamics Phenomena*, Academic Press, New York, 1965.

[4] C. PARK, "On the Convergence of Chemically Reacting Flows", *AIAA Paper 85-0247, AIAA 23rd Aerospace Sciences Meeting, Jan. 14-17, 1985/Reno, Nevada.*

[5] B. VAN LEER, "Flux-Vector Splitting for the Euler Equations", *Lecture Notes in Physics, Vol. 170, 507-512, 1982.*

[6] B. LARROUTUROU, L. FEZOUI, "On the Equation of Multi-Component Perfect or Real Gas Inviscid Flow", *Nonlinear Hyperbolic Problems, Carasso, Charrier, Hanouzet and Joly Eds., Lecture Notes in Mathematics, Springer-Verlag, Heidelberg, 1989*, to appear.

[7] B. STOUFFLET, "Implicit Finite-Element Methods for the Euler Equations", *Proc. of the INRIA Workshop on "Numerical Methods for Compressible Inviscid Fluids", Dec. 7-9, 1983, Rocquencourt (France), Numerical Methods for the Euler Equations of Fluid Dynamics, Angrand F. et al. Eds., SIAM (1985).*

[8] J-A. DESIDERI, A. DERVIEUX, "Compressible Flow Solvers Using Unstructured Grids", *Von Karman Institute for Fluid Dynamics, Lecture Series 1988-05, Computational Fluid Dynamics, March 7-11, 1988.*

[9] B. VAN LEER, "Towards the Ultimate Conservative Difference Scheme III.: Upstream-Centered Finite-Difference Schemes for Ideal Compressible Flow", *J. Comp. Phys.* **23**, *263-275 (1977).*

[10] J. L. MONTAGNE, H. C. YEE, M. VINOKUR, "Comparative Study of High-Resolution Shock-Capturing Schemes for Real Gases", *NASA TM 10004, 1987.*

[11] B. VAN LEER, "Computational Methods fo Ideal Compressible Flow", *Von Karman Institute for Fluid Dynamics, Lecture Series 1983-04, Computational Fluid Dynamics, March 7-11, 1983.*

[12] M.-C. CICCOLI, L. FEZOUI, J-A. DESIDERI, "Efficient Iterative Schemes for Hypersonic Non-Equilibrium Flow", *INRIA report, to appear.*

A STREAMWISE UPWIND ALGORITHM APPLIED TO VORTICAL FLOW OVER A DELTA WING

Peter M. Goorjian and Shigeru Obayashi
NASA Ames Research Center
Moffett Field, California 94035 USA

Summary

Improvements have been made to a streamwise upwind algorithm so that it can be used for calculating flows with vortices. A calculation is shown of flow over a delta wing at an angle of attack. The laminar, thin-layer, Navier-Stokes equations are used for the calculation. The results are compared with another upwind method, a central-differencing method, and experimental data. The present method shows improvements in accuracy and convergence properties.

Introduction

Upwind algorithms are important in computational fluid dynamics for calculations of flows containing shock waves [1]. Some of them are also able to accurately resolve shear layers [2,3]. However, most multidimensional upwind algorithms are first constructed in one dimension and then extended to multidimensions by applying the one-dimensional procedure to each coordinate direction. In comparison, the present method uses the local stream direction and flow velocity to construct the upwinding. Hence the switching of flux evaluations always takes place at sonic values, where the shock waves are located. Therefore this method follows the flow physics more closely and in that respect is analogous to the rotated differencing [4] algorithm developed for the full potential equation.

The present algorithm is an improvement to a streamwise upwind algorithm, which has been applied to steady and unsteady transonic flows over airfoils and wings [5,6,7]. In addition to using rotated differencing to implement upwinding in the streamwise direction, the switching of fluxes across sonic values is smooth and the entropy condition is automatically imposed in a manner similar to Godunov's method. Contact discontinuities are sharply captured [5] and boundary layer profiles are fuller [7] (more accurate) in comparison to a central differencing method for a case of separated flow over a wing.

In the results presented, comparisons are made with the upwind method of Roe [1]. In that method, an entropy correction is needed, which results in a convergence difficulty. The present method does not exhibit that problem. Other features [7] of the present algorithm are that pressure and velocity continuity are enforced in the crossflow direction, and also in the streamwise direction as the velocity approaches zero. These features are adequate for transonic flows without the presence of vortices. However for supersonic flows with vortices, two additional developments were found to be necessary [8]. First, the manner of using the stream direction had to be modified to capture oblique shocks sharply. Second, additional terms were needed to stabilize the calculations. The present algorithm is described in detail.

To demonstrate various capabilities of the present algorithm, the flow was calculated over a delta wing with a leading-edge sweep of 75° at a Mach number of 2.8 and at an angle of attack of 16°. First, a conical flow approximation was used to limit the calculations to two dimensions, so that the influence of grid refinement could be determined. Then full three-dimensional (3D) calculations were performed on a medium-density grid. These computed results were compared with those of other methods and with experimental data.

Governing Equations

The thin-layer Navier-Stokes equations can be written in conservation-law form in a body-conforming, curvilinear, coordinate system (ξ, η, ζ) as follows:

$$\partial_\tau \widehat{Q} + \partial_\xi \widehat{E} + \partial_\eta \widehat{F} + \partial_\zeta \widehat{G} = \frac{1}{Re} \partial_\zeta \widehat{G}^v \,, \tag{1}$$

where Re is the Reynolds number. The vector of conserved quantities $\widehat{Q}$ and the inviscid flux vector $\widehat{F}$ are

$$\widehat{Q} = \frac{1}{J} \begin{bmatrix} \rho \\ \rho u \\ \rho v \\ \rho w \\ e \end{bmatrix} , \qquad \widehat{F} = \frac{1}{J} \begin{bmatrix} \rho \widehat{U} \\ \rho u \widehat{U} + \eta_x p \\ \rho v \widehat{U} + \eta_y p \\ \rho w \widehat{U} + \eta_z p \\ \rho H \widehat{U} \end{bmatrix} ,$$

where J is the transformation Jacobian, ρ is the fluid density, e is total energy per unit volume, and H is the total enthalpy. The contravariant velocity component $\widehat{U}$ is defined as $\widehat{U} = \eta_x\, u + \eta_y\, v + \eta_z\, w$. For the ξ and ζ directions, $\widehat{E}$ and $\widehat{G}$ can be defined similarly. The viscous flux vector $\widehat{G}^v$ is given in reference [8]. The pressure p is related to the conservative flow variables $\widehat{Q}$ through the equation of state for a perfect gas:

$$p = (\gamma - 1)[e - \frac{\rho}{2}(u^2 + v^2 + w^2)] \,. \tag{2}$$

Also, c is the speed of sound, where $c^2 = \gamma p/\rho$. See reference [8] for the form of equation (1) when the conical-flow approximation is imposed.

Numerical Algorithm

The upwind algorithm is applied to the inviscid fluxes $\widehat{E}$, $\widehat{F}$, and $\widehat{G}$ in equation (1). (The viscous term in equation (1) is discretized by a standard procedure [8], which uses second-order, central-differencing.) The upwind algorithm is described by the following formula for the cell interface flux $\widehat{F}$ with a surface vector, $\mathbf{S} = (\eta_x, \eta_y, \eta_z)$,

$$\begin{aligned} \widehat{F}(Q_l, Q_r, \mathbf{S}_{j+\frac{1}{2}}) = \frac{1}{2}\frac{|\nabla\eta|}{J} \times \Big\{ & [\, F_l + F_r] + [\, F_l \text{sign}\ (U_l) + s_l\, \Delta^* F_l] \cos^2\theta_l \\ & -[\, F_r \text{sign}\ (U_r) + s_r\, \Delta^* F_r] \cos^2\theta_r - |A|\, \Delta Q\, \sin^2\theta \Big\} , \end{aligned} \tag{3}$$

where Q_l and Q_r are left and right states, respectively, and the metric terms η_x, η_y, and η_z are normalized by $|\nabla\eta|$ as $k_x = \eta_x/|\nabla\eta|$, $k_y = \eta_y/|\nabla\eta|$, and $k_z = \eta_z/|\nabla\eta|$. The contravariant velocity $\widehat{U}$ is also normalized by $|\nabla\eta|$ and used as $U = k_x u + k_y v + k_z w$. For the first-order-accurate computations, $l = j$ and $r = j+1$. For higher-order extensions, the MUSCL approach [3,9] is used. Sign(U) equals the sign of U and θ is the rotation angle which will be determined later. The symbol * indicates local sonic values [5].

$$\Delta^* F = \Delta^*(\rho q)\mathbf{e_s} = (\rho^* q^* - \rho q)\mathbf{e_s} , \tag{4}$$

$$(q^*)^2 = \left(\frac{2}{\gamma+1}\right)(c^2 + \frac{\gamma-1}{2}q^2) , \quad \rho^* = \rho\left(\frac{(q^*)^2}{c^2}\right)^{\frac{1}{\gamma-1}} , \tag{5}$$

where q is the velocity magnitude and $\mathbf{e_s} = (1, u, v, w, H)^T$ is the sum of the two acoustic wave eigenvectors. Note that equation (4) is based on the rotated difference formula [4,5] for the full potential equation. Equation (4) and the switches s_l and s_r, which will be specified, use the speed q and the Mach number q/c rather than the velocity component U and the Mach number component U/c that many other upwind methods use. With the use of this rotated differencing, the switching of terms at transonic shock waves occurs independently of their alignment to grid lines.

The last term in equation (3) is defined as follows:

$$\begin{aligned} |A|\Delta Q =& |U|\Delta Q + (c - |U|)[\frac{\Delta p}{c^2}\mathbf{e_s} + \rho\Delta U\,\mathbf{e_d}] \\ =& \frac{\Delta p}{c}\mathbf{e_s} + \rho c\Delta U\,\mathbf{e_d} + (\Delta\rho - \frac{\Delta p}{c^2})\;|U|\,\mathbf{e_e} + \rho\,|U|\,\mathbf{e_v} . \end{aligned} \tag{6}$$

The variables in equation (6) are averaged between the left and right states, except when they follow Δ. Then, for example, $\Delta Q = Q_r - Q_l$. Also, $\mathbf{e_d} = (0, k_x, k_y, k_z, U)^T$, which is the difference of the two acoustic wave eigenvectors, $\mathbf{e_e} = (1, u, v, w, q^2/2)^T$ and $\mathbf{e_v} = (0, e_{v2}, e_{v3}, e_{v4}, e_{v5})^T$. Here, $e_{v2} = \Delta u - k_x\,\Delta U$, $e_{v3} = \Delta v - k_y\,\Delta U$, $e_{v4} = \Delta w - k_z\,\Delta U$, and $e_{v5} = u e_{v2} + v e_{v3} + w e_{v4}$. In Cartesian coordinates, $\mathbf{e_e}$ is the entropy wave eigenvector and $\mathbf{e_v}$ is a linear combination of the vorticity-wave eigenvectors.

As originally developed [5,6], equation (3) did not use the terms in equation (6). Next the terms in equation (6) that use $\mathbf{e_s}$ and $\mathbf{e_d}$ were added [7] to enforce pressure continuity in the cross-flow direction and in the streamwise direction as the Mach number approaches zero. Finally, the terms using $\mathbf{e_e}$ and $\mathbf{e_v}$ were added [8] for flows containing vortices.

Following reference [4], the switches s_l and s_r are defined in the manner of Godunov's method as follows: for $U \geq 0$,

$$\begin{aligned} s_l = 1 - \epsilon_m\,\epsilon_l , \quad s_r = (1 - \epsilon_m)(1 - \epsilon_r) , \\ \epsilon_{l,m,r} = \frac{1}{2}[1 + \text{sign}(M^2_{l,m,r} - 1)] , \end{aligned} \tag{7}$$

and M_m denotes the Mach number of the averaged state.

Note that there is current research in improving the sonic point operator [10]. An alternative method to those in reference [10] is: for $U > 0$,

$$s_l = 1 - \epsilon_m(2\epsilon_l - 1) , \quad s_r = (1 - \epsilon_m)(1 - 2\epsilon_r) . \tag{8}$$

This smooth switch is identical to equation (7) except at sonic expansion points. At those points, one- and two dimensional calculations using equation (8) have shown increased accuracy over equation (7). Equation (8) was derived by modeling a transonic expansion wave for Burger's equation. When the sonic value occurs midway between mesh points, this modeling is exact.

Rotated Differencing

As originally developed, the rotation angle θ used the cosine of the velocity as $\cos\theta = U/q$ when the flow was supersonic. However, it is important to detect whether the velocity projected to the grid line is beyond the Mach cone. Thus, U/q is replaced by $M \cdot U/q = U/c$. If U/c becomes larger than one, $\cos\theta$ is set to one. This enhances the ability to capture oblique shock waves [11].

This feature leads to a favorable resolution of bow and crossflow shocks, but it allows the existence of crossflow expansion shocks. To avoid expansion shocks, the rotation angle is determined by a mixture of averaged (m) and pointwise (l, r) values:

$$\cos^2\theta_{l,r} = \min[\ (1-\phi)\frac{U_m^2}{c_m^2} + \phi\frac{U_{l,r}^2}{c_{l,r}^2},\ 1\ ] \tag{9}$$

The following is used for evaluating ϕ in this paper because of the smoothness:

$$\phi = \max\left(\left(\frac{2\gamma}{\gamma+1}\right)\left\{1 - \frac{1}{2\gamma}[(\gamma-1) + (\gamma+1)\frac{p_2}{p_1}]\right\},\ 0\right), \tag{10}$$

where p_1 and p_2 denote upstream and downstream pressures, respectively. The sine is determined by an arithmetic average of the cosines: $\sin^2\theta = 1 - \frac{1}{2}(\cos^2\theta_l + \cos^2\theta_r)$.

Results

The algorithm given by equation (3) has been tested for flow over a delta wing. Both 2D calculations, using the conical flow approximation [8], and 3D calculations have been made. The calculations are compared with Roe's method [1,2] and central differencing [8] (CD).

Computations are carried out in the following manner. The LU-ADI method [12], which can be modified for the conical flow fields [13], is used for testing the two upwind algorithms as well as the CD algorithm. Each of the three algorithms is implemented explicitly into the LU-ADI method, so that steady flows are determined by each of the three algorithms. Laminar flow is also assumed. For third-order accuracy, the MUSCL scheme with Koren's differentiable limiter [9] is used.

Delta-Wing, Conical-Flow Calculations

This test case considers a vortical flow field over a delta wing in order to examine the present formula's capability for computing shear flows. Computations are done for flow past a 75° delta wing at $M_\infty = 2.8$, $\alpha = 16°$, and $Re = 3.565 \times 10^6$, for which experimental data are available [14]. Figure 1 shows the model geometry, and the typical experimental flow field is shown schematically in figure 2. For the computations, the conical approximation is used. Three grids are used for a grid-refinement study. The coarse, medium, and fine grids all use 51 points normal to the body and 27, 51, and 99 points circumferentially, respectively.

Computations were done with the present method, Roe's method, and the CD method. Figure 3 shows a comparison of density contour plots of three numerical solutions on the fine grids. (The density values are on a portion of the sphere of radius equal to one.) Two shock waves can be observed: one is the bow shock wave on the windward side of the delta wing, and the other is the crossflow shock wave on the leeward side. The present method and Roe's method give similar contour plots for those shock waves, but the CD method gives smeared plots. (For the CD method, the smoothing coefficient κ was set to 0.1 in the fine-grid case instead of $\kappa = 0.05$ in the other cases because at convergence the solution had numerical oscillations with $\kappa = 0.05$ in the fine grid.) The low density regions at both the primary and secondary vortices also indicate that both upwind methods give similar solutions, but the CD method gives a smeared one.

A comparison of total pressure contour plots on the fine grids is shown in figure 4. The primary vortex appears similar in both upwind solutions, in respect to its location and the contour level, but not in the CD solution. The primary vortex appears off the boundary layer in both upwind solutions, but the primary vortex and the boundary layer touch each other in the CD solution. The shear-flow region separated from the leading edge also shows differences in the three solutions. The present formula gives a sharper solution than Roe's in this shear-flow region. Again, the CD formula gives the most dissipative solution.

Figure 5 shows comparisons of total pressure profiles normal to the leeward surface of the wing at $y = 0.216$ on the coarse, medium, and fine grids. See reference [8] for comparisons at $y = 0.1$, approximately on the primary vortex, and $y = 0.2$, approximately on the secondary vortex. Figure 5(c) shows the main features of the complicated profiles. The first peak near $z = 0$ indicates an edge of the boundary layer under the secondary vortex. The following local minimum corresponds to the secondary vortex. The next peak near $z = 0.025$ indicates the total pressure recovery between the secondary vortex and the shear layer separated from the leading edge. This shear layer from the leading edge is observed as the next local minimum. Finally, the flow recovers to the free stream. The second peak between the secondary vortex and the leading-edge shear layer appears to be higher with respect to both level and location for all three solutions as the grids are refined. The width between the peak and the region of the recovery to the free stream corresponds to the width of the separated shear layer. The present formula gives the narrowest shear layer in the three, even on the fine grid. This crispness indicates that the present formula computes the shear flow most accurately.

Comparisons of pressure coefficient distributions on the leeward wing surface on the coarse, medium, and fine grids are shown in figure 6. Experimental data [14] are also indicated in figure 6 by upper and lower triangles corresponding to data on the right- and left-hand side of the wing, respectively. Results obtained with the present formula are found to be slightly more accurate than those with Roe's method when compared with experimental data as well as with the fine-grid solution. The CD solution has a large discrepancy between the two upwind solutions on the coarse grid, but the discrepancy decreases as the grids are refined.

Finally, a comparison of convergence histories of calculations, using the present and Roe's methods, is shown in figure 7. The locally varying time stepping was used [12]. Because of the stiffness of conical source term, Δt_0 was set to 0.1 in the first 3000 iterations from the impulsive start. Then, $\Delta t_0 = 0.25$ for the next 1000 iterations, and finally Δt_0 was set to 0.5. The maximum CFL number reached about 20. The present formula shows better convergence in both the L_2 norm and the $J \cdot L_\infty$ norm (rescaled by the transformation Jacobian). On the other hand, Roe's formula reaches a limit cycle. The calculations by both upwind methods were started from uniform free-stream conditions. However, the calculations by the CD method were started from a converged solution by the present formula. The CD method needed a smaller Δt_0: that is, one fifth of the one used for the upwind computations. The present formula converged the best of the three for this shear-flow computation with respect to the convergence rate and the order of magnitude of convergence. The difference in CPU time between the present formula and Roe's formula is less than 1% in the present computations.

Delta-Wing, Three-Dimensional Calculations

A medium-density grid in the curvilinear coordinate system (ξ, η, ζ) was used for calculations for both the present method and Roe's method. There were $25 \times 51 \times 41$ points in the ξ (conical), η (circumferential), and ζ (normal) directions, respectively. The convergence properties were similar to those in the conical-flow calculations. Both methods produce similar results to those obtained in the conical-flow calculations, including the treatment of the bow shock wave and the vortices. Figure 8 shows velocity-magnitude plots at an axial location 90% of the distance from the nose to the trailing edge of the model. First-order accurate results are shown in figure 8(a). The present method (plotted on the right side) shows a slightly larger high-speed region than Roe's method (plotted on the left side). As shown by the third-order results in figure 8(b), this higher speed is more accurate. Hence, the present scheme is less dissipative than Roe's. In this 3D test computation, the CPU time per grid point per iteration is 36.6, 35.7, and 32.1 μsec for the present, Roe's, and CD computations, respectively, on a CRAY X-MP computer. Hence, in 3D, the present method takes about 2.5% more time than Roe's and 14% more time than CD. This confirms that the required arithmetic operations of the present formula are comparable to those of Roe and CD.

Conclusions

An improved streamwise upwind algorithm has been derived and applied to conical flow fields. In comparison with Roe's method, the present formula (1) captures oblique shock waves in the same manner, (2) requires arithmetic operations of the same order, (3) has better convergence properties, i.e., no limit cycle, and (4) has an advantage over Roe's in computing shear flows accurately. The results also indicate that the CD method is more dissipative in the resolution of shock waves and shear layers than upwind methods. In addition, the present method switches differencing at sonic values rather than at values that are dependent on the coordinate system, which is more in accord with the fluid physics.

References

[1] Roe, P. L.: "Characteristics-based schemes for the Euler equations," *Ann. Rev. Fluid Mech.*, 18, 1986, pp. 337-65.

[2] van Leer, B., Thomas, J. L., Roe, P. L., Newsome, R. W.: "A comparison of numerical flux formulas for the Euler and Navier-Stokes equations," AIAA Paper 87-1104-CP, Proc. AIAA 8th Comp. Fluid Dynamics Conf., Honolulu, June 1987.

[3] Vatsa, V. N., Thomas, J. L., Wedan, B. W.: "Navier-Stokes computations of prolate spheroids at angle of attack," AIAA Paper 87-2627-CP, Aug. 1987.

[4] Jameson, A.: "Transonic potential flow calculations in conservation form," Proc. AIAA 2nd Comp. Fluid Dynamics Conf., Hartford, Conn., June 1975, pp. 148-161.

[5] Goorjian, P. M.: "Algorithm development for the Euler equations with calculations of transonic flows," AIAA Paper 87-0536, Jan. 1987.

[6] Goorjian, P. M.: "A new algorithm for the Navier-Stokes equations applied to transonic flows over wing," AIAA Paper 87-1121-CP, Proc. AIAA 8th Comp. Fluid Dynamics Conf., Honolulu, June 1987.

[7] Goorjian, P. M.: "A streamwise upwind algorithm for the Euler and Navier-Stokes equations applied to transonic flows," *Numerical Methods for Fluid Dynamics III,* Inst. Mathematics & Appl. Conf. Series, New Series No. 17, Clarendon Press, Oxford, 1988.

[8] Obayashi, S., Goorjian, P. M.: "Improvements and applications of a streamwise upwind algorithm," AIAA Paper 89-1957-CP, Proc. AIAA 9th Comp. Fluid Dynamics Conf., pp. 292-302, June 1989.
[9] Obayashi, S., Gavali, S.: "Three-dimensional simulation of underexpanded plumes using upwind algorithms," *Supercomputing 88, II: Science and Applications,* pp. 25-33, IEEE Computer Society Press, Washington, D.C., 1989.
[10] van Leer, B., Lee, W.-T., Powell, K. G.: "Sonic-point capturing," AIAA Paper 89-1945-CP, Proc. AIAA 9th Comp. Fluid Dynamics Conf., pp. 176-187, June 1989.
[11] Albone, C. M., Hall, M. G.: "A scheme for the improved capture of shock waves in potential flow calculations," RAE TR 80128, Oct. 1980.
[12] Obayashi, S.: "Numerical simulation of underexpanded plumes using upwind algorithms," AIAA Paper 88-4360-CP, Aug. 1988.
[13] Fujii, K., Obayashi, S.: "Evaluation of Euler and Navier-Stokes solutions for leading-edge and shock-induced separations," AIAA Paper 85-1563, June 1985.
[14] Miller, D. S., Wood, R. M.: "Lee-side flow over delta wings at supersonic speeds," NASA TP-2430, June 1985.

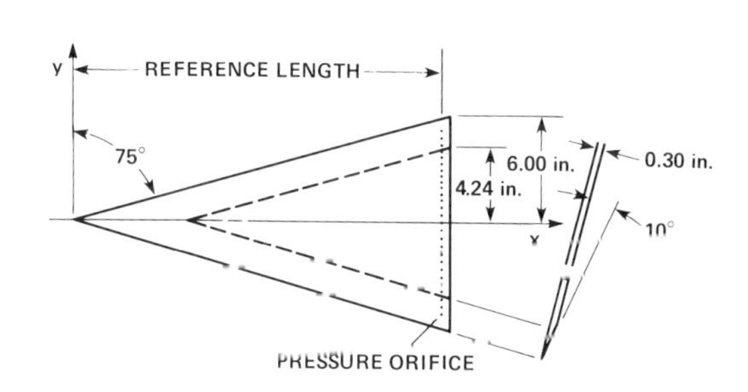

Fig. 1 Model geometry.

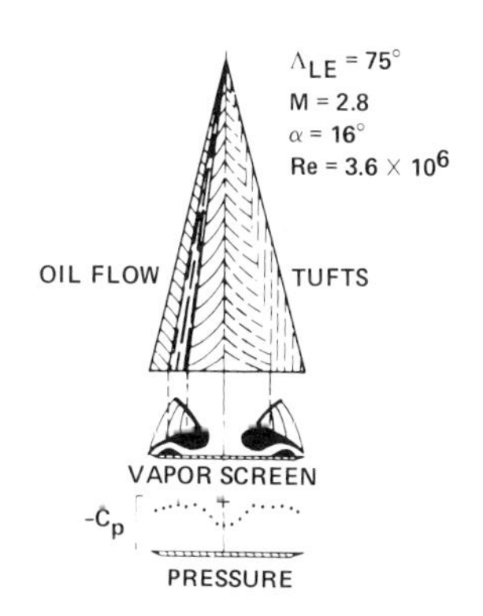

Fig. 2 Experimental flow field of vortex with shock wave (ref. 14).

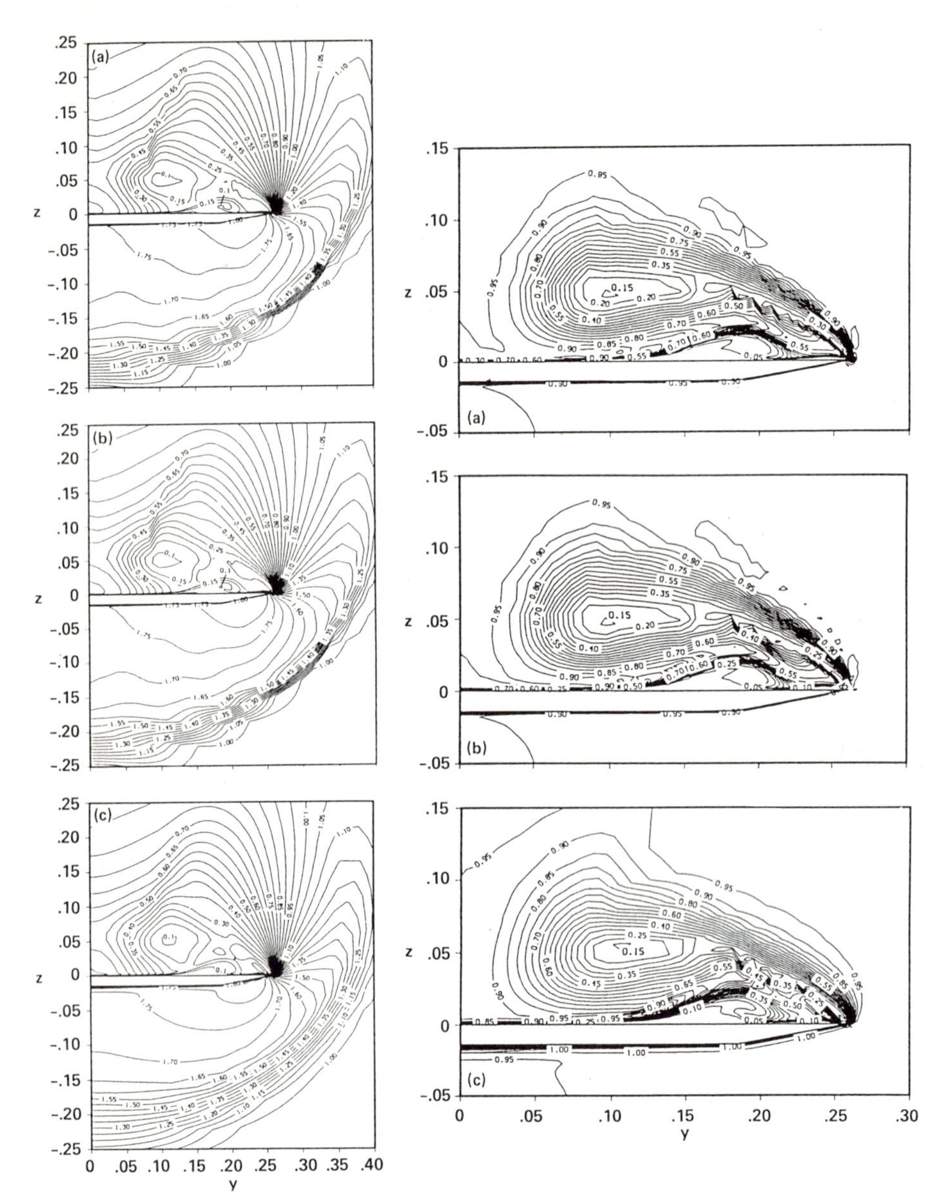

Fig. 3 Comparison of density contour plots on the fine grid; $M_\infty = 2.8$, $Re = 3.565 \times 10^6$, $\alpha = 16°$. a) the present formula; b) Roe's formula; c) the central-difference method.

Fig. 4 Comparison of total pressure contour plots on the fine grid; $M_\infty = 2.8$, $Re = 3.565 \times 10^6$, $\alpha = 16°$. a) the present formula; b) Roe's formula; c) the central-difference method.

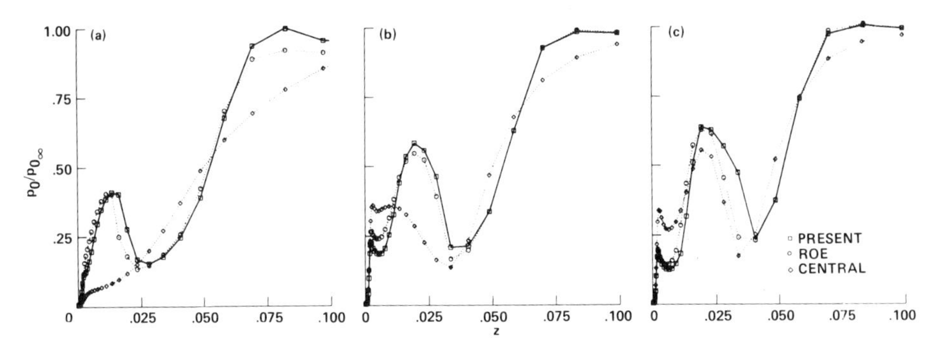

Fig. 5 Comparisons of total pressure profiles at $y = 0.216$. a) coarse grid; b) medium grid; c) fine grid.

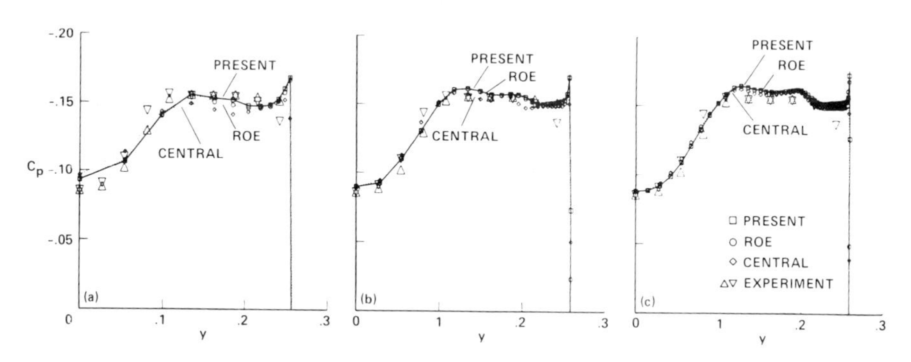

Fig. 6 Comparisons of C_p distributions on the leeward wing surface. a) coarse grid; b) medium grid; c) fine grid.

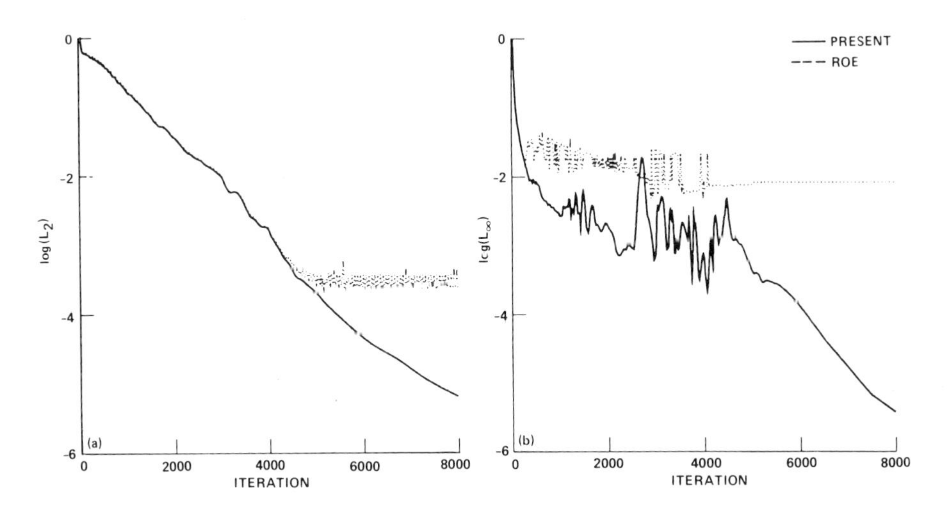

Fig. 7 Comparisons of convergence history on the medium grid. a) L_2 norm; b) $J \cdot L_\infty$ norm.

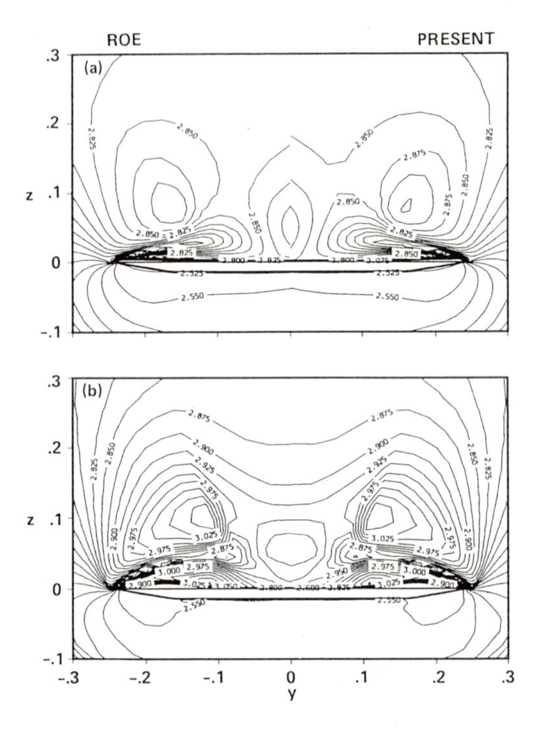

Fig. 8 Comparisons of velocity magnitude contour plots. a) first-order; b) third-order.

A Numerical Study of Hypersonic Stagnation Heat Transfer Predictions at a Coordinate Singularity

Francesco Grasso
Università di Roma "La Sapienza"
Dip. di Meccanica e Aeronautica
Roma, Italia 00184

Peter A. Gnoffo
NASA Langley Research Center
Hampton, Virginia 23665

SUMMARY

The problem of grid induced errors associated with a coordinate singularity on heating predictions in the stagnation region of a three-dimensional body in hypersonic flow is examined. The test problem is for Mach 10 flow over an Aeroassist Flight Experiment configuration. This configuration is composed of an elliptic nose, a raked elliptic cone, and a circular shoulder. Irregularities in the heating predictions in the vicinity of the coordinate singularity, located at the axis of the elliptic nose near the stagnation point, are examined with respect to grid refinement and grid restructuring. The algorithm is derived using a finite-volume formulation. An upwind-biased total-variation diminishing scheme is employed for the inviscid flux contribution, and central differences are used for the viscous terms.

INTRODUCTION

Numerical studies of the convective heating distributions in the stagnation region of blunt three-dimensional and axisymmetric bodies have revealed irregularities in the vicinity of the coordinate axis [1-4]. The heat transfer distributions in the three-dimensional test cases appear discontinuous across the axis in the plane of symmetry [1-2]. The magnitude of the discontinuity has been found to be a function of both grid topology and the location of the stagnation point relative to the axis.

The source of this problem has been identified as a truncation error which jumps from second- to first-order at the axis [3]. In Ref. [3] (and earlier in [4]) a fix for the problem was recommended which expands a derivative in the x–momentum equation. It essentially separated the Jacobian (which goes to zero at the axis) from the finite differences of the primitive variables in the offending term. This procedure successfully removed the irregularity in the heating distribution approaching the axis.

The application of this modification to a three-dimensional finite-volume algorithm is not straightforward. In a finite-volume formulation, derived from the integral form of the conservation laws, the algorithm must define the flux across a cell wall and the shear stress acting on a cell wall. Conservation is assured so long as the flux is uniquely defined as a function of properties at neighboring cells. Various geometrically weighted averages have been considered for defining the flux across cell walls which, in the spirit of [3], diminish the influence of cell volume (Jacobian) in the final finite-volume scheme. The volume weighting discussed in [2] and distance weighting discussed in [5] have improved predictions of heating and pressure as compared with experiment but have not eliminated the irregularities in surface heating near the axis. We do not wish to abandon the conservative finite-volume formulation in physical space in order to cure a problem localized at the axis. Consequently, we concentrate on various aspects of grid topology which may be modified to improve the heating distribution near the axis.

In studying this problem we focus our attention on the Mach 10 flow over a wind tunnel model of the Aeroassist Flight Experiment (AFE) vehicle. The AFE is a flight project designed to obtain critical surface and flowfield data in the Earth's atmosphere at conditions simulating a return from geosynchronous Earth orbit [6]. One important use of this data is for computational fluid dynamic (CFD) code validation for programs which will be used in the design of future aeroassisted space transfer vehicles. The AFE is a blunt body composed of an elliptic nose, a raked elliptic cone, and a circular shoulder. The grid topology in the stagnation region consists of an axis normal to the elliptic nose with pie shaped cells of varying aspect ratio surrounding the axis. This topology, and its effects on truncation error, contribute to the particularly challenging aspects of this problem for the purpose of predicting heating distributions near the axis.

The flowfield algorithm used in this study is the Langley Aerothermodynamic Upwind Relaxation Algorithm (LAURA) [1-2,5]. LAURA is a three-dimensional Navier-Stokes algorithm for the simulation of hypersonic flows about blunt bodies. Its capabilities include both perfect gas flows and flows in thermal and chemical nonequilibrium, though only perfect gas applications are considered here. For hypersonic flows over blunt bodies upwind-biased Total-Variation-Diminishing schemes (UTVD) seem to be superior to the symmetric Total-Variation-Diminishing schemes (STVD), at least as far as discontinuity resolution is concerned [7]. For this reason, to minimize this numerical dissipation intrinsic to the scheme without loss of convergence and stability, a UTVD limiter [8] has been implemented rather than the STVD limiter used in the original algorithm.

NUMERICAL ALGORITHM

The numerical algorithm has been described previously in [1-2,5]. The present algorithm differs from earlier implementations in the use of an upwind Total-Variation-Diminishing scheme as opposed to a Symmetric Total-Variation-Diminishing scheme. An overview of the algorithm and its derivation is presented below.

The integral form of the conservation laws applied to a single cell in the computational domain is written

$$\iiint \frac{\partial \underline{\mathbf{q}}}{\partial t} d\Omega + \iint \underline{\underline{\mathbf{f}}} \cdot \underline{\mathbf{n}} d\sigma = 0 \,. \tag{1}$$

In Equation (1) the first term describes the time rate of change of conserved quantity $\underline{\mathbf{q}}$ in the control volume, and the second term describes convective and dissipative flux through the cell walls.

For viscous flow of a perfect gas, the vectors $\underline{\mathbf{q}}$ and $\underline{\underline{\mathbf{f}}}$ are defined as

$$\underline{\mathbf{q}} = \begin{bmatrix} \rho \\ \rho u \\ \rho v \\ \rho w \\ \rho E \end{bmatrix} \tag{2}$$

$$\underline{\underline{\mathbf{f}}} = \begin{bmatrix} \rho u & \rho v & \rho w \\ \rho uu + p - \tau_{xx} & \rho vu - \tau_{yx} & \rho wu - \tau_{zx} \\ \rho uv - \tau_{xy} & \rho vv + p - \tau_{yy} & \rho wv - \tau_{zy} \\ \rho uw - \tau_{xz} & \rho vw - \tau_{yz} & \rho ww + p - \tau_{zz} \\ \rho uH - \dot{q}_x - \underline{\mathbf{u}} \cdot \underline{\tau}_x & \rho vH - \dot{q}_y - \underline{\mathbf{u}} \cdot \underline{\tau}_y & \rho wH - \dot{q}_z - \underline{\mathbf{u}} \cdot \underline{\tau}_z \end{bmatrix}. \quad (3)$$

In the present work a thin-layer approximation is made, and therefore only the viscous derivatives in the direction normal to the high gradient region are included.

The shorthand notation finite-volume approximation to Equation 1 for a rectangularly ordered, structured grid is

$$\left[\frac{\delta \underline{\mathbf{q}}\Omega}{\delta t}\right]_L + \sum_{l=i,j,k} \left[\underline{\underline{\mathbf{f}}}_{l+1} \cdot \underline{\mathbf{n}}_{l+1}\sigma_{l+1} - \underline{\underline{\mathbf{f}}}_l \cdot \underline{\mathbf{n}}_l \sigma_l\right] = 0 \quad (4)$$

where $\delta \underline{\mathbf{q}} = \underline{\mathbf{q}}^{n+1} - \underline{\mathbf{q}}, \delta t = t^{n+1} - t^n$, L identifies a computational cell having volume Ω, and subscript l indicates a quantity taken on cell face l with surface area σ_l. The quantity $\underline{\mathbf{n}}_l$ is a unit vector normal to cell face l in a direction facing away from the cell center. The dependent variable $\underline{\mathbf{q}}$ is defined at cell centers. The independent variable $\underline{\mathbf{r}} = x\underline{\mathbf{i}} + y\underline{\mathbf{j}} + z\underline{\mathbf{k}}$ is defined at cell corners.

In Equation (4) uppercase integer variables I, J, K and L denote computational coordinates at cell centers, and lowercase integer variables i, j, k and l denote cell faces or cell corners.

Let

$$\underline{\underline{\mathbf{f}}}_l \cdot \underline{\mathbf{n}}_l = \underline{\mathbf{g}}_l + \underline{\mathbf{h}}_l$$

where $\underline{\mathbf{g}}_l$ defines the inviscid terms and $\underline{\mathbf{h}}_l$ defines the viscous terms.

The inviscid flux vector at cell face l is defined as

$$\underline{\mathbf{g}}_l = \left\{a_l' \underline{\mathbf{g}}_{L,l} + b_l' \underline{\mathbf{g}}_{L-1,l}\right\} - \left\{\frac{1}{2(\chi n)_l}\mathbf{R}_l|\mathbf{\Lambda}_l|\left[\underline{\mathbf{s}}_l - \underline{\mathbf{s}}_l^{min}\right]\right\} \quad (5)$$

where $\underline{\mathbf{g}}_{L,l} = [\underline{\underline{\mathbf{f}}}_L \cdot \underline{\mathbf{n}}_l]_{INVISCID}$.

The first term in braces is a second-order accurate volume-weighted interpolation formula for $\underline{\mathbf{g}}_l$, and the factors a_l' and b_l' are geometric weighting functions that account for the relative position of the cell wall with respect to the cell centers. The second term in braces provides the upwind-biased numerical dissipation.

The vector $\underline{\mathbf{s}}_l$ can be thought of as an approximation to the gradients of the characteristic variables in the direction normal to cell face l and is defined as

$$\underline{\mathbf{s}}_l = (\chi n)_l \mathbf{R}_l^{-1}\left(\underline{q}_L - \underline{q}_{L-1}\right). \quad (6)$$

The matrix $\mathbf{R}_l^{-1}$ in Equation (6) and the matrices $\mathbf{R}_l$ and $\mathbf{\Lambda}_l$ in Equation (5) are related to the Jacobian $\mathbf{A}$ of the inviscid flux vector $\underline{\mathbf{g}}$ with respect to $\underline{\mathbf{q}}$. The matrix $\mathbf{R}$ is the matrix of right eigenvectors of $\mathbf{A}$, and $\mathbf{\Lambda}$ is a diagonal matrix containing the eigenvalues of $\mathbf{A}$. The term $(\chi n)_l$ is an approximation to the inverse of the projected distance between cell centers L and $L-1$ on a direction normal to cell face l.

For the scheme to be second-order accurate upwind biased Total Variation Diminishing, the vector $\underline{\mathbf{s}}_l^{min}$ is defined according to

$$\underline{\mathbf{s}}_l^{min} = \frac{1}{2}\psi(a_l)\underline{\mathbf{d}}_l - \psi(a_l + \gamma_l)\underline{\mathbf{s}}_l \quad (7)$$

where $\underline{\mathbf{d}}_l, a_l$ and γ_l represent respectively the anti-diffusive flux and the characteristic speeds of the Jacobian of the inviscid flux vector $\underline{\mathbf{g}}_l$ and the anti-diffusive flux $\underline{\mathbf{d}}_l$. In the present work the use of different formulas for $\underline{\mathbf{d}}_l$ has been tested. In particular

$$\underline{\mathbf{d}}_l = minmod\left(\underline{\mathbf{s}}_{l+1}, \underline{\mathbf{s}}_l\right) + minmod\left(\underline{\mathbf{s}}_{l-1}, \underline{\mathbf{s}}_l\right) \quad (8)$$

where the *minmod* function returns the argument of smallest absolute magnitude when all the arguments are of the same sign or returns 0 if the arguments are of opposite sign;

$$\underline{d}_l = \frac{(\underline{s}_{l+1}\underline{s}_l + |\underline{s}_{l+1}\underline{s}_l|)}{(\underline{s}_{l+1} + \underline{s}_l)} + \frac{(\underline{s}_{l-1}\underline{s}_l + |\underline{s}_{l-1}\underline{s}_l|)}{(\underline{s}_{l-1} + \underline{s}_l)} \tag{9}$$

$$\begin{aligned} \underline{d}_l &= S^+ \max\left[0, \min(2|\underline{s}_{l+1}|, S^+\underline{s}_l), \min(|\underline{s}_{l+1}|, 2S^+\underline{s}_l)\right] \\ &+ S^- \max\left[0, \min(2|\underline{s}_l|, S^-\underline{s}_{l-1}), \min(|\underline{s}_l|, 2S^-\underline{s}_{l-1})\right] \end{aligned} \tag{10}$$

where $S^+ = sgn(\underline{s}_{l+1})$; $S^- = sgn(\underline{s}_l)$ [7].

We define limiter LI as the limiter function given by Equation (8) and LII as the one given by Equation (9) for the nonlinear field and Equation (10) for the linear field.

The numerical viscosity function ψ is defined as discussed in [2]:

$$\psi(z) = \begin{cases} |z| & \text{when } |z| \geq 2\epsilon \\ \dfrac{(z)^2}{4\epsilon} + \epsilon & \phantom{\text{when }} |z| < 2\epsilon \end{cases} .$$

In the present work, ϵ has been defined as

$$\epsilon = \epsilon_0(a_l + |U_l| + |V_l| + |W_l|)$$

where ϵ_0 has been set equal to a value of 0.2. This formula recognizes the fact that ϵ is a dimensional quantity, and for hypersonic flows, it has been found to give enough numerical dissipation for the solution to be stable.

The viscous terms (flux vector $\underline{h}$) are calculated to second-order accuracy in computational space using central differences [5].

A point-implicit relaxation strategy is used to integrate Equation (4). The essence of the strategy is to treat the variables at the cell center of interest implicitly at the advanced iteration level and to use the latest available data from neighbor cells in defining the "left-hand-side" numerics. The success of this approach is made possible by the robust stability characteristics of the underlying upwind difference scheme. The basic algorithm requires only a single level of storage, and numerical experiments show excellent stability characteristics.

The final relaxation equation is defined as

$$\left\{\mathbf{I} + \frac{\delta t}{\Omega_L}\mathbf{M}_L\right\}\delta\underline{q}_L = -\frac{\delta t}{\Omega_L}\sum_{l=i,j,k}\left[\left(\underline{g}_{l+1} + \underline{h}_{l+1}\right)\sigma_{l+1} - \left(\underline{g}_l + \underline{h}_l\right)\sigma_l\right] \tag{11}$$

where $\mathbf{M}_L$ contains the coefficients of the variables that are treated implicitly. The relaxed value of $\underline{q}_L^{n+1}$ is then determined by

$$\underline{q}_L^{n+1} = \underline{q}_L^n - \left[\mathbf{I} + \frac{\delta t}{\Omega_L}\mathbf{M}_L\right]^{-1}\frac{\delta t}{\Omega_L}\left[\sum_{l=i,j,k}\left(\underline{f}_{l+1}\cdot\underline{n}_{l+1}\sigma_{l+1} - \underline{f}_l\cdot\underline{n}_l\sigma_l\right)\right]. \tag{12}$$

Equation (12) involves the inversion of a single 5 by 5 matrix for three-dimensional flow of a perfect gas, and Gauss elimination can be used to solve for $\delta\underline{q}_L$ directly from Equation (11). On computers with large memories, it is more efficient to invert and save the leading matrix in Equation (12) for 10 to 20 relaxation steps before reevaluation.

RESULTS AND DISCUSSION

The flows about a wind tunnel model of the Aeroassist Flight Experiment vehicle at $M_\infty = 10$; $Re = 43,000$ per inch; and three different angles of attack $\alpha = -5°, 0°$, and $5°$ have been computed. Predictions were originally compared with experimental data [9] in [2] and are repeated here (Figure 1) in order to illustrate the numerical problems encountered in the stagnation region near the coordinate singularity. The flux limiter used in all cases is given by Equation (8) unless otherwise specified. The measured heating rate distributions in the stagnation region have been used to assess the effects of grid topology on the quality of the computed solution. Examination of Figure 1 shows a progressively deteriorating heat transfer solution in going from 5° to −5° angle of attack, in terms of the distribution approaching the axis ($s = 0$) from the short side ($s < 0$) and from the long side ($s > 0$) of the vehicle. At −5° angle of attack, the stagnation streamline crosses the axis from the long to short side of the vehicle. In general, the predicted heating levels fall below the experimental data, except for the overshoot in the distribution at −5° angle of attack coming from the long side of the vehicle. The cell Reynolds number for all three cases is approximately equal to 8.0. In contrast, the pressure distributions in the stagnation region for all three angles of attack compare well with the experimental data and show no signs of adverse effects due to the coordinate singularity. The greater sensitivity of the heating distributions on the coordinate singularity is to be expected because of their functional dependence on gradients which are more directly affected by truncation error.

The −5° angle of attack case is chosen to study the effects of grid structure and algorithm modifications on the heating distributions in the stagnation region. This case was chosen because it appeared to be the most challenging with regard to the predictive capabilities of the original algorithm and grid. Several numerical tests were performed and are identified as follows. Case A corresponds to a grid of 39 cells from the axis to the shoulder, 24 cells around the axis, and 64 cells from the body across the shock layer and into the free stream. Case B corresponds to Case A except that a grid restructuring was performed in the stagnation region to remove the axis singularity. Case C employs 128 cells across the shock layer. The grid for case A was derived from this grid shown in Figure 2 by deleting every other mesh point in the direction normal to the body. In like manner, Case D is the fine grid equivalent of Case B, employing 128 cells across the shock layer, as shown in Figure 3.

The surface grid for Cases A and C which involve the axis singularity include cells of varying aspect ratio surrounding the axis. This variation is an artifact of the algebraic grid generation scheme. In order to obtain an approximately equal distribution of cells around the circular shoulder (Figure 2), it was necessary to use small angles at the axis on the long side and large angles on the short side (i.e., $rd\theta = ds$ where $ds \approx$ constant). The restructured grid distributes the points which used to terminate at the axis equidistantly across the plane of symmetry. Even though the axis singularity is removed, a singular jump in the behavior of the metrics is introduced because of the abrupt change in cell definition. The metric coefficients in a finite-volume formulation are equivalent to the ratio of cell wall areas to cell volumes [5]. Consequently, the algorithm, which is formally second-order accurate in computational space, may be only first-order accurate in physical space if the metric coefficients cannot be assumed to vary continuously from cell to cell. The restructuring near the axis is still an interesting option because it is simple to implement and because it opens communication with reflected cells on the opposite side of the symmetry plane. In Cases A and C, for example, the axis singularity represents a wall of zero area, and communication with cells on the opposite side of the singularity is severely restricted, occurring only through the limiter functions in Equation (8) for the cell wall corresponding to $i = 2$.

The predictions for Cases A-D, shown originally in Figure 1c, are presented in Figure 4 using a magnified scale and focusing on the stagnation region. In all cases, the predictions for pressure coefficient are nearly identical and show no signs of adverse affects from the coordinate singularity in Figure 1c. The restructured grid of Case B smoothes out the heating distribution near the axis as compared with Case A. However, the heating level is still almost 20% low. The fine grid results of Case C, which includes the axis singularity, shows a much improved heat transfer distribution in the stagnation region and on the long side of the cone. The maximum difference with respect to the experimental data for heating is 7%. The restructured fine grid results of Case D show no significant improvement over Case C.

The heat transfer distribution is plotted as a function of cell index j around the singularity in Figure 5. The solutions are smoothly varying in all cases. The fine grid results show the minimum variation going around the axis.

The sensitivity of the heat transfer predictions on grid is indicative of significant truncation error effects near the axis singularity. The magnitude of these effects for the $\alpha = -5°$ case is somewhat surprising. Even though the cell Reynolds number ($N_c = \frac{\rho a \Delta z}{\mu}$ is approximately equal to 8 in cases A and B, which is somewhat coarse, the abrupt change in the heating distribution in the vicinity of the axis is larger than expected based on stagnation heating results at other angles of attack (Figures 1a and 1b) and based on heat transfer predictions on the cone flank. The grid restructuring smooths out the irregularity by enhancing communication with cells on the opposite side of the singularity. Still, this is not a satisfactory resolution of the problem, as the computed heating level underpredicts the experimental data by nearly 20%. The improved fine grid results of Case C, which include the axis singularity, demonstrate that the truncation error associated with the normal grid distribution is more important than the truncation error associated with the lateral and circumferential grid distribution. The combination of a somewhat coarse grid in the normal direction and the axis singularity, with varying aspect ratio cells in the circumferential direction, appears to create a nonlinear coupling of truncation error which results in heat transfer predictions that are worse than those obtained away from the singularity. The location of the stagnation streamline relative to the singularity, which determines whether or not the "upwind" direction crosses the axis, also influences the total truncation error contribution.

Algorithm modifications were tested in addition to the grid restructuring and refinement discussed above. These modifications included testing of several different geometrically weighted averages and limiters used in the definition of flux across cell walls. The intention of these modifications was to diminish the influence of cell volume (Jacobian) in the final finite-volume scheme or to diminish the overall truncation error of the scheme. There was some concern, for example, that the problems in the stagnation region may have been compounded by the predominance of near zero eigenvalues and the numerical viscosity function ψ in this region. In all test cases the irregularities across the axis persisted, indicating that the heating distribution problem is predominantly a function of grid-related truncation error. For example, the effect of flux limiters LI (Equation 8) and LII (Equation 9,10) on the heating distribution across the axis is presented in Figure 6 using the restructured grid from Case B.

CONCLUDING REMARKS

The problem of grid-induced errors associated with a coordinate singularity on heating predictions in the stagnation region of a three-dimensional body in hypersonic flow is examined. The test problem is for Mach 10 flow over an Aeroassist Flight Experiment (AFE) configuration. This configuration is composed of an elliptic nose, a raked elliptic cone, and a circular shoulder. The numerical scheme is derived using a finite-volume formulation. An upwind-biased total-variation diminishing scheme is employed for the inviscid flux contribution, and central differences are used for the viscous terms. Irregularities in the heating predictions in the vicinity of the coordinate singularity, located at the axis of the elliptic nose near the stagnation point, are examined with respect to grid refinement and grid restructuring. Attention is focussed on an $\alpha = -5°$ test case which exhibits the greatest sensitivity to grid-induced errors as demonstrated in preliminary numerical tests.

A restructured grid is tested which distributes the points which used to terminate at the axis equidistantly across the plane of symmetry. Even though the axis singularity is removed, a singular jump in the behavior of the metrics is introduced because of the abrupt change in cell definition. Consequently, the algorithm, which is formally second-order accurate in computational space, may be only first-order accurate in physical space if the metric coefficients cannot be assumed to vary continuously from cell to cell. The restructuring near the axis is still an interesting option because it is simple to implement and because it opens communication with reflected cells on the opposite side of the symmetry plane.

Results of the numerical tests show that grid restructuring and grid refinement significantly improve but do not eliminate irregularities in the heating predictions associated with the axis singularity. Grid refinement in the direction normal to the wall is more important than grid restructuring in reducing overall truncation error in the test cases. There appears to be a non-linear coupling of grid-induced truncation errors which exacerbates the heating distributions near the singularity as compared to the quality of the solution away from the singularity. The effects of dissipative properties of different limiters are small compared with the grid-induced truncation errors on the coarsest grid tested with a cell Reynolds number approximately equal to 8.0.

REFERENCES

1. GNOFFO, P.A.: "Upwind-Biased, Point-Implicit Relaxation Strategies For Hypersonic Flowfield Simulations on Supercomputers" 4th International Conference on Supercomputing, April 30 - May 5, 1989, Santa Clara, CA.
2. GNOFFO, P.A., McCANDLESS, R.S., and YEE, H.C.: "Enhancements to Program LAURA for Computation of Three-Dimensional Hypersonic Flow," AIAA Paper 87-0280 (1987).
3. BLOTTNER, F.G.: "Verification of a Navier-Stokes Code for Solving the Hypersonic Blunt Body Problem". 4th Symposium on Numerical and Physical Aspects of Aerodynamic Flows, January 16-19, 1989, California State University, Long Beach, CA.
4. VICTORIA, K.J., WIDHOPF, G.F.: "Numerical Solution of the Unsteady Navier-Stokes Equations in Curvilinear Coordinates: The Hypersonic BluntBody Merged Layer Problem," 3rd International Conference on Numerical Methods in Fluid Dynamics, July 3-7, Paris, France, 1972.
5. GNOFFO, P.A.: "An Upwind-Biased, Point-Implicit Relaxation Algorithm for Viscous, Compressible Perfect-Gas Flows," NASA TP in preparation.
6. JONES, J.J.: "The Rationale for an Aeroassist Flight Experiment", AIAA Paper 87-1508 (1987).
7. YEE, H.C.: "Upwind and Symmetric Shock-Capturing Schemes," NASA TM-89464 (1987).
8. BASSI, F., GRASSO, F., and SAVINI, M.: "Finite Volume TVD Runge Kutta Scheme for Navier Stokes Computations", 11th International Conference on Numerical Methods in Fluid Dynamics, The College of William and Mary, Williamsburg, Virginia, June 27 - July 1, 1988.
9. MICOL, J.R.: AIAA Paper in preparation.

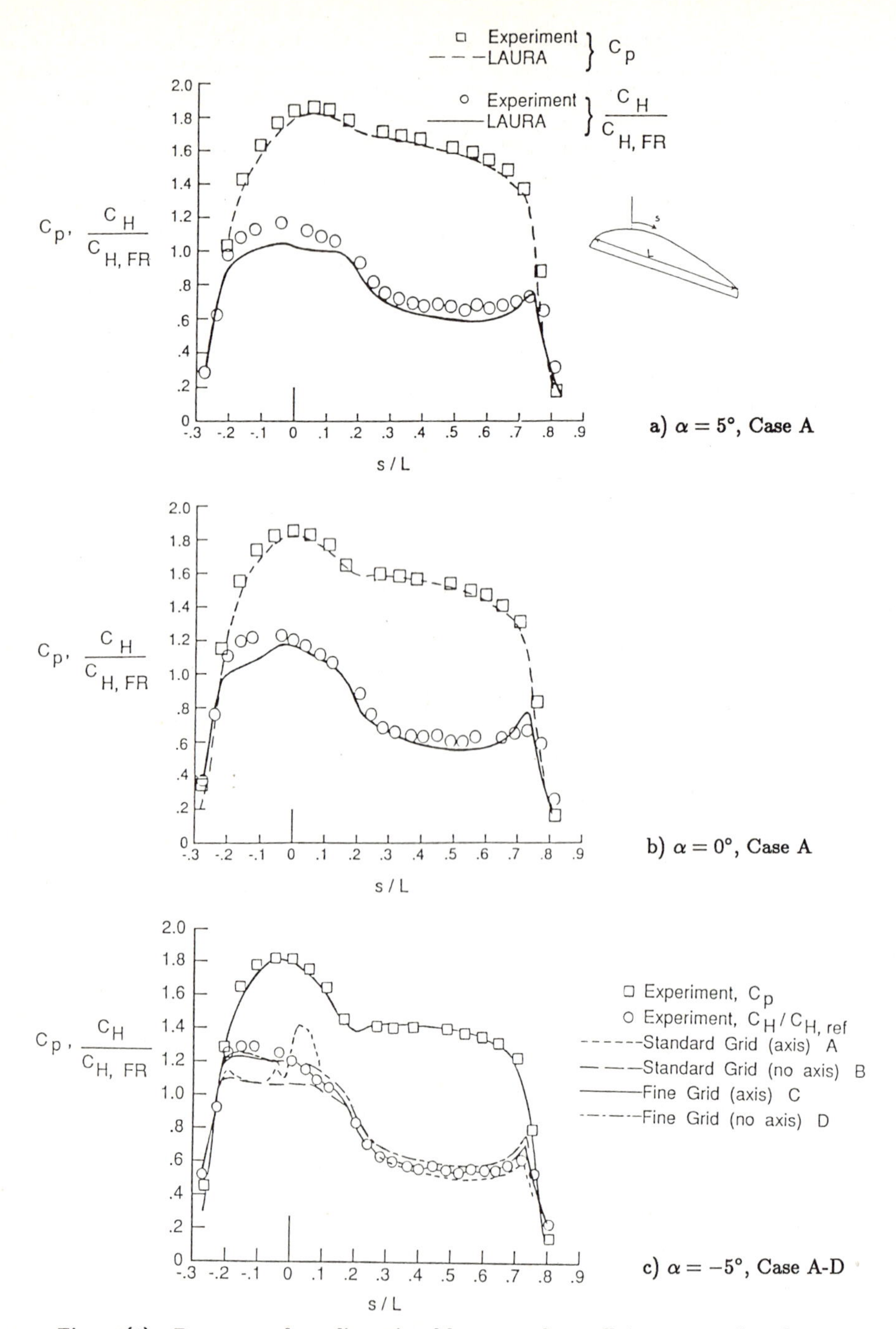

Figure (1) **Pressure and nondimensional heat transfer coefficients vs arc length along the plane of symmetry of the Aeroassist Flight Experiment (AFE) wind tunnel model.**

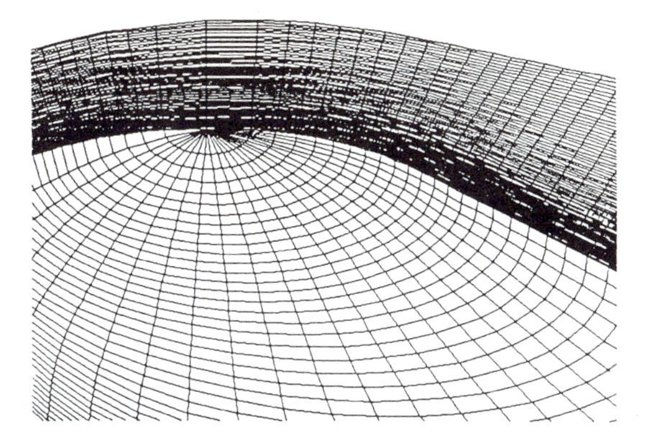

a) Detail of grid in vicinity of axis

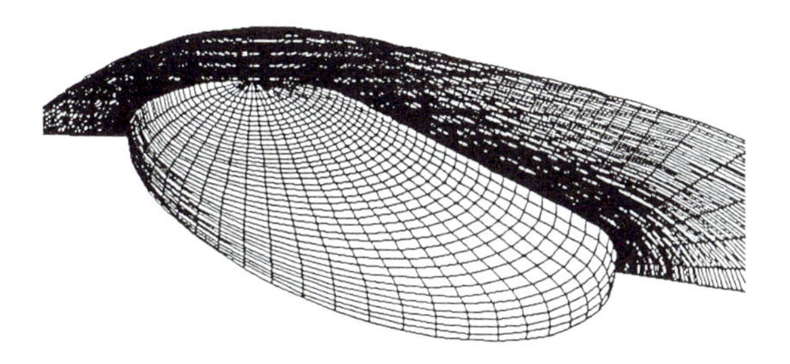

b) Overview of grid on surface and plane of symmetry

Figure (2) AFE grid with axis.

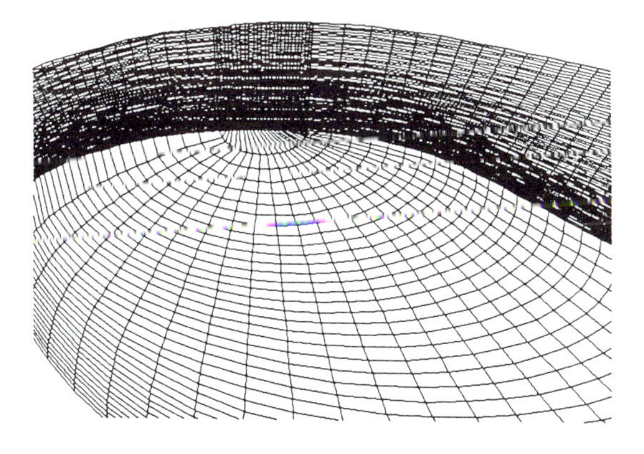

Figure (3) Detail of restructured AFE grid.

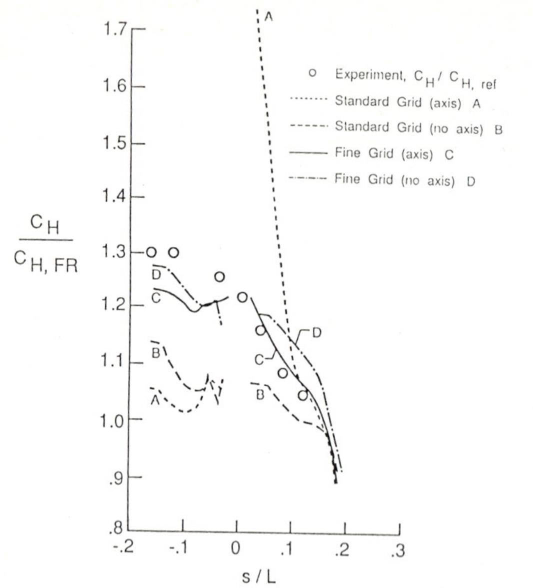

Figure (4) **Heat transfer coefficient distribution vs arc length along plane of symmetry in the stagnation region for Cases A-D**

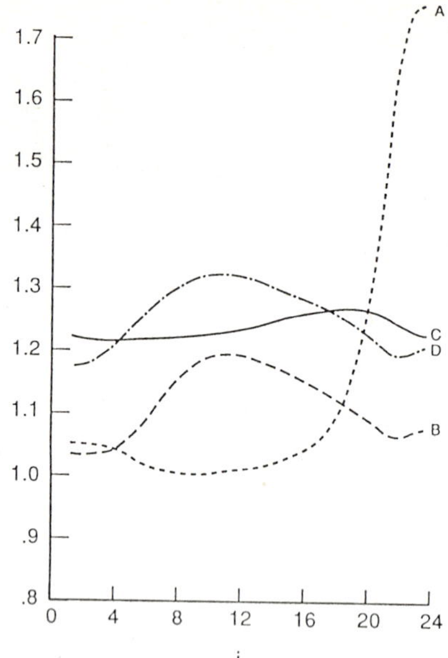

Figure (5) **Heat transfer coefficient distribution vs cell index surrounding the axis for Cases A-D.**

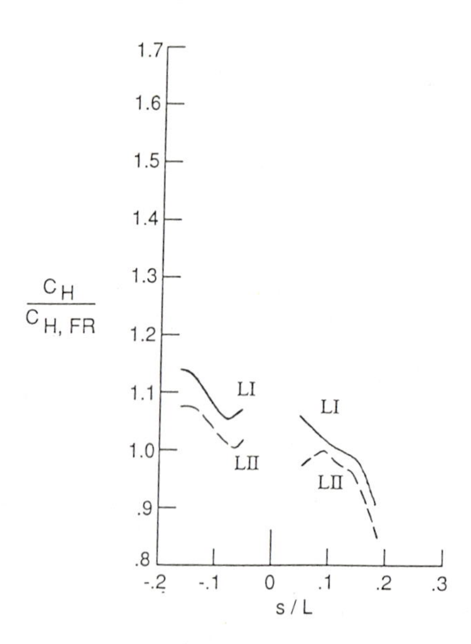

Figure (6) **Heat transfer coefficient distribution along plane of symmetry in the stagnation region for Case B using flux limiters LI and LII.**

VISCOUS, HYPERSONIC FLOWS OVER COMPRESSION RAMPS

Werner Haase
Dornier Luftfahrt GmbH
Postfach 1303, D-7990 Friedrichshafen 1, FRG

SUMMARY

Steady, hypersonic viscous flows over two-dimensional compression ramps are computed. In order to gain insight into the influence of different Mach numbers, Reynolds numbers and ramp angles on the flow structure, results are presented for low and high Mach numbers, for low and high Reynolds numbers as well as for different ramp angles. Furthermore, with respect to high Mach number flows, the influence of real gas effects is investigated. Computed surface pressure, skin friction and heat transfer distributions are compared with available measurements.

INTRODUCTION

In recent years, flows over compression ramps - as illustrated in Fig. 1 - have been investigated as testcases for new or improved numerical methods, e.g. in [1, 2, 3]. This flow problem has received considerable attention because of its importance for the design of control flaps on reentry vehicles and for accurate predictions of pressure and heat loads.

The pressure rise caused by the ramp effects significantly the upstream boundary layer development and results in a complex viscous-inviscid strong interaction situation. If the momentum of the flat plate boundary layer is not efficient to overcome the adverse pressure gradient, separation will occur depending on the combination of Mach number, Reynolds number and ramp angle. Flow separation however, will drastically reduce the flap effectiveness, but on the other hand, aerodynamic surface heating might become less severe [4]. Reattachment will then take place on the ramp due to the interaction of the "free shear" layer with the outer flow. Downstream of reattachment the pressure distribution is still increasing until the flow passes the "neck region" where the boundary layer thickness has its minimum, accompanied by a maximum in skin friction.

All calculations have been run under the assumption that the flow remains laminar - even in the region of strong shock wave boundary layer interaction - as it was reported in the available measurements [5].

NUMERICAL APPROACH

The Navier-Stokes equations describing two-dimensional, unsteady and compressible flows in conservation form are solved by means of a finite volume approach [6] using a Runge-Kutta type time stepping method with multigrid acceleration.
For ideal gas flow the perfect gas equation is used to define the mean static pressure via the internal energy. Real gas effects are included by using density and internal energy as input parameters for a simplified curve fit approach for the thermodynamic properties of equilibrium air [7].

Finite Volume Method

The system of ordinary differential equations in time is solved by a 5-stage Runge-Kutta type time-stepping method:

$$
\begin{aligned}
u^{(0)} &= u^{(n)} \\
u^{(\nu)} &= u^{(0)} - \alpha_\nu P u^{(\nu-1)} \qquad , \qquad \nu = 1,5 \\
u^{(n+1)} &= u^{(5)}
\end{aligned}
\qquad (1)
$$

with the coefficients

$$
\alpha_1 = \frac{1}{4}, \quad \alpha_2 = \frac{1}{6}, \quad \alpha_3 = \frac{3}{8}, \quad \alpha_4 = \frac{1}{2}, \quad \alpha_5 = 1 \qquad (2)
$$

where n denotes the previous time-level and P represents a spatial (central and therefore second order) difference operator. This scheme is stable up to a Courant number of 4.0.

Filtering Technique

To prevent an odd-even decoupling, blended second and fourth order artificial dissipation is used [8]. If the filtering technique is applied only twice, stability analysis indicates the best damping property as well as the largest extension of the stability region to the left of the real axis giving latitude in the introduction of dissipative terms. In practice, the fourth order filter is active throughout the computational domain except in areas with larger pressure gradients where the second order filter takes over.
In order to minimize the numerical dissipation especially in areas with large (geometrical) aspect ratios, filtering has been used taking the eigenvalues λ_i of the x,i- and λ_j of the y,j-direction independently instead of using the sum of these eigenvalues.
However, this approach failed for high hypersonic Mach numbers and/or large ramp angles. Hence, the eigenvalues used for calculating the artificial dissipation have been modified in the following way, leading to stable solutions in all investigated cases:

$$
\begin{aligned}
\lambda_{i,new} &= \lambda_i \times \max(1., \sqrt{Ma}\,) \\
\lambda_{j,new} &= f\,\lambda_{i,new} + (1. - f)\,\lambda_j \ .
\end{aligned}
\qquad (3)
$$

Although f may be a variable factor, it has been chosen to 0.75 for all calculations.

Convergence Acceleration

Introducing the *residual averaging approach*, i.e. collecting the information from residuals implicitly, permits stable calculations beyond the ordinary Courant number limit of the explicit scheme. In the present work however, the Courant number has been chosen to 0.9 for (nearly) all calculations, hence, switching off the residual averaging.
Furthermore, since only the steady state is of interest, a *variable timestep* approach has been used accelerating convergence drastically.
Apart from the 45°-ramp calculations, a *multigrid approach* has been used for all calculations.

Boundary Conditions

The following boundary conditions are valid for all calculations:
At the *solid wall boundary* no-slip conditions are implemented and the flow is assumed to be isothermal ($T_w = 297K$) and with zero normal pressure gradient.
A *symmetry boundary* is used in front of the flat plate for zero incidence.
All properties are fixed at the *farfield boundary*, $T_\infty = 72.2K$.
At the *outflow boundary* (boundary normal to the ramp) linear extrapolation is used for density, mass fluxes and total energy. No distinction has been made between sub- and supersonic outflow areas.

Steady State

The steady state is defined to be reached if an error norm reduction (L2-norm) of approximately 3.5 decades is reached.

Mesh Generation

All meshes are constructed in a way that the resultant shock never hits the farfield boundary, i.e. the domain boundary being "parallel" to the surface. A H-type mesh is chosen for discretization of the computational domains. In all calculations 147 mesh points in streamwise (17 in front of the flat plate, 65 on the flat plate, 65 on the ramp) and 66 mesh points in the wall normal direction are used. The mesh spacing normal to the surface is geometrically stretched with a first meshsize of $y/l = 0.0001$ (l = 0.439m = distance from leading edge of the flat plate to the beginning of the ramp). For the 45°-ramp, a hyperbolic mesh generator [9] is applied in order to avoid a clustering of the mesh near the farfield boundary in the corner region.

RESULTS AND DISCUSSION

Mach Number Influence

For a fixed Reynolds number of $Re_\infty = 2.37 \times 10^5$ and a fixed ramp angle of 15°, Fig. 2 presents the Mach number influence on the pressure coefficient distribution; measurements are only available for $Ma_\infty = 14.1$.
The corresponding skin friction distributions are shown in Fig. 3. For $Ma_\infty = 14.1$ the flow is still unseparated, lower Mach numbers however lead to separation. Moreover, for $Ma_\infty = 2.0$ a second - counterrotating - vortex appears.
The heat transfer distributions, plotted in Fig. 4, show a definitive minimum in the separated region as it was reported in [4].
An examination of the different - Mach number dependent - flowfield structures can be taken from Fig. 5. With increasing Mach numbers an increase in strength of the leading edge shock can be easily detected as well as the formation of the slip line caused by the interaction of the leading edge shock with the induced shock. For Ma = 2 a λ-shaped shock is formed in the corner region due to separation and the corresponding thickening of the viscous layer on the flat plate.

Reynolds Number Influence

The following sequence of Figs. 6, 7, 8 and 9 may now be used to gain a deeper insight into the flow behavior based on Reynolds number variations for the 15°-ramp and a fixed Mach number of $Ma_\infty = 14.1$. Again, results are compared with calculations for $Re_\infty = 2.37 \times 10^5$ and the corresponding measurement. The pressure coefficient distribution is given in Fig. 7, the skin friction in Fig. 8 and the corresponding heat transfer distribution can be taken from Fig. 9.
The influence of the Reynolds number on the global flow field is shown in Fig. 6, indicating a separated area for the high Reynolds number case and an increasing thickness of the viscous (corner region) layer with decreasing Reynolds numbers. The heat transfer distribution is directly proportional to the skin friction, i.e. the heat transfer reaches again a minimum in the region where the flow is separated.

Ramp Angle Influence

With the next sequence of figures computational results for different ramp angles are presented. Figs. 10, 11 and 12 show the pressure coefficient, skin friction and heat transfer distributions, respectively. Note, that for the 45°-ramp, Fig. 13 indicates an attachment shock being almost normal to the undisturbed flow and, therefore, causing an imbedded *subsonic* flow region stretching parallel to the ramp surface. A constant Mach number spacing of $\Delta Ma = 0.1$ for subsonic and $\Delta Ma = 0.5$ for supersonic areas is used in Fig. 13.
In order to demonstrate the complex flow structures for the different ramp angles, streamline plots are presented in Fig. 14. For the 45°-ramp, the separation region contains two counter rotating vortices, a very small vortex can be found directly in the corner region. For ramp angles of 24° and 45°, an unphysical "bump" in the streamlines can be recognized. This bump however, is not related to the calculated flowfield but to the calculation of the stream function itself, i.e. to errors in the integration of the total differential. If particle tracing is applied to the calculated flowfield, the streamlines remain straight in the region where the flow is under free stream conditions.

Incidence Correction and Real Gas Effects

Incidence Correction
The almost constant deviation between the calculation for the 15°-ramp with $Ma_{\infty} = 14.1$ and $Re_{\infty} = 2.37 \times 10^5$ and the corresponding measurement, as it can be derived from Figs. 2-4 may be due to the fact that - in the measurement - the flow approaches the model at a small non-zero angle of attack. This assumption has been numerically investigated by [3]. In Figs. 15, 16 and 17 results - represented by broken lines - with respect to an incidence correction of 0.86° are given for pressure, skin friction and heat transfer. It is remarkable, that the pressure coefficient distribution is now in an excellent agreement with the measurement but only marginal changes in skin friction and heat transfer are visible.

Real Gas Effects
A recalculation of the 15°-ramp case with real gas effects but without an incidence correction should now leave the pressure distribution more or less unchanged, but should lead to a more accurate prediction of the heat transfer. The computed results taking real gas effects into account are given in Figs. 15-17 indicated by the thicker solid lines. Although the skin friction distribution on the flat plate is not completely aligned with the measurements, the heat transfer distribution is now in an excellent agreement with the corresponding measurement.
The global changes for the 15°-ramp in switching from ideal to real gas assumptions are not dramatic due to a relatively low temperature maximum of roughly 1150K. For the 45°-ramp however, a more severe change can be assumed especially with respect to the region where the shock is almost normal to the free stream. The Mach number contours and the corresponding streamlines for ideal and real gas flow can now be taken from Figs. 18 and 19, respectively. Combined with a reduction in size of the recirculated region, the "normal shock" situation disappeared completely for the real gas flow, thus, reducing the maximum temperature from 2700K (ideal) to 2000K (real) and decreasing the heat transfer in this area although the separation zone is reduced.

REFERENCES

[1] Hung, C.M., MacCormack, R.W.: "Numerical Solutions of Supersonic and Hypersonic Laminar Flows over a Two-Dimensional Compression Corner", AIAA Journal, Vol. 14, April 1976.

[2] Stokesberry, D.C., Tannehill, J.C.: "Computation of Separated Flow on a Ramp Using the Space Marching Conservative Supra-Characteristics Method", AIAA Paper 86-0564, Jan. 1986.

[3] Lawrence, S.L., Tannehill, J.C., Chaussee, D.S.: "An Upwind Algorithm for the Parabolized Navier-Stokes Equations", AIAA Paper 86-1117, May 1986.

[4] Ginoux, J.J., "On Some Properties of Reattaching Laminar and Transitional High Speed Flows", von Karman Institute, Technical Note 53, September 1969.

[5] Holden, M.S., Moselle, J.R.: "Theoretical and Experimental Studies of the Shock Wave-Boundary Layer Interaction on Compression Surfaces in Hypersonic Flows", CALSPAN Report AF-2410-A-1, October 1969.

[6] Haase, W., Wagner, B., Jameson, A.: "Development of a Navier-Stokes Method Based on a Finite Volume Technique for the Unsteady Euler Equations", Notes on Numerical Fluid Mechanics, Vol. 7, pp. 99-108, Vieweg Verlag, Braunschweig, 1984.

[7] Tannehill, J.C., Mugge, P.H.: "Improved Curve Fits for the Thermodynamic Properties of Equilibrium Air Suitable for Numerical Computation Using Time-Dependent or Shock-Capturing Methods", NASA CR-2470, October 1974.

[8] Jameson, A., Schmidt, W., Turkel, E.: "Numerical Solutions of the Euler Equations by Finite Volume Methods Using Runge-Kutta Time Stepping Scheme", AIAA Paper 81-1259, 1981.

[9] Seibert, W., Fritz, W., Leicher, S.: "On the Way to an Integrated Mesh Generation System for Industrial Applications", AGARD Fluid Dynamics Panel Specialists' Meeting on "Applications of Mesh Generation to Complex 3-D Configurations, Loen, Norway, May 1989.

FIGURES

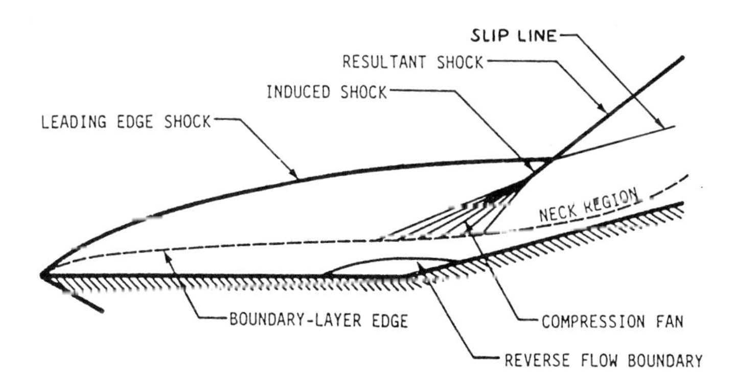

Fig. 1: Supersonic Flow Over a Compression Ramp

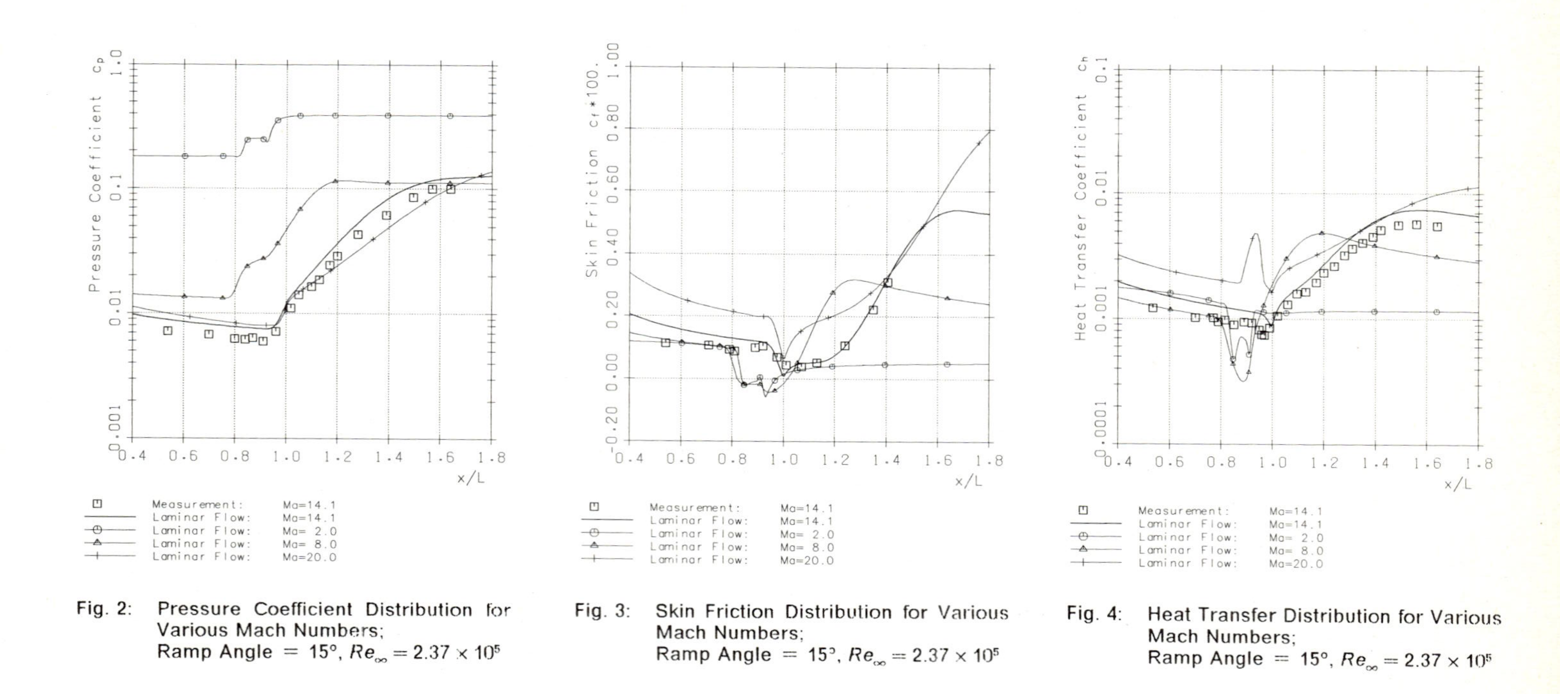

Fig. 2: Pressure Coefficient Distribution for Various Mach Numbers; Ramp Angle = 15°, $Re_\infty = 2.37 \times 10^5$

Fig. 3: Skin Friction Distribution for Various Mach Numbers; Ramp Angle = 15°, $Re_\infty = 2.37 \times 10^5$

Fig. 4: Heat Transfer Distribution for Various Mach Numbers; Ramp Angle = 15°, $Re_\infty = 2.37 \times 10^5$

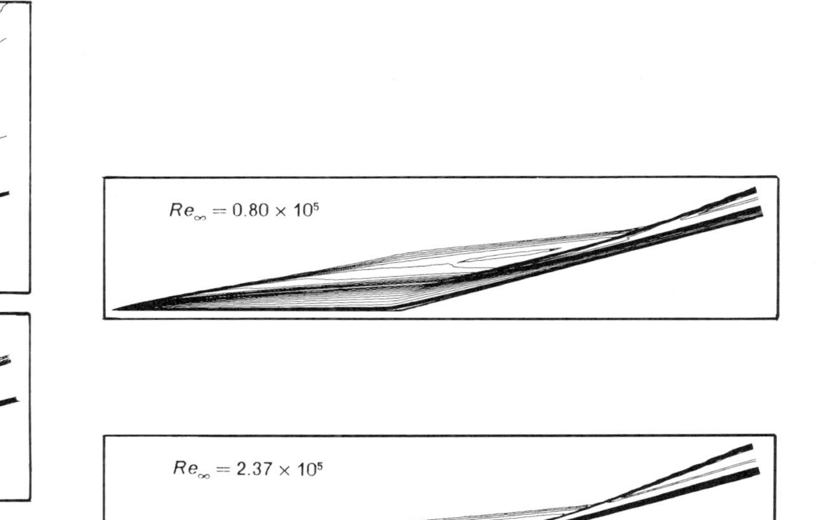

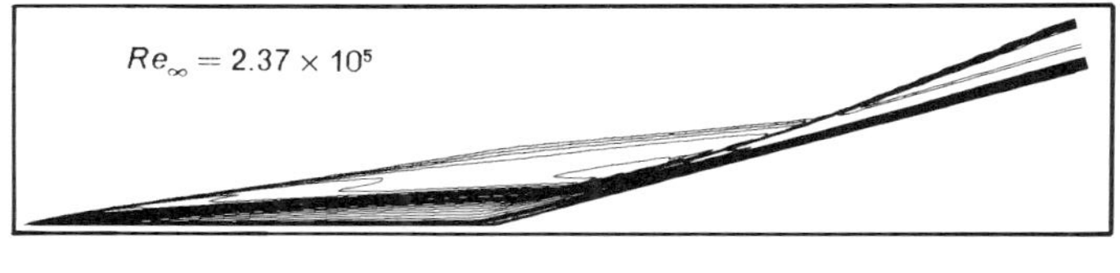

Fig. 5: Mach Number Contours for Various Mach Numbers

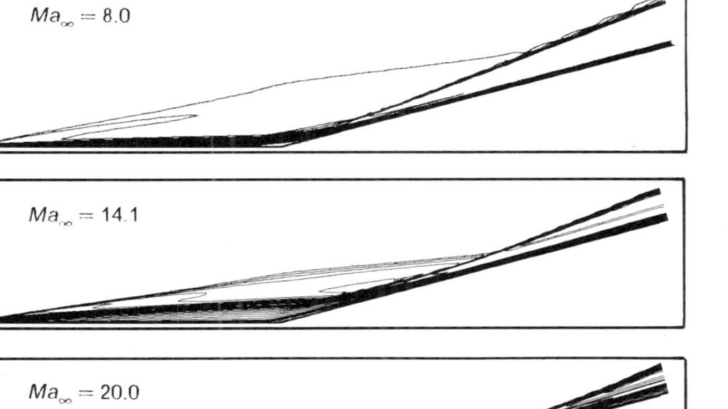

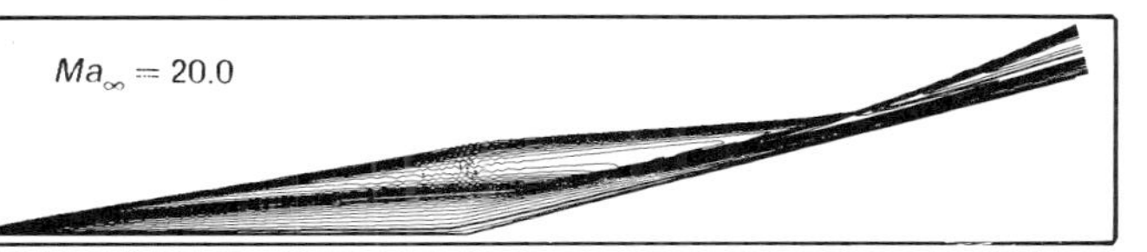

Fig. 6: Mach Number Contours for Various Reynolds Numbers

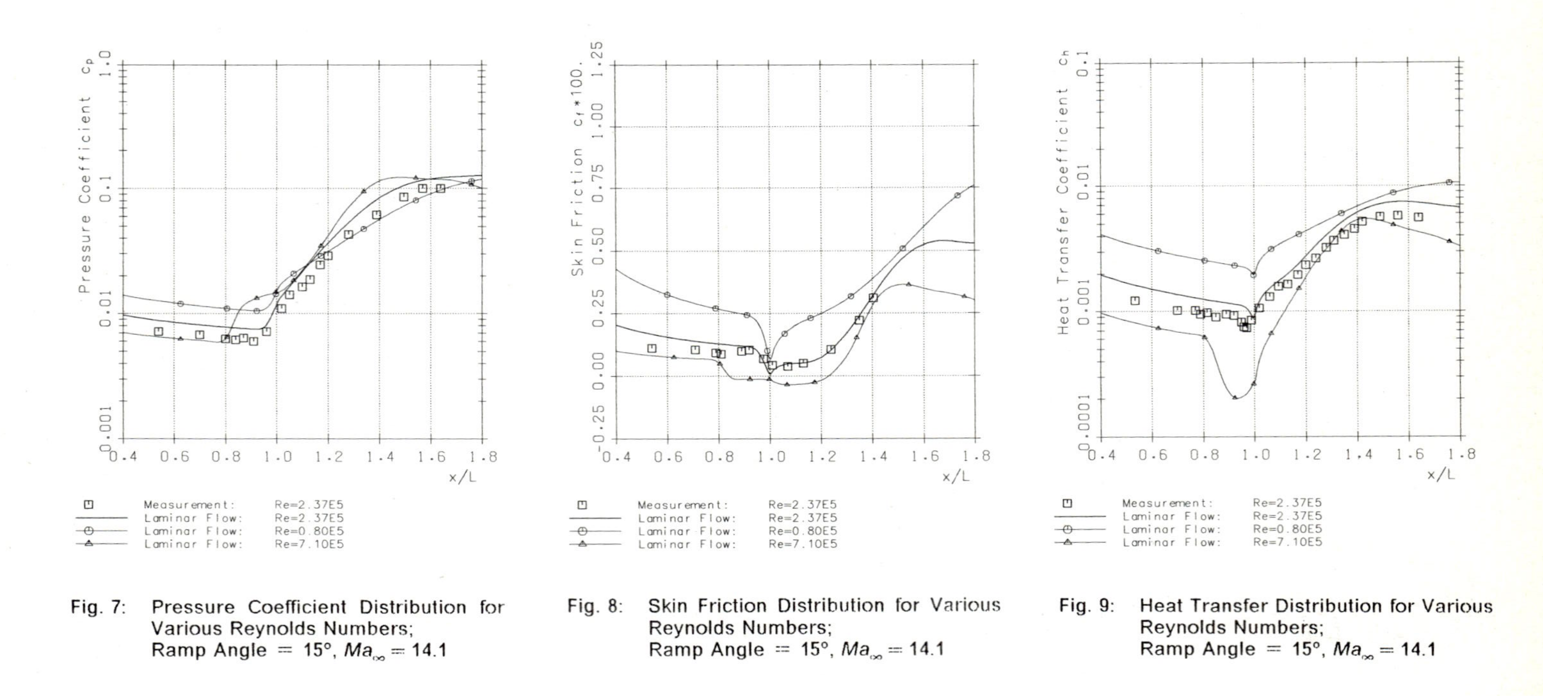

Fig. 7: Pressure Coefficient Distribution for Various Reynolds Numbers; Ramp Angle = 15°, $Ma_\infty = 14.1$

Fig. 8: Skin Friction Distribution for Various Reynolds Numbers; Ramp Angle = 15°, $Ma_\infty = 14.1$

Fig. 9: Heat Transfer Distribution for Various Reynolds Numbers; Ramp Angle = 15°, $Ma_\infty = 14.1$

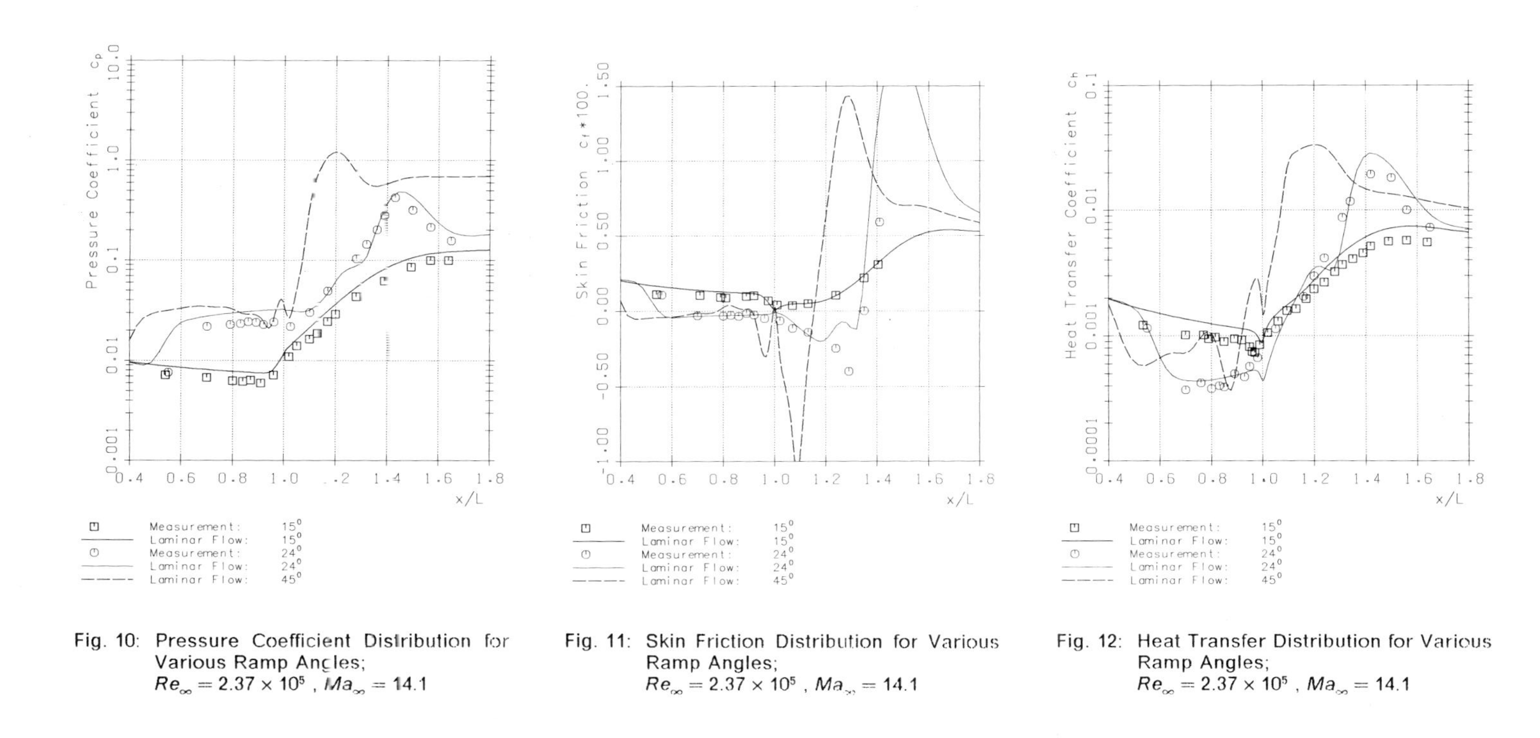

Fig. 10: Pressure Coefficient Distribution for Various Ramp Angles; $Re_\infty = 2.37 \times 10^5$, $Ma_\infty = 14.1$

Fig. 11: Skin Friction Distribution for Various Ramp Angles; $Re_\infty = 2.37 \times 10^5$, $Ma_\infty = 14.1$

Fig. 12: Heat Transfer Distribution for Various Ramp Angles; $Re_\infty = 2.37 \times 10^5$, $Ma_\infty = 14.1$

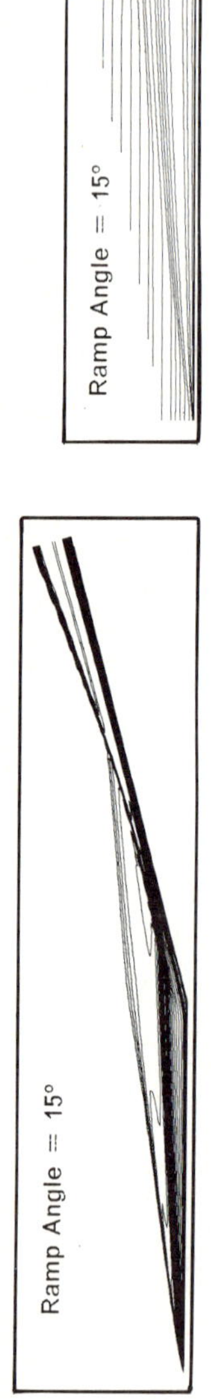

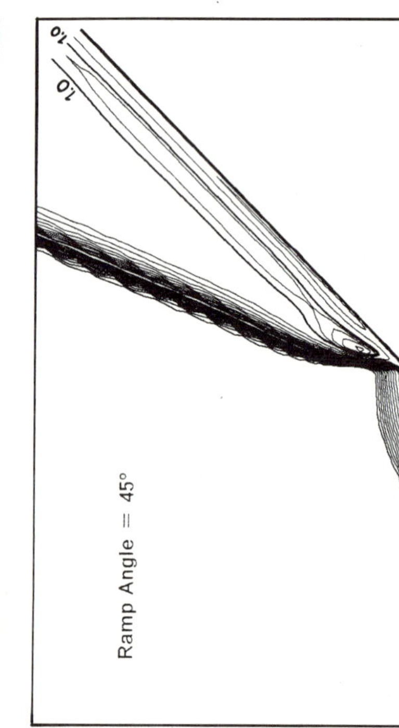

Fig. 13: Mach Number Contours for Various Ramp Angles
$Re_\infty = 2.37 \times 10^5$, $Ma_\infty = 14.1$

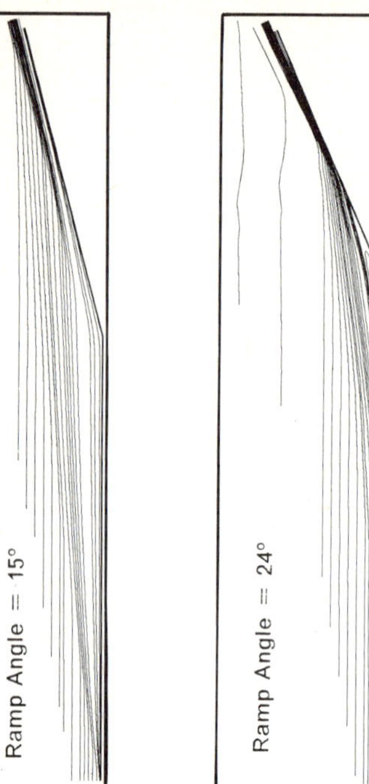

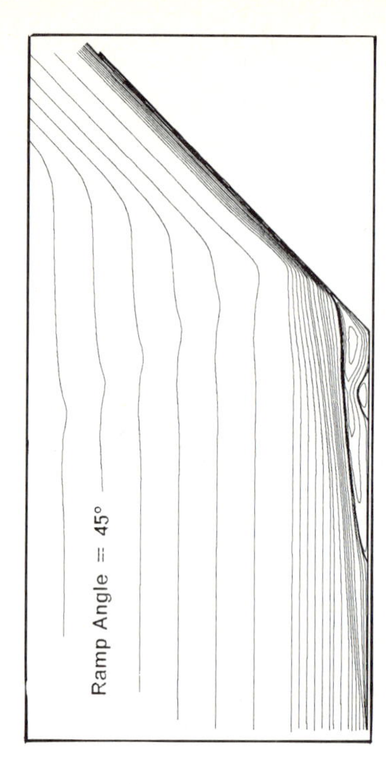

Fig. 14: Streamline Contours for Various Ramp Angles;
$Re_\infty = 2.37 \times 10^5$, $Ma_\infty = 14.1$

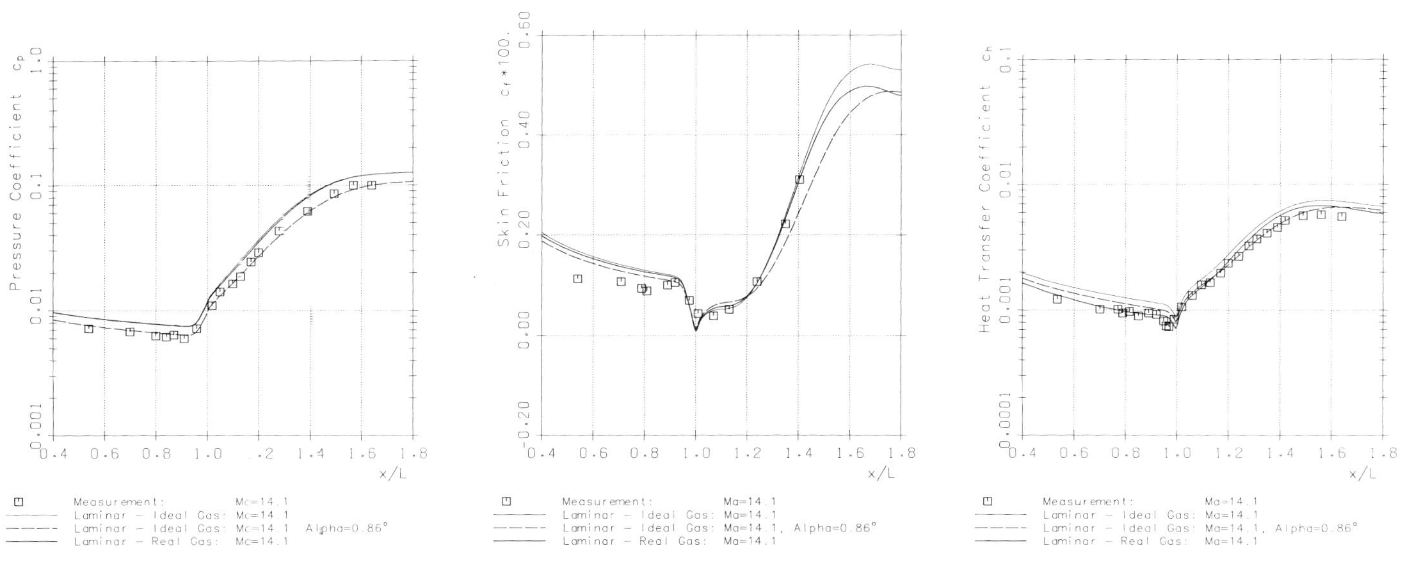

Fig. 15: Pressure Coefficient Distribution for Real Gas Effects and Incidence Correction; Ramp Angle = 15°, $Re_\infty = 2.37 \times 10^5$, $Ma_\infty = 14.1$

Fig. 16: Skin Friction Distribution for Real Gas Effects and Incidence Correction; Ramp Angle = 15°, $Re_\infty = 2.37 \times 10^5$ $Ma_\infty = 14.1$

Fig. 17: Heat Transfer Distribution for Real Gas Effects and Incidence Correction; Ramp Angle = 15°, $Re_\infty = 2.37 \times 10^5$, $Ma_\infty = 14.1$

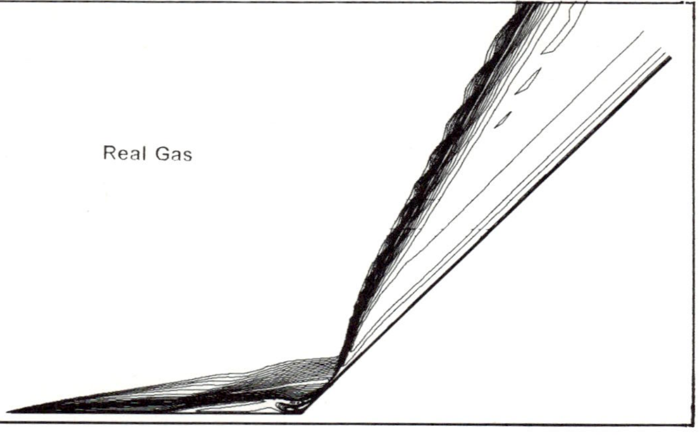

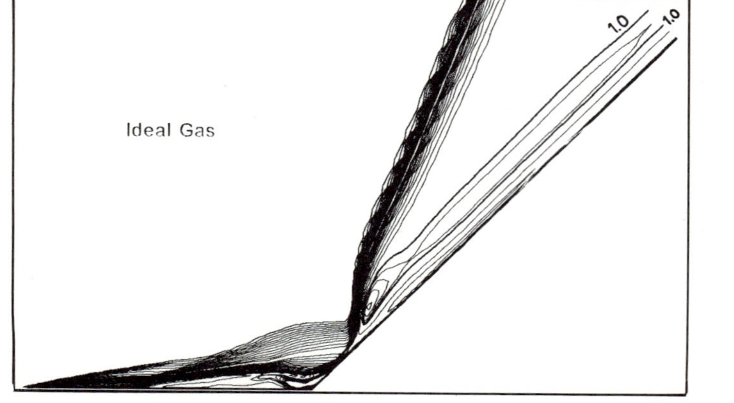

Fig. 18: Mach Number Contours for Ideal and Real Gas Flow. Ramp Angle = 45°, Re_{∞} = 2.37 × 10^5 , Ma_{∞} = 14.1

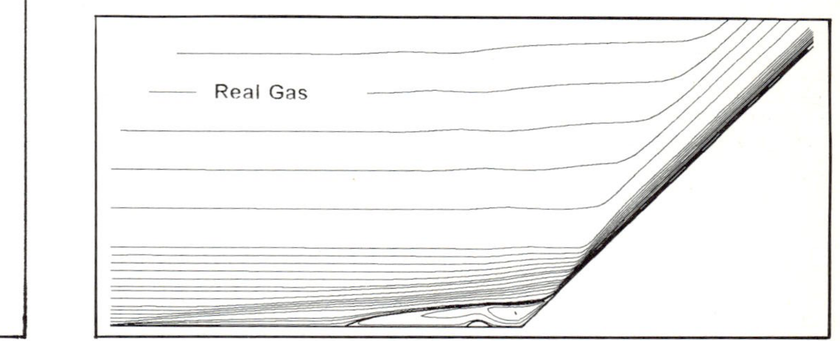

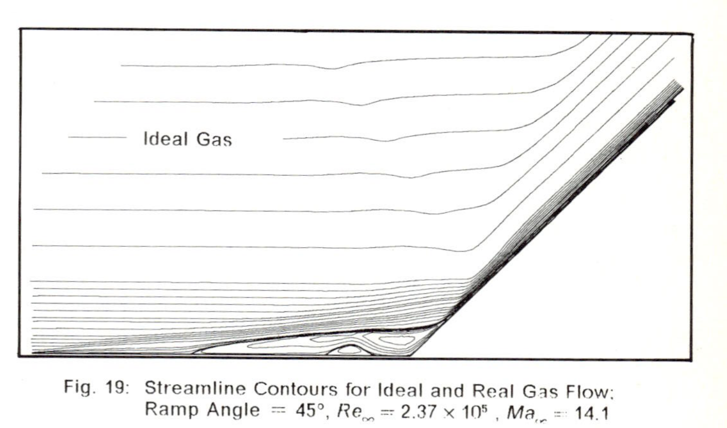

Fig. 19: Streamline Contours for Ideal and Real Gas Flow; Ramp Angle = 45°, Re_{∞} = 2.37 × 10^5 , Ma_{∞} = 14.1

A Fluid—Structure Interaction Model for Heart Valves with a Single Degree of Freedom

J.B.A.M. Horsten, A.A. van Steenhoven and D.H. van Campen

Department of Mechanical Engineering
Eindhoven University of Technology
P.O.Box 513, 5600 MB Eindhoven, The Netherlands

Summary

As a preparatory study for the analysis of the opening and closing behavior of aortic valve prostheses in a viscous fluid flow, a rigid two—dimensional valve, which can rotate around its point of attachment is analyzed. The valve motion and the fluid velocity field are computed. The equations of motion for fluid and structure are iteratively coupled. For the fluid the full unsteady 2D incompressible Navier—Stokes equations are solved. The valve is assumed to be rigid and its inertia is neglected. The equilibrium position of the valve at a point of time is found iteratively, using the Van Wijngaarden—Dekker—Brent method for the root—finding of a nonlinear equation. The numerical method is validated by means of measurement of the interaction forces and the valve displacement in an experimental model. A comparison of numerical and experimental results show that a proper method is developed.

Introduction

Frequently, natural human aortic valves do not function properly and need to be replaced by valve prostheses. An important feature of the natural valve is its partial closure during flow deceleration, such that the reverse flow back into the heart is minimized [1]. This condition is not satisfied by the commonly used disc—type valve prostheses, which consist of one or two rigid plates, mounted in frame [2]. For the design of an improved disc—type aortic valve prosthesis the valvular opening and closing need to be investigated.

Several models for this problem have been developed. Most of them are analytical ones, based on a quasi—one—dimensional approach [1,3,4]. These models give a fair global description of the valve motion, notwithstanding a strong simplification of the fluid flow phenomena, but they cannot give an analysis of the effect of small variations of the valve geometry. A a much more detailed numerical model, specially developed for heart valves, is given by Peskin [5]. It incorporates the full Navier—Stokes equations and flexible boundaries. A major drawback of this model is its limited stability, due to an explicit estimation of the fluid—valve interaction terms. Furthermore, the Reynolds number and spatial accuracy are limited and rigid boundaries can only be incorporated in an indirect way. Somewhat more general fluid—structure interaction models, of which overviews are given in [6] and [7], are not suitable for our purpose, since they generally simplify the fluid flow too much and the inertia of the structure is essential for those models. In our case the situation is opposite: the inertia of natural heart valves is negligible compared to that of surrounding blood.

The object of this study is the development of a numerical fluid—structure interaction model, allowing for a full description of laminar flow patterns in complex domains and for general rigid body structures. In the present model the structure is restricted to have only one degree of freedom and a negligible inertia. However, inertial effects can easily be incorporated. The model valve is shown in figure 1. It is rigid and can rotate around its point of attachment. Behind the disc a cylindrical cavity is present, which is a modeled version of the physiological sinus of Valsalva. The valve is attached to the rigid channel by a membranous hinge. This hinge causes a bending moment in the direction of the closed position. The valve is partly hollow and the average density is 0.95 times that of water, the fluid which is used, so a buoyancy force is acting to the fully opened position. The magnitude of these forces is chosen such that they are in equilibrium for $\varphi \approx 13^o$. The channel width is six times its height in order to minimize three dimensional effects. The fluid flow is assumed to be Newtonian, incompressible, two—dimensional and laminar. Although physiological Reynolds numbers are larger (Re=4000), for computational reasons the maximum Reynolds number was chosen to be 800. The interaction will be taken fully into account. The equations of motion for fluid and structure are iteratively coupled. The subsystems are evaluated separately and an iteration procedure is applied until equilibrium is achieved.

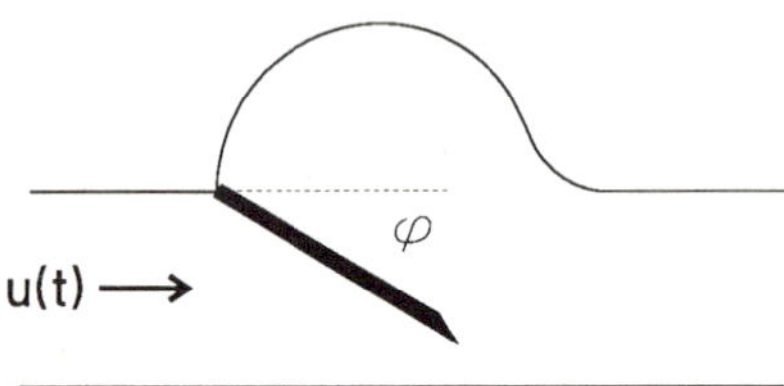

fig.1 Two—dimensional model of the aortic valve

Fluid model

Flow of an incompressible, Newtonian and isothermal fluid must satisfy the Navier—Stokes and continuity equations. In dimensionless form these equations read:

$$St\,\dot{\vec{u}} + \vec{u}\cdot\nabla\vec{u} - \frac{1}{Re}\nabla^2\vec{u} + \nabla p - \vec{f} = \vec{0} \qquad (1a)$$
$$\nabla\cdot\vec{u} = 0 \qquad (1b)$$

with $\vec{u}$ the velocity vector, p the pressure, $\vec{f}$ the body force per unit mass, ∇ the gradient vector operator and the superscript dot $\cdot$ denotes the local time derivative. St denotes the Strouhal number defined as $St = h/u\tau$ with τ a characteristic time scale, h the channel height and u the mean axial velocity. Re is the Reynolds number, defined as $Re = uh/\nu$ with ν the kinematic viscosity.

To obtain an approximation of the velocity and the pressure field within a domain Ω, a standard Galerkin finite element method is applied. Here, only a brief synopsis of the method is given. More details can be found elsewhere [8,10]. A 7—noded triangular Crouzeix—Raviart element, as showed in figure 2, is used. The basis functions of the velocity are extended quadratic functions. Velocity unknowns are defined in all the nodal points. Pressure unknowns are only defined in the center of the element, which are the pressure itself and its three spatial derivatives. The basis functions of the pressure are linear and discontinuous over the element boundaries. The velocity unknowns and the pressure derivatives within the centroid of the element are

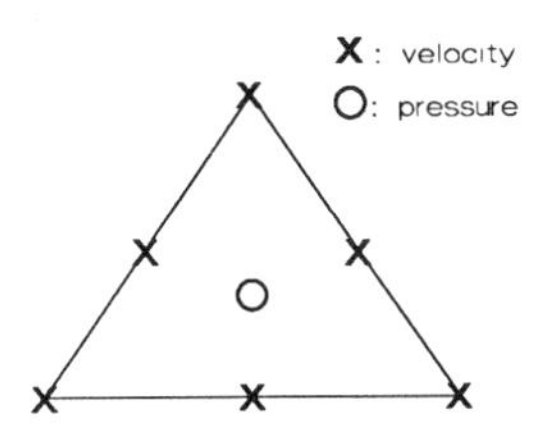

fig.2 Modified $p_2^+ - p_1$ Crouzeix—Raviart element for the spatial discretisation of the Navier—Stokes equations

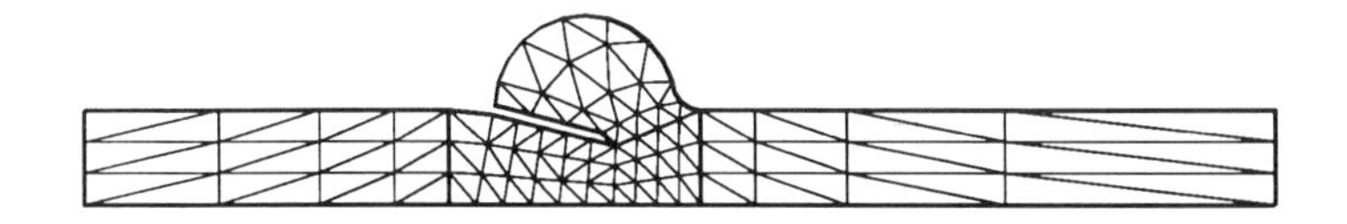

fig.3 Finite element mesh for the fluid flow analysis in the aorta model

eliminated by consideration of the Navier—Stokes and continuity equations elementwise. This leads to a total amount of unknowns of 13. It can be shown that the element satisfies the Babuska—Brezzi condition. The accuracy of the velocity is of order Δx^3 and that of the pressure of order Δx^2, with Δx a characteristic element size. An example of a typical element mesh is given in figure 3.

The spatial discretization leads to a set of matrix equations:

$$\underline{M}\,\dot{\underset{\sim}{u}} + [\underline{S} + \underline{N}(\underset{\sim}{u})]\underset{\sim}{u} + \underline{L}^T\underset{\sim}{p} = \underset{\sim}{f} + \underset{\sim}{b} \tag{2a}$$

$$\underline{L}\,\underset{\sim}{u} = \underset{\sim}{0} \tag{2b}$$

where $\underline{M}\,\dot{\underset{\sim}{u}}$ represents the local acceleration term, $\underline{L}^T\underset{\sim}{p}$ the pressure gradient term, Su the viscous term, $N(\underset{\sim}{u})\,\underset{\sim}{u}$ the convective acceleration term and $\underline{L}\,\underset{\sim}{u}$ the velocity divergence term. $\underset{\sim}{f}$ and $\underset{\sim}{b}$ represent the volume and boundary forces respectively. $\underset{\sim}{u}$ contains the velocity and $\underset{\sim}{p}$ the pressure unknowns in the nodal points. In the set of equations (2) the pressure unknowns do not occur in the continuity equation This leads to zero diagonal elements. In order to avoid partial pivoting and to reduce the number of unknowns, a penalty function method is applied. The discretized continuity equation is replaced by:

$$L\,\underset{\sim}{u} = \varepsilon\, M_p\, \underset{\sim}{p} \tag{3}$$

with $\underline{M}_p$ the pressure matrix and ε a very small parameter. If the right hand side of eq. (3) is small enough, then the incompressibility of the fluid will be sufficiently approximated. For the problems solved in this study the value of ε was chosen to be $\varepsilon = 10^{-6}$, which leads to values of $\varepsilon\,\underline{M}_p\,p$ of $O(10^{-6})$ in the dimensionless formulation.

The local time derivative in (2) is approximated by the Euler—implicit scheme:

$$\dot{\underset{\sim}{u}}^{n+1} = \frac{\underset{\sim}{u}^{n+1} - \underset{\sim}{u}^{n}}{\Delta t} \tag{4}$$

in which $\underset{\sim}{u}^n$ is an abbreviation for $\underset{\sim}{u}(n\Delta t)$ with Δt the time step. This scheme is unconditionally stable and first order accurate in time. Although a Crank—Nicolson scheme would be more accurate than the Euler—implicit scheme, the latter will be used in this study. The motivation is found in the observation that, for reasons which are not completely clear yet, the computed pressure shows small oscillations round the correct values if a Crank—Nicolson scheme is used [10]. The restriction to a first order scheme is not inherent to the fluid—structure interaction algorithm itself (there are no objections for a higher order scheme), but it is a result of a limitation of the fluid solver.

The non—linear convective term $\underline{N}(\underset{\sim}{u})\underset{\sim}{u}$ in (2a) is linearized by a Newton—Raphson iteration scheme:

$$\underline{N}(\underset{\sim}{u}^{i+1})\, \underset{\sim}{u}^{i+1} = \underline{J}(\underset{\sim}{u}^{i})\, \underset{\sim}{u}^{i+1} - \underline{N}(\underset{\sim}{u}^{i})\, \underset{\sim}{u}^{i} \tag{5}$$

with i the index of the iteration step and $\underline{J}(\underset{\sim}{u})$ the Jacobian matrix of $\underline{N}(\underset{\sim}{u})\underset{\sim}{u}$. Substituting equations (3), (4) and (5) into (2) leads to the final set of equations:

$$\begin{aligned}[\underline{M}/\Delta t + \underline{S} + \underline{J}(\underset{\sim}{u}^{n+1,i}) + \varepsilon \underline{L}^T \underline{M}_p^{-1} \underline{L}]\, \underset{\sim}{u}^{n+1,i+1} = \\ \underline{M}/\Delta t\, \underset{\sim}{u}^{n} + \underline{N}(\underset{\sim}{u}^{n+1,i})\, \underset{\sim}{u}^{n+1,i} + \underset{\sim}{f}^{n+1} + \underset{\sim}{b}^{n+1}\,.\end{aligned} \tag{6}$$

At every time step a full Newton—Raphson iteration is carried out until

$$\max_j \|u_j^{i+1} - u_j^{i}\| < \delta$$

with $\delta = 10^{-4}$. At every iteration step the system (6) is built and a LU—decomposition is used to solve it. Experiments with simplified Newton schemes, when the Jacobian matrix is not updated every iteration step, did not result in a decreasing computing time.

After convergence the pressure at t_{n+1} is found with eq. (3) and the normal and tangential stresses are interpolated to the element vertices. The total fluid moment acting on the valve is obtained by integrating the local stress over the valve. A trapezoidal rule is used with the element vertices as integration points. The elements on the valve are chosen of equal size, since this results in the optimal accuracy for the integration of the stresses.

As initial condition, the steady state solution for a fully opened valve is taken. As contact condition on the valve, the normal fluid velocity is set equal to the local valve velocity, determined from the actual valve position and that at the previous point of time. The tangential fluid velocity satisfies a no—slip condition. At the entrance, a fully developed parabolic axial velocity profile is prescribed and the radial flow is set to zero. The entrance channel long enough to garantuee a full development of the unsteady velocity profile. At the outflow a stress free flow condition is prescribed.

Structure model

Since its inertia is assumed to be negligible, the structure is in equilibrium if it satisfies the condition:

$$\Sigma m = m_f + m_g + m_b = 0 \tag{7}$$

where m_f is the fluid moment and m_g the moment due to the buoyancy force

$$m_g = g \cos(\varphi) \int_0^l \Delta\rho(s)\, b(s)\, s\, ds \qquad (8)$$

with l and b the length and thickness of the valve respectively, s the coordinate along the valve, g the acceleration of gravity and $\Delta\rho$ the density difference between valve and fluid. m_b is the bending moment in the point of attachment

$$m_b = -k(\varphi_0 - \varphi) \qquad (9)$$

with k a bending stiffness parameter and φ_0 the angle of fixation of the membranous hinge. At a given point of time, with given inflow conditions and known fluid history, the resulting moment is a function of the unknown valve position φ only. The equilibrium position is the root of the nonlinear equation (7). The derivatives of (7) with respect to φ cannot be computed in a simple way. Therefore, an iterative method is necessary, which does not need the evaluations of the derivatives. Here, the Van Wijngaarden—Dekker—Brent method is used [12,13]. This method assumes that a root is known to be bracketed in a given interval. Then it locates the root, within a given accuracy, by a combination of successive inverse quadratic interpolation and bisection. The method converges never much slower than bisection does and converges superlinear near roots of well—behaved functions.

The bracketing of the root is performed by a restricted extrapolation. As first estimate the solution of the previous time step is taken. The sign of the moment is tested and the second estimate is a trial step in the direction in which the valve is being pushed. Then a linear extrapolation is performed and the next estimate is taken somewhat further than the zero crossing of the extrapolating line. Not the zero crossing is taken, because in this stage only the bracketing is important. Both the extrapolating step and the extrapolated position are restricted to predefined limits. Successive linear extrapolation is performed until the root is bracketed. Always, the two most recent estimates are used. Since the moment appears to be a monotonic function of φ and only extrapolation takes place, they are the closest to the root.

In every timestep, a bracketing and iteration procedure is performed. For each valve position estimate, the mesh is updated and the fluid flow field is computed. The resulting algorithm is fully implicit and hence unconditional numerical stability is achieved. The model is implemented in the SEPRAN software package [9]. The computations are carried out on an Apollo—DN3000 minicomputer and on an Alliant FX/4 mini—supercomputer with two parallel vector processors. The number of position estimates per timestep varies from five to eight. Per position estimate between two and eight Newton iteration steps are necessary. For the typical number of velocity unknowns of 721, this results in about 30 minutes computing time per time step on the Apollo system and about 2 minutes on the Alliant system,

Experiments

The numerical method is validated by means of measurement of the fluid moment acting on the valve and the displacement of the valve as a function of Reynolds number and time. The experimental setup is shown in figure 4. The fluid moment is measured with a force transducer (LVDT) mounted on the upper side of the valve. The valve motion is recorded with a standard video system. The accuracy of the computation of the fluid velocity field has been extensively tested earlier [10,11].

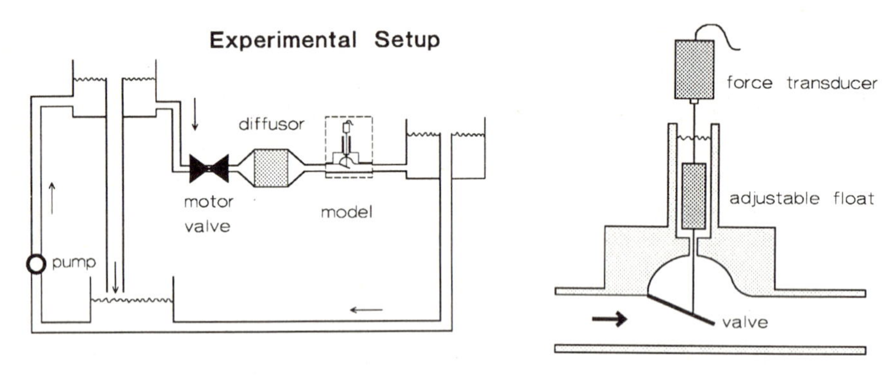

fig.4 Experimental setup: (a) flow channel (b) detail of model

Results

With this method the fluid flow and the valve rotation under steady and pulsating inflow conditions are simulated. A typical result for the steady fluid flow is presented in figure 5. The flow separates at the valve tip and reattaches again at a distance of about the sinus diameter. Behind the valve, a weak vortex is present. The pressure in the sinus is almost constant and equal to that at the valve tip. Figure 6 shows the flow in case of a valve moving towards its closed position under steady inflow conditions. The contents of the sinus rotates together with the valve. This observation supports the negligence of the valve inertia, since it is much smaller than that of the moving fluid.

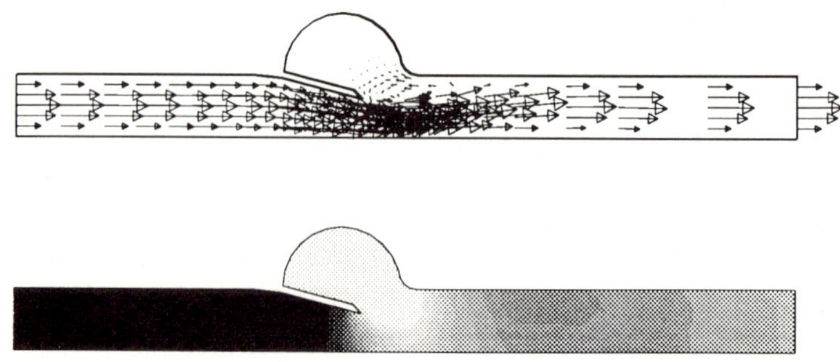

fig.5 Velocity field and pressure distribution for a fixed valve in a steady flow (Re= 200)

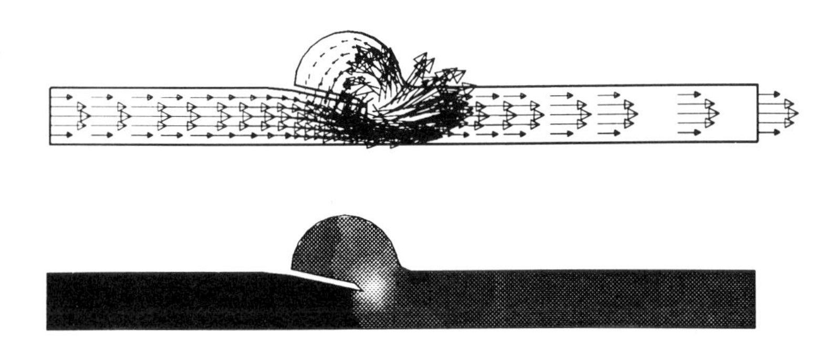

fig.6 Velocity field and pressure distribution for moving valve in a steady flow (Re= 200,St= 1.0)

The computation of the fluid moment is verified in the case that the valve is held in a fixed position, the force transducer being mounted on it. A comparison of numerical and experimental data is given in figure 7. The agreement is close: differences are well within the experimental accuracy limits. Next, the equilibrium position of a freely moving valve in a steady flow is studied as function of the Reynolds number. A preliminary result is given in figure 8. The agreement is not as close as might be expected from figure 7. This deviation is caused by the bending moment in the point of attachment. As stated in eq. (9), it is assumed that the bending parameter is constant. This appears not to hold exactly. More detailed experiments are planned to determine k as a function of the valve position. Also the valve motion due to gravity and bending force in case of a steady inflow is studied. The valve is fixed in its fully opened position till t=0 and then released. The valve rotation is studied at various values of Reynolds number. An example is given in figure 9. Both experimental and numerical results show a second order subcritically damped behavior. The agreement between experimental and numerical results of both the valve motion and the steady equilibrium position is fair. The steady equilibrium positions are the same is in figure 8. The agreement for the valve motion improves if the timestep of the integration is decreased. This indicates that a part of the deviations between theory and experiment is caused by numerical damping, resulting from the first order implicit time integration scheme. Furthermore, the earlier mentioned deviations in the bending stiffness parameter of the membranous hinge may explain some of the observed differences.

Next, valve motion due to volume flow variations at the inlet is studied numerically. A result is given in figure 10. It appears that during flow deceleration the model valve moves already towards its closed position, just like the natural aortic valve does. This early state closure is not observed for commercially available disc—valve prostheses [2]. Experimental verification of this pulsating flow situation is planned.

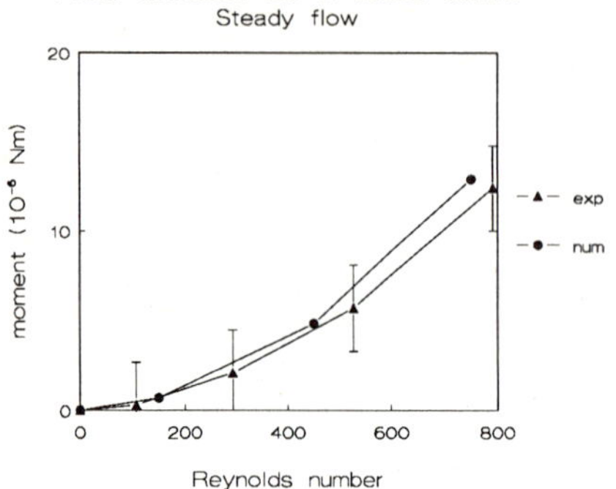

fig.7

Comparison of experimental and numerical moments on a fixed valve due to a steady fluid flow as a function of the Reynolds number ($\varphi = 10^0$)

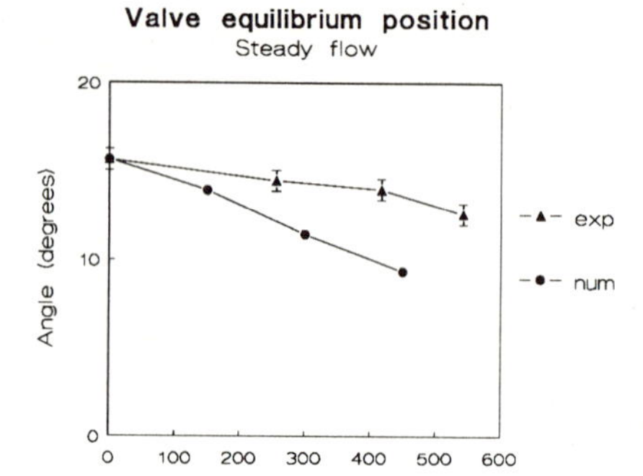

fig.8

Comparison of experimental and numerical valve equilibrium position in a steady flow as a function of the Reynolds number

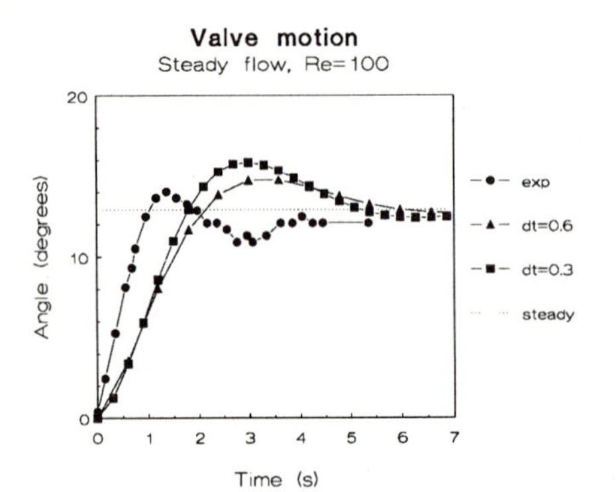

fig.9

Comparison of experimental and numerical valve motion due to gravity and bending force in a steady flow (Re= 100). At t=0 the valve is released from the fully opened position.

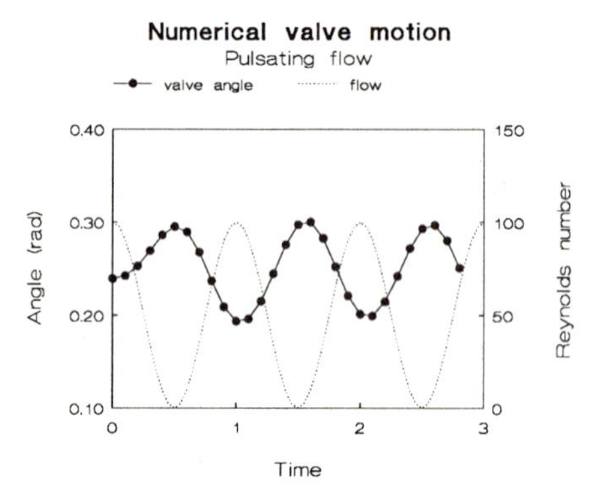

fig.10

Results of a numerical simulation of the valve motion due to a harmonic volume flow variation at the inlet (Re= 0—100, St= 1))

Concluding discussion

In conclusion, it is stated that a proper method is developed for the numerical analysis of the fluid flow around and the motion of a rigid body with one degree of freedom. The agreement between numerical and experimental results is fair. The method is very flexible, since the structure equilibrium condition (3) can easily be modified and any fluid solver can be used, as long as the velocity and pressure fields are evaluated with sufficient accuracy. For numerical reasons, the study has been performed for a maximum Reynolds number of 800. A qualitative comparison of the obtained results with experimental ones at a higher Reynolds number [1] shows the same characteristics of the flow field. Hence, the present method is valuable for the analysis of the actual situation. On the other hand, some increase of the Reynolds number is possible with the fluid solver used.

In this study inertial effects of the structure are neglected, but they can easily be incorporated by extending of the equilibrium condition (7) to the familiar

$$\sum m + I \ddot{\varphi} = 0 \qquad (10)$$

with I the moment of inertia. The second order time derivative can be discretized by either

$$\ddot{\varphi}^{n+1} = [\varphi^{n+1} - 2\varphi^{n} + \varphi^{n-1}]/\Delta t^2 + O(\Delta t) \qquad (11a)$$

or, if a second order accurate fluid solver is used:

$$\ddot{\varphi}^{n+1} = [2\varphi^{n+1} - 5\varphi^{n} + 4\varphi^{n-1} - \varphi^{n-2}]/\Delta t^2 + O(\Delta t^2) \qquad (11b)$$

This can be verified by using a Taylor series expansion.

Further experimental verification, the improvement of the accuracy, the increasing of the Reynolds number and the extension to valves with more degrees of freedom and to flexible valves are subjects of current research.

Acknowledgment

We wish to thank D. Palmen for her valuable contributions. This research is supported by the Dutch Technology Foundation (STW), grant nr. EWT58.0857

References

[1] Steenhoven, A.A. van & M.E.H. van Dongen, "Model studies of the closing behaviour of the aortic valve", *J. Fluid Mechanics vol 90*, pp 21—32, 1979

[2] Steenhoven, A.A. van, Van Duppen, Th.J.A.G., Cauwenberg, J.W.G. and Van Renterghem, R.J., "In vitro closing behaviour of Bjork—Shiley, St Jude and Hancock heart valve prostheses in relation to the in vivo recorded aortic valve closure", *J.Biomechanics vol 15*, 1982, pp841—848

[3] Bellhouse, B.J. & L. Talbot, "The fluid mechanics of the aortic valve", *J. Fluid Mechanics vol 35*, pp 721—735

[4] Wippermann, F.K., "On the fluid dynamics of the aortic valve", *J. Fluid Mechanics vol 159*, pp 487—501, 1985

[5] Peskin, C.S., "Numerical analysis of blood flow in the heart", *J. of Comp. Physics vol 25*, pp220—252, 1977

[6] Belytschko, T, "Fluid—structure interaction", *Computers and structures vol 12*, pp459—470,1980

[7] Park, K.C. & C.A. Felippa, "Partitioned analysis of coupled systems", in: T. Belytschko & T.J.R. Hughes (eds.), *Computational methods for transient analysis*, North—Holland, Amsterdam, 1983

[8] Cuvelier, C., A. Segal & A.A. van Steenhoven, *Finite element methods and Navier—Stokes equations*, D. Reidel Publishing Company, Dordrecht, The Netherlands, 1986

[9] Segal, A., *SEPRAN Manual*, Ingenieursbureau SEPRA, Delft, The Netherlands, 1989

[10] Vosse, F.N. van de, A.A. van Steenhoven, A. Segal & J.D. Janssen, " A finite element analysis of the unsteady two—dimensional Navier—Stokes equations", *Int. J. of Numerical Methods in Fluids vol 6*, pp427—443,1986

[11] Vosse, F.N. van de, A.A. van Steenhoven, A. Segal & J.D. Janssen, " A finite element element analysis of the steady laminar entrance flow in a 90^{0} curved tube", *Int. J. of Numerical Methods in Fluids vol 9*, pp275—287,1989

[12] Press, W.H., B.P. Flannery, S.A. Teukolsky & W.T. Vetterling, *Numerical Recipes*, Cambridge University Press, New York, 1986

[13] Brent, R.P., *Algorithms for minimization without derivatives*, Prentice—Hall, Englewood Cliffs, N.J., USA, 1973

Simulation of Glancing Shock Wave and Boundary Layer Interaction

CHING-MAO HUNG

NASA Ames Research Center, Moffett Field, CA

After months of fishing with a net and carefully studying, a marine professor concluded that 1) all fishes have gills, and 2) no fish is smaller than two inches.

Abstract

Shock waves generated by sharp fins, glancing across a laminar boundary layer growing over a flat plate, are simulated numerically. Several basic issues concerning the resultant three-dimensional flow separation are studied. Using the same number of grid points, different grid spacings are employed to investigate the effects of grid resolution on the origin of the line of separation. Various shock strengths (generated by different fin angles) are used to study the so-called separated and unseparated boundary layer and to establish the existence or absence of the secondary separation. The usual interpretations of the flow field from previous studies and new interpretations arising from the present simulation are discussed.

I. INTRODUCTION

In the past, fluid dynamics has been divided into two branches, theoretical and experimental. Continuing advances in numerical methods and in computer capabilities have at some point qualified computation as a separate branch of fluid dynamics. Using the computer as a tool, computational fluid dynamics (CFD) is able to supplement the other two branches and to carry out its own research, development, and further advancement in fluid dynamics as a field of physical science.

A shock wave generated by one body, glancing across a boundary layer on another body, is a problem well suited to CFD capabilities. This is one of the most common and important three-dimensional (3-D) inviscid/viscous interaction problems. The problem is so complicated that a purely theoretical approach is almost impossible. In early studies, investigators have relied mainly on experimental observations of surface properties, such as static pressure and oil flow, to deduce the basic interaction features of the flow. With the advent of the supercomputer, we are now on the threshold of definitive explorations, finding the details of the flow-field structure and the underlying physical processes.

There are various kinds of glancing shock wave problems. Instead of dealing problems involving complicated geometries, the present paper will focus on the problem of a supersonic flow over a sharp fin mounted on a flat plate, shown in Fig. 1. Recently this simple geometry has attracted a substantial amount of interest and has been studied extensively, (for example, Refs. 1 - 5). However, there are still many flow-field features that need better explanations.

The primary purpose of this paper is to study several basic issues regarding the resultant 3-D flow separation. The separation of the boundary layer on the flat plate, (the sidewall), under the influence of the glancing shock wave will be specially emphasized. To avoid the uncertainty associated with turbulent modeling in our study, the flow is assumed to be laminar. The compressible Navier-Stokes equations will be solved. Cases of sharp fins with various wedge angles will be investigated numerically.

The results are first compared with the experimental and computational results of Degrez [3] for $M_\infty = 2.25, Re = 10^5$ and a wedge angle of 6°. Results for wedge angles of 2° and 12° will also be discussed. Using the same number of grid points, different grid spacings are employed to study the effects of grid resolution on the origin of the line of separation. Various shock strengths generated by different fin angles are used to study the so-called separated and unseparated boundary layer and the existence or absence of a secondary separation. The usual interpretations from previous studies and new interpretations arising from the present simulation will be discussed.

It should be mentioned that this type of flow field, when turbulent, is observed to be unsteady (see discussion in Ref. 6). However, in the present study, the flow is assumed laminar and steady. The applicability of the present discussion to the turbulent cases is based upon the assumption that the random turbulence fluctuations are suppressed and the turbulent flow is steady in a "mean" sense.

II. NUMERICAL PROCEDURE

The governing equations of the present analysis are the time-dependent, compressible Navier-Stokes equations incorporating the concept of the thin-layer approximation in all three directions (Baldwin and Lomax [7]). The flow is assumed laminar and the wall is adiabatic. A numerical procedure developed by Hung and Kordulla [8] without time-splitting is used. The basic numerical scheme is MacCormack's [9] explicit-implicit predictor-corrector algorithm. The solution is carried out until it converges to a steady state. Details of the numerical technique and boundary conditions are discussed in Ref. 8.

Figure 2 shows a typical mesh system of 57x45x27 points for a sharp fin on a flat plate. The apex of the fin (at x = 0.0) is placed at a distance L = 9 cm from the flat plate leading edge and this distance L is used in the characteristic length in the present study. Here (x, y, z) and (I, J, K) are used in the conventional sense of streamwise, crossflow, and vertical directions. The domain of computation lies in the intervals $-1.0 \leq x \leq 3.5$, $0.0 \leq y \leq 4.5$, and $0.0 \leq z \leq 1.5$. Using the same number of grid points, three different grid spacing are employed. The coarse grid is uniform in the streamwise direction and geometrically stretched from the fin and plate (in the J- and K-directions). The medium grid has additional geometric stretching in the streamwise direction from the apex of the fin. The fine grid has finer spacing, (compared to the medium grid), near the fin in the J-direction and near the apex of the fin in the streamwise direction. To avoid over-stretching in the outer region, several zones with different stretching factors are used in the I- and J-directions. Smooth transition in grid spacing is ensured from one zone to another. The grid spacing parameters are listed in Table 1.

III. RESULTS AND DISCUSSIONS

The flow to be simulated has free-stream Mach number M_∞ = 2.25, Reynolds number $Re = 10^5$, and free-stream temperature T_∞ = 263°R. The results are at first compared with the experimental and computational results of Degrez for a wedge angle $\theta = 6^o$. Figure 3 shows a comparison of surface pressures at y = 5 cm. All results are in good agreement, except that the computational result of Degrez shows a "dip" after the separation. The present results and the experimental data do not show the appearance of the dip in surface pressure. The fine grid result (not shown in Fig. 3) is very close to the medium grid result. This indicates that our grid refinement does not affect the prediction of surface pressure.

1. Origin of the Line of Separation

Reference 4 contains an extensive study of the flow-field structure of this geometry for turbulent boundary layer. It is well accepted that the primary separation is a consequence of the high pressure, recovered from the shock system, which induces flow from the fin surface and forces the boundary layer off the sidewall. The question is, where is the origin of the line of primary separation? Figure 4 shows particle traces of the result based on a two-equation model as described in Ref. 4 for the first mesh points above the sidewall ($K = 2$). (The figure is provided by Horstman [10].) This particle trace is constructed by a time integration of velocity components restricted to the plane of $K = 2$. Since the plane of $K = 2$ is very close to the flat plate, (normally it would have a resolution smaller than the size of an oil particle), the particle integrations are treated as surface particle traces and are considered equivalently as a simulation of oil flow in the experiment and as a simulation of skin-friction lines in the theoretical approach. The "oil flow" in Fig. 4 indicates that the line of separation originates somewhere in the plate away from the apex of the fin and that this feature is an open-type separation. (An open-type separation contains only regular points, while a closed-type separation originates from a saddle point, as discussed in Refs. 11 and 12.)

Figures 5a - 5c show surface particle traces for the sequence of three grid refinements. For the coarse grid, the separation is an open type. As the grid spacing near the leading edge is refined, the starting point of the open-type separation moves and eventually the separation becomes a closed type. This clearly demonstrates that the grid resolution can affect the "calculated" topology. The coarse grid simply cannot resolve the vortex structure, while the fine grid can. As the wedge angle increases to 12° the vortex structure is large enough that the medium grid (not shown here) is able to reveal a closed-type separation.

While the existence or not of an open-type separation is still an unanswered question, we believe that some of the numerically observed open-type separations (for instance, Fig. 4) result from insufficient grid resolution. Similarly, every experiment also has resolution problems, such as the size of oil droplets. Some of the experimentally observed open-type separations may be the result of low resolution of the device and facility.

We would like to point out that the concept of a closed 3-D separated region being inaccessible (see Ref. 11) is valid only in the limit of particles moving near/on body surface. The upstream flow particles above the surface are able to access the separated region behind the line of separation through the spiral nature of the separation. Indeed, there is no 3-D separation which is totally closed by a separation surface, as a closed bubble; there must be some fluid flowing in and some fluid flowing out. All 3-D separation surfaces are a kind of vortex sheet.

Hereafter the fine-grid result will be used for discussion, except for cases specially mentioned.

2. Secondary Separation

The second question is the existence of a secondary separation. Experiments (for instance Ref. 13) very often show the appearance of another oil-accumulation line behind the line of primary separation. This has been interpreted as indicating the existence of a secondary separation, (see Fig. 6). The plots of velocity at the first mesh point above the flat plate, $K = 2$, (Fig. 7a) also show that, in addition to the outermost primary separation line, (not clearly visible on the figure), there is a second "line" of clustering or coalescence of velocity vectors. However, based on the result of surface particle traces (Fig. 7b) this is not a line of separation - it is merely a demarcation between regions of high and low surface skin friction. One might imagine that, in a transient stage of an experiment, comparatively more "oil" can be driven in by the high skin friction from the right and less oil carried out to the left near the region of strong variation in skin friction. (For convenience of discussion, here left or right refers to the orientation of one facing the streamwise direction.) Hence the surface may show a temporary accumulation of oil around this second line. Even as the wedge angle increases up to 12° (Fig. 8), our calculations still show no evidence of the existence of secondary separation. Indeed for the 12° case, behind the obvious primary separation line, there is a region that the particle traces show strongly convergent from one side and slowly divergent away from the other side. (These traces eventually converge to the primary separation line.) Hence a high clustering of particle traces in that region occurs. From the plot of velocity vectors in the plane $K = 2$ (Fig. 9a), one can see that this region is associated with the drastic change of the surface skin friction. The flow features near the wall are strongly affected by the surface pressure. Their relations and connections can be seen from Fig. 9b. A strong surface pressure gradient induces a high skin friction and a high velocity near the sidewall. This results in a divergence of particle traces and leads to an appearance of the so-called attachment line (Fig. 9b). After the strong pressure gradient, there is a region of drastic change of skin friction and hence a resulting appearance of clustering of velocity vectors near the wall. The strong pressure gradient appears mainly on the right of the inviscid shock location (Fig. 9b) and the clustering of velocity on the left (Fig. 9a). Note that the appearance of the clustering of particle traces (Fig. 8) does not coincide with the clustering of the velocity vectors near the wall. Instead, showing as a band with the clustering of the velocity vectors on its right, it is close to the pressure plateau region (see below). It is this pressure plateau region that causes the particle traces to run almost parallel to each other before they finally converge to the primary separation line. We would suspect that variations in surface-flow-visualization techniques would also result in different locations of temporary clustering of surface-streak lines.

Note that Degrez's calculation showed a noticeable 'dip' of pressure for the 6° case (see Fig. 4). When it is strong enough, this drop in pressure can significantly retard the primary separated flow (passing beneath the shock system in the opposite y-direction to the main flow) and lead to a secondary separation (see Fig. 12 of Ref. 3). In the present calculation, the pressure shows a plateau region and there is no secondary separation for either the 6° or 12° case. (The 12° case has a large plateau region with a little dip of pressure.) As discussed above a drastic change of velocity leads to a substantial change of skin friction which might also result in a temporary accumulation of oil flow on the surface. Therefore, it is possible that an accumulation of oil flow on the surface in an experiment is not necessarily a line of separation contrary to the usually inference. This argument has also arisen in previous experiments and calculations for other geometries [14, 15].

A note of caution should be given here. We don't know whether the appearance of a secondary oil-accumulation in an experiment is caused by a secondary separation or not. One possible alternative is suggested here. Furthermore, there might be other mechanisms in an experiment, especially for the turbulent case, that could lead to an oil accumulation on the surface. Further detailed and careful studies are needed to answer these questions.

3. Absence of Separation

Whether the boundary layer on the plate is separated or not is usually determined by comparing the turning angle of the limiting streamline with the glancing shock angle in the interaction region (Fig. 10). As shown in Figs. 5 and 8, there is clearly a line, with clustering of particle traces, that originates from a saddle point with turning angle greater than the angle of the glancing shock and the flows are separated. As the wedge angle decreases, one would expect that the turning angle of the skin friction line will decrease and eventually become smaller than the angle of the glancing shock. The flow then will be classified as attached What is the change of the flowfield topology from attached to separated flow? This question is addressed in the following section.

In the present paper a case with 2° wedge angle was calculated with the fine grid distribution. Fig. 11a shows traces of particles for which the origins are almost the same as those of the 6° case (Fig. 5). In contrast to Figs. 5 and 8, there is no obvious line of convergence of particle traces and the turning of skin-friction lines is smaller than the glancing shock angle; this would conventionally be interpreted as an attached case. However, a close examination of the particle traces near the apex (Fig. 11b) shows that actually the flow is separated. Even though it is very small, the separation also is a closed type, and the structure is topologically the same as that for the previous 6° and 12° cases. Actually, all three cases are topologically the same as the structure of a blunt-fin flow field.

One may imagine that, under certain conditions, this type of flow may not separate. However, the conventional method of interpretation using the turning angle of the boundary layer compared to the glancing shock angle as a criterion for separation is not uniformly valid, as demonstrated above. To date, the simplest and most general definition of a 3-D separation line, in the opinion of the author is that of Legendre [16]; "a line of separation . . . has no local property. Its only characteristic is to pass through a saddle point" (There may be doubt as to whether or not there is a local property as pointed out in Refs. 17 and 18.)

Another point should be mentioned here. Figs. 12 and 21 of Ref. 5 showed that, at a low wedge angle, based on the oil-flow picture the boundary layer on the sidewall was separated from a saddle point, as a closed type of separation, on the line of symmetry near the fin apex, and then gradually became attached away from the fin. (This is not an open-type separation as claimed in Ref. 5.) The question arises as to where and how the separation ends. In the opinion of the present author the simplest explanation is that this is case in which the definition of separation based on turning angle and oil-accumulation in experimental observation fails. Because of boundary layer growth on the fin, the shock wave is stronger and hence the pressure rise is higher near the leading edge of the fin than at a position downstream. The difference in pressure rise changes the turning angle and degree of oil accumulation. Based on the concept of continuity, limiting streamlines would not join together except at a singular point. A line of separation, once it originates from a saddle point, will either continue going downstream to infinity or terminate at a singular point. In the other words, once it is separated, the flow can not gradually be reattached without a singular point, according to topological imperatives.

4. Separation on Fin Surface

As sketched in Fig. 12 of Ref. 5, even at low wedge angle there is a separation on the fin surface, as the high pressure flow near the fin surface tries to flow across onto the sidewall. The question arises as to where and how the separation starts. For a boundary layer to separate, an adverse pressure gradient is necessary. Plots of particle traces restricted to the plane of J = 2 and surface pressure are shown in Figs. 12a and 12b for the 12° case. In Fig. 12a, in addition to the vortex spiral nature of the primary separation, we can see that the separation starts from a saddle point (a closed type). Correspondingly, there is a low pressure region and an adverse pressure gradient that triggers the separation (Fig. 12b). By overlaying Fig. 12a with Fig. 12b, one can see that the separation line is downstream of (or on the righthand side of), not coincident with, the pressure minimum along each limiting streamline. The existence of a low pressure region can be attributed to the gradual decrease of total pressure in the incoming boundary layer and hence a decease of the pressure rise after the shock (see Fig. 12b). The small increase in pressure near the plate results from flow stagnation. Topologically, there is even a 'hoseshoe vortex' (Fig. 12a) as in the blunt-fin solution, but it is too weak to induce a significant pressure gradient.

IV. CONCLUSION AND REMARKS

A laminar supersonic flow over a sharp fin mounted on a flat plate has been numerically simulated. Separation of the boundary layer on the flat plate was investigated for various grid refinements and fin wedge angles. Several basic issues concerning 3-D steady separation have been discussed.

The results of the solution have demonstrated that grid resolution can affect the "calculated" topology. For the coarse grid, the separation is an open type. As the grid spacing near the leading edge is refined, the starting point of the open-type separation moves and eventually the separation becomes a closed type. In the opinion of the author, some of the numerically observed open-type separations (for example, Fig. 4) results from insufficient grid resolution. Similarly, experiments have resolution problems, and some of the experimentally observed open-type separations may well be due to low resolution of the device and facility.

Based on the computation of surface particle traces, no secondary separation is found in the present study. A secondary oil-accumulation line has been conjectured to be a demarcation between regions of high and low surface skin friction.

In a calculation with a 2° wedge angle, there is no obvious line of convergent particle traces and the turning angles of skin-friction lines are smaller than the glancing shock angle. This is conventionally interpreted as an attached flow. However, a close examination of the particle traces near the apex has shown that actually the flow is separated, and the structure is the same topologically as that for the blunt-fin flow field.

REFERENCES

[1] Hung, C. M. and MacCormack, R. W.,"Numerical solution of three-dimensional shock-wave and turbulent boundary-layer interaction," AIAA J., Vol. 16, No. 12 Dec. (1978), pp. 1090-1096.

[2] Horstman, C. C. and Hung, C. M., "Computation of 3-D turbulent separated flow at supersonic speed," AIAA Paper No. 79-0002, (1979).

[3] Degrez, G.,"Computation of a three-dimensional skewed shock wave laminar boundary layer interaction," AIAA Paper No. 85-1565, (1985).

[4] Knight, D. D., Horstman, C. C., Shapey, B., and Bogdonoff, S., "The flowfield structure of the 3-D shock wave-boundary layer interaction generated by a 20° sharp fin at Mach 3," AIAA Paper No. 86-0343, (1986).

[5] Fomison, N. R. and Stollery, J. L.,"The effects of sweep and bluntness on a glancing shock wave turbulent boundary layer interaction," AGARD CP 428, paper 8, (1987).

[6] Dolling, D.S.,"Unsteadiness of supersonic and hypersonic shock-induced turbulent boundary layer separation," AGARD-FDP/VKI Special Course on "Three-Dimensional Supersonic and Hypersonic Flows Including Separation", May 8-12, 1989.

[7] Baldwin, B. S. and Lomax, H.,"Thin-layer approximation and algebraic model for separated turbulent flows," AIAA paper No. 78-257, Jan. (1978).

[8] Hung, C. M. and Kordulla, W.,"A time-split finite-volume algorithm for three-dimensional flowfield simulation," AIAA J., Vol. 22, No. 11, (1984), pp. 1564-1572.

[9] MacCormack, R.W. "A numerical method for solving the equations of viscous flow," AIAA J., Vol. 20, No. 9, (1982), pp. 1275-1281.

[10] Horstman, C. C., private communication.

[11] Wang, K.C.,"Boundary layer separation in three dimensions," Reviews in Viscous Flow, Proc of the Lockheed - Georgia Company Viscous Flow Symposium, June 1976, pp. 341-414.

[12] Tobak, M. and Peake, D. J.,"Topology of three-dimensional separation flows," NASA TM-81294, April 1981.

[13] Aso, S., Hayashi, M., and Tan, A. Z.,"The structure of aerodynamic heating in three-dimensional shock wave /turbulent boundary layer induced by sharp and blunt fins," AIAA paper no. 89-1854, June (1989).

[14] Hung, C.M.,"Computation of three-dimensional shock wave and boundary layer interactions," NASA TM-86780, Aug. 1985.

[15] Hung, C.M.,"Computation of separation ahead of blunt fin in supersonic turbulent flow," NASA TM-89416, Dec. 1986.

[16] Legendre, R., "Regular or catastrophic evolution of steady flows depending on parameters," Rech. Aerosp. Vol. 1982-4, pp. 41-49, (1982).

[17] Zhang, H. X.,"The separation criteria and flow behavior for three dimensional steady separated flow," translated from ACTA Aerodynamica Sinica, March (1985), pp. 1-12.

[18] Wu, J.Z., Gu, J.W., and Wu, J.M.,"Steady three-dimensional fluid particle separation from arbitrary smooth surface and formation of free vortex layers," AIAA Paper No. 87-2348,July (1987).

Table 1. Three Grid Systems (57x45x27)

	coarse	medium	fine
Δx min	0.10	0.02	.01
Δy min	0.0025	0.0025	0.0015
Δz min	0.001	0.001	0.001
plate leading edge	I = 4	I = 3	I = 3
fin leading edge	I = 14	I = 22	I = 22

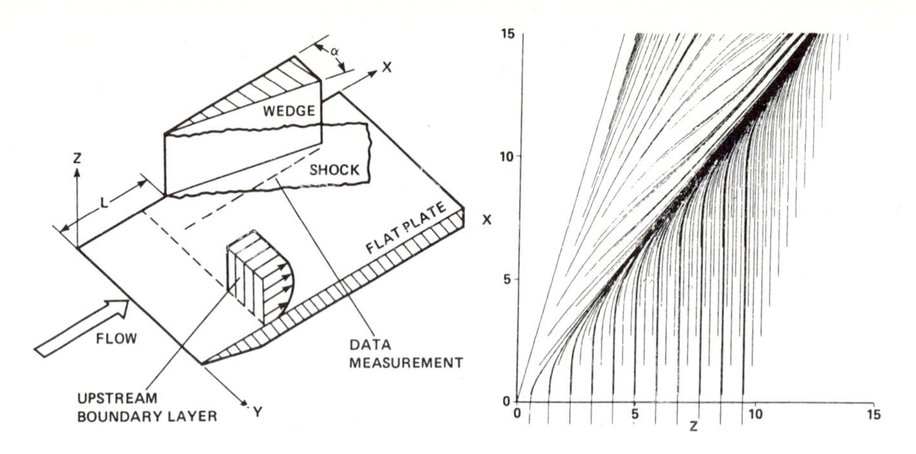

Fig. 1 Glancing shock wave over a boundary layer.

Fig. 4 Surface particle traces from Horstman (Ref. 4)

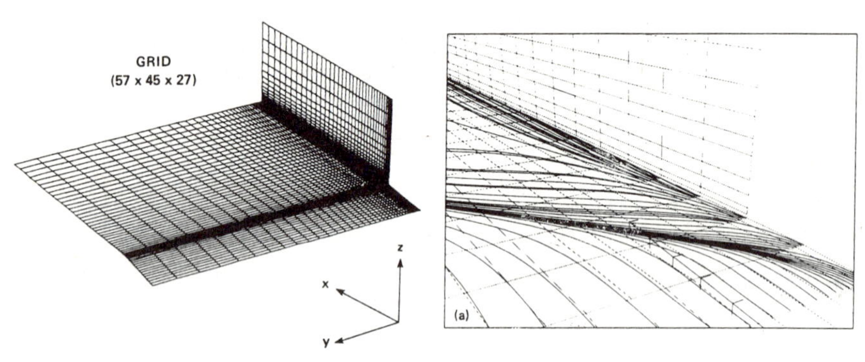

Fig. 2 Mesh system in fin and flat plate.

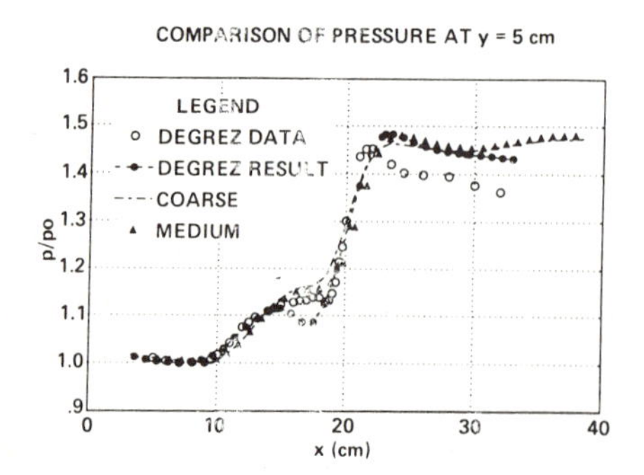

Fig. 3 Comparison of surface pressure at y = 5 cm.

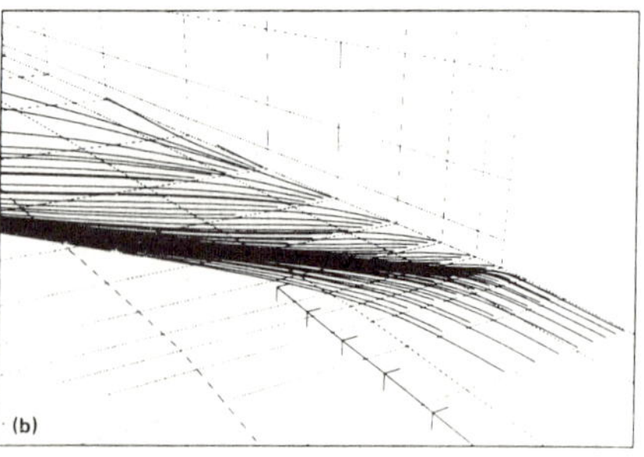

Fig. 5 Surface particle traces (K = 2) for α = 6°: (a) Coarse grid; (b) medium grid.

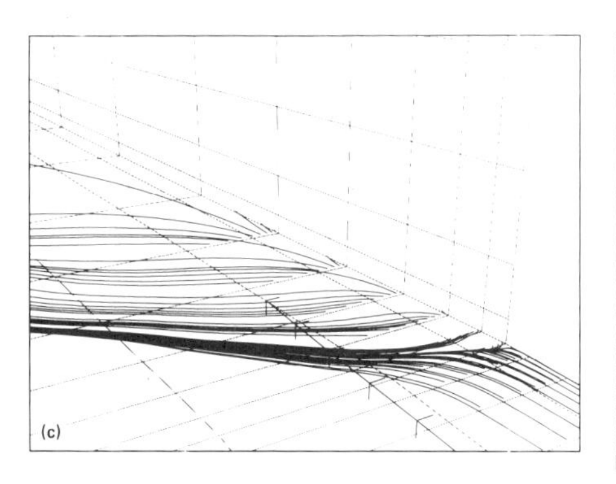

Fig. 5 Concluded. (c) Fine grid.

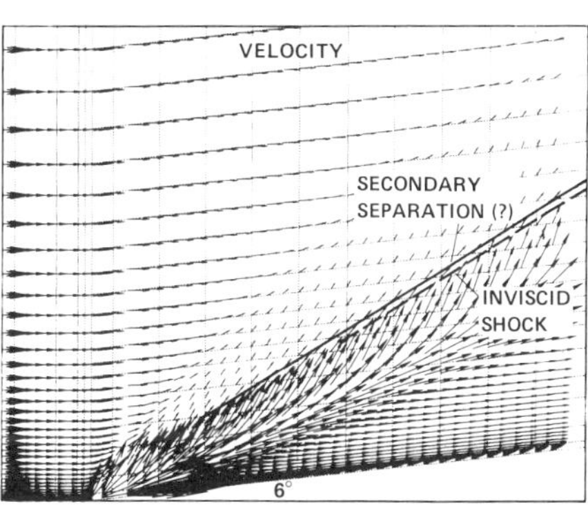

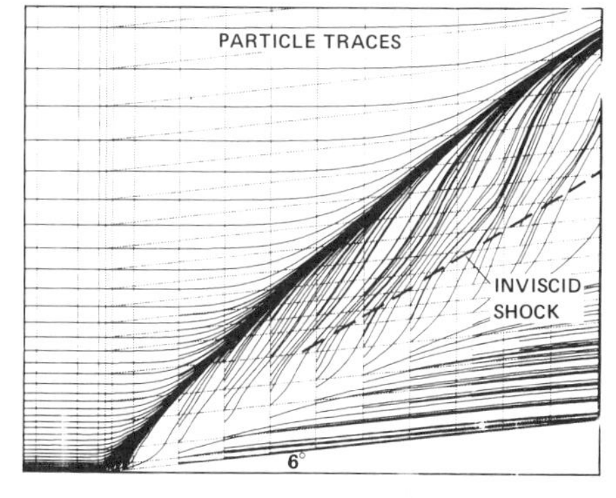

Fig. 7 Velocity vectors and particle traces for K = 2 (α = 6°): (a) Velocity vectors; (b) particle traces.

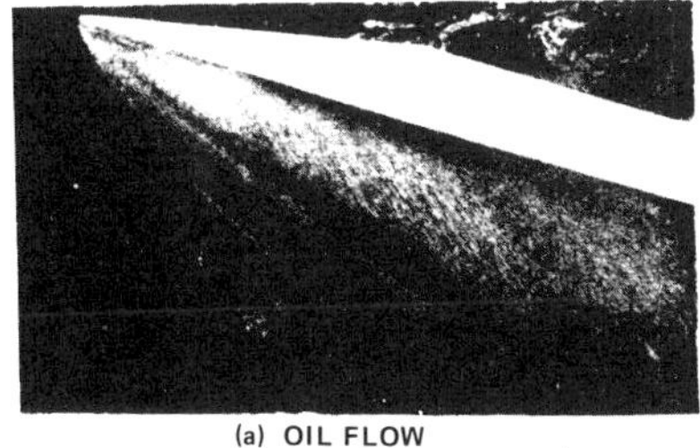

(a) OIL FLOW

0 α = 16° X

M_∞

SHARP FIN

PRIMARY REATTACHMENT

INVISCID SHOCK WAVE

Y n 20 mm

PRIMARY SEPARATION

SECONDARY SEPARATION

(b) SKETCH OF OIL STREAK LINES

Fig. 6 Surface oil flow by Aso, etc. (Ref. 13).

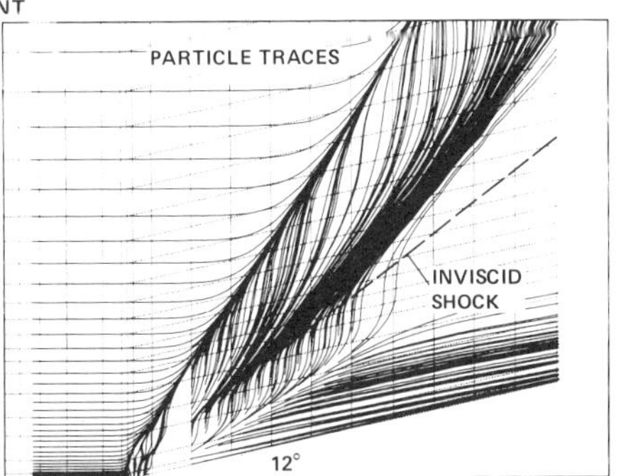

Fig. 8 Surface particle traces (K = 2) for α = 12°.

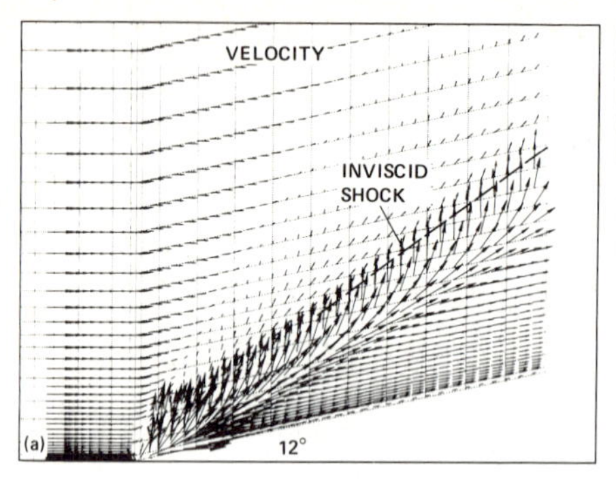

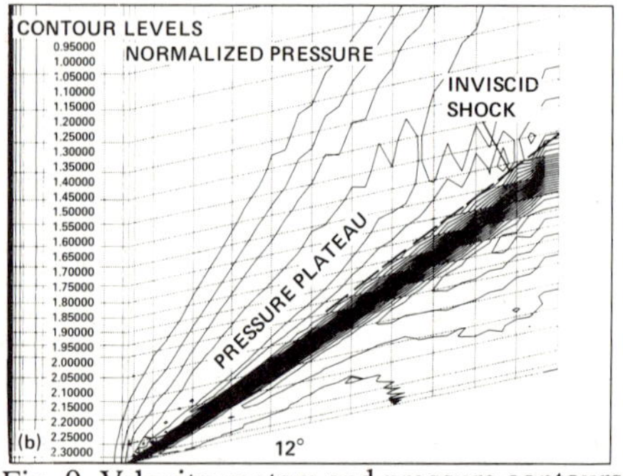

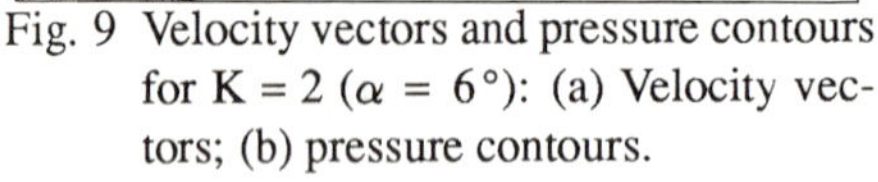

Fig. 9 Velocity vectors and pressure contours for K = 2 (α = 6°): (a) Velocity vectors; (b) pressure contours.

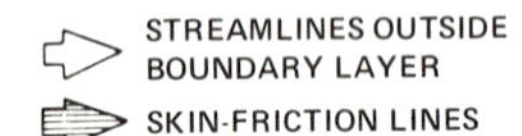

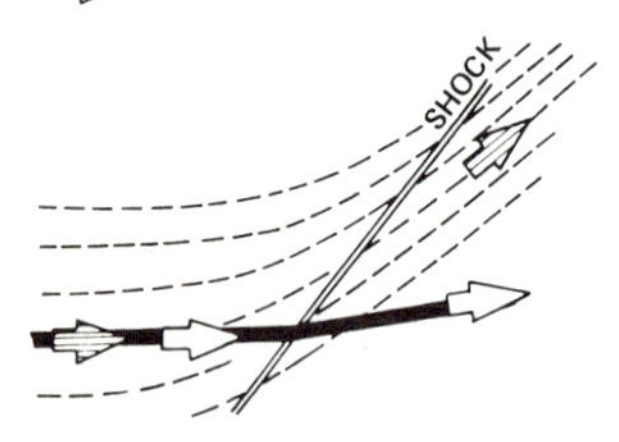

(a) ATTACHED FLOW

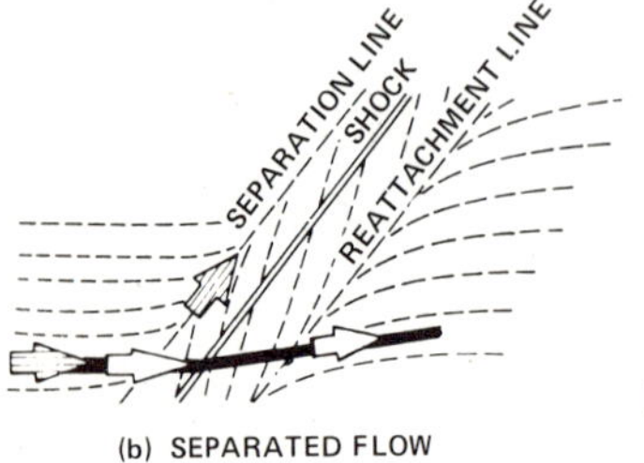

(b) SEPARATED FLOW

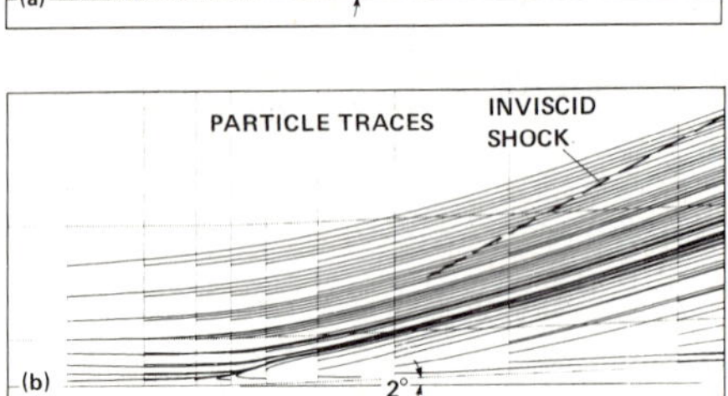

Fig. 11 Surface particle traces (K = 2) for α = 2°: (a) General feature; (b) near apex

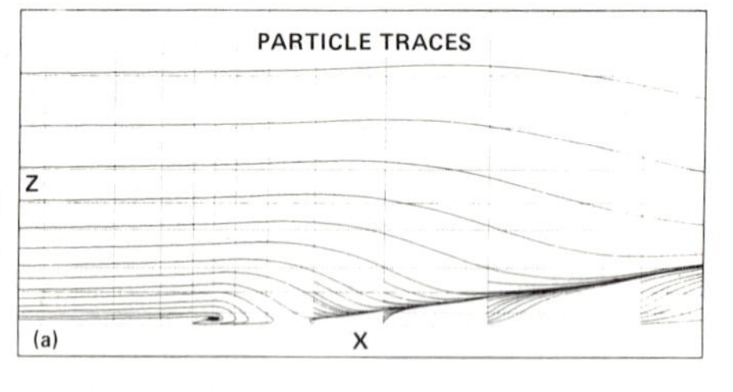

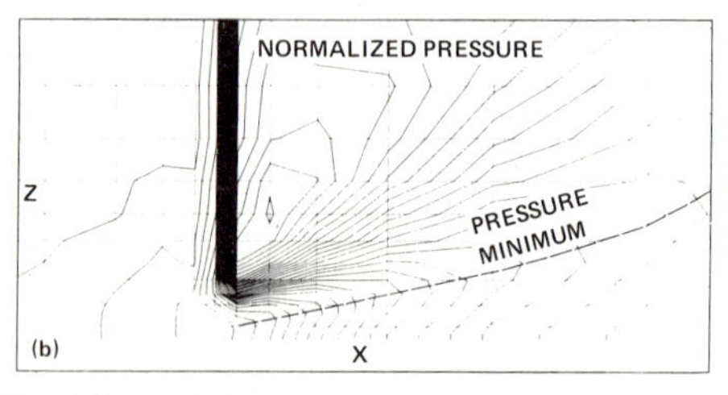

Fig. 12 Particle traces and pressure contours on the fin surface (J = 2) for α = 2°: (a) Particle traces; (b) pressure contours

Fig. 10 Sketches of skin-friction line patterns beneath glancing shock

ASPECTS OF THE APPLICATION OF AN EULER-EQUATION METHOD TO THE SIMULATION OF LEADING-EDGE VORTEX FLOW*)

by

J.M.J.W. Jacobs and H.W.M. Hoeijmakers

National Aerospace Laboratory NLR

Anthony Fokkerweg 2, Amsterdam, The Netherlands

SUMMARY

The flow about a 65-deg sharp-edged cropped delta wing is simulated by solving the Euler equations. Solutions are obtained for transonic vortex flow with and without shocks including cases with strong vortices. This paper concentrates on the assessment of the capability of the Euler method to simulate the details of the flow field, especially with respect to the influence of artificial dissipation and the influence of the computational mesh.

INTRODUCTION

Vortex flow associated with flow separation from wings with highly swept leading edges plays an important role in the high-angle-of-attack aerodynamics of various modern fighter aircraft. An accurate numerical simulation of the detailed flow field is of paramount importance to advance the understanding and utilization of this nonlinear type of flow. For wings with aerodynamically sharp leading edges it is known that at high Reynolds numbers the formation of the leading-edge vortex and the effect of the vortex on the flow over the upper surface of the wing is only slightly dependent on Reynolds number. This implies, that this high-Reynolds-numbers flow can be simulated using inviscid flow models. The Euler equations allow for rotational flow everywhere in the flow field and numerical methods based on them are able to capture along with shock waves vortical flow regions as an integral part of the discrete solution. This renders computational methods based on Euler's equations rather attractive for cases with a complex vortex flow pattern. This is in particular true for transonic flow with vortices and with (strong) shock waves, where there is the additional complexity of the mutual interaction of these two non-linear flow phenomena. Viscous effects, which are responsible for the occurrence of secondary separation and which affect the development of the flow near the center of the primary vortex core, will have to be assessed through a more elaborate mathematical model of the flow.

In the present paper the Euler method under development at NLR is applied to the simulation of the flow about a 65-deg sharp-edged cropped delta

*) This investigation has been carried out under contract with the Netherlands Agency for Aerospace Programs (NIVR) for the Netherlands Ministry of Defence.

wing at sub- and transonic speed. For this configuration experimental data is available and the flow structure is reasonably well known. The investigation concentrates on the assessment of the capability of the method to simulate the details of the vortex flow field. Especially two aspects of the numerical simulation are considered:
i) influence of the artifical dissipation, ii) influence of the computational mesh.

OUTLINE OF METHODS USED

For numerical flow simulation based on the Euler equations a system of computer programs is developed at NLR [1]. It includes a grid generator, a flow solver and a post-processing "flow visualization" facility.

The grid generator produces multi-blocked surface-fitted grids initialized through transfinite interpolation and tuned through elliptic methods. A description of the grid generator can be found in [2].

The numerical algorithm used to solve the time-dependent Euler equations is based on the fully-conservative finite-volume cell-centered scheme of Jameson, Schmidt and Turkel [3]. The conservation equations of mass, momentum and total energy are discretized using a central-difference scheme. The addition of artificial dissipative terms, as a blend of second and fourth differences, damps oscillations in regions with high pressure gradients e.g. near shock waves and stagnation points and suppresses the tendency for odd-even point decoupling. To obtain a steady-state solution, integration in time is carried out by a four-stage Runge-Kutta scheme. Convergence to a steady state is accelerated by the application of local time-stepping, enthalpy damping and residual averaging. For the conditions on the solid surface the method employs a linear extrapolation of the pressure to the wall. The numerical flow simulation system can handle a grid with a multi-block structure, implementing a special boundary condition at internal block interfaces to accommodate slope-discontinuous grid lines and jumps in cell size. It is permitted to have a grid with degenerated cells, i.e. cells with faces and/or edges collapsed to a line or a point.

Because solutions with different levels of artificial dissipation will be considered in this paper, a more detailed description of the artificial dissipation terms is given below. The dissipative term $\vec{D}_{i,j,k}$ added to the discrete approximation of the Euler equations is

$$\vec{D}_{i,j,k} = \vec{d}_{i+\frac{1}{2},j,k} - \vec{d}_{i-\frac{1}{2},j,k} + \vec{d}_{i,j+\frac{1}{2},k} - \vec{d}_{i,j-\frac{1}{2},k} + \vec{d}_{i,j,k+\frac{1}{2}} - \vec{d}_{i,j,k-\frac{1}{2}} \tag{1}$$

with for example

$$\vec{d}_{i+\frac{1}{2},j,k} = \frac{h_{i+\frac{1}{2},j,k}}{\Delta t} \{ \varepsilon^{(2)}_{i+\frac{1}{2},j,k} (\vec{w}_{i+1,j,k} - \vec{w}_{i,j,k}) - \varepsilon^{(4)}_{i+\frac{1}{2},j,k} (\vec{w}_{i+2,j,k} - 3\vec{w}_{i+1,j,k} + 3\vec{w}_{i,j,k} - \vec{w}_{i-1,j,k}) \} \tag{2}$$

with $\vec{w}$ the vector of the flow variables and $\frac{1}{\Delta t} h_{i+\frac{1}{2},j,k}$ a non-linear scaling factor depending on cell dimensions, the velocity vector, the speed of sound and the CFL-number.

Define

$$\nu_{i,j,k} = \frac{|p_{i+1,j,k} - 2p_{i,j,k} + p_{i-1,j,k}|}{|p_{i+1,j,k}| + 2|p_{i,j,k}| + |p_{i-1,j,k}|} \tag{3}$$

where p denotes the static pressure, then

$$\varepsilon^{(2)}_{i+\frac{1}{2},j,k} = \text{Min}\ (\tfrac{1}{2},\ K^{(2)}\nu_{i+\frac{1}{2},j,k})$$

$$\varepsilon^{(4)}_{i+\frac{1}{2},j,k} = K^{(4)}\text{Max}\ (0, \frac{1}{64} - \hat{\alpha}\nu_{i+\frac{1}{2},j,k}) \tag{5}$$

with $K^{(2)}$, $K^{(4)}$ and $\hat{\alpha}$ constants that control the amount of artificial dissipation in the solution and

$$\nu_{i+\frac{1}{2},j,k} = 0.5\ (\nu_{i+1,j,k} + \nu_{i,j,k}). \tag{6}$$

The second difference is used to damp oscillations near shocks, the fourth difference to suppress the tendency for odd-even point decoupling. Near shock waves the fourth difference tends to induce overshoots and therefore it can be switched off by setting the parameter $\hat{\alpha}$ at a positive value.

GRIDS

Three grids with different dimensions and point distributions have been used to compute the flow about the cropped delta wing. A "fine" grid, O-O topology, has been generated around the starboard half of the wing. The grid has grid dimensions 288x76x56 (1,225,728 cells), that is 144 cells on both the wing upper and lower surface in chordwise direction, 76 cells in spanwise direction and 56 cells between the wing surface and the outer boundary of the computational domain. By halving the number of cells in each direction a "medium" grid with dimensions 144x38x28 (153,216 cells) has been subtracted from the fine grid. The grid on the wing surface is "conical" on the forward portion of the configuration, as can be seen in Fig. 1 for the medium grid. The third grid is obtained from W.R. Marchbank of BAe Warton. This grid has a C-H topology, C wrapped around the chord and H in spanwise direction. The dimensions are 192x48x32 (294,912 cells), that is 192 cells in chordwise direction, 48 cells in spanwise direction and 32 cells in the direction normal to the wing. The wing upper and lower surface grids are non-conical, see Fig. 2, and consist each of 72 cells in chordwise direction and 34 cells in spanwise direction.

FLOW COMPUTATIONS

Calculations have been carried out to simulate the flow about the cropped delta wing for the flow conditions (M_∞ = 0.85, α = 10°), (M_∞ = 0.85, α = 20°) and (M_∞ = 0.50, α = 20°). For M_∞ = 0.85 experimental data are available obtained in the NLR high speed wind-tunnel during the International Vortex Flow Experiment [4]. For all calculations the solver converged in 1500 to 2500 iterations, dropping the root mean square norm of the time-like variation of the density by four orders of magnitude. For M_∞ = 0.85, α = 10° the influence of different levels of artificial dissipation on the flow solutions has been studied on the C-H grid. The influence of the computational mesh on the solution has been investigated for all three flow conditions.

Influence of artificial dissipation

For the flow condition of $M_\infty = 0.85$, $\alpha = 10°$, converged solutions have been obtained on the C-H grid for three different sets of values of the artificial dissipation parameters, see table 1. For solution I the parameters have the default values for transonic flow computations. As for the present free-stream condition shocks do not form in the flow domain, a second solution has been computed without the second difference contributing to the artificial dissipative term $\vec{D}_{i,j,k}$. For this solution II the parameters are at their default setting for sub-critical flow. With the parameters set at the minimum values still giving convergence, solution III is computed. For the three different solutions the spanwise pressure distributions at $x/c_R = 0.6$ are depicted in Fig. 3a, while the corresponding spanwise distributions of total pressure loss are presented in Fig. 3b. Fig. 3a indicates that the pressure peak on the upper wing surface becomes steeper and higher when the second-order dissipation is switched off. Reducing the parameter controlling the fourth-order dissipation has a similar effect but of much smaller magnitude. It is shown in Fig. 3b that on part of the wing the total pressure losses turn into total pressure gains when the second-order dissipation is switched off. Total pressure losses are highest at the leading edge.
Fig. 4 compares for solutions I and III, in the plane $x/c_R = 0.6$, the flow field values of the total pressure coefficient. This figure shows that with decreasing artificial dissipation the area where total pressure losses occur becomes somewhat smaller, but that the maximum value, which occurs at the center of the core remains the same. It is concluded that for the present case the solution for default values of the artificial dissipation parameters for transonic flow computations is sufficiently close to the solution with the minimum amount of artificial dissipation needed for convergence. So it is warranted to use these default values in future calculations.

Influence of computational mesh

$M_\infty = 0.85$, $\alpha = 10°$

For the transonic case of $M_\infty = 0.85$, $\alpha = 10°$ the result computed on the medium grid is compared with the result computed on the C-H grid, both for the default values of the artificial dissipation parameters. Comparing the upper-wing-surface isobar pattern of the medium grid solution with the pattern of the C-H grid result, Fig. 5, one observes that the qualitative flow picture is very similar. Both results show the footprint of the leading-edge vortex indicating that the flow starts separating close to the apex. However, due to the better grid point distribution in the apex region the vortex is stronger in the result obtained on the "quasi-conical" medium grid. Comparing for the station $x/c_R = 0.6$ the spanwise pressure distributions computed on the medium grid and the C-H grid with each other and with experimental data, Fig. 6, one observes that the suction peak of the medium grid solution is higher and steeper than of the C-H grid solution. Also the re-attachment point (the local pressure maximum inboard of the peak) is more outboard for the medium grid solution than for the C-H grid solution. The numerical results exhibit a spike in the pressure distribution at the leading edge. The lower-wing-surface pressure distributions of both calculations agree excellently with each other and with the experimental data. The comparison of the computed upper-wing-surface spanwise pressure distributions and experimental data is confirms earlier findings [5], that the main difference between the Euler solution and experimental data is due to the omission of secondary separation modeling in the present calculations. The secondary separation causes the primary vortex to move inboard and upward. The effect on the upper surface pressure distribution is that the suction peak moves inboard and decreases in height, while outboard of the suction peak there is a region of about constant pressure.

In Figs 7 a-b the contours of constant total pressure are presented for the cross-flow plane $x/c_R = 0.6$ as computed on the medium and the C-H grid, respectively. Comparison of these two figures shows that the maximum total pressure loss is slightly higher on the C-H grid while the position of maximum total pressure loss, assumed to coincide with the center of the vortex core, is about the same for the two solutions. It may be concluded that the medium grid solution is at least as good as the C-H solution (in the neighbourhood of the apex even better) although the medium grid has only half the number of cells of the C-H grid (but about the same number of grid points on the surface of the configuration).

$M_\infty = 0.85$, $\alpha = 20°$

For this high angle-of-attack transonic flow case, calculations have been carried out on the medium and the fine grid. In Fig. 8 the isobar pattern on the upper wing surface is shown. The left-hand side of the picture presents the result for the medium grid and the right-hand side the result for the fine grid. Both show the formation of a leading-edge vortex which starts very close to the apex. Furthermore, on both sides the closely-spaced isobars outboard of the pressure minimum indicate the presence of a "cross-flow shock". This shock is strongest on the rear part of the wing. On this part of the wing a second shock appears, which at about 95% root chord merges with the cross-flow shock, forming a Y-shaped shock system. On the central part of the wing at about 83% root chord a so-called "rear shock" appears. This latter shock is normal to the plane of symmetry and extends some distance in the direction of the leading-edge. Comparison of the upper wing surface isobar pattern of the fine with that of the medium mesh result shows that the shocks are steeper, but more importantly, that no new flow features evolve. In Fig. 9, the spanwise pressure distributions at the station $x/c_R = 0.6$ computed on the fine and the medium grid are compared with each other and with experimental data. At about 60% semi-span the experimental data indicate a cross-flow shock provoking an early and strong secondary separation. In the computed results the cross-flow shock is located further outboard and the pressure peak is higher. The differences are primarily due to the absence of a secondary separation model in the computational method. For the lower wing surface the computed results agree excellently with the measured data. Comparison of the results
obtained on the medium and the fine grid learns that increasing the grid density: i) results in a steeper and also somewhat stronger "cross-flow shock", ii) shifts the suction peak slightly inboard, iii) steepens the "rear shock", iv) has no effect on the lower wing surface pressure distribution. The cross-flow-plane contours of the total pressure at the chord-

wise station x/c_R = 0.6, Figs. 10 a-b, show that on both grids at the center of the vortex core the total pressure drops to 45% of its free-stream value. However, compared to the medium grid the region with total pressure losses has shrunk on the fine grid. Furthermore, on the fine grid the position of maximum total pressure loss is slightly more inboard than on the medium grid.

$M_\infty = 0.50$, $\alpha = 20°$

For this case of strong vortex flow converged solutions were obtained on the medium and on the fine grid. The upper wing surface isobars as computed on the medium grid are presented in the left-hand side of Fig. 11, while the right-hand side of the same figure shows the fine grid isobars. For both levels of the grid the low-pressure region on the wing, which is the footprint of the leading-edge vortex on that wing, indicates that flow separation from the leading-edge starts very close to the apex. The lowest values of the surface pressure coefficient are found near the apex underneath the vortex, typical for subsonic flow cases where there is a relatively large effect of the singularity at the apex as well as a strong upstream influence of the trailing edge. Comparison of the left-hand side of the picture with the right-hand side shows that the character of the solution on the medium grid and the one on the fine grid do not differ.
However, on the fine grid the pressures in the peak are substantially lower than found on the medium grid. In Fig. 12 the spanwise pressure distributions as computed on the medium and on the fine grid are compared with each other for the station x/c_R = 0.6. When the grid is refined the upper-wing-surface pressure peak becomes steeper and higher while its position shifts slightly in outboard direction. There is an excellent agreement of the spanwise pressure distributions on the lower wing surface as well as on the upper wing surface in the regions close to the plane of symmetry and near the leading edge. The maximum value of total pressure loss in the cross-flow plane x/c_R = 0.6, Fig. 13, is somewhat higher in the fine grid solution than in the medium grid solution. Furthermore, on the fine grid the area in which there are total pressure losses is smaller, i.e. it is more compact that on the medium grid. Apparently the reduction of the errors in total pressure due to grid refinement cannot balance the increase of the errors due to resolution of larger gradients. The global characteristics computed for this case are presented in table 2. The differences between the forces and moments for the medium and fine grid solutions are small, which might be inferred from the spanwise pressure distributions which show that the area underneath the higher and steeper peak is about equal to the area underneath the lower but broader peak.

CONCLUDING REMARKS

The NLR flow simulation system based on Euler's equations has been employed to simulate the flow about a sharp-edged cropped delta wing. For $M_\infty = 0.85$, $\alpha = 10°$ the solution computed on the C-H type grid with the parameters controlling the amount of artificial dissipation set at the lowest possible values for convergence agrees satisfactorily with the solution with the parameters set at their standard values.
On the O-O type medium grid fully converged solutions were obtained for two transonic cases of compressible vortex flow at $M_\infty = 0.85$, namely $\alpha = 10°$ and $\alpha = 20°$ both with and without a "cross-flow shock" and a "rear

shock", as well as for a case of strong vortex flow, i.e. $M_\infty = 0.50$, $\alpha = 20°$. All cases showed the formation of a leading-edge vortex, none of the cases showed any sign of vortex burst.
On the 0-0 type fine grid fully converged solutions were obtained for $M_\infty = 0.85$, $\alpha = 20°$ showing a strong "cross-flow shock" and a weak "rear shock" and for $M_\infty = 0.50$, $\alpha = 20°$. The solution on the fine grid did not reveal other flow phenomena than the ones already observed on the medium grid, but the flow is better resolved in regions with large gradients.
The differences between the computed surface pressure distributions and experimental data are primarily due to the effect of secondary separation, which is not modeled in the present numerical simulation. To obtain a better agreement between theory and experiment it is necessary to include some modeling of the effect of secondary separation in the simulation.
Further investigation is required to explain to occurrence of the spike in the pressure distribution at the sharp leading edge, this in relation with the investigation of the process of flow separation at such edge.

ACKNOWLEDGEMENT

The authors thank W.R. Marchbank of BAe-Warton for providing the C-H grid

REFERENCES

[1] Boerstoel, J.W., "Progress Report of the Development of a System for the Numerical Simulation of Euler FLows, with Results of Preliminary 3D Propeller-Slipstream/Exhaust-Jet Calculations", NLR TR 88008 U (1988).

[2] Jacobs, J.M.J.W., Kassies, A. Boerstoel, J.W., Buijsen, F., "Numerical Interactive Grid Generation for 3D Flow Calculations", In: Numerical Grid Generation in Computational Fluid Mechanics '88, Pineridge Press, 1988, pp 925-943.

[3] Jameson, A., Schmidt, W., Turkel, E., "Numerical Solution of Euler Equations by Finite Volume Methods Using Runge-Kutta Time-Stepping Scheme", AIAA paper 81-1259 (1981).

[4] Hirdes, R.H.C.M., "U.S./European Vortex Flow Experiment - Test Report of Wind-Tunnel Measurements on the 65° Wing in the NLR High Speed Tunnel HST", NLR Technical Report 85046 L (1985).

[5] Jacobs, J.M.J.W., Hoeijmakers, H.W.M., Van den Berg, J.I., Boerstoel, J.W., "Numerical Simulation of the Flow about a Wing with Leading-Edge Vortex Flow", Proc. 11th Int. Conf. Num. Meth. in Fl. Dyn., Williamsburg, 1988.

TABLE 1 Characteristics of three solutions on the BAe Grid; $M_\infty = 0.85$, $\alpha = 10°$

SOLUTION	$K^{(2)}$	$K^{(4)}$	$\hat{\alpha}$	C_L	C_D
I	1.0	2.0	0.5	0.529	0.0867
II	0.0	2.0	0.0	0.537	0.0880
III	0.0	1.75	0.0	0.538	0.0882

TABLE 2 Characteristics of two solutions on different grids; $M_\infty = 0.50$, $\alpha = 20°$

GRID	C_L	C_D	C_{MY} (apex)	C_{MX} (root)
MEDIUM	1.078	0.3703	-0.664	0.1945
FINE	1.071	0.3676	-0.660	0.1939

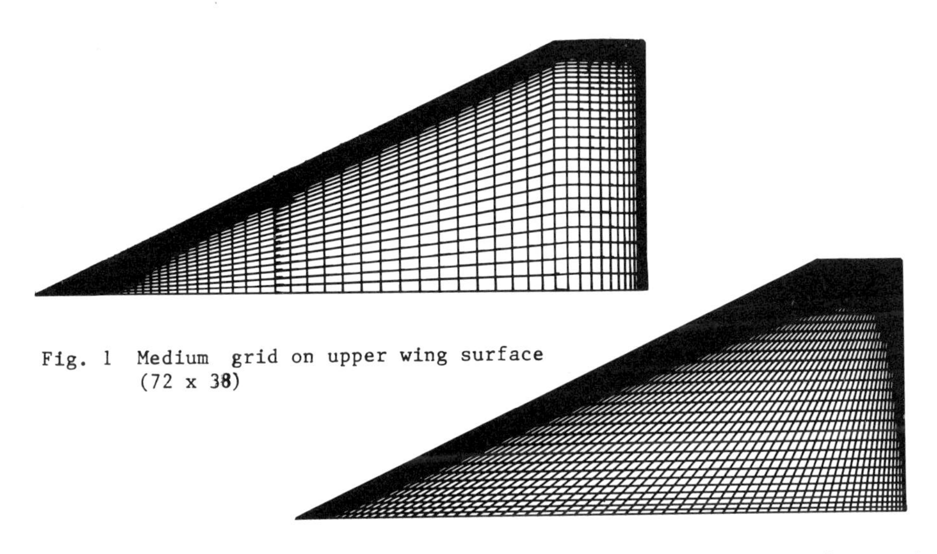

Fig. 1 Medium grid on upper wing surface (72 x 38)

Fig. 2 C-H grid on upper wing surface (72 x 34)

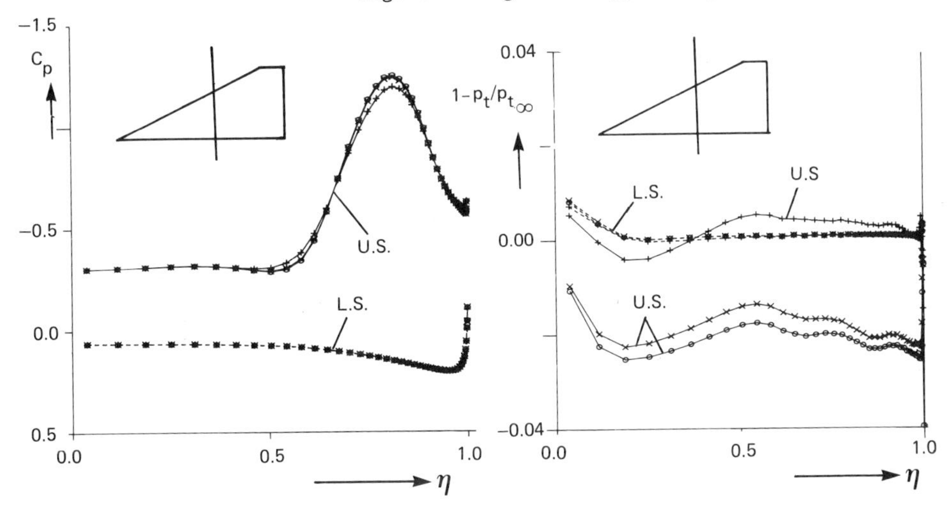

		$K^{(2)}$	$K^{(4)}$	$\hat{\alpha}$
I	+	1.0	2.0	0.5
II	×	0.0	2.0	0.0
III	o	0.0	1.75	0.0

MACH = 0.85
ALPHA = 10 DEG

U.S. : UPPER SURFACE
L.S. : LOWER SURFACE

Fig. 3a Spanwise pressure distribution at $x/c_R = 0.6$, different levels of artificial dissipation

Fig. 3b Spanwise distribution of total pressure at $x/c_R = 0.6$, different levels of artificial dissipation

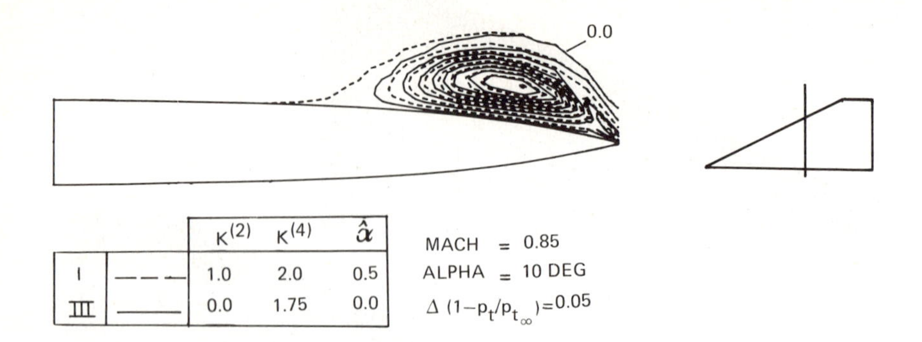

		$K^{(2)}$	$K^{(4)}$	$\hat{\alpha}$
I	----	1.0	2.0	0.5
III	——	0.0	1.75	0.0

Fig. 4 Comparison of contours of constant total pressure loss at x/c_R = 0.6. Different levels of artificial dissipation.

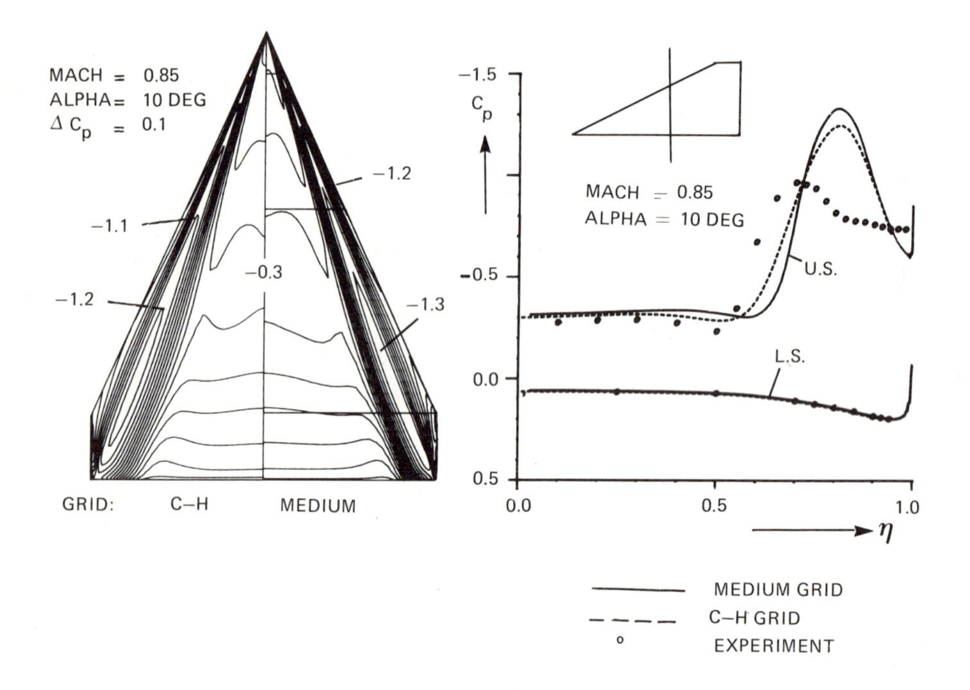

Fig. 5 Isobars on wing upper surface, different grid topologies

Fig. 6 Spanwise pressure distribution at x/c_R = 0.6, different grid topologies

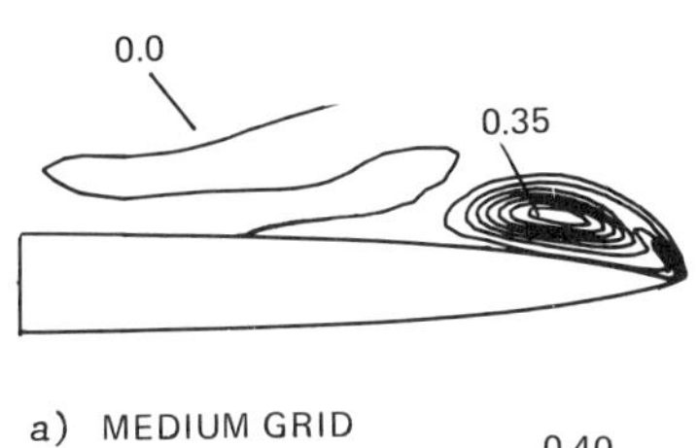

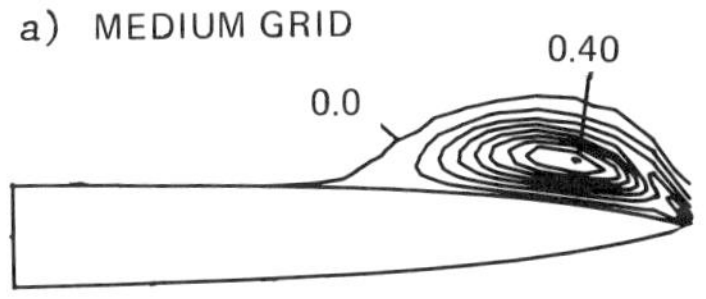

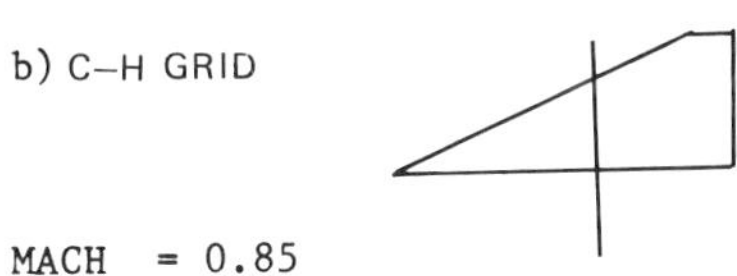

MACH = 0.85
ALPHA = 10 DEG
$\Delta(1-p_t/p_{t\infty}) = 0.05$

Fig. 7 Contours of constant total pressure loss in plane $x/c_R = 0.6$, different grid topologies

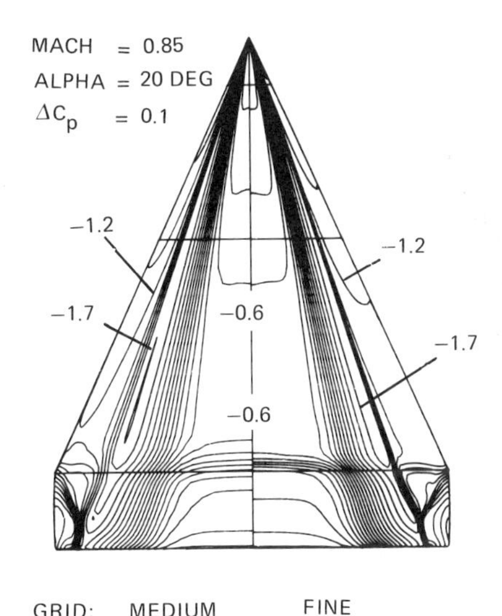

Fig. 8 Isobars on wing upper surface, different grid densities

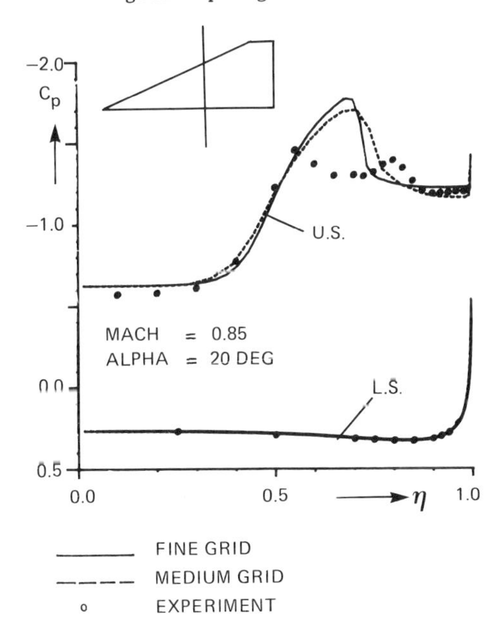

Fig. 9 Spanwise pressure distribution at $x/x_R = 0.6$, different grid densities

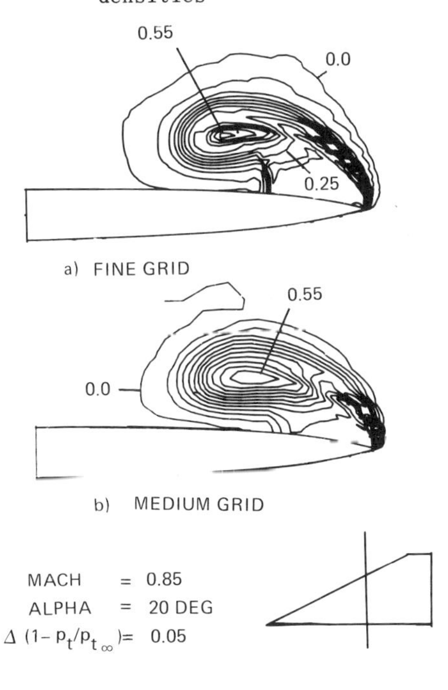

Fig. 10 Contours of constant total pressure loss at $x/c_R = 0.6$, different grid densities

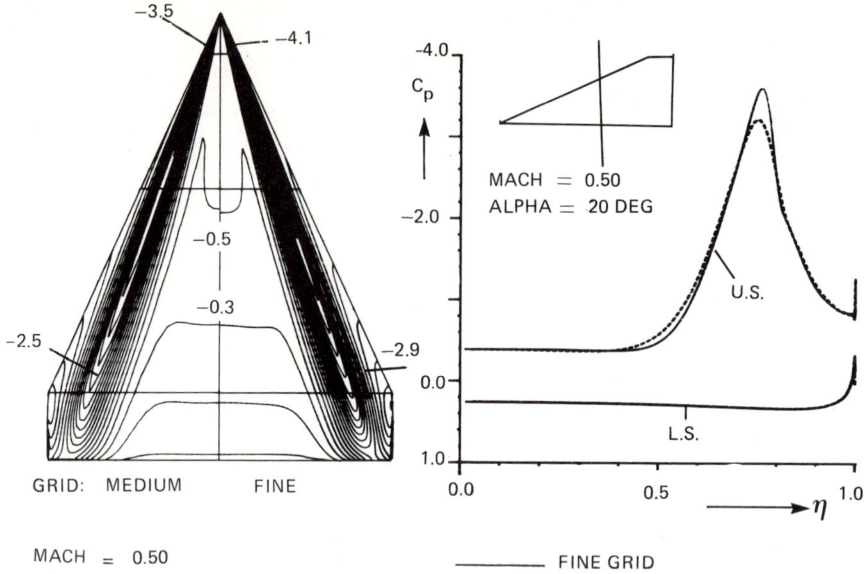

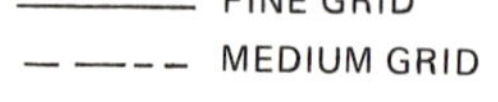

Fig. 11 Isobars on wing upper surface, different grid densities

Fig. 12 Spanwise pressure distribution at x/c_R = 0.6, different grid densities

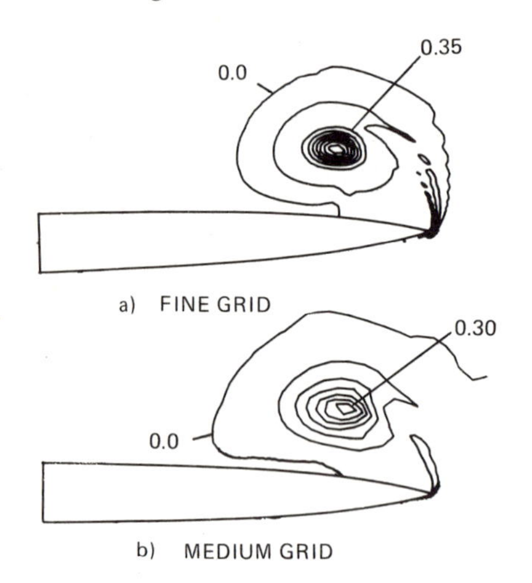

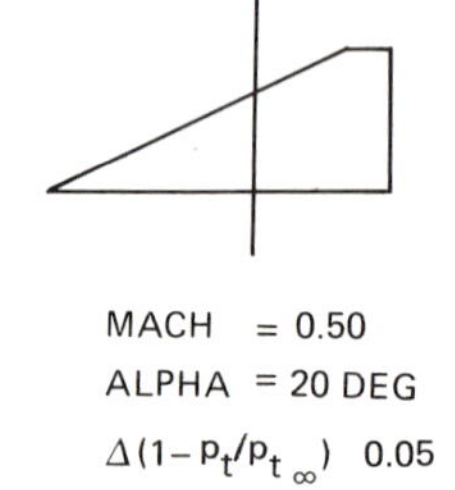

Fig. 13 Contours of constant total pressure loss at x/c_R = 0.6, different grid densities

An Incompressible Inner Flow Analysis by Absolute Differential Form of Navier-Stokes Equation on a Curvilinear Coordinate System

K.Katsuragi and O.Ukai

System Engineering Department, Mitsubishi Heavy Industries,Ltd.
1-1,1-Chome,Wadasaki-cho,Hyogo-ku, Kobe 652 Japan

Summary

In this paper, an incompressible inner flow analysis scheme which is based on the absolute differential form for the conventional equations of the incompressible viscous flow is presented. The physical components of the contravariant vector are taken as velocity variables and the equations are discretized on a staggered mesh system. Consequently, the calculation is performed stably and the good agreement with the experimental results has been shown in the prediction of the natural convection in the horizontal eccentric cylindrical annuli.

Introduction

In the field of a compressible outer flow analysis, most of the applications uses a general curvilinear coordinate system, which is so called the Body-Fitted Method [1,2]. In these applications, the governing equations are transformed from the Cartesian coordinate by the chain rule. So the equations are written with the partial differential form (hereafter PD form) and there are three characteristics in these PD form equations.

(1) Two kinds of velocity vectors are used, the one is the physical velocity vectors in the Cartesian coordinate which is the covariant vectors u_i, and the other is the contravariant velocity vectors U^i.

(2) The unknown velocity variables are the covariant vector u_i.

(3) The continuity equation is expressed by the contravariant vector U^i.

Unlike the expression on the Cartesian coordinate, the continuity equation is not expressed by the unknown velocity variables explicitly. Therefore when the governing equations of the the PD form are solved, it is necessary to convert the unknown velocity variables from the covariant vector to the contravariant vector. If all variables are arranged at the same grid point, the interpolation of the velocity with respect to the space is not necessary. Because of this simplicity, the regular grid system is mainly used on a compressible outer flow analysis.

However, if we apply the same technique to an incompressible inner flow analysis, it is difficult to achieve the exact satisfaction of the continuity equation and to develop the algorithm which has the good accuracy and numerical stability[3].

Particulary, when the thermal hydraulic phenomena that the buoyancy affects are analyzed, the error of the continuity equation induces the fictitious temperature distribution. And then, the boundary conditions at the outlet and on the wall become very complicated.

Shyy et al.[4] have arranged the variables staggered to satisfy the continuity equation exactly. But the velocity interpolation has become necessary somewhere and the treatment of the boundary conditions have become more complicated.

These troubles are arised from the existence of two kinds of the velocity vectors in the governing equations. Recently, Daiguji et al.[5] showed the capability of the absolute differential in a subsonic outer flow analysis to estimate a no-slip boundary condition precsiely. He transformed the governing equations by the tensor calculus and obtained the absolute differential form (hereafter AD form) equations. The absolute differential is indpendent of the coordinate system, so that the usual FDM scheme on the Cartesian coordiante can be applied to the general curvilinear coordinate.

But the contravariant vector is sensitively effected by the geometric quantities. Furthermore, as it has not always the physical dimension of the velocity, the velocity differential with respect to the space has no physical meaning.

In this paper we would like to present the Body-Fitted Method of the AD form. The physical components[6] of the contravariant vectors are taken as the velocity variables[7].

The characteristics of this method are following two points.

(a) In order to exactly satisfy the continuity equation, the equations are discretized on the staggered mesh system. This is very important to the thermal hydraulic analyses.

(b) In contrast to the conventional method using the tensor components, the physical components of the contravariant vector are used as the velocity varibles. Therefore, the differential maintains its physical meaning.

As the examples, the natural convection problems in the annuli between horizontal eccentric cylinders[8] are adopted here. The applicability of the present scheme to calculate an incompressible inner flow with the buoyant effect is examined and is verified by the comparison with the experimental results.

Governing Equations

As the governing equations of an incompressible inner flow in the Cartesian coordinate, we use the continuity, Navier-Stokes and energy equations as follows :

$$\frac{\partial u_j}{\partial x^j} = 0 , \qquad (1)$$

$$\frac{\partial u_i}{\partial t} + u_j \frac{\partial u_i}{\partial x^j} = - \frac{1}{\rho_0} \frac{\partial P}{\partial x^i} + \frac{\partial \sigma^{ij}}{\partial x^j} + f_i , \qquad (2)$$

$$\frac{\partial T}{\partial t} + u_j \frac{\partial T}{\partial x^j} = \frac{\partial e^j}{\partial x^j} . \qquad (3)$$

Here, u_i, t , P , ρ_0 and T are the covariant velocity vector, time, pressure a representative density and temperature, respectively. The stress tensor σ^{ij}, the heat flux e^j and the buoyant term f_i are described as :

$$\sigma^{ij} = \nu \left[\frac{\partial u_i}{\partial x^j} + \frac{\partial u_j}{\partial x^i} \right] , \qquad (4)$$

$$e^j = a \frac{\partial T}{\partial x^j} , \qquad (5)$$

$$f_i = -g_i \beta (T - T_{ref}) , \qquad (6)$$

where ν, a, g_i , β and T_{ref} are the kinematic viscosity, thermal diffusivity, the gravity acceleration in x^i-direction, the coefficient of the thermal expansion and the reference temperature, respectively.

According to the tensor calculus, let replace the partial derivative operators in Eqs.(1)-(3) with the absolute ones, and then we introduce the equations of the AD form, which are written with the contravariant vector U^i, as follows ·

$$\nabla_j U^j = 0 , \qquad (7)$$

$$\frac{\partial U^i}{\partial t} + U^j \nabla_j U^i = - \frac{1}{\rho} (g^{ij} \nabla_j P) + \nabla_j \sigma^{ij} + F^i , \qquad (8)$$

$$\frac{\partial T}{\partial t} + U^j \nabla_j T = \nabla_j e^j , \qquad (9)$$

where

$$\nabla_j U^i = \frac{\partial U^i}{\partial \xi^j} + \left\{ {i \atop j\ k} \right\} U^k . \qquad (10)$$

Eqs.(4)-(6) are also denoted as the following way :

$$\sigma^{ij} = \nu[g^{jk}\nabla_k U^i + g^{ik}\nabla_k U^j] , \tag{11}$$

$$e^j = a[g^{jk}\nabla_k T] , \tag{12}$$

$$F^i = \frac{\partial \xi^i}{\partial x^j}\{-g_j\beta(T-T_{ref})\} . \tag{13}$$

Using the basic metric tensor g_{jk}, g^{jk} is defined by

$$g^{jk} = \frac{G^{jk}}{g} , \tag{14}$$

where G^{jk} and g represent the cofactors and the determinant of the matrix which has g_{jk} as the (j,k)th component, respectively.

$\left\{{i \atop j\ k}\right\}$ is called the Christoffel symbol and defined by

$$\left\{{i \atop j\ k}\right\} = \frac{1}{2} g^{im}\left[\frac{\partial g_{km}}{\partial \xi^j} + \frac{\partial g_{jm}}{\partial \xi^k} - \frac{\partial g_{jk}}{\partial \xi^m}\right] . \tag{15}$$

In Eqs.(7)-(9), the velocity variables are non-physical contravariant vectors , which depend on the grid configuration. Therefore the spatial derivatives may have non-zero values on the curvilinear coordinate even if the flow velocity is uniform on the Cartesian coordinate[7]. Thus we transform the velocity variables in Eqs. (7)-(9) into the physical components, which is defined by Eringen[5].

The physical component of the contravariant vector is defined as follows :

$$U^{(i)} = \sqrt{g_{\underline{ii}}}\,U^i , \tag{16}$$

Here the underscore represents that the summation convention is suspended for these indices.

By substituting Eq.(16) into Eqs(7)-(9), we introduce the governing equations written with the physical components of the contravariant vector.

$$\frac{1}{\sqrt{g}}\left[\frac{\partial(\sqrt{g}\,U^{(j)}/\sqrt{g_{\underline{jj}}})}{\partial \xi^j}\right] = 0 , \tag{17}$$

$$\frac{\partial U^{(i)}}{\partial t} + \frac{U^{(j)}}{\sqrt{g_{\underline{jj}}}}\left[\frac{\partial U^{(i)}}{\partial \xi^j} - \frac{U^{(i)}}{\sqrt{g_{\underline{ii}}}}\frac{\partial \sqrt{g_{\underline{ii}}}}{\partial \xi^j} + \frac{\sqrt{g_{\underline{ii}}}}{\sqrt{g_{\underline{kk}}}}\left\{{i \atop j\ k}\right\} U^{(k)}\right]$$

$$= -\frac{\sqrt{g_{\underline{ii}}}}{\rho}\left[g^{ij}\frac{\partial P}{\partial \xi^j}\right] + \sqrt{g_{\underline{ii}}}\left[\frac{1}{\sqrt{g}}\frac{\partial(\sqrt{g}\,\sigma^{(ij)})}{\partial \xi^j} + \left\{{i \atop j\,k}\right\}\sigma^{(kj)}\right] + F^{(i)}, \tag{18}$$

$$\frac{\partial T}{\partial t} + \frac{U^{(j)}}{\sqrt{g_{\underline{jj}}}}\frac{\partial T}{\partial \xi^j} = \frac{1}{\sqrt{g}}\left[\frac{\partial(\sqrt{g}\,e^{(j)})}{\partial \xi^j}\right], \tag{19}$$

where

$$\sigma^{(ij)} = \nu\left[g^{jk}\frac{\partial(U^{(i)}/\sqrt{g_{\underline{ii}}})}{\partial \xi^k}\right], \tag{20}$$

$$e^{(j)} = a\left[g^{jk}\frac{\partial T}{\partial \xi^k}\right], \tag{21}$$

$$F^{(i)} = \sqrt{g_{\underline{ii}}}\left[\frac{\partial \xi^i}{\partial x^j}\{-g_j\beta(T-T_{ref})\}\right]. \tag{22}$$

The definition of the Christoffel symbol described as Eq.(15) is very complicated. We employed the following simple expressions of the Christoffel symbol in two-dimension using the basic geometrical quantities [6]:

$$\left\{{1 \atop 1\,1}\right\} = \frac{y_\eta x_{\xi\xi} - x_\eta y_{\xi\xi}}{J}, \tag{23-a}$$

$$\left\{{1 \atop 1\,2}\right\} = \left\{{1 \atop 2\,1}\right\} = \frac{y_\eta x_{\xi\eta} - x_\eta y_{\xi\eta}}{J}, \tag{23-b}$$

$$\left\{{1 \atop 2\,2}\right\} = \frac{y_\eta x_{\eta\eta} - x_\eta y_{\eta\eta}}{J}, \tag{23-c}$$

$$\left\{{2 \atop 1\,1}\right\} = \frac{x_\xi y_{\xi\xi} - y_\xi x_{\xi\xi}}{J}, \tag{23-d}$$

$$\left\{{2 \atop 1\,2}\right\} = \left\{{2 \atop 2\,1}\right\} = \frac{x_\xi y_{\xi\eta} - y_\xi x_{\xi\eta}}{J}, \tag{23-e}$$

$$\left\{{2 \atop 2\,2}\right\} = \frac{x_\xi y_{\eta\eta} - y_\xi x_{\eta\eta}}{J}, \tag{23-f}$$

$$J = x_\xi y_\eta - x_\eta y_\xi \, . \tag{24}$$

For notational simplicity, we shift from (x^1, x^2), (ξ^1, ξ^2) to (x, y), (ξ, η).

Numerical Procedure

The governing equations are discretised in the two-dimensional curvilinear coordinate using a staggered mesh system. In the staggered mesh system, here, all scalar variables (i.e. pressure, temperature and so on) are located at the cell center, while the velocity variables $U^{(1)}$ and $U^{(2)}$ are arranged on the middle of the cell face in the ξ^1 and ξ^2 directions, respectively, shown in Fig.1.

The SMAC algorithm presented by Harlow and Amsden[8] is employed.

In the first step, the Navier-Stokes equations(Eq.(18)) are solved for an estimated pressure field to yield a provisional velocity field.

In the second step, a Poisson equation of pressure is solved from the divergence of a provisional velocity field as a source term. Here, owing to the cross derivative term of pressure in Eq.(18), a Poisson equation is discretised by the surrounding 9-points pressure. In this study, this equation is solved by the general band matrix method, whose matrix is regularly sparse but not symmetric.

After the Poisson equation is solved, the velocity is updated as the results of the pressure correction.

Finally, the temperature equation is solved explicitly.

The boundary conditions are treated as almost same as those in the usual finite difference method in the Cartesian coordinate. For example, when no-slip condition will be set at the wall, the opposite velocity will be assumed at the neighboring mesh against the boundary line.

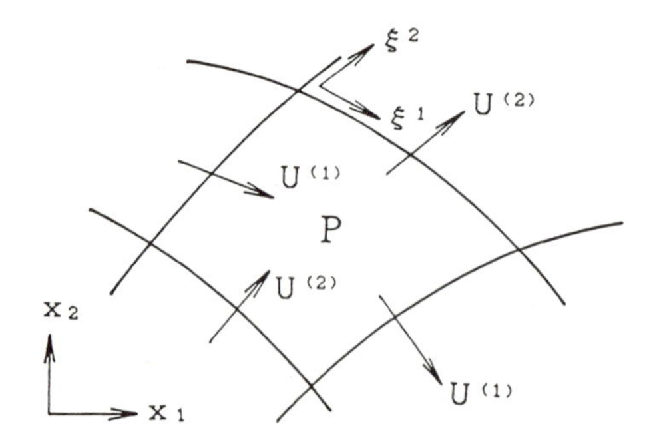

Fig.1 Configuration of a staggered mesh system

Examples

The Kuehn-Goldstein experiments[8] were chosen to examine this present study. We simulated that the fluid in the gap between the inner and outer cylinder was circulating steadily when the inner cylinder is heated and the outer is cooled. The Rayleigh number Ra based on the gap width is 5×10^4 and the Prandtl number Pr is 0.7. The computational grid and the boundary conditions are shown in Fig.2. Because of the annulus eccentricity, the grid system is non-orthogonal. The isotherms and the velocity vectors that are calculated by the above-mentioned scheme are shown in Fig.3(a) and (b), respectively. The comparison between the numerical and experimental results in the radial temperature profiles is plotted in Fig.4. Fig.4 shows the good agreement between the experiments and our numerical solutions. Moreover, the exact satisfaction of the continuity equation is confirmed by the reasonable temperature distribution in Fig.3(a).

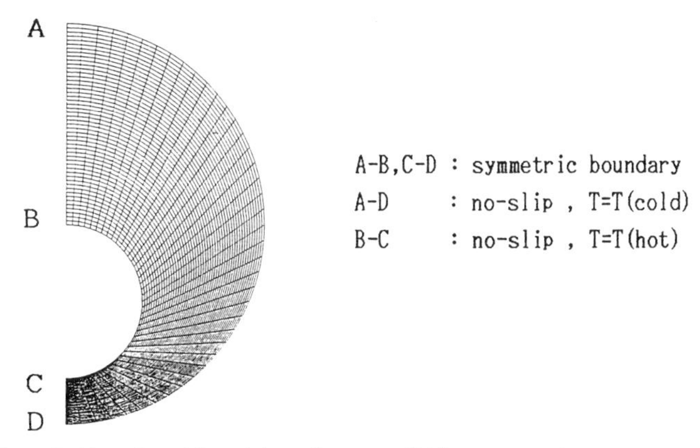

Fig.2 Computational grid and boundary conditions

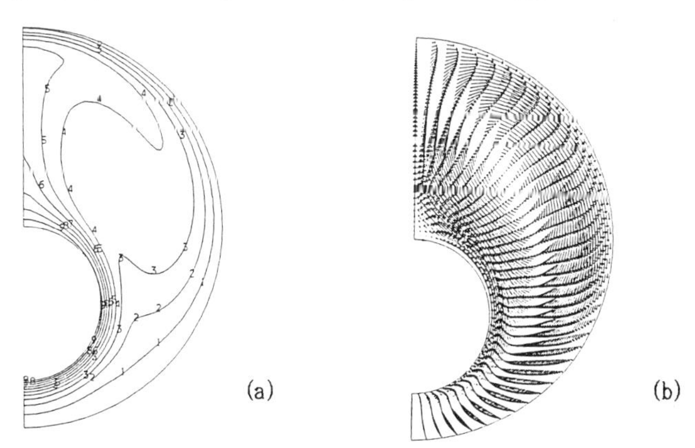

Fig.3 Prediction of the natural convection, where $Ra=5\times10^4$ and Pr=0.7. (a) the isotherms, and (b) the velocity vectors.

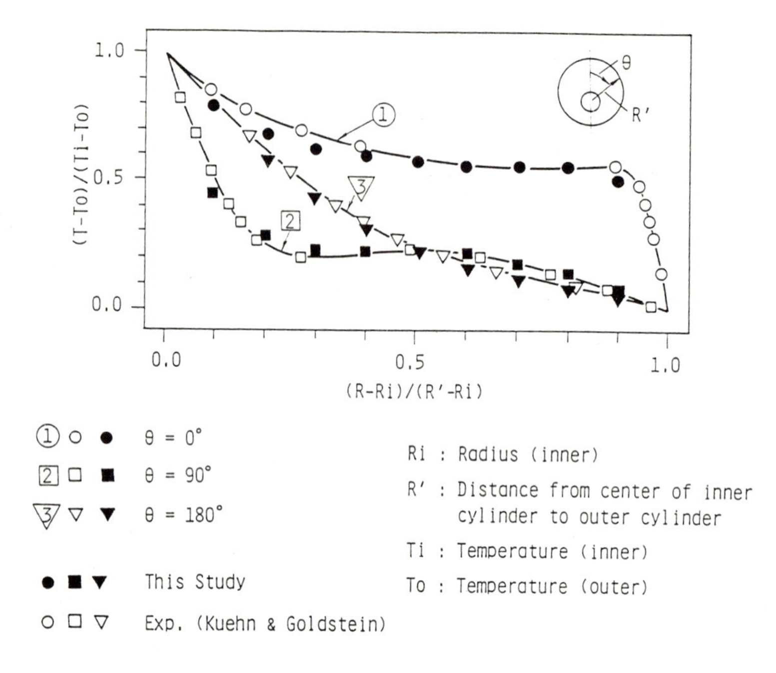

Fig.4 Dimensionless radial temperature profiles

Conclusions

The incompressible inner flow analysis scheme on the curvilinear coordinate system has been presented. The features of this method is following three points.

(1) the governing equations are written with the absolute differential form.

(2) the physical components of the contravariant véctor are taken as the velocity variables.

(3) the equations are discretized on the staggered mesh system.

In contrast to the conventional Body-Fitted Method, this method is effective to achieve the exact satisfaction of the continuity equation and calculate stably. As the interpolation of the velocities is not used, the calculated results become more accurate than those of the conventional method. Moreover, the direction of the velocity components are consist with the coordinate system, the boundary conditions are estimated precisely.

An applicability to analyze the incompressible viscous flow in an arbitrary shaped region is confirmed by some numerical results.

Generally, one of the most important validities of the computational fluid dynamics is the capability of the three-dimensional analyses. From this point of view, this method is promising and easy to be extended to a three-dimensional flow field.

Acknowledgements

I would like to express my gratitude to Dr. S.Koshizuka(Tokyo Univ.) for his useful suggestions, most of which has been incorpolated.

References

1. J.F.Thompson, F.C.Thames and C.W.Mastin "Automatic Numerical Generation of Body-Fitted Curvilinear Coordinate System for Field Containing Any Number of Arbitrary Two-Dimensional Bodies", J.Comp.Phys. 15 (1974) 299-319
2. J.L.Steger "Implicit Finite-Difference Simulation of Flow about Arbitrary Two-Dimensional Geometries", AIAA J. 16 (1978) 679-686
3. S.V.Patankar "Numerical Heat Transfer and Fluid Flow", Hemisphere (1980)
4. W.Shyy "A Numerical Study of Annular Dump Diffuser Flows", Comput.Meths. Appl.Mech.Engrg. 53 (1985) 47-65
5. H.Daiguji and S.Yamamoto , Int.Symp.Comput.Fluid Dyn. , Tokyo, 1985
6. A.C.Eringen "Nonlinear Theory of Continuous Media", McGraw-Hill (1962)
7. S.Koshizuka and Y.Oka "Absolute Staggered Differencing Technique on a Curvilinear Grid", IAHR LMFBR W/G, JAPAN, 1988
8. F.H.Harlow and A.A.Amsden "A Numerical Fluid Dynamic Calculation Method for All Flow Speeds", J.Comp.Phys. 8 (1971) 197-213
9. T.H.Kuehn and R.J.Goldstein "An Experimental Study of Natural Convection Heat Transfer in Concentric and Eccentric Horizontal Cylindrical Annuli", J.Heat Trans. 100 (1978) 635-640

A Numerical Study of Interfacial Instabilities at High Mach Numbers

R. Klein
Princeton University, Department of Mathematics
Fine Hall-Washington Road, Princeton, NJ 08544, USA

C. D. Munz
Kernforschungszentrum Karlsruhe, Institut für Neutronenphysik
und Reaktortechnik, Postfach 36 40, D-7500 Karlsruhe, FRG

L. Schmidt
Universität Karlsruhe, Institut für Angewandte Mathematik,
Englerstr. 2, D-7500 Karlsruhe, FRG

SUMMARY

In the present study we perform direct numerical simulations of high Mach number shear flow instabilities based on the two-dimensional Navier-Stokes equations. To deal with the steep gradients and shocks which emerge after the growth of instabilities we employ modern high resolution shock capturing schemes. We point out some peculiarities arising generally when shock capturing schemes are applied to shear flow stability problems in the limit of vanishing viscosity. Streaklines are calculated to visualize the development of the instabilities.

INTRODUCTION

The stability of compressible high Mach number shear flow has recently found wide interest. This area of problems is not only important for immediate applications such as mixing layers in supersonic combusters or astrophysical jets but also for the understanding of local structures in flows with complicated shock patterns. An example is the vortex sheet behind a triple point in a Mach-reflected shock front. Direct numerical simulations based on the Euler or Navier-Stokes equations allow to follow the build-up of linear instabilites in the laminar background flow up to the fully nonlinear regime. Such calculations are reliable only if the numerical approximation errors do not influence or interfere with the stability properties of the basic flow. Thus, a scheme should avoid unphysical osciallations near steep gradients which can uncontrollably excite the physical instabilities. In Navier-Stokes calculations it should not use up any amount of physical viscosity to stabilize the numerics. In our two-dimensional simulations we employ high resolution shock capturing schemes and fine grids to obey these requirements. First we obtain, for finite and vanishing physical viscosity, the quasisteady parallel flows which establish after long times in the unperturbed case. We discuss their importance for the predicted stability properties in the small amplitude regime. Long time calculations with perturbed initial data then show the growth of instablities and the development of large amplitude flow structures. Strong oblique shocks in the high speed streams and vorticies in the shear regions produce relatively stable large scale structures. These patterns are oscillating on a free shear layer and travel with a certain convection speed in a jet.

GOVERNING EQUATIONS, NUMERICAL METHOD AND FLOW VISUALIZATION

The direct numerical simulations are based on the two-dimensional compressible Navier-Stokes equations which may be written in the conservation form

$$U_t + F(U)_x + G(U)_y = R(U)_x + S(U)_y. \tag{1}$$

Here U is the vector of conserved variables $U = (\rho, \rho u, \rho v, e)^T$. The density is ρ, the velocity components in the x and y directions are u and v, respectively, and e is the total energy per unit volume. F and G are the Euler fluxes, while R and S are the fluxes due to molecular transport:

$$F = \begin{pmatrix} \rho u \\ \rho u^2 + p \\ \rho uv \\ u(e+p) \end{pmatrix}, \quad G = \begin{pmatrix} \rho v \\ \rho uv \\ \rho v^2 + p \\ v(e+p) \end{pmatrix}, \quad R = \begin{pmatrix} 0 \\ \tau_1 \\ \tau_2 \\ \tau_4 + q^x \end{pmatrix}, \quad S = \begin{pmatrix} 0 \\ \tau_2 \\ \tau_3 \\ \tau_5 + q^y \end{pmatrix}, \tag{2}$$

$$\begin{aligned} \tau_1 &= (2\mu + \lambda)u_x + \lambda v_y, \quad \tau_2 = \mu(u_y + v_x), \quad \tau_3 = (2\mu + \lambda)v_y + \lambda u_x, \\ \tau_4 &= u\tau_1 + v\tau_2, \quad \tau_5 = u\tau_2 + v\tau_3, \quad q^x = \gamma\frac{\mu}{Pr}\epsilon_x, \quad q^y = \gamma\frac{\mu}{Pr}\epsilon_y. \end{aligned}$$

Here, p denotes the pressure and ϵ the internal energy per unit mass. The coefficients of shear viscosity and bulk viscosity are denoted by μ and λ, respectively, γ is the adiabatic exponent and Pr the Prandtl number. All calculations have been performed with the equation of state of a perfect gas. The viscosity coefficients are constant and we assume $\lambda = -2\mu/3$. We consider fluid flow of air with Pr $= 0.71$ and $\gamma = 1.4$.

For high Reynolds and Mach number flow the Euler terms dominate the dissipative terms and strong gradients in shock waves and shear layers occur. Hence, in numerical calculations it is important to approximate the Euler terms with high accuracy. Equations (1) should be considered as an inviscid hyperbolic system with a small parabolic perturbation rather than a parabolic system. We use a difference scheme based on a high resolution shock capturing scheme for the Euler terms combined with the usual central differencing of the dissipative terms. The high resolution scheme sharply resolves discontinuities without generating spurious oscillations and is second-order accurate on smooth parts of the flow. In our calculation we use a MUSCL-type scheme introduced by van Leer [6]. The MUSCL-type scheme is designed to solve the one-dimensional Euler equations. A two-dimensional numerical scheme is obtained by dimensional splitting. Equations (1) are split into two one-dimensional problems in x- and y-direction, respectively. These problems are then solved successively in each time step. We will shortly describe the main one-dimensional algorithms; a more detailed description may be found in [10]. In a first step boundary values in each cell of the grid are calculated - U_{i+}^n on the right and U_{i-}^n on the left side - by

$$U_{i\pm}^n = U_i^n \pm \frac{\Delta x}{2} S_i^n \tag{3}$$

where S_i^n is a vector of slopes in the i-th grid zone. These slopes are calculated by a monotonicity preserving interpolation from the integral approximative values U_i^n. These boundary values are updated at half a time step to obtain 2nd order accuracy in time

$$U_{i\pm}^{n+1/2} = U_{i\pm}^{n} - \frac{\lambda}{2}[F(U_{i+}^{n}) - F(U_{i-}^{n})], \quad \lambda = \frac{\Delta t}{\Delta x}. \tag{4}$$

New approximative integral values are obtained at the next time level by a scheme in conservation form

$$U_i^{n+1} = U_i^n - \lambda(h_{i+1/2}^{n+1/2} - h_{i-1/2}^{n+1/2}) \tag{5}$$

with $h_{i+1/2} = h(U_{i+}, U_{(i+1)-})$. The function h approximates the flux between the grid zones and is calculated in an upwind way. Any upwind scheme as reviewed in [4], [5] may be used for this purpose. In the calculations presented below we used the flux vector splitting of van Leer (see [5]). The slope calculation must satisfy accuracy conditions as well as some monotonicity constraints to avoid spurious oscillations near strong gradients. A review of different appropriate slope calculations is given in [11]. We calculate the slopes in terms of characteristic values using the local linearization technique of Roe. The advantage is that different slope calculations may be applied to the nonlinear and linearly degenerate characteristic fields. This enables us to use very compressive slopes in the linear fields which strongly reduce the numerical dissipation of contacts [11]. This is very important for large scale computations because the numerical damping of contacts increases with time. These compressive slopes should not be applied to the nonlinear fields, because there they may be over-compressive and may introduce non-physical discontinuities. In the calculations presented here we determine the slopes by Sweby's formula (see [11]) with the compression parameter k = 1.4 and k = 1.8 for the nonlinear and linear characteristic fields, respectively.

In this paper we present and discuss results for time developing single mixing shear layers and jets. While practical flows develop streamwise, we assume infinite interfaces which are periodically disturbed in space. The numerical results show the development of the interfaces in time. The main advantage of this relatively simple flow problem is that large scale computations with a good resolution can be performed because the solution cannot leave the computational domain. There is no simple transformation between space and time developing mixing layers: besides a Galilei transformation with the streamwise mean velocity we need the assumption that the fluid flow becomes periodic in space. Dunn and Lin [3] performed a boundary layer type analysis for the hydrodynamic stability equations of a spatially evolving mixing layer. They showed that the local inviscid stability problem for the layer at any position in streamwise direction is equivalent to that of an exactly parallel shear layer with the same background velocity profile. Thus results for the temporal mixing layer provide a lot of physical insight into the basic mechanisms of the instabilities. Figure 1 shows vortex sheet initial data, where the tangential velocity u changes discontinuously. We perturb the vicinity of the interfaces by small pressure disturbances of mode k

$$p = p_0 + ae^{-y^2/b^2} sin(2k\pi x). \tag{6}$$

Reynolds and Mach numbers are based on the relative initial velocity of the flow

$$R = 2u_0\ell/\mu \ , \qquad M = 2u_0/c_0 \tag{7}$$

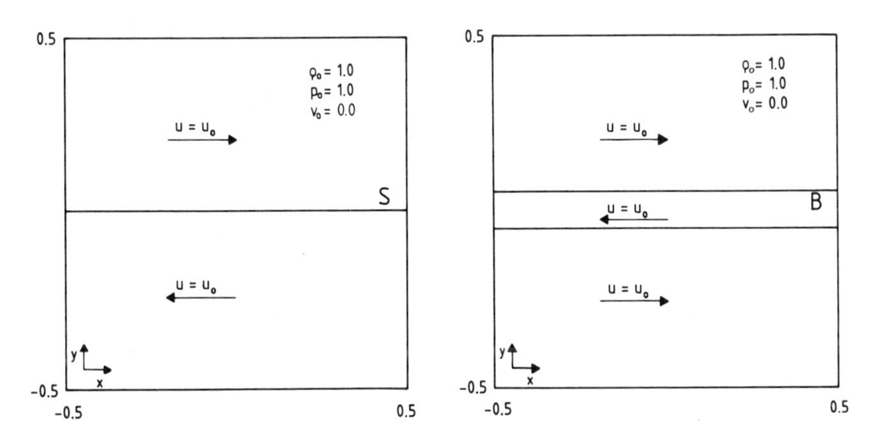

Figure 1: Initial values

where ℓ is the system dimension and c_0 the initial sound velocity ($\ell = 0.5, c_0 = \sqrt{\gamma}$). For the computational region we prescribe periodic conditions at the left and right boundary and outflow conditions at the upper and lower boundary.

To see the temporal development of the interfaces we adopt a visualization technique used in experiments: interfaces are marked by ink or smoke. In our calculations we introduce a number of massless marker particles along the interfaces and advect them according to the flow field without any collisional effect. By graphically displaying these particles we get streaklines which visualize the movement of the interfaces. Figure 2

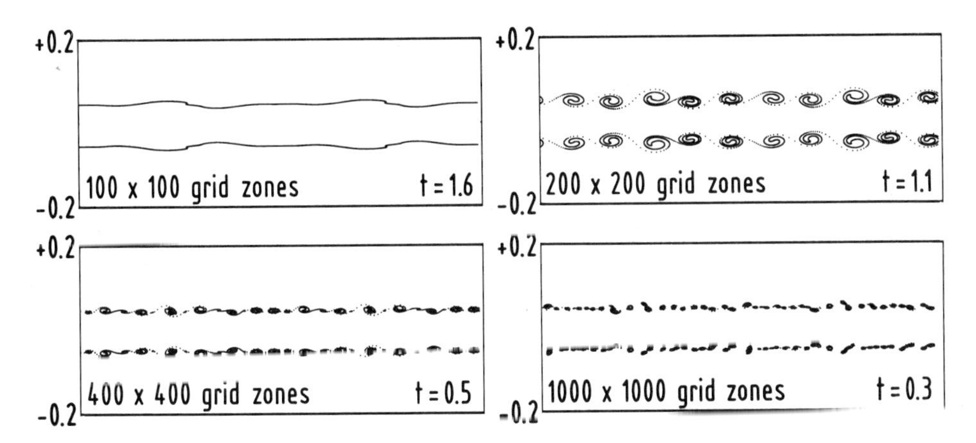

Figure 2: Jet flow at M = 1.0 on different grids

shows a visualization of jet flow at Mach number M = 1.0 for the inviscid equations, calculated on different grids. It also shows the difficulties of numerical simulations of instabilities. The inviscid problem is highly unstable due to Kelvin-Helmholtz instability. Linear stability analysis predicts instability for all wavelengths, with faster growth for the smaller wavelengths. This becomes quite obvious in Figure 2. There we introduced a mode $k = 2$ perturbation according to (6) with amplitude $a = 0.01$ and $b = 0.01$. For a grid with 100 x 100 grid zones the mode 2 perturbation leads to a symmetric roll-up of the vortex sheets into four vortices. On a refined grid smaller structures appear, introduced by approximation errors. These small errors increase and roll up into small vortices due to physical instability - faster than the long wavelength disturbance of mode 2. Due to different vortex interactions such as vortex pairing and merging quite different solutions will appear after a short time. If the physical problem is stable according to fine grained turbulence the asymptotic solutions may have similar large scale structures. In the same way small perturbations (e.g., white noise) of the initial values will also rapidly grow and produce quite different solutions. Of course, the used numerical method also strongly influences the development of the instabilities in the high Reynolds number case. In our calculations we choose for a given Reynolds number the number of grid points sufficiently large so that by further grid refinement no additional small scale structures can be found. Under these conditions we observed that slightly different numerical methods (e.g., using different upwind schemes) produce quite similar or identical results.

THE INSTABILITY OF A SHEAR LAYER

The simplest way to approach the problem is at first to examine the instability of a vortex sheet as given in Figure 1. Any shock capturing scheme in conservation form will introduce some numerical dissipation even in the inviscid case. A contact discontinuity is spread in the first time steps over a few grid zones. This process inevitably leads to numerical entropy production and very much resembles physical dissipation. For small Mach numbers it may reduce the initial growth rate of the instability, but does not lead to severe difficulties. Under strong shear any amount of dissipation leads to large amplitude variations of density and temperature. Due to inertial confinement the first time steps produce a large excess pressure peak which subsequently decomposes into a pair of outward traveling transient acoustic waves. Numerical diffusion establishes a distribution of density, temperature and velocity over a narrow region similar to that of a Navier-Stokes large Reynolds number solution. An asymptotic shear layer solution may be obtained by means of the compressible boundary layer approximation because the initial transients leave the domain after a short time and pressure variations across the layer become negligible. In the special case of unit Prandtl number and with the viscosity proportional to temperature, i.e. $\mathrm{Pr} = 1$ and $\mu = \mu^* T, \mu^* = const$, respectively, there is a direct relationship between velocity and temperature (see [12])

$$T = \frac{1}{\rho} = 1 + \frac{\gamma - 1}{2} \frac{u_0^2 - u^2}{c_0^2}. \tag{8}$$

Furthermore, the compressible boundary layer equations reduce to the *incompressible* ones under the Howarth-Dorodnitzyn transformation. They have the similarity solution

$$u = u_0 \ \ erf(\eta) \ , \ \ \eta = \Big(\int_0^y \rho dy\Big) / 2\sqrt{\mu^* t} \tag{9}$$

for the velocity profile in the main flow direction and describe the thickening of the layer in time. A change in Reynolds number R or in μ^* merely changes the time scale of the outward diffusion of momentum and heat, but does not affect the amplitudes of temperature, density and velocity across the layer. Thus, in the limit of infinite R there is no uniform approach to a contact discontinuity. For fixed time and $R \to \infty$ the layer keeps its internal structure while its thickness shrinks to zero. The numerical dissipation of a shock capturing scheme produces a similar mechanism. Since in the numerical algorithm the grid sizes $\Delta x, \Delta y$ do appear only in form of the ratios $\frac{\Delta t}{\Delta x}, \frac{\Delta t}{\Delta y}$, a mesh refinement with fixed Courant number cannot remedy this behavior. It merely shrinks the width of the "numerical shear layer".

The results of stability analysis of vortex sheets and finite thickness layers are different. The normal mode analysis for a vortex sheet leads to the following results ([1],[9]). For subsonic relative flow ($M < 2$) there exists exponential unconditional Kelvin-Helmholtz type instability. For the range $2 < M < 2\sqrt{2}$ this instability remains, but in addition there is one neutrally stable mode which consists of a pair of downwind facing acoustic waves in either streams. For $M > 2\sqrt{2}$ there are no unstable modes, but three neutrally stable supersonic "kink modes" which all have different phase speeds. The vortex sheet becomes linearly neutrally stable. However, Artola and Majda [2] show that interactions between these kink modes are responsible for inherently nonlinear instabilities. The vortex sheet eigenmodes are independent of the wave number and the unstable and neutrally stable waves can have any structure in x-direction. There is no dispersion so that these perturbations travel without change of shape in the linearized model. The situation is different for finite thickness shear layers. Lessen, Fox and Zien [7] show that for any flow with $M_{rel} < 2$ there is a critical wavelength for neutral stability and longer waves are unstable. Generally higher relative velocities increase this critical wavelength and reduce growth rates in the unstable range. In a companion paper [8] they find for any relative flow Mach number a stability threshold of wavelengths above which the flow becomes unstable. The unstable modes are of the subsonic/supersonic type and neutrally stable purely supersonic disturbances do not exist. Furthermore there is a dispersion relation for the eigenmodes. All these properties are in obvious contradiction to the instabiltity of a vortex sheet. We conclude that the reliability of shock-capturing simulations of vortex sheet instabilities is, at least, questionable, since the numerical background flow introduces different stability characteristics. For shear layers in Navier-Stokes flow one has to fully resolve the viscous structure. Only if the numerical viscosity is an order of magnitude smaller than the imposed physical one, the simulation will accurately present the structure of the layer and show the physical stability characteristics These considerations also give more insight in the phenomenon observed in Figure 2. Since the unstable wavelengths of a finite thickness layer scale with the layer thickness, a mesh refinement will yield increase of smaller distortions of the numerical layer.

The shear layer in our problem grows in time as in the similarity solution (8) in the unperturbed case. Thus in contrast to a spatially growing layer, there is no constant well defined shear layer thickness δ. The wavelength in flow direction in a normal mode analysis of hydrodynamic stability (see, e.g., [3]) scales with the thickness of the shear layer. Out of the continous spectrum of modes included in such a theory our numerical system selects a discrete spectrum compatible with the length of the computational domain, $\lambda_i = (2\ell)/i$. For a given shear layer thickness these wavelengths are associated with particular growth rates $\alpha_i = \alpha(\lambda_i^*)$ which are functions of the scaled wavelengths $\lambda_i^* = \lambda_i/\delta$. As the layer broadens due to viscosity, the nondimensional wavelengths λ_i^* change, since $\delta = \delta(t)$, and so do the associated growth rates. In this way both the growth of the layer in time and variations of the Reynolds number - which also influences the layer thickness at a given time - change the growth rates of the allowed modes in a peculiar way. They make the discrete spectrum of wavelengths in our system move relative to the continous spectrum of an infinitely long layer of constant thickness. For very large or infinite Reynolds number numerical viscosity determines the properties of the shear layer. Predictions of hydrodynamic stability are not directly applicable. We expect, nevertheless, similar effects as described above, because the instabilities are of gasdynamic nature and a high quality code should accurately represent such properties. Since the code resolves a discontinuity over only a few gridpoints (at least for not too long times), the axial wavelengths of the unstable modes become very short, down eventually to the typical length scale of numerical approximation errors. In this case the instabilities are excited uncontrollably by discretization errors in the initial data.

We compared calculations with three different types of initial data: vortex sheet, continuous transition of the velocity by a tanh-profile of small thickness and a "numerical shear layer". In the case of the tanh-profile we assume constant pressure, the density profile is determined by the simple equation (10) and the equation of state. This profile of the initial layer slightly reduces the strength of the outwards travelling transient waves. To obtain the initial data named "numerical shear layer" we start the calculation without perturbation and only after the solution reaches a pseudo-stationary profile we introduce the perturbations. We find that all three types lead to similar results, only the growth rates are different as could have been expected from the above discussion. Figure

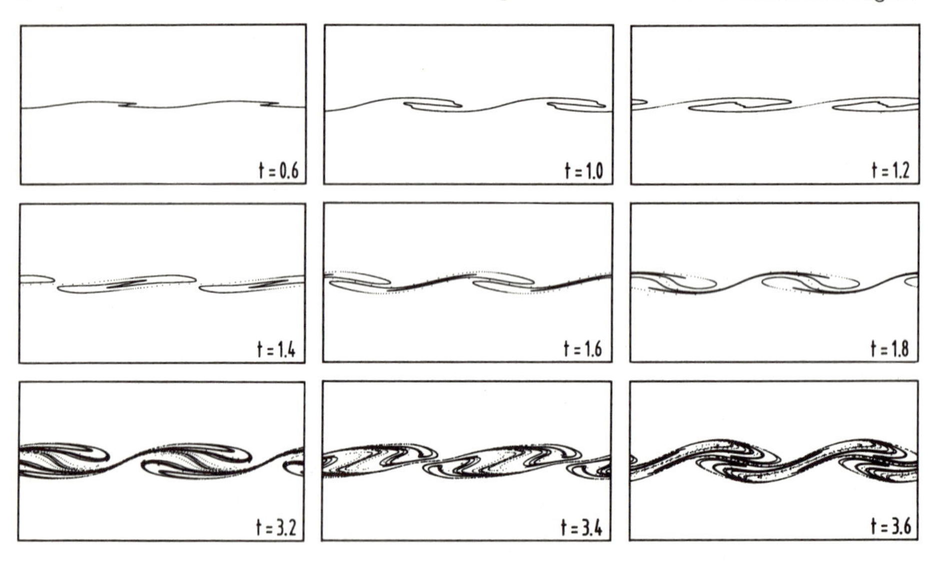

Figure 3: Instability of a shear layer at M = 2.0, R = 11800

3 shows the temporal development of the mixing layer for $M = 2, R = 11800$, visualized by streaklines. The parameters are $a = 0.1, b = 0.05$. Initially the pressure perturbation leads to sinusoidal displacement of the marker particles which grows in the linear regime. At $t = 0.6$ kinks appear and it seems that the shear layer starts to roll up. But the typical Kelvin-Helmholtz roll-up (Figure 2) into spirals does not take place. Instead the blob of vortical layer material that has accumulated around the initial roll-up separates and is

entrained into a new vortical structure. This process repeats several times and is clearly visible in the sequence of plots corresponding to times $t = 1.6$ to $t = 2.6$. At time $t = 2.2$ the vortical blobs are torn apart and at $t = 2.4$ the upper and lower parts of neighboring cores join at intermediate positions to form new blobs. The according pressure contours

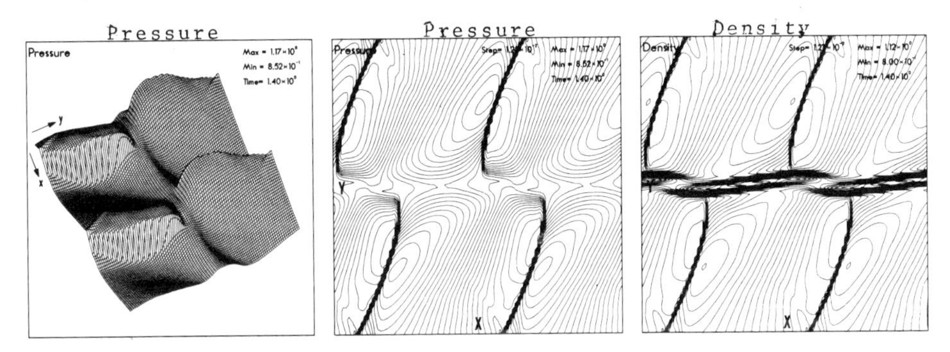

Figure 4: Pressure and density at time t = 1.4

in Figure 4 reveal the intriguing mechanism that produces this pattern. At time $t = 1.6$ we observe pairs of curved shock fronts embracing the vortical blobs, which are visible in the marker particle plot. The shock waves are large amplitude versions of Artola and Majda's stationary kink modes for supersonic vortex sheets. As more and more layer material is entrained and the vortices grow, the shock braces are pushed apart ($t = 1.8$) until they suddenly, between $t = 1.8$ and $t = 2.0$, snap over the stagnation points between the clusters. The pressure peaks holding the stagnation point flows break down and give way to the layer material to go along in flow direction and form new vortical blobs together with material from a neighboring cluster. The new vortices are again embraced by the kink shocks and thus the cycle begins anew. The mechanism described here inherently relies on the interaction of large amplitude, purely gasdynamical structures in the high speed outer flow with the subsonic vortical flow in the finite thickness layer. In a simulation with infinite Reynolds number these patterns can only be established through numerical dissipation, however small it may be.

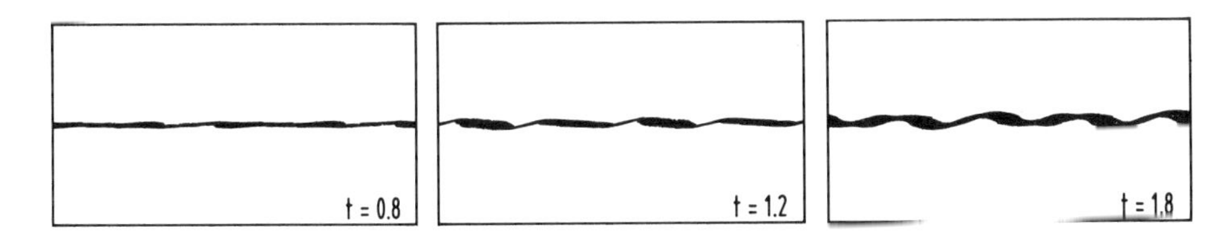

Figure 5: Instability of a shear layer at M = 4.0, R = 237000

As a second example of an unstable shear layer we consider the case of Mach number 4 and of Reynolds number increased to 237000. The perturbation parameters are a = 0.1, b = 0.05, k = 2 and the number of grid zones is 500 x 500. The streaklines pattern for M = 4 in Figure 5 shows only a narrow bounded folding of the layer as expected. A roll-up is missing. The interactions of the layer flow with bow shocks in the outer flow

- as already observed in the case M = 2 - are still present and become the dominant mechanism for the growth of the layer. At time t = 1.2 the streak band exhibits moving distortions, indicating the points of incident of the outer shock waves.

THE INSTABILITY OF A SUPERSONIC JET

In a subsonic jet the flow which is initially parallel and in pressure equilibrium grows in width with time. This is because the shear layers between jet and outer flow are unstable to Kelvin-Helmholtz modes (see Figure 2). The shear layers roll up and eddies are generated which entrain outer flow into the jet. The jet breaks down due to the interaction of these eddies. As outlined in the previous chapter the Kelvin-Helmholtz instability vanishes for high Mach numbers. In contrast to the subsonic case nonlinear gas dynamical effects dominate the supersonic jet flow. Figure 6 shows numerical results

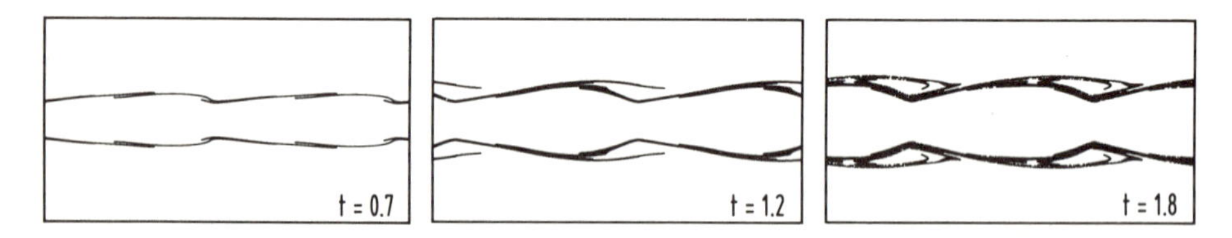

Figure 6: Instability of a jet at M = 5.0, R = 29600

of the evolution of a Mach 5 jet flow. In this case, the parameters for the perturbation (6) are k = 2, a = 0.05 and b = 0.05, and the number of grid zones is 500 x 500. The vortex sheet initial data are used. The results show that as expected the streakline particles do not roll up in the manner of a Kelvin-Helmholtz instability. At t = 0.4 the streaklines start to contract at two locations within the domain. Two nozzle-like throats form out of these contractions. The pattern that has established by time t = 1.2 is relativly stable,

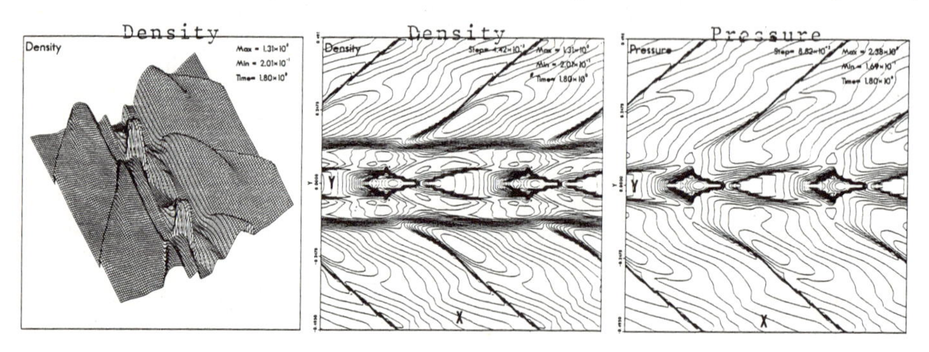

Figure 7: Pressure and density at time t = 1.8

but not stationary. From t = 1.2 to t = 1.8 we observe traces of marker particles being carried outside the main streak band and entrained in two vortices which embrace the throats. A comparison with the density contour lines shows a cross-shock pattern in the jet core between these vortices (Figure 7). Since in this shock structure there is a continuous entropy production the overall entropy of the jet core increases and the whole structure cannot be stationary even in a frame of reference moving with adapted speed. In the outer flow there is a pair of oblique outgoing shock waves which couple to the vortices and form an angle with the jet axis of 41 degrees. Since these waves decay to moderate strength far away from the jet we may interpret them as Mach waves and compute from the angle their speed relative to the outer flow. The pair of vortices travels with about Mach 3.5 upstream relative to the jet core and with Mach 1.5 relative to the outer flow.

The pattern shown here is only one of two stable structures which established in a series of runs for different Reynolds numbers and different initial representations of the shear layers. The second pattern has a more complicated shock structure in the jet core consisting of a strong and a weak cross shock per period. Its phase speed is nearly stationary in our frame of reference. It only appears for very large Reynolds numbers and vortex sheet initial data (or a very narrow initial layer).

REFERENCES

[1] Artola, M., Majda, A.J.: *Nonlinear develompent of instabilities in supersonic vortex sheets I: The basic kink modes*, Physica 28D (1987), 253 - 281

[2] Artola, M., Majda, A.J.: *Nonlinear development of instabilities in supersonic vortex sheets II: Resonant interaction among kink modes*, to appear in Stud. Appl. Math.

[3] Dunn, D.W., Lin, C.C.: *On the stability of the laminar boundary layer in a compressible fluid*, J. Aero. Sci., 22(1955), 455-477

[4] Einfeldt, B.: *On Godunov-type methods for gas dynamics*, SIAM J. Numer. Anal. 25 (1988), 294 - 318

[5] Harten, A., Lax, P.D., Leer, B. van: *On upstream differencing and Godunov-type schemes for hyperbolic conservation laws*, SIAM Rev. 25 (1983), 35 - 62

[6] Leer, B. van: *On the relation between the upwind-differencing schemes of Godunov, Engquist-Osher and Roe*, SIAM J. Sci. Stat. Comput. 5(1984), 1 - 21

[7] Lessen, M., Fox, J.A., Zien, H.M.: *On the inviscid stability of the laminar mixing of two streams of a compressible fluid*, J. Fluid Mech. 23 (1965), 355 - 367

[8] Lessen, M., Fox, J.A., Zien, H.M.: *Stability of laminar mixing of two parallel streams with respect to supersonic disturbances*, J. Fluid Mech. 25(1966), 737 - 742

[9] Miles, J.W.: *On the disturbed motion of a plane vortex sheet,* J. Fluid Mech. 4 (1958), 538 - 552

[10] Munz, C.D., Schmidt, L.: *Numerical simulations of hydrodynamic instabilities with high resolution schemes*, Notes on Numerical Fluid Mechanics Vol. 24, (J. Ballmann, R. Jeltsch, Eds.) Vieweg 1989, 456 - 466

[11] Munz, C. D.: *On the numerical dissipation of high resolution schemes for hyperbolic conservation laws*, J. Comput. Phys. 77 (1988), 18 - 39

[12] Schlichting, H.: *Grenzschichttheorie*, Verlag G. Braun, Karlsruhe 1958

EXPERIENCES WITH EXPLICIT TIME-STEPPING SCHEMES FOR SUPERSONIC FLOW FIELDS

N. Kroll, R. Radespiel, C.-C. Rossow
DLR, Institute for Design Aerodynamics
Flughafen, D-3300 Braunschweig, F.R.G.

SUMMARY

The computation of inviscid and viscous one- and two-dimensional supersonic and hypersonic flows using explicit Runge-Kutta time-stepping schemes is investigated. The spatial discretization is based on a cell-vertex finite-volume scheme with central differencing and various time-stepping schemes. Two artificial dissipation models are discussed with respect to accuracy and convergence behaviour. These are the well-known dissipation model based on fourth and second differences of the flow variables and a flux-limited dissipation model with TVD properties. Furthermore, the application of a multigrid scheme for supersonic flows is discussed. It is shown that substantial CPU-time savings are obtained using multigrid for the computation of viscous high-speed flows.

INTRODUCTION

Vehicles moving with supersonic speed will play an increasingly important role in future aeronautics. In order to predict the flow fields around such configurations, both, experimental examinations as well as numerical calculations are necessary. During the last decade a wide variety of numerical methods simulating the flow fields around complex configurations has been developed. These methods were mainly designed for sub- and transonic flows, and they have been proven to be an efficient tool for the design engineer. Now, when interest is focused on supersonic and hypersonic flows, the efficiency and robustness of the above mentioned schemes have to be investigated again in those flow regimes. Especially the capturing of the strong shocks and the arising shock systems becomes a critical issue.

Among others Runge-Kutta schemes together with central differences using the concept of artificial viscosity [1] reached maturity for sub- and transonic flows. Together with multigrid algorithms convergence rates were achieved which allow to use such schemes in a design process [2, 3]. Concerning supersonic flows the robustness and efficiency of central differencing schemes is not a priori the same as for sub- and transonic flows. Computing flow fields with strong bow shocks oscillations of the flow variables occur during the transient phase, which might inhibit convergence of the time-stepping scheme, even with a careful adjustment of all parameters of the numerical scheme. The objective of the paper is to investigate the robustness of a central differencing Euler code applied to supersonic onflow Mach numbers. The basic method is a finite volume scheme using a cell vertex formulation [4] and an artificial dissipation model given by Jameson et al. [1]. The present investigation concerns different explicit Runge-Kutta time-stepping schemes. The influence of the number of stages, the choice of parameters and number of evaluations of the artificial dissipation are investigated for inviscid supersonic flows. Furthermore, in order to develop a robust method for blunt bodies at high supersonic Mach numbers, a second approach based on TVD-properties is implemented. Here, the dissipation model of Jameson is replaced by a limited, antidiffusive flux, which has been introduced by Yee et al. [5]. Finally, the application of the multigrid algorithm to supersonic flows is discussed. It is shown that significant CPU-time savings are obtained using multigrid at Mach numbers as high as 5.

OUTLINE OF THE METHOD

The governing equations are the unsteady Reynolds-averaged Navier-Stokes equations in integral form where the medium is treated as a Newtonian fluid and an ideal gas is assumed. For the description of inviscid flows, this set reduces to the Euler equations when neglecting the viscous part of the flux density tensor.

The solution scheme for inviscid flows is the DLR code CEVCATS described in [4]. The basic scheme is a finite volume spatial discretization with multi-stage Runge-Kutta integration in time, as outlined in [1]. This has been implemented as a cell-vertex discretization [6], in which the solution vector is evaluated at the vertices of the mesh cells, i.e. at the grid points, with a distribution formula for the cell fluxes as proposed by Hall [7]. The scheme can be shown to be first-order accurate on arbitrary meshes [6]. On smoothly varying meshes it is second-order accurate.

Since the equations are discretized using central differences, additional dissipative terms must be added in order to damp out high-frequency oscillations in the solution. Here, two dissipation models are considered. In the well-known dissipation model of Jameson et al. [1] a blend of fourth and second differences is used to form dissipative terms. The formulation preserves conservation. The second difference terms are multiplied with the nondimensional second difference of the pressure so that the resulting dissipation is of third-order in smooth regions of the flow and first-order terms are added at discontinuities. The implementation of the dissipative terms is described in [8].

The central finite volume scheme described above has been modified towards a scheme with TVD properties. Using a flux-limited dissipation as described in [5] it can be shown that for a properly chosen time differencing this approach leads to a TVD scheme for one-dimensional hyperbolic nonlinear scalar equations and linear constant coefficient systems. Multidimensional schemes are designed by applying the one-dimensional scheme in each coordinate direction. In the present work, we use Yee's symmetric form in which the numerical dissipation terms are independent of the sign of characteristic speed. Roe's average [9] is used to evaluate the dissipative fluxes [5]. For problems containing strong shock waves the minmod limiter is used to obtain solutions without overshoots. Details of the implementation of the scheme may be found in [10].

Recently, the CEVCATS code has been extended to treat viscous flows [3, 11]. Central differencing is used for the viscous terms. In order to prevent excessively large dissipation levels for cells with large aspect ratios, the original scaling factor of the dissipation [1, 8] is replaced by carefully chosen functions of the individual eigenvalues of the Jacobian matrices in the different coordinate directions according to Martinelli [12]. Due to the explicit time-stepping scheme, the convergence of the basic scheme slows down considerably because of the time-step limitation associated with the small mesh cells which are necessary to resolve the thin shear layers. This drawback of the explicit scheme has been removed by applying local time stepping [1], implicit residual averaging with varying coefficients, which depend on the cell aspect ratios [3], and a multigrid algorithm. The multigrid scheme is based on the work of Jameson [2]. For the multigrid process, coarser meshes are obtained by doubling the mesh spacing. Injection is used to transfer the solution to coarser meshes. Residuals are transferred by a weighted average. A forcing function is constructed so that the solution on a coarse mesh is driven by the residuals collected on the next finer mesh. The corrections obtained on the coarse meshes are transferred to the next finer mesh by bilinear interpolation. For the solution of the Navier-Stokes equations at high Reynolds numbers Full Multigrid is applied to provide a well conditioned starting solution for the finest mesh being considered.

SINGLE GRID COMPUTATIONS OF INVISCID FLOWS

In this chapter an examination of robustness and efficiency of explicit Runge-Kutta type time-stepping schemes for supersonic inviscid flows is given. Various multistage schemes are applied to both spatial discretizations outlined above, the central differencing with the artificial dissipation according to Jameson and the TVD type discretization. The influence of the number of stages, the choice of coefficients and the number of evaluations of the dissipative flux is investigated with the aim to show the limits of the various time-stepping schemes with respect to an increase of the free stream Mach number. The investigated schemes are summarized in Table 1. All schemes except the classical Heuns' scheme are hybrid multistage schemes and are described in [8, 13, 14]. As a test problem the one-dimensional flow through a divergent nozzle has been chosen. For all supersonic inflow Mach numbers the outflow conditions have been chosen such that the outflow is subsonic and a normal shock wave occurs at approximately 60% of the nozzle length.

Figure 1 shows the effect of the inflow Mach number on the local Mach number distribution calculated with both the Jameson dissipation model and the TVD scheme. For comparison also the exact solution is given. It is important to note that in all cases if a converged solution could be obtained all time-stepping schemes listed in Table 1 have given the same result. For this study the coefficients of the Jameson dissipation ($k^{(2)} = 0.8$ and $k^{(4)} = 1/64$) as well as the parameter of the entropy function of the TVD scheme ($\delta = 0.05$) has been held fixed for all inflow Mach numbers. The comparison of the results obtained with the two different spatial discretizations shows that the TVD scheme captures the shock more accurately. Increasing the inflow Mach number to M = 10 the TVD scheme shows a similar well resolved shock within two cells. In contrast to that with the dissipation model according to Jameson the shock is more and more smeared as the inflow Mach number is increased.

In Figure 2 the effect of the inflow Mach number on the efficiency of the various time-stepping schemes is shown. Here, the required CPU time to obtain $\|\partial\rho/\partial t\|_2 < 10^{-6}$ as a function of the inflow Mach number is plotted. Using the Jameson dissipation (Fig. 2a) all the time-stepping schemes provide the same efficiency for the low supersonic inflow Mach number M = 2. As the Mach number increases, however, the efficiency of all schemes decreases at different rates. For all schemes there is a limit of the inflow Mach number for which a converged solution could be obtained. It turns out that for the Jameson dissipation model the most robust time-stepping schemes are the 3,2- and 5,2-scheme as well as the classical Heuns' scheme. The most efficient one is the 3,2-scheme. If the various time-stepping schemes are applied to the TVD type discretization (Fig. 2b) the efficiency of each scheme is independent of the inflow Mach number. No upper limit to the inflow Mach number for obtaining a converged solution has been found. The most efficient time-stepping scheme is the 3,2-scheme.

Figure 3 compares the convergence history of the time-stepping schemes in the case of the TVD discretization for the inflow Mach number M = 8. The convergence histories of all schemes show an oscillatory behaviour. The 3,3-scheme requires the minimum of time steps. However, since the dissipative flux is evaluated in each stage, the 3,2-scheme is more efficient with respect to computational time (see Fig. 2b). Note, that the 2,1- and 3,1-scheme show nearly the same convergence history.

It should be mentioned that all one-dimensional calculations have been carried out without implicit residual averaging. It has been found that for strong shocks ($M > 4$) that the residual averaging applied to any of the investigated time-stepping schemes does not improve the efficiency of the scheme for both spatial discretizations. In particular, for higher inflow Mach numbers no converged solution could be obtained if this technique has been applied.

The result of the one-dimensional investigations can be used to form a basis for two- and three-dimensional calculations. Figure 4 shows a comparison of the Iso-Mach lines calculated with the two different spatial discretizations for a circular cylinder at $M = 10$. With both schemes a well resolved bow shock is obtained. In contrast to the TVD scheme the Mach lines calculated with the Jameson dissipation show some oscillations in the vicinity of the shock. For this case the 5,2 and 3,2 time-stepping scheme turned out to be the most efficient time integration schemes.

MULTIGRID COMPUTATIONS OF INVISCID AND VISCOUS FLOWS

Although the original development of the multigrid time-stepping scheme outlined above was directed towards the efficient computation of transonic flows [2], the Mach number range of the multigrid scheme has been extended to values around 7 using the following modifications.

- Smoother

  It has already been mentioned above, that for higher Mach numbers, the Courant number has to be reduced to avoid negative pressures and densities in the transient phase. The multigrid scheme, however, does rely on good high-frequency damping of the Runge-Kutta scheme. Hence, appropriate smoothers for supersonic flows have smaller numbers of Runge-Kutta stages than those for transonic flows. To this point the two-stage scheme of Table 1 has been found to give satisfactory smoothing properties for the Mach number range mentioned above. Using an implicit smoothing of the explicit residuals with variable coefficients according to [3], the convergence rates are further improved. With this technique Courant numbers around CFL = 2.0 are used and the ratio CFL/CFL* = 2.0 where CFL* is the stability limit of the explicit scheme.

  On coarse meshes the original multigrid method proposed by Jameson and coworkers involves the use of linear second-difference dissipative terms which are first-order accurate. For flows with strong shocks the dissipation model on the coarse meshes has to be extended to avoid an oscillatory behaviour of the coarse-mesh corrections near discontinuities. For this purpose the nondimensional second difference of the pressure is used to increase the coefficient of the dissipative fluxes near shock waves.

• Multigrid strategy

For transonic flows with moderate shock strengths, a W-type multigrid cycle with multiple time steps on the coarse meshes has been found more efficient than a single V-cycle. This result changes for flows with strong shocks, where too large corrections from the coarse meshes may lead to negative pressure/density and subsequent divergence of the solution. Therefore, a V-cycle is preferred for supersonic flows. The robustness of the start-up process is improved, if a few, that is 5-10 single-mesh time steps are executed at the beginning to precondition the solution. The robustness of the multigrid scheme for higher Mach numbers can be further improved if two time steps instead of one are executed on the finest mesh level.

• Update of the corrections

The robustness of the multigrid scheme for flows with strong shocks is further enhanced by smoothing of the corrections before they are added to the solution on the fine mesh. In the present scheme we simply use the same factored implicit smoothing operator as for the smoothing of the residuals in the Runge-Kutta scheme, but with constant smoothing coefficients. The magnitude of the smoothing coefficients ε is in the range $0 < \varepsilon < 0.2$ for Mach numbers upto 7.

Computations of two-dimensional inviscid and viscous high-speed flows have been made with the present multigrid scheme. First, the inviscid flow around the NACA 0012 airfoil is used to compare the efficiencies of the single-grid and the multigrid schemes. An O-type coordinate grid of 160 x 48 cells is used here, which was generated for the computation of transonic flows rather than supersonic flows. In Figure 5 the CPU-seconds on CRAY XMP-216 required to obtain $\|\partial\rho/\partial t\|_2 < 10^{-5}$ (corresponding to an error reduction of 6 to 7 orders of magnitude) are plotted versus the free-stream Mach number. For lower Mach numbers the multigrid scheme is superior to the single-grid scheme, whereas at higher Mach numbers the damping behaviour of high-frequency oscillations at the shock becomes dominant and hence, the multigrid scheme is less effective. Note, that the result would have changed in favour of the multigrid scheme if a less stringent convergence criterion had been used.

The next application is the hypersonic turbulent flow over a 35° compression ramp. The flow conditions correspond to test case 3.3 of the Hypersonic Workshop [15]. A mesh of 224 x 128 cells has been used and the grid spacing at the wall is 10^{-5} times the distance from the leading edge to the corner of the ramp. Figure 6 shows that a residual reduction of three orders of magnitude is obtained within 210 multigrid cycles on the fine mesh which corresponds to 185s total computational time on CRAY XMP-216. With the single-mesh code it takes 4100 time steps and 1075s to obtain the same residual level. Clearly, the multigrid scheme is much more efficient when solving the Navier-Stokes equations even for high Mach numbers. The flow solution is shown in Figure 7. There is a strong interaction of the turbulent boundary layer and the oblique shock with a flow separation in the corner region.

CONCLUDING REMARKS

A conventional fourth- and second-difference Jameson-type dissipation model and a flux-limited TVD dissipation in connection with Runge-Kutta time-stepping schemes have been investigated for supersonic flows with respect to accuracy and convergence. 1-D results indicate, that at high Mach numbers the Jameson dissipation gives smeared shocks and convergence slows down due to CFL-restictions. In contrast to that the TVD scheme yields sharply resolved shocks and better convergence properties. These findings are also valid for 2-D flows. Using the Jameson dissipation model in connection with Runge-Kutta time stepping and a multigrid scheme, it is found that convergence is obtained for Mach numbers upto about 7. This is the same range where the Jameson dissipation worked satisfactorily for the 1-D problem. In particular, the computation of viscous flows is greatly enhanced using multigrid. It still remains an open question whether both, the smoothing scheme with a proper scaling of the artificial dissipation and the multigrid algorithm can be improved to allow an efficient computation of flows with very high Mach numbers.

REFERENCES

[1] JAMESON, A., SCHMIDT, W., TURKEL, E.: "Numerical Solution of the Euler Equations by Finite Volume Methods Using Runge-Kutta Time-Stepping Schemes", AIAA 81-1259, (1981).

[2] JAMESON, A.: "Multigrid Algorithms for Compressible Flow Calculations", MAE Report 1743, Princeton University, Text of Lecture given at 2nd European Conference on Multigrid Methods, Cologne, Oct. 1985, (1985).

[3] RADESPIEL, R., ROSSOW, C.-C., SWANSON, R.C.: "An Efficient Cell-Vertex Multigrid Scheme for the Three-Dimensional Navier-Stokes Equations", AIAA 89-1953, (1989).

[4] ROSSOW, C.-C., KROLL, N., RADESPIEL, R., SCHERR, S.: "Investigation of the Accuracy of Finite Volume Methods for 2- and 3-Dimensional Flows", AGARD-CP-437, Vol. 2, P.14, (1988).

[5] YEE, H.C., KLOPFER, G.H., MONTAGNE, J.L.: "High-Resolution Shock-Capturing Schemes for Inviscid and Viscous Hypersonic Flows", NASA-TM 100097, (1988).

[6] ROSSOW, C.-C.: "Berechnung von Strömungsfeldern durch Lösung der Euler-Gleichungen mit einer erweiterten Finite-Volumen Diskretisierungsmethode", Dissertation, TU Braunschweig, 1988, DLR-FB 89-38, (1989).

[7] HALL, M.G.: "Cell Vertex Multigrid Scheme for the Solution of Euler Equations", Proceedings of the Conference on Numerical Methods for Fluid Dynamics, Reading, U.K., (1985).

[8] KROLL, N., JAIN, R.K.: "Solution of Two-Dimensional Euler Equations - Experiences with a Finite Volume Code", DFVLR-FB 87-41, (1987).

[9] ROE, P.L.: "Characteristic-Based Schemes for the Euler Equations", Ann. Rev. Fluid Mech., Vol. 18, pp. 337-365, (1986).

[10] LI, H., KROLL, N.: "Solution of One- and Two-Dimensional Euler Equations Using a TVD Scheme", DFLVR-IB, (1988).

[11] RADESPIEL, R., SWANSON, R.C.: "An Investigation of Cell Centered and Cell Vertex Multigrid Schemes for the Navier-Stokes Equations", AIAA 89-0548, (1989).

[12] MARTINELLI, L.: "Calculations of Viscous Flows with a Multigrid Method", Ph.D. Dissertation, MAE Department, Princeton University, (1987).

[13] JAMESON, A., BAKER, T.J.: "Solution of the Euler Equations for Complex Configurations", AIAA 83-1929, (1983).

[14] VENKATAKRISHNAN, V.: "Computation of Unsteady Transonic Flows over Moving Airfoils", Ph.D. Dissertation, MAE Department, Princeton University, October 1986.

[15] N.N.: Workshop on Hypersonic Flows for Reentry Problems, Co-organized by INRIA and GAMM-SMAI, Antibes (France), 22-26 January 1990.

Scheme	Number of Stages	Number of Evaluations of Dissipation	Coefficients
(2,1)	2	1	$\alpha_1 = 0.7$, $\alpha_2 = 1.0$
(2,2)	2	2	Classical Heun's Scheme
(3,1)	3	1	$\alpha_1 = 0.6$, $\alpha_2 = 0.6$, $\alpha_3 = 1.0$
(3,2)	3	2	$\alpha_1 = 0.21$, $\alpha_2 = 0.5$, $\alpha_3 = 1.0$
(3,3)	3	3	$\alpha_1 = 0.1666$, $\alpha_2 = 0.5$, $\alpha_3 = 1.0$
(5,2)	5	2	$\alpha_1 = 0.25$, $\alpha_2 = 0.16667$, $\alpha_3 = 0.375$ $\alpha_4 = 0.5$, $\alpha_5 = 1.0$

Table 1: Investigated time stepping schemes.

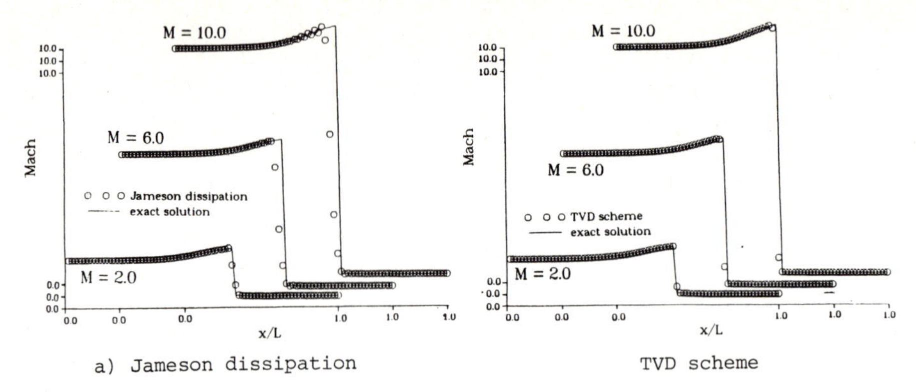

a) Jameson dissipation TVD scheme

Figure 1: Effect of inflow Mach number on local Mach number distribution for a divergent nozzle

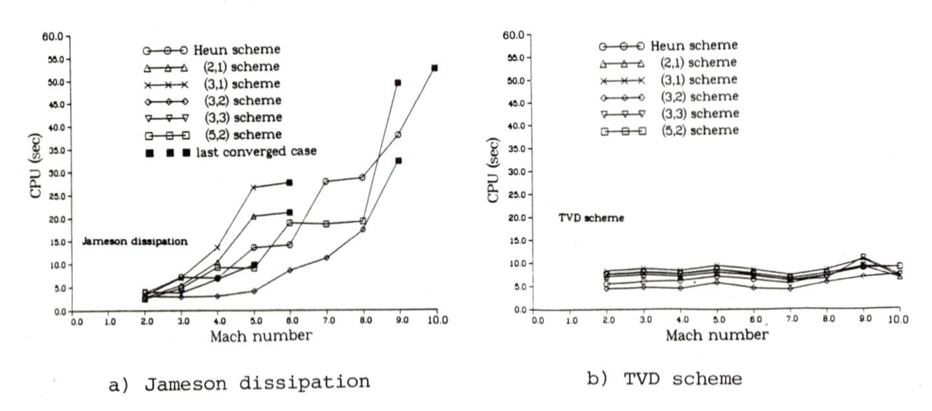

a) Jameson dissipation b) TVD scheme

Figure 2: Effect of inflow Mach number on the efficiency of various time-stepping schemes for a divergent nozzle, convergence criterion $\|\partial\rho/\partial t\|_2 < 10^{-6}$

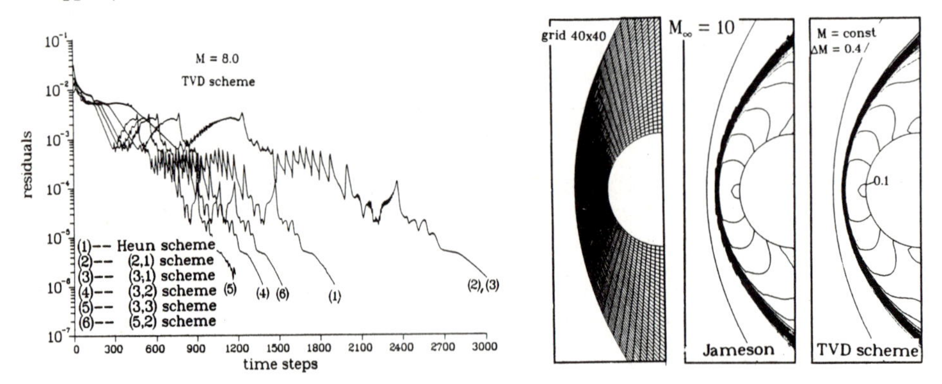

Figure 3: Comparison of convergence histories of various time-stepping schemes for a divergent nozzle (TVD scheme)

Figure 4: Comparison of Mach number contours for a circular cylinder calculated with the Jameson dissipation and the TVD scheme at $M_\infty = 10$

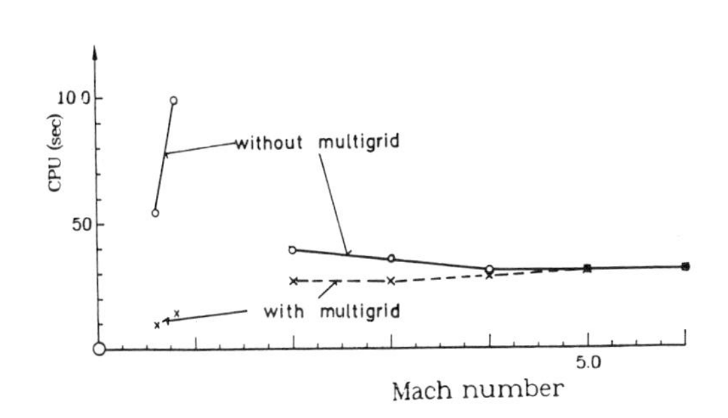

Figure 5: Influence for free-stream Mach number on computing time for NACA 0012 for inviscid flow, mesh 161x49, convergence criterion $\|\partial\rho/\partial t\|_2 < 10^{-5}$, CRAY XMP-216

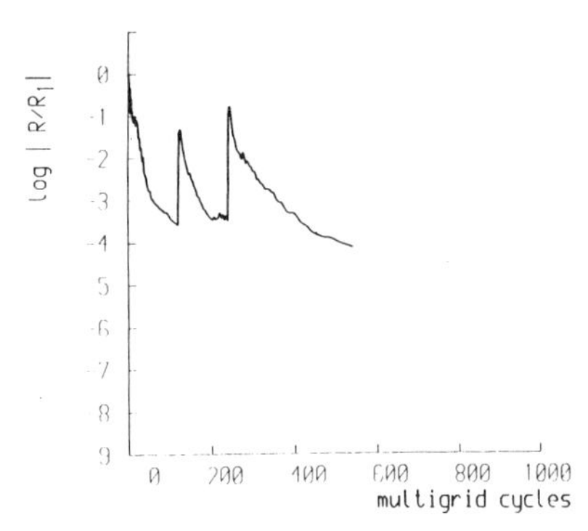

Figure 6: Convergence history of hypersonic turbulent 35° ramp flow, $M_\infty = 5.0$, $Re = 10^7$, $T_W/T_\infty = 3.47$, mesh 225x129

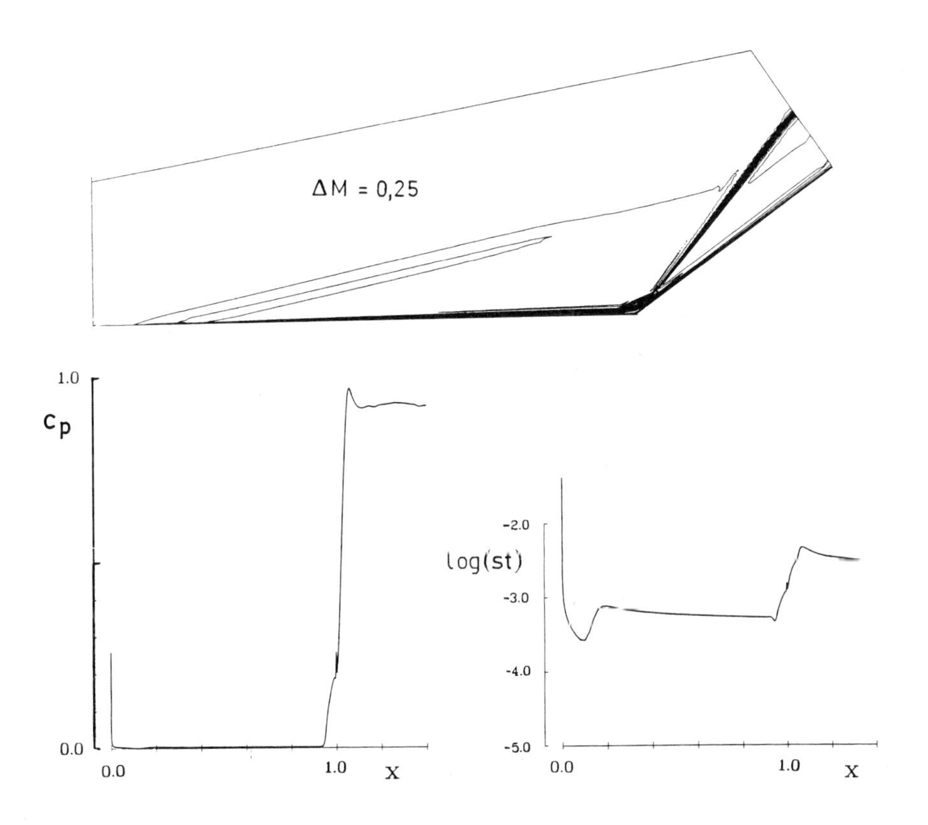

Figure 7: Mach contour and distribution of pressure and Stanton number for hypersonic turbulent 35° ramp flow, $M_\infty = 5.0$, $Re = 10^7$, $T_W/T_\infty = 3.47$, mesh 225x129

About splitting and non-splitting schemes for conservation laws in 2-D

Dietmar Kröner
Universität des Saarlandes, Fachbereich Mathematik
D-6600 Saarbrücken

Abstract

In this paper we shall discuss some problems which may arise in the context of dimensional splitting for conservation laws. We shall study a directionally adapted scheme which has better stability properties at least for anisotropic scalar equations in 2-D. This algorithm can be generalized to the non-linear unsteady Euler equations in 2-D and can be used to reduce the system locally to a scalar equation in 2-D.

1. Introduction

One of the most interesting and important systems of conservation laws are the Euler equations in gas dynamics, in particular in 2-D. In conservative variables they look like

$$\partial_t U + \partial_x F(U) + \partial_y G(U) = 0 \quad in \quad I\!R^2 \times [0,T], \tag{1.1}$$

where

$$U := (\rho, \rho u, \rho v, e)^t,$$
$$F(U) := (u\rho, \rho u^2 + p, \rho uv, u(e+p))^t$$
$$G(U) := (v\rho, \rho uv, \rho v^2 + p, v(e+p))^t.$$

Here ρ denotes the density, u and v the components of the velocity with respect to x and y, e the energy density and p the pressure. For an ideal gas we have the following equation of state

$$p = (\gamma - 1)(e - 0.5\rho(u^2 + v^2)), \tag{1.2}$$

where $\gamma = 1.4$ is the adiabatic exponent.

Systems of this kind, as well as the much simpler scalar hyperbolic conservation laws in 1-D, may have discontinuous solutions even if the data of the problem are smooth. Up to now there are no general, global existence results for the Euler equations in 2-D. On the other hand, there are a lot of numerical schemes, which have been applied to these problems in 2-D. These algorithms can be divided into the following groups: finite volume methods with artificial viscosity, finite volume methods with upwind components, finite element algorithms with different forms of upwinding or artificial viscosity, and dimensional splitting schemes (or fractional step methods). For the last three groups, numerical viscosity is included automatically. We refer to [COL], [DAV], [HLD], [LV1], [LV2], [RO1], [RO2].

In the context of explicit dimensional splitting schemes, there may arise some difficulties concerning the time step restriction for anisotropic problems, for the resolution of shear flows, or for special inhomogeneous problems (see [COO]). These problems will be discussed in §2, §3.

In §2 we shall investigate the stability for an explicit dimensional splitting scheme in 2-D. Applying this to an anisotropic convection equation in 2-D, we obtain a very restrictive bound on the time step. The difficulties of dimensional splitting in the context of the simulation of 2-D shear flows will be discussed in §3 of this paper.

Next, in §4 we shall study the stability analysis for the same problems as in §3 but now for schemes on a locally adapted grid. The choice of the locally adapted coordinate system will depend on the structure of the solution, or more precisely, on a locally preferred direction of propagation. In the following we shall refer to this scheme as the directionally adapted scheme. It turns out that for anisotropic problems the stability analysis allows for a time step which is much larger than for the corresponding dimensional splitting scheme.

In §5 we shall refer to some results, where we have generalized this directionally adapted scheme to the system of the nonlinear unsteady Euler equations in 2-D. This scheme is a generalization of an idea, published by Roe [RO1]. It turns out that for the considered test problems, the results are comparable to those obtained by first order dimensional splitting schemes.

In the last section §6 we shall present two test problems for the directionally adapted scheme: the converging cylindrical shock wave and the Mach 3 wind tunnel with a forward facing step.

2. Stability analysis for dimensional splitting schemes

To derive the stability conditions in the sense of Lax and Richtmyer for the dimensional splitting schemes in 2-D, let us repeat some basic results from the theory. We consider the following linear model problem of first order in two space dimensions:

$$\partial_t u + a\partial_x u + b\partial_y u = 0 \quad in \quad I\!R^2 \times [0,T] \tag{2.1}$$

$$u(.,0) = u_0 \quad in \quad I\!R^2. \tag{2.2}$$

Without restriction we assume that $a > 0,\quad b > 0$. Let us fix some notations. On $I\!R^2$ we define a grid $(i\Delta x, j\Delta y)$ for $i,j \in \mathbb{Z}$ and on $[0,T]$ we set $t_n := n\Delta t$ for $n = 0,1,2,\ldots$. We consider u_{ij}^n as an approximation for $u(i\Delta x, j\Delta y, n\Delta t)$ where u is an exact solution of (2.1), (2.2). Sometimes we omit the subscript n in u_{ij}^n . Furthermore let $u^n := (u_{ij}^n)_{i,j}$. We suppose that we have a finite difference scheme

$$u^{n+1} := Qu^n \tag{2.3}$$

where Q is a polynomial in the shift operators S, S^{-1}, T, T^{-1}. These are defined as follows:

$$Su_{ij} := u_{i+1,j}, \quad Tu_{ij} := u_{i,j+1} \quad .$$

Then it is obvious that

$$S^{-1}u_{ij} := u_{i-1,j}, \quad T^{-1}u_{ij} := u_{i,j-1}.$$

In the following we write Q as a function of S and T. The finite difference scheme (2.3) is called stable in the sense of Lax and Richtmyer with respect to a norm $\| \quad \|$ if there exist constants C and β such that we have for all $\Delta t, \Delta x, \Delta y$, sufficiently small,

$$\| u^n \| \leq ce^{\beta n \Delta t} \| u_0 \|$$

(see [SOD, p. 13]). The symbol or the amplification factor ρ of the difference operator Q is defined as

$$\rho(\xi, \eta) := Q(e^{-i\xi}, e^{-i\eta})$$

for $\xi, \eta \in I\!R$ and $i = \sqrt{-1}$. Now we can formulate the von Neumann condition for stability.

2.1 THEOREM (von Neumann condition) The finite difference scheme (2.3) is stable in the L^2-norm, if there exists a constant C independent of $\Delta x, \Delta y, \Delta t, n, \xi$, and η such that

$$| \rho(\xi, \eta) | \leq 1 + c\Delta t$$

for all $0 \leq \xi, \eta \leq 2\pi$.

For the proof we refer to [RM, p.72]and [SOD, p.18, 37]. As usual if the scheme (2.3) is consistent and stable, the Lax equivalence theorem [RM, p.45]implies the convergence of the scheme. Now let us apply this theorem in order to obtain the maximal time step for the dimensional splitting scheme. Later on in §3 we shall do the same for the directionally adapted scheme.

Let us consider a dimensional splitting scheme based on the simplest one-dimensional upwind scheme. Let $\lambda := \frac{\Delta t}{\Delta x}$ and for simplicity $\Delta x = \Delta y$. Then the scheme for the model equation (2.1), (2.2) is of the form

$$u_{ij}^{n+\frac{1}{2}} := u_{ij}^n - \lambda a(u_{ij}^n - u_{i-1,j}^n) \tag{2.4a}$$

$$u_{ij}^{n+1} := u_{ij}^{n+\frac{1}{2}} - \lambda b(u_{ij}^{n+\frac{1}{2}} - u_{i,j-1}^{n+\frac{1}{2}}). \tag{2.4b}$$

Using the shift operators S and T, the scheme (2.4) can be written as $u^{n+1} := Q(S,T)u^n$ where

$$Q(S,T) = ((1 - \lambda b)Id + \lambda bT^{-1})((1 - \lambda a)Id + \lambda aS^{-1}).$$

For the symbol (or the amplification factor) ρ of Q we obtain

$$\rho(\xi,\eta) = Q(e^{-i\xi}, e^{-i\eta}) = ((1-\lambda b) + \lambda b e^{i\eta})((1-\lambda a) + \lambda a e^{i\xi})$$

and therefore

$$|\,\rho(\xi,\eta)\,|^2 = (1 - 2\lambda b(1-\lambda b)(1-\cos\eta))(1 - 2\lambda a(1-\lambda a)(1-\cos\xi)).$$

Now we can prove the following result.

2.2 LEMMA The dimensional splitting scheme (2.4) is stable with respect to the L^2-norm if and only if

$$1 - \lambda a \geq 0 \quad , \quad 1 - \lambda b \geq 0 \quad . \tag{2.5}$$

Proof: Let us assume (2.5). Since we have the general assumption $a, b > 0$, the difference scheme (2.4) satisfies the von Neumann condition, Theorem 2.1: The function $f(s) := s(1-s)$ assumes its maximal value $\frac{1}{4}$ in $s = \frac{1}{2}$ and therefore we have for all $\xi, \eta \in I\!R$:

$$2\lambda a(1-\lambda a)(1-\cos\xi) \leq 1,$$

$$2\lambda b(1-\lambda b)(1-\cos\eta) \leq 1.$$

Then it is obvious that $|\,\rho(\xi,\eta)\,| \leq 1$.

Now let us assume that the difference scheme (2.4) is stable and hence the symbol ρ satisfies the von Neumann condition, Theorem 2.1. We have to show that (2.5) is valid. Assume that $1 - \lambda a < 0$. The case $1 - \lambda b < 0$ can be treated similarly. Then for $\xi = \frac{\pi}{2}$ and $\eta = 0$ we obtain:

$$|\,\rho(\frac{\pi}{2}, 0)\,|^2 = 1 - 2\lambda a(1-\lambda a)$$

$$= 1 + 2\frac{a}{\Delta x}(\lambda a - 1)\Delta t.$$

The coefficient of Δt will become unbounded if Δx tends to zero, which contradicts the von Neumann condition, Theorem 2.1.

This lemma implies that the maximal time step which guarantees stability of the scheme (2.3), is defined by

$$\Delta t \leq \min(\frac{\Delta x}{a}, \frac{\Delta y}{b}). \tag{2.6}$$

2.3 EXAMPLE Let us consider the initial value problem (2.1), (2.2) for

$$u(x,y,0) := w(x\cos\Theta + y\sin\Theta), \quad x, y \in I\!R \tag{2.7}$$

where Θ is a given constant, $0 < \Theta < \frac{\pi}{2}$ and $w \in L^\infty(I\!R)$ such that $w(s) = 1$ if $s < 0$ and $w(s) = -1$ if $s \geq 0$. This means u_0 depends only on $\xi = x\cos\Theta + y\sin\Theta$ for

$x, y \in I\!R$ and is constant along $\eta = -x \sin\Theta + y\cos\Theta + c$ for any constant $c \in I\!R$. It turns out that the exact solution in the distributional sense of (2.1),(2.7) is

$$u(x, y, t) := w(x\cos\Theta + y\sin\Theta - (a\cos\Theta + b\sin\Theta)t) \tag{2.8}$$

and that u is constant along η for any fixed time t.

Let us approximate the exact solution of (2.1), (2.7) by using the finite difference scheme (2.4) with $\Delta x = 0.01, \Delta y = 0.01, a = 1$, and $b = 1000$. Then the bound on the time step for the dimensional splitting scheme is

$$\Delta t_{\max} = \min(10^{-2}, 10^{-5}) = 10^{-5}. \tag{2.9}$$

We notice that $\Delta t_{\max}$ is independent of the angle Θ, i.e. of the structure of the solution itself. We mention this because this will be different for the directionally adapted scheme (see §4). For this scheme we shall show in §4, using an adapted coordinate system, that we can take a time step which is 1000 times larger than for the dimensional splitting scheme.

3. Dimensional splitting for shear flows

Now we are going to discuss the problems which may arise in the case of a shear flow. We consider the Euler equation (1.1) with the initial conditions

$$U_0(x, y) := W(x\cos\Theta + y\sin\Theta), \tag{3.1}$$

for $x, y \in I\!R$ and some given constant angle Θ, $0 < \Theta < \frac{\pi}{2}$. Here W is a function $W : I\!R \to I\!R^4$ such that $W(\xi) = W_L$ if $\xi := x\cos\Theta + y\sin\Theta < 0$ and $W(\xi) = W_R$ if $\xi \geq 0$ where $W_L = (\rho_0, u_L\rho_0, v_L\rho_0, e_0)$ and $W_R = (\rho_0, u_R\rho_0, v_R\rho_0, e_0)$; $\rho_0, e_0, u_L, v_L, u_R, v_R$ are given constants, such that (u_R, v_R) and (u_L, v_L) are orthogonal to $(\cos\Theta, \sin\Theta)$ and $u_L < 0, v_L > 0, u_R > 0, v_R < 0$ (see Figure 1) . This means, U_0 is piecewise constant in $I\!R^2$: the density ρ_0 and the energy density e_0 are global constants while the velocities (u_L, v_L) and (u_R, v_R) are different on either side of $\xi = 0$ (see also [RM, p.356]).

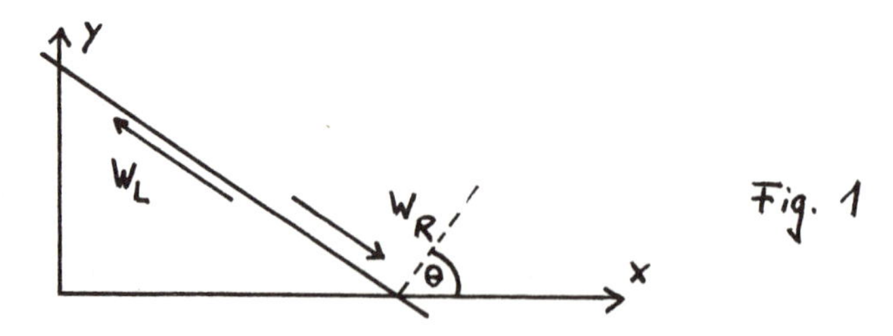

Now let us consider what happens if we apply a dimensional splitting scheme to this initial value problem (1.1), (3.1). The algorithm begins with a one-dimensional step in the $x-$ direction applied to the initial value $U_0 = W_L$ on the left side of $\xi = 0$

and $U_0 = W_R$ on the right side of $\xi = 0$. Actually this means, in the $x-$ step the scheme solves the Riemann problem

(3.2a) $$\partial_t U + \partial_x F(U) = 0 \quad in \quad I\!R^2 \times [0,T]$$

(3.2b) $$U(.,0) = W_L \quad if \quad x > y \tan\Theta$$

$$U(.,0) = W_R \quad if \quad x < y \tan\Theta.$$

This problem differs from the 1-D Riemann problem of gas dynamics only in the equation of state and in the third equation in (3.2a) which describes the transport of momentum ρv in the $x-$direction. For the 1-D Riemann problem in gas dynamics it can be shown that, if the difference $u_R - u_L$ is too large, there may arise a vacuum state [SMO, (18.57)]). Numerical experiments for the shear flow show that a similar effect may happen for the system (3.2). In this case the $x-$ step will produce a wrong solution and the y-step is not able to correct again the wrong solution.

Now let us return to the stability analysis for the scalar model problem in 2-D. In the next section we shall present the directionally adapted scheme and derive its stability condition.

4. Stability analysis for the directionally adapted scheme

We consider again an initial value problem for (2.1). For a discontinuous solution in general a locally preferred direction is well defined in a neighbourhood of the discontinuity. For instance this could be the local direction of propagation of a shock. Instead of a locally defined direction let us consider the global situation as in (2.1),(2.7) (see Figure 2). In the general case one has to carry out the following considerations locally.

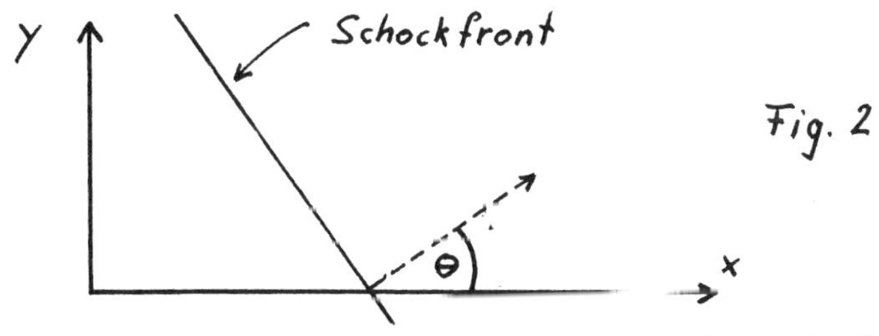

As in (2.8) the exact solution is a travelling wave propagating in the direction given by Θ. Now we would like to adapt the coordinate system to the solution. Let

(4.1) $$\xi = \xi(\Theta) = x\cos\Theta + y\sin\Theta \qquad \eta = \eta(\Theta) = -x\sin\Theta + y\cos\Theta$$

and we define the following transformation:

(4.2) $$u(x,y,t) =: v(\xi,\eta,t).$$

The function v doesn't actually depend on η and satisfies the following differential equation:

$$\partial_t v + (a\cos\Theta + b\sin\Theta)\partial_\xi v = 0 \tag{4.3}$$

$$v(\xi,\eta,0) = w(\xi). \tag{4.4}$$

We use the following one-dimensional upwind scheme for approximating a solution of (4.3):

$$v_i^{n+1} := v_i^n - \frac{\Delta t}{\Delta \xi}\sigma(v_i^n - v_{i-1}^n) \tag{4.5}$$

where v_i^n is considered as an approximation of $v(i\Delta\xi, n\Delta t)$ and $\sigma = a\cos\Theta + b\sin\Theta$. In order to get a condition for the stability, we apply the results of §2 to the initial value problem (4.3),(4.4) with respect to the coordinate system (ξ,η). Then we obtain a bound for the time step in the new coordinate system.

4.1 LEMMA The finite difference scheme (4.5) is stable if and only if

$$1 - (a\cos\Theta + b\sin\Theta)\frac{\Delta t}{\Delta\xi} \geq 0.$$

This result can be proved in the same way as Lemma 2.2. Then the maximal time step for the scheme (4.3) is given by

$$\Delta t \leq \frac{\Delta\xi}{a\cos\Theta + b\sin\Theta}. \tag{4.6}$$

Whereas for the dimensional splitting scheme the maximal time step is independent of the angle Θ, this is not the case for the directionally adapted scheme.

In the following example we shall see that for suitable Θ and a, b as in Example 2.3, the time step for the directionally adapted scheme can be chosen much larger than for the dimensional splitting scheme.

4.2 EXAMPLE We consider the same problem as in Example 2.3 for $a = 1, b = 1000$ and $\Delta\xi = 0.01$. Additionally we assume that $\Theta \ll 1$. Then the largest time step for the directionally adapted scheme (4.5) is

$$\Delta t_{\max} = \frac{0.01}{a\cos\Theta + b\sin\Theta} \sim 0.01. \tag{4.7}$$

This means the bound on the time step of the directionally adapted scheme can be chosen 1000 times larger then for the dimensional splitting scheme. Of course, if Θ

is close to $\frac{\pi}{2}$ the maximal time step of both schemes (2.4), (4.5) are of order 10^{-5}. In between, i.e. if

$$\arcsin\frac{b^2-a^2}{b^2+a^2} < \Theta < \frac{\pi}{2},$$

the time step for the dimensional splitting scheme can be chosen a little bit larger.

5. Directionally adapted scheme for the Euler equations in 2-D

It turns out that the directionally adapted scheme can be used to locally reduce the system (1.1) to a set of scalar equations in 1–D or 2–D (see (5.5), (5.6) below). In this section we shall briefly describe the generalization of this scheme to the Euler equations (1.1) in 2-D. For more details we refer to [KR1], [KR2]. In the first step we obtain an estimate for a locally dominant direction of propagation by approximating a component of the solution locally by a travelling wave. If according to the initial conditions, we expect a large jump e.g. for the density, then this is the component we use for defining a local direction of propagation. In other cases we use the velocity or the energy. Up to now the user of the scheme has to decide this before running the program. If we use the density for defining the main direction of propagation we proceed as follows. Assume that, e.g., $\rho(t,x,y) = \rho_0(x\cos\Theta + y\sin\Theta - St)$ for some unknown function $\rho_0 \in C^1(I\!R, I\!R)$ and $\Theta, S \in I\!R$. Then $\partial_x\rho = \rho_0'\cos\Theta$, $\partial_y\rho = \rho_0'\sin\Theta$ and

$$\tan\Theta = \frac{\partial_y\rho}{\partial_x\rho}. \tag{5.1}$$

This gives an estimate for Θ, if $\partial_x\rho$ is sufficiently far from 0 and if $\partial_y\rho$ is not too large. Otherwise the solution has to be smooth and the scheme should not depend on Θ anyway (see [KR2]).

Now we have to linearize the equations (1.1) locally. Since the angle Θ defines locally a 1-D structure, we take the 1-D Roe mean value $\bar{U}$ (see [KR2, (7)]) and consider instead of (1.1)

$$\partial_t U + F'(\bar{U})\partial_x U + G'(\bar{U})\partial_y U = 0 \quad in \quad I\!R^2 \times [0,T]. \tag{5.2}$$

In order to adapt the scheme locally to the special structure of the solution we rotate the coordinate system by the angle Θ and obtain the new coordinates (ξ,η), where ξ and η are defined as in (3.3). Set

$$V(t,\xi,\eta) := U(t,x,y),$$

and we obtain for V

$$\partial_t V + D(\Theta)\partial_\xi V + D(\Theta + \frac{\pi}{2})\partial_\eta V = 0, \tag{5.3}$$

where $D(\Theta) = F'(\bar{U})\cos\Theta + G'(\bar{U})\sin\Theta$. Let $R_j(\Theta)$ and $\lambda_j(\Theta)$ for $j = 1,2,3,4$ denote the eigenvectors and eigenvalues of $D(\Theta)$ respectively. Then there exist $\alpha_j(t,\xi,\eta), j = 1,...,4$ such that

$$V(t,\xi,\eta) = \sum_{j=1}^{4} \alpha_j(t,\xi,\eta)R_j. \tag{5.4}$$

Due to (5.3) we obtain for α_j:

$$\sum_j (\partial_t \alpha_j + D(\Theta)\partial_\xi \alpha_j + D(\Theta + \frac{\pi}{2})\partial_\eta \alpha_j) R_j = 0.$$

Now we assume that the derivative of V tangential to the discontinuity can be neglected: $\partial_\eta V = 0$. This is exactly satisfied if the states in front of and behind the discontinuity are constant or if the problem is rotationally symmetric. Therefore we obtain for $\alpha_1, ..., \alpha_4$ the following equations:

$$\partial_t \alpha_j + \lambda_j \partial_\xi \alpha_j = 0 \tag{5.5}$$

or in the original coordinate-system

$$\partial_t \alpha_j + \lambda_j \cos\Theta \partial_x \alpha_j + \lambda_j \sin\Theta \partial_y \alpha_j = 0. \tag{5.6}$$

Now assume that U^n is given. Then we discretize (5.1) in order to get an estimate for Θ. After computing the matrix $D(\Theta)$ and its eigenvectors R_j and eigenvalues λ_j we decompose U^n in terms of the R_j with coefficients α_j. These values are used as initial values for (5.6). For solving (5.6) we use an upwind finite volume method (see [KR2]).

6. Test problems

6.1 TEST PROBLEM The converging cylindrical shock wave: As initial conditions we choose radially symmetric values: Within an inner disk of radius 0.25 we prescribe the density $\rho = 1$ and the energy density $e = 0.4$ and in the exterior domain the density $\rho = 4$ and the energy density $e = 1.6$. Initially the velocities are equal to zero everywhere. On the boundaries of the computational domain we use reflecting boundary conditions. Then we should expect a radially symmetric solution.

In our experiments we used a grid of 100×100 points and we chose $\Delta x = 0.01$, $\Delta y = 0.01$ and as the initial time step $\Delta t = 0.05$. The results for the time $t = 0.30$ are shown in Fig. 3, a contour plot of the density.

6.2 TEST PROBLEM Mach 3 wind tunnel with a forward facing step: Initially the tunnel is filled with a gas, which has everywhere density 1.4, pressure 1.0, velocity $u = 3.0$ and $v = 0.0$. On the walls of the tunnel we have reflecting boundary conditions, on the right side homogeneous Neumann and on the left side Dirichlet boundary conditions with values equal to the initial ones. For defining the local angle of propagation we have chosen the velocity. The results of the numerical experiments on a 150×50 grid for $\Delta x = 0.01$, $\Delta y = 0.01$, and $t = 3.0$ are shown in Figure 4. For more details concerning the numerical results and for comparison with other schemes we refer to [KR2].

References

[COL]P.Colella, Multidimensional upwind methods for hyperbolic conservation laws. Report Lawrence Berkeley Laboratory 17023,1984.

[COO]C.H.Cooke, On operator splitting of the Euler equations consistent with Harten's second-order accurate TVD scheme. Numerical methods for partial differential equations. 1(1985), 315-327.

[DAV]S.F.Davis, A rotationally biased upwind difference scheme for the Euler equations. J.of Comp. Physics 56(1984), 65-92.

[HLD]Ch.Hirsch, C.Lacor, H.Deconinck, Convection algorithm on a diagonal procedure for the multidimensional Euler equation. AIAA, Proceedings of the 8-th Computational Fluid Dynamics Conference 1987.

[KR1]D. Kröner. Numerical schemes for the Euler equations in two space dimensions without dimensional splitting. Proceedings of the 2-nd International Conference on Nonlinear Hyperbolic Equations- Theory, Computation Methods, and Applications. Notes on Numerical Fluid Mechanics, Vol. 24, Braunschweig 1989.

[KR2]D. Kröner. Directionally adapted upwind schemes in 2-D for the Euler equations. DFG Priority Research Programme, Results 1986-1988, Notes on Numerical Fluid Mechanics, Vol. 25, Braunschweig 1989.

[LV1]R.J.LeVeque, High resolution finite volume methods on grids via wave propagation. ICASE report 87-68, 1987

[LV2]R.J.LeVeque, Cartesian grid methods for flow in irregular regions. To appear in the Proceedings of the Oxford Conference on Numerical Methods in Fluid Dynamics, 1988.

[RM]R.D. Richtmyer, K.W. Morton, Difference methods for initial-value problems. Second edition, New York 1967.

[RO1]P.Roe, A basis for upwind differentiating of the two dimensional unsteady Euler equations. Numerical methods for fluid dynamics II, Eds.: Morton, Baines, Oxford Univ. Press 1986.

[RO2]P.Roe, Linear advection schemes on triangular meshes. CoA Report No 8720, Cranfield 1987.

[RO3]P.Roe, Discontinuous solutions to hyperbolic systems under operator splitting. Manuscript.

[SMO]J.Smoller, Shockwaves and reaction-diffusion equations. New York Heidelberg Berlin 1983.

[SOD]G.A.Sod, Numerical methods in fluid dynamics. Initial and initial boundary value problems. Cambridge 1985.

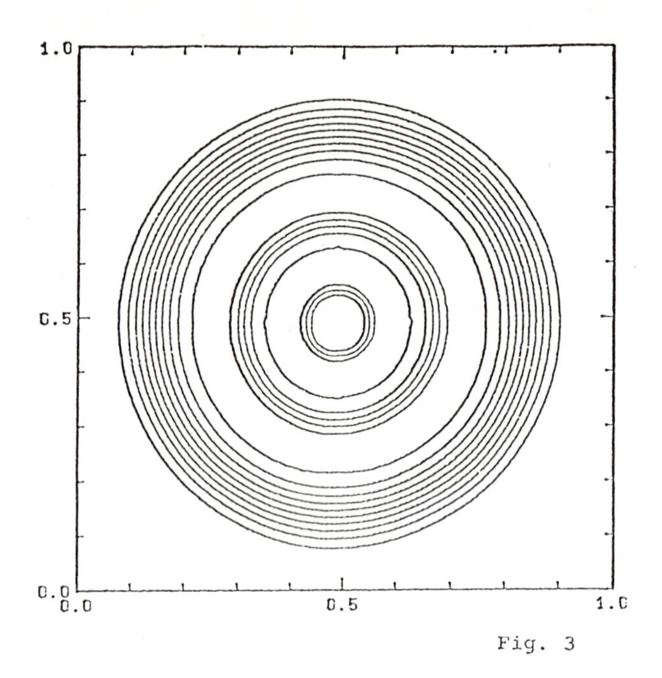

Fig. 3

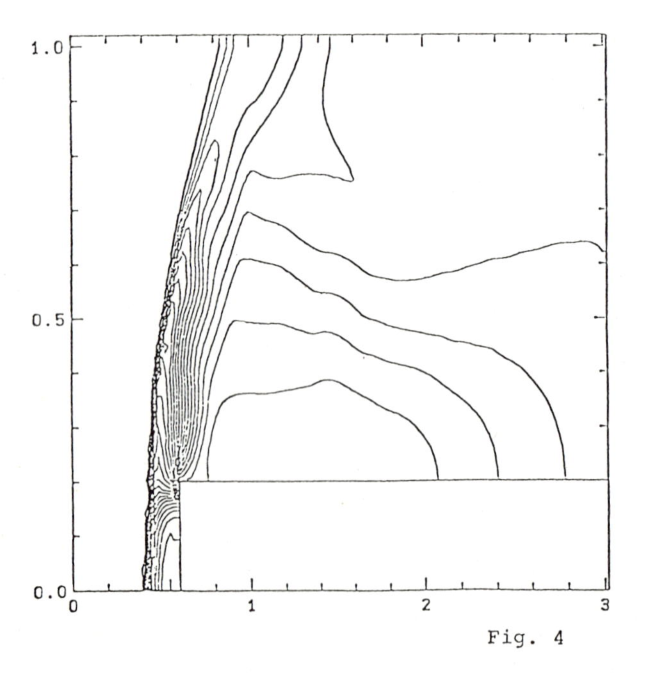

Fig. 4

THE INFLUENCE OF SURFACE-TENSION EFFECTS ON USING VORTEX METHOD IN THE STUDY OF RAYLEIGH-TAYLOR INSTABILITY

Henryk Kudela
Technical University of Wroclaw, 50-370 Wroclaw, Poland

SUMMARY

The effect of surface-tension on the smoothing of irregular motion of vortices in the vortex simulation of Rayleigh-Taylor instability is shown. The irregular motion appears as an effect of short-wave disturbances the source of which is round-off error. Inclusion of surface tension allows the observation of the formation of singularities. The singularities make an infinite jump discontinuity in the curvature of vortex sheet. It is observed that for sufficiently small Atwood number and for initial amplitude perturbation large enough two singularities appear in a half period of the vortex sheet and that only one appears for greater Atwood numbers.

INTRODUCTION

Rayleigh-Taylor (R-T) instability occurs on the interface between two fluids with different densities when the less dense fluid is accelerated in the direction of the denser one [5][14]. With some simplified assumptions, the interface can be regarded as a vortex sheet, and the investigation of its evolution can be reduced to the solution of an initial-value problem. For the solution of this problem we can use a method based on boundary integral techniques. The vortex-sheet formulation described by Baker,Meiron,Orszag (B-M-O) [2] was chosen. The method was successfully used for the case when Atwood number $A=(\rho_1-\rho_2)/(\rho_1+\rho_2),\rho_1>\rho_2$ was equal one. The method failed for $0<A<1$. The linear approximation of R-T, without any stabilizing mechanism, gives an arbitrarily large growth rate of short-wave length perturbations [5]. In computations, the source of short-wave perturbations is round-off error due to finite precision arithmetic [9], and without any stabilizing mechanism it leads to the deterioration of calculations. Similar to Kelvin-Helmholtz instability one may say that the problem is ill-posed in the Hadamard sense [9],[11],[13]. To suppress the chaotic, irregular motion of vortex points in numerical study of vortex sheet evolution many tricks were introduced [12][9]. In this paper it was demonstrated that in the R-T problem the inclusion of surface- tension effects can suppress the irregular motion of vortex points. This permits one to observe the process of singularity formation. The surface tension provides a small-scale stabilizing mechanism which doesn't remove the ill -posedness of the problem. To do this a parameter δ^2 was introduced in the Cauchy-type integral to cut-off the singularity of integrand [10] [8] .In this situation, the surface tension effects caused also the lengthening of time interval in which numerical solutions could be obtained.

STATEMENT OF THE PROBLEM

It is assumed that the motion of the fluids is potential,two-dimensional and that the fluids are inviscid. A rectangular system of coordinates (x,y) is introduced. Before introducing a perturbation of the interface the fluid with density ρ_1 (ρ_2) occupies the upper (lower) half plane and $\rho_1>\rho_2$. A constant gravitational field g acts in the negative y-direction. Both ρ_1 and ρ_2 are constant and there is a density jump at the interface. The interface is described as a complex curve $z(a,t)=x(a,t)+iy(a,t)$, where "a" is regarded a Lagrangian parameter. It is assumed that the interface and the disturbance are periodic in the x-direction with wavelength λ_p. So $a \in [0,a_o]$ and an increment a to a_o gives

$$z(a+a_o,t) = z(a,t) + \lambda_p . \tag{1}$$

The normal velocity component on the interface is continuous and the tangential velocity component on the interface may have jump. So the interface can be regarded as a vortex sheet with intensity:

$$\Gamma(s,t) = (\underline{u}_2 - \underline{u}_1)\underline{s}^o = \frac{\partial\,[\Phi]}{\partial s} \tag{2}$$

where $\underline{u}_1,\underline{u}_2$ are the velocities of the upper and lower fluids at the interface, $\underline{s}^o$=unit tangential vector, s-arclength, $[\Phi]$-jump of the value of the potential across the interface, t= time.
At t=0, the fluid is assumed at rest; then the flat interface y=0 is perturbed and has the form $y(x,t)=\varepsilon(t)\cos Kx$ for $t>0$. In the linear approximation [5] [14] it is known that the amplitude $\varepsilon(t)$ will behave like $\varepsilon(t)=\varepsilon(0)\cosh \omega t$ where

$$\omega^2 = g\,\frac{\rho_1 - \rho_2}{\rho_1 + \rho_2}\,K - \frac{\sigma}{\rho_1 + \rho_2}\,K^3 \tag{3}$$

and where σ is the coefficient of interfacial tension, $K = 2\pi/\lambda$.
From (3) we can see that for $\sigma=0$, then $\omega > 0$ and the interface is unstable. The growth rate for short wavelengths is unbounded. Inclusion of surface -tension effects ($\sigma\neq 0$) stabilizes the perturbations shorter than a critical wavelength

$$\lambda_c = 2\pi\left[\frac{\sigma}{g\,(\rho_1-\rho_2)}\right]^{1/2} . \tag{4}$$

To put the further equations in non-dimensional form we choose λ_p as a unit of length, $\sqrt{\lambda_p/Ag}$ as a unit of time and the surface-tensions coefficient was normalized by $g(\rho_1-\rho_2)\lambda_p^2$.

THE GOVERNING EQUATIONS

By virtue of the Biot-Savart law, the velocity induced by the periodic vortex sheet at a point on the sheet can be expressed as:

$$q^{*} = u(a,t)-iv(a,t)= \frac{1}{2i} \fint_{0}^{1} \gamma\ (a',t)\cot \pi(z(a,t)-z(a',t))da' \qquad (5)$$

where $\gamma(a,t)=\Gamma(a,t)(x_a^2+y_a^2)^{1/2}$ The subscript denotes a differentiation and the dash on the integral sign signifies Cauchy principle value.
The vortex sheet will be replaced by a suitable distribution of vortices (Lagrangian points) each labelled by a parameter "a". The later position of those points allow the determination of the interface shape. We define the Lagrangian point velocity on the vortex sheet as follows [4][3]

$$\frac{\partial z}{\partial t} = \frac{q_1 + q_2}{2} + \alpha \frac{q_2 - q_1}{2} \qquad (6)$$

where q_1, q_2 are the upper and lower limits of the fluid velocities on the interface, and $\alpha \in [-1,1]$ is a weighting factor.Note that for $\alpha=0$ the Lagrangian point is non-material and for $\alpha=-1(1)$ it follows the upper (lower) fluid at interface. In this way we retain some control over the positioning of the Lagrangian points. By virtue of the Sochocky-Plemelji formula [10] (see also [4])

$$q^{*}_{1(2)} = q^{*} (\mp) \frac{\gamma}{2z_a} . \qquad (7)$$

The velocity of a Lagrangian points on the interface can be expressed as:

$$\frac{\partial z^{*}}{\partial t} = q^{*} + \alpha \frac{\gamma}{2z_a} . \qquad (8)$$

To equation (8)(5) there must be added the equation for vortex-sheet strength due to baroclinic generation of vorticity.This equation is [3] :

$$\gamma_t = 2A\ (\mathrm{Re}(q^{*}_t z_a) - \frac{1}{2}\alpha\gamma \frac{u_a x_a + v_a y_a}{z_a z^{*}_a} + \frac{1}{8}\frac{\partial}{\partial a}\frac{\gamma^2}{z_a z^{*}_a} + \frac{1}{2}\alpha\frac{\partial}{\partial a}\frac{\gamma^2}{z_a z^{*}_a}) +$$
$$+2y_a - 2\sigma k_a + \frac{1}{2}\alpha \frac{\partial}{\partial a}\frac{\gamma^2}{z_a z^{*}_a} , \qquad (9)$$

where $k=(x_a y_{aa} - y_a x_{aa})/(x^2_a + y^2_a)^{1.5}$ = the curvature of vortex sheet, σ = the non-dimensional coefficient of surface tension. The time derivatives q^{*}_t are obtained by differentiation of (5). After substitution of these derivative in (9) we obtain:

$$\gamma_t = A \int_0^1 \gamma_t \, B \, da' + r(x,y,u,v,\gamma) \tag{10}$$

where $B = -x_a \dfrac{\sinh 2\pi(y(a)-y(a'))}{\cosh 2\pi(y(a)-y(a'))-\cos 2\pi(x(a)-x(a')} +$

$$y_a \frac{\sin 2\pi(x(a)-x(a'))}{\cosh 2\pi(y(a)-y(a'))-\cos 2\pi(x(a)-x(a'))}$$

and r( ) does not depend on γ_t. Eq. (10) represents a Fredholm integral equation of the second kind and is solved iteratively.
As has already been mentioned in the introduction,the initial value problem for R-T is ill-posed. The solution within a finite time yields a singularity which is an infinite jump in the curvature. In order to regularize the problem the singularity of the integrand of the Cauchy -type integral is cut off. Components of velocity of equation (5) are now:

$$u(a,t) = -\frac{1}{2} \int_0^1 \frac{\sinh 2\pi(y(a)-y(a'))}{\cosh 2\pi(y(a)-y(a'))-\cos 2\pi(x(a)-x(a') + \delta^2} \, da' \tag{5a}$$

$$v(a,t) = \frac{1}{2} \int_0^1 \frac{\sin 2\pi(x(a)-x(a'))}{\cosh 2\pi(y(a)-y(a'))-\cos 2\pi(x(a)-x(a')) + \delta^2} \, da'. \tag{5b}$$

Now the integrals in (5a,b) are not singular and if we calculate with the assistance of (5a), the acceleration u_t, v_t, eq. (10) also will not have a singular integral. In the case of weak stratification when $\rho_1/\rho_2 \longrightarrow 1$, instead of (9) there is often utilized the Boussinesq asymptotic approximation . We take $A \longrightarrow 0$ and $g \longrightarrow \infty$ in such a way that the product Ag goes to a finite limit [15]. Equation (9) takes the form (α=0):

$$\gamma_t = 2y_a \; -2\sigma k_a \, . \tag{9a}$$

Equations (8),(10) constitute the complete description of the evolution for the interface. The numerical procedure goes as follows [3]: for known $x(a), y(a), \gamma(a)$ the interface is marched forward using (8),(5) and next inhomogeneous term r( ) in (10) is calculated. Then the integral equation (10) is solved iteratively for γ_t and finally γ can also be marched in time. When we use Boussinesq approximation ((9a) we do not need to solve integral equation and γ_t is simply calculated from (9a). The time stepping is performed, as in [3] by using a fourth -order Adams-Moulton predictor-corrector scheme. All derivatives with respect to the Lagrangian parameter "a" were computed using cubic spline. All Cauchy type integrals were regularized by subtraction of integral with a known principle value [3] and integrals were calculated using trapezoidal rule on alternate sets of points [4].

NUMERICAL RESULTS

Results presented below were obtained on an IBM PC-AT in double precision (15 decimal digits) arithmetic using RMFORTRAN. The number of vortices accepted for calculations is N=120, time step Δt=0.004. Fig.1a,b shows solutions for A=1 ($\rho_1/\rho_2 = \infty$, the so-called single fluid case). Fig. 1a shows a time sequence of interface profile y(x,t) and fig.1b shows a time sequence of vortex sheet strengths $\Gamma(a,t)$ vs Lagrangian parameter "a". $y(x,0)=\varepsilon \cos 2\pi x$, $\varepsilon=0.5/2\pi$ was the initial condition. Such an amplitude makes possible a comparision results with [2]. These single fluid case results were described precisely by B-M-O [2] and can be regarded as an "exact" solution [15]. They make a good test for the numerical program. The points on the curve indicates the positions of vortices along the interface. For α in (6) we took $\alpha=-1$. This means that vortices move like particles of upper fluid.

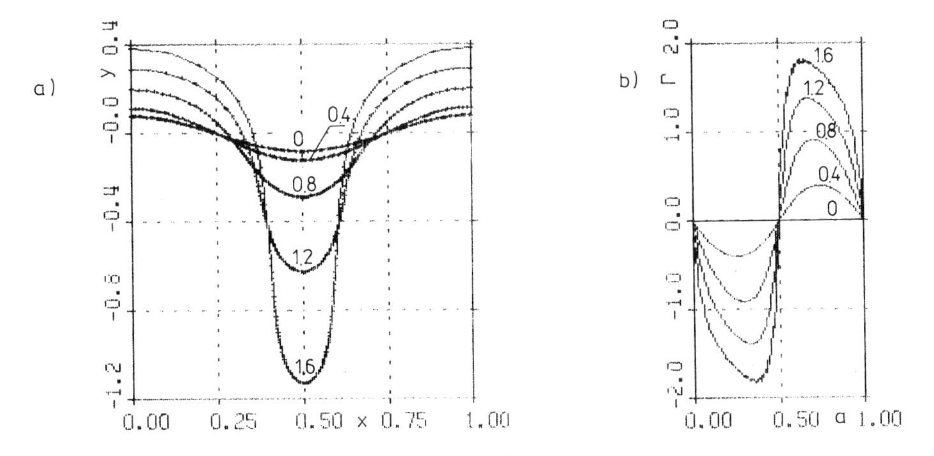

Fig.1. Numerical results for A=1, $\alpha=-1$, $\delta^2=0$, $\sigma=0$. a) interface profile y(x,t), b) vortex sheet strength $\Gamma(a,t)$

Fig.2a)b)c) show numerical results for A=0.0476 ($\rho_1/\rho_2=1.1$), $\sigma=0$ and $\varepsilon=0.1$.

Fig.2a shows the time sequence of interface profiles (only half a period), fig.2b the time sequence of the curvature k(a) of the interface and fig.2c shows the time sequence of the vortex sheet strength $\Gamma(a)$. The inspection of curvature is a good means for checking the smoothness of the solutions and it shows the symptoms of the regularity loss of the solution earlier than one can see on interface profile. In fig.2b for t=0.74 one can see the irregular, noisy changes of curvature at intervals where vortex sheet strength has its maximum approximately. It is due to Helmholtz instability. Those noisy changes start little earlier and their amplitude grows in time

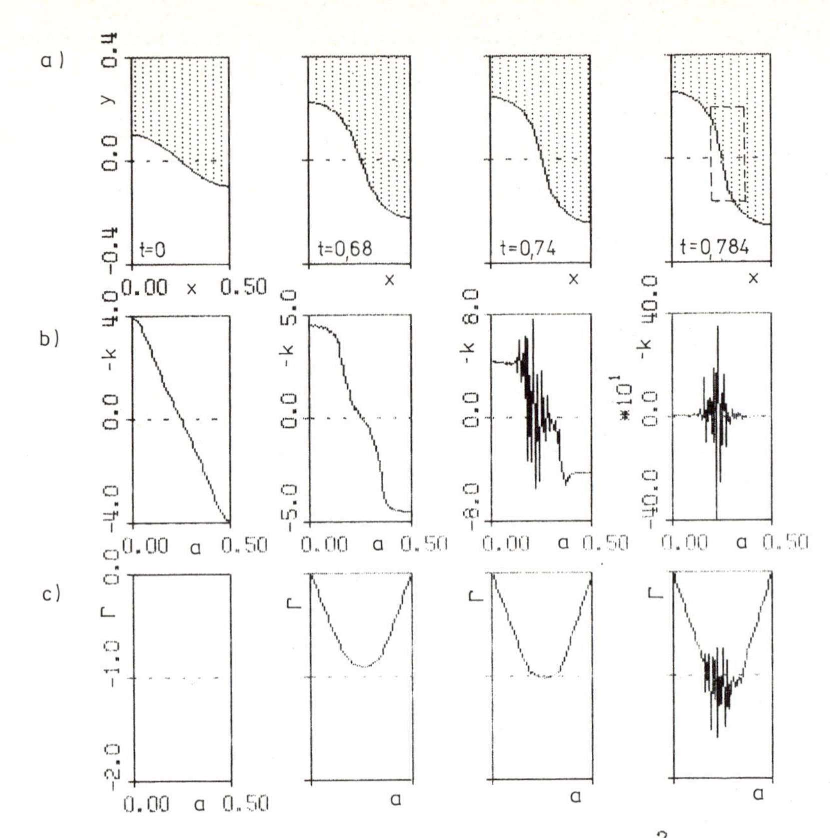

Fig.2 Numerical results for A=0.0476 (ρ_1/ρ_2=1.1), δ^2=0, σ=0, α=0, y(x,0)=0.1cos2πx, a)time sequence of interface profile y(x,t), b) curvature-k(a,t), c) vortex sheet strength Γ(a,t)

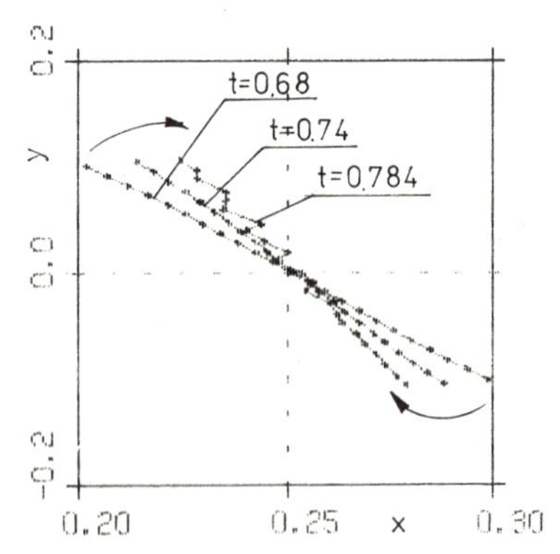

Fig.3 The time sequence of close-up fragments of interface created the same particles for A=0.0476, σ=0.

Fig.3 shows the time sequence of the close-up fragments of the interface matched by the frame in fig.2a t=0.784. These fragments are created by the same particles. For t=0.784 there can be seen distinctly the irregular positions of the vortices. Fig.4a)b) show the time sequence of curvature and strength of vortex sheet in the case when the surface-tension effects are included. For the value of coefficient of interfacial tension we took $\sigma=5*10^{-4}$.Now the noisy ,chaotic changes don´t appear in the curvature.The curvature distribution for t=0.824 has two peaks and typical tips appear in distribution of vortex sheet strength.

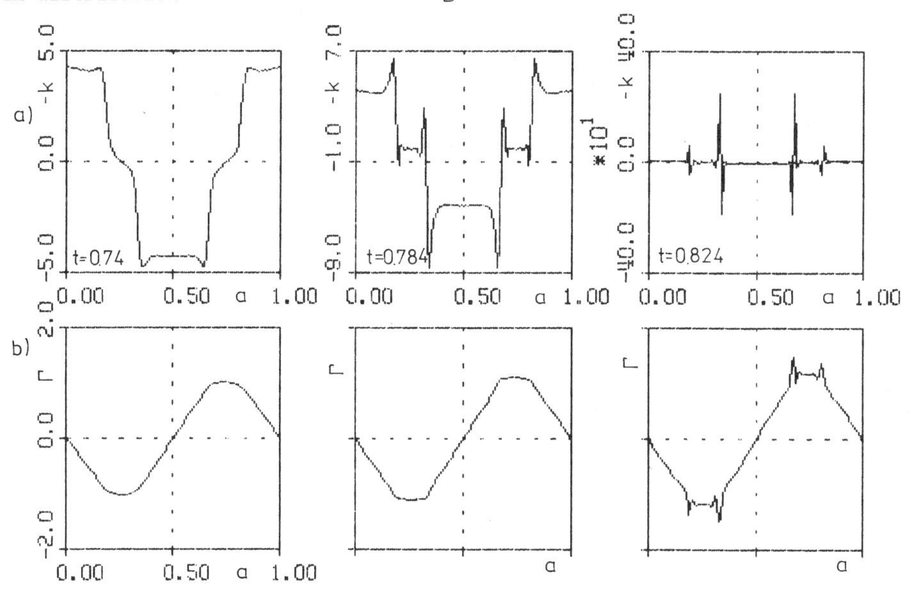

Fig.4 The time sequence a) the curvature k(a,t), b) the vortex sheet strength Γ(a,t) for for the same data as in fig.2 but with surface tension effects σ=0.0005.

Fig.5 shows the time sequence of the close-up fragments of the interface created by the same particle as in fig.3. The origination of cusps is visible and it can be related to the formation of singularities.The suppression of the irregular motion of vortices is evident. The formation of two singular points is connected with large initial amplitudes of perturbation (ε=0.1).For ε=0.01 we can observe only one singularity along half a period of the interface. Fig.6a shows the time sequence of close-up fragments of interface and figs.6b,c show examples of the curvature distribution and vortex sheet strength for time t=1.48.The results which relate to the singularity formation which I present here ,are very similar to the results presented by Krasny [9]. For larger Atwood number in spite of ε=0.1 only one singularity is originated. The singularity point moves towards the point of symmetry x=0.5. Fig.8 shows results for A=0.5. Fig.8a shows the time sequence of close-up of interface fragments and fig.8b,c the example of curvature -k(a) and vortex sheet strength Γ(a) for t=0.8.The singularity formation in solutions is the main reason the B-M-O method failed for 0<A<1. Fig.9 shows the time sequence of the profile interface for the desingularized equations for δ^2=0.1, A=0.0476, σ=0.0005, α=0. The inclusion of surface-tension effects increases the time interval for which the iteration process for γ_t converged.

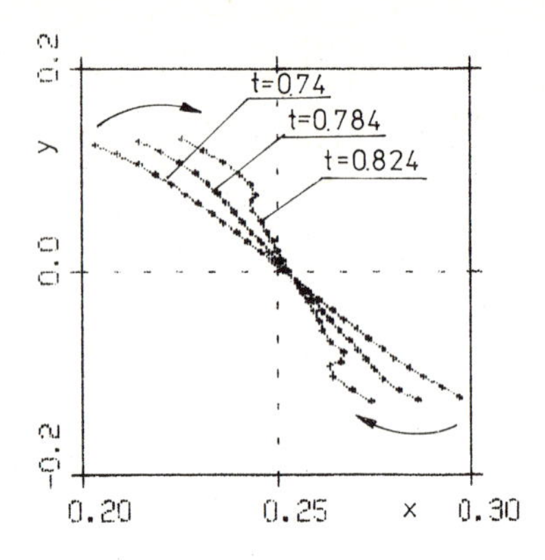

Fig.5 The time sequence of close-up interface fragments for included surface-tension effects σ=0.0005, A=0.0467, α=0, ε=0.1.

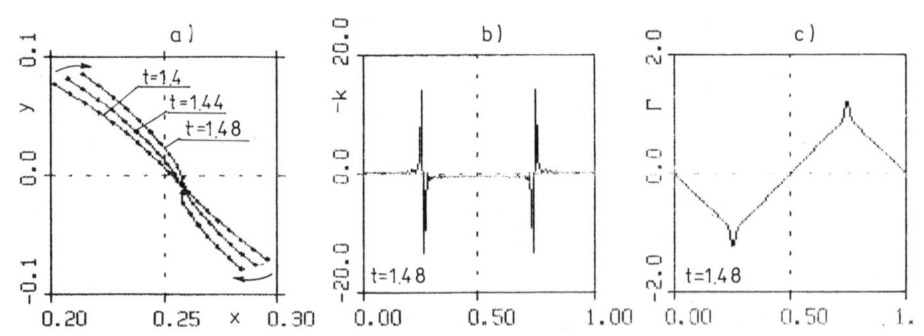

Fig.6 Numerical results for data as in fig 5 but with initial amplitude ε=0.01, a)time sequence of the close-up fragments of interface, b) curvature -k(a) for t=1.48, c)vortex sheet strength Γ(a) for t=1.48

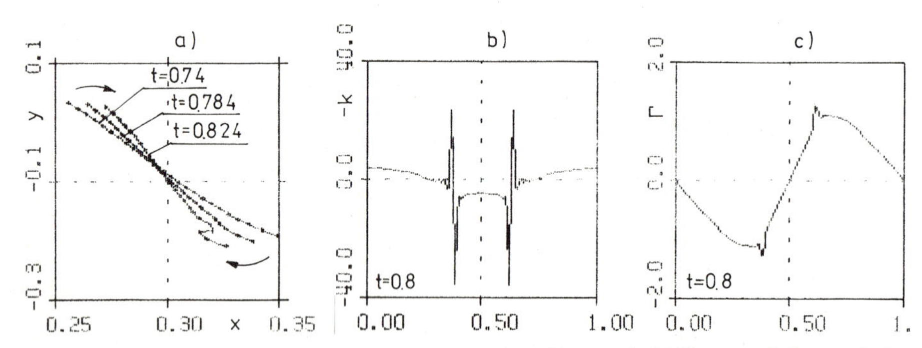

Fig.7 Numerical results for A=0.5 (ρ_1/ρ_2=3), σ=0.0005, α=-0.2, ε=0.1, a)time sequence of close-up fragments of interface b) curvature of vortex sheet -k(a) for t=0.8 c) vortex sheet strength Γ(a) for t=0.8.

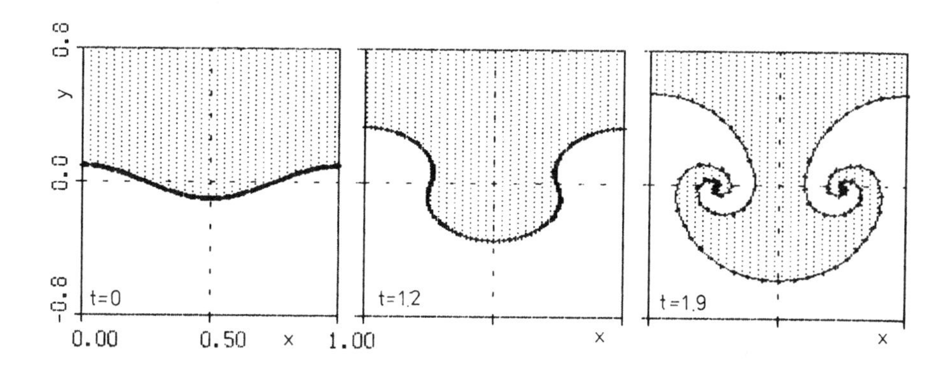

Fig.8 The time sequence of interface profile for desingularized equation $\delta^2=0.1, A=0.0476, \sigma=0.0005, \alpha=0, \varepsilon=0.1$.

Fig.9a,b show the time sequence profiles with and without surface tension for A=0 (Boussinesq approximation), $\delta^2=0.1, \alpha=0$, and for a more complex two-wave initial perturbation as was described by Aref [1]. Viz. $y(x,0)=\varepsilon_1\cos 2\pi x + \varepsilon_2\cos 6\pi x$,where $\varepsilon_1, \varepsilon_2$ were chosen in such a way that the initial profile had six inflection points. As can be seen in fig. 9 a ($\sigma=0$), six vortices grow from these six inflection points. But in fig. 9 b the surface-tension effect prevents the initiation of the vortex in the line y=0. We choose N=200, and the coefficient for interface surface tension σ is two times greater than in previous run viz. $\sigma=0.001$. This smoothing action of surface-tension effects is in good qualitative agreement with the results obtained by Daly in numerical study of the R-T instability by the marker-in-cell method for full Navier-Stokes equations [6].

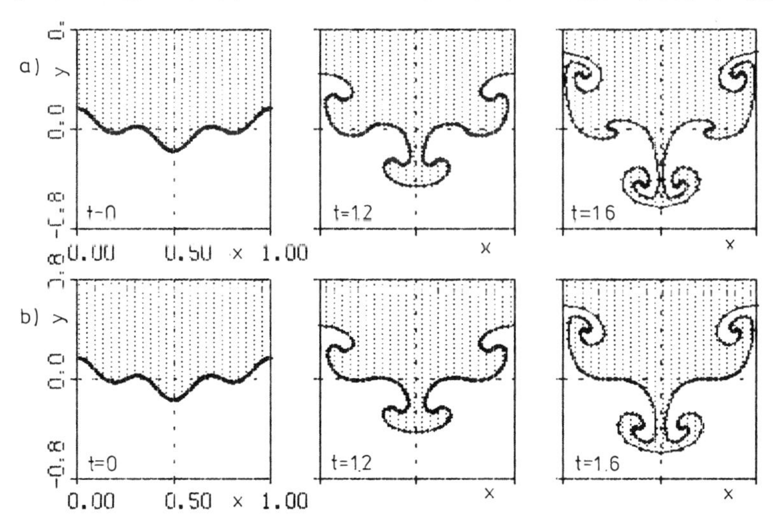

Fig. 9. Numerical results for $y(x,0)= 0.1\cos 2\pi x + 0.07\cos 6\pi x$, A=0, $\alpha=0$, $\delta^2=0.1$, N=200 a) $\sigma=0$, no surface-tension b) $\sigma=0.001$

CONCLUSIONS

The inclusion in the R-T problem of surface-tension effects causes the suppression of irregular motions of vortices. Singularity formations are the main reason why the B-M-O method fails for $0<A<1$. After desingularization of the equation of motion by the δ^2-parameter the surface tension also smoothes the numerical results and causes lengthening of the time interval for which the iteration process for γ_t converges.

Due to nonlinearity of the evolution equation for the vortex sheet strength after a certain time interval we obtained a nearly singular distribution of vortex sheet strength and finally the calculation broke. Alternative to direct summation method,very efficient computationally are vortex-in cell methods described by Tryggvason [15]. But as Tryggvason says "grid free method although considerably more inefficient"...than vortex in cell method.." offers a "cleaner" environment (no grid disturbances) to study such delicate questions as singularity formations and how to apply proper regularizations".

I wish to thank Professor Charles Zemach from Los Alamos Lab. for careful linguistic checking of the manuscript. My work was supported by Central Program of Fundamental Research (CPBP 01.02, X.14).

REFERENCES

[1] Aref H."Finger,Bubble,Tendril,Spike".Fluid Dynamics Transactions, vol.13(1987),PWN, Warsaw,pp.25-54.

[2] Baker G. R. ,Merion D.I.,Orszag S.A. " Vortex simulation of the Rayleigh-Taylor Instability".Phys.Fluids,vol.23, (1980),pp.1485-1490.

[3] Baker G. R.,Merion D.I.,Orszag S.A. " Generalized vortex methods for free surface flow problems". J.Fluid Mech.vol.123, (1982),pp.477-501.

[4] Baker G. R. " Generalized vortex methods for free-surface flows.Proc. Wave on Fluid Interface". Academic Press, (1983),pp.53-81

[5] Bellman R.,Pennington R.H. "Effects of surface Tension and Viscosity on Taylor Instability" .Quart.Appl.Math.,vol.12,(1954),pp.151-162.

[6] Daly B.J. "Numerical Study of the Effect of Surface Tension on Interface Instability",Phys.Fluids ,vol.12,(1969) pp.1340-1354.

[7] Gahov F.G "Boundary problems" (in Russian),1977 ,Moscow,Science

[8] Kerr R.M. "Simulation of Rayleigh-Taylor Flows Using Vortex Blobs", J.Comp.Phys.vol.76 (1988) pp.48-84.

[9] Krasny R. "A study of singularity formation in a vortex sheet by the point-vortex approximation" .J.Fluid Mech.vol.167,(1986) pp.65-93.

[10] Krasny R. "Desingularization of periodic vortex sheet roll-up".J.Comp.Phys.vol.65,(1986) pp.292-313.

[11] Meiron D.I.,Baker G.R.,Orszag S.A. "Analytic structure of vortex sheet dynamics.Part 1.Kelvin-Helmholtz instability".J.Fluid Mech. (1982), vol.114,pp.83-298.

[12] Moore D.W. "On the point vortex method",SIAM J.SCI. STAT.COMPUT., vol.2 (1981) pp.65-84.

[13] Moore D.W."Numerical and Analytical aspects of Helmholtz Instability" ,1985 ,Theoretical and Applied Mechanics ,IUTAM,pp.263-274.

[14] Sharp D.H. "An overview of Rayleigh-Taylor Instability".Physica 12D (1984) pp. 3-18 .

[15] Tryggvason G. "Numerical simulation of the Rayleigh-Taylor instability", J.Comp.Phys.,vol.75,(1988) pp.253-282.

3D EULER FLOWS AROUND MODERN AIRPLANES

J.L. Kuyvenhoven
Fokker Aircraft B.V.
Schipholdijk 231
1117 AB Schiphol-Oost
The Netherlands

J.W. Boerstoel
National Aerospace Laboratory NLR
Anthony Fokkerweg 2
1059 CM Amsterdam
The Netherlands

SUMMARY

The numerical simulation of flows around complex aircraft configurations like the Fokker 50 and Fokker 100, is described. In these simulations multiblock grids and a 3D Euler-flow solver are used.
The major features of the grid-generation procedure and flow solver are outlined.
Results of the validation of the numerical-flow-simulation system for aerodynamic design work are presented.

INTRODUCTION

Until recently, most flow codes used in the aeronautical industry for aerodynamic analysis of complex aircraft configurations were based on simplified physical models, whereas the use of flow codes based on more advanced physical models was restricted to the analysis of geometrically more simple configurations. For example, in the aerodynamic development phases of the Fokker 50 and Fokker 100 aircraft projects, linearized potential-flow solvers were used to analyse complete configurations in cruise, whereas full-potential solvers were used for transonic analysis of wings or wing-body combinations [1].

Rapid advancements in computer technology, together with improvements in flow-solver algorithms and developments in grid-generation techniques have brought realistic aircraft configurations within reach of numerical-flow-analysis codes based on more sophisticated physical models. Design cycle times as well as possible development risks can be reduced by these new developments.

One of the main problems in 3D flow-simulation is the generation of a grid around arbitrary aircraft configurations.
Recently, after recognizing the grid-generation problem, extensive research was put into this area. This research resulted in two major concepts for numerical grid generation, multiblock grids and unstructured grids. Here, multiblock grids are applied, for reasons explained in [2].

The Euler-flow-simulation procedure, encompassing block decomposition, grid generation, and flow solution, is described. Also, results of the flow simulation are presented.
Finally, some improvements of the simulation system are recommended.

SUBTASKS IN NUMERICAL FLOW SIMULATION

The subtasks of the Euler flow simulation are:

- o topological block decomposition,

- o geometrical block decomposition,
- o generation of the grid,
- o flow simulation, and
- o flow visualisation.

The first three subtasks together make up the tasks of the multiblock-grid-generation process. Grid generation is coupled to the flow simulation and flow visualisation by data interfaces which are used for the exchange of information.
The interactive and iterative subtasks (grid generation, visualisation) run on workstations or graphical terminals, while the compute-intensive numerical flow simulation requires supercomputer power.

TOPOLOGICAL AND GEOMETRICAL BLOCK DECOMPOSITION

While simple in principle, considerable practical difficulties are encountered when the multiblock grid for a complete aircraft configuration has to be constructed. Conceptual problems arise where the grid structures from various aircraft components must be integrated. Therefore, the most challenging part of multiblock grid generation is the subdivision of the flow field between the surface of the configuration and some outer far-field boundary into a set of blocks.
The union of the blocks must cover the finite flow field without either gaps or overlaps. Each block is topologically equivalent to a cube (fig.1). An example of a block-decomposed flow domain can be seen in figure 2. In order to allow easy implementation of numerical boundary conditions, blocks are patched to each other face-to-face, and grids are made block- and configuration-boundary conforming.

In principle, the topological subdivision of a flow domain is a highly iterative procedure, whereby interactive graphical methods are used. However, it has been possible to devise rules and strategies for block decomposition and to implement these rules in an automated block decomposition method [3].

The input geometry consists of several aircraft components, each defined separately by an arbitrary number of cross sections with an arbitrary number of points on each section (fig. 3). In general this structured cloud of points does not correspond to the topological subdivision of the aircraft geometry.
So there is a need to easily derive points on the surfaces, other than the definition points, and to make component intersections. This can be done with a commercially available package like CATIA, although a dedicated aerodynamic surface geometry package is preferable.
The output of the topological and geometrical subdivision consists of two files. The topology file specifies the connectivity relations between blocks, faces, edges, and vertices, and the geometry file specifies the geometrical location of vertices and the geometric position and shape of edges and faces.

GRID-GENERATION PROCEDURE

Attractive techniques for the construction of cell partitions in a block are elliptic and algebraic techniques. Algebraic grid generation is based on the application of transfinite interpolation techniques.

Elliptic grid generation is based on the numerical solution of Dirichlet boundary-value problems of the Poisson-like partial-differential equations. The elliptic differential equations used in the grid-generation procedure are somewhat simpler than those of Thompson [4].
The source control functions are replaced by user-specified positive weight functions that should be chosen approximately inversely proportional to desired grid-point distances [5].

By requiring, across block interfaces, only continuity of grid lines (but not slope continuity), the generation of the grid can be done in each block independently in the following 3-step way.

Step 1 First an initial grid in each block-edge interior is constructed. The geometrical shape of each block-edge is specified in numerical analytical form as a sequence of cubic Hermite polynomials, joined together with slope continuity at given points. After grid initialisation, in each edge, the grid points are shifted along each edge curve to desired locations by means of a user-controlled positive weight function in the elliptic equation.

Step 2 The grid interior of each block-face is constructed given the already constructed grid in each block-edge. To obtain an initial grid in the face, transfinite bilinear interpolation is applied. Usually this results in a grid of acceptable quality. If unsatisfactory, again elliptic smoothing can be applied, with two user-controlled positive weight functions in the elliptic equation for grid control.

Step 3 Finally, the initial grid in the interior of each block volume is constructed using transfinite trilinear interpolation in the given grid points on the six block-faces. Although it is possible to make use of the elliptic techniques to improve this initial grid, until now these initial grids were already of sufficient quality, so that the elliptic techniques are not required.

The design of a grid of sufficient quality requires special attention. Usually, the quality of the grid is measured in terms of stretch factors (ratio of two successive mesh lengths), skewness angles (blunt angle between two grid lines, minus 90°) and aspect ratios (ratio of two mesh cell lengths in two different directions).

From experiences [6] it appeared that, in the interior of each block, face, and edge, stretch factors should be preferably in the range 0.95-1.05. Over block faces they can be in the range 0.5-2.0, due to special boundary conditions at internal block-faces that compensate for large stretch factors, slope jumps, etc., over block faces. It is easy to violate these constraints during grid design
Skewness and aspect ratios can vary between wide limits, without affecting numerical flow-simulation accuracy. Skewness angles as large as 50°-60°, and aspect ratios of 100 can be handled by the Euler-flow solver.
Blocks that violate the cuboid form (degenerated blocks) should be avoided in flow regions of primary interest.

Deviations of the above mentioned recommendations result in a non-smooth grid, which in turn results in slower convergence in the flow solver and/or affection of the numerical accuracy of the results.

The flow solver computes the solution of the unsteady 3D Euler equations in an arbitrary multiblock flow domain.
The Euler equations are given by the perfect gas law and five partial differential equations for the conservation of mass, momentum and energy. The equations are suitable for subsonic/transonic inviscid flow, including rotational effects, non-isentropic effects, and non-isenthalpic effects.
The numerical algorithm to solve the Euler equations was developed from the finite-volume cell-centered scheme of Jameson et al [7].
The finite-volume discretization leads to a set of coupled ordinary differential equations, which are integrated to a steady state using a four-stage Runge-Kutta type time-integration scheme.

Numerical dissipative terms are added to suppress the possible tendency for odd-even-point decoupling, and to prevent the appearance of wiggles in regions containing severe pressure gradients (e.g. neighbourhood of shock-waves). To accelerate convergence to a steady-state solution, different devices can be used. Enthalpy damping and residual averaging are the most used devices in the Euler flow solver.

APPLICATIONS OF THE EULER-FLOW SIMULATION SYSTEM

Initially, the Euler-flow simulation system was validated with the 3D wing-nacelle-propeller configuration tested in a NASA Langley windtunnel [8].
After this first 3D test case, the Fokker 100 wing-body configuration was chosen for further validation. The main objective was to analyse the transonic flow around the wing-body configuration at different grid levels, and compare the results with other test data (i.e. other flow codes, flight test data).

The flow domain around half of the configuration was decomposed into 66 blocks, and the size of the grid approximately varied from 20K cells to 1200K cells. To obtain a fully converged solution on the finest grid, four hours CPU time were required on the NEC SX-2 supercomputer of the NLR.

Comparisons were made, for a given value of the lift coefficient, and at a transonic Mach number of 0.78, between the Euler results, full-potential results (non-conservative finite difference: XFLO22) and flight test data.
The agreement in pressure distribution between the various test data results (fig. 4) is quite acceptable.
Other results concerning this validation example can be found in [6,9].

Another important validation test case is the flow analysis of the Fokker 50 aircraft. The Fokker 50 is powered by two Pratt and Whitney PW124 turboprop engines, each driving a Dowty six-bladed, constant speed propeller.
To resolve flow field details near the propeller region, including the interaction of the propeller slipstream with the wing (fig. 5) simulation based on the Euler equations is the natural choice.
The complete flow field is, in general, too complicated for analysis without a certain degree of approximation. The fuselage, wing, and nacelle (without inlet) surfaces are accurately represented by a surface-fitted grid (see fig. 6a-b for details of the grid system).

However, the propeller is simulated by an actuator disk. Here, boundary conditions are imposed to simulate its power loading. Total pressure and swirl angle distributions downstream of the propeller, obtained from experiment or a Dowty Rotol propeller card deck, were used as input for the disk power loading.

It is expected that this simplified propeller model will be found adequate for the evaluation of propeller-power effects on wing loading, so that it is possible to do useful studies of the aerodynamic integration of the airframe propulsion system.

The flow field is covered by 556 blocks, which were filled with approximately 200K cells (coarse grid) or 1600K cells (fine grid).

After some preliminary tests (comparison with linear potential flow, effect of grid refinement on pressure distribution, fig. 7), the propeller power effect on the wing pressure distribution was investigated. The simulated power loading corresponds to a Mach .36 cruise condition and a thrust coefficient of about 0.1. From fig. 8 it can be clearly seen that the upward rotating propeller causes an increase in lift (due to an increase of local angle of attack), while the down-going propeller causes a lift decrease. At this cruise condition, the overall changes are moderate.
More severe distortions of the flow field are encountered when a heavy-loaded propeller is involved. From the spanwise lift distribution in fig. 9 (Mach = .18, thrust coefficient = .87) it can be seen that there are large differences in local lift under the low-speed, high-propeller loading, with maximum local lift increments of 0.5.

A global check (i.e. increase in total lift due to propeller effect) showed that the results, are promising.
More detailed checks will be done when the proposed flight-tests are completed and test-data has become available.

RECOMMENDED IMPROVEMENTS, CONCLUSION

The above mentioned examples showed the capabilities of the simulation system. In general, the results are quite satisfactory. Nevertheless improvements and extensions of the simulation system are necessary to speed up the cycle time of a flow analysis. Some of these are listed below.

- Commercially available geometry packages are generally tailored to CAD/CAM applications rather than the special requirements of grid generation in computational aerodynamic design work. So, there is a need for a dedicated aerodynamic surface geometry package.

- Continuity of grid lines across block-faces made it impossible to refine the grid in only a few blocks where this is desirable. Recently it became possible to apply coarsening/refinement independently in each of the three computational directions, and per block. The only restriction is that the number of grid lines on one side of the block is a power-of-2 multiple of the number of grid lines on the other side.

- The introduction of compound entities (blocks, faces, edges) will make it possible to create local changes in the topology.

Because of dropping the face-to-face coupling, a block-face may now terminate in an interior edge of a compound face of an adjacent block.

- o The design of a grid of sufficient quality requires too much manhour investment. Effort is being put into the research of faster and more direct grid-control means in the grid-generation procedure. Dropping the elliptic techniques and incorporation of grid control in the algebraic techniques might be part of the solution.

- o Investigations must be put into the area of automated block decomposition. The decomposition of a finite flow domain requires much user experience and is directly related to the quality of the flow solution. The various rules and strategies, obtained by experts, must be implemented in an automated block decomposition method, in order to allow non-experts to use the Euler-flow simulation system.

- o As an intermediate step towards the solution of the Navier-Stokes equations, boundary-layer equations should be added to the Euler equations, in order to account for viscous effects.

The Euler-flow simulation system can be seen as a major step forward in the analysis of flows around complex aircraft configurations.
At the moment much attention is paid to the recommendations mentioned above, so that in the near future the simulation system can be a valuable tool in the design and analysis of aircraft configurations.

REFERENCES

[1] Voogt, N., Mol, W.J.A., Stout. J., and Volkers, D.F.: "CFD applications in design and analysis of the Fokker 50 and Fokker 100", AGARD conference, Lisbon, Portugal (1988), pp 19-1 to 19-11.
[2] Boerstoel, J.W.: "Numerical grid generation in 3D Euler flow simulation", in: Numerical methods for Fluid Dynamics, Ed. Morton, K.W., Baines, M.J. Clarendon Press, Oxford (1988).
[3] Allwright, S.: "Multiblock topology specification and grid generation for complete aircraft configurations", AGARD specialists´ meeting, 24-25 May 1989, Norway.
[4] Thompson, J.F., Thames, F.C., Mastin, C.W.: "Automatic Numerical Generation of Body-Fitted Curvilinear Coordinate System for Field Containing any Number of Arbitrary Two Dimensional Bodies", J. Comp. Phys. 15, 1974.
[5] Boerstoel, J.W.: "Problem and solution formulations for the generation of 3D block-structured grids", NLR MP 86020U (1986).
[6] Jacobs, J.M.J.W. et al.: "Numerical interactive grid generation for 3D flow calculations", in: Numerical grid generation in computational fluid dynamics ´88, Ed. Sengupta, S., Peneridge Press (1988).
[7] Jameson, A., Schmidt, W., Turkel, E.: "Numerical solution of Euler equations by finite volume methods using Runge Kutta time stepping schemes", AIAA paper 81-1259.
[8] Amendola, A., Tognaccini, R., Boerstoel, J.W. and Kassies, A.: "Validation of a multiblock Euler flow solver with propeller-slipstream flows", AGARD conference proceedings, Lisbon, Portugal (1988), pp P1-1 to P1-15.
[9] Kuyvenhoven, J.L.: "3D Euler calculations about the Fokker 100", Fokker report L-28-459, 1988.

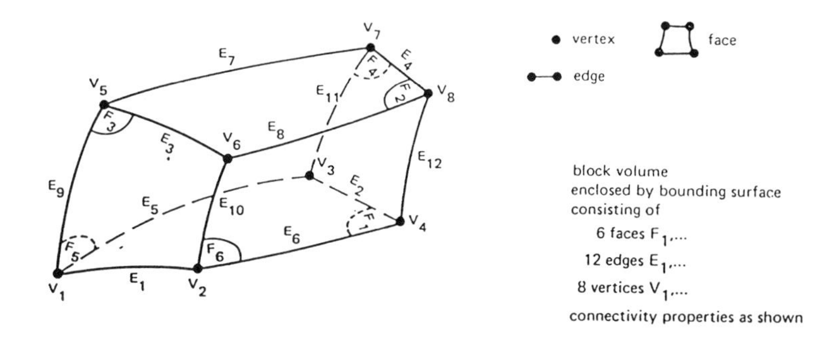

Fig. 1 Topology of a block (hexahedron)

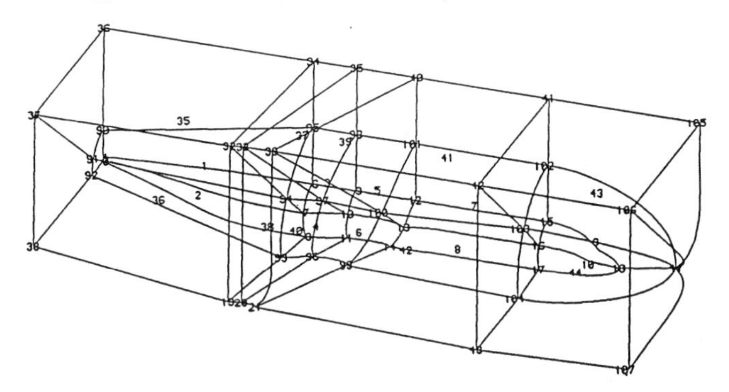

Fig. 2 Upper part of block decomposition around the Fokker 100

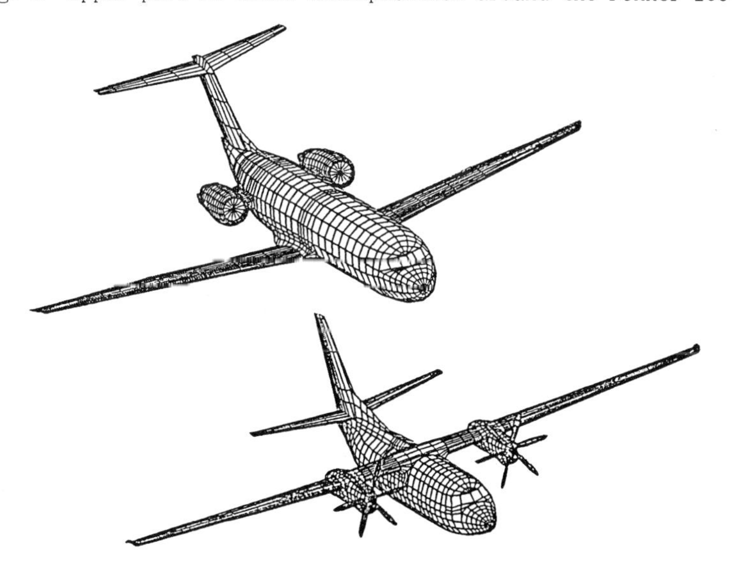

Fig. 3 Geometry definition points of the Fokker 50 and Fokker 100

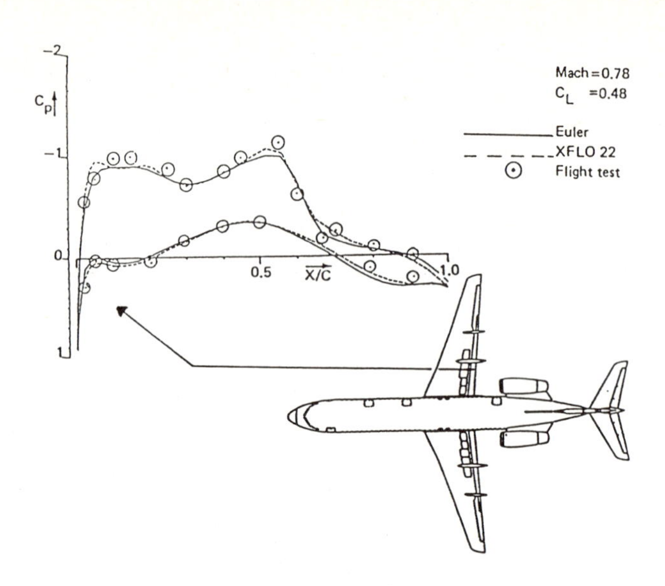

Fig. 4 Comparison of EULER results with full-potential results and flight-test data at the same lift-coefficient

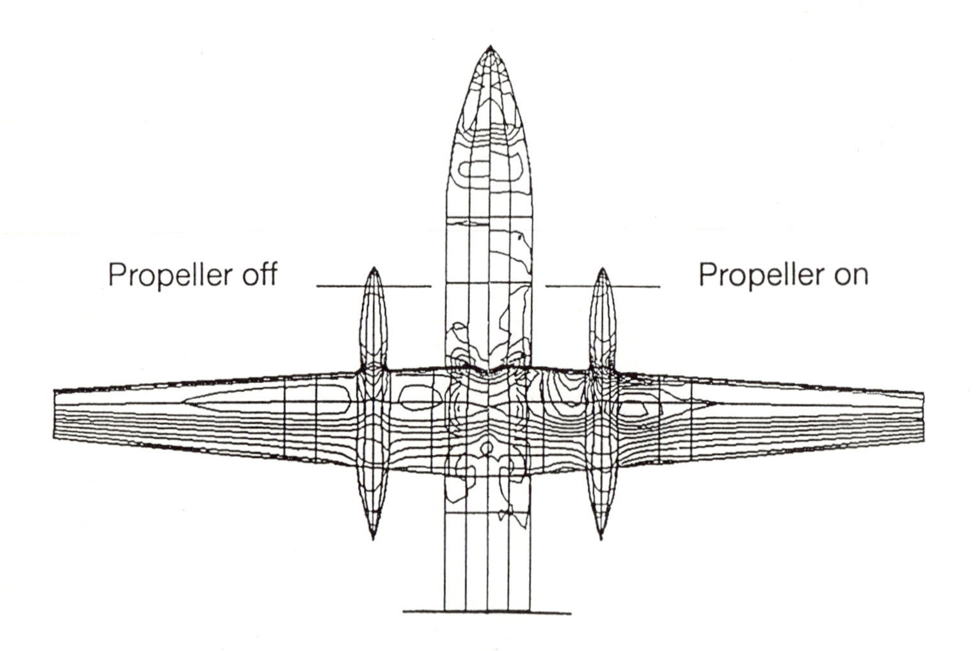

Fig. 5 Differences in computed isobars at Mach=0.18 due to propeller effect (thrust coefficient=0.87)

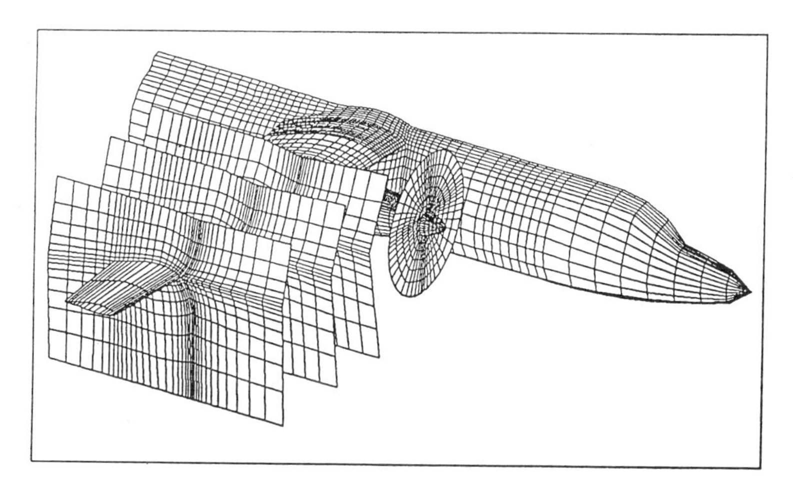

Fig. 6a Grid at three outer wing stations of the Fokker 50

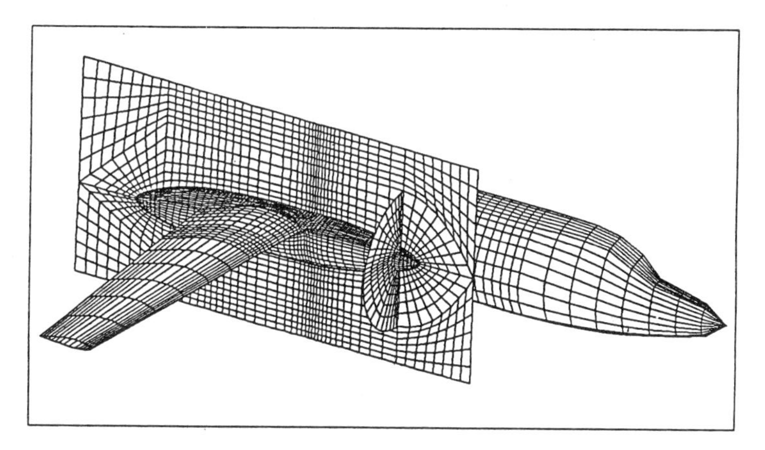

Fig. 6b Grid in symmetry-plane of the nacelle of the Fokker 50

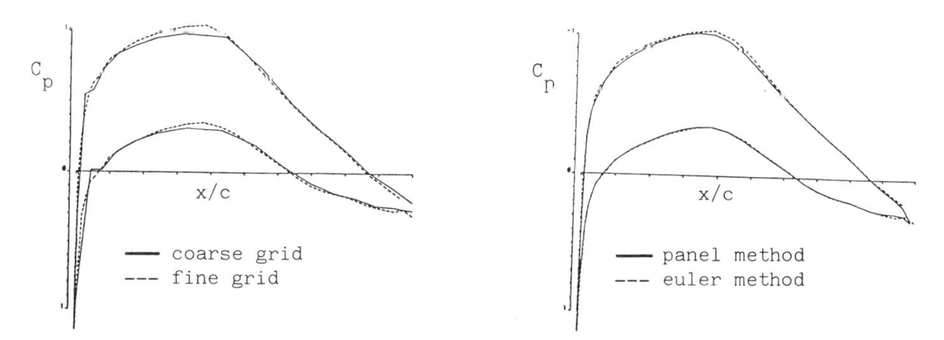

Fig. 7 Accuracy of EULER results at outer wing Fokker 50

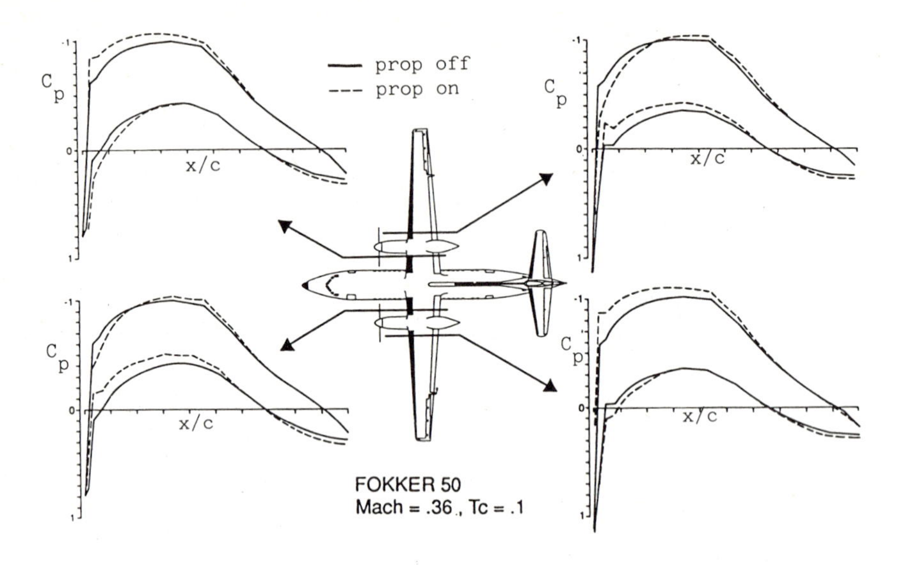

Fig. 8 Differences in pressure distribution due to propeller effect

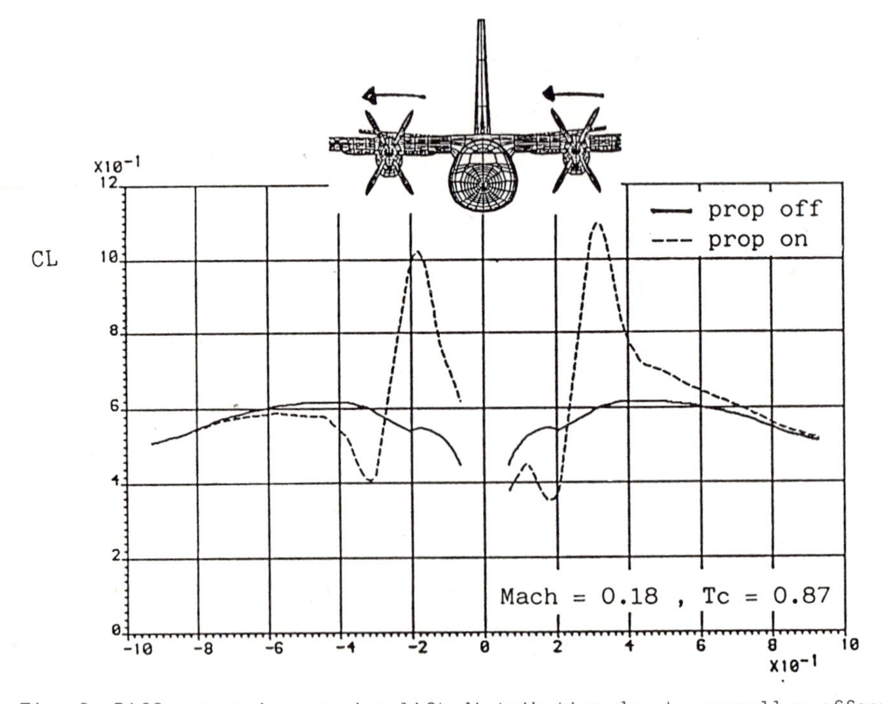

Fig. 9 Differences in spanwise lift distribution due to propeller effect

SIMPLE EXPLICIT UPWIND SCHEMES FOR SOLVING COMPRESSIBLE FLOWS

E. von Lavante
Universität GH Essen, D-4300 Essen, West Germany
A. El-Miligui and F. E. Cannizzaro
Old Dominion University, Norfolk, VA 23508, USA
H. A. Warda
University of Alexandria, Alexandria, Egypt

SUMMARY

Several upwind numerical methods for solving the compressible inviscid and viscous flow equations are discussed. Due to their explicit nature, the schemes are very simple and easy to apply to solutions on multi-block structured grids. Their favourable high frequence damping results in good rates of convergence when combined with multigrid procedures. The schemes are optimized using a simple stability and damping factor analysis. Results for three-dimensional test cases are shown and discussed. Attention is payed to the relative efficiencies of these schemes.

INTRODUCTION

After several decades of existance, computational fluid dynamics finally matured enough to allow the solving of complex, realistic problems with confidence. Several numerical schemes are in widespread use today, solving variety of compressible, viscous and inviscid problems.

The simple central difference schemes are particularly popular and effective in inviscid flow predictions. The necessity of the addition of ambiguous artificial damping makes them difficult to apply to viscous cases. The typical representatives of these numerical methods is for example the implicit ARC3D code by Pulliam and Steger [1] or the prolific Runge-Kutta (R-K) based explicit schemes as introduced by Jameson, Schmidt and Turkel [2].

The upwind schemes are more complex, but also typically more robust and reliable, especially for viscous computations. They are based on several variation of the Riemann problem solver. Typical examples are the CFL3D code developed by Thomas et. al. [3], based on the Roe's approximate Riemann problem solution and implemented as a flux difference scheme, or the Osher scheme as used by Chakravarthy and Osher [4]. Both of these schemes are implicit. The upwind schemes are usually reported to be better suitable for compressible, viscous predictions than the central difference codes.

The above schemes are very effective in converging to steady states on single-block grids of modest complexity. Most of the currently used schemes are implicit. The implicit formulation allows the use of large time steps, decreasing the number of iterations needed to obtain converged steady state. Due to the simplifications made during the development of these methods (linearization) and the frequent use of explicit boundary conditions, the maximum allowable CFL number has been reported as low as ~3.0 for complicated three-dimensional flows. The implicit part of these solvers also provides effective high frequence damping in connection with the application of multigrid procedures.

The application of the above numerical methods to realistic three-dimensional configurations of significant geometric complexity is usually not possible without the use of multi-block structured (zonal) grids. Here, the computational grid is subdivided into a number of blocks of different size and resolution, either overlapping or non-overlapping. A computational grid of this type adapts much easier to the geometric shape of the bodies as well as flow features. The transfer of information between the blocks is typically carried out explicitely by ensuring the conservation of fluxes accross the block interfaces. The consequence of this procedure is a significant reduction of the maximum allowable CFL number to values between 2. and 5. At these CFL numbers, an explicit upwind scheme of good accuracy when applied to viscous problems and suitable for multigrid procedures seems to offer a better choice.

There is a very large number of explicit schemes that have been used or are still in use for solving the compressible flow equations in their inviscid form (Euler) or viscous form (Navier-Stokes). The numerical method should be simple, efficient, robust, have effective damping of high frequence errors (neccesary for multigrid), have low dispersion (low phase error will reduce spurious oscillations and result in faster convergence rates), low levels of numerical dissipation for accurate predictions of viscous effects and maintain high resolution on stretched grids.

BASIC ALGORITHM

The scheme is based on Roe's flux difference splitting [5,3]. This type of upwind scheme is simplified by linearizing the Riemann problem between two cell interfaces about a state obtained by Roe's averaging procedure, Ref. [5]. The present discretization employed the finite volume approach, with the state variables at the cell interfaces determined by the MUSCL interpolation, using mostly the so-called κ scheme.

In the inviscid case, the governing equations are the three-dimensional Euler equations, expressed in nondimensional conservation law form for curvilinear coordinates ξ, η, ζ , using the usual fluid dynamics notation, as:

$$\frac{\partial Q}{\partial \hat{t}} + \frac{\partial F}{\partial \xi} + \frac{\partial G}{\partial \eta} + \frac{\partial H}{\partial \zeta} = 0 \qquad (1)$$

$$Q = \begin{bmatrix} \rho \\ \rho u \\ \rho v \\ \rho w \\ \rho E \end{bmatrix} \qquad F = \begin{bmatrix} \rho \hat{U}_\xi \\ \rho \hat{U}_\xi u + p\hat{\xi}_x \\ \rho \hat{U}_\xi v + p\hat{\xi}_y \\ \rho \hat{U}_\xi w + p\hat{\xi}_z \\ \rho \hat{U}_\xi H \end{bmatrix}$$

$$H = E + \frac{p}{\rho}\ , \qquad E = e + \frac{1}{2}(\ u^2 + v^2 + w^2\ )\ , \qquad \hat{t} = \frac{t}{V}$$

$$U_\xi = \hat{\xi}_x u + \hat{\xi}_y v + \hat{\xi}_z w$$

$$\hat{e}^\xi = \frac{\vec{e}^\xi}{J} = V\ (\vec{\nabla}\xi) = A_\xi \frac{\vec{e}^\xi}{|\vec{e}^\xi|} = (\ \hat{\xi}_x\ ,\ \hat{\xi}_y\ ,\ \hat{\xi}_z\ )$$

where Q is the vector of dependent conservative flow variables and F, G and H are the usual flux vectors in the direction ξ, η and ζ, respect. The semidiscrete form of the governing equations is obtained from (1) by carrying out the spatial discretization, assuming $\Delta\xi = \Delta\eta = \Delta\zeta = 1.0$

$$\frac{\partial F}{\partial \xi} \approx F_{i+\frac{1}{2}} - F_{i-\frac{1}{2}}\ . \tag{2}$$

In equation (2), $F_{i+1/2}$ is the numerical flux vector at the cell interface between cells i,j,k and i+1,j,k; i,j,k are the spatial indices in the directions ξ, η and ζ, respectively. The expressions in the other two directions are similar and will not be repeated here. Then,

$$\frac{\partial Q}{\partial \hat{t}} = R^n_{i,j,k} = (F_{i+\frac{1}{2}} - F_{i-\frac{1}{2}}) + (G_{j+\frac{1}{2}} - G_{j-\frac{1}{2}}) + (H_{k+\frac{1}{2}} - H_{k-\frac{1}{2}})\ . \tag{3}$$

Following Ref. [5], the numerical flux $F_{i+\frac{1}{2}}$, for example, is evaluated as

$$F_{i+\frac{1}{2}} = \frac{1}{2}\left[\ F_R + F_L - \tilde{S}_\xi|\tilde{\Lambda}_\xi|\tilde{S}_\xi^{-1}\ (Q_R - Q_L)\right]. \tag{4}$$

Here, $F_R = F(Q_R)$, $F_L = F(Q_L)$, where Q_R and Q_L are the conservative flow variables describing the states right and left of the cell interface and are obtained by an appropriate inerpolation or extrapolation. The last expression in eq. (4) is a dampimg term due to the upwind character of the present approach; $\tilde{S}_\xi$ and $\tilde{\Lambda}_\xi$ are the eigenvector and eigenvalue matrices of the flux Jacobian matrix $\frac{\partial F}{\partial Q}$, evaluated using the Roe averaged flow variables $\tilde{\rho}$, $\tilde{u}$, $\tilde{v}$, $\tilde{w}$ and $\tilde{h}$, given in [5].

The selection of a particular type of time stepping will determine the characteristics of the numerical method. In the present work, explicit multi-stage schemes were considered; their S-stage version can be written in a somewhat general form as

$$s \leq S:\quad Q^s_{i,j,k} = Q^n_{i,j,k} - \Delta\hat{t}\ \mathfrak{F}(R^{s-1}_{i,j,k}\ ,\ R^{s-2}_{i,j,k}\ ,\ \ldots\ R^1_{i,j,k}\ ,\ R^n_{i,j,k}\ ) \tag{5}$$

$$Q^{n+1}_{i,j,k} = Q^S_{i,j,k}\ .$$

The above formulation of the time stepping procedure applies not only to the modified Runge Kutta method in [2], but also many other existing schemes. From these infinitely many possible schemes, only a few are suitable for the above flux difference splitting spatial discretization. They were selected by considering the much simplified scalar, linear case of the wave equations.

In order to compare some of the time-stepping procedures (5), the standard Fourier stability analysis of several schemes was carried out, yielding the amplification factor g and the phase error ϕ/ϕ_e for the simple wave equation

$$\frac{\partial u}{\partial t} + c\frac{\partial u}{\partial x} = 0 \quad , \quad c > 0 . \tag{6}$$

There are two major blocks of scheme variations that fit eq. (5).

Modified Beam-Warming (BW) Scheme

This scheme was originally introduced by Beam and Warming [6]. It is a two-step method consisting of a first order predictor and second order corrector. It has been slightly modified to make it suitable for the present finite volume formulation:

Predictor:

$$Q_L^1 = Q_i^n \ ; \qquad Q_R^1 = Q_{i+1}^n \tag{7a}$$
$$Q_i^1 = Q_i^n - \Delta\hat{t}\, R^1$$

Corrector:

$$Q_L^2 = Q_i^n + \tfrac{1}{2}(Q_i^1 - Q_{i-1}^n) \ ; \qquad Q_R^2 = Q_{i+1}^n - \tfrac{1}{2}(Q_{i+2}^n - Q_{i+1}^1) . \tag{7b}$$
$$Q_i^2 = Q_i^{n+1} = Q_i^n - \Delta\hat{t}\, R^2$$

In both steps, only the extrapolations (7a) and (7b) are different; the algebra that completes one iteration, consisting of eq. (3) with (4), is identical. The stability analysis of this scheme is given in every good textbook on CFD and will not be repeated here. The resulting plot of the magnitute of the amplification factor |g| can be seen in Fig. 1a, plotted as a function of the spatial wave number β between 0 and π and the CFL-number σ between 0 and 2.

The scheme is still second order accurate in space and time, satisfies the "shift condition" and has a stability limit of σ=2. It has a relatively modest disspertion. It was named the "1-2 BW scheme". It is a very simple and efficient method, consisting of only two steps. It performed well on single grids, in particular when applied to unsteady flow predictions.

There were, however, some problems with this scheme. Since the two steps are different, the steady state result will depend on the time step. This phenomen was observed in only a few cases, represented by convergence to residual that was larger than "machine zero". At this time, none of the flux limitors tested in this scheme converged more than two orders of magnitude. The resulting flow fields agreed well with other, fully converged numerical results. This type of behaviour has been observed by other investigators, but is still disturbing.

A more serious problem is the increase of the damping factor to 1.0 at high frequency and σ=1. In scalar case, the CFL number can be kept at its optimum value (1.7 in the case of the 1-2 BW scheme), but in the case of the Euler equations there are three distinc eigenvalues in each direction. Typically, local time stepping will be implemented, where each cell will be advanced at its optimum time step, corresponding to the maximum eigenvalue at that cell. This means that only the maximum eigenvalue will correspond to the optimum CFL number for high frequence damping and one or more might correspond to the CFL number ranges with very little damping. This was manifested by the lack of convergence of the 1-2 BW scheme when utilized in a Multi-Grid procedure. For multigrid applications, this scheme was modified by making the first step (predictor) also second order accurate:

$$Q_L^1 = Q_i^n + \tfrac{1}{2}(Q_i^n - Q_{i-1}^n) \; ; \qquad Q_R^1 = Q_{i+1}^n - \tfrac{1}{2}(Q_{i+2}^n - Q_{i+1}^n) . \tag{8}$$

The plot of the damping characteristics of this scheme, called 2-2 BW, is shown in Fig. 1b. The maximum stable CFL number is now only 1.0, but the high frequence damping is much better. This scheme worked well with Multi-Grid and will be discussed below.

Modified Runge-Kutta Methods

The modified R-K methods, with the standard set of coefficients, have been rather successfull in combination with the central difference spatial discretization. They have been, however, performing very poorly with upwind differencing. The standard coefficients have to be modified to achieve better performance, resulting in schemes that are in general of reduced accuracy in time. In order to find the optimum sets of the R-K coefficients, a Fourier stability analysis was carried out, similar to the approach in [2]. The extrapolation of the state variables to the cell interfaces was based on the so called κ-scheme:

$$Q_L^s = Q_i^{s-1} + \tfrac{1}{4}\ell_i^{s-1}\{(1 - \kappa\ell_i)\Delta_i^- + (1 + \kappa\ell_i)\Delta_i^+\}^{s-1} \; ; \tag{9a}$$

$$\Delta_i^- = Q_i - Q_{i-1} \; ; \qquad \Delta_i^+ = Q_{i+1} - Q_i \; ; \tag{9b}$$

$$Q_R^s = Q_{i+1}^{s-1} - \tfrac{1}{4}\ell_{i+1}^{s-1}\{(1 + \kappa\ell_{i+1})\Delta_{i+1}^- + (1 - \kappa\ell_{i+1})\Delta_{i+1}^+\}^{s-1} \; ; \tag{9c}$$

where ℓ is one of the possible flux limitors. The value of the parameter κ determines the spatial accuracy of the scheme; κ=-1 is fully upwind, second order accurate; κ=0 is the upwind biased second order Fromm scheme; $\kappa=\frac{1}{3}$ is upwind biased third order and κ=1 is second order central difference scheme. The first order scheme is obtained by setting ℓ=0. In the present stability analysis, the limitor ℓ was set to 1 (no limitor) for simplicity.

The R-K time stepping for equation (6) can be written as

$$u^s = u^n - \alpha_s \, \Delta t \, R^{s-1} \; ; \qquad R^s = \frac{\partial(cu^s)}{\partial x} . \tag{10}$$

Here, α_S=1 for consistancy, with $u^{n+1} = u^S$. Making the assumption that u is harmonic, leads to

$$-R = -c\frac{\partial u}{\partial x} = \frac{\partial u}{\partial t} = u_0 z e^{zt} = zu . \tag{11}$$

Defining P=Δt z, substituting eq. (11) in eq. (10) and some simple algebra yields

$$g = (u^{n+1}/u^n) = 1 + P + \alpha_{S-1}P^2 + ... + \alpha_1\alpha_2...\alpha_{S-1}P^S . \tag{12}$$

Clearly, for stability, $\Re(P)<0$ and $|g|\leq 1$. The stability and damping properties of the scheme are associated with the complex polynomial (12). The damping $|g|$ is a function of the complex P=x+Iy, $I=\sqrt{-1}$, and can be best shown as a plot of contours of constant $|g|$ between 0 and 1. However,

$$u_i = u|_{x=i\Delta x} = \hat{u}_0 e^{I\beta i} \tag{13}$$

where β is the spatial wave number, ranging from 0 to π. Values of β between $\frac{\pi}{2}$ and π are considered high frequencies. P now represents the Fourier transform of the spatial differencing operator and can be superimposed on the damping factor contour plots. Defining

$$\sigma = \frac{c\Delta t}{\Delta x} \tag{14}$$

as the CFL number, and observing that $F_{i+\frac{1}{2}}=cu_{i+\frac{1}{2}}$ gives

$$R = -zu = \frac{1}{\Delta x} c(u_{i+\frac{1}{2}} - u_{i-\frac{1}{2}}) . \tag{15}$$

Eq. (15), combined with eq.(13) and (9), for $\ell=1$, yields finally

$$P = -\sigma(1-e^{-I\beta})\{1 + \frac{1-\kappa}{4}(1-e^{-I\beta}) - \frac{1+\kappa}{4}(1-e^{I\beta})\} . \tag{16}$$

P is a function of σ and β; the plot of $P(\sigma,\beta)$ and $|g|$ can be used to optimize the coefficients α_s. The resulting stability plots are shown only in the upper half of the negative real part of P (fourth quadrant) since they are symmetric with respect to the y=0 coordinate line. The optimization of the coefficients α_s was carried out by displaying these stability plots on a PC-type microcomputer. The changes in the shape of the $|g|$ contours were observed in real time as the coefficients were changed. The "islands" of the low value of $|g|$ correspond to the roots of the polynomial (12). The main purpose of this optimization was to find a combination of the coefficients α_s such that, for as large σ as possible, there would be good high frequency damping (low value of $|g|$) for a large range of σ (maximum size of the "islands" as close to the real axis as possible). The optimization was performed for four different spatial discretizations (1-st order; 2-nd order fully upwind, $\kappa=-1$; 2-nd order Fromm scheme, $\kappa=0$; 3-rd order upwind bias, $\kappa=\frac{1}{3}$) for the two-, three- and four-stage R-K schemes. Even in the case of the four-stage scheme, the actual optimization of the three coefficients ($\alpha_S=1$) was relatively easy and quick, once the influence of the coefficients on the stability plots was understood. Only a few selected stability plots of the most interesting schemes will be presented here. The optimized coefficients are summarized in Table 1.

The simplest schemes to optimize were all the two-stage versions, since only one coefficient is freely selectable. The case of $\kappa=0$ is shown in Fig. 2a. For the optimum coefficient $\alpha_1= 0.42$, the theoretically determined maximum stable cfl number was 1. The Fromm scheme performed very well in most of the test cases due to its very low numerical dispersion.

The four-stage R-K schemes are more interesting. Here, the optimum combination of three coefficients has to be found. The standard R-K coefficients ($\alpha_1=\frac{1}{4}$; $\alpha_2=\frac{1}{3}$; $\alpha_3=\frac{1}{2}$; $\alpha_4=1$), shown in Fig. 2b for the third-order scheme, performend, as expected, relatively poorly. It should be noted that, when multi-grid procedures are implemented, the maximum CFL-number is of little importance. The high frequency damping (or lack of it) will effect the rate of convergence to steady state more significantly than the CFL-number. In this case of standard coefficients, the maximum stable CFL number is relatively low at $\sigma=1.7$, but, more significantly, the damping of high frequencies of error propagation (at $\beta>\frac{\pi}{2}$) is very weak. The result of the present optimization, shown in Fig.2f, is much more promissing.

It should be mentioned here that, in a parallel effort, van Leer et. al. [7] also tried to optimize the R-K coefficients for applications with the upwind methods. Their approach was somewhat different: assuming that a genuine and practical multi-dimensional characteristic formulation of the Euler equations could be found, they optimized the R-K coefficients for only one value of σ, argueing that each wave would be propagated at its optimum CFL-number. Unfortunately, there no such formulation for the three-dimensional case. The advantage of this approach was that the selection process could be automated, eliminating the "guessing game" involved in the present approach. Their results are shown in Fig. 2d and 2h. Generally, their maximum σ was lower, and the damping effective over a narrower range of CFL values.

In the case of the third order scheme, the present optimum α-s are shown in Fig.2f . In the real test cases, however, the van Albada limitors were implemented, shifting the P-curve to the left. The real optimum CFL number was therefore much lower (1.8). A new set of coefficients for the limited case was found. It is shown in Fig.2g (α_1=.11; α_2=.245; α_3=.48); the maximum σ was 2.5.

RESULTS

Most of the above schemes were tested on number of two- and three-dimensional cases. In general, the optimum CFL numbers agreed well with the above theory. In supersonic cases, the real optimum CFL number was usually somewhat higher; in transonic cases, it was mostly the same.

Due to the space limitations of this paper, only one three-dimensional case will be discussed. Here, the supersonic flow in a channel with two 10° compression ramps, at the bottom and left vertical walls, with an inflow Mach number of 3.17, formed a conical shock flow with two Mach tripple points. The resulting flow field, obtained by the Fromm scheme, is shown in Fig. 3. The computations were carried out using a 32x32x32 body fitted grid. The Full-Multi-Grid procedure (FMG) was used with three grid levels with two iterations on each grid level. Considering the results of the above stability analysis, the most promissing schemes of practical importance were the four-stage Fromm scheme (κ=0) and the third order biased scheme ($\kappa=\frac{1}{3}$). The Fromm scheme was preferred due to its low numerical dispersion, demonstrated by results with the least oscillations (no limitors). The convergence of the κ=0 scheme with α_1=.11, α_2=.255, α_3=.46 for different CFL-numbers is shown in Fig. 4. The best rate of convergence was achieved at CFL numbers between 2.0 and 2.3, which was in a surprising agreement with the theory. This scheme, at a σ=2.0, is compared with the simple 2-2 BW scheme at its maximum σ=1.0 in Fig.5. The 2-2 BW scheme seems to be much slower, but it is only a two-stage scheme and thus needs only about half the CPU time per one iteration. Consequently, it is approximately 50% less efficient.

Finally, the present κ=0 scheme is compared with the optimized version of the $\kappa=\frac{1}{3}$ scheme in Ref. [7], used at the optimum CFL number of 1.732. Its performance is much worse, since only one of the three eigenvalues corresponds to the optimum damping case. The application of the van Albada limitor lead to a limit cycle with no apparent convergence. This problem has been reported by other investigators but was still disapointing, since all the two-dimensional cases converged with the limitor in effect. The problem is being further investigated.

CONCLUSIONS

The present study investigated several types of explicit upwind schemes for solving the compressible flow problems. The simple 1-2 BW scheme seems to be very effective for predicting unsteady flows, but it does not work with Multi-Grid due to its insufficient damping of high frequencies at wide range of CFL numbers. The most promissing schemes for Multi-Grid computations are multi-stage R-K methods with optimized sets of coefficients. Here, versions with coefficients optimized for wider range of CFL numbers seem to be more robust than specialized versions. The $\kappa=0$ and $\kappa=\frac{1}{3}$ schemes were successfully tested on several two-dimensional caes 'and subsequently included in a multi-block, multi-grid code used for predictions of complex three-dimensional flows.

ACKNOWLEDGMENTS

This work was partially funded by NASA Langley Research Center Grant NAG-1-633. The authors would like to thank Mr. M. Salas, Mr. R. Gaffney and Dr. B. van Leer for their comments.

REFERENCES

[1] Pulliam, T.H., and Steger, J.L., "Recent Improvements in Efficiency, Accuracy and Convergence for Implicit, Approximate Factorization Algorithms", AIAA paper 85-0360.

[2] Jameson, A., Schmidt, W. and Turkel, E., "Numerical Solutions of the Euler Equations by Finite Volume Methods Using Runge-Kutta Time-Stepping Schemes", AIAA paper 81-1259.

[3] Thomas, J.L., van Leer, B. and Walters, R.W., "Implicit Flux-Split Schemes for the Euler Equations", AIAA paper 85-1680.

[4] Chakravarthy, S.R. and Osher, S., "Numerical Experiments with the Osher Upwind Scheme for the Euler Equations", AIAA Journal, Vol. 21, Sept. 1983, pp. 1241-1248.

[5] Roe, P.L., "Approximate Riemann Solvers, Parameter Vectors, and Difference Schemes", Journal of Comp. Physics, Vol. 43, 1981, pp. 357-372.

[6] Beam, R.M. and Warming, R.F., "An Implicit Finite-Difference Algorithm for Hyperbolic Systems in Conservation Law Form", Journal of Comp. Physics, Vol. 22, 1976, pp. 87-110.

[7] van Leer, B., Tai, Ch. and Powell, K.G., "Design of Optimally Smoothing Multi-Stage Schemes for the Euler Equations", AIAA paper 89-1933.

Table 1.: Optimum R-K coefficients.

	1-σ	κ=-1	κ=0	κ=1/3	
α_1	0.22	0.22	0.42	0.46	2-Stage
α_1	0.105	0.15	0.21	0.22	
α_2	0.325	0.40	0.44	0.48	3-Stage
α_1	0.056	0.091	0.11	0.135	
α_2	0.152	0.24	0.255	0.26	
α_3	0.34	0.42	0.46	0.44	4-Stage

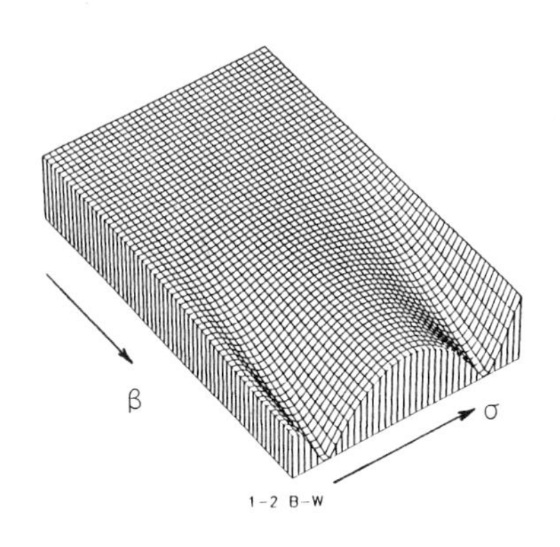

1a) 1-2 BW Scheme

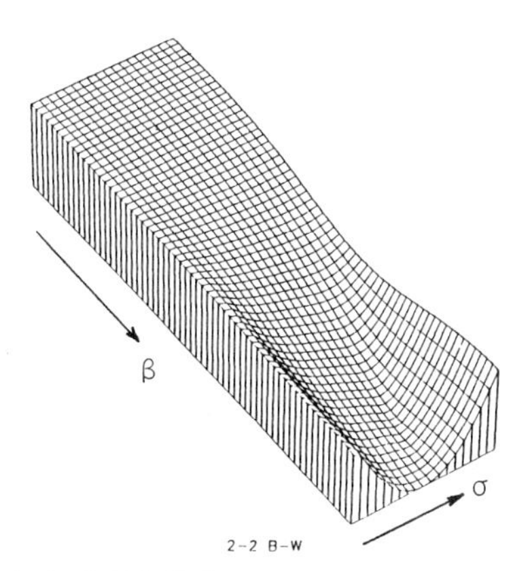

1b) 2-2 BW Scheme

Fig. 1: Amplification factor $|g|$ as a function of CFL-number and wave number.

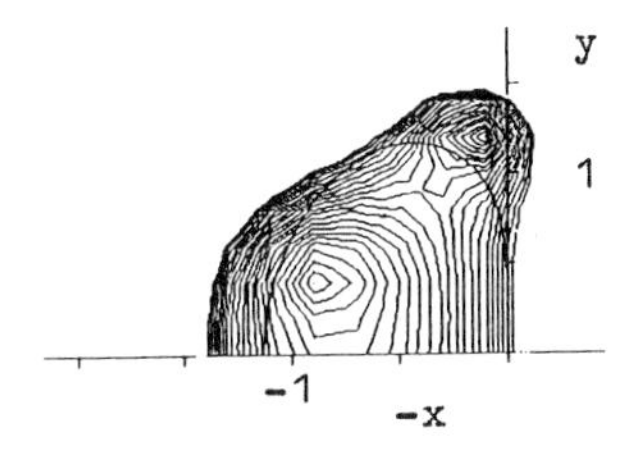

2a) 2-Stage, 2-nd order Fromm, CFL = 1.0

2b) 4-Stage, standard R-K, 3-rd order, CFL = 1.7

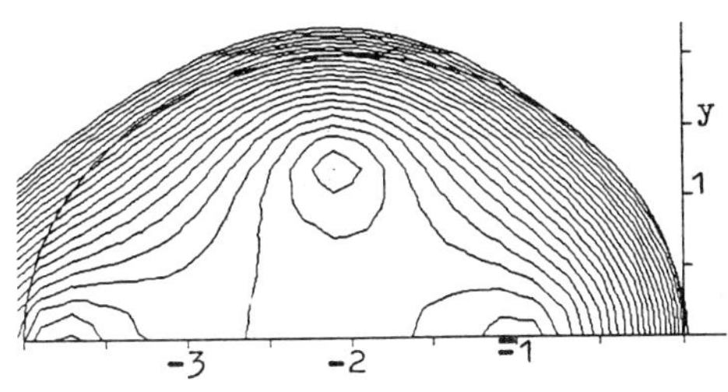

2c) 4-Stage, present optimum, 1-st order, CFL = 4.0

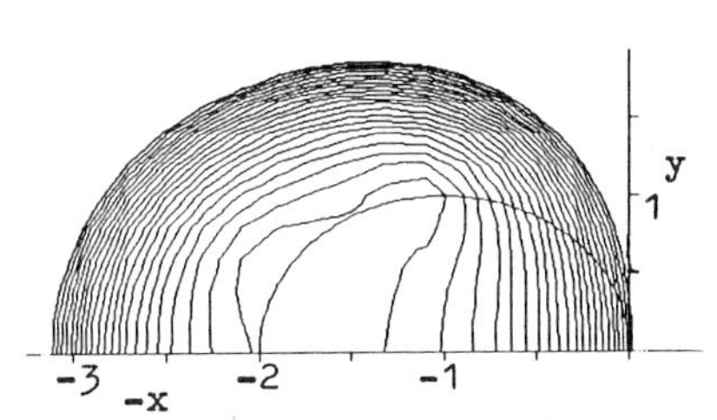

2d) 4-Stage, Ref. 7, optimum, 1-st order, CFL = 2.0

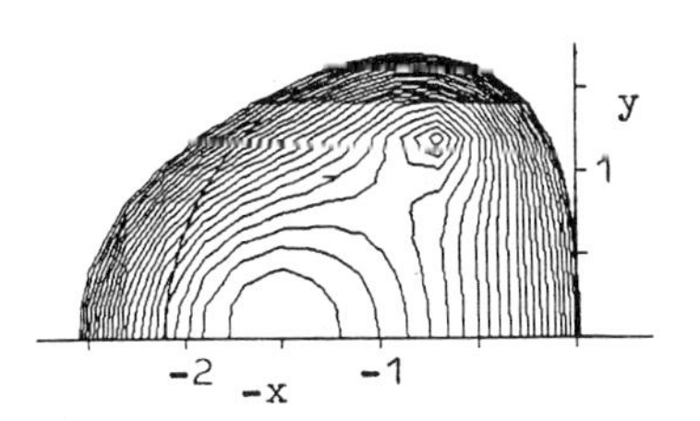

2e) 4-Stage, present optimum, 2-nd order Fromm, CFL = 2.2

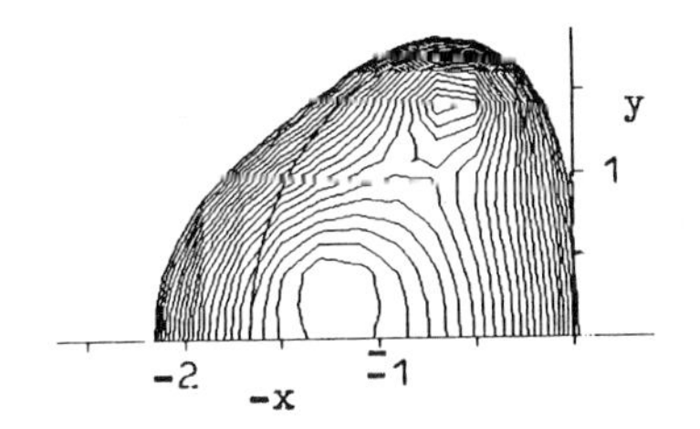

2f) 4-Stage, present optimum, 3-rd order, CFL = 2.5

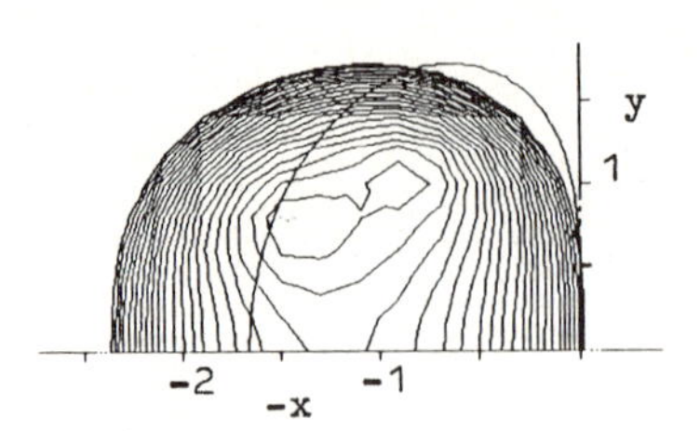

2g) 4-Stage, best w. limitor
3-rd order, CFL = 2.5

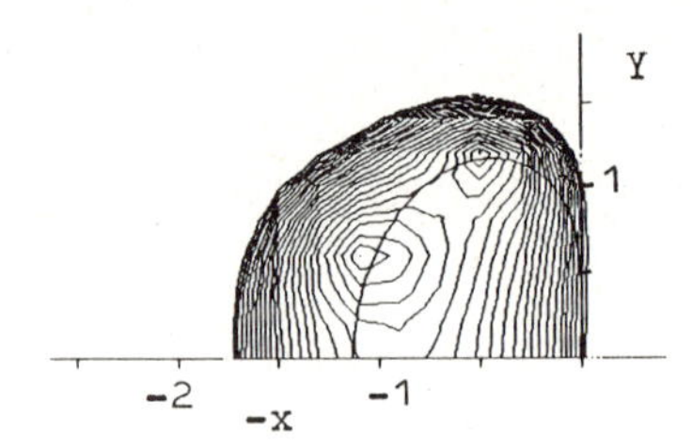

2h) 4-Stage, Ref. 7, optimum
3-rd order, CFL = 1.732

Fig. 2: Stability plots; contours of const. g with a trace of P corresponding to a maximum CFL-number.

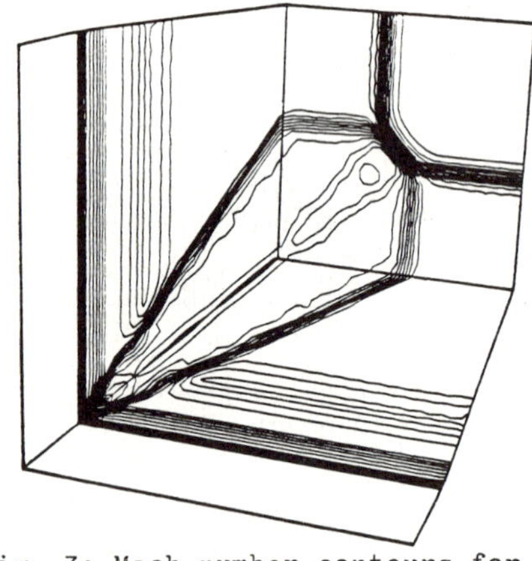

Fig. 3: Mach number contours for corner flow, $M_{in} = 3.17$

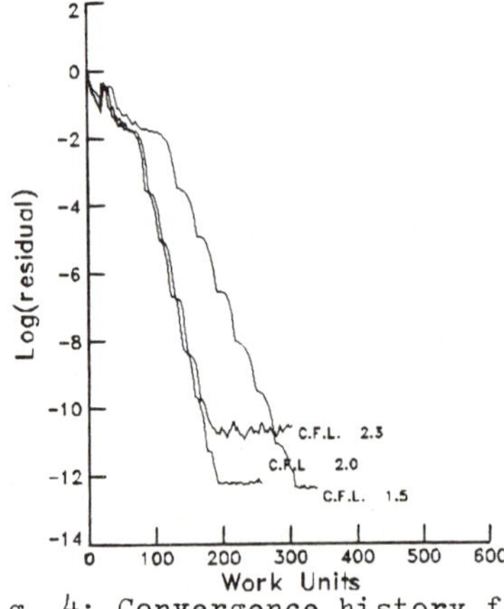

Fig. 4: Convergence history for corner flow, 4-Stage Fromm scheme.

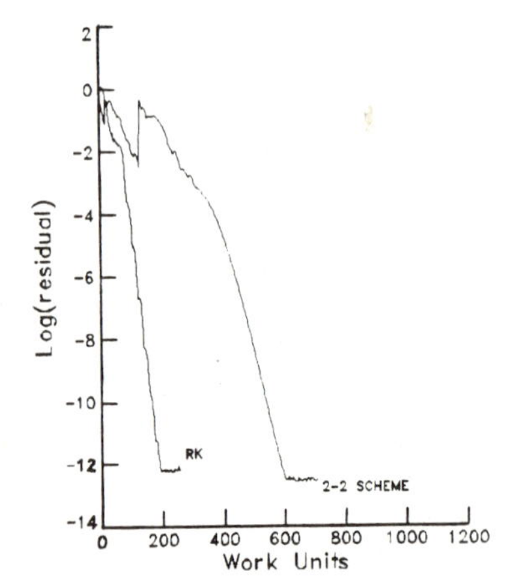

Fig. 5: Comparison of convergence histories for corner flow

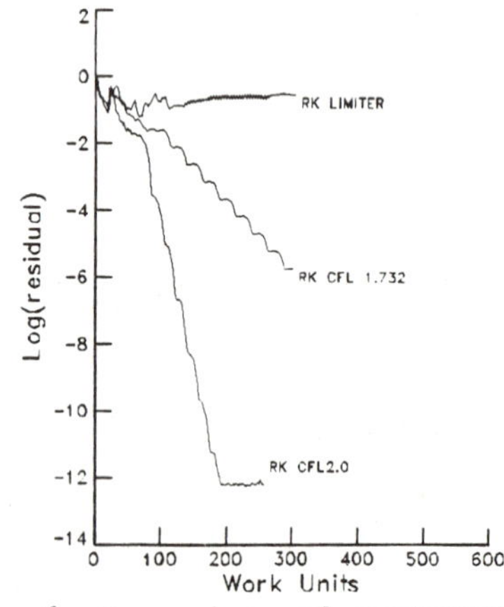

Fig. 6: Comparison of convergence histories for corner flow

Development of a Three-Dimensional Upwind Parabolized Navier-Stokes Code

Scott L. Lawrence, Denny S. Chaussee
NASA Ames Research Center, Moffett Field, California

John C. Tannehill
Iowa State University, Ames, Iowa

Summary

An algorithm for the integration of the parabolized Navier-Stokes (PNS) equations that is based on Roe's flux-difference splitting approach in both crossflow directions has been developed. The algorithm was developed using finite-volumes to ensure accurate conservation of numerical fluxes and modifications have been applied to make the scheme implicit and second-order accurate in the crossflow directions. The resulting PNS code has been applied to hypersonic flow past two simple test geometries and results are presented here. The computed flow-fields for a 10 deg half-angle cone at a wide range of incidence angles are compared with experimental surface pressure and heat transfer as well as lee side pitot pressure profiles. Generally good agreement is observed though high grid density is needed to capture the lee side pitot pressure behavior at moderate to high angles of attack. Computed results are also presented for turbulent flow past a generic elliptic cone-based geometry.

Introduction

The design of future hypersonic flight vehicles will depend heavily on computational fluid dynamics for the prediction of aerodynamic and thermodynamic loads, as well as engine performance. One of the features that characterizes the hypersonic flow regime is the presence of strong shock waves generated by the vehicle and by protuberances from the main body such as wings, canopies, and engine inlets. Thus, a need exists for a robust computational tool that can efficiently and accurately resolve flow-fields containing discontinuities.

The present work describes the extension of an upwind algorithm designed for the integration of the parabolized Navier-Stokes (PNS) equations [1] to three dimensions. Conventional PNS solvers [2,3] are based on the central differencing of crossflow fluxes and, therefore, have difficulty capturing strong embedded shocks. The objective of this work is to mate the efficiency of the space-marching approach with the advantageous shock capturing characteristics of upwind schemes. In addition to being upwind, the new algorithm is implicit (including boundary conditions) and is based on the use of finite-volumes for accurate flux conservation. The resulting code is applied to hypersonic flows past two simple body shapes: a 10° half-angle circular cone and a generic all-body hypersonic vehicle. Cone flow solutions have been computed for three angles of attack and results are compared with experimental surface pressure and heat transfer measurements as well as pitot pressure profiles. Results have also obtained for the flow past the all-body vehicle at an angle-of-attack of 15°.

Method

The PNS equations, obtained from the steady Navier-Stokes equations by neglecting streamwise viscous derivatives, can be written with respect to a generalized coordinate system as

$$\frac{\partial \bar{\mathbf{E}}^*}{\partial \xi} + \frac{\partial \bar{\mathbf{E}}^p}{\partial \xi} + \frac{\partial (\bar{\mathbf{F}}_i - \bar{\mathbf{F}}_v)}{\partial \eta} + \frac{\partial (\bar{\mathbf{G}}_i - \bar{\mathbf{F}}_v)}{\partial \zeta} = 0 \tag{1}$$

where $\bar{\mathbf{F}}$ and $\bar{\mathbf{G}}$ represent fluxes in the η- and ζ-coordinate directions, respectively. The subscripts i and v indicate inviscid and viscous components, respectively. The first two terms of Eq. (1) arise from the use of the Vigneron technique [2] in which the streamwise flux vector is split into two parts: the first part ($\bar{\mathbf{E}}^*$) is the modified streamwise flux, and the remainder ($\bar{\mathbf{E}}^p$) is that part of the streamwise pressure gradient responsible for introducing ellipticity into the equations in subsonic regions. The latter part is usually neglected or treated as a source term so that Eq. (1) becomes hyperbolic-parabolic in nature.

Equation (1) is differenced in a finite-volume manner to yield the following discretized conservation law:

$$\begin{aligned} \hat{A}^{*^n}_{k,l} \delta^{n+1} \mathbf{U}_{k,l} = & - (\hat{A}^{*^n}_{k,l} - \hat{A}^{*^{n-1}}_{k,l}) \mathbf{U}^n_{k,l} \\ & - [(\hat{\mathbf{F}}_i - \hat{\mathbf{F}}_v)^{n+\frac{1}{2}}_{k+\frac{1}{2},l} - (\hat{\mathbf{F}}_i - \hat{\mathbf{F}}_v)^{n+\frac{1}{2}}_{k-\frac{1}{2},l}] \\ & - [(\hat{\mathbf{G}}_i - \hat{\mathbf{G}}_v)^{n+\frac{1}{2}}_{k,l+\frac{1}{2}} - (\hat{\mathbf{G}}_i - \hat{\mathbf{G}}_v)^{n+\frac{1}{2}}_{k,l-\frac{1}{2}}] \\ & - [\hat{\mathbf{E}}^p(d\mathbf{S}^{n+1}_{k,l}, \mathbf{U}^n_{k,l}) - \hat{\mathbf{E}}^p(d\mathbf{S}^n_{k,l}, \mathbf{U}^{n-1}_{k,l})] \end{aligned} \tag{2}$$

where

$$\mathbf{U} = [\rho, \rho u, \rho v, \rho w, E_t]^T \quad \text{and} \quad \hat{A}^{*^{n-1}} = \frac{\partial \hat{\mathbf{E}}^*(d\mathbf{S}^n, \mathbf{U}^{n-1})}{\partial \mathbf{U}^{n-1}} \; .$$

The left-hand-side and the first term on the right-hand-side of Eq. (2) result from the linearization of the streamwise flux vector

$$\hat{\mathbf{E}}^*(d\mathbf{S}^n, \mathbf{U}^n) = \hat{A}^{*^{n-1}} \mathbf{U}^n$$

which is applied in order to avoid the difficulty of extracting flow properties from $\hat{\mathbf{E}}^*$. This linearization also simplifies the application of the implicit algorithm. The indices on the arguments $d\mathbf{S}$ and $\mathbf{U}$ indicate the location where the geometry and the physical variables, respectively, are evaluated.

A conventional central-differencing scheme would be obtained from Eq. (2) by simply averaging flow properties from adjacent cell centers to determine the flux at the cell faces. Upwind schemes derive their superior shock capturing characteristics from the introduction of flow physics at this level of the algorithm. In the present algorithm, the inviscid fluxes in the crossflow directions are evaluated through the solution of the approximate form of the governing equations

$$\frac{\partial \hat{\mathbf{E}}^*}{\partial \xi} + D_{m+\frac{1}{2}} \frac{\partial \hat{\mathbf{E}}^*}{\partial \kappa} = 0$$

with initial conditions

$$\hat{\mathbf{E}}^{*^n}(\kappa) = \begin{cases} \hat{\mathbf{E}}^*(d\mathbf{S}^n_{m+\frac{1}{2}}, \mathbf{U}_m) & \text{where } \kappa < \kappa_{m+\frac{1}{2}}; \\ \hat{\mathbf{E}}^*(d\mathbf{S}^n_{m+\frac{1}{2}}, \mathbf{U}_{m+1}) & \text{where } \kappa > \kappa_{m+\frac{1}{2}}. \end{cases}$$

where

$$D_{m+\frac{1}{2}} = \frac{\partial \hat{\mathbf{F}}_i}{\partial \hat{\mathbf{E}}^*} \quad \text{or} \quad \frac{\partial \hat{\mathbf{G}}_i}{\partial \hat{\mathbf{E}}^*}$$

and κ and m are replaced with η and k or ζ and l for the calculations of $\hat{\mathbf{F}}_i$ or $\hat{\mathbf{G}}_i$, respectively. This problem is a steady form of Roe's approximate Riemann problem [4] and the flow properties making up $D_{m+\frac{1}{2}}$ are averaged between m and $m+1$ using standard Roe averaging.

The solution to the above approximate Riemann problem consists of four constant-property regions separated by three surfaces of discontinuity emanating from the cell edge, $(\xi^n, \kappa_{m+\frac{1}{2}})$, and having slopes given by the eigenvalues of $D_{m+\frac{1}{2}}$. Of particular interest to the numerical algorithm is the resulting flux across the $m+\frac{1}{2}$ cell interface. This first-order-accurate inviscid flux consists of an unbiased component and a first-order upwind dissipation term as follows

$$\mathbf{H}^{\mathrm{I}}_{m+\frac{1}{2}} = \mathbf{H}^{\mathrm{c}} - \frac{1}{2}|D|_{m+\frac{1}{2}} \left[\hat{\mathbf{E}}^*(d\mathbf{S}^n_{m+\frac{1}{2}}, \mathbf{U}_{m+1}) - \hat{\mathbf{E}}^*(d\mathbf{S}^n_{m+\frac{1}{2}}, \mathbf{U}_m)\right] \tag{3}$$

where $\mathbf{H}^{\mathrm{c}}$ is a simply averaged flux and $|D|$ is the matrix which has the same eigenvectors as D and has eigenvalues which are the absolute values of those of D. First-order inviscid numerical fluxes in the η- and ζ-directions are then given by

$$(\hat{\mathbf{F}}^{\mathrm{I}}_i)_{k+\frac{1}{2},l} = \mathbf{H}^{\mathrm{I}}_{k+\frac{1}{2},l} \quad \text{and} \quad (\hat{\mathbf{G}}^{\mathrm{I}}_i)_{k,l+\frac{1}{2}} = \mathbf{H}^{\mathrm{I}}_{k,l+\frac{1}{2}}$$

respectively, where κ in Eq. (3) is replaced by η or ζ accordingly.

The scheme is extended to second-order accuracy following the approach of Chakravarthy and Szema [5] in which a second-order generic numerical flux is defined in terms of the first-order flux and anti-dissipative correction terms. The added terms are limited relative to one another in order to eliminate overshoots and undershoots that are characteristic of second-order schemes. In order to eliminate nonphysical behavior at locations where eigenvalues change sign, local dissipation is added in regions where eigenvalues are small in magnitude.

The algorithm is made implicit by evaluating the first-order numerical flux at the $n+1$ marching station and lagging the second-order correction terms at the n'th level. A straight-forward linearization in ξ is then applied assuming the $|D|$ matrix is locally frozen. The resulting system of algebraic equations is factored in a conventional manner (e.g. see Refs. [2] and [3]) to produce an alternating direction implicit (ADI) scheme of the form

$$\left[\hat{A}^*_{k,l} + \frac{\partial(\delta_\eta\{\hat{\mathbf{F}}^{\mathrm{I}}_i - \hat{\mathbf{F}}_v\})}{\partial \mathbf{U}_{k,l}} + \bar{\delta}_\eta \left(\frac{\partial\{\hat{\mathbf{F}}^{\mathrm{I}}_i - \hat{\mathbf{F}}_v\}}{\partial \mathbf{U}}\right)\right]^n$$
$$\times \left[(\hat{A}^*_{k,l})^{-1}\right]^n \left[\hat{A}^*_{k,l} + \frac{\partial(\delta_\zeta\{\hat{\mathbf{G}}^{\mathrm{I}}_i - \hat{\mathbf{G}}_v\})}{\partial \mathbf{U}_{k,l}} + \right.$$
$$\left. \bar{\delta}_\zeta \left(\frac{\partial\{\hat{\mathbf{G}}^{\mathrm{I}}_i - \hat{\mathbf{G}}_v\}}{\partial \mathbf{U}}\right)\right]^n \delta^{n+1}\mathbf{U}_{k,l} = RHS^n$$

where RHS^n is of the same form as the right-hand-side of Eq. (2) except that the crossflow fluxes are evaluated using flow properties at the n'th marching level.

Due to the Vigneron technique, the eigenvalues of D may obtain values as large as 10^9 in the subsonic region when large stepsizes are being taken. The associated eigenvector matrices are highly ill-conditioned and computer round-off errors can be amplified to the point where they destabilize the calculations. In the present version of the code, this is avoided by simply deactivating the upwinding in the subsonic region. Efforts are in progress which would allow a modified upwind dissipation term to be included near the wall.

Results and Discussion

Test Case I

The first test case consists of Mach 7.95 laminar flow past a 10° half-angle circular cone. The Reynolds number of the flow, chosen to match the experimental conditions used by Tracy [6], was $Re_L = 4.2 \times 10^5$, where the cone length L was 4 inches. Three angles of attack were investigated including low ($\alpha = 8^o$), moderate ($\alpha = 16^o$), and high ($\alpha = 24^o$) incidence cases. Three mesh sizes were used in these calculations: a medium grid containing 60 points circumferentially and 70 radial points was used in the calculation of the low and medium angle-of-attack cases, and coarse and fine grids of dimensions 60×50 and 115×90, respectively, were applied to the high angle-of-attack case. The calculations required approximately 0.08 seconds of CPU time per crossflow grid point on a Cray Y-MP computer.

Figure 1 shows computed Mach number contours at $x/L = 1.$ for the low and moderate angle-of-attack cases. For comparison, flow-field patterns determined experimentally by Tracy are shown as the overlaid symbols: circles represent shocks, and diamonds and triangles show the locations of the outer and inner edges, respectively, of the viscous layer. This figure shows that the grids used for these cases are sufficient to accurately predict the outer shock shapes and positions. When the angle of attack is increased from 8° to 16° a region of crossflow separation develops on the lee side of the cone. The location of the boundary-layer separation is well predicted by the PNS code, but the size of the recirculation region is slightly underpredicted by the computation. The experimental data, deduced from pitot pressure surveys, also indicate the presence of weak crossflow shock stems near the boundary-layer separation locations. These shocks are very weak at this angle of attack and are not observed in the computed Mach contours.

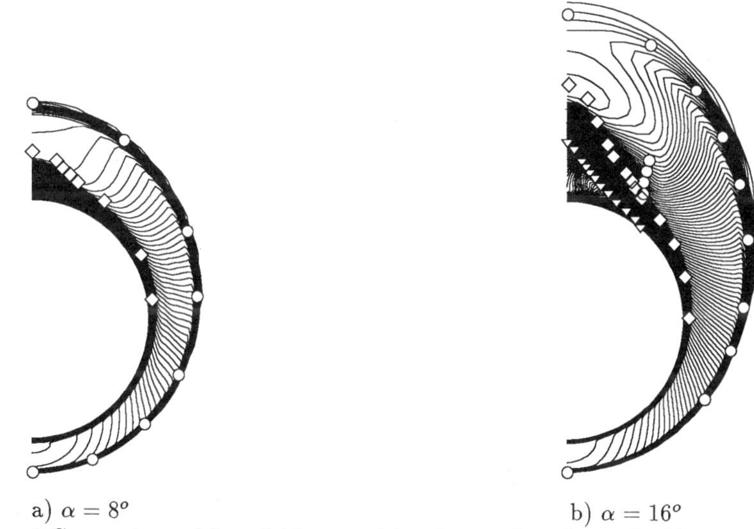

a) $\alpha = 8^o$ b) $\alpha = 16^o$

Fig. 1 Comparison of flow-field geometries: low and moderate incidence.

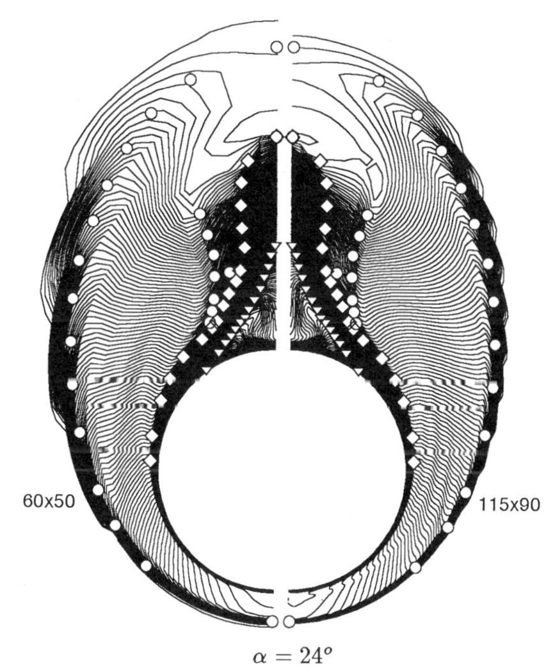

$\alpha = 24^o$

Fig. 2 Comparison of flow-field geometries: high incidence.

At high incidence ($\alpha = 24°$), the crossflow shock is of sufficient strength to produce a Λ-shock pattern where it interacts with the separated shear layer. Figure 2 shows results for this case computed on the coarse and fine grids in comparison with experimental data. This figure indicates that the coarse grid is sufficient to capture the global features of the flow. Of course, the outer shock is more sharply defined by the fine grid as are some of the details of the lee side flow-field. In particular, the additional grid points in the fine grid are necessary to resolve the Λ-shock pattern and the shape of the separated shear layer.

Pitot pressure profiles at the lee-side plane of symmetry are presented in Fig. 3 for all three angle-of-attack cases. At low angle-of-attack, the flow remains attached and the computed results are in good agreement with experiment, though the boundary-layer thickness is slightly underpredicted. At moderate angle-of-attack, the experimental data profile exhibits an inflection associated with the recirculation region beneath the separated boundary layer. The computed results for this case also indicate a recirculation region; however, this region does not extend as far from the cone as in the experiment and the inner edge of the separated shear layer is somewhat smeared compared to the data. This discrepancy is more pronounced at high angle-of-attack when the coarse grid is used. Results from the fine grid calculation show a marked improvement in the resolution of the inner edge of the separated boundary layer. Results that were computed on a 60×90 grid (not shown) indicate that the improvement is due mainly to the additional circumferential rather than radial resolution.

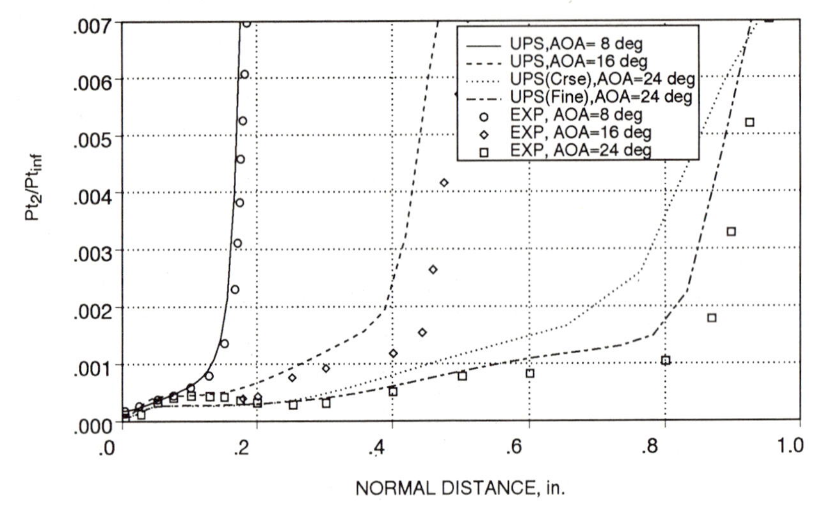

Fig. 3 Pitot pressure profiles.

Surface pressure distributions are compared with experiment in Fig. 4. Good agreement with experiment is observed for each angle of attack and results at high angle-of-attack are insensitive to grid refinement. Figure 5 shows comparison of computed and experimental heat transfer rates normalized by the heat transfer at zero incidence. Excellent agreement is seen for the low and moderate angle-of-attack cases and for the leeward distribution at high angle of attack. The computed results for

$\alpha = 24^o$ fall below the available data on the windward side (data for circumferential angles less than 30° could not be extracted from Ref. [6]). The cause of this discrepancy is unknown, but the agreement is not greatly influenced by the use of the the finer grid.

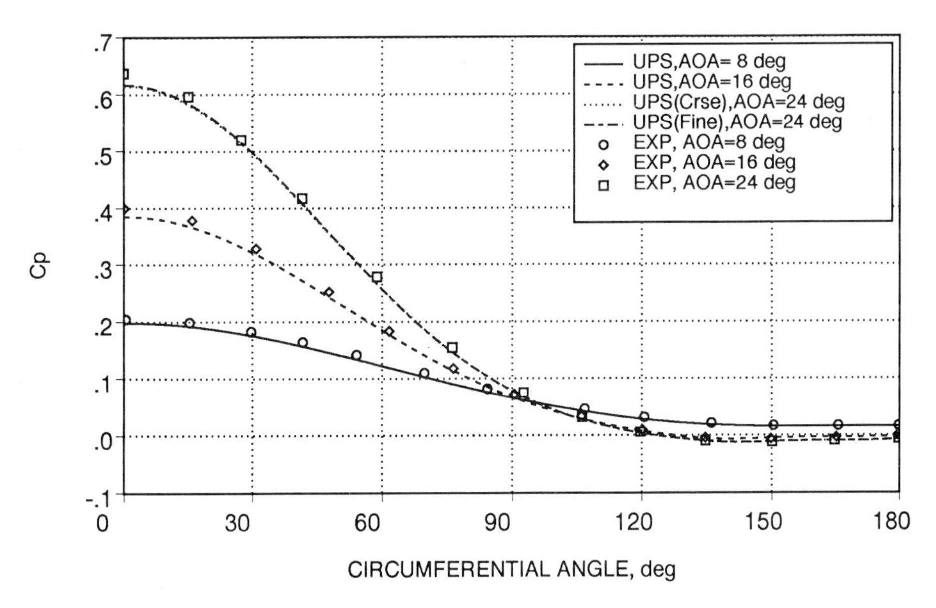

Fig. 4 Circumferential pressure distributions.

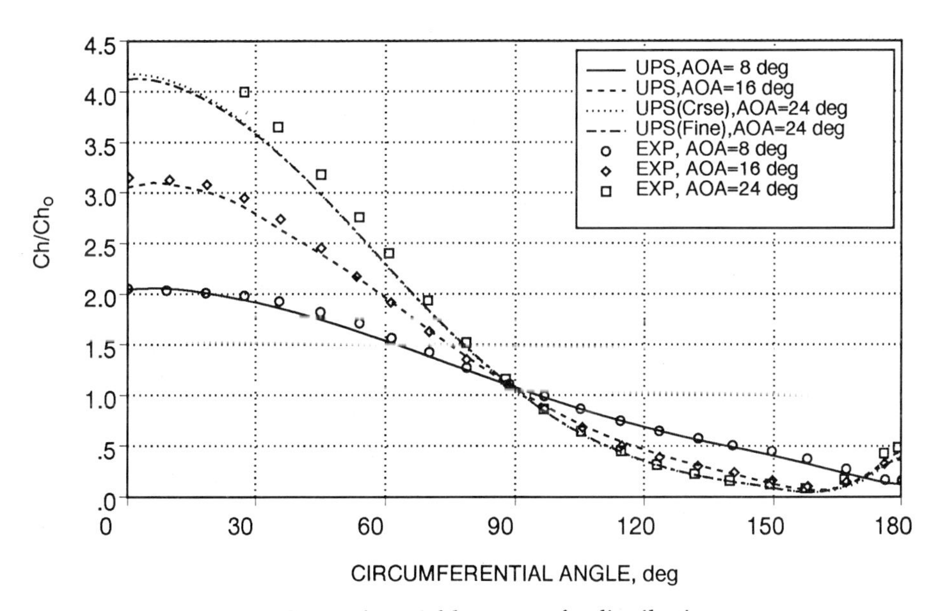

Fig. 5 Circumferential heat transfer distributions.

Test Case II

The second test case consists of Mach 7.4 flow past the simple all-body hypersonic aircraft shown schematically in Fig. 6. Calculations are being performed in conjunction with an experimental investigation of this geometry in the 3.5-ft hypersonic wind tunnel located at NASA Ames Research Center. Publication of the experimental measurements of surface pressure, flow-field pitot pressure, and surface heat transfer is expected in the near future. The grid employed in the present computation contained 90 cells in both the circumferential and normal directions and was generated hyperbolically at each cross section sweeping from the body radially outward [7].

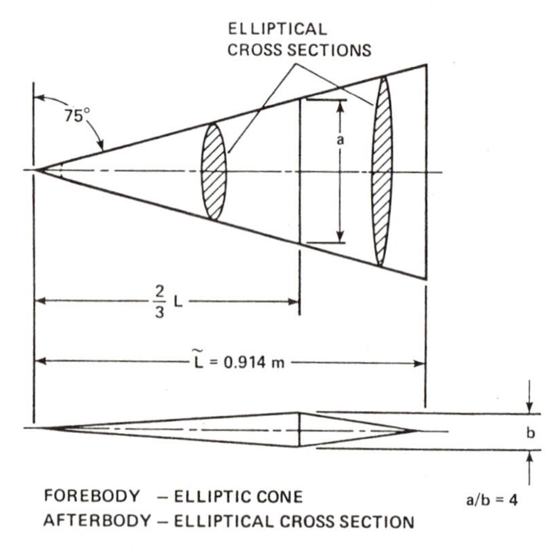

Fig. 6 Schematic of the Ames All-Body vehicle.

The present solution was computed for the region extending from $x/L = 0.05$ to $x/L = 0.9$. Conical flow is assumed from the apex to $x/L = 0.05$. The calculation required approximately 10 minutes of CPU time on the Cray Y-MP computer including the generation of the starting solution. The Reynolds number of the computed flow was 5.6 million based on the body length. Turbulent flow was assumed in the present calculation and the Baldwin-Lomax [8] algebraic turbulence model was used to compute local turbulent viscosities. Mach contours in the plane of symmetry illustrating the dominant features of the computed flow-field are shown in Fig. 7. The windward bow shock is sharply captured on the forebody and is observed to spread somewhat aft of the body "break point". This is due to the grid downstream of the break point which is not as well aligned with the shock as is the forebody grid. Other features illustrated in Fig. 7 are the expansion fan emanating from the break point and the boundary layers at the upper and lower centerlines.

Finally, Mach contours at two cross sections are presented in Fig. 8. The cross section of Fig. 8a is slightly upstream of the break point while that shown in Fig. 8b is at the last computed axial station. A large recirculation region exists on the lee side of the body. This region spreads toward the leading edge as the body thickness decreases because of the decreasing leading edge radius. Also present in these figures is a weak crossflow shock extending outward from the separation bubble qualitatively similar to the cone flow shock seen in Fig. 2.

The present computed results have been compared with preliminary experimental data and the agreement is generally good. At this time, the data is not publicly available.

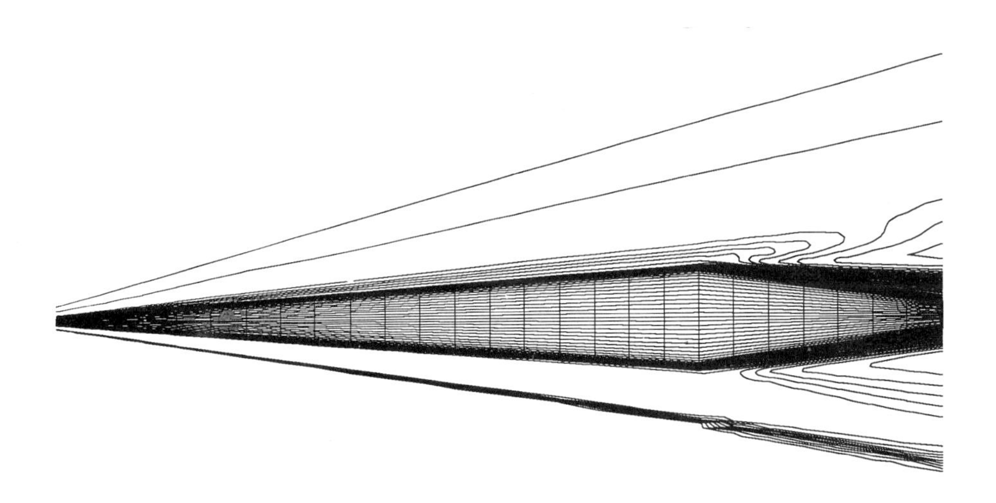

Fig. 7 Mach contours in the plane of symmetry: $\alpha = 15^o$.

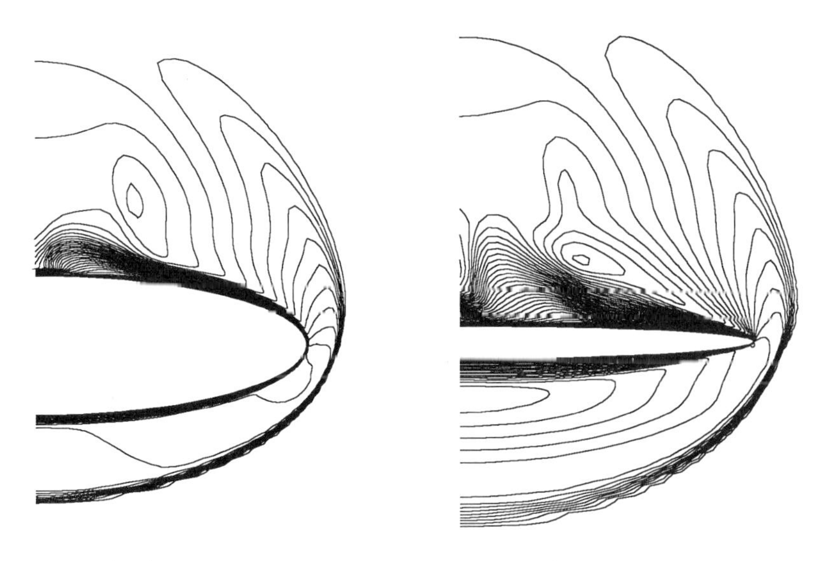

a) x/L = 0.63 b) x/L = 0.9.

Fig. 8 Mach contours: $\alpha = 15^o$.

Conclusions

Development and application of an upwind algorithm for the integration of the three-dimensional parabolized Navier-Stokes equations is described. The algorithm is based on the solution of steady approximate Riemann problems in both crossflow directions, analogous to Roe's method for the time-dependent equations. The scheme has been modified to make it implicit and second-order accurate in the crossflow directions. In application to high speed flows past two test geometries, the upwind character of the method has proved effective in providing sharp shock definition while greatly reducing dependence on user-added numerical dissipation. Comparisons with experimental data for Mach 8 flow past a 10 deg half-angle cone indicate that the numerical solutions do well in predicting surface pressure and heat transfer. At high incidence, accurate prediction of the complex lee side viscous flow-field can be achieved provided a high density grid is used.

References

[1] Lawrence, S. L., Tannehill, J. C., and Chaussee, D. S., "An Upwind Algorithm for the Parabolized Navier-Stokes Equations," AIAA Paper 86-1117, July 1986.

[2] Vigneron, Y. C., Rakich, J. V., and Tannehill, J. C., "Calculation of Supersonic Viscous Flow over Delta Wings with Sharp Subsonic Leading Edges," AIAA Paper 78-1137, July 1978.

[3] Schiff, L. B. and Steger, J. L., "Numerical Simulation of Steady Supersonic Viscous Flow," AIAA Paper 79-0130, Jan. 1979.

[4] Roe, P. L., "Approximate Riemann Solvers, Parameter Vectors, and Difference Schemes," *Journal of Computational Physics* , Vol. 43, 1981, pp. 357-372.

[5] Chakravarthy, S. R. and Szema, K. Y., "An Euler Solver for Three-Dimensional Supersonic Flows with Subsonic Pockets," AIAA Paper 85-1703, July 1985.

[6] Tracy, R. R., "Hypersonic Flow over a Yawed Circular Cone," California Institute of Technology Graduate Aeronautical Laboratories, Pasadena, Calif., Memorandum No. 69, Aug. 1963.

[7] Steger, J. L. and Chaussee, D. S., "Generation of Body Fitted Coordinates Using Hyperbolic Partial Differential Equations," *SIAM Journal Sci. Stat. Comput.* , Vol. 1, No. 4, Dec. 1980.

[8] Baldwin, B. S. and Lomax, H., "Thin Layer Approximation and Algebraic Model for Separated Turbulent Flows," AIAA Paper 78-257, Jan. 1978.

Development of the MZM numerical method for 3D boundary layer with interaction on complex configurations

M. Lazareff and J.C. Le Balleur

O.N.E.R.A. - BP 72 - 92322 Chatillon Cedex (France)

Summary : The viscous 3D steady flow problem at high Reynolds number is decomposed into a "Viscous-Defect" VD problem interacted with a pseudo-inviscid problem, the former VD problem being hyperbolic along the boundaries under thin-layer assumptions. Both problems are solved on the whole 3D field, and coupled by the boundary conditions, thus restoring the global elliptic character of the solution. The "Viscous-Defect" problem is solved here by a hybrid field/integral method, involving modelled 3D parametric velocity profiles, discretized in the normal direction. The MZM numerical method of solution of the resulting hyperbolic system of equations has been extended to complex multi-structured configurations. The new method is here applied to wings in attached flows, to ellipsoids at incidence, and to airplane wing-body configurations.

1 - Introduction

We are interested in the numerical solution of the viscous 3D steady flow problem at high Reynolds number. This allows the use of thin-layer approximations, and leads to viscous-inviscid interaction methods. On ellipsoids and general slender bodies, the present viscous-inviscid methods are applied to the study of open separation [1,2]. On wings, the methods are able to compute the effect of viscosity on lift and drag [1,2,3] in attached flow. The methods are now extended here to complex configurations, such as a wing-body airplane model. The same method can be useful for land- or sea-faring vehicles.

In this paper, we consider a swept wing mounted in low position on a fuselage, as in modern commercial airplanes. We seek to compute the viscous incompressible flow over this wing-body configuration, including effects on the global lift and drag of the global and local interactions between the two bodies. The primary interaction effect is a modification of the wing loading. This is not necessarily limited to the root sections, but the effect is a diminution of the local loading at the root of the wing, with lift transfer to the fuselage, as compared to the same wing in stand-alone symmetrical mounting. This unloading, generally controlled by wing twisting, has a favourable effect on the evolution of the viscous layer, for a given section profile and incidence. This in turn allows the use of thicker wing sections, as necessitated by structural and landing-gear storage constraints.

The resulting rapid evolution of the wing section thickness and twist is a first pecularity for the computation of the viscous layer, and a yawed-cylinder approximation is no longer a precise initialization of the viscous solution near the dividing streamline at the leading-edge. The computation of the viscous layer on the fuselage, on the other hand, combines the difficulties of viscous shear-layer separation on a prolate body with the wing-junction problems.

Apart from the wing-body configuration , results are also given for a prolate spheroid at incidence, and a wing at incipient separation with accumulation of skin-friction lines at the trailing-edge.

2 - Method of solution

The "direct" numerical solution of the Navier-Stokes (NS) equations at high Reynolds number introduces difficult problems of grid refinement and of numerical conditioning. As a numerical alternative, the "Defect-Formulation" of Le Balleur [4,5,6] splits the fluxes of the NS conservation equations into Euler and Viscous-Defect (VD) fluxes. When introducing in addition thin-layer approximations in the Viscous "Defect-Formulation", the VD system of equations is boundary layer-like and of hyperbolic type.

2.1 - Viscous-inviscid coupling

The correct global elliptic behaviour is here however restored, even at supersonic speeds or with massive separation, by the coupling relation (exactly derived from the continuity equation), which links the pseudo-inviscid and VD problems via inviscid normal velocity distributions on solid surfaces and wakes, Le Balleur [9,4,10,11,12,2]. In three dimensions, the corresponding numerical problem of the viscous-inviscid coupling has not yet received a general numerical method of solution, as those designed for two dimensions, Le Balleur

[10,3,11,12], but converged solutions are obtained in cases of practical interest, mainly without separation, with improvement of simple fixed-point coupling methods, Le Balleur, Lazareff [1,2,3].

2.2 - Viscous equations - Turbulent modelling

Then we shall deal first here with the solution method for the Viscous Defect boundary-layer-like problem, in three-dimensions. The viscous grid is composed of the curvilinear (x^1, x^2) trace on the coupling surfaces of the (suitably refined) inviscid grid, and of an x^3 direction locally normal to the (x^1, x^2) coupling surface.

The thin-layer Viscous Defect VD problem is solved by a hybrid field/integral method, involving a rank-3 hyperbolic system of integral equations on (x^1, x^2), and x^3-discretized 3D turbulent mean velocity profiles. These profiles are however parametrically modelled with the 3-parameter model of Le Balleur [6,12,5,7]. The turbulence is modelled algebraically trough a parametric entrainment function, Le Balleur [5,11,12], deduced from the velocity profiles modelling and from mixing-length assumptions. In compact tensor form [2], the (coupling) continuity equation (1) together with the momentum (2) and entrainment momentum (3) of the VD system read:

$$\mathrm{div}\,[\rho q \underline{\delta}] = \rho w - \bar{\rho}\bar{w} \ . \tag{1}$$

$$\mathrm{div}\,[\rho q^2 \underline{\underline{\theta}} + \rho q \underline{u} \otimes \underline{\delta}] = \rho \underline{u} w - \bar{\rho}\bar{\underline{u}}\bar{w} + \tfrac{1}{2} \rho q^2 \underline{C_f} \ . \tag{2}$$

$$\mathrm{div}\,[\rho \underline{u} \delta - \rho q \underline{\delta}] = \bar{\rho}\bar{w} + \rho q E \ . \tag{3}$$

$w = u^3$, $q = \|\underline{u}\|$, E: entrainment function, $\underline{C_f}$: surface friction vector, δ: viscous layer thickness .

$$\underline{\delta} = (\rho q)^{-1} \int_0^\infty (\rho \underline{u} - \bar{\rho}\bar{\underline{u}})\, dx^3 , \quad \underline{\underline{\theta}} = (\rho q)^{-2} \int_0^\infty (\rho \underline{u} \otimes \underline{u} - \bar{\rho}\bar{\underline{u}} \otimes \bar{\underline{u}})\, dx^3 - q^{-1} (\underline{u} \otimes \underline{\delta}) \ . \tag{4}$$

This system is solved in projection on the $(\underline{x_1}, \underline{x_2})$ local holonomic basis associated with (x^1, x^2).

2.3 - Numerical viscous problem

In contrast to the Euler hyperbolic system for steady inviscid supersonic flow, the VD system in 3D cannot generally be solved by a space-marching numerical method along one single main-flow direction. Its local characteristic cone may vary widely in direction and angle, leading to severe CFL limitations or even to local impossibility of integration, with a classical space-marching scheme.

We have developed the MZM ("Multi-Marching Multi-Zonal") numerical method of solution in the (x^1, x^2) plane, which can be used both in integral- and field- methods, to overcome this specific difficulty [1,2].

3 - 'Multi-Zonal' and 'Multi-Marching' MZM method

3.1 - 'Multi-Marching' iterative sweeps

In the suggested MZM method, Le Balleur, Lazareff [1,2], the discretized VD system is first solved with a "Multi-Marching" technique on a given $[i_1, i_2] \times [j_1, j_2]$ zone, using as many alternate sweeps along the four possible marching directions of the grid (+i,-i,+j,-j) as necessary.

For each of these sweeps, for example when marching in the x^1 (or i) direction, the marching step Δx^1 is subject to the linear local CFL stability criterion, and also to a non-linear criterion based on a maximal value of the ratio $(h_1 \Delta x^1/\delta)$ of the step size to the viscous layer thickness $(h_1 = \|\underline{x_1}\|)$. The above equations are written at node (i,j):

$$\left[\frac{\partial}{\partial x^1} F^k(f^m) + \frac{\partial}{\partial x^2} G^k(f^m) \right]_{(i,j)} = b^k_{(i,j)} \quad , \quad k=1,4 \ , \ m=1,3 \ . \tag{5}$$

where the unknowns are the 3 parameters f^m of the turbulent velocity profiles modelling, plus the inviscid transpiration velocity at the walls or wake-cuts (included in $b^k_{(i,j)}$).

Discretization is presently by the explicit MacCormack scheme, which marches in the x^1-direction the unknown $\mathbf{F}^k$, with an explicit x^2-discretization of $\mathbf{G}^k$. A Newton method is used for the non-linear computation of f^m from F^k, the Jacobian matrix $\partial F^k/\partial f^m$ being provided by the velocity profile modelling.

3.2 - 'Multi-Zonal' automatic decomposition and chaining

In the MZM method, the numerical efficiency of this Multi-Marching technique is further augmented by a "Multi-Zonal" treatment of the (i,j) solution domain, automatically adapting to each of several flow topologies, as explained below.

The first step is to define the type of obstacle and associated flow. Second, a specialized routine looks for characteristic features of the flow (i.e. the leading edge attachment line region) and stores the corresponding grid indices in a "remarkable indices" array. Third, based on this array, a decomposition of the computational domain is automatically generated ("Multi-Zonal"). Any other zone may be defined, either by "absolute" or "remarkable" indices, the latter being independent of the grid definition and of the flow parameters (i.e. the incidence angle).

The resulting "zones" globally cover the computational domain, with possible marginal, partial or total overlap. They are sequenced in a way that respects a correct order of dependency, as deduced from the probable viscous flow topology. In this, the method enables a degree of feedback from user's past experience on the same type of flow. The initial "Multi-Marching" idea is represented by successive solution on "oriented zones", each with a different $\pm i, \pm j$ marching direction. More generally, the problem of inter-dependency is solved by iterative solution on the concerned group of oriented zones, not necessarily sharing the same (i, j) rectangle.

3.3 - Optimization

In the limit, the initial "Multi-Marching" idea could be applied on a node-by-node basis, with an explicit scheme as presently used. Sweeping together several lines in the x^2-direction, when marching for example in x^1-direction, allows an increase in computing efficiency, due to vectorization. This is compounded by the need to use an integration step that satisfies the linear (CFL) and non-linear stability criteria for all these lines, while quite often the maximum step in x^1-direction widely varies along the x^2-direction.

The Multi-Zonal decomposition of the computational domain alleviates this difficulty, when based on previous knowledge of the general features of the flow. Inside a given zone, a local optimization limits the re-computation of already accessed nodes (as may occur in multiple sweeps of the zone) to those whose characteristic cone does not lie close to the general direction of a possible sweeping direction. These nodes are accessed (by integration from known nodes) along the two best possible directions (if at all possible) and the two results are weighted according to an empirical law.

This weighting is necessary because different numerical integration errors along two completely different paths leading to neighbouring nodes may end up as a "fracture line" in the solution. In this respect, care has to be taken so as to assure the invariance of intrinsic differential quantities (i.e. components of a gradient on an intrinsic basis) in a permutation of the discretization coordinate directions. The discretization scheme would have to be equivalent to a centered scheme plus a tensorial artificial viscosity. The MacCormack scheme is an example of such a non-rotational-invariant scheme, leading to a dependence on the integration path even in the fine-grid limit. In fact, even permuting the bias at the predictor and corrector stages may lead to different results.

3.4 - Example of 'MZM' strategy on a prolate spheroid at incidence

The MZM computation has enabled the approximate display of "discontinuities" (or weak discontinuities) in the hyperbolic "uncoupled" boudary-layer-like VD problem,, as occur when prescribing an outer potential (non-lifting) velocity field on a prolate spheroid at incidence. The MZM numerical strategy described below allows the computation of the viscous layer starting from the stagnation cell at the nose.

The initial Cauchy data is formed by application of the stagnation point similarity solution on a 9-point "extended stagnation cell". Starting from these 9 points, radial multi-marching-sweeps along the four possible directions $\pm i, \pm j$ of the grid are repeated on a "nose zone", covering the front part of the ellipsoid, down to a section aft of the stagnation point across which the flow is everywhere downwind.

At incidences near 10°, the "discontinuity" cannot be captured in a single sweep, but the azimuthal component of the inviscid flow is sufficient for the azimuthal upward sweep to proceed, using as Cauchy data the result of the longitudinal sweep. The azimuthal downward sweep cannot be used with an inviscid non-lifting outer velocity field because this fictitious non-lifting inviscid field is everywhere upward in azimuth. A number of iterations may be needed, depending on the grid used, after which no further nodes are accessed.

At incidences over 20°, for Reynolds numbers in the 10^6 range, the discontinuities generated by "uncoupled" viscous calculations becomes too stiff for the scheme to compute cleanly, and oscillations appear ("ringing"). These oscillations propagate towards the lower side (a problem with the assymmetrical MacCormack scheme) and completely disrupt the computation when a unique zone is used for covering the whole domain. Here the MZM "Multi-Zonal" capability enables to compute the lower side first, once and for all, thus suppressing any possibility of downward propagation. By the way, this "Multi-Zonal" decomposition is also efficient from the step-size point of view, as explained before.

3.5 - Progress in the prediction of open separation

Numerical progress allows a first weak-interaction tentative approach for predicting the roots of vortex-sheets (separating viscous layers) which characterize "open separation" on smooth slender bodies.

An initial step was the computation of the viscous layer on "both sides" of the accumulation, with friction lines converging towards it and diverging from the upper symmetry line, reference [1]. Thence the friction-lines pattern was essentially different from that of the inviscid streamline pattern from which it had been constructed.

A second step, reference [2], was achieved with the computation of the viscous layer across the accumulation line, exhibiting a sharp evolution of the computed normal velocity and a change of sign in the azimuthal skin-friction component. This feature of the viscous-layer solution is conserved in a grid refinement (see 5.1), although some details may depend on the solution scheme. At the initial step [1] the method was unable to cover this region. This "discontinuity", specific of an uncoupled viscous-layer computation, may be viewed as an indication of the possible location of the vortex-sheet's root in the pseudo-inviscid component of a coupled viscous-inviscid computation.

A third step now in progress tries to involve the two main aspects of the coupling model between the vortex-sheet and the body viscous layer, upstream of and around the vortex sheet's root (locus of the viscous layer separation), namely:

- computation of the viscous layer in the presence of a (prescribed) vortex sheet in the inviscid field
- computation of the quasi-inviscid velocity field with prescibed normal velocities (displacement), deduced from the viscous-layer solution computed with the inviscid field

The first aspect only reminisces of a boundary-layer computation (direct mode), on a lifting wing. The difference here, for the wake separation on a smooth surface, is that the "trailing edge" dihedral is 180°, and that the main direction of the flow is parallel to the "trailing edge" (vortex root).

The second aspect does not have an equivalent in airfoil or wing computations, because here the absence of geometric singularity allows to study the effect of a normal velocity distribution, apart from the Kutta condition. We have used tentatively here, as a preliminary step, the normal velocity distribution deduced from the uncoupled non-lifting computation, by extrapolating downstream the azimuthal distribution obtained at the section of maximal "discontinuity". The new result (Fig. 1a) is that it is then possible to obtain a surface velocity field with the same type of accumulation towards an off-symmetry line as with a Kutta condition, which is not to say that they are equivalent. When using the original, non-extrapolated, distribution the effect is similar but of lesser intensity (Fig. 1b).

3.6 - Striction criterion for the detection of accumulation lines

The convergence (or divergence) S of the envelope lines of a velocity field $\underline{u}$ is computed by applying the divergence operator to the velocity direction vector field $\underline{u}/\|\underline{u}\|$:

$$S = \mathrm{div}\,(\underline{u}/\|\underline{u}\|)\ . \tag{6}$$

On a $(\underline{x_1}, \underline{x_2})$ basis with metric coefficients $(h_1 = \|\underline{x_1}\|,\ h_2 = \|\underline{x_2}\|,\ \lambda = (\underline{x_1}, \underline{x_2}))$, this is expressed by:

$$S = \|\underline{u}\|^{-3}\ [\,(h_2 u^2)^2\, u^1_{,1} + (h_1 u^1)^2\, u^2_{,2}\,]\,/\,J\ ;\ J = \|\underline{x_1} \times \underline{x_2}\| = h_1 h_2 \sin\lambda\ . \tag{7}$$

For a bundle of lines, converging nearly parallel to the x^1-direction :

$$u^2 \ll u^1\ ,\ u^1_{,1} \ll u^2_{,2}\ ,\ S \sim u^2_{,2}/(J\, h_1\, u^1)\ . \tag{8}$$

Alternatively, the "striction" of a bundle of converging envelope lines with local mean tangent and normal $(\underline{s}, \underline{n})$ may be estimated by computing the logarithmic derivative S' of the distance between a pair of neighbouring lines. The distance $d_{\underline{\eta}}$ is computed along a direction $\underline{\eta}$, roughly parallel to $\underline{n}$, while the derivative is taken along $\underline{\sigma}$, roughly parallel to $\underline{s}$:

$$S' = (\mathrm{Log}\, d_{\underline{\eta}})_{,\underline{\sigma}}\ . \tag{9}$$

The logarithmic treatment takes care of the specific value of the line-spacing d, and of the projection cosines, as long as the angles $\boldsymbol{\eta} = (\underline{n}, \underline{\eta})$, $\boldsymbol{\sigma} = (\underline{s}, \underline{\sigma})$ vary slowly with respect to the curvature radius of the bundle. The error $\Delta S'$ on striction S', when computing $d_{\underline{\eta}}$ along the direction $\underline{\eta}$ in lieu of $\underline{n}$ is :

$$\Delta S' = S'_{\underline{\eta}} - S'_{\underline{n}} \sim \eta_{,s}\ \mathrm{tg}\,\eta\ . \tag{10}$$

The use of direction $\underline{\sigma}$ in lieu of $\underline{s}$ roughly introduces a factor $\cos(\boldsymbol{\sigma}) \sim 1$, with the hypotheses of (8). In terms of $\alpha = (\underline{x_1}, \underline{u})$, a simple estimate $\hat{S}'$ of S', consitent with intuition and equivalent to (8), is:

$$\hat{S}' = \alpha_{,y}\ . \tag{11}$$

The "striction" measure of convergence, either by (7) or by (11), provides a numerical detection of an accumulation line. It does not involve the inspection of the pattern of envelope lines, while this is generally quite instructing. It is necessary however to "filter out" other loci of strong convergence, such as the rear stagnation point in the inviscid flow over a smooth body. This can be done following the same topological features as used in the "Multi-Zonal" decomposition (see 3.3).

4 - Multi-structured MZM method for complex configurations

The present MZM method has recently been further generalized to problems defined on several independent structured grids, as occur in complex configurations.

For the wing-body configuration, the present viscous computation is made with a pseudo-inviscid routine based on the panel method developed at Aérospatiale by Rivoire, Eichel [8].

The grid is made of two structured parts. One extends over the fuselage, and the other over the wing surface and its wake. The staggered coupling grid of the viscous solver follows the same arrangement.

The extended MZM method is thus made to work on the wing-body configuration, using the topology information for wings and slender bodies. Work is in progress relative to an acceptable numerical treatment of the wing socket problem.

5 - Results

5.1 - Prolate spheroid at incidence

Results are shown first for a 6:1:1 prolate spheroid at 10° incidence and $Re_{2a} = 6.7\ 10^6$ (AGARD WG10 Working Group test-case). This is an uncoupled solution using the potential outer velocity-field. Good agreement is obtained with experiment on the computed discretized velocity profiles (Fig. 3a), until section x/2a=.64 where viscous-layer separation begins.

For the same case, Fig. 3b shows the comparison of computed results on two grids, and with experiment on the fine grid, for streamwise displacement thickness δ^1 and skin-friction direction γ. This last parameter develops a "quasi-discontinuous" behaviour, evident at section .73 . There the fine-grid solution shows "ringing" at the foot of the discontinuity, owing to the absence of artificial viscosity in the solution scheme (MacCormack scheme).

The progressive development of this azimuthal discontinuity, and its dependence on the incidence, is illustrated on Fig. 2, showing several azimuthal sections of the δ^1 parameter at 15° and 20° incidence. The scheme is unable to compute sections further than .40 at the higher incidence, leading to oscillations and underprediction of the extrema. This limitation is however primarily consequent to the lack of viscous-inviscid coupling in these preliminary computations.

5.2 - Wing-body configuration

Results for the new wing-body capability are shown for an Airbus-like shape (Fig. 4a). At this time converged results have been obtained with viscous strong coupling on the wing and wake and without coupling on the fuselage.

Pseudo-inviscid wall streamlines on the wing and fuselage are shown on Fig. 4b, while Fig. 4c represents the skin-friction lines of the coupled computation on the wing at 7.5° incidence, drawn together with the skin-friction lines on the fuselage, the latter lines being issued from an uncoupled viscous computation. These friction-lines cover the part of the fuselage over the wing upper surface, but are interrupted on the approach of the "root trailing-edge" section, probably due to an insufficient grid resolution with respect to the (unfavourable) pressure gradient. A tendency towards accumulation near the trace (on the fuselage) of the wing wake-sheet is apparent.

Figures 4d and 4e display 3D representations of the fields for two viscous parameters, coded as elevations over the wing planform. Parameter H_i on Fig. 4d is the shape parameter δ^1/θ^{11}, a measure of the tendency to separate. Parameter β on Fig. 4e is the local angle of twist of the viscous 3D velocity vector, between external flow and skin-friction lines.

5.3 - High aspect ratio NACA0012 tapered swept wing with twisted tip

This configuration has been designed to test the capability of the method to progressively approach separation along the upper trailing-edge, leading to accumulation at the outboard sections. The large incidence (11.5°), moderate Reynolds number ($1.1\ 10^6$ based on the smallest chord), sweep (near 27°) and taper (.3) all concur to large 3D viscous effects at the outer wing. The grid has been refined there in the span direction to accomodate this anticipated behaviour. In order to get a flow stabilized near incipient separation ahead of the wing tip, a twist with constant slope has been added, parabolically growing from 0° at 70% span to .8° at 100%. The tip section is thus at 10.7° actual incidence.

Converged coupled results show a definite tendency to accumulation of the skin-friction lines along the upper trailing edge (Fig. 5a), locally deflecting to within 5° of its direction. This 5° margin has been close to 0° in the course of the coupling iterations.

The iso-lines of skin-friction modulus C_f (Fig. 5b) and shape parameter H_i (Fig. 5c) at the outer wing show the emergence near the leading-edge of values typical of the trailing-edge area, near the root sections. This tendency to separate is stabilized by the added twist, as is also evident on the 3D plot of C_f (Fig. 5d) and Reynolds number $R_{\theta''}$ (Fig. 5e). On this last figure, the lower surface is in front, the wake is in the back with the upper surface in between, and the wing tip is on the left. (Notice that the evolutive wing planform (see Fig. 5a) is transformed into a rectangle by the use of chord- and span-fraction parameters in this representation)

6 - Conclusions

A new numerical method for the computation of viscous flows over complex configurations of practical interest, based on the "MZM" numerical technique previously suggested by the authors [1,2] and on the turbulent boudary layer modelling of Le Balleur [12], is in development.

The new method is here applied to slender bodies or ellipsoids at incidence, and to an airplane wing-body configuration, where as a first step, only the inviscid influence of the fuselage on the wing and wake flows has been taken into account. Preliminary results, however, are also obtained, with uncoupled computation of the boundary layer over the fuselage, going round the wing socket and nearing the trailing edge adverse pression region. Converged solutions of the viscous-inviscid interaction along the wing and the viscous wake are obtained, in attached flows. Work is in progress for adding the solution and coupling of the viscous system on the whole fuselage (without wing-body junction problems).

References :

[1] LE BALLEUR J.C., LAZAREFF M. - A Multi-Zonal-Marching integral method for 3D boundary layer with viscous-inviscid interaction. - Proceed. 9th ICNMFD, Saclay, France (1984) Lecture Notes in Physics, 218, Springer Verlag (1985).

[2] LAZAREFF M., LE BALLEUR J.C. - Computation of three-dimensional flows by viscous-inviscid interaction using the "MZM" method - AGARD -CP-412 Paper 25, 1986 (Aix-en-Provence).

[3] LAZAREFF M., LE BALLEUR J.C. - Computation of three-dimensional viscous flows on transonic wings via boundary layer-inviscid flow interaction - La Recherche Arospatiale 1983-3, p. 155-173.

[4] LE BALLEUR J.C. - Computation of flows including strong viscous interactions with coupling methods - AGARD, General introduction, Lecture 1, Colorado-Springs 1980, AGARD-CP-291 (1981).

[5] LE BALLEUR J.C - Strong matching method for computing transonic viscous flows including wakes and separations. Lifting airfoils. - La Recherche Aérospatiale 1981-3, p. 21-45, English and French editions (1981).

[6] LE BALLEUR J.C. - Numerical viscid-inviscid interaction in steady and unsteady flows. - Proceed. 2nd Symp. Numerical and Physical Aspect of Aerodynamic Flows, Long-Beach, (1983), Springer-Verlag, T. Cebeci editor, chapt. 13, p. 259-284 (1984).

[7] LE BALLEUR J.C. - Viscous-Inviscid interaction solvers and computation of highly separated flows - "Studies of Vortex Dominated Flows" - (ICASE-NASA Langley 1985) chap. 3 p. 159-192, Hussaini and Salas ed., Springer-Verlag 1987.

[8] RIVOIRE V., EICHEL P. - Rapport DRET 86/304 Aérospatiale/Div. Avions (1987).

[9] LE BALLEUR J.C. - Viscous-inviscid flow matching : Analysis of the problem including separation and shock waves. - La Recherche Aérospatiale 1977-6, p.349-358 (Nov.1977). French, or English transl. ESA-TT-476.

[10] LE BALLEUR J.C. - Viscous-inviscid flow matching : Numerical method and applications to two-dimensional transonic and supersonic flows. - La Recherche Aérospatiale 1978-2, p. 67-76 (March 1978). French, or English transl. ESA-TT-496.

[11] LE BALLEUR J.C. - Numerical flow calculation and viscous-inviscid interaction techniques. - Recent Advances in Numerical Method in Fluids, Vol 3. : Computational methods in viscous flows, p. 419-450, W. Habashi editor, Pineridge Press, (1984).

[12] LE BALLEUR J.C. - New possibilities of Viscous-Inviscid numerical techniques for Solving Viscous flow equations, with massive separation. - Fourth Symposium "Numerical and Physical Aspects of Aerodynamic Flows", Long Beach, USA (16-19 January 1989). ONERA TP 1989-24 (and Proceed. Springer-Verlag, selected papers, editor T. Cebeci, to appear).

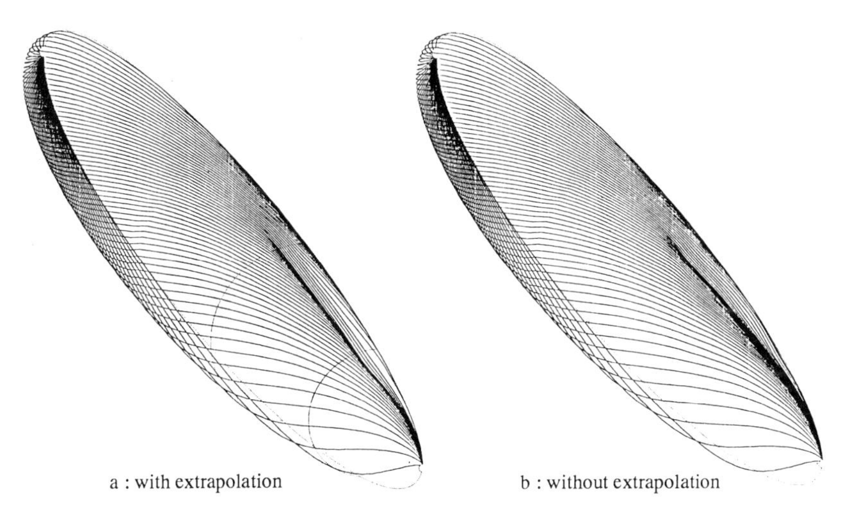

Fig. 1 : 6:1:1 ellipsoid, $\alpha = 10^o$, $Re_{2a} = 6.7\,10^6$.
Pseudo-inviscid surface streamlines (non-zero normal velocity from boundary-layer computation).

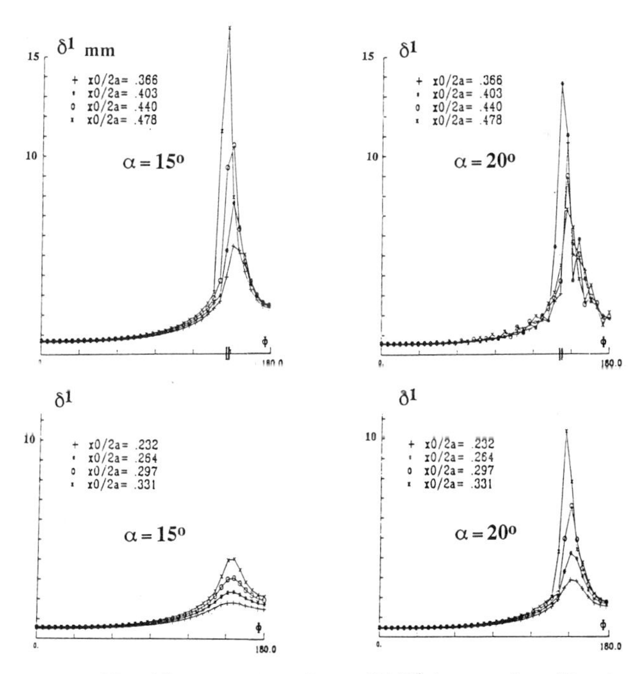

Fig. 2 : 6:1:1 ellipsoid, $\alpha = 15^o$, 20° , $Re_{2a} = 6.7\,10^6$ (computation with outer potential flow)
Streamwise displacement thickness δ^1, azimuthal sections.

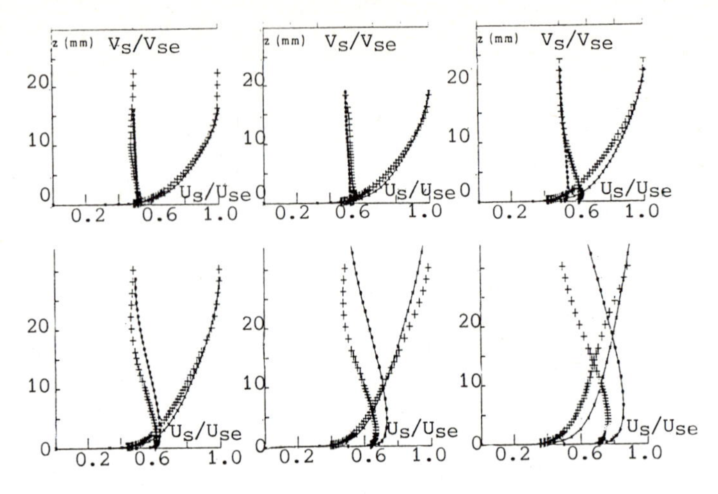

a : Computed discretized viscous velocity profiles.

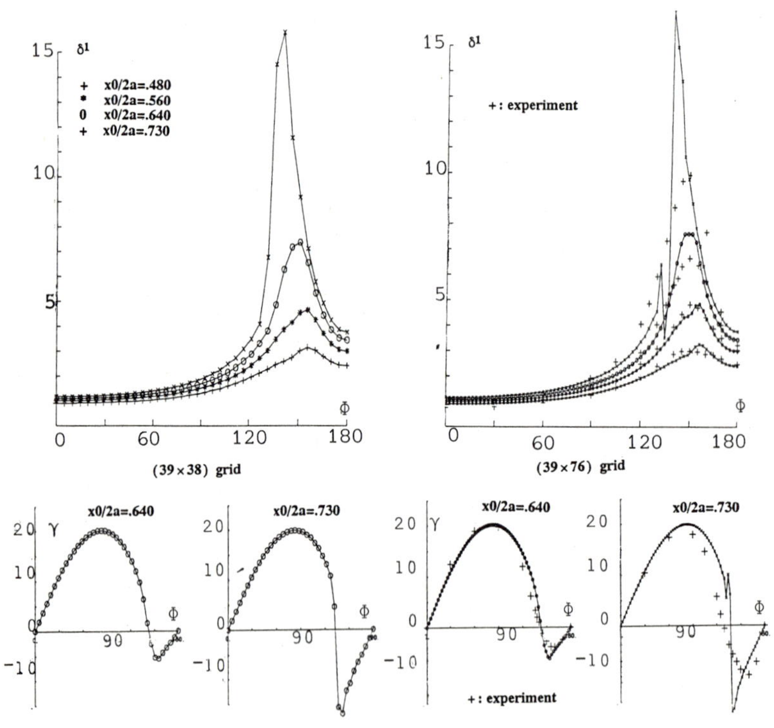

b : Streamwise displacement thickness δ^1 and skin-friction direction γ.
Grid refinement study. Comparison with experiment.

Fig. 3 : 6:1:1 ellipsoid, $\alpha = 10^o$, $Re_{2a} = 6.7\,10^6$ (computation with outer potential flow)

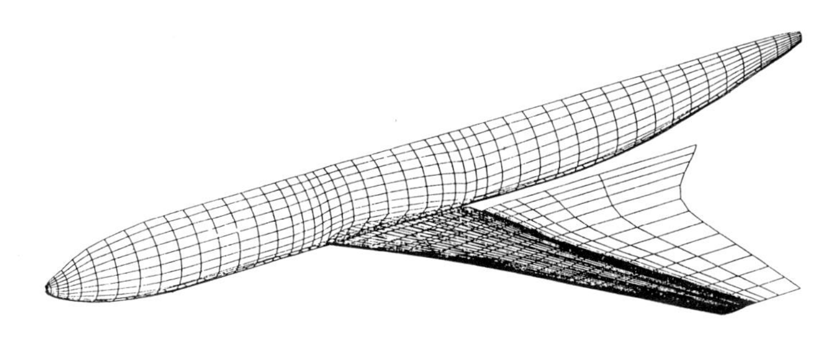

a : Coupling grid for the viscous computation on the skin and wake.

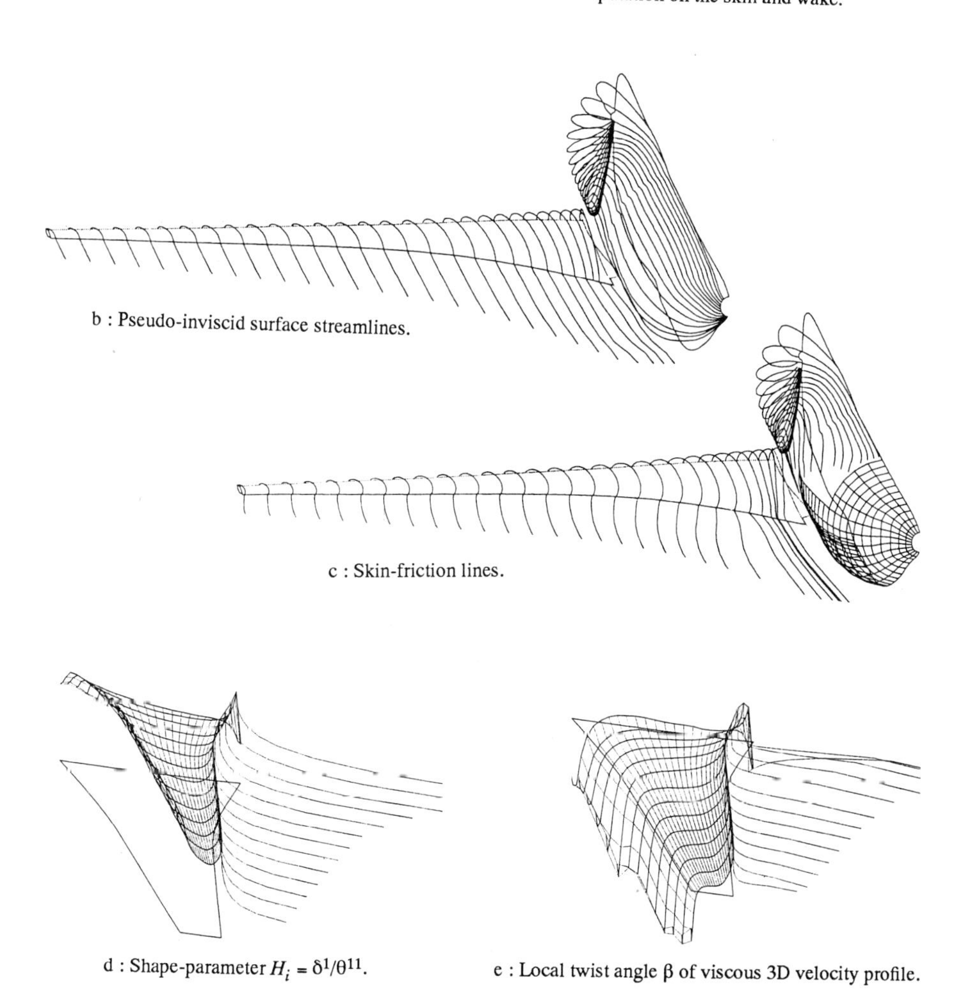

b : Pseudo-inviscid surface streamlines.

c : Skin-friction lines.

d : Shape-parameter $H_i = \delta^1/\theta^{11}$.

e : Local twist angle β of viscous 3D velocity profile.

Fig. 4 : Airbus-like wing-body configuration, $\alpha = 7.5^o$, $\mathrm{Re}_c = 15\,10^6$, $M = .4$

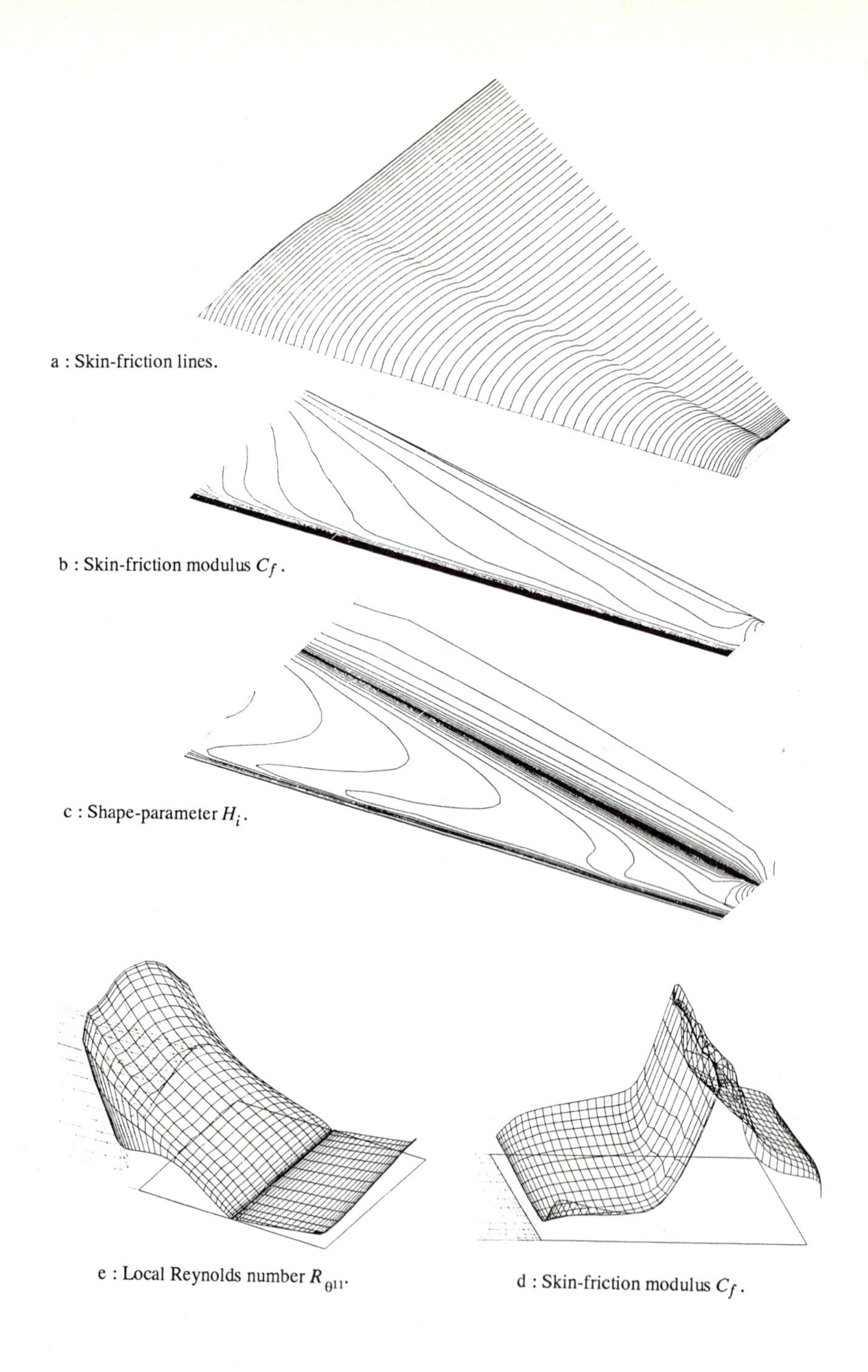

Fig. 5 : NACA0012 tapered swept wing with twisted tip, $\alpha = 11.5^o$, $\mathrm{Re}_c = 3.3\,10^6$, $M = .4$

Numerical Simulation of Turbulent Flows around Airfoil and Wing

Yves P. Marx *

NASA Langley Research Center

Hampton VA 23665-5225

Introduction

During the last years the simulation of compressible viscous flows has received much attention. While the numerical methods were improved drastically — discontinuous flows can now be well resolved with high-order, non-linear, upwind schemes [1]; the computing times have significantly been reduced by implicit or multigrid techniques [2] — a satisfactory modeling of the Reynolds stresses is still missing. Thus, even if with the present generation of supercomputers, the computation of flows around complete aircrafts has become possible [3,4], the computation of realistic separated flows remains unpractical on complex geometries. In this paper, after a short description of the numerical procedure used for solving the Reynolds equations, experiments with a promising simple turbulence model [5] are discussed.

Numerical Method

The three-dimensional Reynolds equations are solved with an upwind cell centered implicit scheme. The MUSCL approach is followed using the κ scheme with the Chakravarthy-Osher limiter [6] for the reconstruction of the flowfield from its cell averages values, and the Roe scheme [7] for the resolution of the discontinuties created at each cell iterfaces by the reconstruction scheme. Steady states computations are accelerated with the combination of the local time stepping procedure and a "diagonal" three factor upwind implicit operator. The numerical procedure is detailed in [8,9].

Turbulence Modeling

It is well known that equilibrium models such as the Baldwin-Lomax model, are not suited for separated flows for which the diffusion and the convection of the turbulence are no more negligible and introduce some imbalance between the production and the dissipation rate of turbulence. While retaining the eddy viscosity assumption, these non equilibrium effects can be taken into account by two-equation models, $K-\epsilon$, $K-\omega$, but it seems that despite their "universality" the two-equation models does not improved significantly the agreement between the computed results and the experimental data for separated flows [10]. A less ambitious approach is to modified two-layers mixing length models in order to extend their successes to separated flows. Such an approach was taken by Johnson and King [5] and the model they derived appeared to be adequate for the computation of separated flows on airfoils and wings [10] [11].

The idea behing the Johnson-King model is (i) to scale the turbulent velocity to the square root of the maximum Reynolds shear stress rather than to a length scale wall vorticity product; (ii) to compute the maximum Reynolds shear stress by solving a differential equation in which non-equilibrium effects are taken into account. As the level of the turbulent shear stress is then determined by the differential equation, the Johnson-King model in contrary to standard mixing length models, neither depends only on local mean flow gradients nor assumed a turbulence in equilibrium. The eddy viscosity distribution in the inner layer used with the Johnson-King model is thus

$$\nu_{ti} = K\eta D^2 u_M \tag{1}$$

where

$$u_M = \left(\frac{\tau_M}{\rho}\right)^{\frac{1}{2}}$$

*This work was done while the author held a National Research Concil-NASA Larc Research Associateship.

$$D = 1 - \exp(-\frac{\eta^+}{17})$$

$$\eta^+ = \frac{\eta \max(u_M, u_\tau)}{\nu}.$$

Here and below the index $_M$ indicates the location where the Reynolds shear stress is maximum.

In the original formulation of Johnson and King the outer eddy viscosity layer was based on the Cebeci-Smith distribution. This formulation was well suited in the boundary-layer context used by Johnson and King for the derivation of their model, but with Navier-Stokes codes, the Baldwin-Lomax formulation is more convenient. The outer eddy viscosity layer is therefore calculated using

$$\nu_{to} = \sigma 1.6\, \mathrm{F}_w \mathrm{Fk} \tag{2}$$

with

$$\begin{aligned}
\mathrm{F}_w &= \min\left[\eta_{\max}\mathrm{F}_{\max}, \frac{\eta_{\max}(|\breve{u}|_{\max} - |\breve{u}|_{\min})^2}{\mathrm{F}_{\max}}\right] \\
\mathrm{F}_{\max} &= \max_j (F(\eta))_{i,j,k} \;;\; F(\eta_{\max}) = \mathrm{F}_{\max} \\
F(\eta) &= \eta|\omega|D \\
\mathrm{Fk} &= 0.0168\left[1 + 5.5(\tfrac{\eta}{\delta})^6\right]^{-1} \\
\delta &= 1.9\,\eta_{\max}.
\end{aligned}$$

The coefficient σ is introduced to force the value of the maximum shear stress $\tilde{\tau}_M = \mu_t|\omega|$ to match the value τ_M obtained by the resolution of the differential equation (6). This coefficient can be computed by solving the equation

$$\tilde{\tau}_M(\sigma) - \tau_M = 0 \tag{3}$$

with a Newton method, or with a procedure proposed by Abid [12],

$$\begin{aligned}
\sigma^{t+\Delta t} &= \sigma^t \frac{\tau_M}{\tilde{\tau}_M(\sigma^t)} \\
\sigma^0 &= 1.
\end{aligned} \tag{4}$$

Knowing the values of ν_{ti} and ν_{to}, the actual value of the turbulent eddy viscosity ν_t is computed with

$$\nu_t = \nu_{to}(1 - \exp(-\frac{\nu_{ti}}{\nu_{to}})). \tag{5}$$

As stated, the level of the turbulent shear stress is obtained through the resolution of a differential equation, see [5] for its derivation. In the three-dimensional case, this equation is

$$\breve{u}_M \frac{\partial}{\partial \xi}(\frac{\tau_M}{\rho})^{-\frac{1}{2}} + \breve{w}_M \frac{\partial}{\partial \zeta}(\frac{\tau_M}{\rho})^{-\frac{1}{2}} = \frac{0.125}{L_M}\left[(1 - (\frac{\tau_M}{\tau_{eq}})^{-\frac{1}{2}}) + \frac{2L_M}{\delta(0.7 - \frac{\eta}{\delta})}\sqrt{(1 - \sigma^{\frac{1}{2}})^+}\right] \tag{6}$$

with

$$L_M = \min(K\eta_M, 0.225K\delta).$$

The left hand side of (6) represents the convection of the turbulent shear stress. The diffusion of the turbulent shear stress being modeled by the last term of the right hand side, the remaining term corresponds then to the imbalance between the production and dissipation of turbulence. This term is consequently approximated by the difference between the actual shear stress and the shear stress that would have been obtained if the turbulence was in equilibrium. This shear stress, τ_{eq}, is computed at $\eta = \eta_M$ using

$$\begin{aligned}
\tau_{eq} &= \mu_{t\,eq}|\omega| \\
\nu_{t\,eq} &= \nu_{to\,eq}\left(1 - \exp(-\frac{\nu_{ti\,eq}}{\nu_{to\,eq}})\right) \\
\nu_{ti\,eq} &= K\eta D^2 u_{eq} \\
\nu_{to\,eq} &= 1.6\,\mathrm{F}_w \mathrm{Fk}
\end{aligned} \tag{7}$$

with

$$D = 1 - \exp(-\frac{\eta u_{\text{eq}}}{17\nu})$$
$$u_{\text{eq}} = \max(\nu_{t\,\text{eq}}|\omega|, \nu|\omega|).$$

The convection terms of (6) are approximated by first-order upwind differences. The equation is then solved with a Point Alternate Symmetric Gauss-Seidel Relaxation with one relaxation performed in each direction. An alternative approach used in [13] and [11], is to add a time dependent term and to solve the equation in the same way as the Navier-Stokes equations.

Results and Discussions

The simulation of two-dimensional flows were first considered. In order to demonstrate the improvements resulting from the use of a non-equilibrium model for separated flows, comparisons between the results obtained with the Johnson-King model and a standard mixing length model [14] were performed. From figures (1,2) it appears clearly that whereas the shock location predicted by the Baldwin-Lomax model on the RAE2822 airfoil, $M_\infty = .75$, $\alpha = 2.81$, $\text{Re} = 6.5 \times 10^6$, is too much downstream of the experimental data of Cook et al. [15], a correct shock location is found with the Johnson-King model. A grid refinement study was performed, fig.(3,4), to establish that on the 257x65 mesh, an almost grid independant solution has been obtained. This grid refinement study shows however, that the Johnson-King model is more sensitive to the grid density than the Baldwin-Lomax model for which, satisfactory results can be obtained on a 81x17 mesh, fig.(5). This sensitivity is in a large part due to the initialization of the maximum shear stresses. The "solution" (unsteady) shown in figure (6) was for instance obtained in the following manner : the Reynolds equations with the Baldwin-Lomax model were first solved on a 81x17 mesh; this solution was then interpolated on the 161x33 mesh and served as the initial condition for this finer mesh; fifty iterations were then performed on the 161x33 mesh with the Baldwin-Lomax model before the original Johnson-King model was turned on. If more iterations had been performed on the 161x33 mesh with the Baldwin-Lomax model, steady but wrong solutions would have been obtained as well. The solution shown in figure (3) was computed by taking as the initial condition, the converged solution obtained with the Baldwin-Lomax model on the 161x33 mesh. In order to enforce a unique solution — independant of the initial condition — the velocity scale used in the inner layer of the equilibrium eddy viscosity, $\nu_{ti\,\text{eq}}$, has to be replaced. Instead of using

$$u_{\text{eq}} = \nu_{t\,\text{eq}}|\omega|$$

as in the original Johnson-King model,

$$u_{\text{eq}} = \max(\nu_{t\,\text{eq}}|\omega|,\ \nu|\omega|)$$

must be employed. This fix was yet not sufficient in the computation on the ONERA M6 wing shown below. The level of the starting maximun shear stress was still too low and in consequence, the shock location was moving upstream without any bound. The fix was then to replace in the computation of $\nu_{ti\,\text{eq}}$, the inner layer formulation of Johnson and King by the Baldwin and Lomax formulation,

$$\nu_{ti\,\text{eq}} = K^2\eta^2 D^2|\omega|.$$

Unfortunately with this inner formulation, the pressure recovery obtained on the RAE2822 airfoil is not predicted as well as with the original formulation and a pressure bump is found, fig.(7). This bump has also been observed by Radespiel [13] and Swanson [16], with Jameson type scheme and with the "original" formulation of Johnson and King. The extreme sensitivity of the Johnson-King model shows up also on the convergence of the method. From figure (8), it is apparent that the lift converges with difficulties, the clear decay of the oscillations nevertheless indicates, that if enough iterations are performed, a steady solution can be expected. It should also be pointed that the computation of σ proposed by Abid, equation (4), introduces some time dependency in the solution. The steady state will slightly depend on the time integration path and depending whether the turbulence quantities were updated at every iterations or only at every five iterations, different convergence histories were found.

As already noticed by Coakley [10], we also observe that whereas the Johnson-King model improves significantly the prediction of separated flows, attached flows are not as well resolved with the non-equilibrium model as with the Baldwin-Lomax model, fig.(9). The reason again lies in the inner layer eddy viscosity formulation of Johnson and King. Recently, Johnson and Coakley [17] have proposed a new formulation which consists of a non-linear blending of the Johnson-King and Baldwin-Lomax formulations. This new law seems to ameliorate the results, but the added non-linearity increases also the convergence problems.

The computations done in three-dimensions on the ONERA M6 Wing, confirm the behavior of the Johnson-King model already observed in two-dimensions. For an attached flow case $M_\infty = .84$, $\alpha = 3.06$, Re $= 11 \times 10^6$, even if the Johnson-King model predicts a shock position slightly upstream of the position computed with the Baldwin-Lomax model, the solutions obtained with both models agree well with the experiments of Schmitt and Charpin [18], fig.(10). For a higher angle of attack $\alpha = 6.06$, a separated region forms on the wing, fig.(11). As non-equilibrium effects are then important, a proper pressure distribution on the wing cannot be found with the Baldwin-Lomax model, fig.(12). On the contrary, with the Johnson-King model a good agreement with the experimental data is found, fig.(12). The pressure plateau, typical of separated flows, is for instance well predicted. No experimental visualizations of the wall streamlines were available on the ONERA M6 wing, but both the computed wall streamlines, fig.(11), and pressure contours, fig.(13), are closed to the one obtained by Abid et al. [11].

Conclusion

Computations with a non-equilibrium model have been performed and the results compared to the results obtained with the equilibrium model of Baldwin and Lomax. The comparison shows that whereas the non-equilibrium model improves significantly the computed solutions for separated flows, for attached flows, the solutions obtained with the Baldwin-Lomax model are more accurate than with the non-equilibrium model. An excessive sensitivity of the non-equilibrium model has also been observed. Therefore, some changes in the non-equilibrium model in order to improve the accuracy of the solutions for attached flows *and* to reduce the sensitivity of the model, have to be made before engineering computations can be done routinely with the non-equilibrium model of Johnson and King.

Acknowledgement

The author is much obliged to Doctors B. Wedan for providing the grids, R. Abid for his help in the implementation of the Johnson-King model, and V. Vatsa for the numerous discussions concerning the computations.

References

[1] A. Harten. On the nonlinearity of modern shock-capturing schemes. *ICASE Report*, 178206, 1986.

[2] A. Jameson. Computational transonics. *Comm. Pure Appl. Math.*, XLI:507–549, 1988.

[3] A. Jameson, T. J. Baker, and N. P. Weatherill. Calculation of inviscid transonic flow over a complete aircraft. *AIAA Paper*, 86-0103, 1986.

[4] J. L. Thomas, R. W. Walters, T. Reu, F. Ghaffari, R. P. Weston, and J. M. Luckring. A patched-grid algorithm for complex configurations directed towards the F/A-18 aircraft. *AIAA Paper*, 89-0121, 1989.

[5] D. A. Johnson and L. S. King. A mathematically simple turbulence closure model for attached and separated turbulent boundary layers. *AIAA Journal*, 23:1684–1692, 1985.

[6] S. Chakravarthy and S. Osher. A new class of high accuracy TVD schemes for hyperbolic conservation laws. *AIAA Paper*, 85-0363, 1985.

[7] P. L. Roe. Approximate Riemann solvers, parameter vectors, and difference schemes. *Journal of Computational Physics*, 43:357–372, 1981.

[8] Y. P. Marx. Computation of turbulent flow on a CAST 10 wing using an upwind scheme. *AIAA Paper*, 89-1836, 1989.

[9] Y. P. Marx. Numerical solution of 3D Navier-Stokes equations with upwind implicit schemes. *NASA Technical Memorandum*, to be published, 1989.

[10] T. J. Coakley. Numerical simulation of viscous transonic airfoil flows. *AIAA Paper*, 87-0416, 1987.

[11] R. Abid, V. N. Vatsa, D.A. Johnson, and B.W. Wedan. Prediction of separated transonic wing flows with a non-equilibrium algebraic model. *AIAA Paper*, 89-0558, 1989.

[12] R. Abid. Extension of the Johnson-King turbulence model to the 3-D flows. *AIAA Paper*, 88-0223, 1988.

[13] R. Radespiel. A cell-vertex multigrid method for the Navier-Stokes equations. *NASA Technical Memorandum*, 101557, 1989.

[14] B. Baldwin and H. Lomax. Thin-layer approximation and algebraic model for separated turbulent flows. *AIAA Paper*, 78-257, 1978.

[15] P. H. Cook, M. A. McDonald, and M. C. P. Firmin. Aerofoil RAE 2822 - pressure distributions and boundary layer and wake measurements. *AGARD-AR*, 138, 1979.

[16] R. C. Swanson. Personal communication. 1989.

[17] D. A. Johnson and T. J. Coakley. Improvements to a nonequilibrium algebraic turbulence model. *AIAA Journal*, to be published, 1989.

[18] V. Schmitt and F. Charpin. Pressure distributions on the ONERA-M6 wing at transonic mach numbers. *AGARD-AR*, 138, 1979.

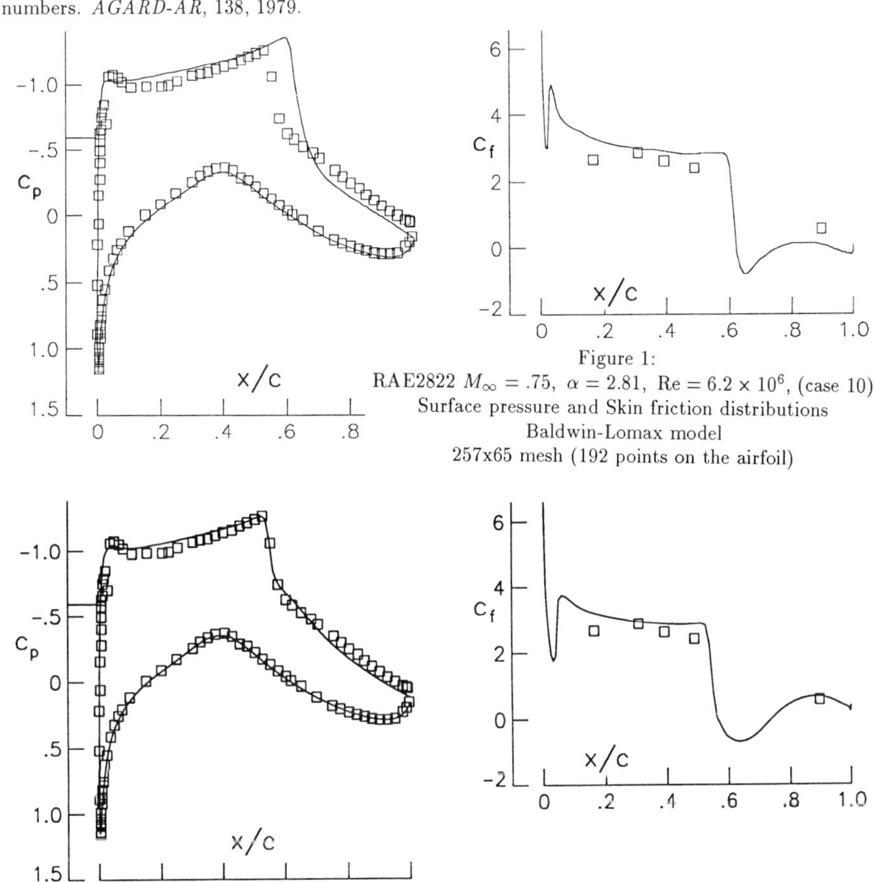

Figure 1:
RAE2822 $M_\infty = .75$, $\alpha = 2.81$, $\mathrm{Re} = 6.2 \times 10^6$, (case 10)
Surface pressure and Skin friction distributions
Baldwin-Lomax model
257x65 mesh (192 points on the airfoil)

Figure 2:
RAE2822 (case 10)
Surface pressure and Skin friction distributions
Johnson-King model
257x65 mesh (192)

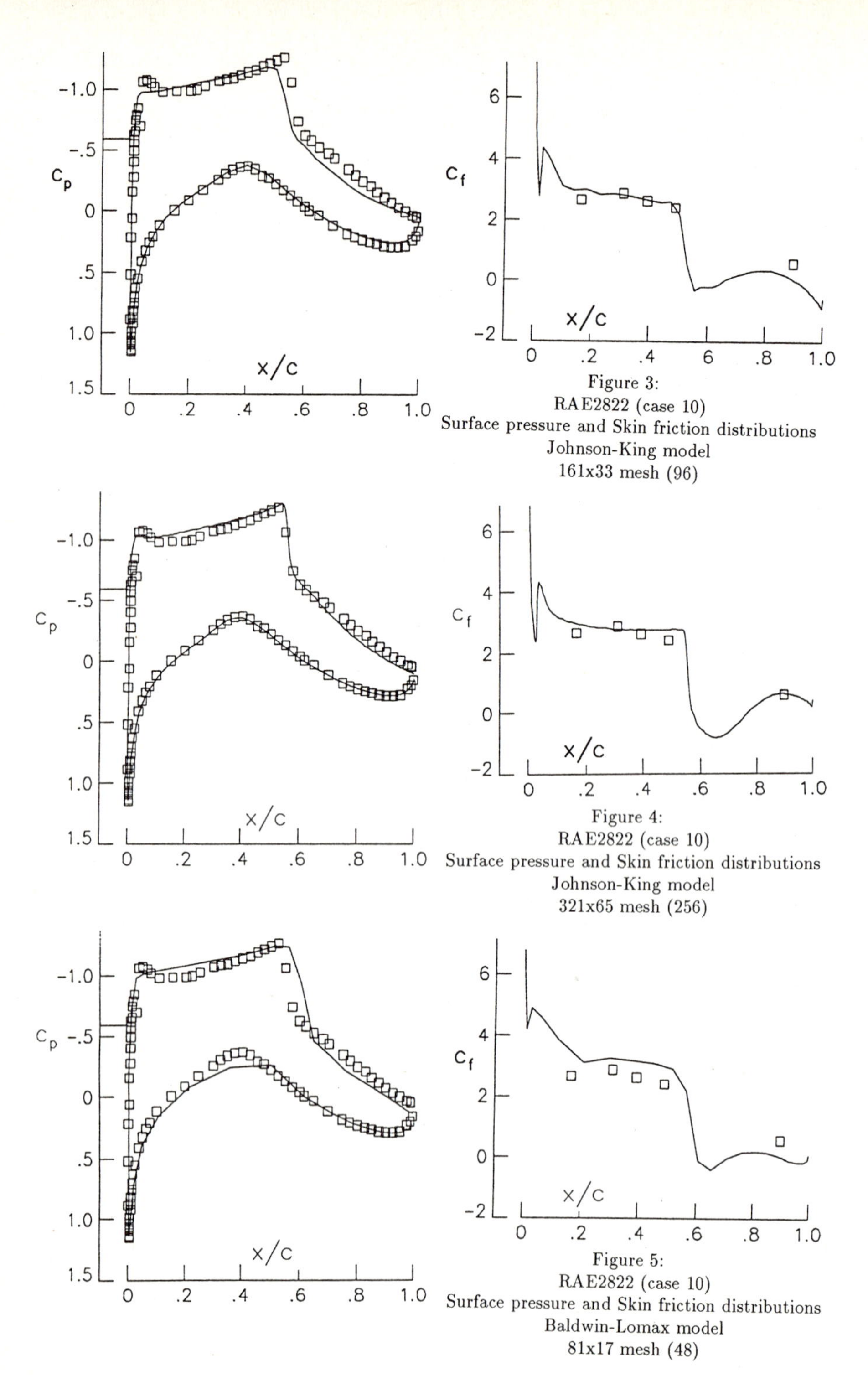

Figure 3:
RAE2822 (case 10)
Surface pressure and Skin friction distributions
Johnson-King model
161x33 mesh (96)

Figure 4:
RAE2822 (case 10)
Surface pressure and Skin friction distributions
Johnson-King model
321x65 mesh (256)

Figure 5:
RAE2822 (case 10)
Surface pressure and Skin friction distributions
Baldwin-Lomax model
81x17 mesh (48)

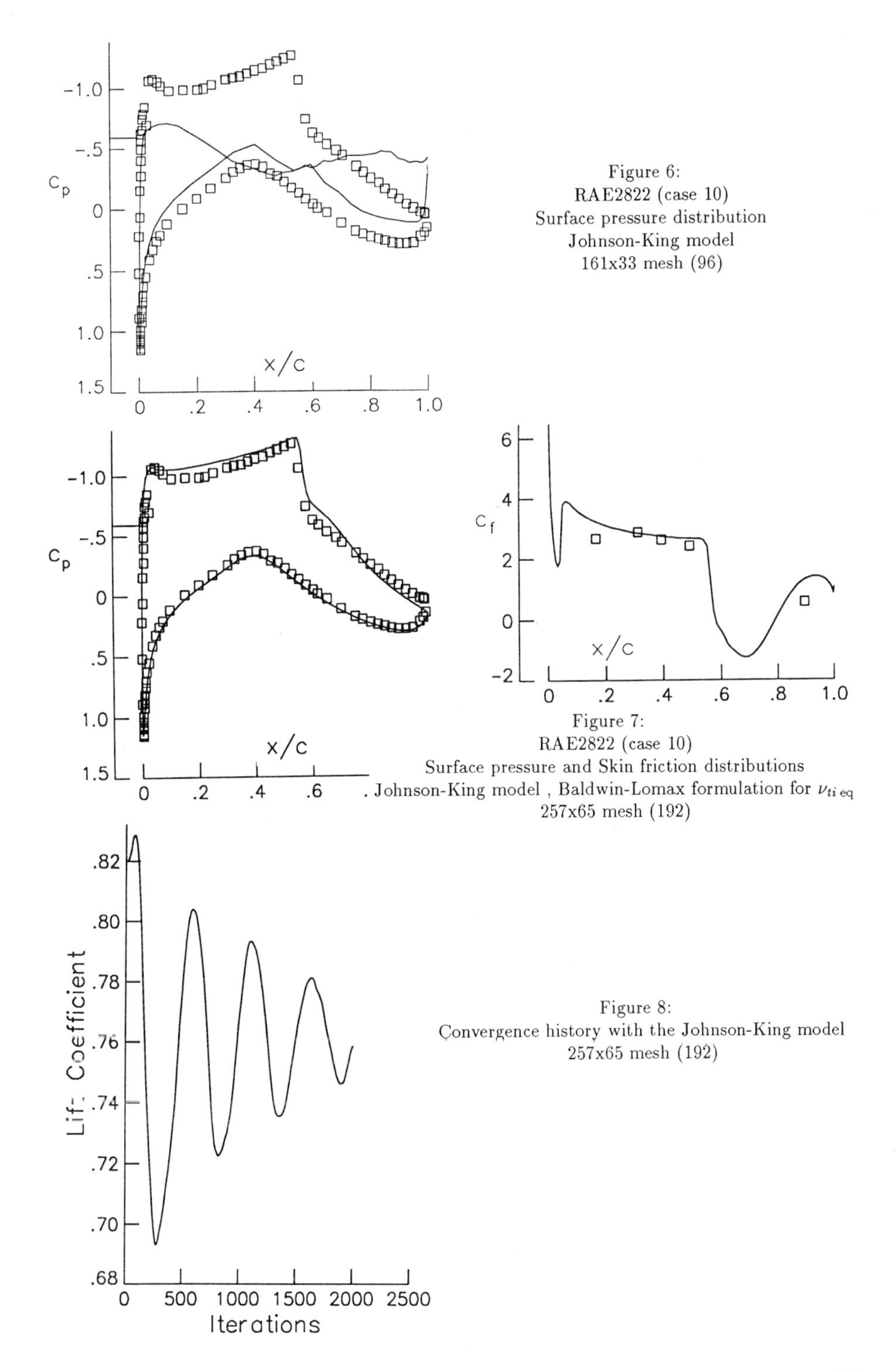

Figure 6:
RAE2822 (case 10)
Surface pressure distribution
Johnson-King model
161x33 mesh (96)

Figure 7:
RAE2822 (case 10)
Surface pressure and Skin friction distributions
Johnson-King model , Baldwin-Lomax formulation for $\nu_{ti\,eq}$
257x65 mesh (192)

Figure 8:
Convergence history with the Johnson-King model
257x65 mesh (192)

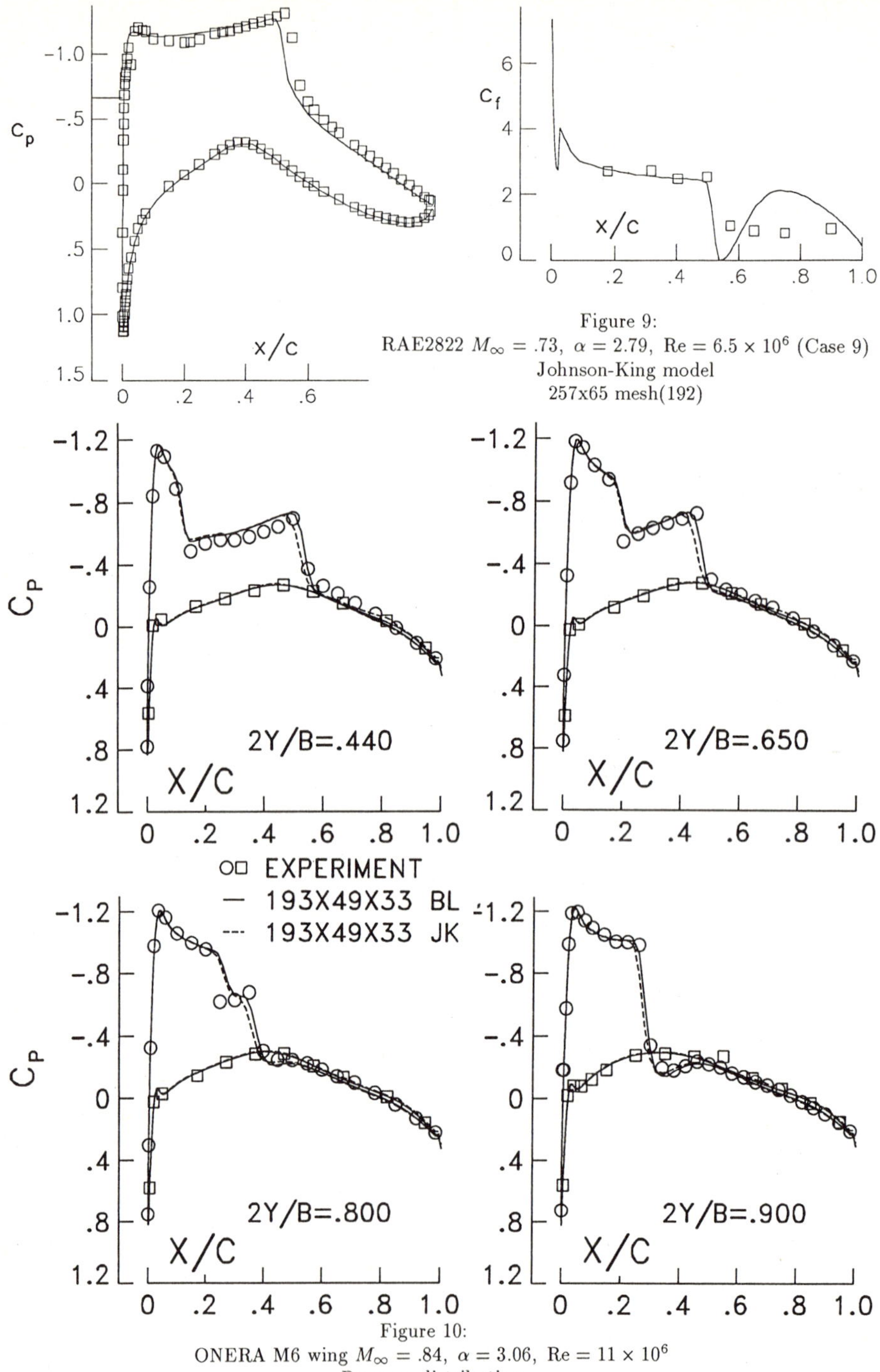

Figure 9:
RAE2822 $M_\infty = .73$, $\alpha = 2.79$, $\mathrm{Re} = 6.5 \times 10^6$ (Case 9)
Johnson-King model
257x65 mesh(192)

Figure 10:
ONERA M6 wing $M_\infty = .84$, $\alpha = 3.06$, $\mathrm{Re} = 11 \times 10^6$
Pressure distribution
Baldwin-Lomax and Johnson-King models

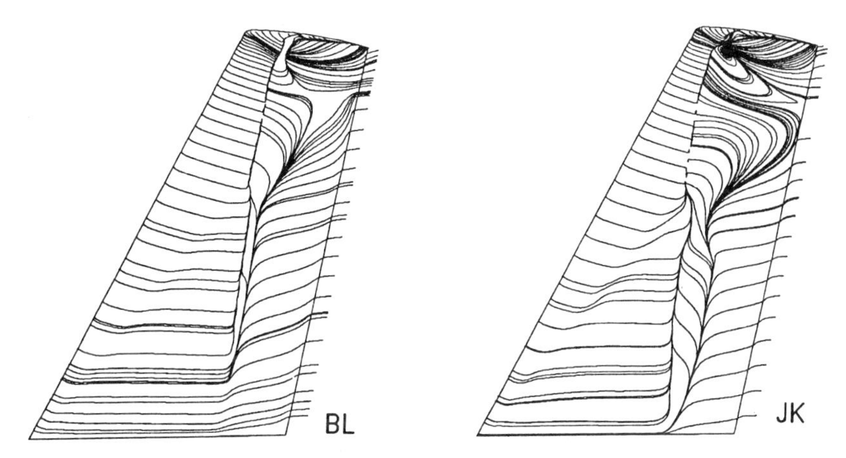

Figure 11:
ONERA M6 wing $M_\infty = .84$, $\alpha = 6.06$, $\text{Re} = 11 \times 10^6$
Wall streamlines
Baldwin-Lomax and Johnson-King models

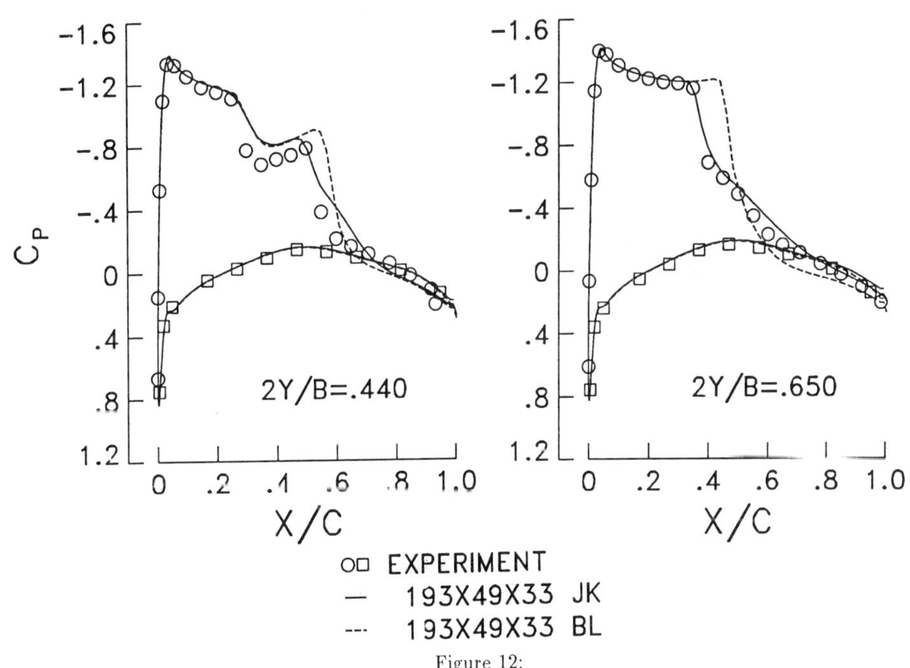

Figure 12:
ONERA M6 wing $M_\infty = .84$, $\alpha = 6.06$, $\text{Re} = 11 \times 10^6$
Pressure distribution
Baldwin-Lomax and Johnson-King models

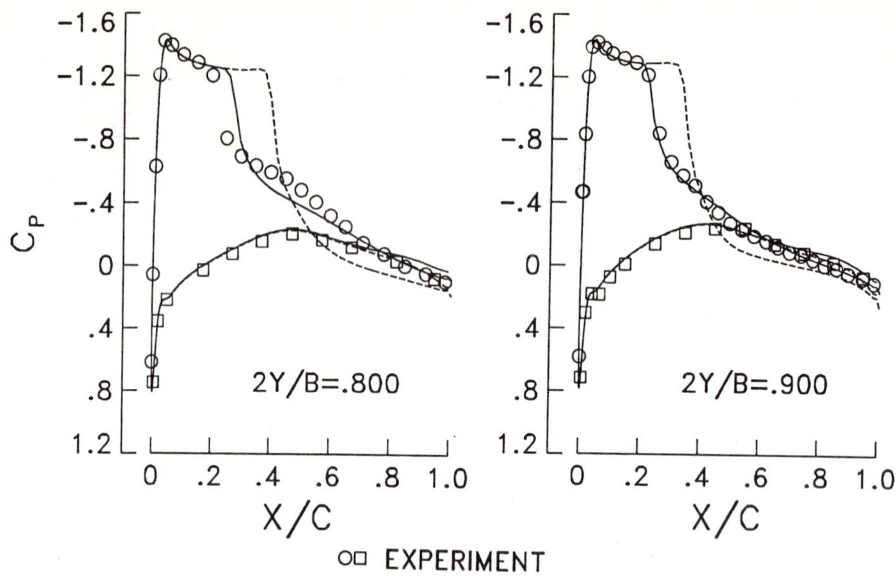

Figure 12:
ONERA M6 wing $M_\infty = .84$, $\alpha = 6.06$, $\mathrm{Re} = 11 \times 10^6$
Pressure distribution
Baldwin-Lomax and Johnson-King models

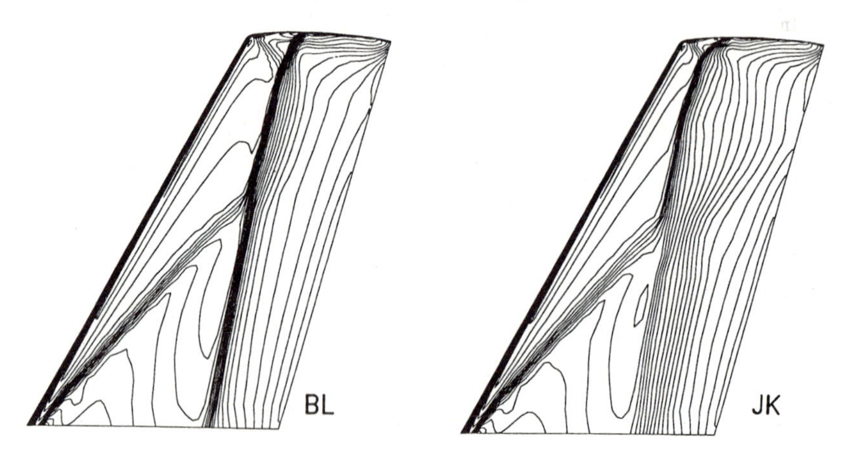

Figure 13:
ONERA M6 wing $M_\infty = .84$, $\alpha = 6.06$, $\mathrm{Re} = 11 \times 10^6$
Pressure contours
Baldwin-Lomax and Johnson-King models

A POSTERIORI ERROR ESTIMATORS FOR ADAPTIVE SPECTRAL ELEMENT TECHNIQUES

CATHERINE MAVRIPLIS
Program in Applied and Computational Mathematics
207 Fine Hall, Princeton University, NJ 08544, USA

SUMMARY

Aspects of adaptive spectral element methods are presented with emphasis on the *a posteriori* error estimators used in the automatic mesh refinement process. The nonconforming formulation of the method is reviewed in an effort to illustrate the various mesh refinement options available. Single mesh *a posteriori* error estimators are developed for the spectral element method. These estimate the actual error incurred by the discretization on a local per element basis and predict the convergence behaviour of the numerical solution as well. As a result the error estimators serve as criteria in the choice of refinement options. The usefulness of the estimators is demonstrated through examples of Navier-Stokes calculations.

INTRODUCTION

The goal of this ongoing work is to provide the spectral element method with adaptive mesh capabilities for flexibility and efficiency, in order to enlarge the scope of problems that can be solved with this method. Spectral element methods are highly accurate weighted residual techniques for the numerical solution of partial differential equations. As their high convergence rate depends on smoothness of the solution, it follows that these methods are appropriate for a class of problems where high order regularity is common. They are currently used for the direct simulation of low Reynolds number ($0 \sim 5000$) incompressible fluid flow.

In the spectral element discretization, first introduced by Patera [1], the computational domain is broken up into K macro-elements, on each of which the independent and dependent variables are approximated by Nth (high) order tensor product polynomial expansions. Although the spectral element method is, by construction, general enough to handle complex geometries [2], there remain some limitations which could be restrictive to the widespread use of the method. Primarily, the limitation of the spectral element method is in flexibility and generality as regards complex geometry mesh generation and locally refined resolution capabilities, as the large brick elements are somewhat impractical. Furthermore, these limitations severely hinder any developments in the areas of automatic mesh generation, adaptive mesh refinement and the treatment of moving boundaries. In this paper we address these limitations with the goal of developing an adaptive spectral element method.

In a first part, we briefly review the standard conforming spectral element formulation and introduce the nonconforming discretization. While the nonconforming discretization is an integral part of the adaptive method, it will only be briefly described as it is the subject of a more detailed paper (see [3,4]). Nonconforming methods have been developed for $h-$type finite element methods [5,6], however they are less vital to the success of the finite element method as the flexibility of the conforming discretization can compensate for irregular geometries and functional discontinuities due to the local quality of the method. In contrast, $p-$type spectral methods are severely limited in scope due to the global quality of the method. Spectral element methods may therefore exploit and maximize the competitive advantages of both of these techniques. In particular, these methods can benefit from the choice of $h-$ and $p-$refinement procedures.

The result is that the method can accept nonconforming mesh geometries of fairly arbitrary type, an example of which is shown in Figure 1b. To arrive at such a mesh, however, it is not obvious *a priori* how to proceed. For this objective, we develop, in this paper, the second key feature in the development of adaptive meshes: error estimators to serve as criteria for refinement decisions.

The first role of the error estimators is, of course, to provide an estimate of the actual error on a per element basis as well as globally. Comparison of the elemental error estimates provides criteria as to which elements must be further refined. The estimate must be efficiently calculated so that it may be used as a postprocessing step as well as in the course of the calculation so that time-dependent mesh refinement can proceed. For this purpose we develop single mesh *a posteriori* error estimates. The use of a single mesh is in contrast to currently used error estimators in high [7] and low order methods [8], as multiple mesh calculations would be too expensive for the spectral element formulation. However our estimators are similar to those of Babuska and Dorr [7] as they are both calculated *a posteriori* and use the decay rate to predict convergence behaviour.

The emphasis in this paper is on the theoretical formulation and implementation of the error estimators for adaptive mesh refinement techniques. The last section provides some illustrations of the validity and application of the error estimators for Navier-Stokes calculations.

DISCRETIZATION

The objective is to solve the incompressible Navier-Stokes equations, stated as:

$$\frac{\partial \vec{u}}{\partial t} + \vec{u} \cdot \nabla \vec{u} = -\nabla p + \frac{1}{R} \nabla^2 \vec{u} + \vec{f} \tag{1a}$$

$$\nabla \cdot \vec{u} = 0, \tag{1b}$$

where $\vec{u}$ is the velocity vector, p is the pressure and R is the Reynolds number. These equations are advanced in time using an explicit/implict fractional time-stepping scheme [9], which results in a set of separate equations for p and $\vec{u}$. The nonlinear terms are treated explicitly as a simple inhomogeneity in the elliptic equation set. The discussion will therefore concentrate on a formulation for the solution of elliptic problems; the extension to Stokes and Navier-Stokes problems is detailed in [10,4]. We consider the two-dimensional Poisson equation on a domain $\Omega \subset \mathbf{R}^2$, with homogeneous boundary conditions: Find $u \in \mathbf{R}$ such that

$$-\nabla^2 u = f \quad \text{in } \Omega, \quad u = 0 \quad \text{on } \partial\Omega.$$

Conforming Spectral Element Method

The variational statement for the Poisson problem is: Find $u \in X = \mathcal{H}_0^1(\Omega)$ (the standard Sobolev space of functions satisfying homogeneous boundary conditions) such that

$$\int_\Omega \nabla u \cdot \nabla v \, d\vec{x} = \int_\Omega f v \, d\vec{x} \quad \forall v \in X. \tag{2}$$

The spectral element discretization proceeds by breaking up the domain Ω into K (for the moment) rectilinear elements Ω^k in (x, y). The domain decomposition has the constraint that the intersection of two adjacent elements is either an entire edge or a vertex. Relaxing this constraint is the subject of the next section.

We then require that the variational statement (2) be satisfied for a polynomial subspace X_h of X defined on the Ω^k. This is a tensor product space of all polynomials of degree less than or equal to N with respect to each space variable x, y on each subdomain Ω^k. Using an affine mapping from (x^k, y^k) to $(r, s) \in [-1; 1]^2$ on each element k, the jacobian of which is denoted by J^k and performing (tensor product) GLL $\times$ GLL quadrature in (r, s), the disrete problem becomes

$$\sum_{k=1}^{K} {}'|J^k|\rho_i\rho_j \nabla u_h(r_i, s_j) \cdot \nabla v_h(r_i, s_j) = \sum_{k=1}^{K} {}'|J^k| f(r_i, s_j) v_h(r_i, s_j) \quad \forall v_h \in X_h \tag{3}$$

where r_i, s_j are the Gauss-Lobatto collocation points in the x, y directions respectively, ρ_i, ρ_j the Gauss-Lobatto weights for $i, j = 0, N$. $\sum'$ denotes direct stiffness summation between elements. It has been shown by Maday and Patera [10] that such an implementation converges exponentially fast to the exact solution for fixed $K, N \to \infty$ for smooth data and solution.

To complete the discretization, we choose a basis for $u_h \in X_h$ to be

$$u_h^k(x, y) = u_{ij}^k h_i(r) h_j(s) \quad x, y \Longrightarrow r, s \tag{4}$$

where the h_i, h_j are the one-dimensional Lagrangian interpolants defined in [10] and $u_{ij}^k = u_h^k(r_i, s_j)$, and where we make use of tensor product forms. We require the polynomials to be $\mathcal{C}^0$ continuous across elemental boundaries and enforce homogeneous Dirichlet boundary conditions for $u_h \in X_h$. Choosing v_h to be nonzero at only one global collocation point, the discrete formulation (3) becomes

$$A\underline{u} = B\underline{f} \tag{5}$$

where A, B are the global multi-dimensional K element Laplacian and mass matrices. The resulting matrix system is solved by conjugate gradient iteration. A typical conforming spectral element mesh for flow past a 60° wedge is shown in Fig. 1a. This mesh has $K = 90$ elements, shown here, and $(N+1) \times (N+1)$ collocation points per element, not shown here.

Nonconforming Spectral Element Method

A nonconforming discretization is one in which the discretization space X_h is not a subset of the proper space X, in which the continuous problem is well posed. An additional consistency error associated with the deviation of X_h from X is therefore incurred. Unlike other nonconforming methods (see [5,6]), this method is based on the explicit construction of the appropriate nonconforming space of approximation to minimize the consistency error. Here, consistency errors are commensurate with the approximation errors, thereby preserving the convergence properties of the spectral element discretization.

To define the approximation space X_h, we first introduce a new structure known as the set of mortars γ^p, which are defined as the intersection of adjacent element edges. Upon this structure, ϕ is defined as the mortar function, which is a polynomial of degree N in the local one-dimensional mortar variable s. The approximation space X_h consists of the functions v in $\mathcal{L}^2(\Omega)$ that are tensor products of polynomials of degree N in each direction of each element k, such that the two following conditions are satisfied:

1 the vertex condition: at each vertex q of each element Ω^k, $v_{|\Omega^k}(q) = \phi(q)$.

2- the $\mathcal{L}^2$ condition: over each elemental edge, $\forall \psi \in P_{N-2} \quad \int_{edge} (v - \phi)\psi ds = 0$.

For a conforming approximation this definition of X_h reduces to the standard spectral element approximation space. The vertex condition ensures exact continuity at cross points while the $\mathcal{L}^2$ condition represents a $\mathcal{L}^2$ minimization of the jump in functions at internal boundaries. The combination of these two conditions ensures the optimality of the discretization, as explained and illustrated in [4].

Using the same choice of basis as in (4) for the functions v and performing the inner products by Gauss-Lobatto quadrature as in (3), the discrete matrix equation set (5) becomes

$$Q^T A Q \underline{u} = Q^T \underline{f}.$$

In practice, the Laplacian operations are performed entirely locally at the elemental level with all transmission coupling through the Q, Q^T operations.

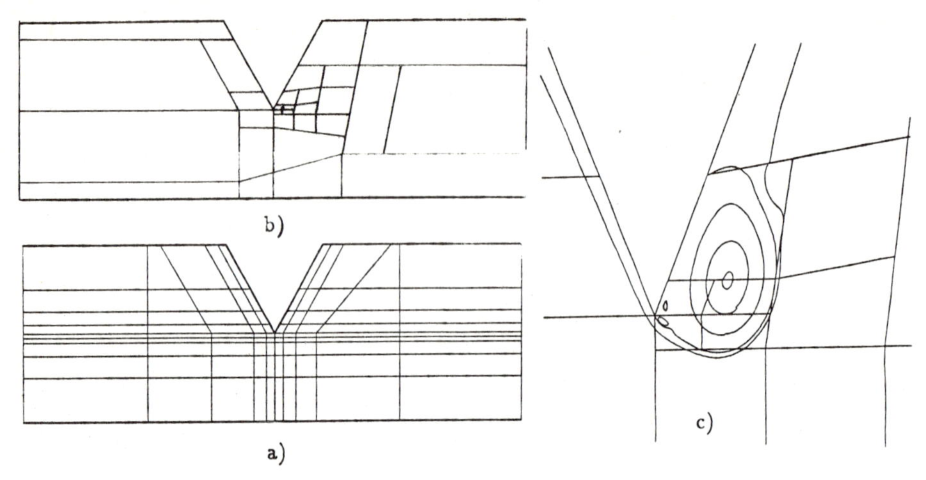

Figure 1: Conforming (a) and nonconforming (b) meshes for the wedge calculation and instantaneous streamlines (c) near the wedge tip based on the nonconforming mesh at t=5s

To illustrate the validity and advantages of the nonconforming formulation, we propose to calculate the startup flow past a 60° wedge-like sharp edge at a Reynolds number of 1560, with both conforming and nonconforming discretizations. The nonconforming mesh ($K = 29$ elements) corresponding to the conforming mesh ($K = 90$ elements) of Figure 1a for this problem is shown in Figure 1b. Both calculations were performed with $N = 6$. In both cases a large number of elements have been placed around the tip of the wedge where the vortex forms. In the conforming case (Fig. 1a) such clustering necessitates a large number of elements in the rest of the field as well. In contrast, the nonconforming approach (Fig. 1b) can afford more flexibility, yielding a globally irregular structure, keeping, however, the same degree of refinement near the wedge tip. The savings in computational effort are significant as they are proportional to the reduction in number of elements.

Instantaneous streamline patterns near the wedge tip at $t = 5s$ are shown in Fig. 1c for the nonconforming calculation. Qualitative agreement with the flow visualization results of [11] and the conforming calculation [9] is good, demonstrating the usefulness of the nonconforming formulation. For a lengthier discussion see [12]. The question is then how to arrive at such nonconforming meshes automatically and how to achieve optimal meshes. For this we turn to error estimators.

ERROR ESTIMATORS

In this section we develop single-mesh "*a posteriori*" error estimators for the spectral element solution u_h to a general problem $G(u) = f$. Knowledge of the error, or at least of its behaviour, is helpful in conducting large scale calculations in complex geometries, but absolutely essential in conducting such calculations efficiently, which includes the use of adaptivity.

Adaptivity is common and effective in low order finite difference and finite element methods. Furthermore, adaptivity is very flexible for low order methods due to the low order coupling throughout the domain. However, the local quality is also restrictive for error estimation: it often requires calculation of multiple solutions on finer and finer meshes to estimate the decay rate or convergence of the solution to estimate the actual error e. g. [8].

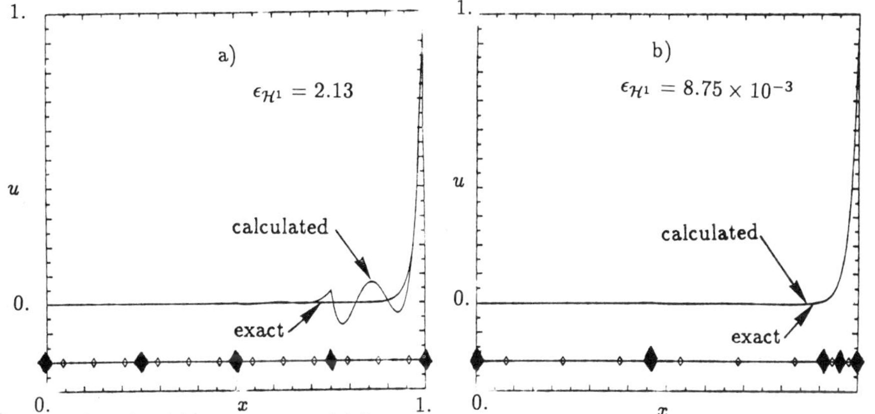

Figure 2: Regular (a) and optimized (b) mesh solutions for the 1D boundary layer problem

High order methods, on the other hand, can make use of the fact that high order polynomials are used on each element/subdomain to model solutions. High order methods thus have a **locally global** quality which provides sufficient information to estimate errors without extensive additional calculation. In $p-$type finite element methods, *a posteriori* error estimates have been developed in the same spirit as those presented here. Babuska and others [9] calculate error estimates based on increased order calculations.

The proposed error estimators rely on the calculation and extrapolation of the spectrum to estimate the error as well as predict convergence rates. They are therefore **single mesh** *a posteriori* error estimators in contrast to those of Babuska *et al.* Furthermore, there exists a spectrum for each element: we therefore obtain **local** per element estimates and convergence properties independent of each other.

As a motivation to develop adaptive meshes, consider a simple one-dimensional example in which error estimates are used in conjunction with an intuitive refinement scheme to demonstrate the power of adaptivity. Consider the problem

$$-u_{xx} + \lambda u_x = 0 \quad \text{on } \Omega =]0;1[, \quad u(0) = 0, \; u(1) = 1.$$

which exhibits a boundary layer near $x = 1$ for large λ. A naïve spectral element discretization using equal-sized elements) yields a poor solution for $\lambda = 50$ as shown in Figure 2a ($K = 4$, $N - 4$) The grid points are indicated by the symbols, the solid symbols representing element boundaries.

The quality of the solution can be drastically improved by a simple dynamical rearrangement of the K (fixed) elements We compare elemental error estimates, taken here to be the Nth order term of the Legendre expansion of u_h, a_N^k, and change element sizes in consequence. The procedure is repeated until the difference in error estimates across all the elements is less than a certain tolerance level. For the same resolution, the overall actual error in the $\mathcal{H}^1$ norm is reduced by two orders of magnitude as illustrated in Fig. 2b.

Though the above scheme is simple and effective in one dimension, in higher dimensions and complex geometries, repositioning of the element boundaries becomes a more complex task. Increasing the order of interpolation (p-refinement), repositioning of element boundaries as well as increasing the number of elements (h-refinement) must all be used in conjunction, in a compatible manner to efficiently refine the calculation. For this purpose, we must provide more detailed information than a simple estimate of the error, namely a "smart" error estimator.

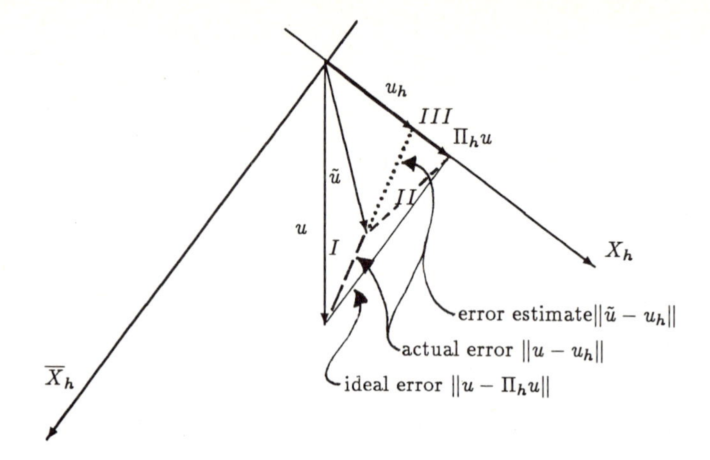

Figure 3: Geometric representation of the approximation error contributions

Derivation

A property of the spectral element discretization is that the error between the exact solution (u) and spectral element approximation (u_h) is at most a constant away from the difference between the solution and the best polynomial approximation to the solution, which can be expressed as [10]

$$||u - u_h|| < C \inf_{v_h \in X_h} ||u - v_h||. \tag{6}$$

The best polynomial approximation to the function u is the projection $\Pi_h u$ of u onto X_h, the function which minimizes the error, as can be seen in Fig. 3. The actual error incurred by the spectral element approximation $||u - u_h||$ is thus bounded by the sum of two contributions: $||u - \Pi_h u||$ which represents the difference between the actual solution and the best polynomial approximation and $||u_h - \Pi_h u||$ which represents the fact that we never achieve the exact projection of the actual solution. For the $\mathcal{H}^1$ norm the constant C in Inequality (6) is close to 1, which implies that the first term dominates in the actual error.

Since u is not known, in order to estimate these contributions we extrapolate u_h to $\tilde{u}$. The approach taken is to make use of the large amount of information available in the truncated sum representation of the solution in this locally global high order method: extrapolation of the coefficients of the sum representation provide a reasonably close estimate of u.

The error between the exact and numerical solutions is therefore bounded by

$$||u - u_h|| \leq ||u - \tilde{u}|| + ||\tilde{u} - \Pi_h \tilde{u}|| + ||u_h - \Pi_h \tilde{u}||. \tag{7}$$

The three contributions to the upper bound (7) are I- the extrapolation error $||u - \tilde{u}||$, II- the approximation error due to truncation $||\tilde{u} - \Pi_h \tilde{u}||$ and III- the approximation error due essentially to quadrature and the fact that one can only get to within a constant of the best polynomial approximation (Ineq. 6), $||u_h - \Pi_h \tilde{u}||$. They are shown geometrically in Figure 3. Assuming the extrapolation error to be small, provided the extrapolation is good, we propose an error estimate consisting of the sum

$$\epsilon_{est} = ||\tilde{u} - \Pi_h \tilde{u}|| + ||u_h - \Pi_h \tilde{u}||.$$

Consider first the error due to truncation only, ($||\tilde{u} - \Pi_h \tilde{u}||$). For this purpose, we assume that $\tilde{u} = u$ is the exact solution sum. In one dimension, the numerical solution is expressed as a Legendre polynomials expansion with coefficients a_n^k, $n = 0, N$ and hence the pointwise error on a per element basis is the equivalent infinite sum starting from $N + 1$. The coefficients a_n^k of the spectrum are easily calculated calculated by numerical quadrature using the orthogonality relation of the Legendre polynomials. It then follows that an estimate of the approximation error due to truncation (in the $\mathcal{L}^2$ norm) is written

$$\|\tilde{u} - \Pi_h \tilde{u}\| = \left(\sum_{n=N+1}^{\infty} \frac{\tilde{a}_n^2}{\frac{2n+1}{2}} \right)^{\frac{1}{2}}, \tag{8}$$

where the $\tilde{a}_n$ are the extrapolated coefficients (for $n > N$). The superscript has been dropped as we assume all calculations are on a per element basis. Derivative coefficients are used for the $\mathcal{H}^1$ norm error estimate.

Second, we consider only the approximation error due to quadrature and the fact that the solution is a constant away from the best polynomial fit, henceforth denoted as the error due to "quadrature". In the numerical solution, the coefficients a_n for $n = 0, N$ are not exact since they are but numerical approximates. Denoting the exact solution as before with the a_n coefficients, the discretized numerical solution similarly with coefficients b_n, this contribution is reduced to

$$\|u_h - \Pi_h \tilde{u}\| = \left(\sum_{n=0}^{N} \frac{(b_n - a_n)^2}{\frac{2n+1}{2}} \right)^{\frac{1}{2}} \simeq \left(\frac{b_N^2}{\frac{2N+1}{2}} \right)^{\frac{1}{2}}. \tag{9}$$

Summing these two contributions, we propose to use as a rigorous error estimate an approximation to the two contributions in (8,9), namely

$$\epsilon_{est} = \left(\frac{b_N^2}{\frac{2N+1}{2}} + \sum_{n=N+1}^{\infty} \frac{\tilde{b}_n^2}{\frac{2n+1}{2}} \right)^{\frac{1}{2}}$$

where the coefficients b refer to the coefficients of the numerical solution for the $\mathcal{L}^2$ error norm or those of the derivative for the $\mathcal{H}^1$ norm. Details of the two-dimensional case are given in [4].

We now detail the method of extrapolation of the spectrum. A typical spectrum is shown in Figure 4a. Using a four point least squares best fit to $b(n)$, we solve for c and σ in the exponential decay approximation,

$$b(n) \sim c e^{-\sigma n}.$$

In a log-linear plot this corresponds to a straight line. As this behaviour is fairly simple we can expect reliable results: we use linear regression to fit the $\log b(n)$. The question of statistics is important in a least squares best fit: for 3 to 5 ($N = 4$) terms, we cannot necessarily trust our results, but for 6 ($N = 5$) and higher numbers of terms, results are more decisive. For the purpose of error estimation, this means that the reliability of the estimate will increase with N. A regression factor is monitored to indicate the quality of fit, to ensure the assumption that extrapolation errors are small.

Now, if we can calculate the coefficients $b(n)$ for $n = 0, N$ and estimate those for $n \geq N+1$ using extrapolation, the elemental $\mathcal{L}^2$ error norm can be estimated as

$$\epsilon_{est} = \frac{b_N^2}{\frac{2N+1}{2}} + \int_{N+1}^{\infty} \frac{(b(n))^2}{\frac{2n+1}{2}} dn,$$

where the infinite sum of Equation (8) has been replaced by an integral.

Behaviour

An important feature of the error estimates is that they provide two pieces of information: the estimate of the error and the rate of decay σ of the spectrum which is related to the convergence rate. This information is in turn used to make decisions on remeshing. We consider two problems to illustrate this point.

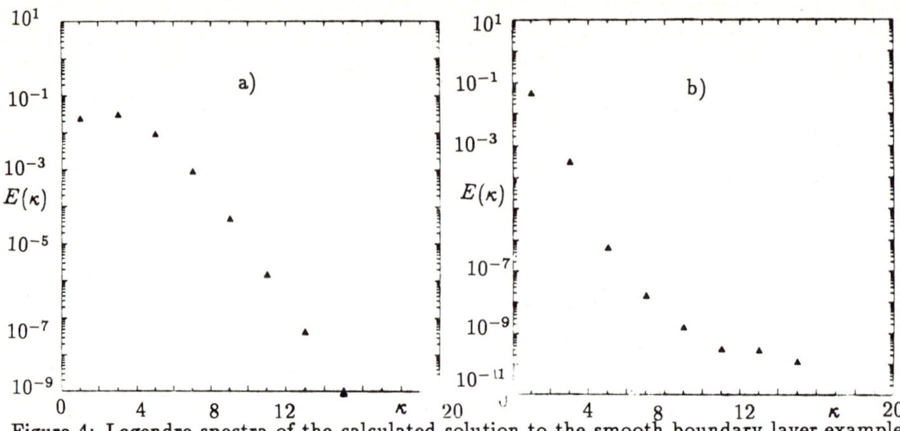

Figure 4: Legendre spectra of the calculated solution to the smooth boundary layer example (a) and the singularity example (b)

First, we solve the Helmholtz problem $\nabla^2 u - \lambda^2 u = 0$ on $I =]0;1[^2$ with Dirichlet boundary conditions $u = u_{\text{exact}} = e^{\frac{\lambda}{\sqrt{2}}(x-1)} e^{\frac{\lambda}{\sqrt{2}}(y-1)}$, which exhibits a sharp but smooth boundary layer near the corner (1,1) for large λ. The two-dimensional spectrum for the element containing the boundary layer, $E(\kappa)$, the calculation of which is detailed in [4], is shown in Figure 4a for $N = 10$, $K = 4$ and $\lambda = 50$. The exponential decay ($\sigma > 1$) indicates that the elemental discretization is adequate and increasing N will refine the solution optimally.

Second, we solve the Poisson problem $\nabla^2 u = 1$ with homogeneous boundary conditions on $I =]-1;1[^2$ which exhibits weak singularities of the form $r^2 \log r$ as $r \to 0$ where r is the radial distance from each corner. The spectrum, $E(\kappa)$, for a corner element, is shown in Figure 4b for $N = 6$, $K = 4$. The slow algebraic(-like) decay ($\sigma < 1$) is due to the singularities. This would encourage remeshing to keep the slow decay rate local to the singularities (e. g. placing small elements around each corner).

These two examples illustrate the advantage of the present error estimators: the capability to predict convergence behaviour is crucial to an efficient mesh refinement process. The important consideration in the refinement decision process is that the work is proportional to KN^3. Hence, increasing the order N of the polynomial ($p-$refinement) should only be used when a large reduction in error is expected for the extra effort, which is the case of large decay rates. Otherwise, for slow decay rates, an increase in N will incur very little improvement in error and is not worth the extra effort. $h-$refinement should be used until higher decay rates are achieved, as little extra effort is required. In general, it is not clear where the limit lies between h- and p-refinement and decisions are difficult. For now we consider h-refinement to be implemented for $\sigma < 1.$ and p-refinement for $\sigma > 1$.

RESULTS

Kovasznay Solution

The success of the formulation and the projected error estimates is illustrated by solving a Navier-Stokes problem (1). Kovasznay [13] has developed an exact solution which may represent low Reynolds number flow past a wire mesh as shown by the streamlines in Figure 5a, calculated with $N = 6$, $K = 8$ for $R = 40$. Spectral accuracy is achieved as the $\mathcal{L}^2$ and $\mathcal{H}^1$ error norms in u and v shown in Figure 5b decay more than three orders of magnitude when N is increased from 4 to 8. $\mathcal{L}^2$ and $\mathcal{H}^1$ error estimates for the velocity components u and v are

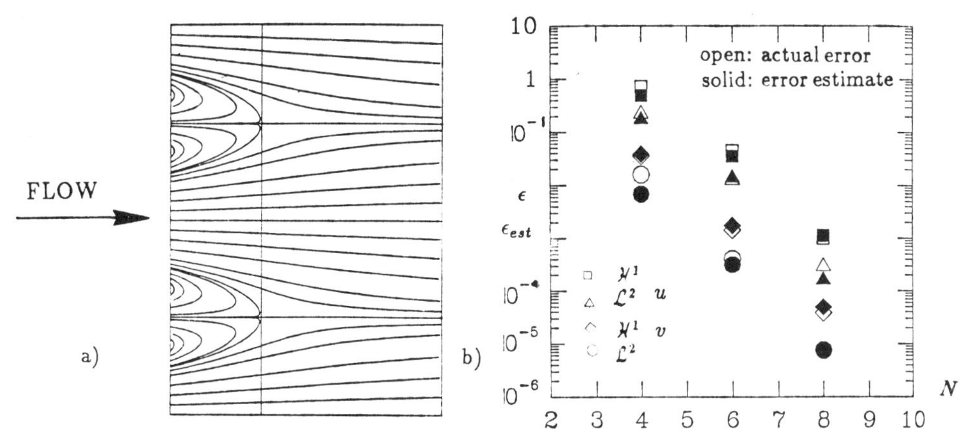

Figure 5: Streamlines (a) and comparison of actual errors and error estimates in u and v (b) for the Kovasznay solution

also given in Figure 5b for the cases $N = 4, 6, 8$. The error estimates are within a factor of 2 from the actual errors for $N = 6$. Though this may seem large, it is the order of magnitude of the error which is important to predict for comparison with other elements and other cases. In general, from experience with many other calculations, we may assert that the error estimators predict errors to within 5-10% of the log value. The exponential decay is accurately captured, which implies that the solution is well resolved. Values of σ from the extrapolation scheme were found to be reasonably high $\sim 4, 5$ to support this fact. Considering the low amount of spectrum data available since $N \sim 4, 6$ the estimates do remarkably well.

Flow in a Channel Expansion

In cases where the problem is more one of inadequate resolution and therefore the area to be refined is unknown, the error estimates become most useful, particularly in the case of automatic refinement (adaptivity). The following example illustrates the point by step-by-step manual adaptation of the mesh, guided by the error estimator information.

Consider the laminar separated flow in a channel expansion, sketched in Figure 6a, at a Reynolds number $R = 109.5$. Intuition tells us that the flow separates at the point $(0, 0)$, and reattaches on the bottom wall a certain distance downstream, leaving a recirculating "dead" zone in the corner. We start the problem guided by intuition with a generic macromesh consisting of 4 elements, as shown in Figure 6a.

The progression of the mesh refinement process is illustrated in Figure 6 where the solution is presented in terms of instantaneous streamlines plots at $t = 10$ superimposed on the elemental mesh, along with tables of values of the error estimators and decay rates for each element. On the first try, a low decay rate in element 3 suggests splitting of this element ($h-$refinement) while elements 1 and 2 are $p-$refined (by increasing N or "equivalently" by shortening the element) due to their high decay rates. Element 4 has an adequate solution: no refinement is needed. We proceed similarly at each step. Ideally we should stop when all errors are of the same order, which is almost achieved in the case of Fig. 6c. Inadequate resolution is evident by the "wiggle" (discontinuity in higher derivatives) in the streamlines near the interface of elements 3 and 4. High errors and low decay rates near the step are probably due to the corner singularity. Though this final discretization is by no means an adequate solution in terms of the error estimators, it is shown to illustrate how the refinement process proceeds with the error estimators. The final results are improved as the wiggle at the 3-4 element interface is

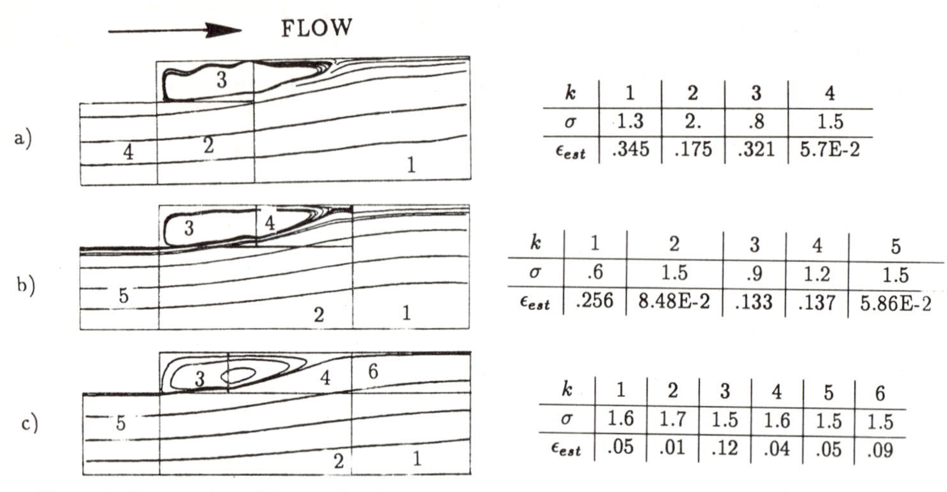

a)

k	1	2	3	4
σ	1.3	2.	.8	1.5
ϵ_{est}	.345	.175	.321	5.7E-2

b)

k	1	2	3	4	5
σ	.6	1.5	.9	1.2	1.5
ϵ_{est}	.256	8.48E-2	.133	.137	5.86E-2

c)

k	1	2	3	4	5	6
σ	1.6	1.7	1.5	1.6	1.5	1.5
ϵ_{est}	.05	.01	.12	.04	.05	.09

Figure 6: Progression of the mesh refinement process for flow in channel expansion: streamlines and error estimates and decay rates

smoother and all error estimates have been reduced substantially. The reattachment length $L \simeq 5$ agrees with Patera's calculation [1] and Armaly *et al.*'s experiments [14].

In this example, the mesh is fairly simple so that we are able to resolve the flow in only a few adaptation steps. In more complex situations, manual repetition of this procedure yields a final mesh which has adequate resolution everywhere. Automatic implementation of this procedure is the subject of work currently underway.

The author would like to thank Anthony Patera for his input in this work.

REFERENCES

1. Patera, A.T., *Journal of Computational Physics*, 54, 468, 1984.
2. Fischer, P.F., Ronquist, E.R. and Patera, A.T., in *Parallel Supercomputing - Methods, Algorithms and Applications*, Wiley and Sons, 1989.
3. Maday, Y., Mavriplis, C. and Patera, A.T., *Proceedings of the Second SIAM Conference on Domain Decomposition Methods*, 1988.
4. Mavriplis, C., Ph. D. thesis, MIT, 1989.
5. Dorr, M., *Numerische Mathematik*, submitted, 1989.
6. Strang, G. and Fix, G., *An Analysis of the Finite Element Method*, Prentice-Hall, 1973.
7. Babuska, I. and Dorr, M., *Numerische Mathematik*, 25, 257, 1981.
8. Berger, M.J. and Oliger, J., *Journal of Computational Physics*, 53, 484, 1984.
9. Karniadakis, G.E., *Numerical Applied Mathematics*, to appear, 1989.
10. Maday, Y. and Patera, A.T., in *State-of-the-art Surveys in Computational Mechanics*, ASME, 1988.
11. Pullin, D.I. and Perry, A.E., *Journal of Fluid Mechanics*, 97, 239, 1980.
12. Mavriplis, C., Fischer, P.F. and Karniadakis, G.E., *Tenth Australian Fluid Mechanics Conference*, to appear, 1989.
13. Kovasznay, L.I.G., *Proceedings of the Cambridge Philosophical Society*, 44, 58, 1948.
14. Armaly, B.F., Durst, J. *et al*, *Journal of Fluid Mechanics*, 127, 473, 1983.

AN ALTERNATE DOMAIN COUPLING PROCEDURE FOR CFD

D. R. McCarthy and D. W. Foutch

Boeing Commercial Airplanes

P. O. Box 3707, M/S 7K-06, Seattle, Washington 98124, USA

SUMMARY

Alternate Domain Coupling is a general procedure, implemented in a modular software system developed by the authors, for the application of zonal methods to computational fluid dynamics. The method has origins in the Schwarz alternating procedure, and is intended to provide fast communication across zone boundaries. The advantages of such a procedure include smooth variation of solutions from zone to zone and enhanced usefulness of strongly coupled implicit solution procedures.

INTRODUCTION

Zonal methods, in which a computational domain is subdivided into a collection of abutting or partially overlapping subdomains, currently enjoy wide popularity in Computational Fluid Dynamics. In addition to supporting coarse-grain parallel calculations, the procedure simplifies body-fitted mesh generation, admits the calculation of problems too large to fit in central memory, and facilitates the treatment of complex geometric configurations. However, there are also two less obvious but potentially important advantages.

First, by standardizing the data structure by which the zones are described, the solution procedure may be divorced from the global geometry of the problem, permitting the construction of general purpose software. The results given in this paper are obtained using such a system, described briefly below.

Second, if zone sizes could be limited without undue penalty, then fast solution procedures which would ordinarily be impractical or costly might be made attractive. For example, because of the nonlinear relationship between problem size and cost for direct Newton-linearized solvers, as well as practical limits on the size of matrix inversions, these methods are often feasible only for small problems. Nevertheless, for some regions of the flowfield, they may be preferable to explicit calculations with small time steps

Aside from the obvious overhead of moving data to and from auxiliary storage, the price usually paid for zonal methods is a slow rate of transmission of information across the zone boundaries. In most implementations, information transfer is accomplished by copying from one zone to another a small band of cells or points whose width is determined by the needs of the computational molecule. For time-dependent or spatially hyperbolic problems, this often represents no special burden, as the method is compatible with and natural to the solution procedure itself. In such cases, the zonal approach serves primarily as a data handling device, and the zones are usually taken as large as storage will permit. But for spatially elliptic or mixed problems, this procedure limits convergence at the interface to the explicit rate supported locally by the molecule itself. A companion argument concludes that implicit procedures on individual zones can only produce expensive and uselessly accurate solutions which are inevitably discarded once adjacent zones have been updated. An obvious corollary finds that explicit algorithms are to be preferred, with little work done on each visit to any zone, so that frequent transfer of data to and from storage or other processors is inevitable.

This paper concerns itself with a means whereby the rate of information transfer between zones can be substantially enhanced. This is accomplished by procedures and software which provide for the actual elimination of each internal boundary at some point during the calculations. We thus provide rapid communication between all portions of the flowfield in spite of the presence of complex geometric obstructions. The actual sequence of calculations can be controlled by the user for compatibility with the flow physics and the geometry at hand. We thus expect to obtain solutions with good continuity and smoothness properties at the zone boundaries, even for problems with elliptic behavior. In addition, we expect that in some cases circumstances will come to favor fast implicit calculations which would have been otherwise computationally intractable. In this paper we explore the use of Alterante Domain Coupling using three simple examples.

A ONE-DIMENSIONAL MODEL PROBLEM

Zonal, or Domain Decomposition, methods are rooted in classical attempts to solve continuum problems on complicated domains by reducing them to a sequence of similar problems on simpler or smaller subdomains. Early in the process, it becomes clear that the boundary conditions posed at zone interfaces, and the degree of continuity across the interfaces, play a central role. For a particularly simple example, one may consider the (elliptic) Dirichlet problem

$$y''(x) = 0, \quad y(a) = A, \quad y(b) = B \tag{1}$$

whose exact solution is the linear function $y(x)$ indicated in Figure 1a. A simple zonal method, representative of procedures common in CFD, would decompose the interval $[a,b]$ into two zones $[a,c]$ and $[c,b]$. Suppose an estimate C is made of the value of the solution at the interface, and the pair of Dirichlet problems

$$y_1''(x) = 0, \quad y_1(a) = A, \quad y_1(c) = C \tag{2}$$

$$y_2''(x) = 0, \quad y_2(c) = C, \quad y_2(b) = B \tag{3}$$

are solved separately but exactly. The result is the spurious "solution" consisting of the functions $y_1(x)$ and $y_2(x)$ shown in Figure 1a. This "solution" actually possesses zero residuals everywhere but at a single point, where the first derivative is discontinuous. If this solution were monitored in a calculation by examining the average residual, it might be concluded that it is quite well converged, although in fact it lies arbitrarily far from the true solution. Attempts to improve the solution by passing Dirichlet data at the interface are useless, as they can only transfer the original guess. A more sophisticated approach might try to remove the discontinuity by solving, e.g., the Neumann problem

$$y_2''(x) = 0, \; y_2'(c) = y_1'(c), \; y_2(b) = B, \tag{4}$$

instead of (3). This procedure, however, even if combined with transfer of Dirichlet data in the opposite direction, can only produce iterates whose graphs lie on the parallelogram indicated in Figure 1a.

Suppose, however, that the interval $[a, b]$ is decomposed instead into overlapping zones $[a, d]$ and $[c, b]$ with

$$a < c < d < b,$$

and that the method then proceeds to solve (exactly, but repeatedly) the sequence of Dirichlet problems

$$y_1''(x) = 0, \; y_1(a) = A, \; y_1(d) = y_2(d) \tag{5}$$

$$y_2''(x) = 0, \; y_2(c) = y_1(c), \; y_2(b) = B. \tag{6}$$

Then the procedure converges nicely as shown in figure 1b. In fact, it is easy to show that the convergence rate for this example is

$$r = \frac{c-a}{d-a} \, \frac{b-d}{b-c} < 1.$$

Each factor represents the proprtion of non-overlap in its respective zone. Thus the rate behaves as expected, improving from 1 to 0 as the overlap increases.

THE MODULAR SYSTEM (MOSYS) AND THE SCHWARZ ALTERNATING PROCEDURE

A similar procedure to that given above was introduced for two-dimensional problems by H. A. Schwarz in 1869 [1, 2], and is now known as the Schwarz Alternating Procedure. The original motivation was to extend constructions of continuum solutions to Laplace's equations from convex to non-convex domains. There has been much investigation, especially in the finite element context, of the convergence properties of the method for various model problems and domain shapes; see, e.g., [3, 4].

Here we present the method from a slightly different point of view which makes it a suitable basis for a rather general software system for Computational Fluid Dynamics. Such a system, known as MOSYS (MOdular SYStem), has been created by the authors for the treatment of zonal problems on body-fitted meshes in two and three dimensions. The system consists of two parts: an executive program, which executes high level instructions written by the user to manipulate the zones and associated data, and a library of solution modules which operate interchangeably on the flowfield to solve the desired equations.

The system accepts as input a collection of non-overlapping zones called fundamental zones. Each zone consists of a grid whose underlying data structure is rectangular, together with any cell-centered, nodal or mixed field data, and a system of patches on the faces of the zone which describe the boundary conditions. Each zone may also possess a "rind" of arbitrary thickness, consisting of auxiliary cells lying physically outside the zone itself. Each zone possesses its own computational coordinate system; no global coordinate system is assumed to exist.

The most basic operation which the system performs is the JOIN, or union. By this operation the user may combine any number of physically adjacent fundamental zones into a single working zone, provided only that a rectangular computational coordinate system is possible for the resulting union. The JOIN operation furnishes this working zone with the same structure as the fundamental zones, including a unified coordinate system, a field which is a composite of the last known values on each of the component zones, and a patch system describing the external environment. Where other zones, not in the current union, are adjacent to the working zone, their grids and fields are copied into the rind. This working zone is then passed, via a standard interface, to the desired solution module, whose job it is to update the flowfield thus assembled. Since working zones are constrained to have rectangular data structure, solution modules may be written without regard for the global problem geometry. Only the solution modules know the meaning of the physical variables or what equations are being solved. (In principle, modules could be written to solve any desired problem in computational physics, e.g., electromagnetics, acoustics, etc.) Upon return from the solution module, the executive SPLITs the working zone into its fundamental components and replaces them in storage, thus releasing the workspace for the next JOIN operation.

A complete description of MOSYS is beyond the scope of this work and will appear elsewhere.

ALTERNATE ZONE COUPLING

Implementation of the Schwarz procedure follows easily and flexibly by what we term Alternate Zone Coupling. In this procedure, one regards the working zones as the overlapping zones of the Schwarz procedure. Their intersections, which provide the overlap, are the fundamental zones themselves. Almost any desired overlap scheme can then be implemented provided that the original fundamental zones are sufficiently small that each desired intersection of working zones is itself a fundamental zone (or union of fundamental zones). In practice, one identifies the desired working zones, and defines the fundamental zones as their various intersections. There is no penalty in dealing with small fundamental zones since the working zones may be as large as core storage will permit. The exact sequence of JOINs to be executed is controlled by the user and in general will depend on the nature of the problem and the equations to be solved. For problems with spatially elliptic components, however, one will in general wish to use the coupling feature to eliminate each internal boundary at some point during the calculations. Conversely, one may wish to limit the sizes of the working zones in order to facilitate the application of implicit procedures in the solution modules. Optimal use of the system is clearly problem dependent.

It is worthwhile to note that all the customary advantages of zonal methods are retained. In particular,the JOIN of a single fundamental zone is legal, so no capability is lost over the usual implementation. Simultaneously, the Alternate Zone Coupling scheme facilitates smooth solutions across zone boundaries, even if fast implicit solvers are used, and provides a mechanism for information to travel between remote locations in the flowfield within a single sweep through the JOIN sequence.

A TWO-DIMENSIONAL MODEL PROBLEM

As a simple two-dimensional demonstration, we solve the Dirichlet problem

$$U_{xx} + U_{yy} = 0 \quad \text{on} \quad [-16, 16] \times [-16, 16]$$

subject to

$$U(-16, y) = 1, \quad U(16, y) = 2, \quad U(x, -16) = 3, \quad U(x, 16) = 4$$

on a uniform 33 x 33 grid, using the usual five-point molecule. The domain is broken into 16 fundamental zones, each 9 x 9, as shown in Figure 2. The calculations are on cell centered data with $U = 0$ as an initial guess. One layer of rind cells is passed, and the solution module solves working zone problems directly by Gaussian elimination. We choose a JOIN sequence designed to eliminate not only the internal boundaries but also the 9 internal zone corners. The sequence

corner joins	side joins	central join
1) A1 U A2 U B1 U B2,	5) A2 U A3 U B2 U B3,	9) B2 U B3 U C2 U C3
2) A3 U A4 U B3 U B4,	6) B1 U B2 U C1 U C2,	
3) C1 U C2 U D1 U D2,	7) B3 U B4 U C3 U C4,	
4) C3 U C4 U D3 U D4,	8) C2 U C3 U D2 U D3,	

illustrated in figure 2, is designed to transmit boundary information from the edges toward the center. A sweep through the JOIN sequence consists of a direct solve on each of these 9 working zones with boundary values inherited from adjacent zones. Figure 3a shows a contour plot of the evolving solution at the end of the first sweep, with discontinuities at zone boundaries clearly visible. The exact (discrete) error is known, since the problem admits direct solution of the discrete problem, obtained from the union of all zones at once. Figure 3b shows a contour plot of the error at the end of the first sweep; as might be expected from the maximum principle, the maximum errors occur along the zone boundaries. Figure 3c shows the evolving solution after 5 sweeps through the sequence. Increasing smoothness of the solution across zone boundaries is evident as the procedure converges. Figure 3d shows the maximum value of the error on exit from the last join in the sequence as a function of the number of sweeps through the sequence. The iteration converges linearly at an observed rate of about .32 per sweep.

APPLICATION TO A BASIC FLOW PROBLEM

To demonstrate application of the system to a realistic flow problem, we chose an axisymmetric choked nozzle for which good data are available [5]. The 4 x 4 zonal structure was identical with that of the preceding example, as shown in figure 4a. For demonstration purposes, coarse grids were sufficient, with each fundamental zone 8 cells by 4 cells for an overall size of 32 x 16.

The Euler equations were solved using two 2-d/axisymmetric modules developed specifically for MOSYS. Owing to the non-linearity of the equations, the nature of the modules is more complex than in the preceding case. These modules both solve the same variant of Roe's flux-difference split equations [6] as modified by Shima [7]. The left and right flow states required by the Riemann solver are obtained from second order upwind extrapolations, and no flux limiters are used. Because of the upwinding, a rind thickness of two cells is required.

With a view toward fast implicit procedures, the resulting non-linear equations are carefully linearized. The linearization is the Newton linearization of the first order contributions, except for the tensor-like terms arising from the derivatives (with respect to the flow variables) of the Roe coefficient matrix itself, which are omitted. Second order contributions are explicit. A module called ROEDIR (ROE DIRect), incorporating a sparse matrix solver, executes a Newton-like procedure by repeatedly forming this linearization and solving directly for the new solution. The number of such linearizations per call is variable; in all calculations given here, ROEDIR solved only a single linearized problem per call.

Because of the storage required by this particular sparse solver, the ROEDIR module is not directly applicable to the union of all 16 zones in the nozzle problem, even on the CRAY XMP-24. Therefore, to obtain a comparison solution, a second module called ROEJAC (ROE JACobi), which solves the individual linearized problems by point-implicit Jacobi iteration, was used to solve the overall problem as a unit. This module is known to converge well for this problem and to agree reasonably with data. A contour plot of the resulting Mach number distribution is shown in figure 4b.

To determine whether direct solution on smaller zones could produce these results for the Euler equations, as it did for Laplace's equation, the same 9-zone join sequence as for the earlier example was executed using ROEDIR as the solution module. In addition, a second calculation was performed with the simpler corner JOIN sequence

1) A1 U A2 U B1 U B2 3) C1 U C2 U D1 U D2
2) A3 U A4 U B3 U B4 4) C3 U C4 U D3 U D4

to determine whether a less expensive scheme would be competitive. Both schemes converged to results indistinguishable from figure 4b; all methods, of course, solve the same difference equations. Convergence histories are given in figure 4c; the upper pair of curves represents the maximum error and the lower pair the rms error, measured by comparison with the single-zone solution described above. Within each pair, the faster convergence is obtained with the 9-zone scheme, although apparently not sufficiently faster to justify 9/4 as much work per sweep. However, because the rind is thick relative to the sizes of the zones, even the 4-join case is very strongly coupled. Thus results such as this on very coarse grids may not be indicative of the situation on denser mesh. Further, one notes that the convergence rates themselves in the 4 and 9 join cases are apparently the same, leading to the conclusion that the outer (or zonal) iteration is not the controlling factor in this example. In fact, close examination reveals a complex interaction of the oblique shock with the axis and outflow boundary, which is likely to impede convergence on zone A4. Experiments with additional work confined to this corner of the flowfield show improved converge rates on a per-sweep basis. Additional work is apparently needed to understand the convergence rates of the method when mixed hyperbolic and elliptic effects are present.

CONCLUSION

We have proposed and implemented in software a quite general philosophy for the application of zonal methods to problems not only in CFD but in computational physics in general. We have demonstrated that the method is capable of providing smooth solutions across zonal boundaries and of rapidly transmitting data throughout the domain, even where elliptic influences are important. Much remains to be done to understand the optimal use of the method, both with respect to convergence rates of the (outer) zonal iterations themselves and the degree of implicitness which is useful within the solution modules. These questions and others will be addressed in future work.

REFERENCES

[1] SCHWARZ, H. A.: Uber Einige Abbildungsaufgaben. Ges. Math. Abh., 11 (1869), p. 65-83.

[2] COURANT, R., and HILBERT, D.: Methods of Mathematical Physics, Vol. 2. New York, Interscience, 1961.

[3] Proceedings of the First International Symposium on Domain Decomposition Methods for Partial Differential Equations. Glowinski, R., Golub, G., Meurant, G., and Periaux, J., eds. Society for Industrial and Applied Mathematics, Philadelphia, 1988.

[4] Proceedings of the Second International Symposium on Domain Decomposition Methods. Chan, T., Glowinski, R., Periaux, J., and Widlund, O,. eds. Society for Industrial and Applied Mathematics, Philadelphia, 1989.

[5] CUFFEL, B., BACK, L., and MASSIER, P.: Transonic Flowfield in a Supersonic Nozzle with a Small Throat Radius of Curvature. AIAA Journal, Vol. 7 (1969), No. 7, pp. 1364-1366.

[6] ROE, P.: Approximate Riemann Solvers, Parameter Vectors, and Difference Schemes. Journal of Computational Physics, Vol. 43 (1981), pp. 357-372.

[7] SHIMA, E.: Numerical Analysis of Multiple Element High Lift Devices Using Implicit TVD Finite Volume Method. AIAA paper No 88-2574-CP, AIAA 6th Applied Aerodynamics Conference Proceedings, June 1988.

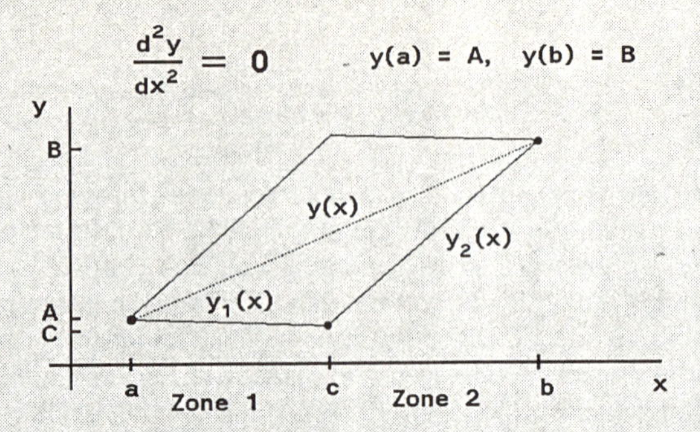

Figure 1a. Without Overlap

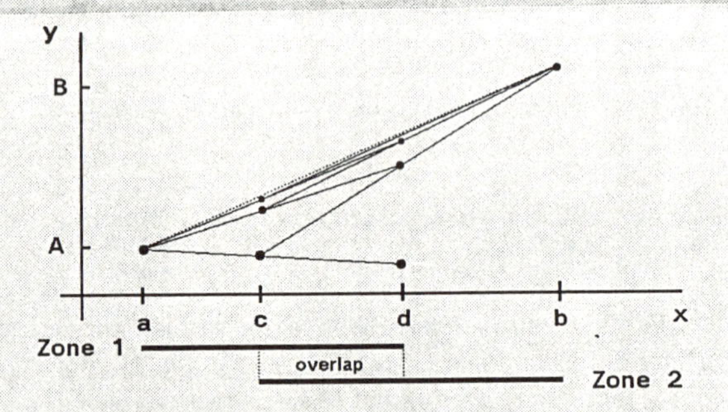

Figure 1b. With Overlap

Figure 1. Comparison of Zonal Data Passing Techniques for the One Dimensional Model Problem

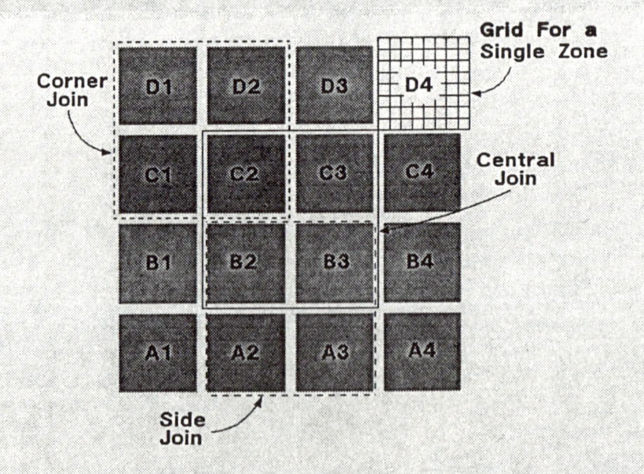

Figure 2. Zonal Structure for the Two Dimensional Model Problem

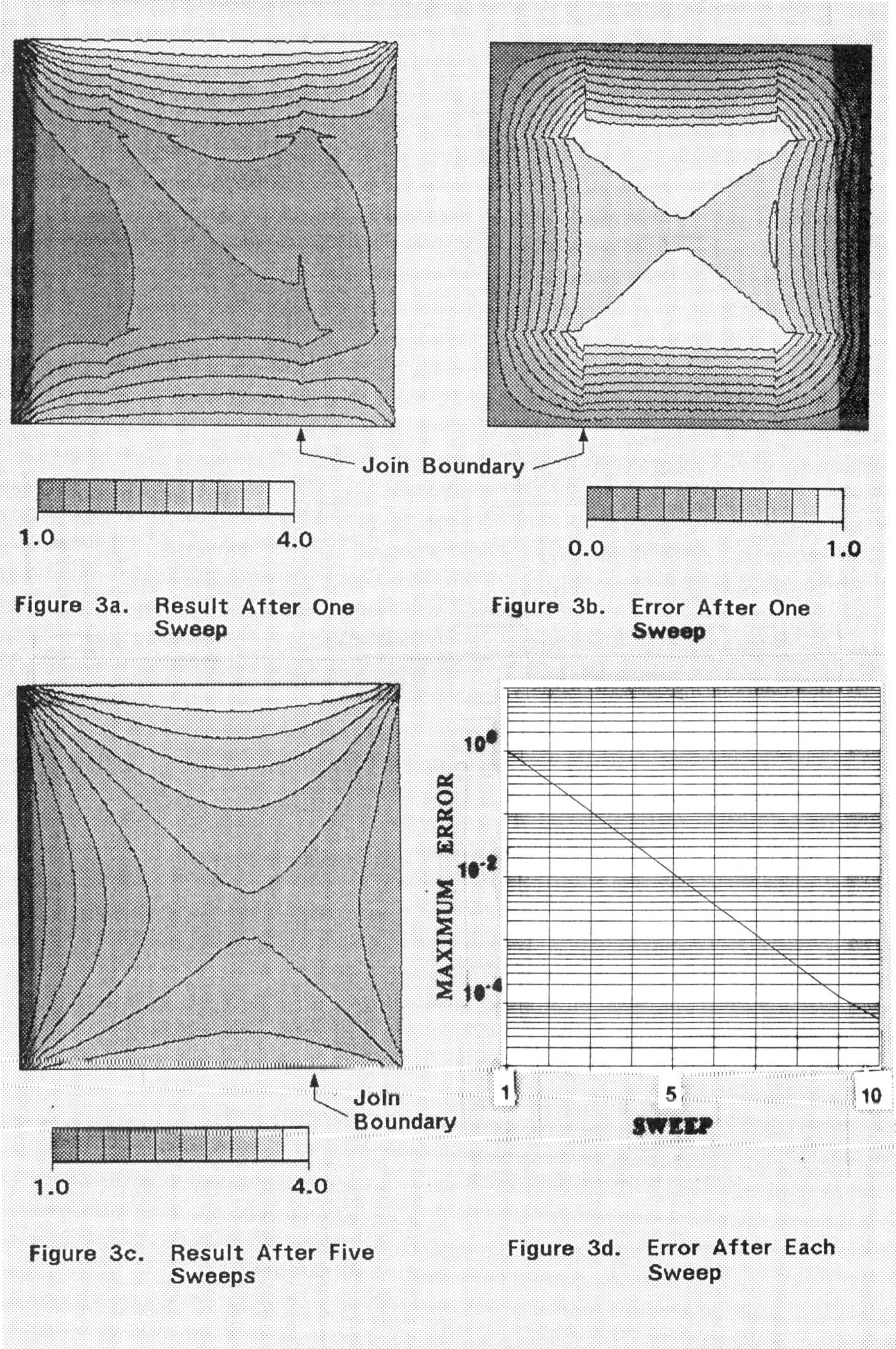

Figure 3a. Result After One Sweep

Figure 3b. Error After One Sweep

Figure 3c. Result After Five Sweeps

Figure 3d. Error After Each Sweep

Figure 3. The Convergence History of the Two Dimensional Model Problem

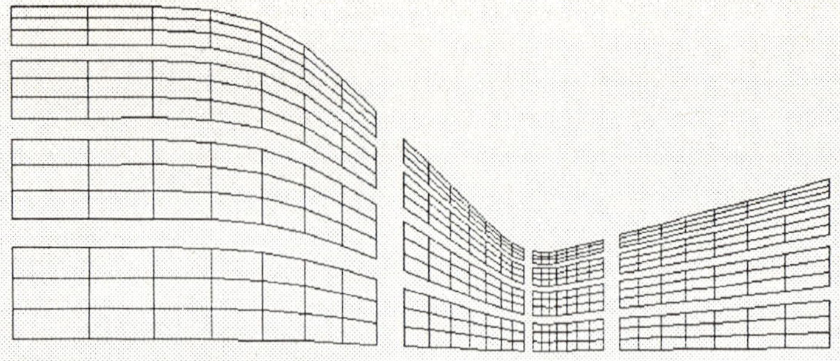

Figure 4a. Zonal Structure

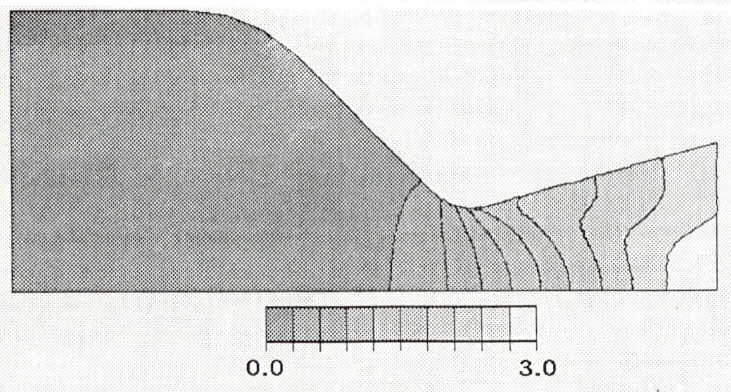

Figure 4b. Mach Number Contours

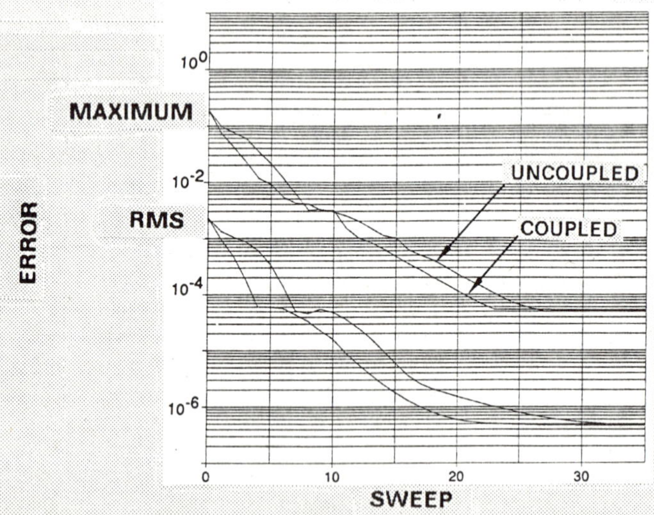

Figure 4c. Convergence History: Comparsion of Coupled and Uncoupled Calculations

Figure 4. The Converging-Diverging Nozzle Calculation

SPLIT-MATRIX MARCHING METHODS FOR THREE-DIMENSIONAL

VISCOUS AND INVISCID HYPERSONIC FLOWS

S. Menne and C. Weiland

Messerschmitt-Bölkow-Blohm GmbH
D-8000 München 80, Fed. Rep. of Germany

SUMMARY

Two split-matrix marching methods for hypersonic flow are presented: a space marching method for inviscid flow and a time-integration marching method for viscous flow. For both methods the solution algorithm is explained. Flow calculations of the hypersonic flow over a two-stage transport vehicle show that the methods allow an economic computation of flows over complex geometries with sharp shocks and no spurious oszillations without using any explicit artificial damping.

I. INTRODUCTION

The initiation of hypersonic research programs for spaceplanes, as, e.g., the National Aerospace Plane in the USA, or the Hermes in Europe, has led to growing efforts in the research of hypersonic flows, especially CFD. For the computation of flows about complex geometries, marching methods show attractive features due to their efficiency and robustness as long as a marching solution exists, i.e., as long as the flow is supersonic in marching direction. Unfortunately, there is always a subsonic sublayer for viscous flow over a body due to the no-slip boundary condition. Two different marching methods have been developed: for supersonic flow in the entire flowfield the Space Marching method (SM) for inviscid flow and for mainly supersonic flow with subsonic boundary-layers and/or subsonic pockets the Time-Integration Marching method (TIM) for both viscous and inviscid flow. These are described in the next section.

II. PROBLEM FORMULATION AND SOLUTION METHOD

The governing equations can be written in strong conservation law form as

$$\Omega Q_{\tau} + E_{\xi} + F_{\eta} + G_{\zeta} = E_{\nu,\xi} + F_{\nu,\eta} + G_{\nu,\zeta} \quad , \tag{1}$$

with the vector of the conservative variables

$$Q = (\rho, \rho v_x, \rho v_y, \rho v_z, e)^T \quad ,$$

the convective fluxes E, F, and G, and the viscous fluxes E_ν, F_ν, and G_ν which can be found in, e.g., [1], and the metric Jacobian $1/\Omega$.

For a steady, inviscid, and supersonic flow, the governing equations reduce to

$$Q_\zeta + K \cdot Q_\xi + J \cdot Q_\eta = 0 \quad , \tag{2}$$

with $K = G_Q^{-1} \cdot E_Q$, $J = G_Q^{-1} \cdot F_Q$ and $F_Q = \partial F / \partial Q$.

Eqs.(1,2) are written in generalized curvilinear coordinates (ξ, η, ζ). Boundary conditions have to be specified at an initial plane in marching direction and along all boundaries in crossflow planes.

II.1 SPACE MARCHING METHOD (SM)

For flows with a Mach number in flow direction ζ greater than one the steady Euler equations are applied. A starting solution has to be given on the body as an initial condition. For bodies with sharp noses and attached bow-shocks, this initial solution can be easily obtained assuming a conical flow. In the crossflow planes, the boundary conditions are the slip boundary condition on the body surface and jump conditions at the outer bow-shock. The implementation of the slip boundary condition can be found in [2]. At the outer bow-shock, a very accurate bow-shock fitting procedure is applied which is consistently imbedded in the Runge--Kutta stepping scheme.

INTEGRATION IN MARCHING DIRECTION

The space marching from level k to level (k+1) is done with a second--order accurate explicit three-stage Runge-Kutta method

$$\begin{aligned} Q^{(1)} &= Q^k - a_1 \, \Delta\zeta \, P(Q^k) \\ Q^{(2)} &= Q^k - a_2 \, \Delta\zeta \, P(Q^{(1)}) \\ Q^{k+1} &= Q^k - a_3 \, \Delta\zeta \, P(Q^{(2)}) \quad , \end{aligned} \tag{3}$$

where the Runge-Kutta coefficients a_1=1/4, a_2=1/2, a_3=1 are optimized for the third-order upwind biased space discretization scheme [3]. CFL numbers between 1 and 1.5 are used in the simulations.

SPATIAL DISCRETIZATION

The steady Euler equations (2) are considered in quasi-conservative form, i.e., the Euler equations are expressed in non-conservative form in terms of the conservative variables Q

$$Q_\zeta + K^+ Q_\xi^+ + K^- Q_\xi^- + J^+ Q_\eta^+ + J^- Q_\eta^- = 0 \quad . \tag{4}$$

The matrices $K^\pm$ and $J^\pm$ are split according to the sign of their eigenvalues [4]. The space derivatives of Q are calculated using a finite-difference scheme with a third order accurate upwind-biased formula [3], e.g.,

$$\begin{aligned} Q_{\xi,i}^{\ +} &= (\ Q_{i-2} - 6Q_{i-1} + 3Q_i + 2Q_{i+1})/6\Delta\xi \\ Q_{\xi,i}^{\ -} &= (-Q_{i+2} + 6Q_{i+1} - 3Q_i - 2Q_{i-1})/6\Delta\xi \quad . \end{aligned} \tag{5}$$

A MacCormack-type artificial diffusion term prevents wiggles near shocks. Details of the space marching scheme can be found in Weiland [4].

II.2 TIME-INTEGRATION MARCHING METHOD (TIM)

A drawback of conventional marching methods is the restriction to supersonic flows. In viscous flows, a boundary-layer develops along the body surface due to the no-slip boundary condition. This subsonic layer prohibits a marching in flow direction, although the type of flow inside the boundary-layer is parabolic in flow direction, i.e., a marching solution is possible in both the outer supersonic part of the flow and in the boundary-layer. The most popular way to overcome this problem is to modify the pressure gradient in streamwise direction in order to obtain a modified system which is hyperbolic in the entire domain. This technique was proposed by Vigneron [5].

A different way is followed here. The general solution algorithm (as described below) is modified in the sense that the downstream states in marching direction are obtained by extrapolating the upstream states, namely first- or zeroth-order extrapolation. Therefore, the solution itself in each marching plane is identical to the global solution system for the whole domain. Since the downstream states are totally dependent on the upstream states, flow perturbations from downstream do not enter the marching plane allowing a marching in flow direction. The neglection of downstream influences in marching direction is possible as long as no strong interaction is present, e.g., a detached shock, incident shock - boundary-layer interaction etc. .

For viscous flow, the complete Navier-Stokes equations are considered. Due to the extrapolation in (ζ) marching direction, a correct treatment of the viscous forces in marching direction is not possible. However, the flow in the crossflow planes is correctly described as long as the spatial resolution is sufficient to resolve all flow details. Especially, strong interactions (shock - shock, shock - boundary-layer, vortex - boundary-layer) occuring in crossflow planes are accounted for. Moreover, the solution can be started with freestream values with the initial plane located in front of the body. The initial solution is therefore well defined. In crossflow planes boundary conditions have to be given at all boundaries. The body boundary conditions are the slip condition for inviscid flow and the no-slip condition for viscous flow. Due to the shock capturing capabilities of the TVD algorithm, there is no need to fit the outer bow-shock. The outer boundary is therefore taken some distance outside the outer bow-shock, where freestream conditions can be prescribed.

TIME INTEGRATION

An explicit one-step time integration (Euler explicit) is used here. Since the time integration is used only as a driver to converge the solution to a time-independent state, this time integration is equivalent to an iteration process.

$$
\begin{aligned}
\Omega/\Delta\tau \; \Delta_n Q_{i,j,k} &= -\left(\Delta_i(E-E_\nu) + \Delta_j(F-F_\nu) + \Delta_k(G-G_\nu)\right)^n{}_{i,j,k} \\
Q^{n+1} &= Q^n + \Delta_n Q \; .
\end{aligned} \tag{6}
$$

The one-step Euler implicit scheme was also employed with point Gauss--Seidel and line Gauss-Seidel relaxation. So far, numerical tests showed no advantage of the implicit scheme over the explicit one. The advantage of larger time steps in the implicit version was counterbalanced by the larger necessary computational work per time step compared to the explicit version. Moreover, the implicit solution showed a stagnation in the convergence for hypersonic test cases. Therefore, only the explicit version is described in the following. Local time steps are used in order to overcome the problem of slow convergence due to stability restrictions.

SPATIAL DISCRETIZATION

The governing equations (1) are solved using a finite volume discretization, i.e., coordinates are defined at cell vertices and flow variables at cell centers. The flux derivatives are expressed by a symmetric TVD scheme [6]. In the following, the numerical fluxes are briefly described. The numerical inviscid flux $\Delta_i E$ can be written as

$$
\begin{aligned}
\Delta_i E_{i,j,k} &= E_{i+1/2} - E_{i-1/2} \\
E_{i+1/2} &= 1/2\,(E_i + E_{i+1} - \\
&\qquad (T/c^2\,\Psi(\Lambda))_{i+1/2}\,(S_{i+1/2} - R(S_{i-1/2}, S_{i+1/2}, S_{i+3/2}))) \\
E_Q &= T\cdot\Lambda\cdot T^{-1}
\end{aligned} \tag{7}
$$

$$S_{i+1/2} = c^2 \, T^{-1}{}_{i+1/2} \, (Q_{i+1} - Q_i)$$

$$R(s^-,s,s^+) = 2 \, \text{minmod}(s^-,s,s^+,(s^-+s^+)/4) \quad ,$$

with the right eigenvector matrix T, the diagonal eigenvalue matrix Λ, the entropy correction function Ψ, and the speed of sound c. Details of the numerical flux formulation are described in Pfitzner et al. [7].

The numerical viscous flux $\Delta_i E_\nu$ can be written as

$$\Delta_i E_\nu = E_{\nu,i+1/2} - E_{\nu,i-1/2}$$

$$E_{\nu,i+1/2} = E_\nu(Q_{i+1/2}, Q_{r,i+1/2}), \quad r = \xi,\eta,\zeta \quad , \tag{8}$$

where $Q_{i+1/2}$ denotes the arithmetic mean of Q_i and Q_{i+1} and $Q_{r,i+1/2}$ stands for a centrally differenced first-order derivative in ξ, η, and ζ direction, respectively. An extrapolation of the metric terms is necessary for the evaluation of the viscous fluxes near boundaries. A second order accurate four-point extrapolation is used here.

Five marching planes have to be stored for the computation of the solution in one marching plane. The first two planes are known from previous marching steps. The third plane is the actual computation plane. The flow variables in the fourth and fifth plane are either extrapolated from the second and the third plane (first-order extrapolation) or transferred from the third plane (zeroth-order extrapolation). Due to the reduced number of grid points in each marching plane compared to the complete domain, the number of time-steps required to attain steady state is about the K-th part of that of a global solution (with K being the number of marching planes). Contrary to the gain in computation time, there is a loss in input/ output time for the marching scheme since each plane has to be read/written separately. This problem is removed by using direct access files which are very effective in handling large amounts of data with fast input/output.

III. RESULTS

Results are presented for the flow over a two-stage transport vehicle (TSTV) and for the flow over a cone. For the space marching (SM) results to the TSTV, the flowfield in each marching halfplane is resolved by 73 grid points along the body contour and 33 grid points from body surface to bow-shock (finite difference scheme; shock-fitting). Due to the explicit marching, the number of marching planes depends on the local CFL number. Here, the marching proceeds in planes normal to the z-direction, i.e., $\zeta \equiv z$. For the time-integration marching (TIM) results, the TSTV flowfield in each halfplane consists of 70 cells along the body contour and 40 cells from body surface to outer boundary. The number of marching planes is 40.

Fig. 1 shows isobars on the lower (a) and upper (b) surface of the TSTV for inviscid flow (SM) with a Mach number Ma=6.8 and an angle of attack

$\alpha=8°$. The canopy shock is clearly indicated. Fig. 2 presents the flowfield in the symmetry plane. Fig. 2a displays the static pressure distribution for inviscid flow (SM) with Ma=6.8 and $\alpha=8°$. The canopy shock and a subsequent expansion appears in the upper part. On the lower side, a compression is formed followed by an expansion fan due to the convex-concave form of the lower body contour. Figs. 2b and 2c show isobars and iso-Mach lines for viscous flow (TIM) with a Reynolds number $Re=8\ 10^6$. Compared to the inviscid results, the canopy shock is weaker and the shock angle is smaller because the shock position is moved further downstream due to the boundary layer. The outer bow-shock appears only on the lower, windward side. The static pressure distribution on the upper TSTV surface is shown in Fig. 3 for inviscid flow (SM). Additionally to Fig. 2, an intersection of the outer bow-shock with the wing is indicated in the rear part of the wing.

The hypersonic flow at Ma=7.95 around a cone at an angle of attack of $\alpha=24°$ is considered as a test case for the TIM method. The local Reynolds number is $Re_x=3.6\ 10^5$ and the Prandtl number is set to Pr=0.72 for air. The cone semivertex angle is $\alpha=24°$. The freestream temperature is $T_\infty=310$ K and the wall temperatue is set constant to $T_{wall}=310$ K. Fig. 4 displays the flow picture obtained experimentally by Tracy [8] (left half) with the outer bow-shock (points), the edge of the viscous layer (squares), and the location of minimum pitot pressure (triangles). The numerical are calculated on crossflow plane grids with 50X50 grid points and 10 marching planes. The right half of Fig. 4 presents iso-Mach lines in a crossflow plane with $Re_x=3.6\ 10^5$. The comparison shows a larger extension of the calculated outer bow-shock and of the detached viscous layer on the leeside in the numerical data. The recompression shock indicated by a kink in the Mach number contours cannot be resolved accurately.

Fig. 5 presents isobars in TSTV crossflow planes at z=3.5 (a), and z=4.2 (b) for inviscid flow (SM) with Ma=4.5 and $\alpha=6°$. Fig. 5a shows the formation of an internal wing shock with an expansion above the wing. The canopy shock is already weak. In Fig. 5b, the wing shock intersects with the outer bow-shock which is correctly fitted. Fig. 6 shows viscous (TIM) isobars for the same flow case with $Re=5.3\ 10^6$ at z=2.05 (a), z=2.95 (b), z=3.55 (c), and z=4.15 (d). The flow behaviour is very similar to the inviscid one except for the boundary-layer region. The canopy shock distance to the body surface is smaller in the viscous case (Fig. 6b). The location and shape of the outer bow-shock of both inviscid and viscous case are nearly identical. There are no perturbations outside the outer bow-shock. Note that the pressure levels in Figs. 6a-d are different, because 30 equally distributed contour levels are presented with different extremal values.

In Fig. 7, isobars (a-f) and iso-Mach lines (g-i) are compared for Ma=6.8 and $\alpha=8°$ for inviscid flow (SM) at z=3.0 (a), z=3.5 (b), z=4.2 (c), and for viscous flow (TIM) with $Re=8\ 10^6$ at z=2.95 (d,g), z=3.55 (e,h), and z=4.15 (f,i). Note that the contour levels are different. The region outside the outer bow-shock is not shown here. Fig. 7 displays the same characteristics as Figs. 5 and 6. The location and shape of the outer bow-shock is nearly identical for both viscous and inviscid flow. The boundary-layer on the windward side is much thinner than on the leeward (upper) side where two compressions induced by the body contouring lead to a massive boundary layer thickening. The grid spacing from body to outer edge is chosen such that around 10 grid points are located inside the thin windward boundary-layer.

REFERENCES

[1] Bird, R.B., Stewart, W.E., and Lightfoot, E.N.: "Transport Phenomena", Wiley, Inc., New York 1960.

[2] Weiland, C.: "A Split-Matrix Method for the Integration of the Quasi-Conservative Euler Equations", Notes on Numerical Fluid Mechanics, Vol. 13, Vieweg 1986.

[3] Weiland, C., Pfitzner, M.: "3-D and 2-D Solutions of the Quasi-Conservative Euler Equations", Lecture Notes in Physics, Springer Verlag, Vol. 264 (1986).

[4] Weiland, C.: "A Split-Matrix Runge-Kutta Type Space Marching Procedure", to be published.

[5] Vigneron, Y.C., Rakich, J.V. and Tannehill, J.C.: "Calculation of Supersonic Viscous Flow over Delta Wings with Sharp Subsonic Leading Edges", AIAA paper 78-1137 (1978).

[6] Yee, H.C.: "Upwind and Symmetric Shock Capturing Schemes", NASA TM 89464 (1987).

[7] Pfitzner, M., Schröder, W., Menne, S., and Weiland, C.: "Three-Dimensional Simulations of Hypersonic Flows", Int. Conf. on Hypersonic Aerodynamics, Royal Aeronaut. Soc., Sept. 4-6, Manchester 1988.

[8] Tracy, R.R.: "Hypersonic Flow over a Yawed Circular Cone", Ph.D-Thesis, Cal. Inst. of Technology, Graduate Aeronaut. Lab. Memo 69, Aug. 1963.

Fig. 1a Isobars on TSTV lower surface, inviscid (SM), Ma=6.8, $\alpha=8°$.

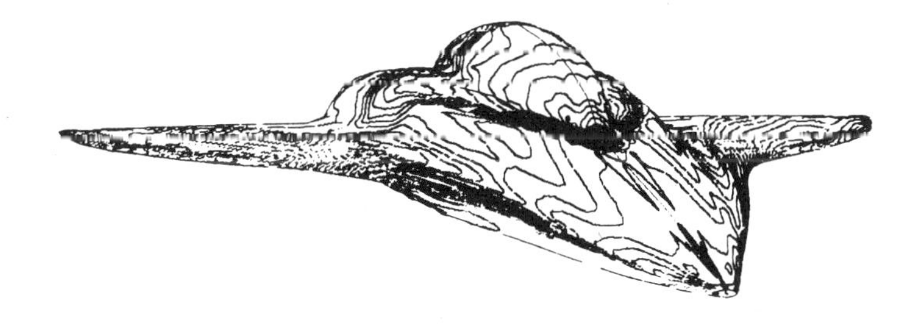

Fig. 1b Isobars on TSTV upper surface, inviscid (SM), Ma=6.8, $\alpha=8°$.

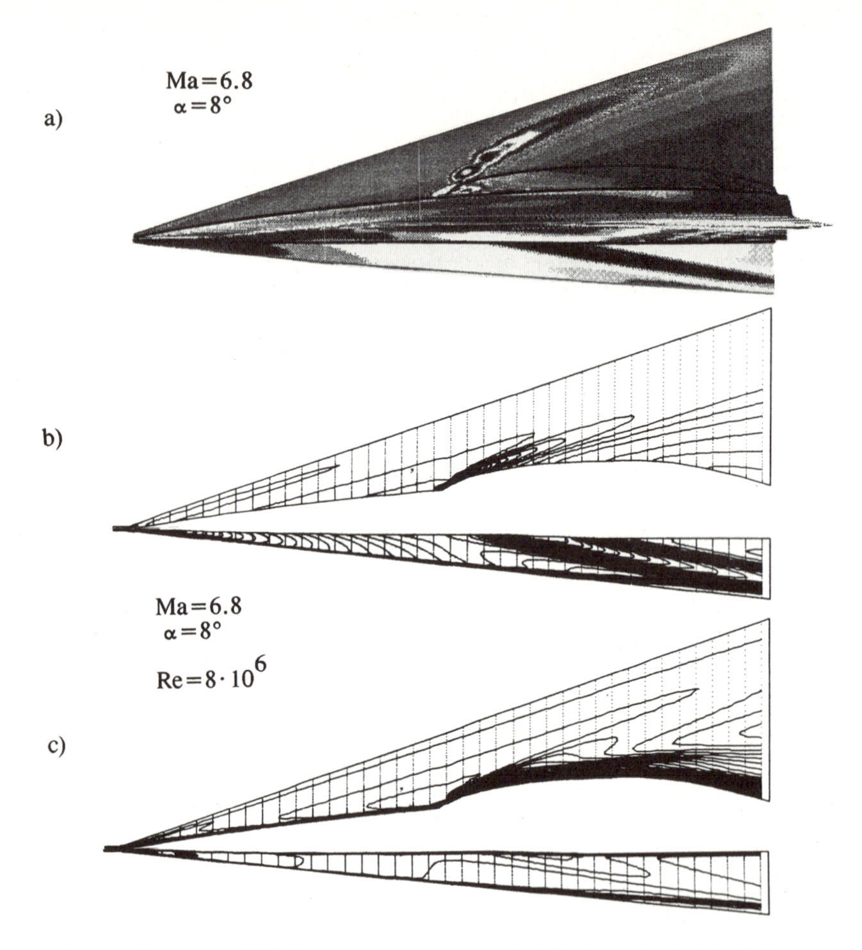

Fig. 2 Flow over TSTV in symmetry plane. (a) Isobars, inviscid (SM), (b) isobars, viscous (TIM), (c) iso-Machlines, viscous (TIM).

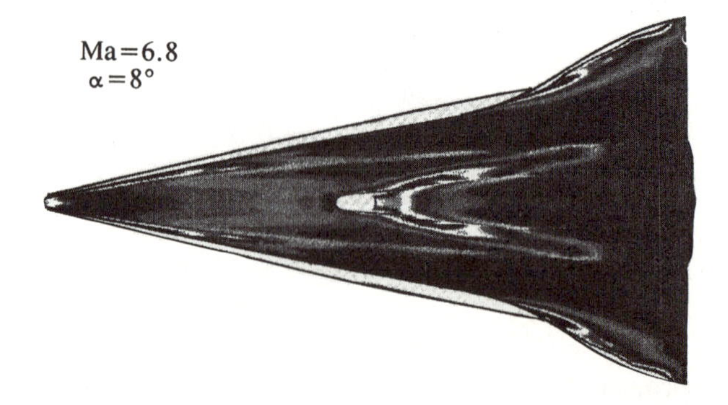

Fig. 3 Static pressure distribution on upper TSTV surface, inviscid (SM).

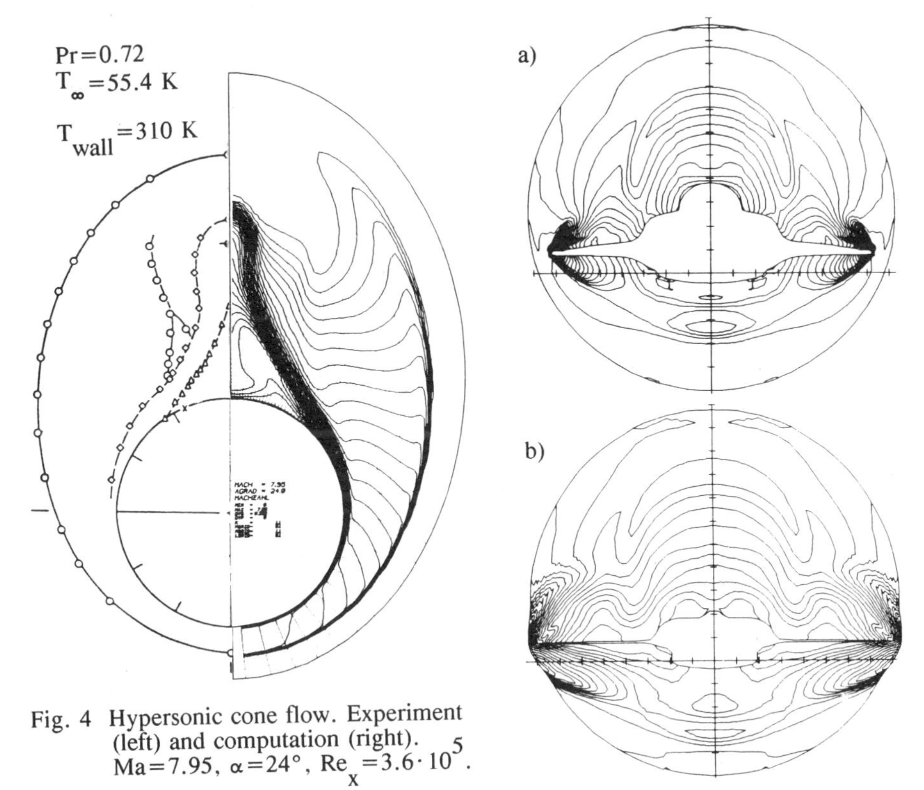

Fig. 4 Hypersonic cone flow. Experiment (left) and computation (right). $Ma=7.95$, $\alpha=24°$, $Re_x=3.6\cdot 10^5$.

Fig. 5 Isobars in crossflow plane of TSTV, inviscid (SM) at $z=3.5$ (a), and $z=4.2$ (b). $Ma=4.5$, $\alpha=6°$.

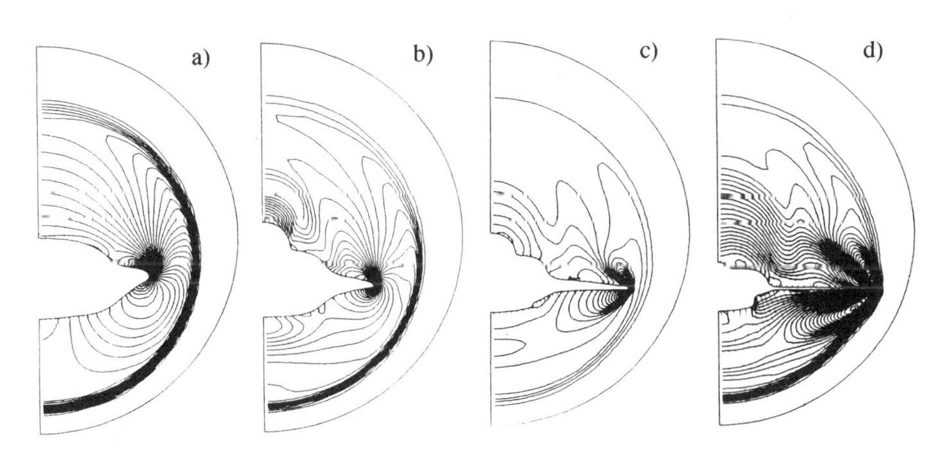

Fig. 6 Isobars in crossflow plane of TSTV, viscous (TIM) at $z=2.05$ (a), $z=2.95$ (b), $z=3.55$ (c), and $z=4.15$ (d). $Ma=4.5$, $\alpha=6°$, $Re=5.3\cdot 10^6$.

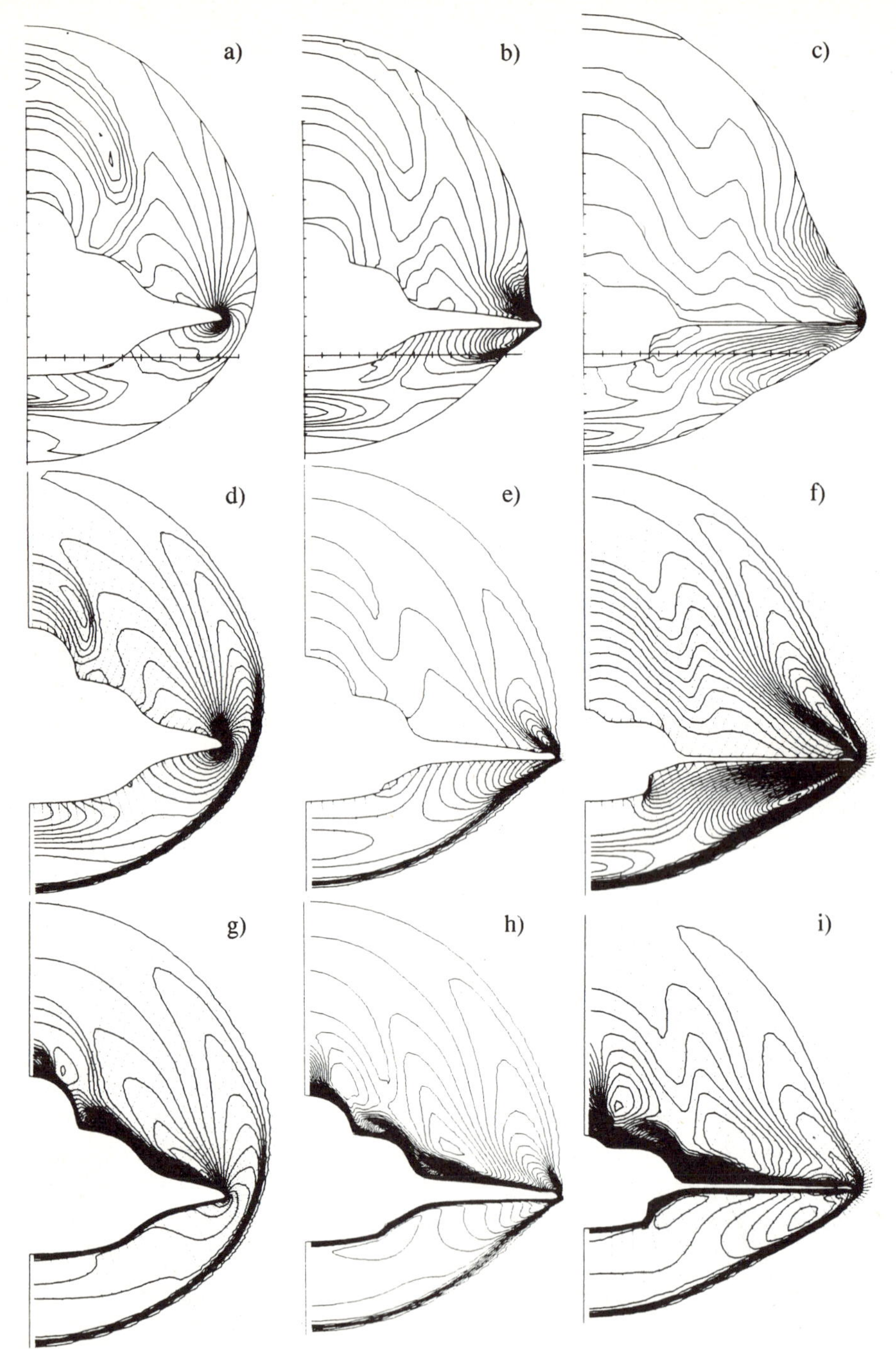

Fig. 7 Isobars (a-f) and iso-Machlines (g-i) in crossflow plane of TSTV for Ma=6.8, $\alpha=8°$. Inviscid (SM) at z=3.0 (a), z=3.5 (b), z=4.2 (c). Viscous, $Re=8\cdot10^6$ (TIM) at z=2.95 (d,g), z=3.55 (e,h), and z=4.15 (f,i).

NUMERICAL SIMULATION OF THE NEAR FIELD OF A PLANE JET

J. MEYER, A. SEVRAIN, H.C. BOISSON, H. HA MINH

Institut de Mécanique des Fluides de Toulouse
Avenue du Professeur Camille Soula, 31400 Toulouse, France

SUMMARY

The near field of a plane jet is simulated by direct resolution of 2-D incompressible unsteady Navier-Stokes equations, using primitive variables, for Reynolds numbers Rd= 1000, 1700. This simulation predicts correctly the physical flow, in the potential region, including the instability process, the vortex apparition and its evolution and the jet parameters variation. It enables to investigate the flow by simultaneous analysis of the streaklines pattern and the temporal signals.

INTRODUCTION

The near field of a plane jet is less documented and less understood than its far field, which is an affinity region. It is generally admitted that the near field consists of a pair of mixing layers, growing independently on both sides of the potential core, and interacting at the end of the potential region [2]. In this potential region, the flow is mainly two-dimensional [3], by contrast with the three-dimensional flow in the affinity region [4].

Having shown in an experimental study the existence in the near field of a pair of symmetrical and highly correlated vortices, undergoing one pairing, then growing, before bursting [5], a numerical simulation of the flow was realized.

Previous numerical simulations have generated vortices in the near field of the plane jet [6], but they did not predict the pairing mechanism as in the physical experiments.

This simulation was realized for Reynolds numbers R_d= 1000 and 1700, and its results may be directly compared with physical experiments.

NUMERICAL METHOD

The near field of a plane jet is simulated by direct resolution of 2-D incompressible unsteady Navier-Stokes equations, using a control volume method in primitive variables(pressure- velocity). The numerical code was previously applied to transitional free shear flows: the near wake of a circular cylinder by BRAZA et al. [7] and the plane mixing layer by KOURTA et al. [8].

A time marching predictor-corrector procedure is used to solve sequentially the linearized equations of momentum and mass conservation. An auxiliary function Φ is introduced such as:

$$\vec{V}^{n+1} = \vec{V}^{*} - \overrightarrow{\text{grad}}\,\Phi \quad \text{and} \quad P^{n+1} = P^{n} + \Phi/\delta t \tag{1}$$

with $\vec{V}^{n+1}$ the velocity field at time t^{n}, $\overrightarrow{V^{*}}$ the first guess of the velocity using the pressure P^{n} in the momentum equation at the intermediate time station t^{*}.

Φ is then obtained by solving the Poisson equation satisfying the mass conservation constraint:

$$\Delta\Phi = \text{div}\ V^{*} \tag{2}$$

and the pressure and velocity fields are updated.

The discretization is made by a control volume method, on a staggered grid arrangment. A forward time centered space scheme is used. An A.D.I. method is implemented to solve both the momentum equation and the Poisson equation for the auxiliary function. The method provides a second order precision in space.

In this simulation, an unsteadiness of the solution may be observed. It is not introduced artificially but generated by the code itself [9].

In most of the test cases, the flow is solved for a half plane jet with a symmetry condition in the median plane, using the experimental evidence of symmetrical flow in the near field. Nevertheless the numerical simulation of the whole jet has also confirmed the complete plane symmetry of the flow pattern at these Reynolds numbers.

BOUNDARY AND INITIAL CONDITIONS

The problem of fixing the adequate boundary and initial conditions is essential for the jet simulation in many requests. It remains an open question. The choice retained here is sketched on figure 1.

The upward difference schemes are adopted to express the continuous boundary conditions set are given in Table 1.

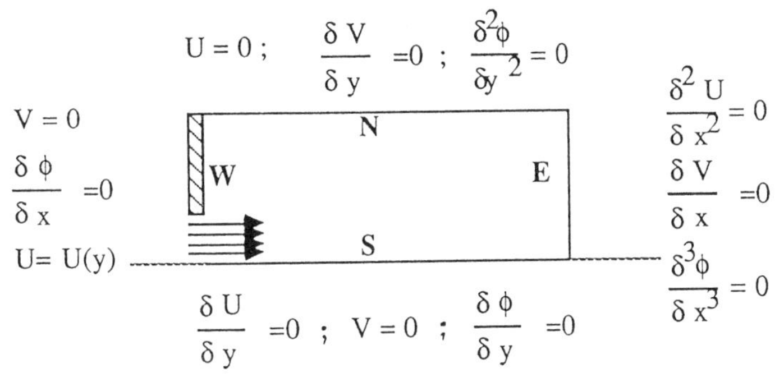

1 a   1a Domain and boundary conditions

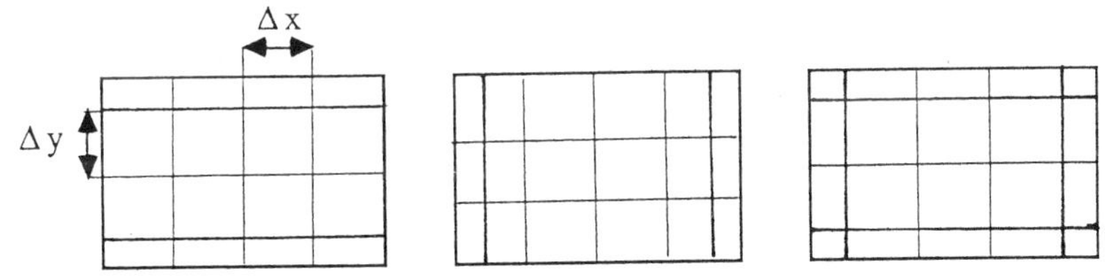

1 b   Meshing for U,V, and ϕ at boundaries

Fig .1 Domain staggered grid and boundary conditions

Table 1. Discrete formulae for Boundary Conditions (the suffix 1 is taken on the boundary , 2 and 3 being the next points insiside the domain in the considered direction).

$\frac{\partial v}{\partial y}$	$\frac{1}{2\delta y}(3V_1 - 4V_2 + V_3)$	$\frac{\partial^2 U}{\partial x^2}$	$\frac{1}{\Delta x^2}(U_1 - 2U_2 + U_3)$
$\frac{\partial u}{\partial y}$	$\frac{1}{3\delta y}(8U_1 - 9U_2 + U_3)$	$\frac{\partial^2 \Phi}{\partial y^2}$	$\frac{4}{3.\delta y^2}(2\Phi_1 - 3\Phi_2 + \Phi_3)$
$\frac{\partial v}{\partial x}$	$\frac{1}{3\delta x}(8V_1 - 9V_2 + V_3)$	$\frac{\partial^3 \Phi}{\partial x^3}$	$\frac{20}{\delta x^3}\left(\Phi_1 - \frac{15}{8}\Phi_2 + \frac{5}{4}\Phi_3 + \frac{3}{8}\Phi_4\right)$

$\frac{\partial \Phi}{\partial x} = 0$ and $\frac{\partial \Phi}{\partial y} = 0$ are expressed with centered schemes on a fictitious frontier.

The inflow boundary conditions correspond to a prescribed velocity profile. In this simulation the cases of parabolic and trapezoidal profiles are taken. As in the experiments, the momentum thickness θ_o attached to the profile is a key parameter of the developping instabilities.

The condition on the top of the domain, simulating the incomming flow from the external medium at rest was not found of great importance and it is taken very simple in order to allow the lateral entrainment by the jet.

Of major importance is the influence of the outflow boundary condition which have been chosen to not disturb the downstream convection of the vorticity shedded by the intake velocity profile. The choice of higher order conditions is similar to the one proposed by [6], despite the fact that the discretization of the boundary relationship is not the same. Comparisons with a control volume approach using the same schemes as theses authors and performed by [10] show that the results are not very sensitive to the differences in both formulations.

Related to this choice of boundary conditions, it is found that the initial conditions in the time marching process are also very important. In the case of the flow at rest and of an abrupt application of the initial velocity profile, a large vortex is created, convected downstream and followed by other smaller ones. But the boundary conditions then act as a miror and deflect the vortices towards the top surface. The computation finally diverges. Conversely, starting with an initial inviscid flow in the domain allow all the vortices to flow out the domain and at a Reynolds number of 500, this transient leads to a steady laminar jet.

A detailed set of tests on the initial conditions as well as the sizes of the domain and the meshes has been conducted. It has led to the following compromise between computational cost and numerical relevance for the purpose of a comparison with physical experiments:

Size of the domain : 20. x 3.
Number of grid points : 200 x 50
Time Step : 0.01
Time range : 0 to 120 .

The velocity reference (U_o=1) is the axis exit velocity, the length reference (d=1) is the initial jet width and the viscosity is fixed through the overall Reynolds number R_d ($U_o = U_o d/\nu$).

GENERAL RESULTS

The flow was found to reach a steady state for a parabolic velocity profile at the entry section but it remains unsteady for the other profiles. This simulation was realized for the following three cases by adjusting the slope of the trapezoidal velocity profile :

- case A, with R_d = 1000, and θ_o = 0.02 ;
- case B, with R_d = 1000, and θ_o = 0.01 ;
- case C, with R_d = 1700, and θ_o = 0.02 ;

Typical streaklines pattern of this simulation is presented in figure 2 for case B, at different time steps : t= 65 - 80.

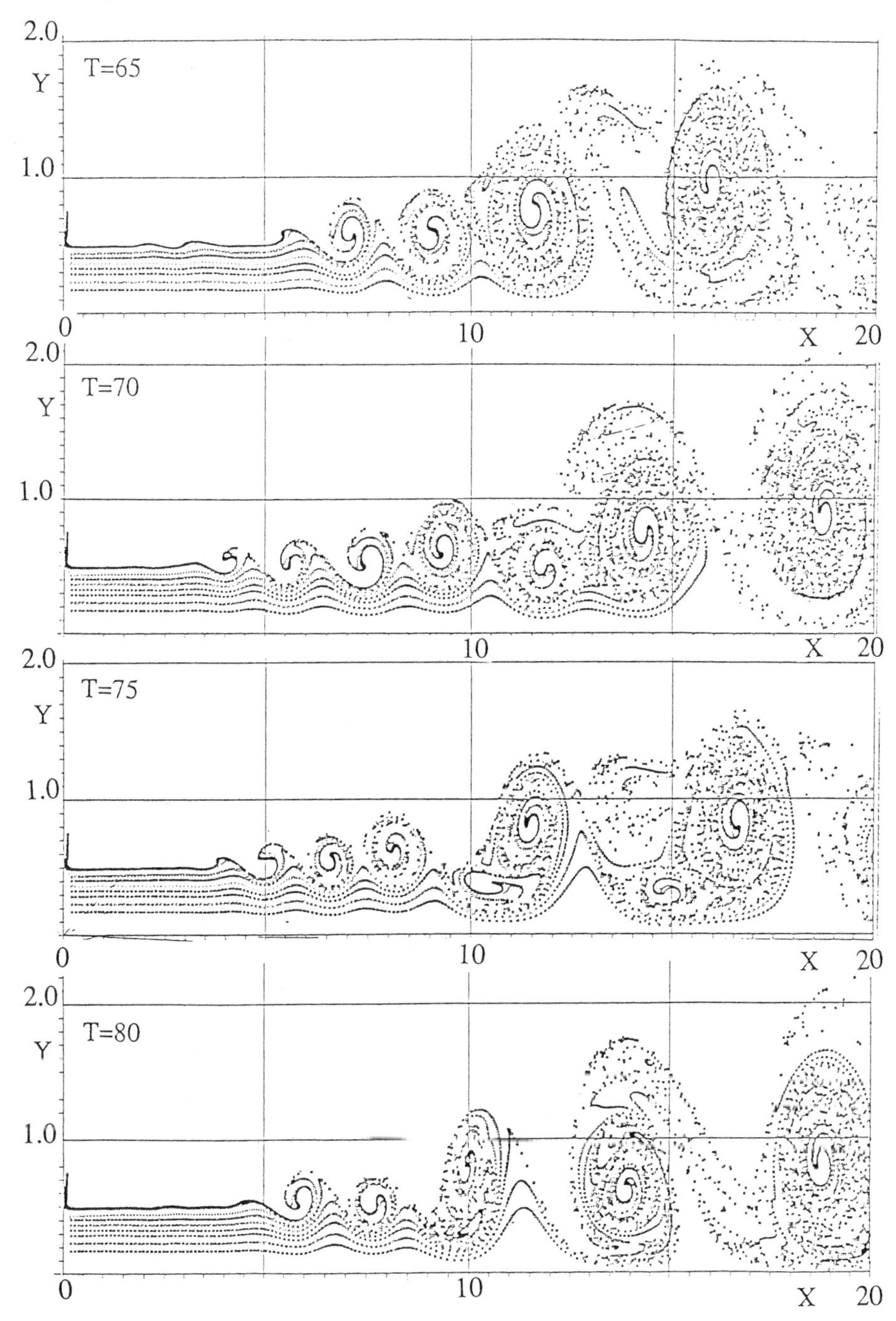

Fig . 2 Streklines pattern , Rd=1000 , $\theta_o = 0.01$, t = 65-80

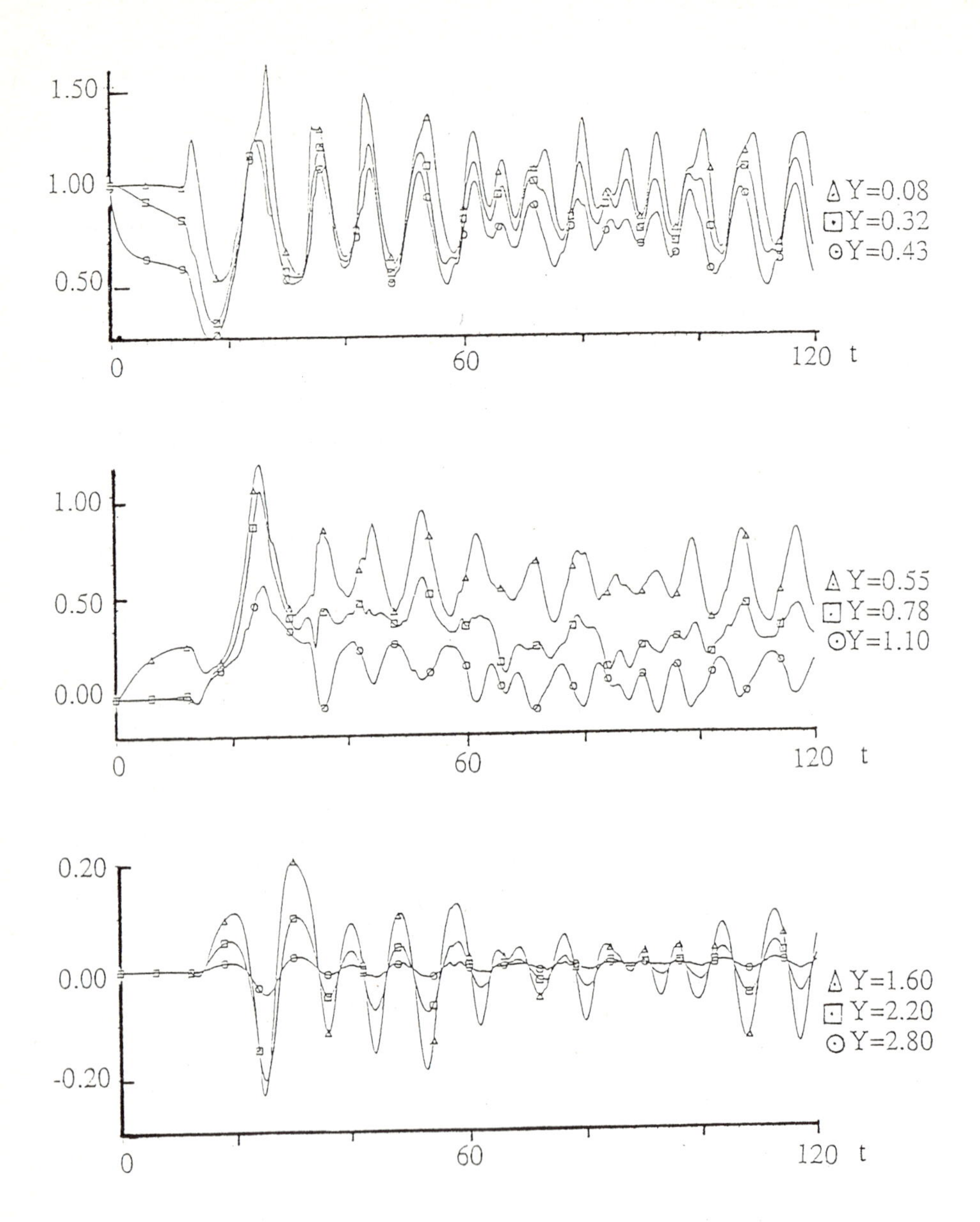

Fig . 3 Longitudinal velocity temporal signal, for Rd=1000
$\theta_{\circ}$ =0.01 , x=10

One observes successively waves, then a train of vortices convected downstream at velocity $Uo/2$, then a pairing between consecutive vortices, and finally widening of the resulting vortices. Diagonal lines are seen to convect fluid particles from the vortices region in the shearline downstream to the centerline.

All these features are in complete agreement with visualization and spectral analysis of the physical flow [12].

Typical temporal signals of longitudinal velocity $u(t)$ are presented in figure 3, for case B, at position $x=10$, and multiple lateral positions. Different signals and different frequencies are obtained in the shearline ($y=b$, where $U=Uo/2$) and the centerline, in agreement with physical results. The signal is more irregular in the shearline, as it is sensitive to the details of the moving vortices.

Simultaneous analysis of streaklines and temporal signals enables to find the signature of vortices in the temporal signal. As an example, one observes, at $y = b = d/2$, a double peak in $u(t)$ at some time-stations. This double peak corresponds to a vortex pairing, where the two paired vortices have separate traces. For instance, the double peak in the temporal signal at $x= 10$, $y= .43$, $.55$, at $t= 80$ (figure 3), corresponds to the vortex pairing, visible in the streaklines (figure 2) at same time and positions. Such double peaks have been observed in physical measurements, along the shearlines [5], but could not been explained without simultaneous analysis of streaklines and temporal signals, made possible by this numerical simulation.

INFLUENCE OF THE INITIAL VELOCITY PROFILE

Computation of mean and r.m.s. velocities enables to measure the individual influence of the initial conditions on flow development, which is almost impossible to obtain in physical experiments. Such an analysis is presented in figure 4, where the mid-plane mean velocity Um, the fluctuating intensity in the shearline $u'b$, mean and fluctuating momentum fluxes J and J' are plotted as a function of longitudinal distance x, for the 3 simulated cases :

$$J = \int_{\infty}^{+\infty} \rho_o \, U^2 dy \quad \text{and} \quad J' = \int_{-\infty}^{+\infty} \rho u'^2 dy \, , \tag{3}$$

An increase of the Reynolds number leads to a more rapid mean velocity decrease and turbulence build-up, in accordance with physical experiments [5], while change in the initial momentum thickness has little effect on these parameters.

On the other hand, a decrease of the initial momentum thickness has significant effect on the longitudinal evolution of the mean momentum flux, while increase of the Reynolds number has no effect on this parameter. The observed continuous decrease of J when $\theta_o = 0.01$, in opposition to an increase, followed by a decrease for $\theta_o = 0.02$, may be attributed to the very sharp initial velocity profile [11].

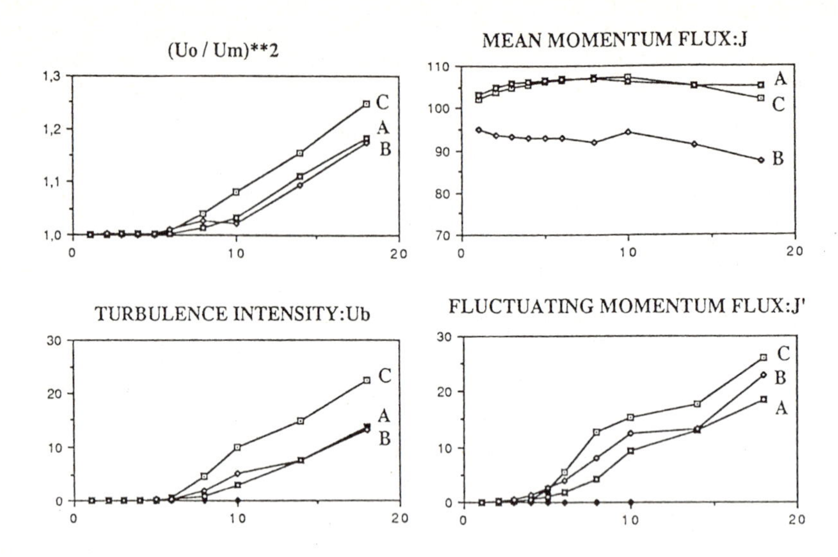

Fig . 4 Longitudinal evolution of jet parameters,
case A (Rd= 1000, θ_o=0.02) case B (Rd= 1000 , θ_o=0.01)
case C (Rd=1700 , θ_o =0.02)

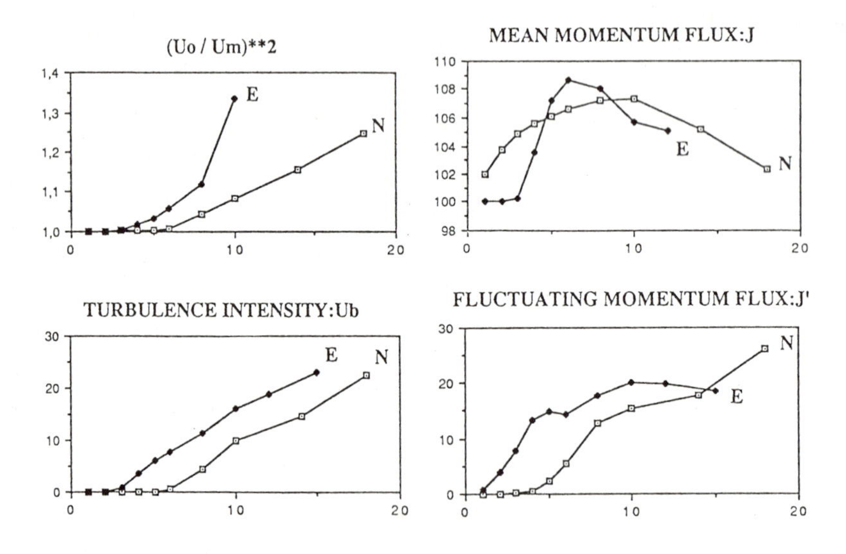

Fig . 5 Longitudinal evolution of jet parameters,
E:Experiment, N:Numerical Simulation :
case C (Rd= 1700 θ_o =0.02)

COMPARISON WITH PHYSICAL EXPERIMENTS

A detailed comparison of the jet parameters, obtained by the numerical simulation, with those obtained by physical experiments, is presented in figure 5, for case C (with similar initial conditions : $R_d = 1700$, $\theta_o = 0.02$). It is based also on the mean velocity Um, the fluctuating intensity u'_b, and mean and fluctuating momentum fluxes J and J', as a function of the distance x.

This comparison shows good agreement for the longitudinal evolution of these parameters, except for a longitudinal shift, probably because of the lack of turbulence in the entry section for the numerical simulation, while it exists in the physical experiment (measured intensity = 5.10^{-3}). It is to be noted than an exponential increase for fluctuating intensities in the mid-plane, as well as along the shearline, is get with the simulation, in accordance with physical experiment and theory of instability. A good agrement is get for momentum fluxes J and J'. For mean velocity Um, the agreement is not as good as for fluctuating intensity, probably because of still non- perfect boundary conditions at the outflow boundary.

CONCLUSION

This numerical simulation stops to be physically correct at the end of the potential region, because it does not predict the vortex bursting, occuring there, due to merging and interaction of the two mixing layers [5].

It has been shown [3] that in this interaction region, a sharp transition occurs from 2-D to 3-D flow. It is therefore clear that this present 2-D simulation cannot predict these 3-D features, and cannot be valid beyond than the potential region.

But, in this potential region, this numerical simulation predicts correctly the physical flow, including : the instability process, the vortex apparition, growing, and pairing, and the longitudinal evolution of mean and fluctuating velocities and momentum fluxes.

Moreover, simultaneous analysis of streaklines pattern and temporal signals, allows an investigation of the flow, which is generally not possible in physical experiments.

ACKNOWLEGMENTS

The authors are grateful to Professor CHASSAING for having initiated this study. They thank Dr. BRAZA and Dr. KOURTA for their numerical contribution and Mr. PONS, BONNEFONT and MRS. NGUYEN for their technical suppport.

This work was supported by the Ministery of Defence Research Department (D.R.E.T.) under contract number 87/131.

REFERENCES

[1] ABRAMOVICH, G.N., 1963, The Theory of Turbulent Jets, M.I.T. Press, Cambridge, Mass., U.S.A.

[2] ZAMAN, K.B.M.Q., HUSSAIN, A.K.M.F., 1981, J.Fluid Mech., 103, 133-159.

[3] THOMAS, F.O., GOLDSCHMIDT, V.W., 1986, J.Fluid Mech., 163, 227-256.

[4] EVERITT, K.W., ROBINS, A.G., 1978, J.Fluid Mech., 88, 563-583.

[5] MEYER, J., SEVRAIN, A., BOISSON, H.C., HA MINH, H., 1989, 2nd IUTAM Symp.Struc.Turb.Drag Reduct., Zurich.

[6] SHIMIZU, A., WADA, T., 1985, Computers & Fluids, 13, 83-97.

[7] BRAZA, M., CHASSAING, P., HA MINH, H., 1986, J.Fluid Mech., 165, 79-130.

[8] KOURTA, A., BRAZA, M., CHASSAING, P., HA MINH, H., 1987, AIAA J., 25, 279-286.

[9] GRANT, A.J., 1974, J.Fluid Mech., 66, 707-724.

[10] ESTIVALEZES J.L., 1989, Thèse de Docteur de l'INP de Toulouse.

[11] GUTMARK, E., WYGNANSKI, I., 1976, J.Fluid Mech., 73, 465-495.

[12] MEYER, J., 1989, Thèse de docteur de l'I.N.P.T., Toulouse.

EFFICIENT SOLUTION OF TURBULENT INCOMPRESSIBLE SEPARATED FLOWS

Vittorio Michelassi and Francesco Martelli
Energy Engineering Department
University of Florence, Italy

SUMMARY

A computational method for incompressible separated flows based on two-dimensional approximate factorization is presented. Turbulence effects are accounted for by low-Reynolds number forms of the k-ϵ model. Mass conservation is enforced by the artificial compressibility method. Decoupling and coupling of the equations of motions with the turbulence model equations are investigated. Testing of the coupled solver showed no improvement in convergence or accuracy in comparison to the classical decoupled approach. The solver was then applied to several large-recirculation flows using a modified version of the low-Reynolds-number form of the k-ϵ model proposed by Chien and a two-layer version of the k-ϵ model introduced by Rodi. Both versions gave fast convergence rates and good agreement with experiments.

INTRODUCTION

The need for accurate and efficient solvers is especially pressing when turbomachinery designers must compute turbulent internal flows in the presence of large separated regions frequently encountered at off-design conditions. For compressible flows, good results have been obtained by Coakley [1] with two equations turbulence models and the approximate factorization solver. The same solution technique originally developed for compressible flows, was applied to laminar incompressible separated flows by Michelassi and Benocci [2],[3] who used the pressure-velocity formulation and artificial compressibility concept in order to be able to march in time avoiding the Poisson equation for the pressure or the well known pressure correction method.

The artificial compressibility method allows overcoming the pressure-density decoupling typical of compressible solvers used at very low Mach number flows. On the basis of the fast and accurate results obtained by Michelassi and Benocci in the laminar flow regime, the possibility of extending the solver to turbulent flows was examined. Following the work of Patel et al.[4], the low-Reynolds-number form of the k-ϵ model - which has been used mainly for not predominantly elliptic flows - were selected to enhance accuracy and efficiency. The fastest and most accurate solver was then determined by investigation of the decoupled and coupled solutions of the Navier-Stokes (N-S) and turbulence model equations.

TURBULENCE MODELLING

Turbulent flows have been successfully computed over a wide range of flow regimes with Reynolds averaged N-S equations using the high-Reynolds-number form of the k-ϵ model [5]. However, this formulation has the drawback of requiring "wall functions" to bridge the viscous and buffer layers in proximity to solid walls. This approach, in which the influence of molecular viscosity is not modelled, is strictly valid only for attached shear flows and can give rise to inaccuracies for more complex or recirculating flows. With the low-Reynolds-number forms of the k-ϵ model, the k-ϵ equations are integrated down to the wall accounting for the various viscous effects (or viscosity modelled even if physically not viscous) that are generally neglected in the high-Re form. Unlike the standard k-ϵ model, the low-Reynolds-number forms are tuned to fit the experimental data in the viscous and buffer layers. From this standpoint, there are no theoretical restrictions to applying the low-Re forms to elliptic flows. The standard nondimensional form of the k-ϵ model is

$$k_{,t} + (u_i\, k)_{,i} = (\Gamma_k\, k_{,i})_{,i}/Re + \mathbf{v}_t\, P/Re - \epsilon + D \tag{1}$$

$$\epsilon_{,t} + (u_i\, \epsilon)_{,i} = (\Gamma_\epsilon\, \epsilon_{,i})_{,i}/Re + C_1 \epsilon \mathbf{v}_t\, P/(k\, Re) - C_2\, \epsilon^2/k + E \tag{2}$$

in which

$$Re = \text{Reynolds number}$$
$$u_i = \text{velocity}$$
$$P = u_{i,j}\,(u_{i,j} + u_{j,i})$$
$$\mathbf{v}_t = Re\, C_\mu k^2/\epsilon \qquad \Gamma_k = \mathbf{v}_t/\sigma_k \qquad \Gamma_\epsilon = \mathbf{v}_t/\sigma_\epsilon$$
$$C_1 = 1.44 \quad C_2 = 1.92 \quad C_\mu = 0.09 \quad \sigma_k = 1.0 \quad \sigma_\epsilon = 1.3$$

The extra terms D and E are set equal to zero only in the high-Re form. The equations are written in tensor form so that, for example, $u_{i,j}$ means derivative of the ith component of the velocity vector in the jth coordinate direction. For high-Re flows under the hypothesis of homogeneous turbulence, the total dissipation rate ϵ and its isotropic part ϵ' coincide and may be expressed as;

$$\epsilon = \epsilon' = \mathbf{v}_l\, \overline{u_{i,j}\, u_{j,i}}$$

in which the overbar denotes a time average. Whenever the assumption of isotropic dissipation loses validity - for example, very close to walls - modifications must be made to account for the nonisotropic part of the dissipation rate. This requires either solving directly for the total dissipation rate, ϵ, that includes the non-isotropic effects or else introducing the explicit additional term D in (1) that represents the nonisotropic part of the dissipation, $\epsilon = \epsilon' + D$. Both the approaches will be considered here.

In the comprehensive overview of low-Reynolds-number k-ϵ models by Patel et al. [4] only three models were found to posses the degree of generality usually associated with a two-equation turbulence model. Among the more general forms, the Chien model [6] was selected since it showed good agreement with measurements for boundary layers.

A two-layer version of the k-ϵ model, introduced and tested by Rodi and Cordes [7], in which viscous effects are accounted for, was also considered.

In the author's investigation of these and other models [8] emphasizing quality of results and numerical behavior of the formulation, the Chien and the two-layer models were found to be the best suited for the computation of incompressible separated flows. A brief description of the two models follows.

The Chien Low-Reynolds-number k-ϵ model (CH)

In the Chien's formulation [6] the two transport equations (1) and (2) for k and ϵ' are valid with the definitions of the additional terms, D and E, expressed by

$$
\begin{aligned}
D &= -2\, k / (Re\, d^2) \\
E &= -2\, \epsilon'\, f_3 / (Re\, d^2) \\
u_* &= \sqrt{v_l\, u_{,j}\ wall} \\
v_{eff} &= v_t + v_l \qquad \Gamma_k = v_l + v_t/\sigma_k \qquad \Gamma_\epsilon = v_l + v_t/\sigma_\epsilon \\
C_\mu &= C_{\mu\infty}\, (1 - \exp(-C_3\, Re\, d\, u_*)) \\
C_2 &= C_{2\infty}\, (1 - 0.022 \exp(-(Re\, k^2 / (6\, \epsilon'))^2)) \\
f_3 &= \exp(-C_4\, Re\, d\, u_*) \\
d &= \text{distance from the closest wall} \\
v_l &= \text{laminar kinematic viscosity .}
\end{aligned}
$$

The empirical constants C_1, $C_{2\infty}$, $C_{\mu\infty}$ are the high Re values given for the standard k-ϵ formulation, while C_3=0.0115 and C_4=0.5 are additional constants. The total dissipation rate is introduced in the k equation observing that

$$\epsilon = \epsilon' + D = \epsilon' - 2\, k / Re / d^2$$

in which D represents the extra near-wall destruction term. Note that the wall boundary condition is $k=\epsilon'=0$. The extra term D tends to a finite value at the wall, since $k \alpha d^2$ holds in the viscous sublayer. Also, an extra term, E, is added to the ϵ equation to ensure that $\epsilon' \approx d^2$ holds in the wall region.

The choice of the friction velocity u_* as a scale parameter in the exponential damping functions, valid only for attached flows, may result in prediction of an unphysical growth of the boundary layer thickness at the separation and reattachment points. After intense numerical testing, we found convenient to compute u_* via the wall function expression for the turbulent kinetic energy k

$$u_* = \sqrt{k_{max}\ \sqrt{C_{\mu\infty}}}\ . \tag{3}$$

The value of k_{max} in (3) was chosen as the turbulent kinetic energy peak closest to the wall for every cross section.

The Two-Layer k-ϵ Model (TL)

In testing lower order turbulence models Rodi and Scheuerer [9] found that the high-Re k-ϵ failed to reproduce the skin friction coefficient over a flat plate, while the simple Norris and Reynolds [10] formulation generally gave better results in terms of fit to measurements. In the Norris and Reynolds model [10], the following ramp distribution for the turbulence length scale L is assumed

$$L = C_D \min (K\, d\ ,\ \Phi\, \delta) \tag{4}$$

in which C_D and Φ are empirical constants, K=0.41 is the Von Karman constant, δ is the boundary layer thickness. Since the computation of δ is not straightforward when flow conditions are not fully developed or in presence of recirculations, Rodi and Cordes applied (4) in its linear variation region close to the wall, ($y^+ \leq 60$-80), thereby obtaining

$$L = C_D \, K \, d \; . \tag{5}$$

The damping function f_μ, in which $R_y = k^{\frac{1}{2}}$ (d Re) is implemented as

$$f_\mu = 1 - \exp\left(- 0.0198 \, R_y \right) \; . \tag{6}$$

Rodi and Cordes proposed solving the k-equation, with v_t damped according to (6), and then computing the total dissipation rate ϵ according to the Norris and Reynolds one-equation model in the inner layer ($y^+ \leq 60$-80), while maintaining the standard k-ϵ model in $y^+ \geq 60$-80 thereby setting D=E=0 (Table 1). The original values of the Norris and Reynolds model constants were modified to be able to retain the f_μ in the entire computational domain and not only in the inner layer [11] obtaining

$$C_\epsilon = 12.9858 \qquad C_D = 6.0858 \; .$$

With this form of the k-ϵ model, it is possible to avoid the cumbersome solution of the ϵ equation very near the wall that often gives rise to numerical problems. Note that because of the implementation of (5) in place of (4) it is necessary to ensure that matching between the two layers takes place in the region of linear variation of L where the wall effect on ϵ nearly vanishes. Cordes and Rodi [7] found it convenient to use the ratio $v_t/v_l \approx 36$ as switching criterion, which is like saying $f_\mu \approx 0.6 - 0.9$ (for fully developed flow this corresponds to $y^+ \approx 60$-90). In the present calculations, f_μ was used as the switching criterion to pass from one set of equations to the other between the inner and outer layers.

Table 1: Two-Layer Model Formulation.

Wall Distance	L	ϵ	k
$y^+ \leq 60$-80	$C_D \, K \, d$	$k^{3/2}(1+C_\epsilon/(\mathrm{Re}\, L\, k^{\frac{1}{2}}))/L$	k-eq. with f_μ
$y^+ \geq 60$-80	$k^{3/2}/\epsilon$	Standard ϵ equation	k-eq. with f_μ

THE SOLUTION METHOD

The Artificial Compressibility Method

The proposed solver is designed for steady incompressible flows. As long as the primitive variable formulation (pressure and velocity) is retained for incompressible flows, neither pressure nor density appears in the continuity equation with the result that the equation cannot be directly used to compute the pressure field.

A pseudo-unsteady form of the continuity equation was

obtained by Chorin [12] by a fictitious state equation, $P=\varrho \mathbf{E}$, where $\mathbf{E}>0$ is an artificial compressibility parameter, ϱ a fictitious density, placing the pressure in the time-dependent form of the continuity equation that reduces to

$$p_{,t} + \varrho \, \mathbf{E} \, u_{i,i} = 0 \; . \tag{7}$$

Equation (7) can be utilized to compute the pressure field, p. However, since mass conservation is enforced only at the steady state, it is impossible to follow a physical time transient. The most convenient values for the artificial compressibility parameter $\mathbf{E}$ was found to range between 1 to 10 times the average velocity module, which keeps it constant over the entire computational domain.

The Reynolds averaged N-S equations (8) under the Boussinesq assumption, are then coupled to (7)

$$u_{i,t} + (u_i \, u_j)_{,j} + p_{,i}/\varrho = (\mathbf{v}_{eff} \, (u_{i,j}+u_{j,i}))_{,j}/Re \; . \tag{8}$$

The artificial compressibility parameter should not be given excessive physical meaning, since the momentum equations are written in incompressible form and the whole set of differential equation is physically unbalanced until the steady state solution is reached. Accordingly, $\mathbf{E}$ can be seen as the relative weight of the physical part of the artificial compressibility equation with respect to the unsteady and nonphysical term.

The Conservative Form of the Basic Differential Equations

Strongly stretched grids and, in the case of an irregular physical domain, a curvilinear nonorthogonal coordinate system (α,β) are introduced for solving viscous turbulent flows. The partial differential equations are discretized with centered finite differences adopting Pulliam and Steger's [13] strong conservative form to minimize the errors caused by undifferentiated metrics: the following metric relations hold

$$\alpha_{,x} = J \, y_{,\beta} \qquad \alpha_{,x} = - J \, x_{,\beta} \qquad \beta_{,x} = - J \, y_{,\alpha} \qquad \alpha_{,x} = J \, x_{,\alpha}$$
$$J^{-1} = x_{,\alpha} \, y_{,\beta} - x_{,\beta} \, y_{,\alpha} \; . \tag{9}$$

With the introduction of the artificial compressibility equation, the differential system can be written in vector form

$$q_{,t} + F_{,\alpha} + G_{,\beta} = F_{v,\alpha} + G_{v,\beta} + H \tag{10}$$

in which the flux vectors are

$$q = J^{-1} \, [\, p \, , \, u \, , \, v \, , \, \epsilon \, , \, k \,]^{t}$$

$$F = J^{-1} \left| \begin{array}{l} \mathbf{E} \, U \\ u \, U+\alpha_{,x} \, (p/\varrho - 2/3 \, k) \\ v \, U+\alpha_{,y} \, (p/\varrho - 2/3 \, k) \\ \epsilon \, U \\ k \, U \end{array} \right| \qquad G = J^{-1} \left| \begin{array}{l} \mathbf{E} \, V \\ u \, V+\beta_{,x} \, (p/\varrho - 2/3 \, k) \\ v \, V+\beta_{,y} \, (p/\varrho - 2/3 \, k) \\ \epsilon \, V \\ k \, V \end{array} \right|$$

$$F_V=J^{-1}\begin{vmatrix} 0 \\ \alpha_{,x}\,\tau_{,xx} + \alpha_{,y}\,\tau_{,xy} \\ \alpha_{,x}\,\tau_{,yx} + \alpha_{,y}\,\tau_{,yy} \\ \alpha_{,x}\,e_{,x} + \alpha_{,y}\,e_{,y} \\ \alpha_{,x}\,h_{,x} + \alpha_{,y}\,h_{,y} \end{vmatrix} \qquad G_V=J^{-1}\begin{vmatrix} 0 \\ \beta_{,x}\,\tau_{,xx} + \beta_{,y}\,\tau_{,xy} \\ \beta_{,x}\,\tau_{,yx} + \beta_{,y}\,\tau_{,yy} \\ \beta_{,x}\,e_{,x} + \beta_{,y}\,e_{,y} \\ \beta_{,x}\,h_{,x} + \beta_{,y}\,h_{,y} \end{vmatrix}$$

$$H = J^{-1}\begin{vmatrix} 0 \\ 0 \\ 0 \\ C_1\,C_\mu\,k\,P - C_1\,2/3\,\epsilon\,P_D - C_2\,\epsilon^2/k + E \\ C_\mu\,k^2\,/\,\epsilon\,P - 2/3\,k\,P_D - \epsilon + D \end{vmatrix}$$

in which

$$P_D = u_{i,i}$$

$$\tau_{,xx}=2\mathbf{v}_{eff}\,u_{,x} \quad \tau_{,yy}=2\mathbf{v}_{eff}\,u_{,y} \quad \tau_{,xy}=\tau_{,yx}=\mathbf{v}_{eff}(u_{,y} + v_{,x})$$

$$e_{,x} = \Gamma_\epsilon\,\epsilon_{,x} \qquad e_{,y} = \Gamma_\epsilon\,\epsilon_{,y} \qquad U = \alpha_{,x}\,u + \alpha_{,y}\,v$$

$$h_{,x} = \Gamma_k\,k_{,x} \qquad h_{,y} = \Gamma_k\,k_{,y} \qquad V = \beta_{,x}\,u + \beta_{,y}\,v$$

and, for example, the x-derivative of u is given by

$$u_{,x} = u_{,\alpha}\,\alpha_{,x} + u_{,\beta}\,\beta_{,x}.$$

Note that the system is solved with respect to the cartesian velocities u_i and that the turbulent normal stress contribution, $\delta_{ij}\ 2/3\ k$, is explicitly retained; (δ_{ij} is the Kronecker delta).

The production term appearing in the k-ϵ model is normally written as

$$P = \overline{u'_i\,u'_k}\,(u_{j,k}) + \overline{u'_j\,u'_k}\,(u_{i,k}) \tag{11}$$

where u' are the fluctuating velocity components. When a proper closure is performed under the Boussinesq assumption, the Reynolds stresses can be written in general tensor form

$$\overline{u'_i\,u'_j} = \mathbf{v}_t\,(\,u_{i,j} + u_{j,i}\,) - 2/3\,\delta_{ij}\,k\,. \tag{12}$$

Introduction of (12) in (11) gives the following expression for the production term

$$P=[\mathbf{v}_t\,(u_{i,j}+u_{j,i})u_{i,j} - 2/3\,u_{i,i}\,k] = [\mathbf{v}_t P - 2/3 P_D\,k]\,. \tag{13}$$

In the second term of (13), $u_{i,i}$ is zero, according to the continuity equation, only at the steady state. In the numerical transient large mass conservation error may occur so that the neglection of $u_{i,i}$ in (13) can bring to values of k that differ considerably from $\Sigma_i\,\overline{u'^2}_i$ with the possible result of negative turbulent kinetic energies. So, the term P_D was kept in the production of both k and ϵ in vector H.

<u>The Numerical Algorithm</u>

The approximate factorization method proposed by Beam and Warming [14] splits an N-dimensional operator into the product of N 1-dimensional operators that can be solved in sequence. For a two-dimensional problem, the linear system resulting from (10) can be written

$$[I + \theta\delta t(\theta_\alpha H_j + (A+R_{,\alpha})_{,\alpha} - R_{,\alpha\alpha} - \Omega_i J^{-1}D_{,\alpha\alpha}(J))] \delta q^* = RHS$$
$$[I + \theta\delta t(\theta_\beta H_j + (B+S_{,\beta})_{,\beta} - S_{,\beta\beta} - \Omega_i J^{-1}D_{,\beta\beta}(J))] \delta q = \delta q^* \quad \textbf{(14)}$$

in which
$$RHS = \delta t \; (-F_{,\alpha}-G_{,\beta}+F_{v,\alpha}+G_{v,\beta}+H - \Omega_e(D_{,\alpha\alpha\alpha\alpha}(q) + D_{,\beta\beta\beta\beta}(q))).$$

δt is the time step, q^n is the solution at time $t= n\,\delta t$, δq equals $q^{n+1} - q^n$; the θ parameter weights the implicit and explicit contribution to the space operators (with θ=1 used for all the calculations). The jacobians of convective, diffusive and source terms are defined as

$$A = F_{,q} \quad B = G_{,q} \quad R = F_{v,(q,\alpha)} \quad S = G_{v,(q,\beta)} \quad H_j= H_{,q} \,.$$

The resulting linear system is solved by block tridiagonal matrix inversion. System (14) serves to advance the solution in time until the change δq between two consecutive solutions is small enough (typically $\delta q \approx 10^{-7}$). H_j and the two weighting functions related to it,

$$\theta_\alpha= |V| \;/\; (U^2 + V^2)^{\frac{1}{2}} \qquad\qquad \theta_\beta= |U| \;/(U^2 + V^2)^{\frac{1}{2}}$$

will be discussed in the next section.

The implicit formulation may be fully exploited by using a local time step based on an approximation of the Courant condition in which the diffusive terms contribution to δt is accounted for to avoid stability problems in viscosity dominated regions. The following expression for the local time step was used

$$\delta t=\frac{CFL}{|U|+|V|+C(\alpha_{,x}^2+\alpha_{,y}^2+\beta_{,x}^2+\beta_{,y}^2)^{\frac{1}{2}}+\mathbf{v}_{ef}(\alpha_{,x}^2+\alpha_{,y}^2+\beta_{,x}^2+\beta_{,y}^2)/Re}$$

in which $C= 1 \;/\; (\mathbf{E}^{\frac{1}{2}})$ according to the artificial equation of state associated with (7).

Fourth-order-difference explicit ($D_{,\alpha\alpha\alpha\alpha}$, $D_{,\beta\beta\beta\beta}$) and second-order-difference implicit ($D_{,\alpha\alpha}$, $D_{,\beta\beta}$) numerical dissipation terms are introduced in equation (14) to damp the numerical modes with the highest wave numbers (Beam and Warming, [14]) with the local weights (Ω_e and Ω_i) definition given by Michelassi and Benocci [3]. The use of artificial damping was necessary only in the artificial compressibility equation which contains no physical damping term and might consequently cause a decrease in convergence speed; moreover, they allowed avoiding pressure-velocity decoupling that may occur because of the adopted centered discretization. While the introduction of these artificial terms only in the continuity equation prevents undesired momentum diffusion, it nevertheless generates an inlet/outlet mass error ranging from 0.01% to 1% in highly distorted grids. Typical weigths are Ω_e=0.01 and Ω_i=2Ω_e.

Boundary Conditions

Boundary conditions were imposed explicitly by setting δq equal to zero on the computational boundaries when solving implicitly the internal domain: at the end of every iteration the unknowns were updated to match the boundary condition on every specific side. The no-slip condition was applied on solid walls, where the static pressure was computed imposing zero normal to the wall derivative. On walls, k was set equal to zero for both CH and TL, while $\epsilon'=0$ and $\epsilon_{,y}=0$ were the ϵ condition for CH and TL. At the domain exit, the static pressure was fixed, while all the other variables were extrapolated assuming fully developed flow. At the inlet section, the velocities and turbulence quantities were given for a fully developed channel flow.

<u>The Decoupled and Coupled Approaches</u>

Despite their structural differences, for both CH and TL models the implicit treatment of the k-ϵ sink and source terms was compulsory in the decoupled and coupled modes. The jacobian of the sink and source terms of the Chien's model, H_j, which is obtained by differentiating the vector H with respect to k and ϵ, is given by:

$$H_j = \begin{vmatrix} 0 & 0 & 0 & 0 & 0 \\ 0 & 0 & 0 & 0 & 0 \\ 0 & 0 & 0 & 0 & 0 \\ 0 & 0 & 0 & \begin{array}{l} -2/3\ C_1\ P_D - 2C_2\ \epsilon/k + \\ (C_{2\infty}-C_2)\ Re^2\ k^{3/2}/(18\ \epsilon) \\ -2\ f_3\ /(Re\ d^2) \end{array} & \begin{array}{l} C_1\ C_\mu\ P + C_2\ \epsilon^2\ /\ k^2 - \\ (C_{2\infty}-C_2)\ Re^2\ k^2\ /\ 9 \end{array} \\ 0 & 0 & 0 & -C_\mu\ k^2\ /\ \epsilon^2\ P - 1 & \begin{array}{l} 2\ C_\mu\ k/\epsilon\ P - \\ 2/3\ P_D - 2\ /\ (Re\ d^2) \end{array} \end{vmatrix} .$$

The jacobian matrix was added in both the sweeps according to the Θ_α and Θ_β weighting funtions used in (14); these expressions allows to introduce H_j in the sweep where the largest gradients take place enhanching the dominance of the jacobian matrix diagonal.

In applying two equations models, it is common practice to solve the partial differential equations with the turbulence model decoupled from the equations of motion. A solver in which the turbulence model source terms are treated implicitly and the N-S equations are solved coupled with the k and ϵ transport equations was compared with a the standard decoupled method. In the decoupled approach, the N-S equations are implicitly solved to update the pressure and the velocity field: then, the k-ϵ equations are solved together with the new given velocities to compute a new turbulent viscosity field. The coupled approach consists of the coupled solution of the N-S plus k-ϵ equations yielding a 5x5-block tridiagonal inversion in which the turbulent viscosities in the diffusive terms are treated explicitly. The following matrix gives, for example, the α-direction convection term jacobian for the coupled solver;

$$A = \begin{vmatrix} 0 & \alpha_{,x} & \alpha_{,y} & 0 & 0 \\ \alpha_{,x} & (2\alpha_{,x}u+\alpha_{,y}v) & \alpha_{,y}u & 0 & -2/3\ \alpha_{,x} \\ \alpha_{,y} & \alpha_{,x}v & (\alpha_{,x}u+2\alpha_{,y}v) & 0 & -2/3\ \alpha_{,y} \\ 0 & \alpha_{,x}\epsilon & \alpha_{,y}\epsilon & (\alpha_{,x}u+\alpha_{,y}v) & 0 \\ 0 & \alpha_{,x}k & \alpha_{,y}k & 0 & (\alpha_{,x}u+\alpha_{,y}v) \end{vmatrix}$$

while the two following matrices refer to the decoupled

solver, the first one of which, A_{N-S}, for the N-S sweep, the second one, $A_{k-\epsilon}$, for the k-ϵ sweep;

$$A_{N-S}=\begin{vmatrix} 0 & \alpha_{,x} & \alpha_{,y} \\ \alpha_{,x} & 2\alpha_{,x}u+\alpha_{,y}v & \alpha_{,y}u \\ \alpha_{,y} & \alpha_{,x}v & \alpha_{,x}u+2\alpha_{,y}v \end{vmatrix} \quad A_{k-\epsilon}=\begin{vmatrix} \alpha_{,x}u+\alpha_{,y}v & 0 \\ 0 & \alpha_{,x}u+\alpha_{,y}v \end{vmatrix}.$$

In the first approach the convection of k and ϵ is treated implicitly coupled with the fluid motion equations; moreover, the normal stress contribution to the static pressure is implicitly accouted for.

RESULTS

Comparison of Decoupled and Coupled Approaches

The coupled and decoupled solvers were compared using the CH model on a typical bump geometry with a 70x31 computational grid. The Reynolds number, based on the inlet section width and maximum velocity, was 12,300. The flow field was initialized with the fully developed flow conditions specified at the inlet section. CFL was set equal to 5, **E** = 5 for both the coupled and decoupled approaches. The implicit-explicit artificial damping weigth was $\Omega_e = 0.0025$.

The two approaches gave identical flow patterns (Figure 1). The convergence histories (Figure 2) show no appreciable differences for the maximum change δq with machine accuracy (10^{-7}) reached in fewer than 500 iterations. Moreover, all the points on the computational domain converged at the same rate for both the solvers.

With the convergence histories so alike, it is clear that the implicit treatment of the k and ϵ convective terms provided by the coupled solver hardly plays a significant role. Nevertheless, the flow field investigated here is largely dominated by diffusive processes, as demostrated by the presence of the large separated region; this is probably the reason why the coupling of the turbulence model with the N-S equations via the convective transport of k and ϵ does not bring about any gains in convergence. Still, it would be interesting to verify this result for convection dominated flows.

The decoupled solver was applied for all the further tests since it brought a 15% reduction in computational time with respect to the coupled solver in which inversion of a 5x5 block tridiagonal matrix is required.

Testing on a 180-Deg Planar Turnaround Bend

After prior code validation [8] the Chien low-Re form of the k-ϵ model was applied to the flow in a planar turnaround duct. The Reynolds number, based on the inlet section width and averaged velocity was 80,000. The test geometry, as described by Sandborn and Shin [15], consisted of a 180-deg bend in a constant cross-section channel with smooth walls.

In low-Re k-ϵ models, a strong clustering of points at the wall is needed in order to capture the shape and peaks of velocities and turbulent kinetic energy. The three grids obtained by a simple algebraic mesh generator are characterized in Table 2, with the 90 x 65 grid used for the first run shown in Figure 3.

Averaged residuals of the order of 10^{-6} were obtained in approximately 700 iterations with the decoupled solver. Weak pressure wiggles were detectable only at the bend exit section of grid # 1. These local instabilities could easily be removed incresing the fourth order damping weigths (Ω_e = 0.0025), but it was found that the inlet-outlet mass error could rise to 1% should Ω_e = 0.1 be exceeded, while it would average around 0.01% if the weights were not increased. At any rate, the wiggles vanished with the mesh refinement afforded by the 151 x 121 grid.

Table 2. Grids for Planar Turnaround Bend

Grid Id	Number of Grid Points	Cross-Flow Expansion Ratio
1	90 x 65	1.15
2	110 x 95	1.20
3	151 x 121	1.18

Figure 4a shows the bend inlet profiles of the streamwise velocities for the the three grids. Agreement with measurements is generaly good. Velocity gradients are slightly underestimated close to the walls indicating that the computed boundary layer is thicker than the experimental one, regardless of grid.

As visible in Figure 4b, fit deteriorates at the bend exit section where the velocity profiles predicted by the model are smoother than experimental values due to a strong momentum transfer in the crossflow direction. The poor fit common to all grids is probably due to the behavior of the k-ϵ model in presence of strong streamline curvature and should not be imputed to mesh quality. The model remarkably predicts the same lack of separation indicated by the experimental velocity profile, despite the uncertainty in measurements caused by flow unsteadiness.

The calculated value of the static pressure coefficient C_p defined as

$$C_p = p \ / \ (1/2 \ \rho \ u^2{}_{in})$$

and plotted in Figure 4c compares favorably with measured values: Only a slight underestimation of the pressure minimum is detectable along the inner wall. This underestimation is related to the smooth velocity profile predicted at the bend exit section that causes a loss redistribution in the cross-stream direction. Mesh dependence is nevertheless quite remarkable, with no sizeable changes found between grid # 1 and grid # 3 which has approximately 3 times as many points.

Testing on Cylindrical Diffusers

Tests of greater significance were carried out in 45- and 15-deg cylindrical diffusers. The Reynolds number, based on the inlet section diameter and mean velocity, was 200,000. The computation of cylindrical diffusers was regarded as expecially valuable in verifying the capability of the model to detect large separated regions at high Re and in presence of strong adverse pressure gradients. Results calculated by CH and TL models were compared with the measured values reported by Chaturvedi [16].

A word is in order regarding the differences between CH and TL models. The turbulence length scale and turbulent kinetic energy profiles as predicted by the models for a fully developed channel flow are shown in Figure 5 (a and b). The turbulence length scale profile (Figure 5a) highlights the region of linear variation of L within $y^+ < 80$ where TL switches to equation (5). The underestimation of k given by TL is especially noticeable in the range $0 < y^+ < 100$. The differences fade in the core of the flow region ($y^+ > 500$). The differences are clearly located only in the viscous and buffer layers, at least for this simple flow condition. For further details about the models, the reader can refer to [8].

<u>45-Deg Flow</u>. The 80 x 45 grid for the 45-deg model (Figure 6) was selected on the basis of the mesh dependence tests performed on the 180-deg turnaround duct. The grid expansion ratio in the crossflow direction was 1.2 - a point-clustering that allowed 20 points to be placed within $y^+ < 40$, but that produced a very coarse mesh in the core flow region. A symmetry boundary condition was applied at the top boundary, while the standard solid wall condition was specified at the bottom wall.

Fully converged results were obtained with CFL=5 in approximately 1000 iterations. With the decoupled solver, the CH and TL models required approximately the same number of iterations to converge, although TL with only one exponential damping expression proved slightly less costly in computational terms than CH in which three functions are introduced. The choice of ϵ' as unknown made in CH as opposed to ϵ used in TL proved not to be critical; the sharper gradients exhibited by ϵ' as compared to ϵ in the wall region did not cause convergence or stability problems.

A qualitative indication of the recirculation length is visible in Figure 7 (a, b) where the computed and measured streamline patterns are compared. CH predicts the reattachment position at $Z \approx 8.25$, while the TL model predicts $Z \approx 8.90$. Both results are in good agreement with the experimental value of $Z \approx 9.50$. A secondary counter-rotating vortex was detected by both models at the lower corner of the pipe expansion (Figure 8). This feature is of notable interest, since the vortex would probably be lost using the standard high-Re form of the k-ϵ, since it is located inside the wall function application field.

The three crossflow velocity profiles in Figure 9 show close agreement with measurements, with the TL model predicting a slightly sharper velocity gradient at the wall.

Qualitative information on k is given by the isoline plots in Figure 10 which shows underestimation of the turbulent kinetic energy in the symmetry axis proximity for both the models. This discrepancy disappears as the exit section is approached and may be attributed to poor mesh refinement. For this flow pattern, TL predicts approximately the same k values as CH and these are always in good agreement with measurements (Figure 11). This is likely due to the fact that the k peak is located far from the wall so that the damping function effect is lost and the models behave like a high-Re k-ϵ.

The axial pressure and velocity distribution, allowing

evaluation of diffuser perfomance are shown in Figure 12 (a, b). While agreement for both models is good, the closer agreement of CH pressure rise prediction to experiments is proof of correct loss reproduction.

15-Deg Flow. Results were analogous for the 15-deg configuration investigated on an 80 x 45 grid. To test the influence of matching position on TL, two computations were carried out, with the first matching at f_μ = 0.55 and the second at f_μ = 0.65.

The computed and measured streamlines patterns are compared in Figure 13. The predicted reattachment positions ranging between Z = 7.5 - 8.0 are in good agreement with the experimental value of Z = 7.6.

The axial pressure plot in Figure 14 provides an inside view of the TL model. If matching occurs at f_μ = 0.65, the pressure profile approaches the one predicted by CH; if matching occurs at f_μ = 0.55 (closer to the wall), the pressure growth is stronger. Since only viscous and buffer layers are affected, we can conclude that matching position does influence boundary layer thickness and consequently energy dissipation, although only slightly.

CONCLUSIONS

The artificial compressibility method gave accurate prediction of internal turbulent incompressible steady flows. While the coupled solver used with the approximate factorization gave the same results of the decoupled solver, there was no gain in terms of accuracy and efficiency for the diffusion dominated flow considered here.

The proposed implicit treatment of the sink-source terms yields a stable algorithm for a broad range of flow situations. The two low-Reynolds-number forms of the k-ϵ model give fast and moderately accurate flow prediction in a 180-deg planar bend and in 45- and 15-deg configurations of a cylindrical diffuser at high Re provided that the artificial compressibility formulation is accounted for in the production term of both the k and ϵ transport equations.

Although the TL model proved to be a good compromise between accuracy and computational efficiency, it is necessary to improve the matching criterion for the two fluid layers governed by different sets of equations.

REFERENCES

[1] Coakley, T.J., "Turbulence Modelling Methods for the Compressible Navier - Stokes Equations", AIAA Paper 83-1693 (1983).

[3] Michelassi, V. and Benocci, C., "Solution of the Steady State Incompressible Navier-Stokes Equations in Curvilinear Nonorthogonal Coordinates", Von Karman Institute Technical Note 158 (1986).

[3] Michelassi, V. and Benocci, C., "Prediction of Incompressible Flow Separation with the Approximate Factoriza-

tion Technique", International Journal of Numerical Methods in Fluids, Vol. 7, (1987).

[4] Patel, V.C., Rodi, W. and Sheuerer, G., "Turbulence Models for Near-Wall and Low Reynolds Number Flows: A Review", AIAA Journal, Vol.23, No. 9 (1985).

[5] Rodi, W., "Turbulence Models and their Applications in Hydraulics", I.A.H.R., second revised edition, (1984).

[6] Chien, K.Y., "Predictions of Channel and Boundary Layer Flows with a Low-Reynolds-Number Turbulence Model", AIAA Journal, Vol. 20, No. 1 (1982).

[7] Cordes, J. and Rodi, W., "Arbeitsbereicht zum Forshungsvorhaben Ro558/5-2. Berechnung abgeloster Stromungen an Tragflugenprofilen", University of Karlsruhe, Karlsruhe, West Germany (1988).

[8] Michelassi, V., "Testing of Turbulence Models with an Artificial Compressibility Solution Method", Report SFB 210/T/49, University of Karlsruhe, Karlsruhe, West Germany, (1988).

[9] Rodi, W. and Sheuerer, G., "Scrutinizing the k-ϵ Turbulence Model under Adverse Pressure Gradient Conditions", ASME Transactions, Journal of Fluids Engineering, Vol. 18 (1986).

[10] Norris, L.H. and Reynolds, W.C., "Turbulent Channel Flow with a Moving Wavy Boundary", Stanford University Mechanical Engineering Report FM10, Stanford, California (1975).

[11] Schönung, B., Private Communication, University of Karlsruhe, Karlsruhe, West Germany (1988).

[12] Chorin, A.J., "A Numerical Method for Solving Incompressible Viscous Flow Problems", Journal of Computational Physics, Vol. 2, No. 1, pp.12-26 (1967).

[13] Pulliam, T.H. and Steger, J.L., "Implicit Finite Difference Simulations of Three-Dimensional Compressible Flow", AIAA Journal, Vol. 18, No. 2, (1980).

[14] Beam, R.M. and Warming, R.F., "Implicit Numerical Methods for the Compressible Navier-Stokes and Euler Equations", Von Karman Institute LS 1982-04 Computational Fluid Dynamics, Brussels, Belgium, (1982).

[15] Sandborn, V.A. and Shin J.C., "Evaluation of Turbulent Flow in a 180-Degree Bend or Bulk Reynolds Numbers from 70,000 to 160,000", NASA CR NAS8-36354 (1987).

[16] Chaturvedi, M.C., "Flow Characteristics of Axisymmetric Expansion", Journal of Hydraulic Division of the Americal Society of Civil Engineers, pp.61-92 (1963).

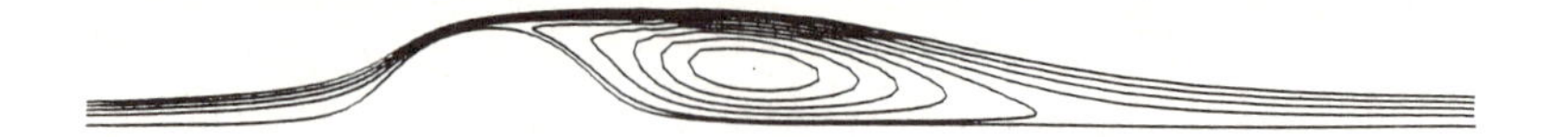

Figure 1. Bump: Streamline Pattern.

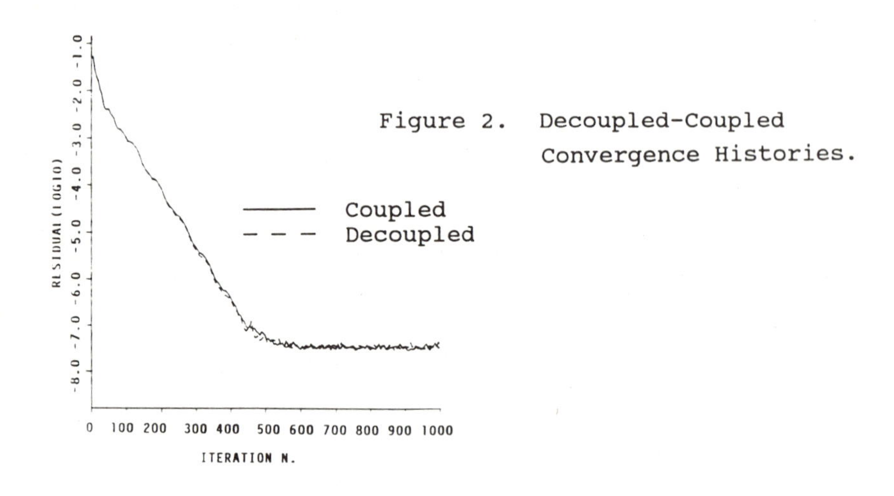

Figure 2. Decoupled-Coupled Convergence Histories.

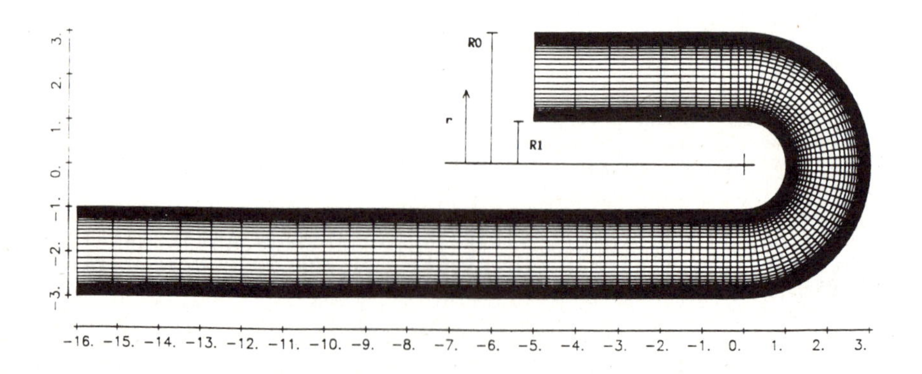

Figure 3. 180-deg Planar Turnaround Duct: 90x65 Grid.

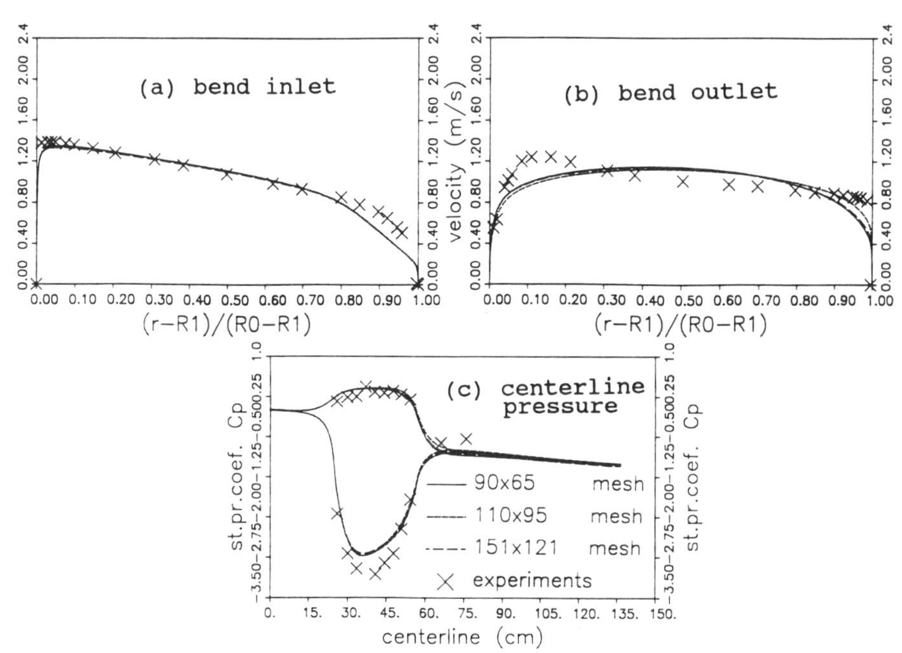

Figure 4. 180-deg Planar Turnaround Duct: Bend Inlet-Outlet Velocities and Axial Pressure Profile.

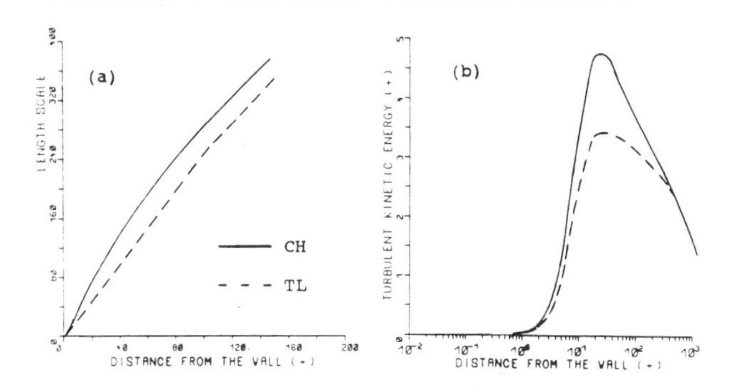

Figure 5. Fully Developed Flow: Turbulent Kinetic Energy and Mixing Length.

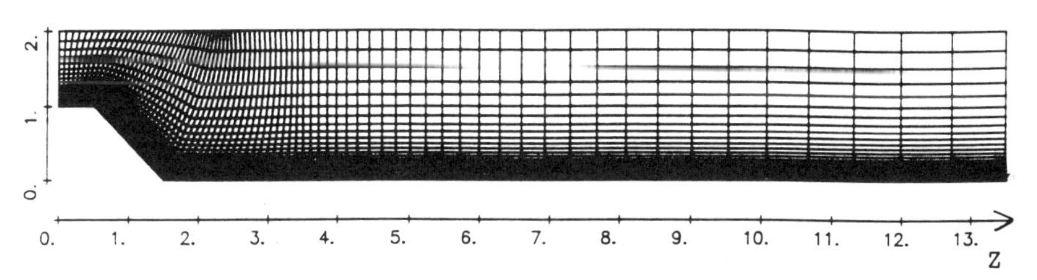

Figure 6. 45-deg Diffuser: 80x45 Grid.

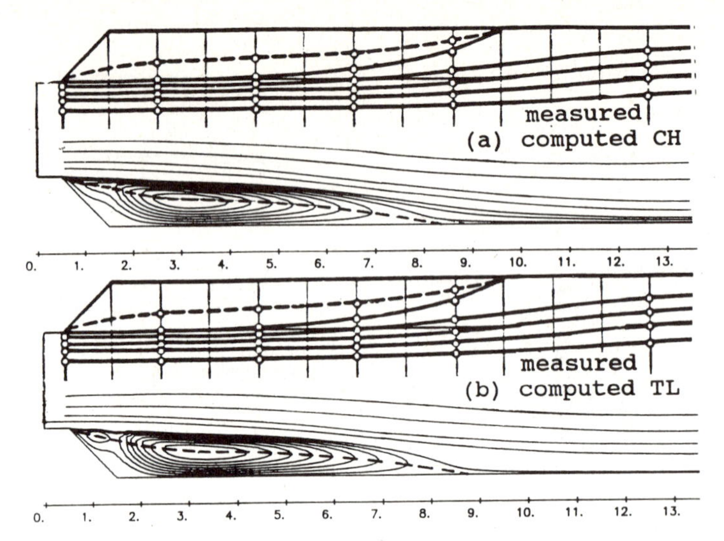

Figure 7. 45-deg Diffuser: Streamline Pattern.

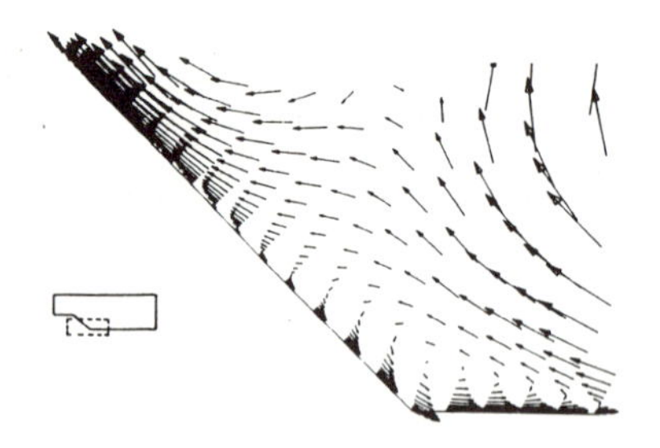

Figure 8. 45-deg Diffuser: Lower Corner Velocity Field.

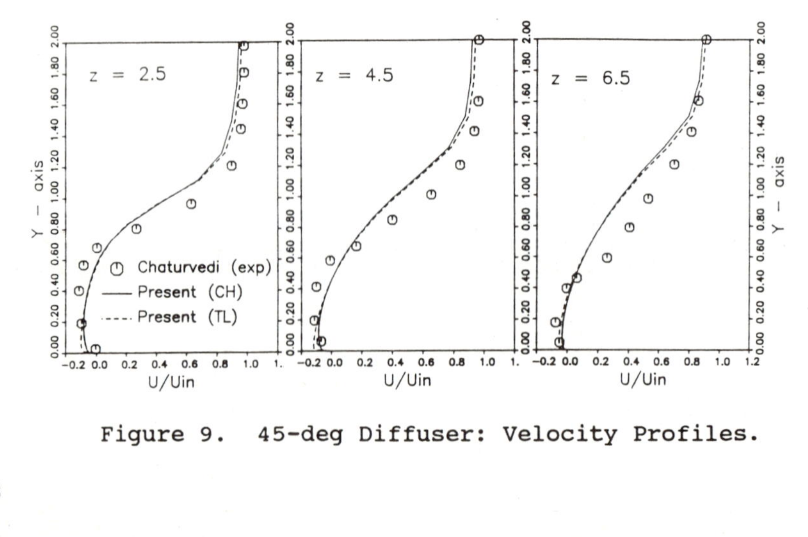

Figure 9. 45-deg Diffuser: Velocity Profiles.

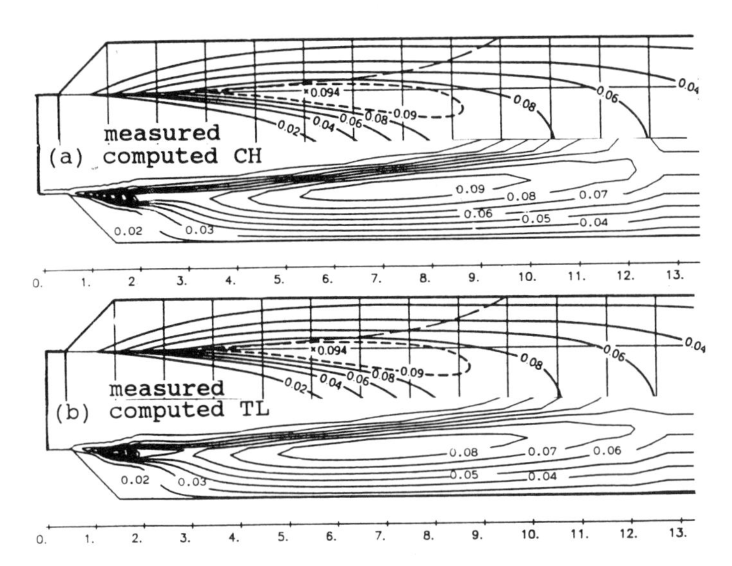

Figure 10. 45-deg Diffuser: k Isolines.

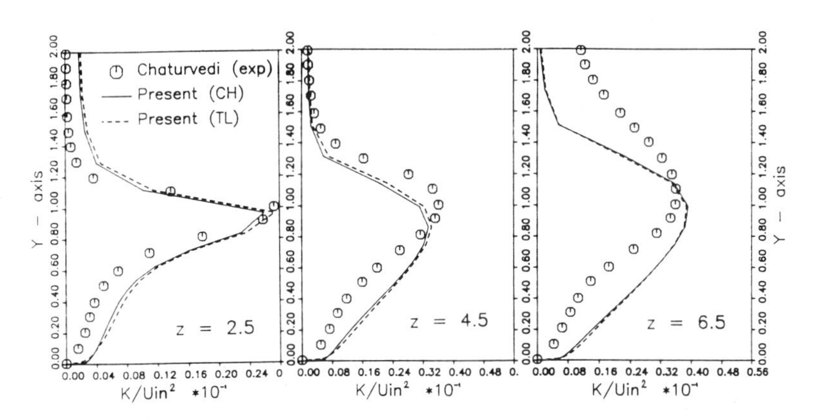

Figure 11. 45-deg Diffuser: k Profiles.

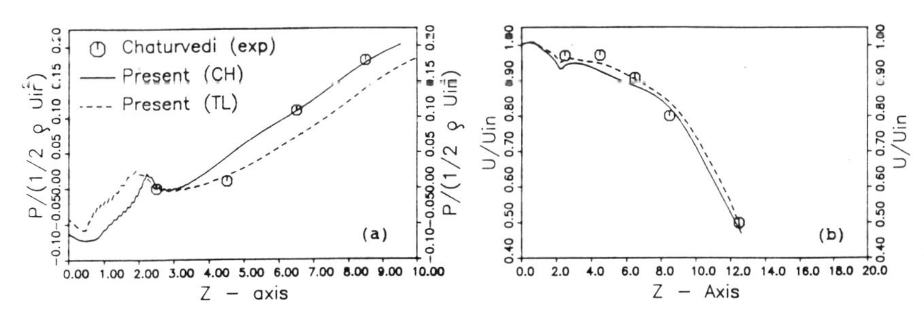

Figure 12. 45-deg Diffuser: Axial Velocity and Pressure.

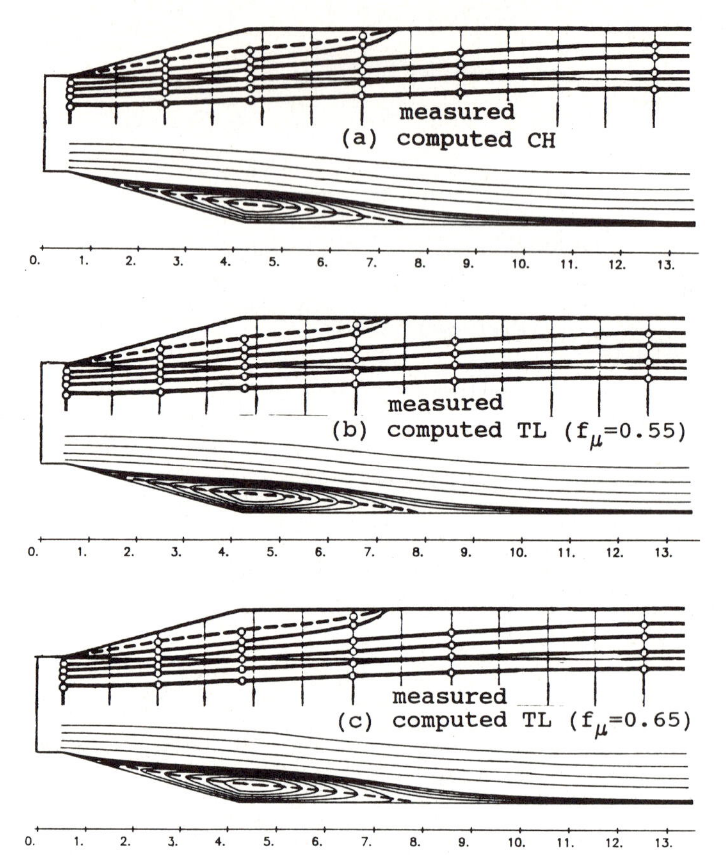

Figure 13. 15-deg Diffuser: Streamline Pattern.

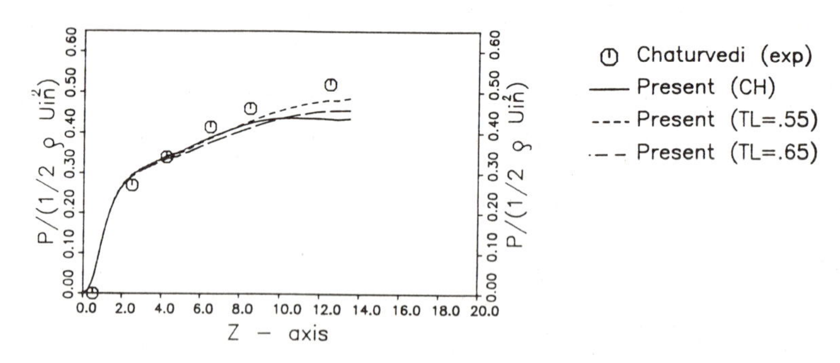

Figure 14. 15-deg Diffuser: axial pressure.

MULTIGRID-BASED GRID-ADAPTIVE SOLUTION OF THE NAVIER-STOKES EQUATIONS

Jess Michelsen

DTH, Dept. of Fluid Mech., Build. 404, DK-2800, Lyngby, Denmark

SUMMARY

A finite volume scheme for solution of the incompressible Navier-Stokes equations in 2D and axisymmetry is described.

Solutions are obtained on non-orthogonal, solution adaptive BFC grids, based on the Brackbill-Saltzman generator. Adaptivity is achieved by the use of a single control function based on the local kinetic energy production.

Non-staggered allocation of pressure and Cartesian velocity components avoids the introduction of curvature terms associated with the use of a grid-direction vector-base. A special interpolation of the pressure correction equation in the SIMPLE algorithm ensures firm coupling between velocity and pressure field.

Steady-state solutions are accellerated by a full approximation (FAS) multigrid scheme working on the decoupled grid-flow problem, while an algebraic multigrid scheme is employed for the pressure correction equation.

INTRODUCTION

Reliable calculations of practical flow problems generally require very fine discretizations, this especially being true for high-*Re* flow. Timestep criteria and convergence properties of standard iterative schemes will often preclude the solution of such problems within a meaningful turnaround time, leading to the acceptance of less qualified results.

On a given hardware, the task of speeding up fine-grid calculations can be approached in two ways. One can either improve the convergence properties of the solution algorithm, or seek to attain the same level of precision on a smaller number of grid-points.

The first approach often leads to the construction of multigrid schemes, while the second approach may lead to the use of solution adaptive grid methods.

Multigrid schemes for incompressible Navier-Stokes problems have been reported by a number of authors. Ghia, Ghia & Shin [1] proposed an FAS scheme for the streamfunction-vorticity formulation, while Barcus [2] and Barcus, Peric & Scheuerer [3] preferred a correction grid (CG) scheme for the primitive variables. Vanka [4] proposed a block-implicit FAS scheme for the primitive variables.

For all of the above references, orders of magnitude convergence accellerations were achieved.

Solution adaptive grids are being constructed in a number of ways. Here, only block-structured elliptic grid-generators are considered. The Poisson generator of Thompson et al. [5] has two control functions, one controlling the grid density in each of the two local grid directions. Brackbill & Saltzmans [6] Euler system employs only one function, which controls the grid density distribution. Moreover, the Euler system has two user-specified parameters, which shift the emphasis between grid smoothness, orthogonality, and density.

The objective of the present work is to economize fine-grid computations by simultaneous employment of both of the above mentioned approaches, i.e to construct a multigrid scheme for the solution of the dynamically–adaptive Navier–Stokes problem.

In order to facilitate the treatment of non–linear source terms, viz. the Coriolis and centrifugal forces, an FAS scheme will be preferred. The Brackbill–Saltzman generator will be preferred for its orthogonality control.

CONSERVATION LAWS

The conservation of mass and the Cartesian momentum components is written in a general transformed space (ξ,η), in which the physical positions of the grid points move in time

$$\frac{\partial Q}{\partial t}+\frac{\partial(F+F_p)}{\partial\xi}+\frac{\partial(G+G_p)}{\partial\eta}=\frac{\partial(V_1+V_2)}{\partial\xi}+\frac{\partial(W_1+W_2)}{\partial\eta} \tag{1}$$

where the transfomation

$$\frac{\partial}{\partial x}=\frac{1}{J}\cdot(y_\eta\cdot\frac{\partial}{\partial\xi}-y_\xi\cdot\frac{\partial}{\partial\eta}),\ \frac{\partial}{\partial y}=\frac{1}{J}\cdot(-x_\eta\cdot\frac{\partial}{\partial\xi}+x_\xi\cdot\frac{\partial}{\partial\eta}) \tag{2}$$

is employed. The cross–product of the base vectors, $J=x_\xi y_\eta-x_\eta y_\xi$, represents the cell volume.

The conserved quantities and the fluxes are written

$$Q=J\cdot\begin{Bmatrix}\rho\\ \rho u\\ \rho v\end{Bmatrix} \tag{3}$$

$$F=[y_\eta(u-\dot{x})-x_\eta(v-\dot{y})]\cdot\begin{Bmatrix}\rho\\ \rho u\\ \rho v\end{Bmatrix},\quad F_p=\begin{Bmatrix}0\\ y_\eta P\\ -x_\eta P\end{Bmatrix}, \tag{4}$$

$$G=[-y_\xi(u-\dot{x})+x_\xi(v-\dot{y})]\cdot\begin{Bmatrix}\rho\\ \rho u\\ \rho v\end{Bmatrix},\quad G_p=\begin{Bmatrix}0\\ -y_\xi P\\ x_\xi P\end{Bmatrix} \tag{5}$$

where $(\dot{x},\dot{y})$ is the grid speed. The viscous forces are written

$$V_1=\frac{\mu}{J}\cdot(x_\eta^2+y_\eta^2)\cdot\begin{Bmatrix}0\\ \frac{\partial u}{\partial\xi}\\ \frac{\partial v}{\partial\xi}\end{Bmatrix},\quad V_2=-\frac{\mu}{J}\cdot(x_\eta x_\xi+y_\eta y_\xi)\cdot\begin{Bmatrix}0\\ \frac{\partial u}{\partial\eta}\\ \frac{\partial v}{\partial\eta}\end{Bmatrix}, \tag{6}$$

$$W_1=\frac{\mu}{J}\cdot(x_\xi^2+y_\xi^2)\cdot\begin{Bmatrix}0\\ \frac{\partial u}{\partial\eta}\\ \frac{\partial v}{\partial\eta}\end{Bmatrix},\quad W_2=-\frac{\mu}{J}\cdot(x_\eta x_\xi+y_\eta y_\xi)\cdot\begin{Bmatrix}0\\ \frac{\partial u}{\partial\xi}\\ \frac{\partial v}{\partial\xi}\end{Bmatrix}, \tag{7}$$

where the V_2 and W_2 terms arise with grid skewness. Additional stress terms arise if the viscosity varies over the region, or if the velocity field is not divergence–free. These terms are not of prime interest for the presentation, and are omitted.

ADAPTIVE GRID GENERATION

The Brackbill–Saltzman system, which is preferred for the present use, comprises a set of quasi–linear Euler equations for the minimization of a weighted sum of measures for global grid smoothness, orthogonality, and density distribution

$$b_1 x_{\xi\xi} + b_2 x_{\xi\eta} + b_3 x_{\eta\eta} + a_1 y_{\xi\xi} + a_2 y_{\xi\eta} + a_3 y_{\eta\eta} = \lambda_c w J^2 \cdot \frac{\partial w}{\partial x}, \tag{8}$$

$$a_1 x_{\xi\xi} + a_2 x_{\xi\eta} + a_3 x_{\eta\eta} + c_1 y_{\xi\xi} + c_2 y_{\xi\eta} + c_3 y_{\eta\eta} = \lambda_c w J^2 \cdot \frac{\partial w}{\partial y}, \tag{9}$$

with coefficients given by

$$a_i = a_{si} + \lambda_c w^2 a_{ci} + \lambda_o a_{oi}, \text{ etc.,} \tag{10}$$

where s,c, and o refer to *smoothness*, *concentration*, and *orthogonality*. λ_c and λ_o are user–specified parameters, which shift the emphasis from smoothness to concentration, respectively orthogonality.

The majority of the boundary points in the grid are assigned a Cauchy–like boundary condition, which specifies the points to follow the internal points, while sliding along a curve, always preserving grid orthogonality on the boundary Γ

$$\forall(\xi,\eta)\in\Gamma : f[\underline{x}(\xi,\eta)]=0 \wedge x_\xi x_\eta + y_\xi y_\eta = 0 . \tag{11}$$

In order to render the grid problem well–posed, a limited number of boundary grid–points are assigned Dirichlet conditions. In order to ensure the presence of grid points on eventual sharp corners, such grid–points are always assigned Dirichlet conditions.

ADAPTIVE CONTROL FUNCTION

Some examples of dynamically–adaptive grids controlled by a function of derivatives of a scalar flow variable have been reported, notably Saltzman & Brackbill [7], who based the grid control on the pressure gradient. For incompressible viscous flow, solution adaptation would be based on the velocity field rather than on the pressure field. Hence, an invariant of the velocity vector field is sought. Being the simplest invariant of the strain tensor, the production of kinetic energy

$$\nu \cdot U_{i,j} \cdot (U_{i,j} + U_{j,i}) \tag{12}$$

is preferred as control function. However, the viscosity will be omitted.

In high velocity–gradient zones, (10) shows that emphasis is almost totally on grid concentration. Subsequently, grid smoothness is ultimately controlled by the distribution of the control function w in those zones. Thus, for practical purposes, the control function must be smoothed. This is commonly done by performing a few relaxation sweeps on the control function, using a Laplace operator. In order to ensure early detection of wakes etc., a convective–diffusive operator for a cell *Pe* between zero and unity is preferred. Thus, the need for fine cells in wake zones is recognized somewhat in advance.

BASIC SMOOTHING ALGORITHM

The conservation laws (1) are put in finite volume form using standard central difference expressions for the viscous forces. The pressure at cell faces is found by linear interpolation between cell centers. The convective part is discretized by the power–law upwind form [8].

The momentum equations are now on the form

$$a_P u_P^{n+1} + \frac{\rho J^n}{\Delta t}\cdot(u_P^{n+1} - u_P^n) = \sum_{nb=E,W,N,S} a_{nb} u_{nb}^{n+1} + S_u , \tag{13}$$

$$a_P = \sum_{nb=E,W,N,S} a_{nb} \quad . \tag{14}$$

S_u here represents source terms including the pressure forces and the cross derivative terms V_2 and W_2. The present evaluation of the diagonal coefficient a_P includes the grid speeds at the cell faces. Thus, the cell volume update during the timestep will be consistent, whichever upwind scheme is used, provided the old rather than then the new cell volume is used at the end of the timestep in (13). In the next timestep, the cell volume is of course recalculated on basis of the updated coordinates.

In order to avoid velocity–pressure uncoupling in the non–staggered arrangement, a special interpolation practice, first proposed by Rhie [9], is employed for calculation of the cell face fluxes entering the mass conservation

$$C_{\xi e} - C_{\xi w} + C_{\eta n} - C_{\eta s} = S_m . \tag{15}$$

Here, e,w,n,s refer to the cell faces. The fluxes are written

$$C_\xi = \rho\cdot[y_\eta u - x_\eta v] \ , \ C_\eta = \rho\cdot[-y_\xi u + x_\xi v] \ . \tag{16}$$

where the appropriate momentum equations are now substituted for the cell face velocities

$$u_e^{n+1} = \frac{\frac{\rho J}{\Delta t}\cdot u_e^n + \sum a_{nb} u_{nb}^{n+1} + S_u + [y_\eta\cdot(P_P - P_E) - y_\xi\cdot(P_{se} - P_{ne})]}{a_P + \frac{\rho J}{\Delta t}} . \tag{17}$$

The cell face velocities are found by linear interpolation of the terms in (17) evaluated at the cell centers. For the pressure part, however, it is the coefficient that is interpolated. Overbars denoting linear interpolation, the cell face velocity is written

$$u_e^{n+1} = \left(\frac{\frac{\rho J}{\Delta t}\cdot u_e^n + \sum a_{nb} u_{nb}^{n+1} + S_u}{a_P + \frac{\rho J}{\Delta t}}\right)_e + \left(\frac{1}{a_P + \frac{\rho J}{\Delta t}}\right)_e \cdot [y_\eta \cdot (P_P - P_E) - y_\xi \cdot (P_{se} - P_{ne})] , \tag{18}$$

where the coupling between the velocity and pressure fields is clearly seen. The righthand side of (18), and the corresponding expressions for the remaining velocity components are finally substituted into (16), yielding a Poisson equation for the pressure, here written for an orthogonal grid

$$\frac{1}{a_e}(P_P - P_E) - \frac{1}{a_w}(P_P - P_W) + \frac{1}{a_n}(P_P - P_N) - \frac{1}{a_s}(P_P - P_S) = S_m . \tag{19}$$

A total sweep of the basic smoother now commences by evaluation and smoothing of the control function, based on the latest available velocity field. One sweep on the grid system is then performed, here using a zebra line–implicit smoother. The old and new grids define the grid speeds. Finally, a timestep in the SIMPLE algorithm is carried out.

For multigrid applications, time–fidelity is normally irrevant. Hence, a local timestep is used, changing the SIMPLE algorithm into its more common underrelaxed version.

MULTIGRID ALGORITHM

Multigrid technique is employed at two different levels of the present scheme. On the outer level, the solution is advanced in time towards steady–state using an FAS scheme, while on the inner level, the Poisson equation is solved by an algebraic scheme.

Multigrid timestepping

Assume the following form of the problem

$$\underline{L}^k \cdot \underline{X}^k(\underline{\xi}^k) = \underline{F}^k(\underline{\xi}^k) \text{ for } \underline{\xi}^k \in \Omega , \tag{20}$$

$$\underline{\Lambda}^k \cdot \underline{X}^k(\underline{\xi}^k) = \underline{\Psi}^k(\underline{\xi}^k) \text{ for } \underline{\xi}^k \in \Gamma , \tag{21}$$

where underscore denotes vector. $\underline{\xi}^k$ is the set of independent variables and $\underline{X}^k$ the exact solution to the G^k–problem. $\underline{L}^k$ and $\underline{\Lambda}^k$ are discretized operators and $\underline{F}^k$, $\underline{\Psi}^k$ fine–grid residual vectors. The following analysis applies for the grid problem as well as for the flow problem, i.e. the flow variables are treated as dependent on $\underline{\xi}^k$ rather than on the grid coordinates.

Let $\underline{x}_0^1$ be an approximation to the fine–grid problem, the residuals are then

$$\underline{f}^1(\underline{\xi}) = - \underline{L}^1 \cdot \underline{x}_0^1(\underline{\xi}) \text{ for } \underline{\xi} \in \Omega, \tag{22}$$

$$\underline{\varphi}^1(\underline{\xi}) = -\underline{\Lambda}^1 \cdot \underline{x}_0^1(\underline{\xi}) \text{ for } \underline{\xi} \in \Gamma. \tag{23}$$

The fine–grid solution is now restricted along with its residuals

$$\underline{x}_0^2 = I_1^2 \underline{x}_l^1, \tag{24}$$

$$\underline{F}^2 = R_1^2 \underline{f}^1 + \underline{L}^2 \underline{x}_0^2, \tag{25}$$

$$\underline{\Psi}^2 = R_1^2 \underline{\varphi}^1 + \underline{\Lambda}^2 \underline{x}_0^2. \tag{26}$$

The grid coordinates and residuals are injected. Flow variables are restricted by linear interpolation, while restriction of momentum residuals is performed as simple sums. The cell face fluxes are restricted in much the same ways as the flow residuals. This is necessary due to the special interpolation used in the pressure correction. This restriction procedure ensures correct mass conservation on the coarse grid.

Both corrections of grid coordinates and flow variables calculated on the coarse grid are prolongated using linear interpolation.

The described multigrid components are used in a full multigrid scheme, using a fixed V–cycle. The number of pre–smoothings increase with grid level, while one post–smoothing is performed on each level.

Algebraic MG Poisson solver

The pressure equation is solved in a fixed V–cycle correction grid (CG) scheme. Restriction and prolongation are identical to the ones used for the flow variables in the outer loop. An alternating direction zebra solver is employed as basic smoother.

Calculation of the coarse–grid coefficients is best explained by a resistance analogy, where the massflux change δC in fig. 1 acts as current, while the change in pressure drop acts as voltage

$$\delta C_{e1}^h = \frac{\delta(P_{P1}^h - P_{E1}^h)}{\Delta\xi^h \cdot a_{e1}^h}. \tag{27}$$

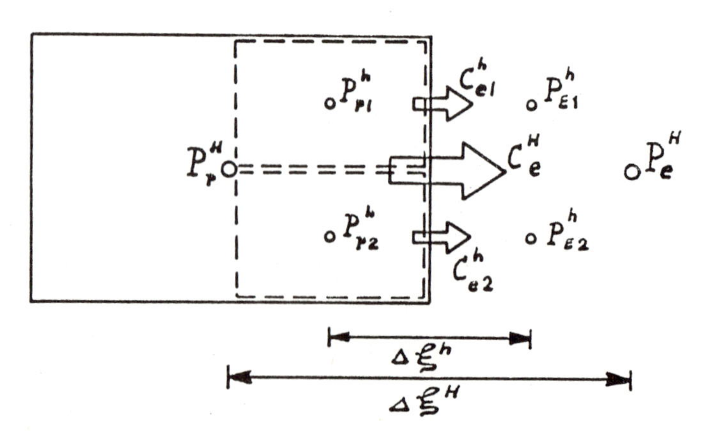

Fig.1 Calculation of coefficients for pressure correction on coarser levels.

The expression for massflux change on the coarse grid is now approximated by

$$\delta C_e^H = \left(\frac{1}{a_{e1}^h} + \frac{1}{a_{e2}^h} \right) \cdot \frac{\delta(P_P^H - P_E^H)}{\Delta \xi^H} . \tag{28}$$

Finally, assuming standard coarsening, the coarse coefficients in (19) are calculated as

$$\frac{1}{a_E^H} = \frac{1}{2} \cdot \left(\frac{1}{a_{e1}^h} + \frac{1}{a_{e2}^h} \right) . \tag{29}$$

EXAMPLES

Two test cases are considered,

—— the flow in a cylindrical lid–driven cavity at $Re = 2492$, $H/R = 2.5$, and

—— flow around a NACA 64_2A015 airfoil at 5^0 incidence and $Re = 7000$.

Fig. 2a describes the cylindrical lid–driven cavity problem. Also, the finite volume solution obtained on 192×96 equidistant control volumes is compared to the visualization of Escudier [10]. The flow details of the separating zones compare excellent with the experimental evidence. The calculations were done on a 4–level scheme, using a PISO [11] algorithm, which is similar to the SIMPLE reported herein. The convergence history is shown in fig. 2b.

An adaptive grid was employed in the airfoil example. Three grid–levels were employed in the outer loop, while four grid levels were used for pressure correction. The finest grid comprised 40×208 control volumes. The SIMPLE algorithm was employed, using underrelaxation paramters $\alpha_u = 0.6$, $\alpha_p = 0.4$. Fig. 3 shows the final grid, velocity vectors, and streamlines compared to the photo of Werlé [12]. Convergence history is shown in fig. 4.

The grid is seen to reproduce the high–gradient zones very nicely. An interesting flow–detail is seen on the streamline plot, where a second recirculating zone is rotating in clockwise direction, partly driven by the pressure–side flow. The author believes to see the same phenomenon in the photo.

For the algebraic multigrid solution of the Poisson system, convergence rates between 0.31 and 0.45 were recorded. Stüben & Linden [13] report convergence rates of 0.23 for isotropic coefficients and 0.33 for 1:100 anisotropy (1:10 sidelength ratio). As the grid skewness is not treated implicitly in the zebra smoother, the values recorded here are reasonable.

CONCLUSION

The cylindrical driven cavity problem showed the multigrid scheme preforming at near–optimum, while convergence was somewhat slower in the adaptive case. While the grid curvature and stretching is not believed to affect the convergence substantially, the *Re* number is known to influence multigrid convergence rates adversely. Moreover, the dynamical grid affects the convergence.

In order to obtain convergence, the grid movements had to be underrelaxed. In the near future, a red–black basic smoother will be employed for the grid instead of the present zebra smoother. It will be faster in CPU–time, while yielding lower grid–speeds.

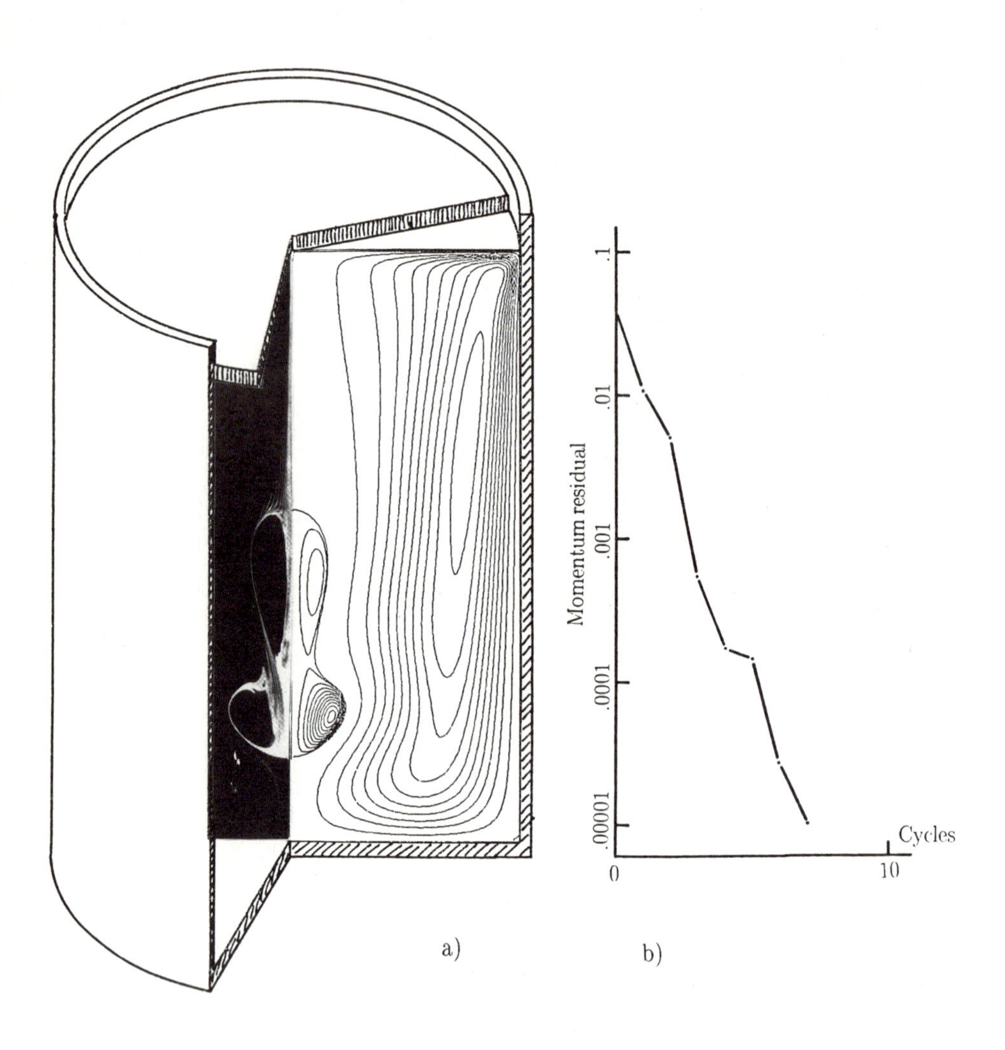

Fig.2 Cylindrical lid driven cavity. a) FV solution on 192×96 VC compared to visualisation of Escudier. b) Convergence history of 4–level scheme.

ACKNOWLEDGEMENT

The present work is supported by the Danish Technical Science Research Council under grant 95 83 02.

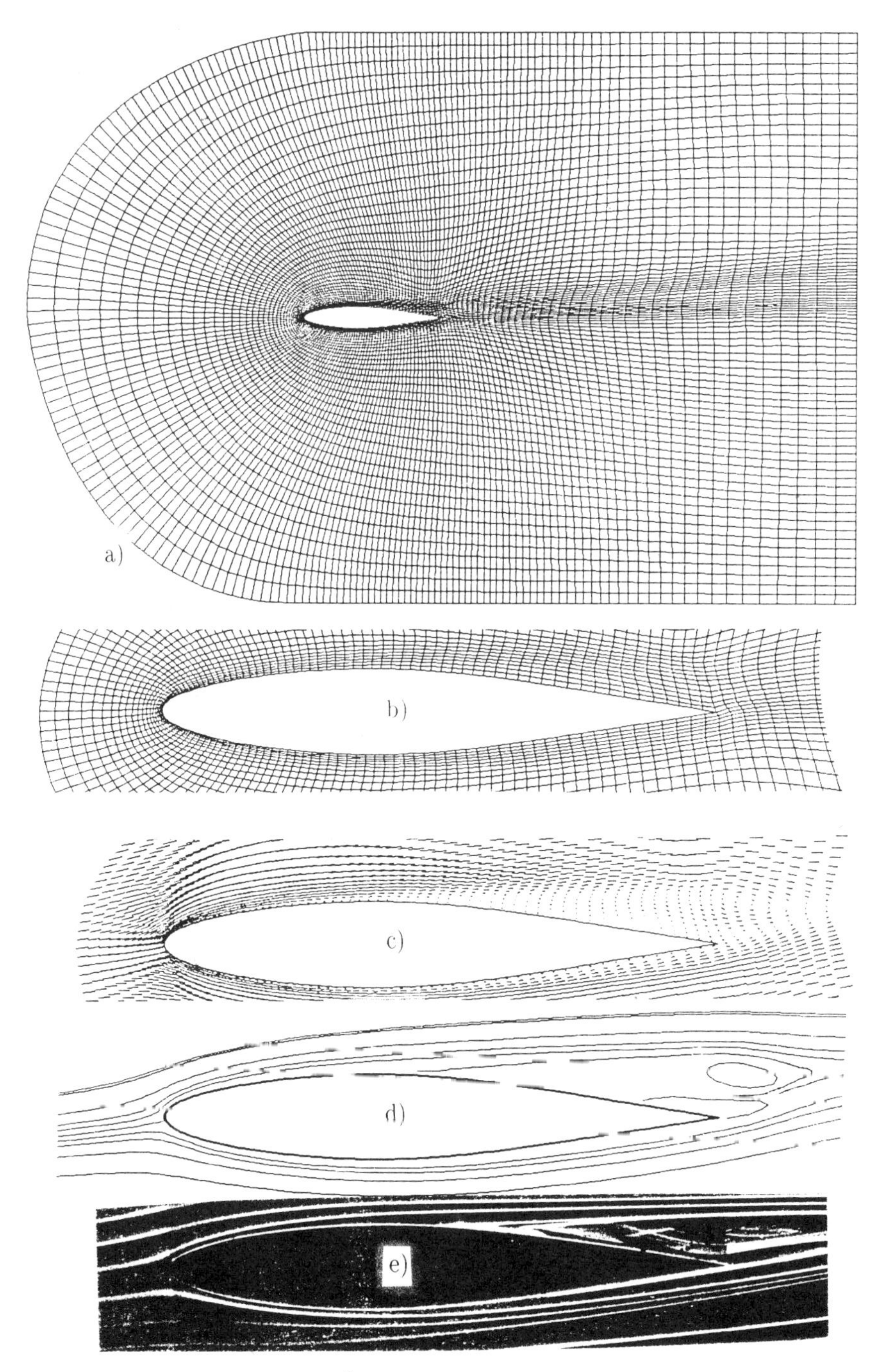

Fig.3 NACA 64_2A015 airfoil at 5^0 incidence, $Re = 7000$. a) 40×208 grid, b) Grid detail, c) Velocity vectors, d) Streamlines, e) Visualization of Werlé.

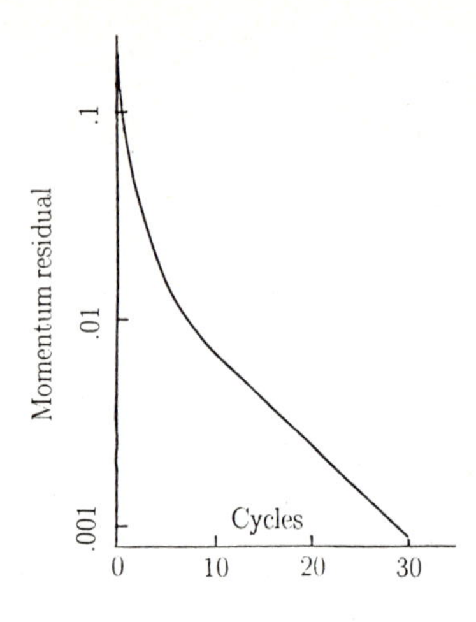

Fig.4 NACA 64_2A015 airfoil, convergence history of 3–level grid–adaptive scheme.

REFERENCES

[1] Ghia U, Ghia K.N, Shin C.T., *J. Comp. Phys.*, 48, p387, 1982.

[2] Barcus M, '*Berechnung zweidimensionaler strömungsprobleme mit mehrgitterverfahren*', M.Sc. Thesis, LSTM, Erlangen, 1987.

[3] Barcus M, Peric M., Scheuerer G., 'A control volume based full multigrid procedure for the prediction of two–dimensional laminar incompressible flows', Proc. 7th GAMM Conf. Num. Meth. Fluid Mech., Louvain–la–Neuve, 1987.

[4] Vanka S.P., *Comp. Meth. Appl. Mech. Eng.*, 55, p321, 1986.

[5] Thompson, J.F., Thames F.C., Mastin C.W., *J. Comp. Phys.*, 15, p299, 1974.

[6] Brackbill J.U., Saltzman J.S., *J. Comp. Phys.*, 46, p342, 1982.

[7] Saltzman J.S., Brackbill J.U., *Appl. Math. Comp.*, 10, 1982.

[8] Patankar S.V., *Num. Heat Transfer*, 2, 1979.

[9] Rhie C.M., '*A numerical study of the flow past an isolated airfoil with separation*', Ph.D. Thesis, Univ. of Illinois, Urbane–Champaign., 1981.

[10] Escudier M.P., *Experiments in Fluids*, 2, p189, 1984.

[11] Issa R.I., *J. Comp. Phys.* 61, 1985.

[12] Werlé H., Publ. no. 156, ONERA, France, 1974.

[13] Stüben K., Linden J., 'Multigrid methods: an overview with emphasis on grid generation processes', *Numerical Grid generation in Comp. Fluid Dynamics*, eds. Häuser, Taylor, 1986.

INVISCID FLOW ABOUT A DOUBLE ELLIPSE

Gino Moretti
GMAF, Inc., P.O.Box 184, Freeport, NY, USA

Mauro Valorani
Dpt. of Mech. and Aeron., via Eudossiana 18
University of Rome, Italy

"Authors, on the whole, limit themselves to explanations of their way of doing things, apparently with total success (it would be refreshing to read a paper where attempts were not successful)."

G.J. Hancock [1]

Summary

The λ-scheme and shock-fitting are applied to analyze the flow about a double ellipse on a variety of free-stream Mach numbers and angles of attack, and accuracy and reliability of the results are discussed.

1. Introduction

With computers assuming an increasingly larger (almost exclusive) role in design, accuracy and reliability are major requirements of a computational code in fluid mechanics. In recent times, the flow field about a double ellipse (simulating the bow of a vehicle with a canopy) has been the subject of numerical exercises, prompted by the industrial needs of the Hermes project. Even the simplest (not so realistic) two-dimensional, inviscid problem has shown the need for a benchmark providing a standard of accuracy, short of which no attempt should be made to compute more complicated, three-dimensional flows.

We have tried to generate such a benchmark using the computational techniques that we consider most accurate. Basically, we used the λ-scheme [2] to integrate the Euler equations at grid nodes, and shock-fitting techniques. We used two different approaches to handle the shock generated by the canopy: in the first, the flow field is divided into two regions, using the imbedded shock as a boundary; in the second, a single region is considered, with the imbedded shock fitted across mesh lines. Suitable grids were generated for each case, using basic conformal mappings plus linear normalizations to produce rectangular computational meshes and, in certain cases, stretchings of both coordinates. Computations performed on strictly Cartesian meshes will not be reported here. The reason for using different approaches is obviously to provide elements of comparison, in an attempt to find standards of accuracy.

On the whole, the results obtained are consistent on a wide gamut of values of free stream Mach number (M_∞) and angle of attack (α). Pictures of the entire flow field obtained with different calculations and different number of grid points look exactly the same for each set of M_∞ and α.

Therefore, we could accept such results as the sought for benchmark. On a closer inspection, however, we find certain discrepancies that still make us wonder how much numerical error may affect the apparently outstanding results. In the spirit of the quotation above, and because of space limitations, we decided to minimize the description of the numerical techniques (that we consider right) and to emphasize the analysis of what may go wrong. We believe that our analysis highlights some relevant connections between the physics of the problem and its numerical interpretation (or distortion), and we hope it will inspire some further study. We have seen plenty of calculations of double-ellipse flow fields in recent times, but we have not seen yet an analysis of results such as the one we present here. We would like to urge analysts working on other techniques to repeat our effort and draw their own conclusions.

2. The physical challenge

The geometry of Hermes concern and the range of M_∞ (4 through 25) and α (0° through 30°) may produce flow fields with local difficulties, that are not clear in all their physical details. In general, regardless of the grid being used, high vales of M_∞ make the calculation much harder than low values. The subsonic region in front of the ellipse becomes very thin. The mesh, however, has to maintain a certain resolution. Consequently, the number of steps needed to reach convergence may exceed reasonable limits. If the calculation stops, say, after 4000 steps and the standoff distance is found to be 0.02 (all lengths taken as multiples for the semi-axis of the main ellipse), with residuals of the order of $1/10^4$, and we find, by stopping the calculation after 40000 steps, with residuals of the order of $1/10^6$, that the standoff distance is 0.015, what can we conclude about the real standoff distance? What would it be after 400000 steps or 4 million steps, when the residuals get much closer to "machine zero"? We should admit that questions of this kind have a pure academic interest. In practice, the Euler equations do not account for viscous effects; moreover, at high Mach numbers real gas (equilibrium or non-equilibrium) effects play a dominant role. Nevertheless, a better understanding of the capacity of a numerical technique to solve the Euler equations is imperative as the backbone of reliability for its extension to more complicated physical structures.

The same remarks apply at low Mach numbers but high angles of attack. Obviously, the combination of high Mach numbers and high angles of attack is the riskiest and may lead into real trouble.

The region surrounding the intersection of the main ellipse with the canopy enhances the difficulty described above, with a twist of its own. The corner produces a deflection of about 41°. If the imbedded shock is detached, the velocity must vanish at the corner, but we know that the influence of the singularity is confined to a very small region, generally on a scale smaller than the size of a mesh interval. In other words, the effect of the singularity is not felt on the nodes bracketing the corner and derivatives can be taken bypassing the singularity. This, of course, may be a matter of common practice and is amply justified, except if the size of the grid interval itself tends to vanish. Now, for increasing M_∞ and/or α the root of the imbedded shock gets closer and closer to the corner (we are talking again of standoff distances of less than 0.01). A numerical description of the flow field requires a few intervals between the shock and the corner, with an individual length of the order of 0.001.

This not only makes convergence extremely slow again, but it brings in a need for a better handling of the corner. If the corner is not properly handled, total temperature losses occur, that propagate along the body surface. The velocity downstream of the corner tends to become too slow; the sonic line tends to reach the body too far downstream. On the contrary, the sonic point on the shock tends to get closer and closer to its root, and this seems to be correct because it should lead continuously to the transition from a detached shock followed by a subsonic region to the attached shock followed by a supersonic region. It seems that a perfect numerical technique should be able to show such a transition, with an ever diminishing subsonic bubble; so far, however, we cannot understand whether the bubble should maintain the shape that it has at lower Mach numbers or should really become elongated as the numerical results show. We believe that an improper treatment of the corner is what produces the elongation.

It is important to state that, for the Hermes project geometry, within the ranges of M_∞ and α mentioned above, <u>no configuration with attached shocks is possible</u> (the Mach number in front of the shock should be higher than 5, but it grows at most a little above 3).

3. An outline of computational techniques

In all cases, the grids are generated using a basic conformal mapping between the physical $z=x+iy$ plane and an auxiliary plane, $\zeta=\xi+i\eta$. If the imbedded shock is considered as a boundary between two regions, a classic Joukowski mapping is used for the first ellipse; the same mapping, followed by a power to eliminate the corner and a bilinear mapping to straighten the image of the body, is used for the second region. When the imbedded shock is fitted between mesh points, the Joukowski mapping is followed by a Karman-Trefftz mapping to eliminate the corner. In general, neither the image of the body (b) nor the image of the shock (c) in the ζ-plane is a straight line. The computational region is reduced to a rectangle by the non-conformal transformation: $X=\xi$, $Y=(\eta-b)/(c-b)$; equal intervals are taken along the X- and Y-axes, but stretchings of the X- or Y-coordinate can be used to cluster grid lines where needed. The principles and equations presented in [2] are used here; the codes are based on a combination of the formulas for orthogonal grids and for H-grids (because of the non-conformal step in the mappings). Briefly, the local time derivatives of speed of sound, velocity components and entropy are expressed as sums of contributions (f-terms) carrying signals from the proper domains of dependence. The integrations at each step are made on two levels to achieve second-order accuracy; the space derivatives are approximated by two-point differences, and their coefficients are averaged between such points, to provide non-linear second-order accuracy. Some exceptions, reducing accuracy to first-order locally, are necessary in the vicinity of body, shock and sonic lines. Boundary conditions are applied as mentioned in [2]. The fitting of shocks is performed as outlined in [2] and [3].

Some features of the code are worth mentioning. For both approaches:

(1) At the grid rows next to the shocks, the contributions in the directions normal to them are evaluated only to a first-order accuracy.

(2) All geometric differences needed to evaluate the shock slope and its curvature are taken backwards with respect to the tangential component of the impinging velocity.

For the calculations on two regions, separated by the imbedded shock:

(3) We may (i) alternate the calculations at every step, or (ii) compute the first region in its entirety, ignoring the existence of the canopy, until convergence is reached, and then compute the second region only.

(4) In either case, the upstream conditions for the imbedded shock are obtained by proper interpolations of the values obtained in the first region in the vicinity of the imbedded shock. Bilinear interpolations on the computational grid (that is rectangular) are used.

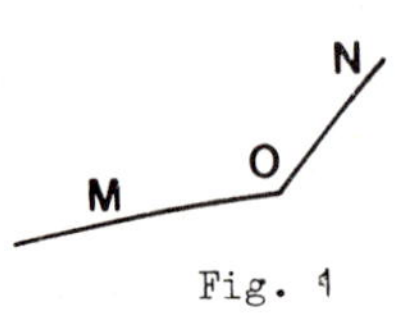

Fig. 1

(5) At the intersection of the two ellipses (O, Fig. 1), we use linear combinations of the moduli of **q** and a at M and N to define the modulus of **q** and a, weighing their contributions according to the lenghts OM and ON. The modulus of **q** is then split in two equal parts to define u and -v.

(6) Some calculations have been made using a global Δt; others using local Δt's. The steady state results are the same; local Δt's accelerate the convergence, particularly when the standoff distance of the imbedded shock is very small. It is important to note that local Δt's can be used only when the shocks do not move too fast. However, the displacements of the shocks at each step must be reduced in size to maintain stability.

(7) We have not attempted to compute the second region with its shock merging with the bow shock; therefore, we always tried to cut it to have it all contained within the first region.

For the single region calculation, with the imbedded shock fitted between mesh points:

(8) To achieve resolution near the intersection of the two ellipses, independent stretchings are provided in both directions; the stretching functions are continuous with their first and second derivatives to avoid additional causes of inaccuracy.

(9) The computational boundary is displaced by a minimal amount from the actual body surface to skirt the singularity at the intersection of the ellipses.

(10) The first and second derivatives of b and c are approximated by formulas of the type: $(b_n-b_{n-1})/(\xi_n-\xi_{n-1})$ and $2(b_{\xi n}-b_{\xi,n-1})/(\xi_n-\xi_{n-2})$ to maintain accuracy.

(11) Special care must be taken to avoid minimal vibrations of the imbedded shock in reaching a steady state. When a fine computational mesh is used, a fitted shock is well defined, and its motion can be well described, by assuming that the upstream and downstream values at any shock point are simply the values at the grid nodes bracketing it. The location of a shock point is merely needed to determine whether such a point remains in the same cell or moves to an adjacent cell. For all practical purposes, this is not a major shortcoming. Nevertheless, if a shock point is practically in a steady position but very close to a grid node, it may oscillate back and forth from one cell to the next; sizeable jumps may thus occur that prevent the residuals from reaching such low levels as required as a proof of convergence. The inconvenience is eliminated by introducing fictitious, but reasonable, values on either side of the shock point, obtained by carefully extrapolating values from the flow field in such a way that they are continuous, together with their first derivatives, when the shock point crosses a grid line [4]. In addition, the value at the first node downstream of the shock is interpolated from the updated downstream shock value and the next node.

(12) Derivatives cannot be approximated by differences taken across a shock. Therefore, the technique outlined in [2] is applied.

4. Summary of results

A) Parametric studies have been made, to determine the overall order of accuracy of the code and the rate of convergence. The flow field about a 4:1 ellipse at no incidence, with $M_\infty=4$, was evaluated using 7 grids of increasing fineness. Local Δt's were used in all cases. In the following Table, we show the normalized standoff of the bow shock, the maximum excursion of total temperature in the field (in percents of the theoretical value) and the percent error in the mass flow at the exit of the shock layer.

GRID	12x4	18x6	24x8	36x12	48x16	72x24	96x36
STF.DIST	.0543	.0495	.0470	.0446	.0437	.0431	.0429
Δt_0	8	6	3.2	1.5	1	.4	.24
Δm	.0095	-.0086	-.0085	-.0057	-.0041	-.0026	-.0018

The trends of the standoff distance and of the total temperature show an overall second-order accuracy. The trend of mass flow error is only slightly better than for first-order accuracy, but this is no surprise since the mass flow is obtained via a crude trapezoidal integration. On

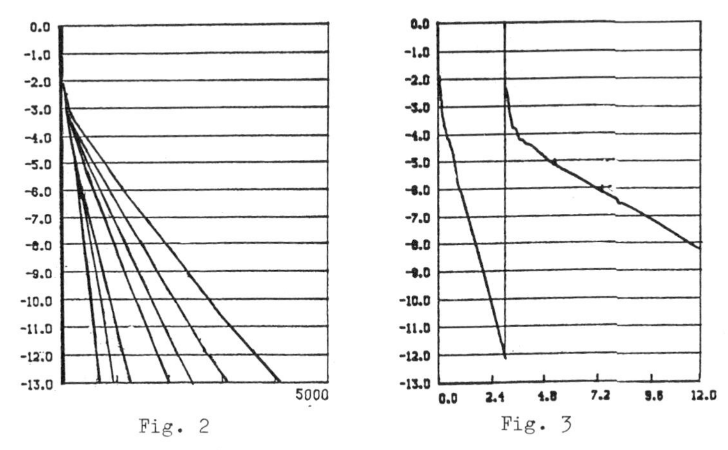

Fig. 2 Fig. 3

the whole, we consider the tests positive. The convergence histories for all cases are shown in Fig. 2 (mean square values of Δa between successive steps vs number of steps). Another important result of the test above and all the other tests made at no incidence is the perfect symmetry of the computed flow field (to within 14 digits). It should be noted that the grid used for the main ellipse itself is symmetric; more about results for a non symmetric grid, below.

B) The second parametric study was made, using the following free stream Mach numbers: 4, 6, 8, 12, and 25, at no incidence, for a 4:1 ellipse topped by a 1.4:1 ellipse. The calculation was made on two regions, separated by the imbedded shock. We used an 80x15 grid for the first ellipse and a 25x10 grid for the second ellipse. The initial shape of both shocks is a parabola. We began by computing at least 8000 steps for each case, letting the first region evolve for 3000 steps and then introducing the second region. In these runs, we used a local Δt (with a global Δt, at a Courant number of 1.6, convergence is rather slow). A typical history of the residual is shown in Fig. 3 for the case $M_\infty=4$, $\alpha=0$

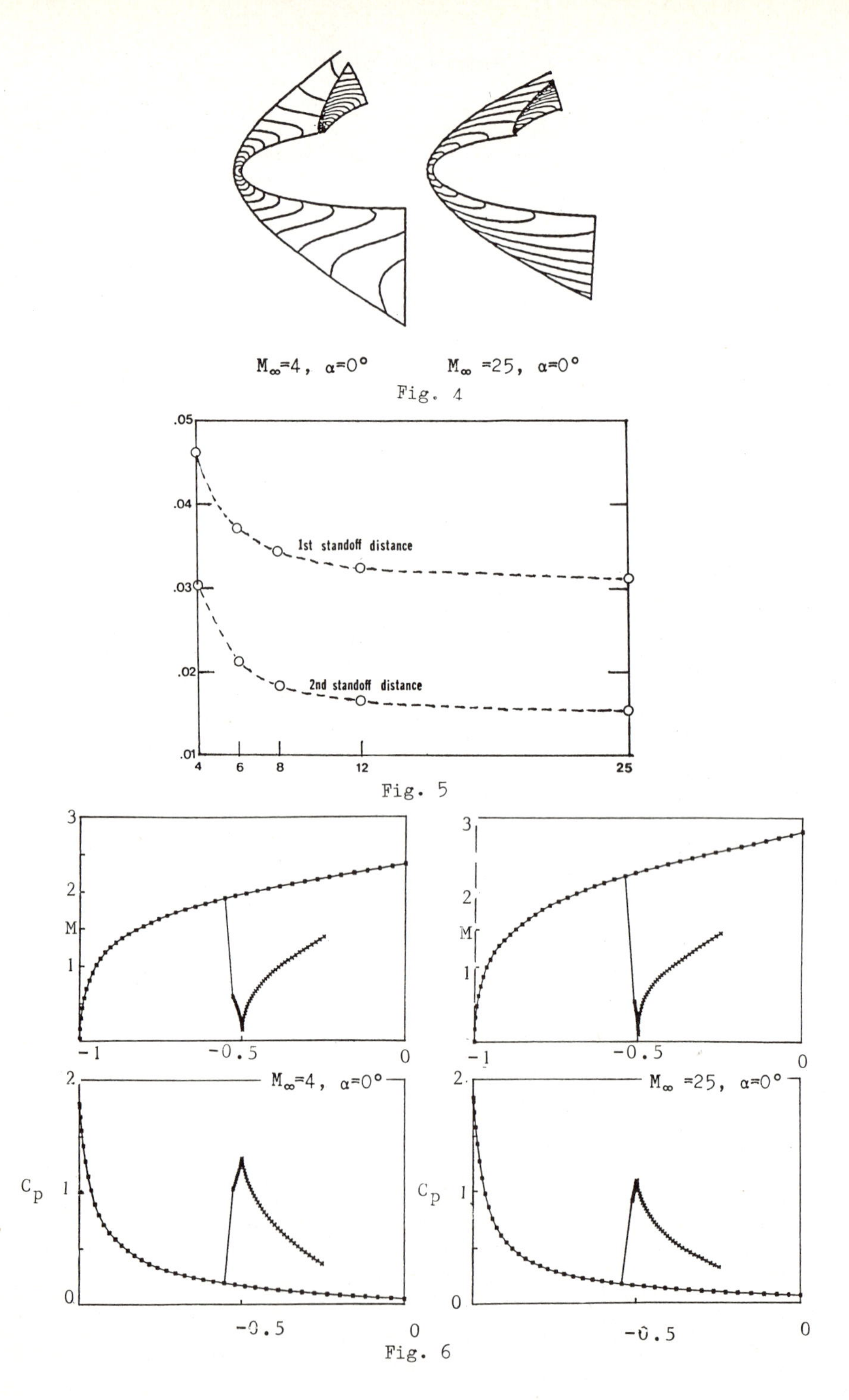

Fig. 4

Fig. 5

Fig. 6

(12000 steps). The jolt at step 3000 occurs when the computation on the second region starts. A complete description of the results involves a minimum of 40 figures; here we must limit ourselves to a few samples. Fig. 4 shows shocks and isomachs for $M_\infty=4$ and 25, at $\alpha=0°$. Fig. 5 shows the interesting trend of the standoff of the bow shock and of the imbedded shock as a function of M_∞ at no incidence. Limits for $M_\infty=\infty$ are clearly visible. The imbedded shock can never become attached for this configuration. Fig. 6 shows body values of C_p and M for the same cases as in Fig. 4. Again, the symmetry of results in the lower and upper portion of the first ellipse at no incidence is remarkable.

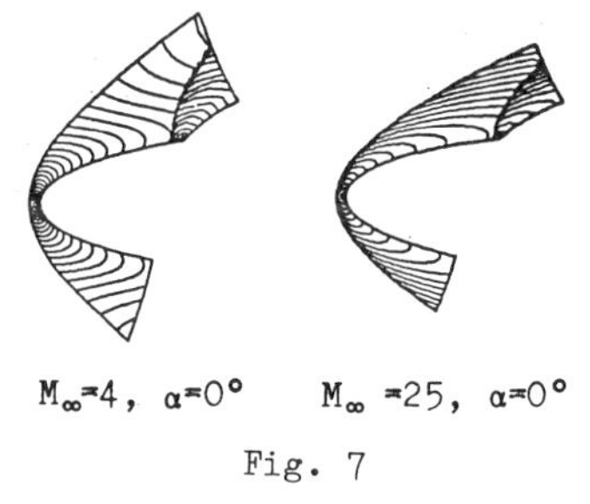

Fig. 7

C) The third parametric study was made, for the same cases as in B), computing the flow on a single region. We expect some departure from the previous results, for three reasons: (i) the grid is not symmetric; therefore the results at no incidence for the first ellipse may lose some of the symmetry; (ii) despite the stretchings, the resolution in the vicinity of the corner is not as high as when two regions are used, and (iii) the treatment of the corner itself is different. Nevertheless, we expect such differences to be irrelevant, in the context of the most important question: Does the fitting of a shock between mesh points provide an accuracy comparable with the one above?

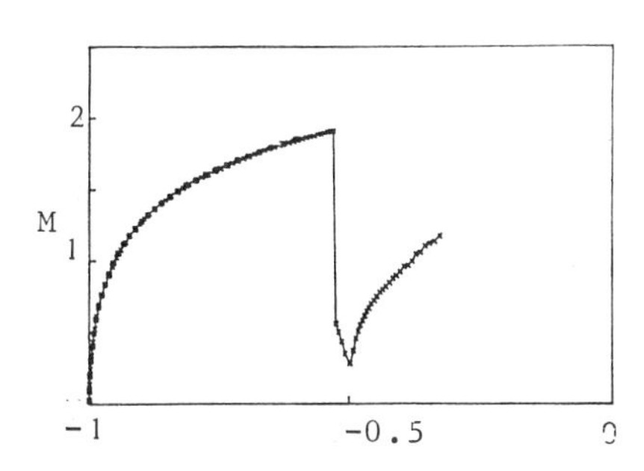

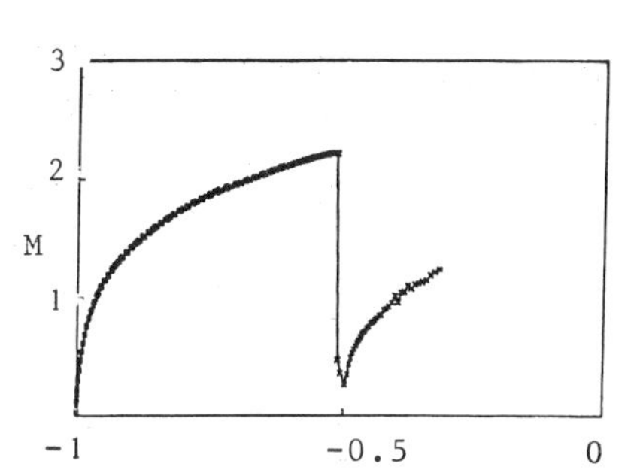

Fig. 8

As we said above, the results obtained with the two techniques are consistent. For example, compare Fig. 7 with Fig. 4 and Fig. 8 with Fig. 6; Figs. 7 and 8 are obtained using a single grid. There is no appreciable difference between the two sets. It is also important to see that the curves of M for the upper and lower portions of the first ellipse do overlap exactly, despite the lack of symmetry of the grid; this supports our statement about the accuracy of the λ-scheme (obviously, one cannot compare any point on the upper surface with a corresponding point on the lower surface, as we did in A), because no two points lie on the same vertical line). On a closer inspection, one can observe a certain loss of accuracy on the body, downstream of the imbedded shock; this is due to the lesser resolution of the grid and to the different treatment of the corner. The standoff distances of bow and imbedded shocks, however, are very close (for example, for the bow shock at $M_\infty=4$ and 25, we find 0.046 vs 0.043 and 0.029

vs 0.028, respectively; for the imbedded shock, in the same order, 0.032 vs 0.034 and 0.014 vs 0.018)

D) The fourth parametric study deals with changes in the angle of attack. The initial shape of the bow shock is still a parabola, but no longer symmetric. First, we used two regions. Fig. 9 shows how the standoff distance of the imbedded shock decreases with increasing incidence, at $M_\infty=4$. A closer analysis of the values on which Figs. 5 and 9 are based leads us to investigate the consistency between standoff distance and Mach number on the body in front of the imbedded shock (M_1), regardless of the incidence or free-stream Mach number. The answer is found in Fig. 10, where the standoff distance of the imbedded shock is plotted vs M_1. All plotted values, and all others computed for different cases or with other grids and not shown, fall on a curve tending to the value, slightly larger than 5, at which the shock would become attached. The consistency of results obtained using two regions or a single region is, once more, remarkable. For lack of space, we must limit our presentation to Fig. 11, showing the isomach pattern for $M_\infty=4$ and $25, \alpha=30°$, and Fig. 12, that shows corresponding plots of M on the body surface.

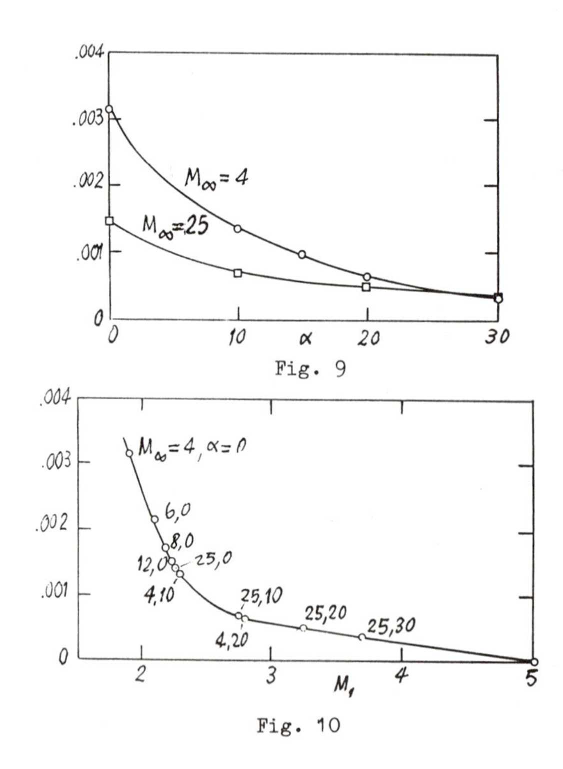

Fig. 9

Fig. 10

5. Sobering thoughts and suggestion for future work

An overall glance to our vast collection of cases suggests that, for practical purposes, the results obtained computing on two regions are reliable, correct to at least one tenth of one percent. If a single region is used, some departure is noted in the standoff distance of the imbedded

$M_\infty=4,\ \alpha=30°$ $M_\infty=25,\ \alpha=30°$

Fig. 11

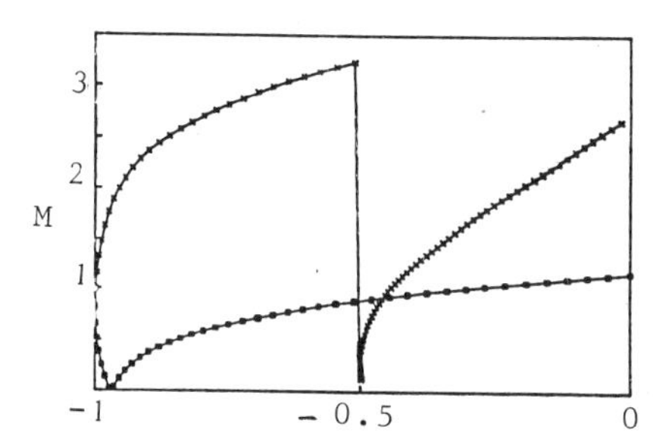

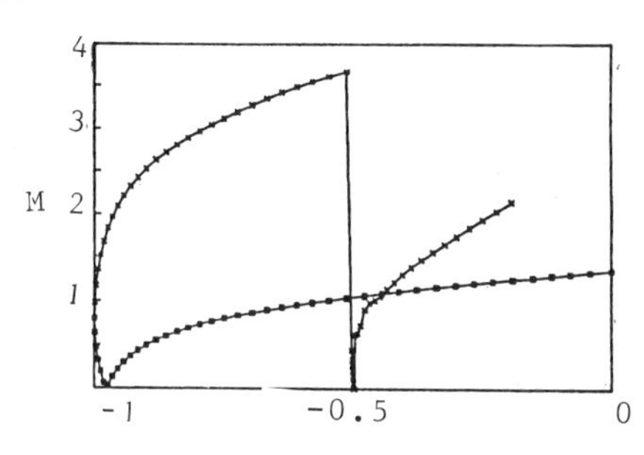

Fig. 12

shock, for example, but still within a good range of accuracy. The flow patterns obtained with the two types of calculations are very close to each other and there is no reason for distrusting them.

Nevertheless, there are still open questions, that we consider very relevant, both for a better understanding of numerics and for a better insight in the physics of the problem.

1) The highest difficulty lies where it is hardly detectable, that is, in the neighborhood of the corner. It is, indeed, commonly understated or plainly overlooked. We have tried to use our techniques for an exploration in depth, but we are unsatisfied. Under circumstances that, in the framework of our technique, are trivial (such as low free-stream Mach numbers at no incidence), the computed patterns are smooth and reasonable; convergence is reached steadily, even if slowly at times. When the Mach number is high the velocities are very high, the imbedded shock is close to the corner, all gradients become large and accuracy is lost. A disturbing fact is that higher resolution apparently worsens the results. Note that, with a standoff distance of the order of 0.001, twenty intervals between the shock and the corner produce an interval length of the order of 0.00005, so that all differences in the calculation must be multiplied by 20000. Any small error in such differences is enhanced, producing round-off errors that may overtake the truncation errors. If our conjectures are correct, no adaptive grid may help.

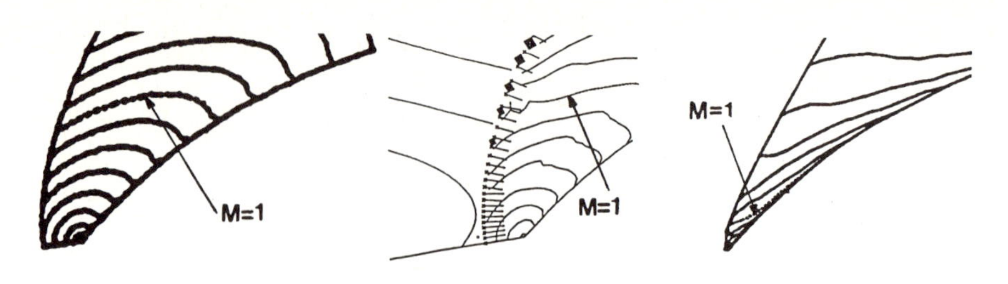

Fig. 13

2) The second puzzling question arose by inspecting our results. We cannot state that the isomach patterns are perfect on the body surface, because of total temperature losses at the corner; but we cannot overlook the appearance of a consistent pattern, that changes drastically as the standoff distance decreases. Physically, a clear picture of the neighborhood of the corner as a function of the standoff distance is important to understand how the flow evolves when a detached shock tends to become attached. In the first case, there is a subsonic zone behind the shock (that is normal to the surface); the subsonic zone does not exist when the shock is attached and oblique. How does the subsonic zone shrink as the standoff distance decreases? So far, our conjecture is that the transition is not self-similar but rather as shown in Fig. 13. On the left, the pattern for large standoff distances is shown twice (once as obtained computing with two regions, and then using a single region). On the right, the pattern for small standoff distances is shown, as obtained with the two-region calculation. It is evident that the sonic point on the shock tends to get closer to the corner than the sonic point on the wall. Again, an exhaustive analysis of the problem requires more sophisticated numerical tools.

3) Finally, once the problem of the corner is understood, it would be advisable to use the transonic region as a starting condition for a supersonic, march-on calculation to describe the entire region behind the shock, till its interaction with the bow shock and further on. A simple question arises, for which we did not see an answer yet (of course, conventional shock-capturing codes are not the best choice for answering it): At low Mach numbers and no incidence, the imbedded shock reaches the bow shock in a short distance. Does the imbedded shock ever reach the bow shock at high angles of attack?

We would like to see the above questions answered in the near future. Should we, perhaps, spend more time on our basic tools before undertaking complicated analyses of viscous, three-dimensional, non-equilibrium flows?

References

1. Hancock, G.J., A review in The Aeronautical Journal, 91:45, 1987.
2. Moretti, G., A technique for integrating two-dimensional Euler equations, Comp. Fl. 15:59-75, 1987.
3. Moretti, G., Efficient Euler solver with many applications, AIAA J. 26:655-60, 1988.
4. Moretti, G., Thoughts and afterthoughts about shock computations, Pol. Inst. of Brooklyn PIBAL Rep. 72-37, 1972.

IMPLICIT UPWIND FINITE-DIFFERENCE SIMULATION OF LAMINAR HYPERSONIC FLOW OVER FLARED CONES

B. Müller
DLR, Institute for Theoretical Fluid Mechanics,
Bunsenstr. 10, D-3400 Göttingen, FR Germany

SUMMARY

Axisymmetric laminar hypersonic flow of perfect gas over flared cones is simulated by solving the thin-layer Navier-Stokes equations for the steady state using an implicit upwind scheme based on approximate factorization and the total variation diminishing (TVD) method of Harten and Yee. The algorithm is outlined, and applications are discussed for Mach 25 flow in 77 km altitude over a 30° cone with 20° flare and a 30° cone with convex rear part and a 27° flare.

INTRODUCTION

During the reentry of a spacecraft, the body flap at the lower rear is deflected for maneuvering from high to low angles of attack. Therefore, the aerothermodynamic efficiency of this control device has to be guaranteed for the trajectory of a hypersonic lifting vehicle. Such investigations have quite often been carried out based on two-dimensional ramp flow simulations (cf. e.g. [1]). Here, axisymmetric laminar hypersonic flow of perfect gas over 30° cones with straight and convex rear parts and 20° and 27° straight flares (Figs. 1a, 2a) is investigated. The body contours correspond to two simplified geometries of the European spacecraft Hermes in the windward symmetry plane at an angle of attack of 30°. The flow conditions assume perfect gas at reentry conditions of Mach 25 in 77 km altitude. The geometry and flow conditions were proposed by G. Durand, CNES, as a first step to investigate the effects of straight and curved body contours on the aerodynamic efficiency and thermal loads of flaps for the design of Hermes.

The objective of the present paper is to outline the implicit upwind total variation diminishing (TVD) scheme, which was presented for the two-dimensional Navier-Stokes equations in [2], for the axisymmetric thin-layer Navier-Stokes equations and to discuss the strong shock-shock and shock-boundary layer interactions for the simulation of laminar hypersonic flow over flared cones.

The governing axisymmetric thin-layer Navier-Stokes equations are stated in section 2. The implicit upwind TVD finite-difference scheme based on approximate factorization and the modified flux approach of Harten and Yee [3] is outlined in section 3. In section 4, calculated results for Mach 25 flow of perfect gas in 77 km altitude over flared cones are discussed. The conclusion are given in section 5.

GOVERNING EQUATIONS

For axisymmetric Newtonian fluid flow of perfect gas, the time-dependent compressible Navier-Stokes equations read in dimensionless conservation-law form and in general coordinates as follows (cf. e.g. [4]):

$$\frac{\partial q}{\partial \tau} + \frac{\partial E^{(\xi)}}{\partial \xi} + \frac{\partial E^{(\eta)}}{\partial \eta} + H = Re_{\infty}^{-1}\left[\frac{\partial G}{\partial \eta} + S\right] \tag{1}$$

where $q = J^{-1}(\rho,\ \rho u,\ \rho v,\ e)^T$ is the vector of the conservative variables scaled by the Jacobian J of the transformation of the independent variables

$$\tau = t\,,\quad \xi = \xi\,(t, x, y)\,,\quad \eta = \eta\,(t, x, y)\ . \tag{2}$$

$E^{(\xi)}$ and $E^{(\eta)}$ are the inviscid fluxes in the ξ- and η-directions, respectively. H denotes the inviscid axisymmetric source term. The thin-layer viscous flux and source term are represented by G and S, respectively. The freestream Reynolds, Prandtl and Mach numbers Re_∞, Pr_∞ and M_∞, resp., are defined in terms of freestream values denoted by subscript ∞ and a characteristic length L. Stokes' hypothesis and the Sutherland law are employed. The Prandtl number and the ratio of specific heats are constant, namely Pr = 0.72 and γ = 1.4.

ALGORITHM

The axisymmetric thin-layer Navier-Stokes equations (1) are solved at the interior grid points of a boundary-fitted structured mesh. The approximate factorization scheme of Beam and Warming [5] is employed as basic implicit algorithm [4]:

$$\left[I + \frac{\vartheta \Delta\tau}{1+\psi}\,\mu_\xi \delta_\xi\,\frac{\partial E^{(\xi)n}}{\partial q} + D_I^{(\xi)n}\right]\Delta q^{*n}$$
$$= \frac{\psi}{1+\psi}\,\Delta q^{n-1} - \frac{\Delta\tau}{1+\psi}\left[\mu_\xi\,\delta_\xi\,E^{(\xi)n} + \mu_\eta\,\delta_\eta\,E^{(\eta)n} - \right. \tag{3a}$$
$$\left. - Re_\infty^{-1}\left(\delta_\eta\,G^n + S^n\right)\right] + D_E^n\,,$$

$$\left[I + \frac{\vartheta\Delta\tau}{1+\psi}\left(\mu_\eta\delta_\eta\,\frac{\partial E^{(\eta)n}}{\partial q} + \frac{\partial H^n}{\partial q} - \right.\right.$$
$$\left.\left. - Re_\infty^{-1}\left(\delta_\eta\frac{\partial G^n}{\partial q} + \frac{\partial S^n}{\partial q}\right)\right) + D_I^{(\eta)n}\right]\Delta q^n = \Delta q^{*n}\,, \tag{3b}$$

$$q^{n+1} = q^n + \Delta q^n\ . \tag{3c}$$

I is the 4 x 4 identity matrix. $\Delta\tau$ denotes the time step, and n the time level.

The first-order Euler implicit time differencing formula, i.e. $\psi = 0$ and $\vartheta = 1$ in (3), is used for steady-state calculations. The second-order three-point-backward formula, i.e. ψ = 1/2 and ϑ = 1 in (3), is preferred for unsteady flow simulations.

The classical finite-difference operators are defined by

$$\delta_\xi\,a_{i,j} = a_{i+1/2,j} - a_{i-1/2,j}\,,\quad \mu_\xi\,a_{i,j} = (a_{i+1/2,j} + a_{i-1/2,j})/2\,, \tag{4}$$

etc. For convenience, $\Delta\xi = \Delta\eta = 1$ is assumed.

The upwind scheme based on the modified flux approach of Harten and Yee [3] is used here. For one-dimensional hyperbolic systems of conservation laws with constant coefficients or with nonlinear fluxes in the scalar case, Harten showed that his high-resolution scheme is total variation diminishing (TVD) in the sense of total variation non-increasing and second-order accurate except for extrema where the accuracy is reduced to first-order [6]. The upwind TVD scheme can be implemented into the basic implicit algorithm (3) by choosing [7,8].

$$D_E = -\frac{\Delta\tau}{1+\psi}\ \frac{1}{2}\left[\delta_\xi R^{(\xi)}\phi^{(\xi)} + \delta_\eta R^{(\eta)}\phi^{(\eta)}\right] , \tag{5a}$$

$$D_I^{(k)} = -\frac{\vartheta\Delta\tau}{1+\psi}\ \frac{1}{2}\left[\delta_k \lambda^{(k)} I \delta_k J\right] \tag{5b}$$

where the components of $\phi^{(\xi)}$ are defined by (supressing the j-index):

$$\phi^{(\xi)l}_{i+1/2} = Q\left(\lambda^{(\xi)l}_{i+1/2}\right)\mu_\xi g^{(\xi)l}_{i+1/2} - Q\left(\lambda^{(\xi)l}_{i+1/2} + \gamma^{(\xi)l}_{i+1/2}\right)\alpha^{(\xi)l}_{i+1/2} \tag{6}$$

with $\alpha^{(\xi)}_{i+1/2} = R^{(\xi)^{-1}}_{i+1/2}\ \delta_\xi (Jq)_{i+1/2}$,

$$g^{(\xi)l}_i = \text{minmod}\left(\alpha^{(\xi)l}_{i-1/2} , \alpha^{(\xi)l}_{i+1/2}\right) ,$$

$$\gamma^{(\xi)l}_{i+1/2} = \frac{1}{2}\ Q\left(\lambda^{(\xi)l}_{i+1/2}\right)\left(\delta_\xi\, g^{(\xi)l}_{i+1/2}\right)\alpha^{(\xi)l}_{i+1/2}\ /\left(\max\left\{\left|\alpha^{(\xi)l}_{i+1/2}\right| , \tilde{\varepsilon}\right\}\right)^2 ,$$

$\tilde{\varepsilon} = 10^{-12}$ for 64-bit computer words.

The minimum modulus limiter is defined by

$$\text{minmod}\,(a, b) = s \max\left\{0,\ \min\left\{|a| , s\,b\right\}\right\} \tag{7}$$

where s = sign (a) .

The function Q is equal to the modulus for the eigenvalue of $\partial E^{(\xi)} / \partial (Jq)$ of multiplicity two. For the other two eigenvalues, Q is the entropy function [6] defined by

$$Q(\lambda) = \begin{cases} |\lambda| & if|\lambda| \geq \delta \\ \dfrac{\lambda^2+\delta^2}{2\delta} & if|\lambda| < \delta \end{cases} \tag{8}$$

where the entropy parameter δ is an anisotropic function of the spectral radii $\lambda^{(k)}$ of $\partial E^{(k)}/\partial (Jq)$ [2]:

$$\delta^{(\xi)} = \tilde{\delta}\,\lambda^{(\xi)}\left[1 + \left(\frac{\lambda^{(\eta)}}{\lambda^{(\xi)}}\right)^{2/3}\right] \tag{9}$$

with $\tilde{\delta} = 0(10^{-1})$. $\lambda^{(\xi)l}$ denotes the l-th eigenvalue of $\partial E^{(\xi)} / \partial(Jq)$. The columns of $R^{(\xi)}_{i+1/2}$ [9] are the right eigenvectors of $\partial E^{(\xi)} / \partial(Jq)$ evaluated with the Roe average [10] and the arithmetically averaged metric terms. The second eigenvector corresponding to the eigenvalue of multiplicity two is scaled by the speed of sound c, because dimensional arguments suggest a velocity and c is strictly positive and easily available. Fictitious values of α outside the boundaries are obtained by zeroth-order extrapolation, e.g. $\alpha^{(\xi)}_{1/2} = \alpha^{(\xi)}_{1+1/2}$. The simplified implicit treatment (5b) of the explicit upwind TVD terms (5a) is chosen because of its low computing cost. However, the linearization of the first-order upwind TVD terms, i.e. (5a) with $g^{(k)} \equiv 0$, improves the convergence of the scheme in some cases significantly [8], but takes about 38 % more CPU time per time step than (5b) with the present code on the IBM 3090 computer.

If the grid lines are not aligned with flow discontinuities, the wave propagation is not normal to cell interfaces. Thus, the operator splitting employed by TVD methods in multi-dimensions will break up a flow discontinuity into staircases along grid lines because the basic 1D Riemann solver will recognize a discontinuity only normal to a cell interface. Staircasing leads to a loss of accuracy and may cause failure of the computation, if the flow discontinuity which is oblique to the mesh is a strong shock.

Therefore, if the grid lines are not aligned with strong shocks, a small amount of numerical damping needs to be added to (5). Here, scalar nonlinear second differences in the conservative variables sensed by the second difference of the pressure and multiplied by the ratio of local and freestream Mach numbers are chosen [2]. The coefficient of the added numerical damping terms is 0 (100) times smaller than the value used for central discretizations [1]. Instead of adding numerical damping, the robustness of the TVD scheme can be maintained for hypersonic flow simulations by making the arguments of the limiter proportional to the pressure instead of density differences [8].

RESULTS

Assuming laminar axisymmetric flow of perfect gas in 77 km altitude calculations are carried out for the two flared cones depicted in Figs. 1a, 2a with γ = 1.4, $M_\infty = 25$, $Re_\infty = 17364$, L = 1m, Pr = 0.72, $T_\infty = 192.34\,K$, $T_w = 800\,K$ [11]. The half angles of both cones are 30°, and the flare angles are 20° and 27°, respectively. The convex part of the second part corresponds to the Hermes 94 geometry supplied by D. Devezeaux, ONERA.

The region of interest is bounded by the flared cone, the farfield and the outflow boundary (cf. Figs. 1a, 2a). On the body contour, the no-slip condition and wall temperature are imposed. The wall pressure is calculated from the boundary layer approximation of the wall normal momentum equation. The uniform freestream flow is fixed at the farfield. At the outflow boundary, the conservative variables are linearly extrapolated. The conical approximation is used at the leading edge, i.e. the conservative variables are multi-valued in the computational plane and the first ξ-derivative is approximated by the second-order three-point-forward formula.

The flow field is initialized for a 30° cone with a 5° flare on a coarse 45 x 33 mesh by prescribing the oblique shock relations for $M_\infty = \infty$ along grid lines, except for the body contour where the no-slip condition and the wall temperature $T_w = 800\,K$ are prescribed from the very beginning. The flare is deflected at increments of 5° (for the 27° flare, the last deflection increment is 7°) taking the previous result as initial condition. The result is interpolated on a fine 89 x 65 mesh and advanced 3600 time levels at a constant time step $\Delta\tau$ = 0.0025. δ in (9) is 1/80. In order to improve the resolution, the result is interpolated on a manually adapted 53 x 33 mesh shown in Figs. 1a, 2a for the two bodies.

The result on the coarse adapted mesh is interpolated on the fine 105 x 65 mesh (not shown) and advanced 4800 and 3600 time levels for the bodies with straight and curved contours, respectively. The latter computations of the axisymmetric thin-layer Navier-Stokes equations take 209 and 138 CPU minutes, resp., on the IBM 3090 computer. The maximum cell Reynolds numbers adjacent to the body contour are smaller than 2 and the maximum CFL numbers in the ξ- and the η-directions are $O(10^{-1})$ and $O(10^2)$, respectively. Using α^k in (6) proportional to the pressure difference [8] instead of the second-order numerical damping for the last 1200 time levels yields deviations of less than 1 % in the sensitive peak values of pressure, density and temperature.

The results for the flow over the 30° cone with straight contour and 20° flare were reported in [2]. Only the pressure and temperature contours are shown here for comparison purposes. The results (Figs. 1b, c) indicate the interaction of the weak oblique cone shock with the strong flare induced shock. The pressure contours show that the curved resultant shock is weaker than the flare induced shock but stronger than the oblique cone shock. A contact discontinuity and an expansion fan emanate from the shock-shock interaction.

Similar features are found for the flow over the 30° cone with convex contour and 27° flare (Figs. 2b, c). Whereas the temperature levels for the two flow cases are similar, the flare induced pressure rise is about 50 % larger than for the straight cone with 20° flare. The beginning and end of the compression region in the corner approximately indicate the locations of the separation and reattachment points. The separation induced shock interacts with the oblique shock. Further downstream, the outer shock becomes curved due to the interaction with a smeared nearly straight shock (cf. also Fig. 2d), which interacts with the flare induced shock. As for the flared straight cone, a contact discontinuity (Fig. 2c) and an expansion fan (Fig. 2b) are generated.

The velocity vectors given at every other point of the 105 x 65 mesh show the flow deceleration downstream of the shock-shock interaction, a velocity overshoot close to the flare and separated flow in the corner (Fig. 2e). The wall tangential velocity profiles at $x/L = 11.865$, i.e. at the corner, and at $x/L = 13$ give a more detailed picture and illustrate the good shock resolution (Fig. 2f). The flow separates at $x/L \approx 9.5$ and reattaches at $x/L \approx 12.9$. Secondary separation and reattachment occur at $x/L \approx 11.5$ and 10.0, respectively, and also in the corner (Fig. 2g). Between the corner and the primary reattachment point, the wall pressure increases about 50% more than for the straight cone with 20° flare (Fig. 2h). Downstream, the flow expands quickly. The peak heat transfer at the primary reattachment point is about three times larger than for the straight cone with 20° flare (Fig. 2i).

CONCLUSIONS

An implicit upwind TVD finite-difference scheme is outlined to solve the axisymmetric thin-layer Navier-Stokes equations for hypersonic flow. Strong shock-shock and shock-boundary layer interactions are found in simulations of hypersonic flow over flared cones. The results indicate that the convex Hermes 94 geometry exerts only a small impact on the surface pressure, whereas a 7° higher flare angle leads to a 50% larger pressure rise and a three times larger peak heat transfer on the flare.

ACKNOWLEDGEMENTS

Part of the research was supported by the Centre National d'Etudes Spatiales, France. The help of K. Droste and H. Holthoff in producing the present results is gratefully acknowledged.

REFERENCES

[1] MÜLLER, B.: "Calculation of Laminar Supersonic Flows over Ramps and Cylinder-Flares", DFVLR-IB 221-88 A 14, 1988.

[2] MÜLLER, B.: "Simple Improvements of an Upwind TVD Scheme for Hypersonic Flow", AIAA Paper 89-1977-CP, 1989.

[3] YEE. H.C., HARTEN, A.: "Implicit TVD Schemes for Hyperbolic Conservation Laws in Curvilinear Coordinates", AIAA J., Vol. 25 (1987), pp. 266-274.

[4] MÜLLER, B.: "Navier-Stokes Solution for Hypersonic Flow over an Indented Nosetip", AIAA Paper 85-1504-CP, 1985.

[5] BEAM, R.M., WARMING, R.F.: "An Implicit Factored Scheme for the Compressible Navier-Stokes Equations II: The Numerical ODE Connection", AIAA Paper 79-1446, 1979.

[6] HARTEN, A.: "High Resolution Schemes for Hypersonic Conservation Laws", J. Comp. Physics, Vol. 49 (1983), pp.357-393.

[7] YEE, H.C.: "A Class of High-Resolution Explicit and Implicit Shock-Capturing Methods", NASA TM 101088, 1989.

[8] YEE, H.C., KLOPFER, G.H., MONTAGNE, J.-L: "High-Resolution Shock-Capturing Schemes for Inviscid and Viscous Hypersonic Flows", NASA TM 100097, 1988.

[9] VINOKUR, M., LIU, Y.: "Equilibrium Gas Flow Computations II: An Analysis of Numerical Formulations of Conservation Laws", AIAA Paper 88-0127, 1988.

[10] ROE, P.L.: "Approximate Riemann Solvers, Parameter Vectors, and Difference Schemes", J. Comp. Physics, Vol. 43 (1981), pp. 357-372.

[11] MÜLLER, B., DROSTE, K., HOLTHOFF, H.: "Calculation of Laminar Hypersonic Flow over Flared Cones", DLR-IB 221-89 A 17, 1989.

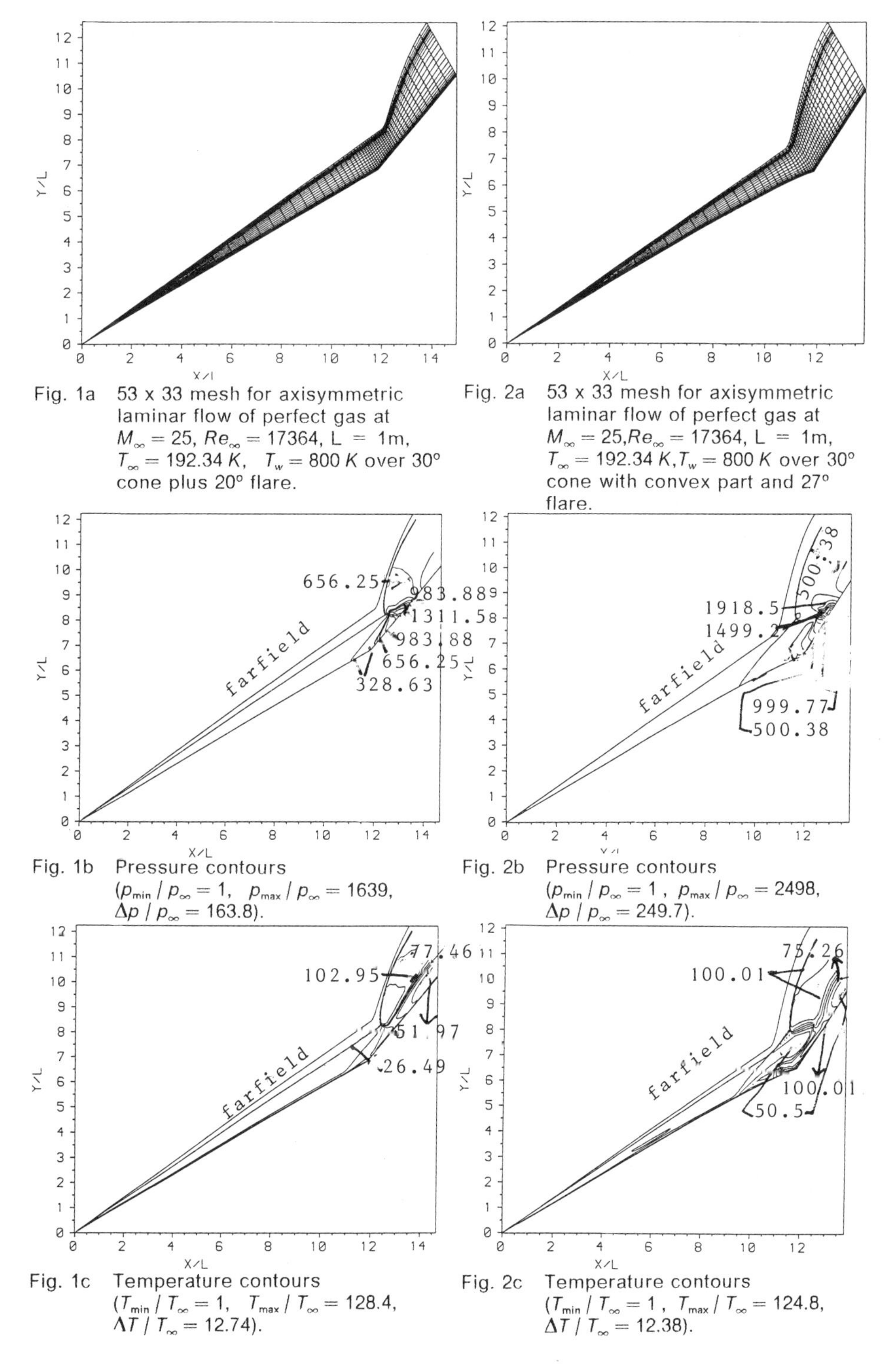

Fig. 1a 53 x 33 mesh for axisymmetric laminar flow of perfect gas at $M_\infty = 25$, $Re_\infty = 17364$, L = 1m, $T_\infty = 192.34\ K$, $T_w = 800\ K$ over 30° cone plus 20° flare.

Fig. 2a 53 x 33 mesh for axisymmetric laminar flow of perfect gas at $M_\infty = 25$, $Re_\infty = 17364$, L = 1m, $T_\infty = 192.34\ K$, $T_w = 800\ K$ over 30° cone with convex part and 27° flare.

Fig. 1b Pressure contours ($p_{min} / p_\infty = 1$, $p_{max} / p_\infty = 1639$, $\Delta p / p_\infty = 163.8$).

Fig. 2b Pressure contours ($p_{min} / p_\infty = 1$, $p_{max} / p_\infty = 2498$, $\Delta p / p_\infty = 249.7$).

Fig. 1c Temperature contours ($T_{min} / T_\infty = 1$, $T_{max} / T_\infty = 128.4$, $\Delta T / T_\infty = 12.74$).

Fig. 2c Temperature contours ($T_{min} / T_\infty = 1$, $T_{max} / T_\infty = 124.8$, $\Delta T / T_\infty = 12.38$).

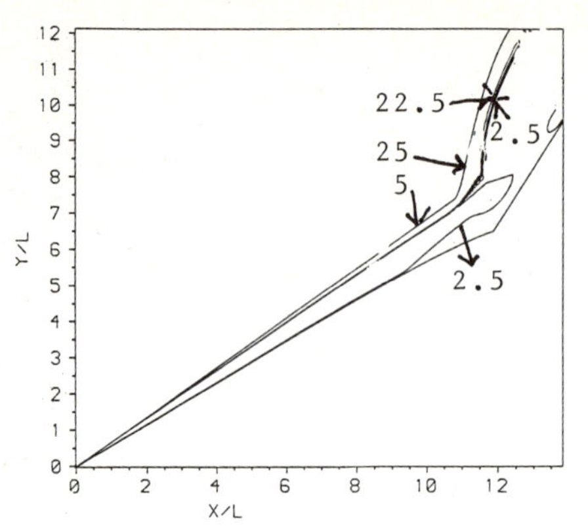

Fig. 2d Mach number contours ($\Delta M = 2.5$).

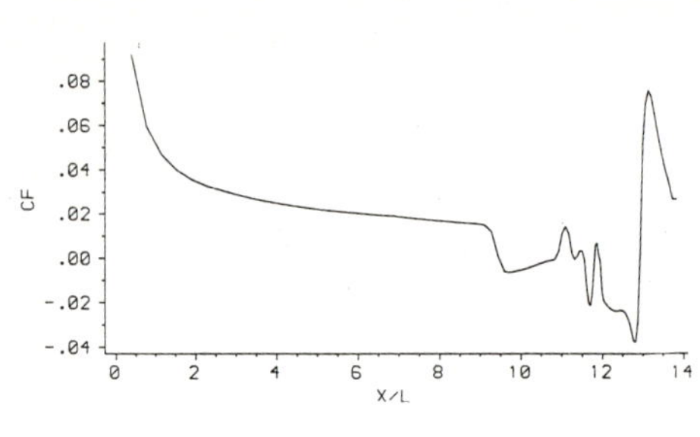

Fig. 2g Skin friction coefficient

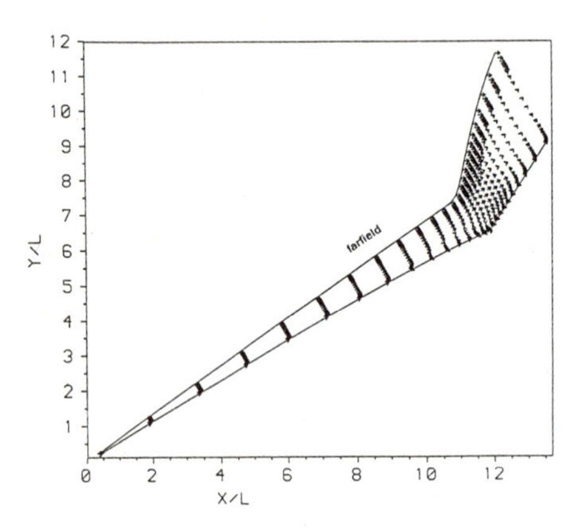

Fig. 2e Velocity vectors

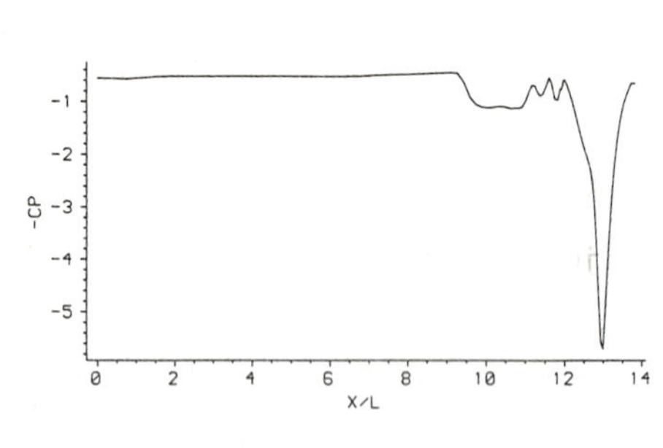

Fig. 2h Pressure coefficient

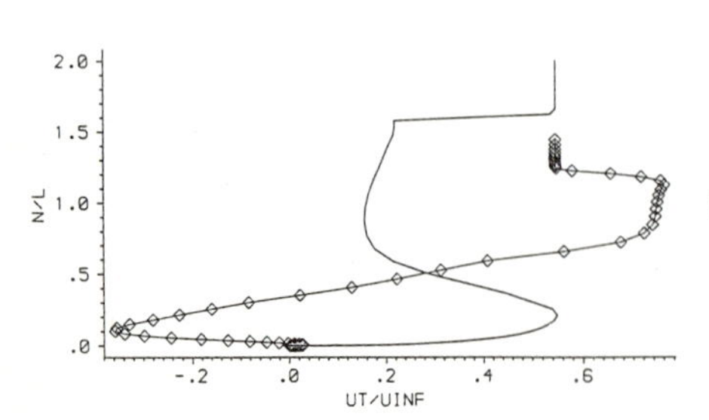

Fig. 2f Wall tangential velocity profiles normal to flare (□:$x/L = 11.865$, ——:$x/L = 13$).

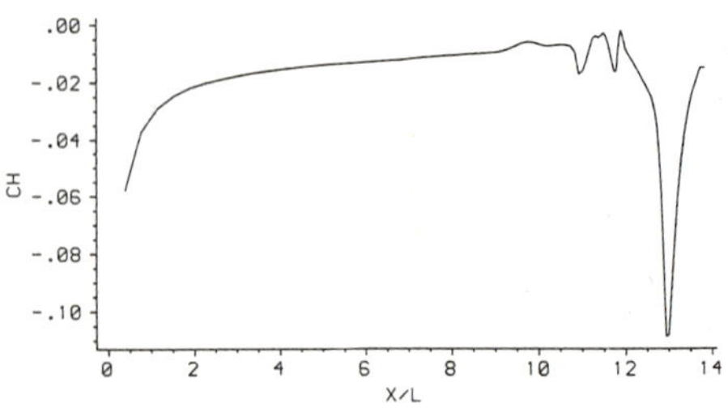

Fig. 2i Heat flux coefficient

Calculation of viscous hypersonic flows using a coupled Euler / second order boundary layer method

Ch. Mundt, M. Pfitzner, M.A. Schmatz

Messerschmitt-Bölkow-Blohm GmbH, Postfach 80 11 60,
D - 8000 München 80, Germany

Summary

A method is presented which allows the calculation of flows with weak viscous interaction by solving the second-order boundary-layer equations in the viscous and the Euler equations in the inviscid regions. Both parts of the solution are coupled. Ideal and equilibrium real gas results are presented for laminar flows. A comparison with Navier-Stokes results shows good agreement.

Introduction

The determination of viscous hypersonic flowfields remains a major challenge and is of great importance for hypersonic projects. A variety of physical and chemical phenomena not observed in sub-, trans- and supersonic flows has to be accounted for to make an optimized, efficient configuration possible. The evaluation of the characteristic properties of the flow can be done using three tools: free flight experiments, wind tunnel experiments and computational fluid dynamics. Free flight experiments are very expensive, and wind tunnels fail to simulate the important parameters of the free stream simultaneously. Both tools have large turn-around times. On the other hand, calculations do not have these drawbacks. However, the accuracy of the modelling has to be accessed and the results have to be validated. The aim is to provide methods which are efficient to use and which have the required accuracy.

For flowfields or parts of them showing no strong viscous-inviscid interaction, an efficient calculation is possible by solving the boundary layer equations and the Euler equations for the viscous and the inviscid parts, respectively. Appropriate boundary conditions have to be prescribed to couple both equation systems. Thus, similar results as obtained from a Navier-Stokes calculation can be expected. This approach, together with a coupling to the Navier-Stokes equations for regions of strong interaction, has been developed for sub- and transonic flows, see /1,2/.

Considering hypersonic flows, new effects have to be accounted for. The strong bow shock, the thin shock layer and the development of an entropy layer have to be addressed on the inviscid level. For the viscous part the development of thick boundary layers and the entropy layer swallowing are important phenomena which have to be accounted for. For both parts thermodynamic and chemical processes are of major importance due to the high temperatures and enthalpies involved.

The method developed to account for the phenomena sketched above is presented in the following. Based on the underlying methods SOBOL and EULSPLIT /3,4/, the solution of the governing equations is briefly discussed. The coupling process is described and results from this method are shown. Comparisons between different gas models (ideal gas and equilibrium real gas /5/) and with results from Navier-Stokes (NSFLEX) calculations /6/ validate the approach for a broad range of applications. Finally an outlook on future work is given.

The method of solution

Inviscid part /4/

The Euler equations are solved using a split-matrix algorithm with Runge-Kutta time stepping. The basic equations read, in a Cartesian coordinate system using conservative variables:

$$\mathbf{Q}_{,t} + \mathbf{F}_{,x} + \mathbf{G}_{,y} + \mathbf{H}_{,z} = 0, \qquad (1)$$

with

$$\mathbf{Q} = (\rho, \rho u, \rho v, \rho w, \rho e + 1/2\rho(u^2+v^2+w^2))^T,$$
$$\mathbf{F} = (\rho u, \rho u^2+p, \rho uv, \rho uw, u(\rho e+1/2\rho(u^2+v^2+w^2)+p))^T,$$
$$\mathbf{G} = (\rho v, \rho uv, \rho v^2+p, \rho vw, v(\rho e+1/2\rho(u^2+v^2+w^2)+p))^T,$$
$$\mathbf{H} = (\rho w, \rho uw, \rho vw, \rho w^2+p, w(\rho e+1/2\rho(u^2+v^2+w^2)+p))^T,$$

and express the conservation of mass, momentum and energy. ρ, e, p, u, v and w denote the density, mass specific internal energy, pressure and the three velocity components, respectively. Eq.(1) is transformed to a time dependent, locally monoclinic coordinate system

$$\mathbf{Q}_{,\tau} + \mathbf{A}\mathbf{Q}_{,\xi} + \mathbf{B}\mathbf{Q}_{,\eta} + \mathbf{C}\mathbf{Q}_{,\zeta} = 0. \qquad (2)$$

Note that the variables are not transformed, i.e. Cartesian velocity components are retained. $\mathbf{A}$, $\mathbf{B}$ and $\mathbf{C}$ denote the Jacobians of the flux matrices and are generalized as $\mathbf{K}$, in the following.

The eigenvalues of $\mathbf{K}$ are determined and the corresponding eigenvectors are calculated. Then, $\mathbf{K}$ can be diagonalized

$$\mathbf{\Lambda}_K = \mathbf{T}_K^{-1}\ \mathbf{K}\ \mathbf{T}_K, \qquad (3)$$

where $\mathbf{\Lambda}_K$ denotes the matrix of the eigenvalues and $\mathbf{T}_K$ the matrix of the eigenvectors. The matrices $\mathbf{K}$ are split according to the sign of their eigenvalues, and are replaced in eq.(2).

Thus, the equation to be discretized reads

$$Q_{,\tau} + T_A \Lambda_A{}^+ T_A{}^{-1} Q_{,\xi}{}^+ + T_A \Lambda_A{}^- T_A{}^{-1} Q_{,\xi}{}^- + T_B \Lambda_B{}^+ T_B{}^{-1} Q_{,\eta}{}^+ + T_B \Lambda_B{}^- T_B{}^{-1} Q_{,\eta}{}^- + T_C \Lambda_C{}^+ T_C{}^{-1} Q_{,\zeta}{}^+ + T_C \Lambda_C{}^- T_C{}^{-1} Q_{,\zeta}{}^- = 0. \quad (4)$$

For the discretization the third-order accurate upwind biased formula

$$Q^{\pm}{}_{,\xi\ i} = \pm\ (Q_{i\mp 2} - 6\ Q_{i\mp 1} + 3\ Q_i + 2\ Q_{i\pm 1})\ /\ 6\Delta\xi \quad (5)$$

is used. The equations are integrated in time using a three step Runge-Kutta procedure

$$\begin{aligned} Q^{(1)} &= Q^n - a_1\ \Delta t\ P\ (Q^n), \\ Q^{(2)} &= Q^n - a_2\ \Delta t\ P\ (Q^{(1)}), \\ Q^{n+1} &= Q^n - a_3\ \Delta t\ P\ (Q^{(2)}) \end{aligned} \quad (6)$$

to enhance stability. The operator P contains the space derivatives.

A bow-shock fitting approach is used. The grid extends in the ξ - direction from the body surface to the shock and is therefore time dependent. Appropriate boundary conditions have to be prescribed. At the shock, the Rankine-Hugoniot equations are applied. A normal mass flux is prescribed as the result of the solution of the boundary layer equations at the body surface.

Viscous part /3/

The second-order boundary-layer equations are derived from the non-dimensionalized Navier-Stokes equations by an order of magnitude analysis. Whereas in first-order theory only terms of order one are retained, in second-order theory terms of order $Re^{-1/2}$ are kept, too. A transformation of the variable boundary-layer thickness to the outer boundary of the computational domain is done to ensure a constant resolution of the boundary layer. Using tensorial concepts and thus a different notation than for the formulation of the Euler equations, the equations read in a locally monoclinic coordinate system, which is necessary to account for the special properties of the boundary layer normal to the wall:

Continuity equation (greek indices = 1,2)

$$(\sqrt{a}\ M\ \rho\ v^{\alpha})_{,\alpha} + \sqrt{a}(M\ \rho\ v^3)_{,3} = 0, \quad (7)$$

x^1- and x^2- momentum equations

$$\rho(v^{\beta} v^{\alpha}_{,\beta} + v^3 v^{\alpha}_{,3} + \underline{\Lambda^{\alpha}_{\varepsilon} M^{\varepsilon}_{\beta,\gamma} v^{\gamma} v^{\beta}} + \Lambda^{\alpha}_{\varepsilon} \Lambda^{\delta}_{\beta} \Gamma^{\varepsilon}_{\delta\gamma} v^{\beta} v^{\gamma} -$$

$$- \underline{2\ Re^{-1/2}\ \Lambda^{\alpha}_{\delta} b^{\delta}_{\beta} v^{\beta} v^3}) = - \Lambda^{\alpha}_{\gamma} \Lambda^{\beta}_{\delta} a^{\gamma\delta} p_{,\beta} + (\mu\ v^{\alpha}_{,3})_{,3} -$$

$$- \underline{Re^{-1/2} (2 \Lambda^{\alpha}_{\gamma} b^{\gamma}_{\delta} \mu v^{\delta}_{,3} + \Lambda^{\delta}_{\gamma} b^{\gamma}_{\delta} \mu v^{\alpha}_{,3})} , \quad (8)$$

x^3- momentum equation

$$\underline{Re^{-1/2} M^{\delta}_{\alpha} b_{\delta\gamma} v^{\alpha} v^{\gamma}} = - \rho^{-1} p_{,3} , \quad (9)$$

energy equation

$$\rho (v^{\alpha} e_{,\alpha} + v^3 e_{,3}) - Ec \gamma p \rho^{-1} (v^{\alpha} \rho_{,\alpha} + v^3 \rho_{,3}) =$$

$$= Ec \gamma M^{\delta}_{\alpha} M^{\varepsilon}_{\beta} a_{\delta\varepsilon} \mu v^{\alpha}_{,3} v^{\beta}_{,3} +$$

$$+ \gamma Pr^{-1} (k (T_{,e} e_{,33} + T_{,\rho} \rho_{,33} + T_{,e3} e_{,3} + \quad (10)$$

$$+ T_{,\rho 3} \rho_{,3} - \underline{Re^{-1/2} \Lambda^{\alpha}_{\varepsilon} b^{\varepsilon}_{\alpha} k T_{,3}}) + k_{,3} T_{,3}) .$$

In addition to the variables explained before, T, v^{α}, v^3, μ, and k denote temperature, surface tangential contravariant velocities, surface normal velocity, viscosity and heat conductivity. Free stream reference values occuring are the ratio of specific heats, γ, the Eckert, Prandtl and Reynolds numbers Ec, Pr and Re. The quantities unexplained up to now describe the geometry and the terms underlined are special second order terms omitted in first-order theory. The Einstein summation convention is used.

The equations are of parabolic type and are therefore solved using a finite-difference space-marching method. The equations are discretized using the second-order Crank-Nicolson scheme. The equations containing second derivatives are solved simultaneously using a standard implicit elimination method. The continuity and x^3-momentum equations are solved decoupled.

At the wall, the no-slip condition is used for the velocity. Concerning the temperature, an adiabatic wall, a wall with fixed temperature and a wall in radiation equilibrium can be considered. Boundary conditions corresponding to the second-order theory have to be prescribed at the outer edge of the boundary layer. All variables are interpolated according to the boundary-layer thickness from the inviscid profiles delivered after the solution of the Euler equations.

Coupling procedure

Some details of the coupling process have been mentioned before when discussing the boundary conditions for both parts of the method. The results of the boundary layer and the Euler solution are coupled as follows: The inviscid profiles of the velocities and the thermodynamic quantities are provided as boundary conditions for the boundary layer solution. As these profiles are evaluated at the outer edge of the boundary layer, the effects of the entropy layer swallowing are taken into account, as will be shown by the results.

After the solution of the boundary layer equations, the equivalent source distribution is calculated by

$$\sqrt{a}\,(\rho\, v^3)_0 = \left(\sqrt{a}\int_0^\delta M\,(\rho_e v_e^\alpha - \rho\, v^\alpha)\, dx^3\right)_{,\alpha} . \tag{11}$$

Here, δ denotes the boundary layer thickness. The use of the sources is equivalent to the thickening of the body by the displacement thickness. After the boundary-layer calculation the source distribution is impressed on the Euler calculation as a boundary condition at the body, filling the region in-between the body and the displacement thickness with fluid. The resulting equivalent inviscid flow is then coupled to the boundary layer again.

As far as the coupling strategy is concerned, at first it was thought from the experience gained in sub- and transonic calculations, that an iterative coupling with many boundary layer calculations is best suited (close coupling). Therefore, a relaxation procedure was developed to eliminate the disadvantageous effect of travelling pressure waves on the convergence. These pressure waves are developing because in supersonic regions of the flow they are the only way to propagate information upstream. During the iteration, the equivalent source distribution was diminished compared to the source distribution of the first boundary-layer calculation. This effect is avoided by using only one coupling iteration step (perturbation coupling). However, the differences on the final result are small.

This observation is also confirmed by the perturbation theory, see /7,8/. Following these theoretical considerations, it would make sense to continue the iteration process only if additional, higher order terms were introduced.

Although the coupling is working quite well, the question needs further investigation whether it is sufficient from the theoretical point of view to prescribe a mass flux source only. It is possible to calculate momentum or energy sources from the boundary layer and inviscid profiles as well. It has to be shown whether a coupling of these sources to the inviscid flow is possible and neccessary.

Modelization of the gas behaviour

Considering hypersonic flows with high kinetic energies, it is necessary in general to deal with high total enthalpies. Accordingly, the ideal gas assumption breaks down in most cases and the excitation of vibration, dissociation and ionization as well as chemical reactions have to be accounted for. In the present method this is possible for equilibrium air. The thermodynamic and transport properties routines of /5/ are used, which are vectorized and twice differentiable. The governing equations shown above are written in the formulation of the real-gas method. Of course, simplifications are possible for the ideal gas assumption. There, the equation system is closed by using the perfect gas state equation and the Sutherland law. For real-gas calculations, the system is closed by the real-gas routines. A comparison of the CPU time shows that the real-gas calculation consumes only about 15 per cent more CPU-time than the ideal gas calculation.

Results

In this section, results of the method described above are presented and the importance of several physical phenomena is shown. Most calculations have been done using hyperbolas or hyperboloids as test bodies. For all calculations laminar flow is considered. In Fig. 1 a hyperboloid is shown with a mesh of 56*17 points for the inviscid and an overlapping grid of 56*51 points for the viscous calculation, respectively. For the conditions $M_\infty=10$, $Re_r=2.47\cdot10^3$, adiabatic wall, the outer grid line represents the bow shock, and the thick line in the computational domain the boundary-layer thickness. Lines of constant pressure show the smooth transition as well as the importance of using second-order boundary-layer theory ($\partial p/\partial x^3 \neq 0$!).

Fig. 2 shows the pressure coefficient c_p along the x-axis of a hyperbola at $M_\infty=10$, $Re=2.47\cdot10^3$, adiabatic wall, ideal gas assumption. The lower curve is the uncoupled result and clearly underpredicts the c_p, because the displacement effect of the viscous flow is not present. This effect tends to make the body blunter and thus raises the c_p. The upper curve shows the nearly coinciding pertubation coupled and Navier-Stokes (NS) results, and the close-coupled result, which underpredicts the pressure very slightly.

The effect of the entropy layer swallowing is exemplified by Fig. 3, showing the skin-friction coefficient for a hyperboloid at the same free-stream conditions as before, but using the real-gas equilibrium model. The lower curve shows the first-order boundary-layer calculation, which uses the the inviscid wall conditions as the boundary value at the boundary-layer thickness. The upper line corresponds to the uncoupled second-order calculation. In between lies the pertubation coupled second-order result, which uses the correct boundary conditions. Downstream, it is approached by the first-order result, but near the nose, an important difference remains.
It should be mentioned that the temperature is overpredicted by the first-order theory over the complete surface.

Concerning the real-gas effects, Fig.4 shows the important change in temperature at the wall along the hyperbola with the same conditions as before in comparison to the ideal gas results. The difference is due to the energy consumption caused by the excitation of vibration and dissociation as well as by other chemical reactions. Assuming an emissivity coefficient of $\varepsilon=0.8$, the wall temperature for radiation equilibrium at the solid surface is shown, too. It causes the temperature to fall further. It should be mentioned that the decrease of temperature is Reynolds-number dependent /9/. Note that the temperature distributions are not similar. The real-gas case with $\varepsilon=0$ has the lowest difference between the stagnation point and the end of the body due to the de-exitation of internal degrees of freedom. For the result with radiation the permanent energy loss leads to the highest temperature difference. Also, the flowfield is strongly affected, as depicted in Fig. 5 for the skin friction coefficient.

Comparison and validation of the results

To assess the accuracy of the coupled Euler/second-order boundary-layer method, the results have been extensively compared to the Navier-Stokes (NS) method /6/. For both methods much experience was gained during the comparison and some improvements were made. The validation is very encouraging, as the results are very similar, as will be seen in the following.

The shock shape and shock standoff distance for both the real-gas (lower part) and the ideal-gas (upper part) assumption is shown in Fig. 6. A denser mesh is used with 110 points distributed along the body surface. The mesh for both methods is identical with respect to the wall-tangential direction. At the left hand side the shock-fitted result of the coupled calculation is shown with lines of constant pressure, and at the right hand side the shock-captured NS result. The agreement is very good. Note the different shock shape and standoff distance which is due to the lower temperature in the real-gas case.

The temperature distribution at the wall for the ideal gas case (Fig. 7) shows a small difference for the values at the stagnation point, which agree with the theoretical value within a range of about one per mille. Further downstream, the temperature falls at first slower and than faster for the coupled than for the NS result. Note, however, that the scale is such that the maximum deviations are of the order of one percent.

For the real-gas cases, the results compares well, too. For example, the skin friction coefficient is predicted with very small deviations, as depicted in Fig. 8.

Also, the profiles normal to the wall have been checked. The temperature profile is given at the station I=100 (see Fig. 6) for ideal gas and the velocity is given at the station I=50 for the real gas in Fig. 9. Both profiles are normalized with the boundary layer thickness and the value of the temperature and the velocity at the boundary layer thickness. Only minor differences are evident.

Finally, the wall temperature is compared for a wall in radiation equilibrium. Fig. 10 compares the wall temperature for a hyperboloid. The emissivity coefficient is set to ε=0.8. In the stagnation region the deviation is negligible. Downstream, the temperature distribution follows the same trends as before and finally leads to a slightly lower temperature resulting from the coupled method than from the Navier-Stokes method.

Conclusions

The coupled solution of the second-order boundary-layer and Euler equations provides an efficient tool for calculating viscous hypersonic flows. It is restricted, however, to flows with weak viscous interactions, i.e. no shock/ boundary layer interaction or streamwise separation can be allowed.

Results have been presented which show that the important effects in hypersonic flow can be accounted for. A comparison with Navier-Stokes results validates the method which has been presented above. However, further validation will be done by comparing the results of the method to shock tunnel results /10/. The application to a reentry body configuration is now in progress and will be reported in /11/.

In the future, the method will be extended to account for non-equilibrium chemistry and thermodynamic effects.

Acknowlegement

The authors wish to thank Dr. F. Monnoyer for providing the SOBOL method and the IABG-Ottobrunn VP200 supercomputer group for their support.

References

/1/ E.H. Hirschel, M.A. Schmatz: Zonal solutions for viscous flow problems. In: Finite approxiamtions in fluid mechanics, DFG Priority Research Program, Results 1983-1985, E.H. Hirschel (ed.), Notes on numerical fluid mechanics, vol. 14, pp. 99-112, Wiesbaden, Vieweg Verlag, 1986.

/2/ K.M. Wanie, M.A. Schmatz, F. Monnoyer: A close coupling procedure for zonal solutions of the Navier-Stokes, Euler and boundary-layer equations. ZFW Vol.11,1987,p.347-359.

/3/ F. Monnoyer:Calculation of three-dimensional attached viscous flow on general configurations with 2nd order boundary layer theory. MBB-S-PUB-345, 1988, to appear in ZFW.

/4/ C. Weiland, M. Pfitzner: 3-D and 2-D solutions of the quasi - conservative Euler equations. Lecture Notes in Physics No. 264, 1986, pp. 654 - 659.

/5/ Ch. Mundt, R. Keraus, J. Fischer: New, accurate, vectorized approximations of state surfaces for the thermodynamic and transport properties of equilibrium air. Paper submitted to the ZFW, 1989.

/6/ M.A. Schmatz: Hypersonic three-dimensional Navier-Stokes calculations for equilibrium gas. AIAA paper 89-2183, 1989.

/7/ W. Schneider: Mathematische Methoden der Strömungsmechanik. Vieweg Verlag, Braunschweig, 1978.

/8/ M. van Dyke: Perturbation methods in fluid mechanics. The parabolic press, Stanford, 1975.

/9/ E.H. Hirschel, Ch. Mundt, F. Monnoyer, M.A. Schmatz: Reynolds-number dependancy of radiation-adiabatic wall temperature. Paper submitted to the AIAA Journal, 1990.

/10/ R.R. Boyce, Ch. Mundt, A.F. Houwing, R.J. Sandeman: Initial comparisons of CFD calculations and shock tunnel experiments for hypersonic argon flow. National space symposium, Canberra, 1989.

/11/ F. Monnoyer, Ch. Mundt, M. Pfitzner: Calculation of the hypersonic viscous flow past reentry vehicles with an Euler-boundary layer coupling method. AIAA paper 90-0417, 1990.

Figures

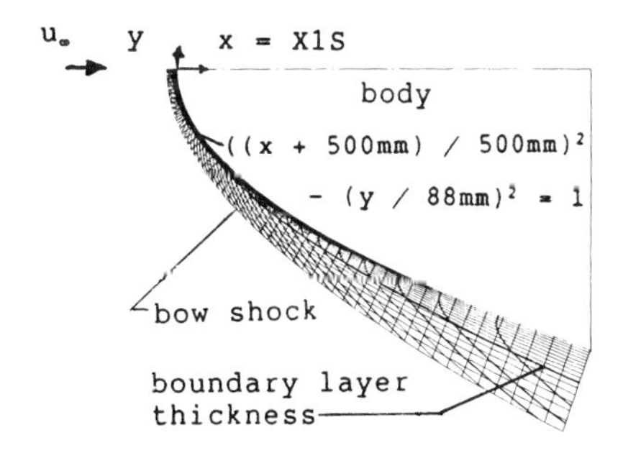

Fig. 1:
mesh and isobars for a coupled solution at the hyperboloid, $M_\infty=10$, $Re=2.47\cdot10^3$, $T_{ref}=220K$, $\rho_{ref}=7.7\cdot10^{-4}$ kg/m³, adiabatic wall, real gas

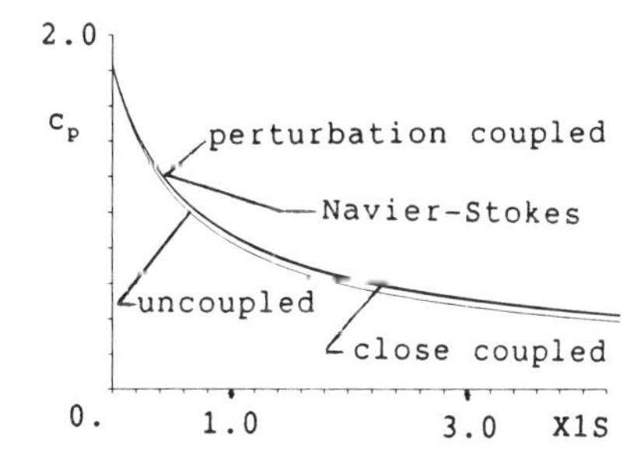

Fig. 2:
Pressure coefficient at the hyperbola, $M_\infty=10$, $Re=2.47\cdot10^3$, $T_{ref}=220K$, adiabatic wall, ideal gas

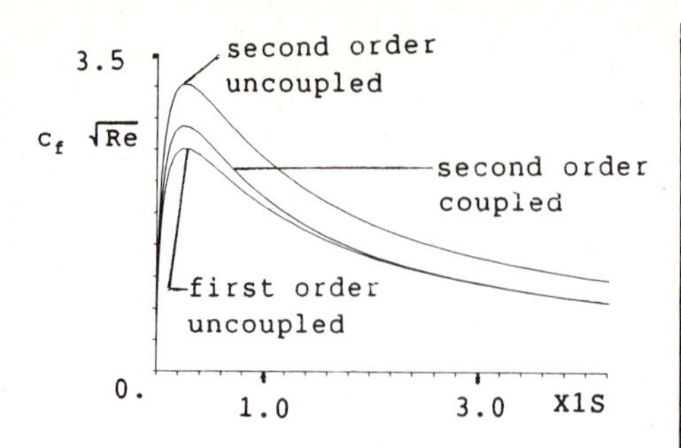

Fig. 3:
Skin friction coefficient at the hyperboloid, $M_\infty=10$, $Re=2.47\cdot10^3$, $T_{ref}=220K$, $\rho_{ref}=7.7\cdot10^{-4}$ kg/m^3, adiabatic wall, real gas

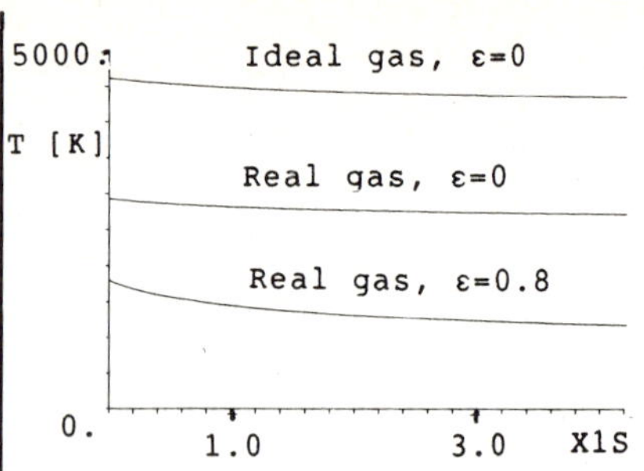

Fig. 4:
Temperature at the hyperbola, $M_\infty=10$, $Re=2.47\cdot10^3$, $T_{ref}=220K$, $\rho_{ref}=7.7\cdot10^{-4}$ kg/m^3

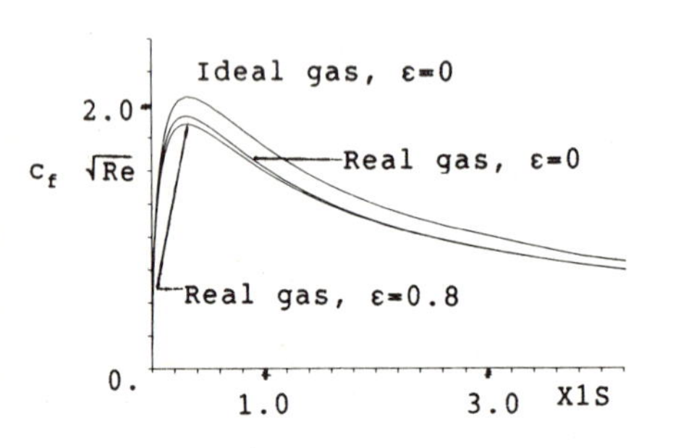

Fig. 5:
Skin friction coefficient at the hyperbola, $M_\infty=10$, $Re=2.47\cdot10^3$, $T_{ref}=220K$, $\rho_{ref}=7.7\cdot10^{-4}$ kg/m^3

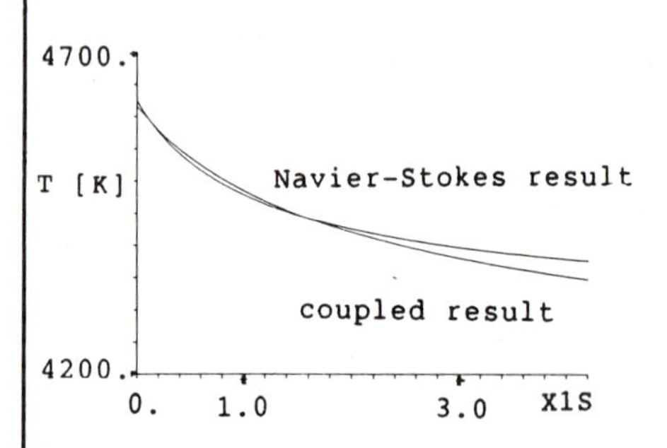

Fig. 7:
Comparison of the temperature at the hyperbola, $M_\infty=10$, $Re=2.47\cdot10^3$, $T_{ref}=220K$, ideal gas

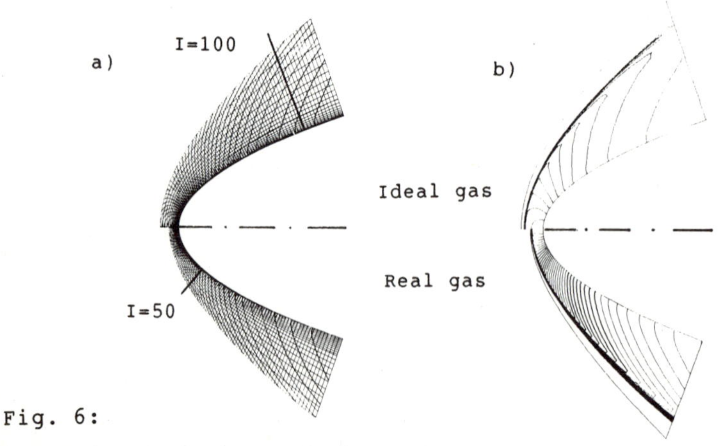

Fig. 6:

Comparison of a) coupled result and b) NS result for the hyperbola, $M_\infty=10$, $Re=2.47\cdot10^3$, $T_{ref}=220K$, $\rho_{ref}=7.7\cdot10^{-4}$ kg/m^3, adiabatic wall

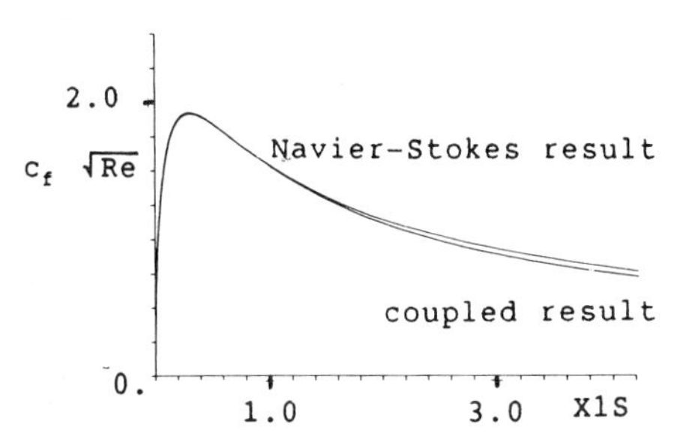

Fig. 8:

Comparison of the skin friction coefficient at the hyperbola, $M_\infty=10$, $Re=2.47 \cdot 10^3$, $T_{ref}=220K$, $\rho_{ref}=7.7 \cdot 10^{-4}$ kg/m³, adiabatic wall, real gas

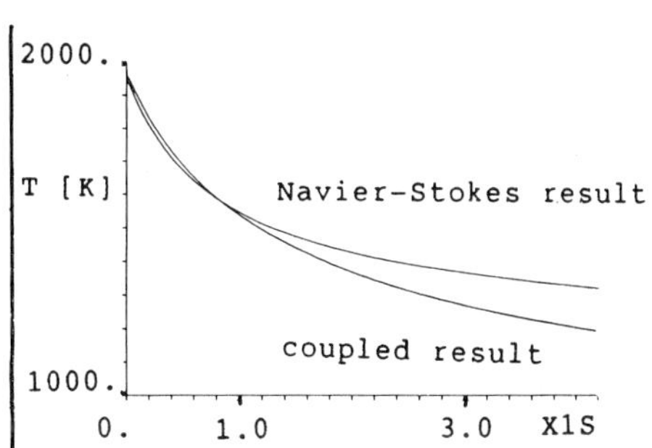

Fig. 10:

Comparison of the temperature for a hyperboloid with radiating solid surface, $M_\infty=10$, $Re=2.47 \cdot 10^3$, $T_{ref}=220K$, $\varepsilon=0.8$, $\rho_{ref}=7.7 \cdot 10^{-4}$ kg/m³, real gas

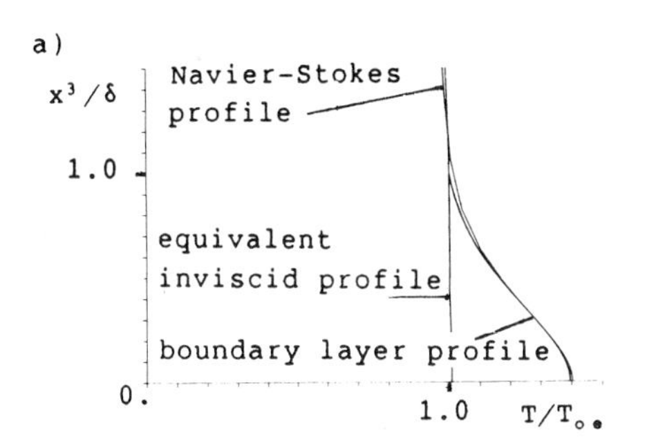

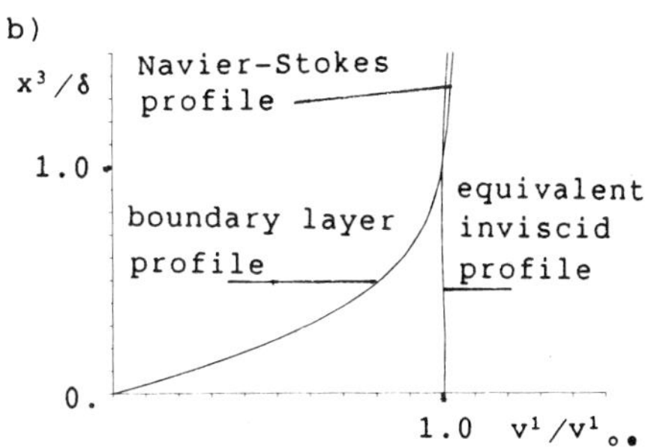

Fig. 9:

Normalized profiles normal to the wall
a) temperature for ideal gas, b) velocity for real gas, $M_\infty=10$, $Re=2.47 \cdot 10^3$, $T_{ref}=220K$, $\rho_{ref}=7.7 \cdot 10^{-4}$ kg/m³

A MULTIGRID LGS METHOD FOR THE VORTICITY-VELOCITY NAVIER-STOKES EQUATIONS

M. NAPOLITANO and G. PASCAZIO

Istituto di Macchine ed Energetica, Università di Bari, via Re David, 200, 70125 Bari, Italy

Introduction.

In the last years robust and efficient numerical schemes have been developed by one of the authors for solving viscous steady flows in two dimensions |1-4|. All of these schemes use the vorticity-stream function Navier-Stokes equations discretized in space by means of central differences to achieve solutions free of numerical viscosity. A clever use of a deferred correction strategy |5|, made possible by the delta form of the equations |6|, combined with a robust multigrid block-line-Gauss-Seidel relaxation procedure, allows to achieve steady solutions to difficult problems at a very reasonable computer cost |7|. A new effort has now been undertaken, aimed at developing a similar approach for three-dimensional flows. The vorticity-vector potential equations |8|, which are the extension to three dimensions of the vorticity-stream function formulation, have been discarded because of the difficulty in the treatment of the boundary conditions. The primitive variable (pressure-velocity) formulation has also been discarded in order to avoid the cumbersome treatment of the correct integral conditions for the pressure |9|. Therefore, the vorticity-velocity equations are being considered.

For such a formulation, several studies have already appeared in the scientific literature, see, e.g., |10-14|. In particular, Orlandi |13| and Guj and Stella |14| use a staggered grid to enforce continuity, whereas References 10-12, which all use node-centred schemes, employ an exponential scheme |10| or some kind of upwinding for the advection terms |11,12| in order to avoid instabilities at moderate-to-high values of the Reynolds number. More recently, a node-centred incremental block-line-Gauss-Seidel method has been developed for the two-dimensional vorticity-velocity equations |15|, very similar to that of |2|. The scheme was shown to be robust and efficient for the two-dimensional driven-cavity-flow problem |16| at Reynolds number of 100 and 1000. However, the steady solutions were found to be extremely sensitive to the treatment of the "boundary conditions" for the vorticity and did not satisfy mass conservation, neither exactly, nor within the second-order-accuracy of the numerical scheme |15|. Actually, for the Re = 1000 flow case, using two different (both second-order-accurate) discretizations for the vorticity condition at the wall causes the results to change by almost

100% |15|. It was thus concluded that, unless one is able to find appropriate boundary conditions which enforce the discrete form of the continuity equation, exactly or at least to second-order-accuracy, the use of a staggered grid is an essential item in the development of a numerical method for solving the vorticity-velocity Navier-Stokes equations.

In the present paper, an incremental multigrid scalar line-Gauss-Seidel (LGS) method is developed for solving the vorticity-velocity Navier-Stokes equations on a staggered grid, which is more accurate and reliable than those using a node cented discretization (e. g., |11| and |12|) and more efficient than those using a staggered-grid discretization (e. g., |13-14|).

Governing equations and numerical method.

For 2-D flows and Cartesian coordinates, the nondimensional vorticity-velocity Navier-Stokes equations can be written as:

$$\omega_t + u\,\omega_x + v\,\omega_y = (\omega_{xx} + \omega_{yy})/Re \qquad (1)$$

$$u_{xx} + u_{yy} = -\,\omega_y \qquad (2)$$

$$v_{xx} + v_{yy} = \omega_x \qquad (3)$$

where u and v are the longitudinal and vertical velocity components (which are prescribed at the boundary), subscripts indicate partial derivatives and ω is the vorticity, which has to satisfy the following boundary condition:

$$\omega = v_x - u_y\,. \qquad (4)$$

Equations (1-3) are discretized and linearized in time using a two-level implicit Euler scheme and the delta form of Beam and Warming |6| to give:

$$\Delta\omega^H/\Delta t + u^h\,\Delta\omega^H_x + v^h\,\Delta\omega^H_y - (\Delta\omega^H_{xx} + \Delta\omega^H_{yy})/Re$$

$$= C^H_h\left[-(u\,\omega)^h_x - (v\,\omega)^h_y + (\omega^h_{xx} + \omega^h_{yy})/Re\right] \qquad (5)$$

$$\Delta u^H/\Delta t - \Delta u^H_{xx} - \Delta u^H_{yy} = C^H_h\left[u^h_{xx} + u^h_{yy} + \omega^h_y\right] \qquad (6)$$

$$\Delta v^H/\Delta t - \Delta v^H_{xx} - \Delta v^H_{yy} = C^H_h\left[v^h_{xx} + v^h_{yy} - \omega^h_x\right]. \qquad (7)$$

In Eqs (5-7), u, v and their derivatives refer to the solution at any given time (iteration) level and the Δ's indicate the unknowns, namely the variations of the dependent variables (ω, u, v) and of their derivatives from the old to the new time (iteration) level.

Also, superscripts H and h indicate the current and finest grids, respectively and C_h^H indicate suitable collection operators from the finest grid h to the current grid H (a different operator is to be used for each equation, due to the staggered grid). Notice that a relaxation-like time derivative has been added to Eqs (6) and (7), to parabolize them |16|. Starting from an arbitrary initial condition (i. e., flow at rest), Eqs (5-7) are solved on the finest grid h by means of a two sweep alternating direction LGS iteration, to provide $\Delta\omega^h$, Δu^h, Δv^h; the solution (ω^h, u^h, v^h) is updated and Eqs (5-7) are solved on successively coarser grids (H = 2h, 4h, and 8h); the entire process is repeated until the finest-grid residual is reduced to a suitably small value. In more detail, at every grid level H, the following steps are required: a) the coefficients in the LHS of Eqs (5-7) are evaluated at the H-mesh gridpoints using the finest-grid solution (ω^h, u^h, v^h) locally, whereas the RHS steady state residuals are evaluated on the finest grid h and collected up to the current grid H; b) Eqs (5-7) are then solved approximately, using a single sweep of the aforementioned LGS smoother and homogeneous boundary conditions, to provide $\Delta\omega^H$, Δu^H, Δv^H; c) $\Delta\omega^h$, Δu^h, Δv^h are evaluated as

$$(\Delta\omega^h, \Delta u^h, \Delta v^h) = I_H^h (\Delta\omega^H, \Delta u^H, \Delta v^H) \tag{8}$$

where I_H^h are standard bilinear interpolation operators from the current grid H to the finest grid h; d) the finest-grid solution is updated as

$$(\omega^h, u^h, v^h) = (\omega^h, u^h, v^h) + (\Delta\omega^h, \Delta u^h, \Delta v^h) \tag{9}$$

e) the vorticity at the boundary is finally corrected so as to satisfy Eq (4) on the finest mesh. All of these steps are performed twice, with the LGS solution method marching from left to right and from top to bottom of the computational domain, respectively. It is noteworthy that the proposed methodology is very simple, since it does not require any logical choice to be made and employs a single free parameter, namely, the time step Δt. Furthermore, it does not need any additional storage with respect to the basic smoother, insofar as only the finest-grid solution is computed and a single array is used for the deltas at all grid levels. However, its work per iteration is slightly greater than that required by most current multigrid methods -- due to the additional interpolations and collections needed to visit and update the finest-grid solution after every coarse-grid calculation -- and it is likely to be less efficient than more sophisticated multigrid methods.

Results.

The proposed multigrid method has been applied to solve the rather difficult driven-cavity-flow problem |16| at Re = 1000. Figure 1 provides the convergence history of the method when using 1, 2, 3, and 4 grid levels, the finest grid containing 97x97 nodal points. The logarithm of the residual of the vorticity equation is plotted versus the work, one work unit being the work required to perform a two-sweep iteration on the finest grid. The improvement in the convergence rate due to multiple grids is quite clear: machine zero on an HP 840S computer using single precision arithmetic is reached after about 3000 work units, when using one grid level, but after only about 300 work units (about 100 global iterations), when using 4 grid levels. It is noteworthy that the work required to perform a global multigrid cycle when using 2, 3, and 4 grid levels is equal to 1.59, 2.10, and 2.62, respectively; also, one work unit corresponds to 4 CPU seconds on the aforementioned computer.

The numerical solution is given in figure 2 as the u and v velocity profiles at the vertical and horizontal centerlines of the cavity. The very accurate results of Ghia et al. |17| are also given (as circles) for comparison. The two solutions coincide within plotting accuracy. Also, as already shown by Guj and Stella |14|, the present results, obtained using a staggered grid, second-order-accurate central differences and a conservative discretization of the advection terms, satisfy the discrete form of the continuity equation exactly, i. e., to machine zero.

In order to test the proposed method versus a very difficult problem, the Re = 3200 flow case was computed using a finest grid containing 129x129 nodal points. The convergence history and the numerical solution are given in figures 3 and 4, where the very accurate results of |17| are also given (as circles) for comparison. It is noteworthy that the method has some difficulty to converge when using 4 grids. This is not surprising insofar as the present multigrid method is extremely simple and does not employ any sophisticated means for optimizing the convergence rate, such as full multigrid cycling or Reynolds number continuation. Incidentally, both of these improvements could be implemented without any difficulty. Also, the time step for the vorticity equation had to be reduced from 1 to 0.2.

Acknowledgement.

This research has been supported by MPI (60%) and C.N.R., grants n. 86.2478.07 and 87.2763.07.

References.

1. M. Napolitano, "Efficient ADI and Spline ADI Methods for the

Steady-State Navier-Stokes Equations", Intern. J. for Numerical Methods in Fluids, **4**, 1101-1115, 1984.

2. M. Napolitano and R.W. Walters, "An Incremental Block-Line-Gauss-Seidel Method for the Navier-Stokes Equations", AIAA J., **24**, 770-776, 1986.
3. M. Napolitano, "An Incremental Multigrid Strategy for the Fluid Dynamic Equations", AIAA J., **24**, 2040-2042, 1986.
4. J. H. Morrison and M. Napolitano, "Efficient Solutions of Two-Dimensional Incompressible Steady Viscous Flows", Comput, Fluids, **16**, 119-132, 1988.
5. P. K. Khosla and S.G. Rubin, "A Diagonally Dominant Second-Order-Accurate Implicit Scheme", Comput. Fluids, **2**, 207-209, 1974.
6. R. M. Beam and R. F. Warming, "An Implicit Factored Scheme for the Compressible Navier-Stokes Equations", AIAA J., **16**, 393-402, 1978.
7. M. Napolitano, "Efficient Solution of Two-Dimensional Steady Separated Flows", Invited Lecture, Int. Symp. Comp. Fluid Dynamics, Sydney, August, 1987.
8. S. M. Richardson and A. R. H. Cornish, "Solution of Three Dimensional Incompressible Flow Problems", J. Fluid Mechanics, **82**, 309, 1977.
9. L. Quartapelle and M. Napolitano, "Integral Conditions for the Pressure in the Computation of Incompressible Viscous Flows", J. Comput. Physics, **62**, 340-348, 1986.
10. S. C. R. Dennis, D. B. Ingham and R. N. Cook, "Finite-Difference Methods for Calculating Steady Incompressible Flows in Three Dimensions", J. Comput. Physics, **33**, 325-339, 1979.
11. R. K. Argarwal, "A Third-Order-Accurate Upwind Scheme for Navier-Stokes Solutions in Three Dimensions", Presented at the ASME Winter Annual Meeting, Washington, D.C., 15-20 November 1981.
12. B. Farouk and T. Fusegi, "A Coupled Solution of the Vorticity-Velocity Formulation of the Incompressible Navier-Stokes Equations", Intern. J. for Numerical Methods in Fluids, **5**, 1017-1034, 1985.
13. P. Orlandi, "Vorticity-Velocity Formulation for High Re Flows", Comput. Fluids, **15**, 137-150, 1987.
14. G. Guj and T. Stella, "Numerical Solutions of High Re Recirculating Flows in Vorticity-Velocity Form", Intern. J. for Numerical Methods in Fluids, to appear.
15. P. Giannattasio and M. Napolitano, "Numerical solutions to the Navier-Stokes equations in vorticity-velocity form", Proceedings of the Third Italian Meeting of Computational Mechanics, Palermo, Italy, 7-10 June 1988, pp. 95-101.
16. O. R. Burggraf, "Analytical and Numerical Studies of the Structure of Steady Separated Flows", J. Fluid Mechanics, **24**, 113-151, 1966.
17. U. Ghia, K. N. Ghia and C.T. Shin, "High-Re Solutions for Incompressible Flow Using the Navier-Stokes Equations and a Multigrid Method", J. Comput. Physics, **48**. 387-411, 1982.

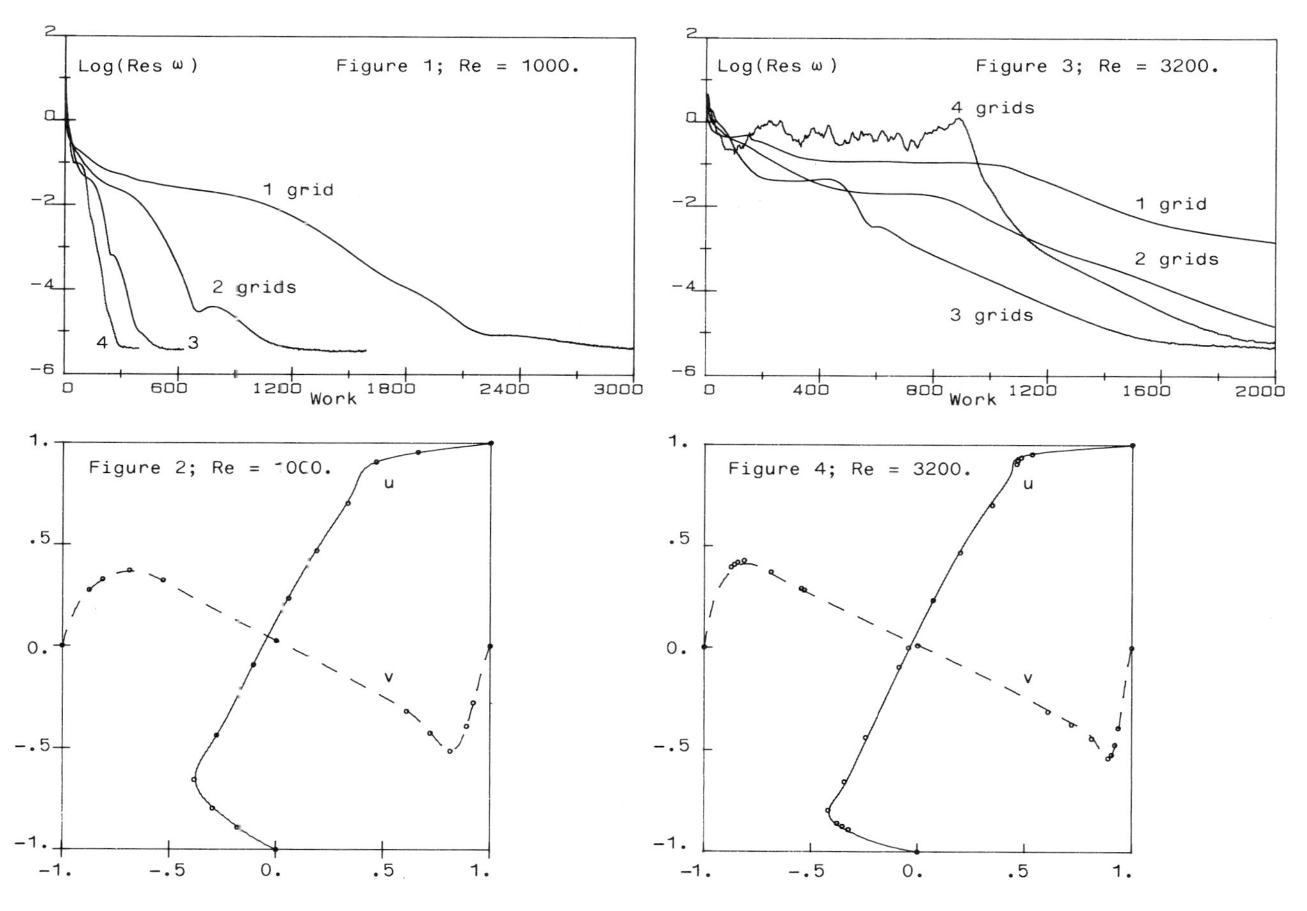

Figures 1–4: Convergence histories and numerical results for Re = 1000 and Re = 3200.

NUMERICAL SIMULATION OF VORTICES MOTION IN PRESENCE OF SOLID BOUNDARIES

Paolo Orlandi

Dipartimento di Meccanica e Aeronautica
Università di Roma, "La Sapienza"

Introduction

The important role of coherent structures on the comprehension of turbulence has been emphasized, recently, by experimental and numerical studies. Interactions of these large structures at different scales among themselves and interactions with the random field of the small scales is the mechanism which make turbulence a field not completely known. In the present study a numerical simulation has been done to describe, how large structures behave in presence of bodies of different shape. Although the calculation is limited to 2-D flowfields, this is a first step to build a numerical simulation for 3-D flowfields.

A numerical experiment requires a very accurate discretization in time and space. Dealing with homogenous or very simple domains spectral methods are very accurate but in presence of complex domains they are very difficult to implement then the traditional finite difference scheme is more suitable. In the two dimensional case the N-S system of equations, in the vorticity-stream function formulation allows to reduce the number of operations, while in the 3-D case the system in primitive variables is more efficient. Our final goal is to develop a numerical method for 3-D flows in complex domains, thus the ω, ψ formulation has been developed to transfer its qualities to the primitive variable formulation. As an intermediate step, a very accurate method to treat high Re flows has been obtained.

Equations and Numerical scheme

In the 2-D case the N-S equations written in general coordinates are

$$\frac{\partial \omega}{\partial t} + \frac{1}{\sqrt{a}} \frac{\partial q^i \omega}{\partial x_i} = \frac{1}{Re} \frac{1}{\sqrt{a}} \left(\frac{\partial}{\partial x_k} \alpha^{kl} \frac{\partial \omega}{\partial x_l} \right) . \tag{1}$$

$\frac{1}{\sqrt{a}}$ is the Jacobian of the coordinate trasformation. $\alpha^{kl} = \sqrt{a} a^{kl}$ are metric coefficients with the contravariant coefficients a^{kl} evaluated from the covariant coefficients

$$a_{ij} = \frac{\partial y_k}{\partial x_i} \frac{\partial y_k}{\partial x_j} .$$

q^i are the fluxes related to the stream function by

$$q^1 = \partial \psi / \partial x_2 \quad , \quad q^2 = - \partial \psi / \partial x_1 \tag{2}$$

The stream function ψ is related to the vorticity ω by

$$\omega = -\frac{1}{\sqrt{a}} \left(\frac{\partial}{\partial x_k} \alpha^{kl} \frac{\partial \psi}{\partial x_l} \right) . \tag{3}$$

This system of equations is very simple and requires a limited number of metric quantities, due to the fact that in 2-D, ω and ψ are scalar quantities. In the case of 3-D flows it is convenient to solve a system of equations with unknown the fluxes q^i. The derivation of the system of equations is rather complicate and requires a large number of metric quantities evaluated at different grid positions [1]. Furthermore on the derivatives discretization appear fluxes and metric quantities at grid locations where these are not defined. Several choices on averaging can be done, then the evaluation of discrete differential operators is not unique as for Cartesian or polar coordinates. As an example the analysis of the discrete operator $\nabla \times (\nabla \phi)$ is enlightening. The relation $\mathbf{q} = -grad(\phi)$ in general coordinates is

$$q^k = -\alpha^{kl} \frac{\delta \phi}{\delta x_l} \tag{4}$$

and the relation $\omega = curl(\mathbf{q})$ is

$$\omega = \frac{1}{\sqrt{a}} \left(-\frac{\delta q^1 \alpha^{k2}}{\delta x_k} + \frac{\delta q^2 \alpha^{k1}}{\delta x_k} \right) . \tag{5}$$

In 2-D the fluxes are located at the sides of the cell, the pressure at the center and the vorticity at the grid points. The metric quantities in Eq. (4) are defined in a unique way, instead the derivatives in Eq. (5) can be expressed combining the fluxes and the metric quantities in several ways. At the moment, we were not able to show that the discretized form of $\nabla \times (\nabla \phi)$ is identically null, as in Cartesian or polar coordinates. This is one among the several difficulties encountered, in the primitive variable formulation when general curvilinear coordinates are used.

For the system of Eq.(1) the discretization of the RHS does not, usually, present any difficulty when the metric quantities are appropriately located. More critical is the discretization of the non-linear terms and the derivation of vorticity boundary conditions. With regard to the nonlinear terms, Arakawa in a fundamental paper [2], presented a scheme, which, in the inviscid and periodic case, conserves total energy, total enstrophy and skew symmetry. This scheme has often been used in meteorological applications where periodic boundary conditions are imposed. On the contrary in aeronautical applications Arakawa's scheme has had a limited spreading. Eq. (1) together with Eq. (2) shows that the nonlinear terms, multiplied by $\sqrt{a}$, have a form equal to the form in a Cartesian coordinate system, then the Arakawa's scheme can be straightforwardly extended to general curvilinear coordinates. In this case ω, ψ, with respect to the primitive variable formulation, has the further quality that the nonlinear terms discretization brings to a scheme which conserves total energy, total enstrophy and skew symmetry. The extension of the Arakawa's scheme to the primitive variables formulation, also in the simple case of Cartesian coordinates, requires a very large number of mathematical operations to maintain global conservation properties [3]. The task is more onerous or even impossible in the case of general coordinates. Then, before we pursue this effort, it is worth to verify how the results by the ω, ψ formulation are sensitive to the global conservation properties, especially for wall bounded flows. In this paper the expression of the the discretized form of the nonlinear terms by the Arakawa's scheme (Ref.[1] Eq.(46)) is not given and it is indicated by $H(\psi,\omega)$.

With regard to the role of wall boundary conditions the primitive variable formulation presents advantages respect to the ω,ψ formulation. With staggered fluxes particular attention must be used on evaluating

$$\frac{1}{\sqrt{a}}\frac{\partial}{\partial x_l}\alpha^{ll}\frac{\partial q^m}{\partial x_l} \qquad l \neq m$$

at the first grid point near the boundary. In the the ω,ψ formulation the wall vorticity is usually evaluated by Eq. (5). In the present paper Eq. (5) has been differentiated with respect to x_k, for boundaries at $x_k = const$. This operation has the merit to relate the wall vorticity to the inner vorticity field. For the x_2 direction, when orthogonal coordinates are used, follows

$$\frac{\partial \omega}{\partial x_2} = \frac{\partial}{\partial x_2}\frac{1}{\sqrt{a}}\left(\frac{\partial \alpha^{11}q^1}{\partial x_1} - \frac{\partial \alpha^{22}q^2}{\partial x_2}\right). \tag{6}$$

This equation is evaluated at the position $i, j_w + 1/2$, with j_w indicating the boundary. The discretization of the first term on the right hand side is straightforward. The term with q^2 has been discretized by the same scheme used on the primitive variable formulation achieving an accuracy $O(\Delta x^2)$. This expression is slightly different from the Woods [4] formulation and does not present the incovenience that at high Re and with non-uniform grids the solution diverges as reported in Ref. [5]. Eq. (6) becomes more complex when general non-orthogonal coordinates are used. In the present paper a simulation was done with general curvilinear non-orthogonal coordinates, but free-slip boundary conditions have been considered. In this case the wall vorticity is

$$\omega = -\frac{q^1\Gamma^1_{12}}{a_{22}}$$

where

$$\Gamma^n_{li} = a_{nj}\frac{\partial y_k}{\partial x_j}\frac{\partial y_k}{\partial x_i \partial x_l} .$$

A further crucial aspect of the solution of unsteady flows is the time advancement scheme, which requires at least second order accuracy and not too restrictive stability conditions. A three-step time advancement for Eq. (1) can be written as [6].

$$\begin{aligned}(\omega^k - \omega^{k-1}) - \alpha_k\frac{\Delta t}{\sqrt{a}Re}\nabla^2(\omega^k - \omega^{k-1}) &= \Delta t(\gamma_k H^k(\psi,\omega) + \rho_k H^{k-1}(\psi,\omega)) \\ &+ \alpha_k\frac{2\Delta t}{\sqrt{a}Re}\nabla^2(\omega^k)\end{aligned}$$

where $k = 1,2,3$ denotes the substep number; ω^0 and ω^3 are the vorticities at time step n and $n+1$ the ∇^2 operator is the second order finite difference operator in the right hand side of Eq.(1) which retains only the $i = j$ derivatives. The cross derivatives ($i \neq j$) have been included in the $H(\psi,\omega)$ operator. The coefficients γ_k and ρ_k, $k = 1,2,3$ are constants selected such that the total time advancement between t^n and t^{n+1} is nearly third-order accurate in time. These coefficients are

$$\gamma_1 = 8/15\ ,\ \gamma_2 = 5/12\ ,\ \gamma_3 = 3/4\ ,\ \rho_1 = 0.\ \ ,\rho_2 = -17/60\ ,\ \rho_3 = -5/12\ ,$$

$$\alpha_l = \gamma_l + \rho_l \quad , \quad \sum \alpha_l = 1.$$

At each substep $H(\psi, \omega)$ is advanced explicitly and the viscous term implicitly.

The use of the unknown $\Delta\omega^k = \omega^k - \omega^{k-1}$ yields, at each substep k, for the boundary conditions the expression

$$\frac{\delta \Delta\omega^k}{\delta x_2} = \frac{\delta\omega^k}{\delta x_2} - \frac{\delta\omega^{(k-1)}}{\delta x_2} + O(\Delta t^2)$$

where the right hand sides are calculated by Eq.(6). Boundary conditions second order accurate in time are necessary for unsteady flows. This time advancement scheme has time step restrictions due to the explicit treatment of the non linear terms. The stability limit of the third order Runge-Kutta is larger than the one used with Adams-Bashfort scheme [6]. A reduction of the CPU time is thus achieved.

Results and Conclusions

In order to emphasize the qualities of a numerical scheme, significant physical cases must be considered, and, to our opinion, it is mandatory to chose flows for which analytical solutions or solutions obtained by other methods exist. As a first case the motion of a dipolar vortex has been considered: this flow structure, in absence of viscous effects, translates with constant velocity preserving its shape. This case has been considered to test the importance of global conservation properties.

The vortex dipole used here has an initial distribution $\omega = k^2\psi$ throughout the recirculation domain. In the inviscid case this dipole moves with a constant velocity U_c. The stream function distribution is given by

$$\psi = -\frac{2U_c J_1(kr)}{kJ_0(kr_0)} sin\theta$$

in the region $r < r_0$ and $\psi = 0$ elsewhere [7]. In Fig. 1 the dipolar vortex velocity at each time is given for the three Arakawa's schemes. The solution was obtained by a 128×128 with 1000 time steps by using a CPU of 122 secs on a CRAY XMP48. The scheme which conserves both kinetic energy and enstrophy and preserves the skew symmetry ($c_1 = c_2 = c_3 = 1/3$) reduces the dispersive errors and predicts a constant translation velocity. The scheme which conserves energy and vorticity ($c_1 = c_2 = 1/2, c_3 = 0$) gives better predictions than the scheme conserving only the vorticity ($c_1 = 1, c_2 = c_3 = 0$). The third order Runge-Kutta allows much larger time steps than those used by a leap-frog scheme.

The second case considered is the time evolution of small perturbations starting from the plane Poiseuille flow at Re=7500. The initial perturbation comes from solutions to the Orr-Sommerfeld eigenvalue problem. This is the problem Canuto et al. [8] at pg. 21 have chosen to emphasize the qualities of spectral methods with respect to finite differences. In Ref.[8] the authors presented finite difference solutions in a very poor agreement with theoretical results. Our opinion is that the inaccuracy of the finite difference depends on several reasons. The first reason is that they used a coordinate trasformation optimal for the Chebyshev spectral collocation method but not appropriate for finite difference. This is not the main reason which, instead resides on the treatment of the near wall second derivative and on the discretization of the convective terms. The same consideration has been drawn by Rai and Moin [9] who obtained accurate results using special upwinding in a primitive variable formulation. This case has been considered to test the wall vorticity assumption of Eq. (6). Fig. (2) shows how important are also the effects of the convective terms. All the schemes used in the present paper, however give much better results than those presented in [8]. In this case, again the Arakawa's scheme with ($c_1 = c_2 = c_3 = 1/3$) gives very good results.

As an interesting and difficult case to simulate the case of a vortex dipole impinging a small cylinder has been chosen. The same geometrical parameters, used in the clear and well documented experiment of Homa et al. [10], have been assumed. From the computational point of view, it is necessary to introduce a very enhanced coordinate stretching near the small cilinder to accurately represent the high vorticity gradients near the wall. Far enough from the cylinder a free-slip boundary has been assumed. The case of a symmetrical and a weak non-symmetrical interactions was considered and the trajectories of primary and secondary vortices were compared with the experimental trajectories. A more detailed comparison with the experimental results, the dependence of the numerical results on the outer boundary location and on the number of grid points will be published elsewere.

Fig. (3) shows that, in the case of symmetric impingment, the present numerical scheme predicts the trajectories of vortex centers agree with the experimental observations. The Reynolds in the numerical scheme is higher than the Re used in the experiment, but as pointed out in Ref. [10] this flow, in a short time scale, is not high Re dependent. In fact immediately after the secondary vortices formation, interactions among primary and secondary vortices are of inviscid nature. The viscosity acts in a subsequent period of time in a similar manner to the interaction of a vortex dipole with flat walls [11]. When the interaction is non-symmetric the flow changes in a large measure as shown by the trajectories in Fig. (4). In this case discrepancies between numerical and real experiments occur. The reason in a smaller measure can be also attributed to the difficulties, in the real experiment, to estimate the trajectories of vortices center from ink visualizations.

As final case the case of a dipole encountering a constriction in a 2-D channel has been simulated. For this case experimental studies are not available, however it has been considered in order to show the present numerical scheme handles satisfactory flows in general curvilinear coordinates. The simulation done with a very coarse grid 40×30, shows in Fig. (5) that the dipole conserves its shape when the constriction is wide enough to allow the dipole passage. Fig. (6) shows that a small reduction on the constriction stops the straight motion of the dipole, and the dipole brakes into two structures which travel backward along the free-slip walls.

As conclusion, we started this research with the aim to develop an accurate vorticity stream function numerical scheme and to transfer some of the qualities in the primitive variable formulation. We found this to be a very difficult task, especially for general curvilinear coordinates. The present numerical scheme based on general conservation properties is very promising on the achievement of 2-D numerical experiments. Recently true 2-D flows received a renewed interest, especially because real 2-D experiments have been designed. When instead simulations of 3-D flows are achieved, the use of primitive variables is worth. The use of schemes conserving energy are mandatory. In a 3-D flow the enstrophy is not any longer conserved, then numerical schemes conserving other high order quantities must be developed. Helicity, that recently has attracted a large interest, is one of the candidates for conservation properties.

References

1. Orlandi P., Moin P., Kim J. "Numerical Solutions of 3-D Flows Periodic in One Direction and with Complex Geometries in 2-D", 1988, Unpublished.
2. Arakawa A., J. of Comp. Physics, 1, 119, 1966.
3. Grammeltvedt A. ,Montly Weather Review., 97, 384, 1969.
4. Woods L.C. Aero. Q., 5, 176, 1954.
5. Roache P. Computational Fluid Dinamics Hermosa Publishing, 1976.
6. Wray A., "Minimal storage Time advancement Schemes for Spectral Methods" Submitted to J. of Comp. Physics, 1988.
7. Lamb H. Hydrodynamics Cambridge University Press, 1932.
8. Canuto C., Hussaini M.Y., Quarteroni A. and Zang .T.A., Spectral Methods in Fluid Dynamycs Springer Series in Computational Physics, 1987.
9. Rai M.M., Moin P. AIAA paper 89-0369 27th Aerospace Sciences Meeting Reno, 1988.
10. Homa J., Lucas M. and Rockwell D., J. Fluid Mech., 197, 571, 1988.
11. Orlandi P. "Vortex Dipole Rebound From a Wall", Submitted to Phys. of Fluids, 1989.

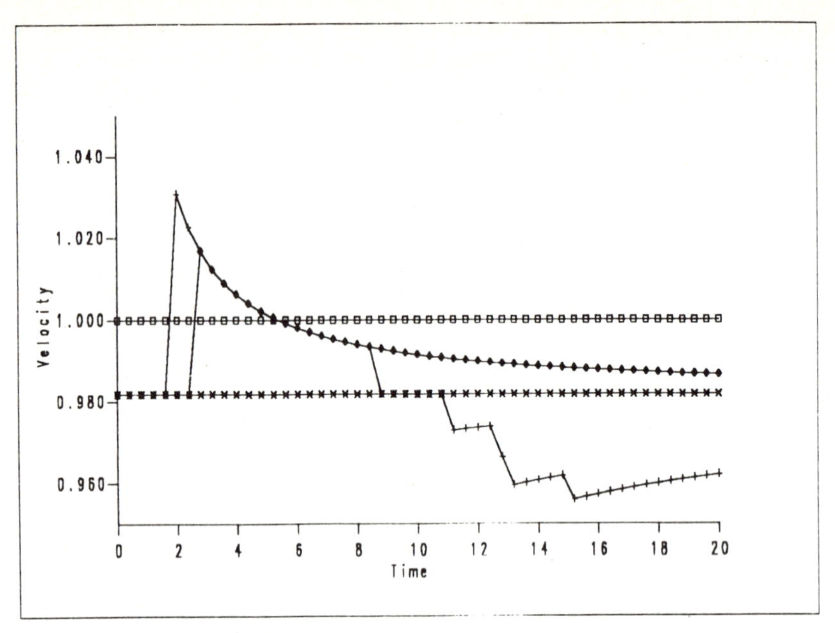

Figure 1 Translation modon velocity.
$-\square-\square-$ Theoretical; $-\times-\times-\times-$ $c_1 = c_2 = c_3 = 1/3$; $-\diamond-\diamond-$ $c_1 = c_2 = 1/2$, $c_3 = 0$; $-+-+-$ $c_1 = 1, c_2 = c_3 = 0$.

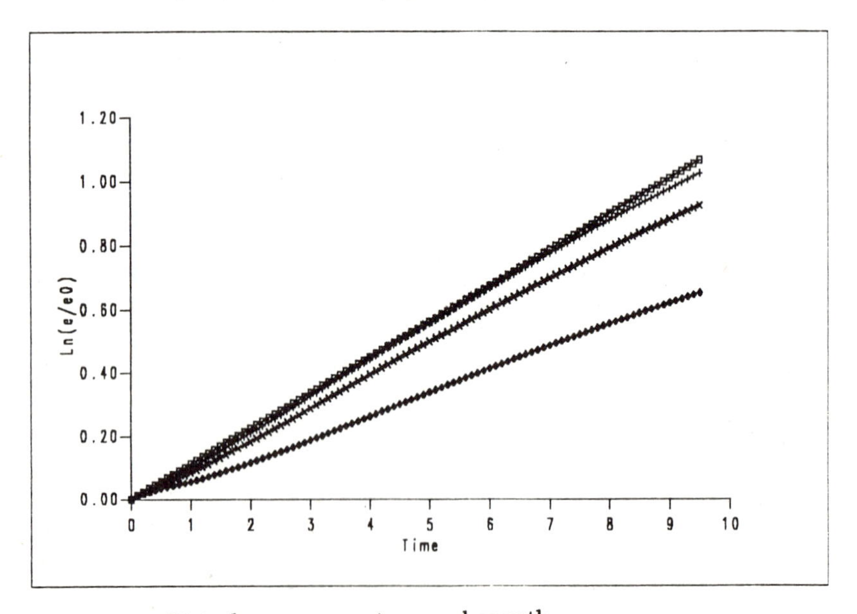

Figure 2 Disturbance energy temporal growth.
$-\square-\square-$ Theoretical; $-+-+-$ $c_1 = c_2 = c_3 = 1/3$; $-\times-\times-$ $c_1 = c_2 = 1/2, c_3 = 0$; $-\diamond-\diamond-$ $c_1 = 1, c_2 = c_3 = 0$.

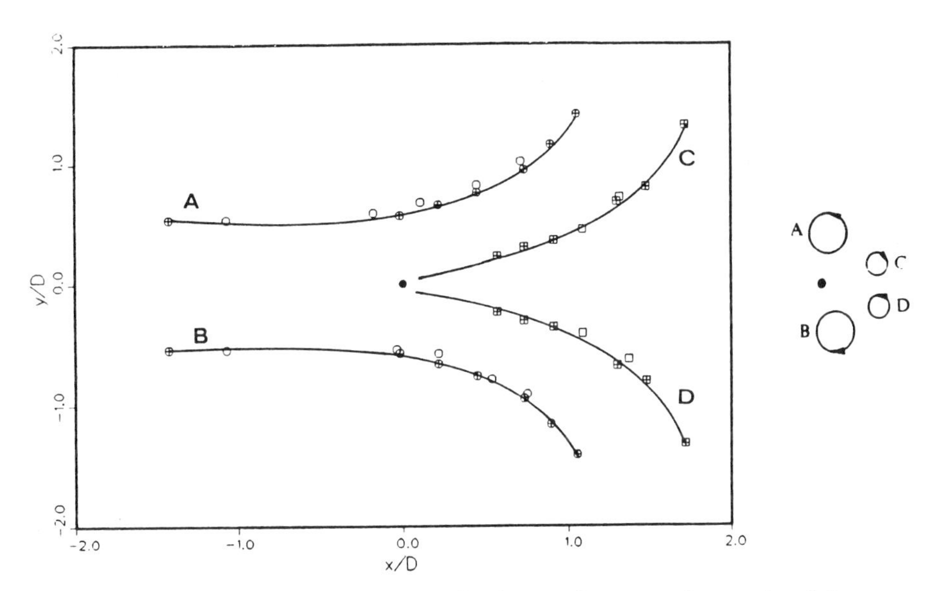

Figure 3 Trajectory of vortices and scheme of vortices interaction [1]. Symmetric case $e/D = 0$. ○, □, Experiment of Ref.[1]; $-\oplus-\oplus-$, $-\Box-\Box-$ Present simulation.

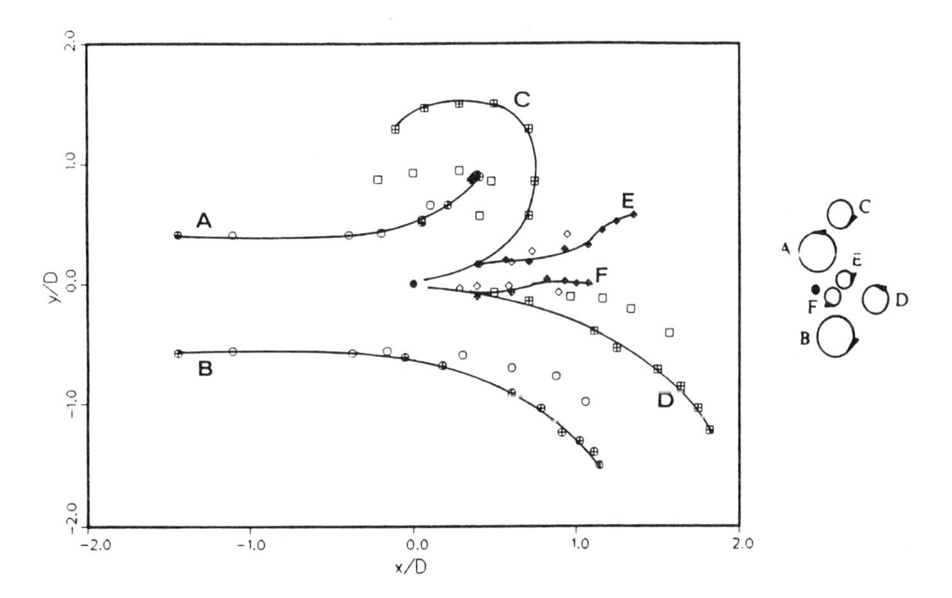

Figure 4 Trajectory of vortices and scheme of vortices interaction [1]. Non symmetric case $e/D = 0.12$. ○, □, Experiment of Ref.[1]; $-\oplus-\oplus-$, $-\Box-\Box-$ Present simulation.

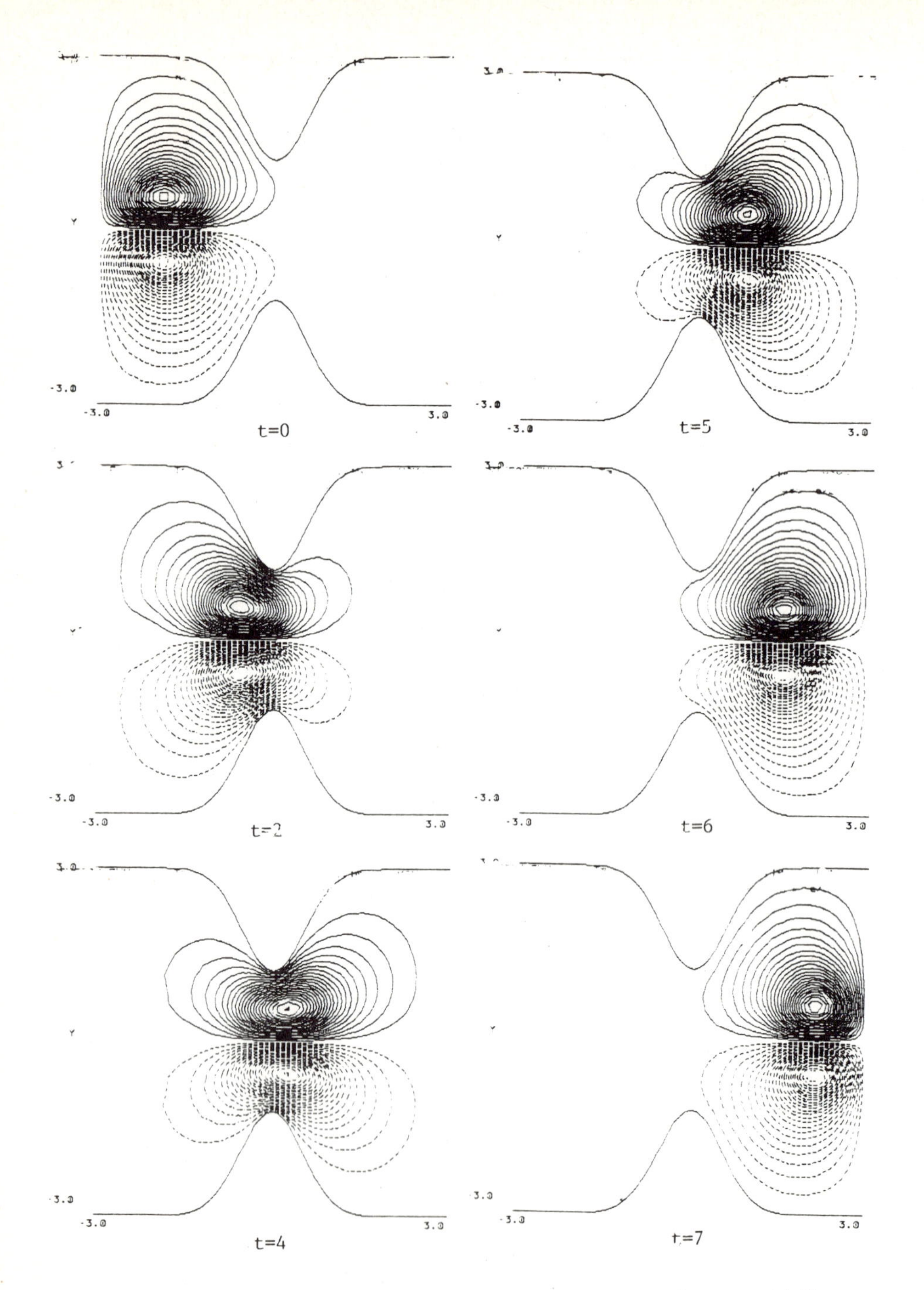

Figure 5 Stream function of a dipole passing through a constriction (width = .75).

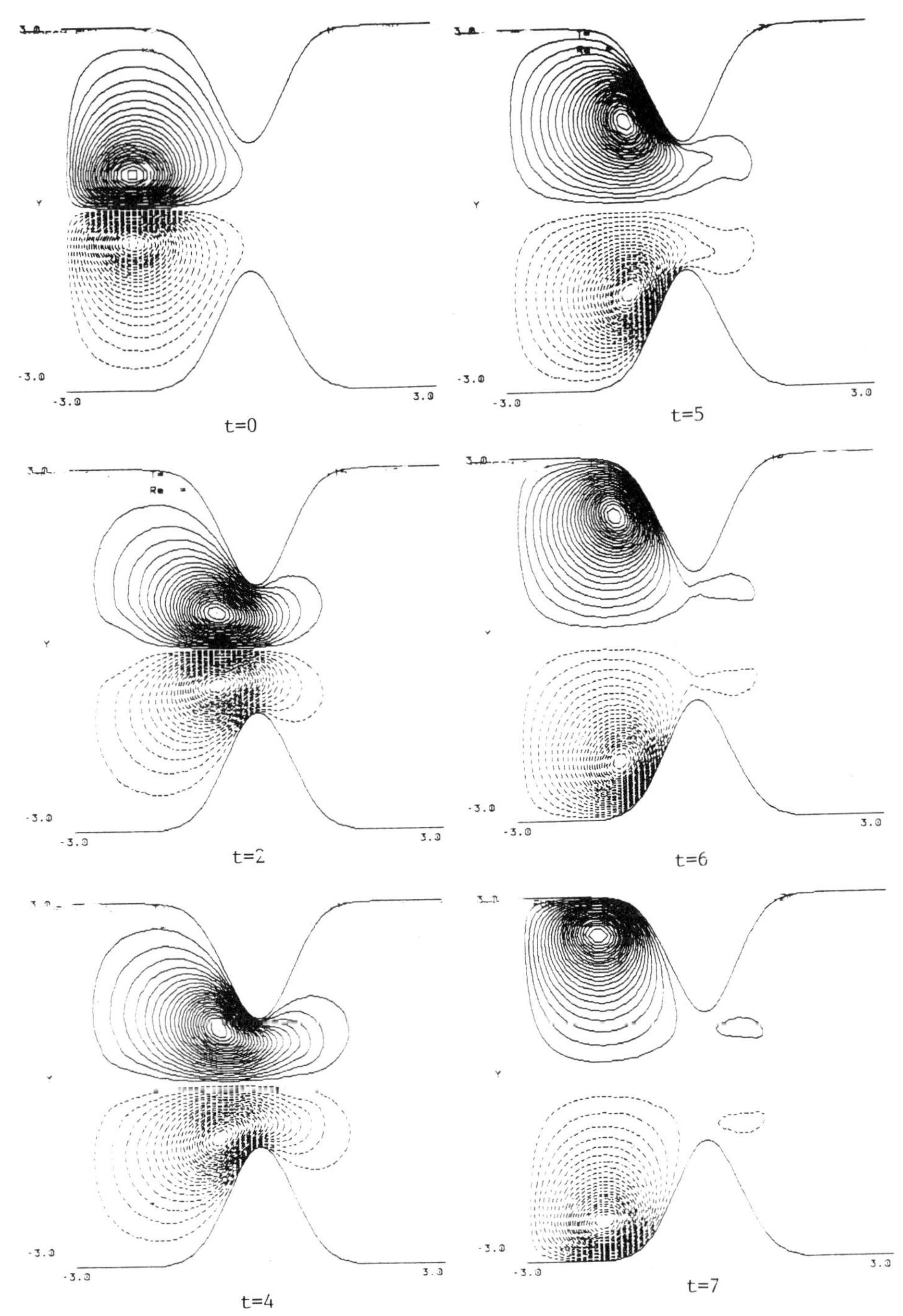

Figure 6 Stream function of a dipole passing through a constriction (width = .65).

CALCULATION OF 3-D LAMINAR FLOWS WITH COMPLEX BOUNDARIES USING A MULTIGRID METHOD

A. Orth, B. Schönung
Institute for Hydromechanics
University of Karlsruhe, Kaiserstr.12, D-7500 Karlsruhe, F.R.Germany

SUMMARY

The paper reports on the use of a multigrid method for calculating 3-D incompressible, laminar flows with complex geometries. The flow solver, which is briefly described, is based on a finite-volume method using non-staggered grid arrangement and general curvilinear coordinates. Following the SIMPLE-algorithm of Patankar and Spalding the discretized momentum equations are solved sequentially. The applied multigrid method is a modified Full-Approximation Scheme (FAS), updating the variable-dependent components in the source terms on coarser grids too. For the different flow configurations the multigrid performance is described and the predicted results are presented.

INTRODUCTION

Originally multigrid methods have been developed for the solution of linear elliptic equations as for example the Laplace-equation. They have been adopted more and more for the calculation of problems in fluid mechanics (see Hackbusch and Trottenberg [1]). However the solution of the Navier-Stokes equations has been mainly performed for simple flow geometries and cartesian coordinates. For such configurations excellent speed-up factors could be obtained with various multigrid methods (see e.g. Vanka [2], Barcus et al. [3] and Arakawa et al. [4]). It is of course of interest to apply multigrid methods for the calculation of 3-D problems of practical importance. Compared with the ideal Laplace-equation, the 3-D-Navier-Stokes equations can however cause some problems, which may reduce the efficiency of the multigrid method: For the solution of problems of practical interest general curvilinear coordinates have to be used, which result in cross-derivative terms in the partial differential equations. Large aspect and expansion ratios may occur in the generated body-fitted grids, which lead to strongly non-isotropic operators in the difference equations. A coupled system of nonlinear equations has to be solved. Due to limitations in the storage-capacity, the number of grid points per space direction is in general smaller than for 2D-problems as well as the number of grid levels, which can be used. In addition the problem occurs, that even with the coarsest grid of a multigrid cycle the complex boundaries have to be prescribed accurate enough. All these properties of the equations to be solved will influence the performance of the used multigrid-method.

MATHEMATICAL MODEL

The mathematical model as well as the numerical calculation procedure are based on the idea of Peric [5] and are described in detail in Majumdar et al. [6], [7]. The equations for steady, three-dimensional flows in non-orthogonal coordinates are solved, using cartesian velocity components. They may be written in a generalised form as follows:

$$\frac{\partial}{\partial x_i}(C_i\Phi + D_i^{\Phi}) = JS_{\Phi} \ . \tag{1}$$

The convective and diffusive coefficients C_i and D_i^{Φ} as well as the source term S_{Φ} are given in table 1 for the continuity and momentum equations. x_i are the curvilinear and y_i the cartesian coordinates and J is the Jacobian.

NUMERICAL CALCULATION PROCEDURE

Grid Generation. In connection with the finite-volume method body-fitted numerical grids are used, which are generated by algebraic (Zhu et al. [8]) as well as differential [Naar and Schönung [9]] grid generation methods. The latter allows the distribution of grid points along boundaries as well as the specification of grid intersection angles at the boundary and the mesh size normal to the boundary.

Finite-Volume Method. The kind of control volume and the variable arrangement used in the present method are shown in fig.1. All the variables are stored at the geometric center (P) of the control volume (collocated variables).

The difference equations representing the flux balance are derived by integrating the governing differential equations over the control volumes with the aid of the Gauß-theorem. The resulting balance equation for each control volume may be expressed as follows

$$I_e - I_w + I_n - I_s + I_t - I_b = \int\limits_{\Delta V} S_{\Phi} dV \ . \tag{2}$$

Each of the surface fluxes I_e, I_w at the east, west, north, south, top and bottom controlvolume faces consist of three distinct parts, namely a convective distribution I^C, a normal diffusive contribution I^{DN} and a cross-derivative diffusive part I^{DC}.

For the approximation of the normal diffusive fluxes I^{DN} central differences and for the convective fluxes I^{C} hybrid upwind/central differences are used (see Patankar [10]). The cross-derivative diffusive fluxes I^{DC} are treated explicitly and linear interpolations are used for their approximations.

To satisfy the continuity equation the SIMPLE-algorithm of Patankar and Spalding [11] has been adapted for the finite-volume method with collocated variables. In this method the discretized forms of the differential equations are solved sequentially. The extension for the collocated variable-arrangement uses special interpolations to avoid checkerboard splitting.

For solving the system of linear difference equations, which is obtained after the discretization of the flux-balance equation (2) the nine diagonal-solver of Peric [12], which is based on the "Strongly-Implicit-Procedure" of Stone [13], is used.

Multigrid Method. A neccessary condition for using a multigrid method is a good smoothing algorithm. As shown by Shaw and Sivaloganathan [14] the SIMPLE algorithm of Patankar and Spalding [11] is a possible choice.
The basic discretized equations on a given grid with meshsize h for the three cartesian velocity components read:

$$-a_p^h U_i^h + \sum_{nb} a_{nb}^h U_{nb}^h - S^h(U_i^h) = 0 \qquad (3)$$

nb runs through the directions east, west, north, south, top and bottom of the controlvolume. The source term $S^h(U_i^h)$ contains the pressure difference terms in each cartesian direction i and the cross derivative fluxes I^{DC}. In general S^h depends on all velocity components U_i.
An approximation $\tilde{U}_i^h$ for the exact solution U_i may be written as

$$\tilde{U}_i^h = U_i^h - e_i^h \qquad (4)$$

where e_i^h is a *correction*. Writing equation (3) for the approximation $\tilde{U}_i^h$ and introducing equation (4) leads to

$$-a_p^h(U_i^h - e_i^h) + \sum_{nb} a_{nb}^h(U_{nb}^h - e_{nb}^h) - S^h(U_i^h - e_i^h) = r_i^h \qquad (5)$$

with r_i^h as *residuum*. By using the Full-Approximation Scheme equation (5) is approximated on a coarser grid with meshsize H as follows:

$$-a_p^H(U_i^H - e_i^H) + \sum_{nb} a_{nb}^H(U_{nb}^H - e_{nb}^H) - S^H(U_i^H - e_i^H) = \qquad (6)$$

$$I_h^H(r_i^h) - a_p^H(I_h^H(\tilde{U}_i^h)) + \sum_{nb} a_{nb}^H(I_h^H(\tilde{U}_{nb}^h)) - S^H(I_h^H(\tilde{U}_i^h)) .$$

The RHS of (6) is constant on the coarser grid; starting values for the terms $(U_i^H - e_i^H)$ are choosen as $I_h^H(\tilde{U}_i^h)$. The operator of *restriction* I_h^H is based on a full-weighting interpolation for the variables U_i and a simple addition of the fine grid residuals. The coefficient a_{nb}^H on the coarser grid depend on the fluxes in and out the controlvolume. These fluxes are restricted from the finer grids by summation over corresponding sides. On the coarser grid a seperate pressure-correction equation based on the velocity corrections is solved to fullfill the continuity equation. The resulting pressure-corrections are used to update the pressure field on the fine grid level. The new fine grid velocity values $\tilde{U}_i^{h\ new}$ are obtained by

$$\tilde{U}_i^{h\,new} = \tilde{U}_i^{h\,old} + I_H^h\left((U_i^H - e_i^H) - I_h^H(\tilde{U}_i^h)\right). \qquad (7)$$

For the *prolongation* I_h^H a bilinear interpolation is used. To get a better convergence rate for cases, where the cross-derivative fluxes in the source term are of importance, $S^H(U_i^h - e_i^h)$ is taken into account on coarser grids too, as shown in equation (6). A **F**ull**M**ulti**G**rid method with fixed V-cycles is used. Five iterations for pre-smoothing have been found enough for most cases. At the coarsest grid the residuals are diminished to 20%. Different tests have shown that this is enough to get a good convergence rate without wasting to much time for an exact solution. For post-smoothing one iteration is performed to rub out high error components rising from the interpolation procedure.

RESULTS

All calculations have been performed on a Siemens/Fujitsu VP400-EX computer with a peak performance rate of 1.7 GFLOPS, and a maximum main storage of 512 MByte. The used computer code is vectorized up to 90%.

3-D Cavity. As a test case for a 3D flow with simple boundaries the flow in a lid-driven cavity has been investigated. Figure 2 shows the flow configuration and figure 4a-b the velocity vectors in two planes, calculated with a 42*42*42 grid.
In figure 4a the position of the main vortex near the symmetry-plane (z=0.5) are displayed. Small vortices can be seen at the lower corners. In figure 4b, which show a y-z plane, a complex system of vortices can be regarded. The flow structure is in good agreement with the results of Vanka [15]. Table 2 gives the number of iterations at the finest grid level and the computing times on different grids, comparing single grid and multigrid algorithm. For these calculations the residuals based on the euclidian norm have been reduced to 0.005%. The residua history is given in fig.3. The computing time by using the multigrid method increases linearily with the number of gridpoints, whereas the singlegrid algorithm does to the power of 1.76.

S-shaped Diffusor. In practice S-shaped diffusors are used to connect different components of fluid mechanics facilities. Figure 5 shows the grid used in the

calculations. Figure 6a displays velocity vectors at the symmetry plane (z/H=0.). 6b give the velocity vectors in the y-z plane at B-B. Two vortices of secundary motion indicate the complex flow behaviour. Results have been compared with the calculations of Majumdar [6] and the experiments of Rojas et al. [16] and show good agreement.
In table 3 iterations and computing time for reducing residuals to 0.6% are compared for two grids at two Reynoldsnumbers. Single grid calculations at low Re-number show a very bad convergence. The convective parts in the equations are very small and the matrix of the pressure-correction equation is very bad conditioned. Because multigrid methods are better for elliptic equations than for parabolic ones, this problem doesn´t occur there. The computing time increases linearily with the gridpoints used, but the 32*30*14 grid is to coarse to get a multigrid speedup.

3-D Cylinder Flow. The flow over a cylinder between two plates is still steady at Re-number 67. The problem is of practical interest for heat-exchanges. Fig.7 shows the numerical grid. In fig.8a and 8b x-y planes, one at the symmetry plane (z/H=0.5), the other at z/H=0.12, are displayed. Behind the cylinder a recirculating zone occurs. In table 4 iterations and computing time for single grid and multigrid calculations are given. Although only two grid levels could be used a speed-up factor over six has been reached for reducing residuals to 0.03%.

CONCLUSIONS

The multigrid calculations of laminar flow in a 3-D Cavity show the same speed-up factors for computing time as previous 2-D ones, when the same number of grid-points per direction is used. For practical 3-D problems these number is often limited by available storage capacity, so that very large factors cannot be expected. At higher Reynoldsnumbers it is observed that the multigrid method becomes less efficient, but it is still working. It is therefor only a question of grid refinement to get larger speed-up factors. By using carefully generated, smooth grids, multigrid methods can be used even in complex flow situations and geometries.

REFERENCES

[1] Hackbusch, W., Trottenberg, U.: Multigrid Methods, Lecture Notes in Mathematics, Vol.960, Springer-Verlag, Berlin, 1982

[2] Vanka, S.P.: Block-Implicit Multigrid Calculation of 2-Dim. Recirculating Flows, Comp. Meth. Appl. Mech. Eng., Vol.59, 1986, pp.29-48

[3] Barcus, M., Peric, M., Scheuerer, G.: A Control Volume Based Full Multigrid Procedure for the Prediction of Two-Dimensional, Laminar, Incompressible Flows,7th GAMM Conf. Num. Meth. Fluid Mech., Louvain-la-Neuve, Belgium, 1987

[4] Arakawa, C., Demuren, A.O., Rodi,W., Schönung, B.: Application of multigrid Methods for the Coupled and Decoupled Solution of the Incompressible Navier-

Stokes Equations, 7th GAMM Conf. Num. Meth. Fluid Mech., Louvain-la-Neuve, Belgium, 1987

[5] Peric, M.: A Finite-Volume Method for the Prediction of Three-Dim. Fluid Flow in Complex Ducts, PHD-Thesis, Imp. Coll. Sci. and Techn., Univ. London, 1985

[6] Majumdar, S., Schönung, B., Rodi, W.: A Finite Volume Method for Steady Two-Dimensional Incompressible Flows Using Non-staggered Non-Orthogonal Grids, Proceedings of the Seventh GAMM Conference on Num. Meth. in Fluid Mech., Notes on Num. Fluid Mech., Volume 20, Vieweg-Verlag, Braunschweig, 1988, pp.191-198

[7] Majumdar, S., Rodi, W., Schönung, B.: Calculation Procedure for Incompressible Three-Dimensional Flows With Complex Boundaries, to be published by Vieweg-Verlag

[8] Zhu, J., Rodi, W., Schönung, B.: Algebraic Generation of Smooth Grids, Num. Grid Gen. in Comp. Fluid Mech., Ed. Sengupta et al., Pineridge Press, Swansea, UK, 1988, pp.217-226

[9] Naar, M., Schönung, B.: Numerische Gittererzeugung bei Vorgabe von Randwinkeln und Randmaschenweiten, Inst. f. Hydromechanik, Univ. Karlsruhe, Bericht Nr.644, 1986

[10] Patankar, S.V.: Numerical Heat Transfer and Fluid Flow, Hemisphere Publishing Corporation, McGraw-Hill, New York, 1980

[11] Patankar, S.V., Spalding, D.B.: A Calculation Procedure for Heat, Mass and Momentum Transfer in Three-Dimensional Parabolic Flow, Int. J. Heat Mass Transfer, Vol.15, 1972, pp.1787-1806

[12] Peric, M.: Efficient Semi-Implicit Solving Algorithm for Nine-Diagonal Coefficient Matrix, Numerical Heat Transfer,Vol.11, 1987, pp.251-279

[13] Stone, H.L.: Iterative Solution of Implicit Approximations of Multidimensional, SIAM J. Numer. Anal., Vol.5, No.3, 1968, pp.530-558

[14] Shaw, G.J., Sivaloganathan, S.: On The Smoothing Properties Of The SIMPLE Pressure Correction Algorithm, Int. J. Num. Meth. Fluids, 8, 1988, pp.441-461

[15] Vanka, S.P.: A Calculation Procedure for Three-Dimensional Steady Recirculating Flows using Multigrid Methods, Comp. Meth. Appl. Mech. Eng., 55, 1986, pp.321-338

[16] Rojas, J., Whitelaw, J.H., Yianneskis, M.: Developing Flow in S-shaped Diffusors, Part 1, Fluids Sect. Rep. FS/83/12, 1983, Mech. Eng. Dep., Imp. Coll. London

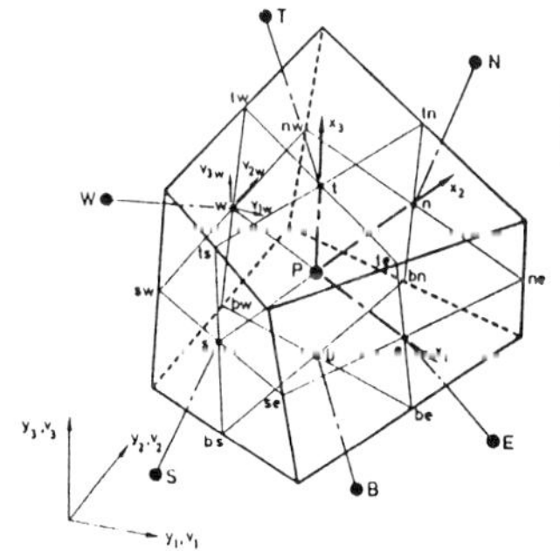

Fig.1: 3-D Controlvolume

Table 1: Coefficients and sourceterms for equation (1).

	Φ_i	C_i	D_i^Φ	S_Φ
a)	1	U_i	0	0
b)	v_k	U_i	$-\frac{\mu}{J}\left(B_j^i \frac{\partial v_k}{\partial x_j} + \beta_j^i \omega_k^j\right)$	$-\frac{1}{J}\frac{\partial}{\partial x_j}(p\beta_k^j)$

$J = \left|\frac{\partial y_i}{\partial x_i}\right|$ $\quad \beta_j^i =$ cofactors of $\frac{\partial y_i}{\partial x_j}$ in J $\quad B_j^i = \beta_l^i \beta_l^j$

$U_i = \rho v_j \beta_j^i$ $\quad \omega_k^j = \frac{\partial v_j}{\partial x_l}\beta_k^l$

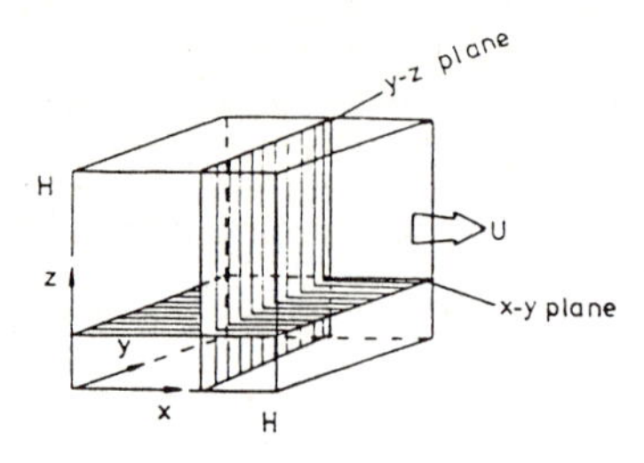

Fig.2: Geometry of 3-D Cavity

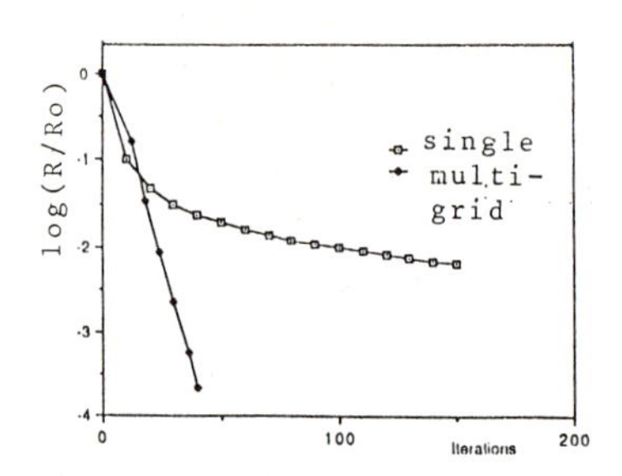

Fig.3: Residua history for 3-D Cavity flow calculation

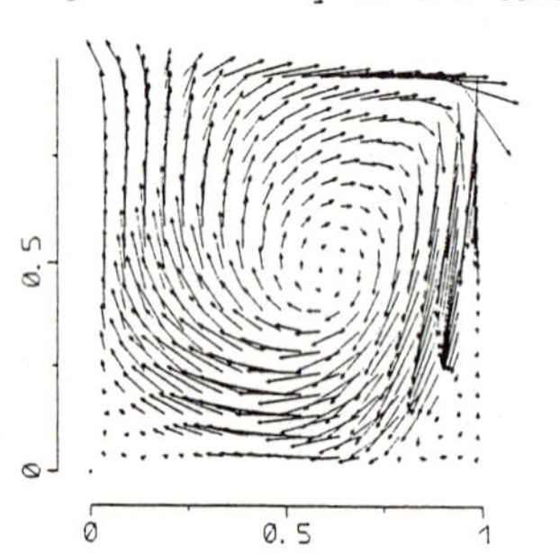

Fig.4a:x-y plane z/H = 0.4875

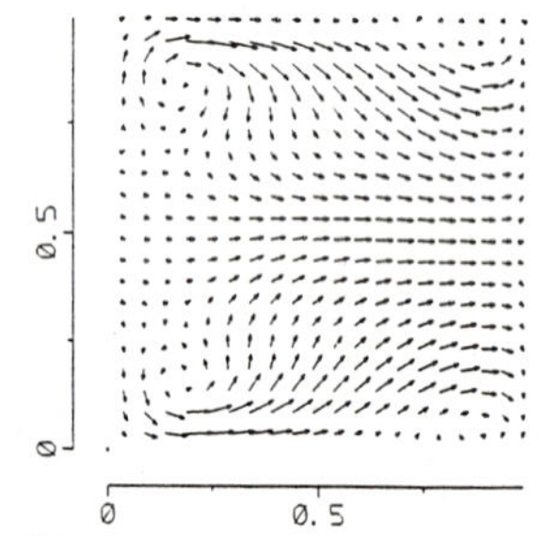

Fig.4b:y-z plane x/H = 0.4875

Fig.4 a-b: Flow in a 3-D Cavity at Re=1000

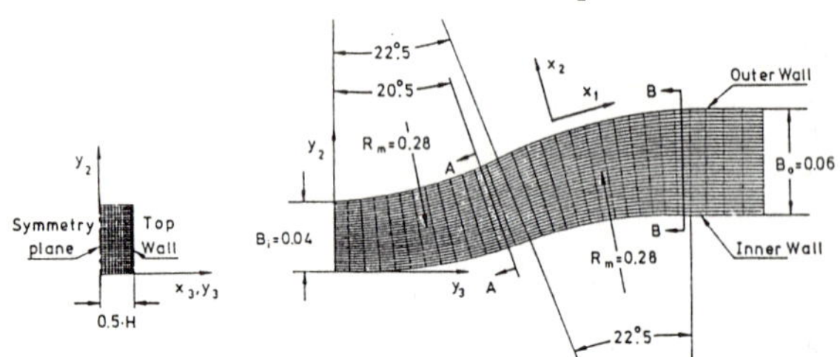

Fig.5: Grid and Geometry of S-shaped diffusor

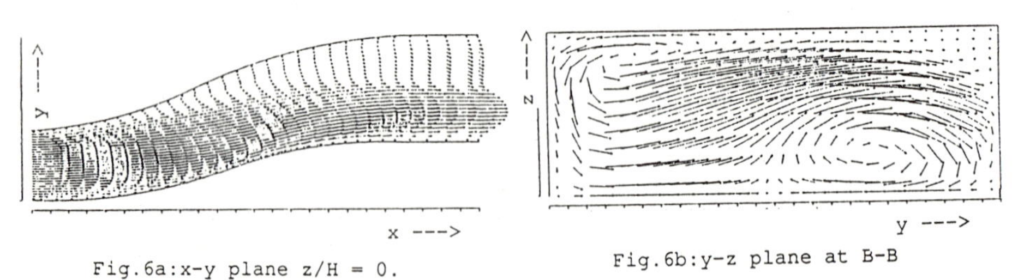

Fig.6a:x-y plane z/H = 0.

Fig.6b:y-z plane at B-B

Fig.6a-b: Flow in a S-shaped diffusor at Re=790

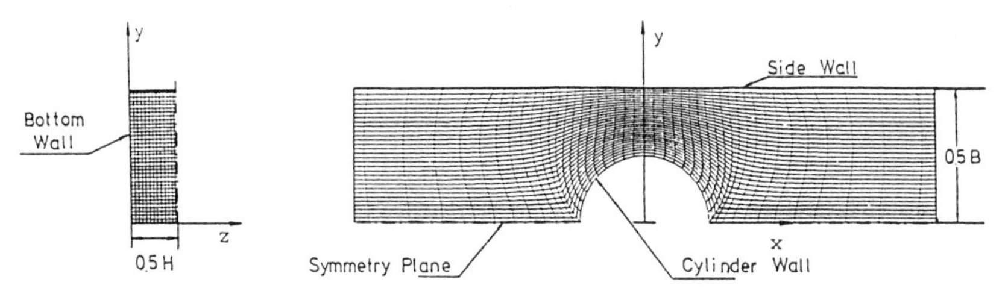

Fig. 7: Grid and geometry of cylinder between two walls

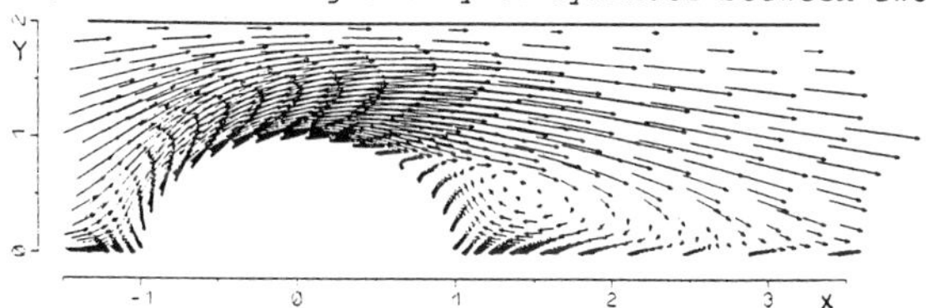

Fig.8a: x-y plane z/H=0.5

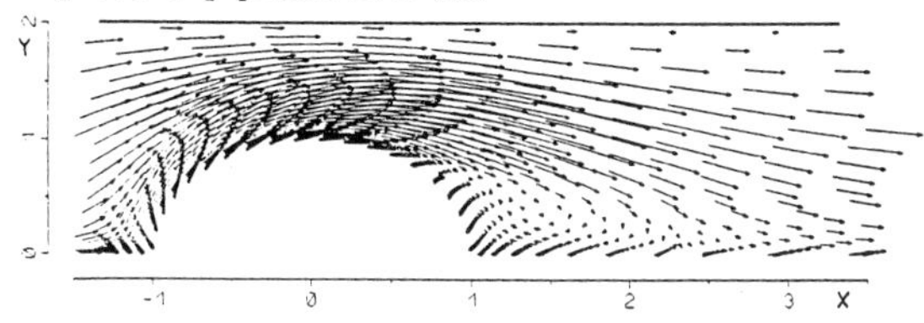

Fig.8b: x-y plane z/H=0.12

Fig.8 a-b: Flow over cylinder between two walls at Re=67

Table 2: Iterations and computing time for flow in a 3D cavity

Gridpoints		22*22*22	42*42*42	66*66*66
SG	Iterations	171	560	1300*
	CPU(sec.)	66.2	2037.6	22500.*
MG	Iterations	46	42	40
	CPU(sec.)	26.3	171.4	655.5
* estimated				

Table 3: Iterations and computing time for flow in a S-shaped diffusor

Gridpoints		32*30*14		62*58*26	
		Re=79	Re=790	Re=79	Re=790
SG	Iterations	146	100	335*	191
	CPU(sec.)	140.8	70.7	4000*	1775.4
MG	Iterations	44	113	47	130
	CPU(sec.)	35.0	77.8	258.1	646.0
* estimated					

Table 4: Iterations and computing time for flow over a cylinder

Gridpoints		62*32*14
SG	Iterations	1099
	CPU(sec.)	2845.9
MG	Iterations	346
	CPU(sec.)	442.3

ON THE RELATION BETWEEN TVD AND MESH ADAPTION AND APPLICATION TO NAVIER-STOKES CALCULATIONS

B. PALMERIO (*), C. OLIVIER (**), A. DERVIEUX (**)
(*) University of Nice and INRIA-Sophia-Antipolis
(**) INRIA-Sophia-Antipolis, B.P.109, 06560 VALBONNE, FRANCE

ABSTRACT

Considering the TVD methods as accuracy-adaption methods, we propose a strategy for deciding mesh refinements for TVD-approximated compressible flows. Both mesh enrichment and deformation are applied to two typical test cases involving boundary layers: a flow past a flat plate and a flow around an airfoil.

INTRODUCTION

From many years, the occuring of oscillations in numerical solutions of viscous layers have been an important problem in CFD. The CFD community where advised by Ph. Gresho and R. Lee [1] not to "suppress the wiggles" because they were "telling us something". In this paper, we suggest to listen what they are telling and then to anyway suppress them.

They are telling "the mesh is not fine enough here". Then we can:

- either adapt the approximation by degrading it to a first-order accurate diffusive one (this point of view is exactly the opposite of the p-method!),
- or adapt the mesh (applying a h-method).

We propose to apply both strategies: indeed, if the mesh is not adapted, the accuracy is not good; if the first-order diffusive approximation is not applied, the preliminary solution oscillates, and this may result in divergence of the solution algorithm, particularly in hyperbolic compressible models.

In the TVD-mesh-adaption method that we propose, when the possible arising of oscillation is detected by applying the TVD principle, an ad hoc diffusion is added and mesh refinement is decided. An ideal situation would be obtained when all TVD-limited regions of the computational domain are refined until the approximation is second-order accurate everywhere. This is the ultimate goal of the work presented.

1. THE TVD SCHEME

We first shortly define the scheme that is used for the calculation of the flow; this kind of scheme has been introduced in [5]; a more complete description can be found in [4].

The 2-D Navier-Stokes equations is written in short as follows:

$$\frac{\partial W}{\partial t}+\frac{\partial F(W)}{\partial x}+\frac{\partial G(W)}{\partial y}=\frac{1}{Re}\left(\frac{\partial R(W)}{\partial x}+\frac{\partial S(W)}{\partial y}\right), \tag{1}$$

in which $W(x,y,t)$ is a vector function in R^4 :

$$W=\begin{pmatrix}\rho\\ \rho u\\ \rho v\\ E\end{pmatrix},$$

where ρ is the density, u and v are the components of the velocity and E the total energy per volume unit. $F(W)$ and $G(W)$ are given by:

$$F(W)=\begin{pmatrix}\rho u\\ \rho u^2+p\\ \rho uv\\ (E+p)u\end{pmatrix}, \quad G(W)=\begin{pmatrix}\rho v\\ \rho uv\\ \rho v^2+p\\ (E+p)v\end{pmatrix},$$

where p is the pressure defined by the perfect gas law ($\gamma=1.4$):

$$p=(\gamma-1)\left(E-\frac{1}{2}\rho(u^2+v^2)\right).$$

$R(W)$ and $S(W)$ are the diffusive flux functions, defined as follows:

$$R(W)=\begin{pmatrix}0\\ \tau_{xx}\\ \tau_{xy}\\ u\tau_{xx}+v\tau_{xy}+\frac{\gamma k}{Pr}\frac{\partial\varepsilon}{\partial x}\end{pmatrix}, \quad S(W)=\begin{pmatrix}0\\ \tau_{xy}\\ \tau_{yy}\\ u\tau_{xy}+v\tau_{yy}+\frac{\gamma k}{Pr}\frac{\partial\varepsilon}{\partial y}\end{pmatrix}.$$

ε is the specific internal energy; T is the temperature, k is a normalised thermal conductivity, τ_{xx}, τ_{xy} and τ_{yy} are the components of the stress tensor.

Numerical integration

The integration domain is assumed to be a polygon, divided in a classical finite-element triangulation. The degrees of freedom are the values of W on each vertex. The dual finite-volume mesh is defined from cells C_i around each vertex i drawn with sections of the surrounding medians (Fig.1). These volumes will be used for the integration of the time derivative and of the advection terms.

The viscous terms are discretised by applying the usual Galerkin variational method. Then we can write the approximation as follows:

$$
\begin{aligned}
&aire\,(C_i)\,\frac{W_i^{n+1}-W_i^n}{\Delta t^n}+\int_{\partial C_i}\mathcal{F}(W,\vec{\eta})\,d\sigma = \\
&-\frac{1}{Re}\sum_{T,i\epsilon T}\iint_T\left(R\frac{\partial\varphi_i}{\partial x}+S\frac{\partial\varphi_i}{\partial y}\right)dxdy + boundary\ integral\ ,
\end{aligned}
\tag{2}
$$

in which φ_i hold for the P_1 Galerkin basis functions, the sum is taken over the triangles having i as a vertex, and

$$
\mathcal{F}(W,\vec{\eta}) \;=\; F\,\eta_x \;+\; G\,\eta_y\ . \tag{3}
$$

The advection terms are discretised by a finite-volume scheme that is compatible with the Galerkin finite-element for advection [9]. The flux are computed through the boundary of cells defined by centroids and mid-sides (Figure 1)

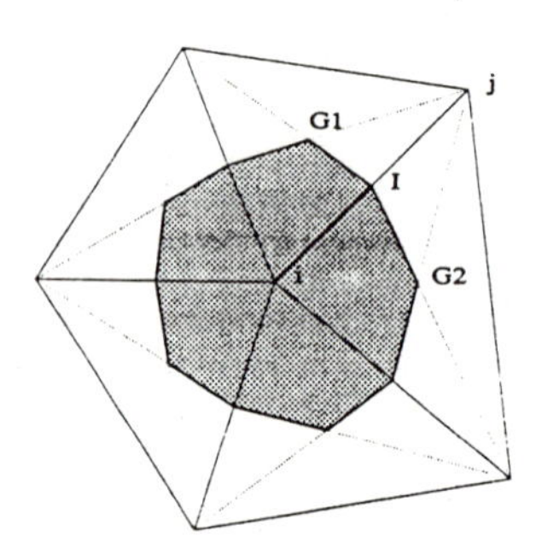

Figure 1 : barycentric cell around a vertex

$$
\int_{\partial C_i}\mathcal{F}(W,\vec{\eta})\,d\sigma=\sum_{j\epsilon K(i)}\int_{\partial C_{ij}}\mathcal{F}(W,\vec{\nu})\,d\sigma+\int_{\partial C_i\cap\Gamma_h}\mathcal{F}(W,\vec{\eta})\,d\sigma\ ,
$$

where $K(i)$ holds for the set of vertices that are neighbors of vertex i, $\Gamma_h=\partial\Omega_h$, et $\partial C_{ij}=\partial C_i\cap\partial C_j=[G_{1,ij},I_{ij}]\cup[I_{ij},G_{2,ij}]$ (see Figure 2).

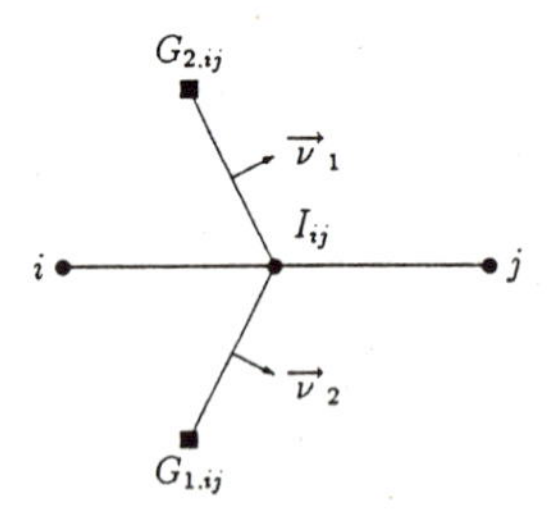

Figure 2 : Flux integration between two cells

where the sum is taken over the neighbouring vertices j around vertex i. Then each flux between cell i and cell j is computed through an approximate Riemann solver which define a flux function:

$$\int_{\partial C_{ij}} \mathcal{F}(W,\vec{\nu})\,d\sigma = \Phi\left(W_{ij},W_{ji},\vec{\nu}_{ij}\right) \ , \tag{4}$$

with:

$$\vec{\nu}_{ij} = \int_{\partial C_{ij}} \vec{\nu} d\sigma \ .$$

in which the metrics is taken into account in the normal mean vector; the upwinding is obtained by using Ph. Roe 's flux splitting [8], written in short:

$$\Phi_{ROE}\left(U,V,\vec{\nu}\right) = \frac{\mathcal{F}(U,\vec{\nu}) + \mathcal{F}(V,\vec{\nu})}{2} - \mid \mathcal{R}\left(U,V,\vec{\nu}\right) \mid \frac{(V-U)}{2} \ , \tag{5}$$

to complete the description of the scheme, W_{ij} and W_{ji} will be defined through interpolation involving "limited slopes" [3] which provide the basic first-order accurate upwind scheme ($W_{ij} = W_i$ and $W_{ji} = W_j$) where extrema are detected.

Solution algorithm

In this presentation the choice of the resolution algorithm will not be discussed since we focus on the accuracy point of view.

In short, the algorithm [4] is an pseudo-unsteady one with local time stepping. The time-stepping is a linearised unfactored implicit one with Jacobi linear iteration. Time steps are increased in function of the inverse of the residual for better convergence, as in the SER method of Mulder and van Leer [7].

2. MESH ADAPTION

Mesh Criterion

The above scheme is considered as a perturbation of a central differenced one obtained by using, instead of ROE's flux splitting, the following function:

$$\Phi_C\left(U,V,\vec{\nu}\right) = \mathcal{F}\left(\frac{U+V}{2},\vec{\nu}\right) \tag{6}$$

then the following numerical viscosity estimate is considered:

$$E_{ij} = ||\Phi_{ROE}(W_{ij},W_{ji},\vec{\nu}_{ij}) - \Phi_C(W_i,W_j,\vec{\nu}_{ij})|| \tag{7}$$

where the norm is the Euclidean one.

Mesh Enrichment

One first way to locally adapt a mesh is to divide the elements of a well chosen region of the computational domain.

We can choose here to divide any segment ij for which E_{ij} is larger than a given parameter. In a smoother option, every triangle in which one at least side ij satisfies the above inequality is divided in four subtriangles.

Mesh Deformation System

In order to refine the mesh more efficiently in one direction (layers), we consider a mesh deformation method, relying on a spring analogy. The mesh iteration is applied in alternation with the flow solution. See [6] for details.

The TVD criterion E_{ij} is introduced in the strength of each attraction spring between vertices i and j.

3. NUMERICAL EXPERIMENTS

Capture of shocks: A numerical shock profiles generally consists in one region B with high gradient but more or less linear behaviour between two regions A and C with less high gradients, higher further derivatives and likely the arising of extremas. In most TVD schemes, the behaviour of the TVD sensor will in short produce first-order accuracy in region A and C and second-order accuracy in region B. With a TVD scheme that is central central-differenced for the second-accurate component, this may result in a small numerical viscosity in region B, and therefore unfortunately, in no refinement of this central zone of the shock.

Conversely, when using a MUSCL-TVD formulation, the second-order accurate component carries a numerical viscosity (fourth-order derivatives that bring a third order error) that can be still large in region B, mainly because fourth-order derivatives are computed on larger molecules while region B has only a one to three elements width; we shall illustrate this point in the experiments below.

Flow past a flat plate

For a demonstration of the basic qualities of the method we consider the flow past a plate at Reynolds number 1000 and Mach number at farfield 3.

A uniform initial mesh (1138 vertices) is enriched twice using the TVD criterion, resulting in a 6480-vertices mesh (Fig.1), and then deformed using the same criterion (Fig.2).

While an intuitive idea of the overall improvement is given by several contours (Figs. 3 and 4), the more striking comparison is related to the friction coefficient (Figs. 5 and 6).

Flow around a NACA0012 airfoil

We consider a more complex flow around a NACA0012 airfoil, with farfield Mach number of 2 an angle of attack of 10 degrees and a Reynolds number of 1000. This case was proposed in a workshop [10] in 1985. A very thin detached shock is present. The same strategy (enrichment and then deformation) is applied. The final mesh contains 7987 nodes (Fig. 7). We present the corresponding Mach contours (Fig. 8): some

extra refinement seems needed for a thin capture of the shock; the boundary layer is rather well computed.

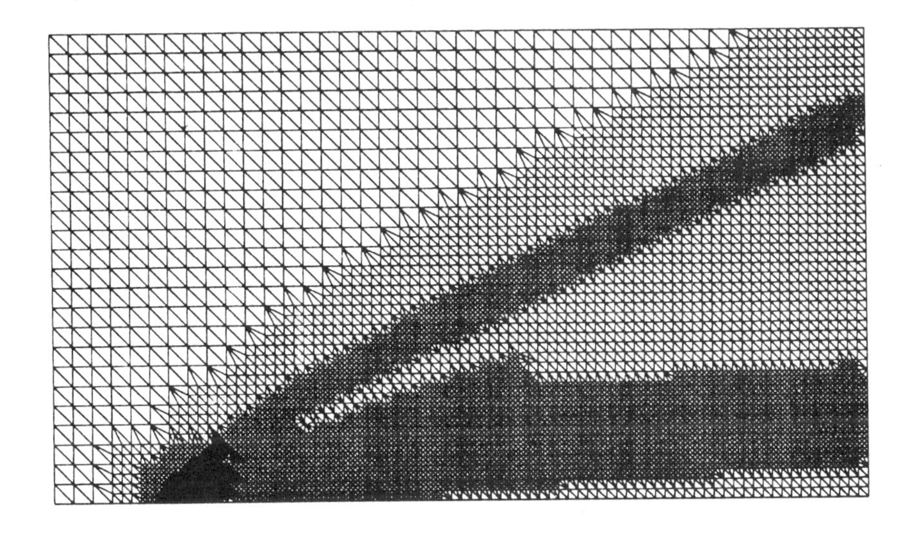

Figure 1: Enriched mesh for the plate problem

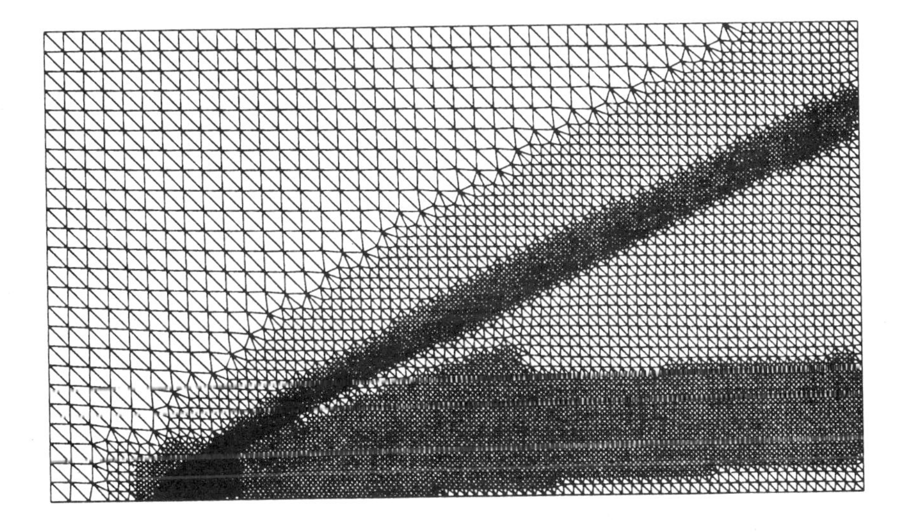

Figure 2: Enriched and deformed mesh for the plate problem

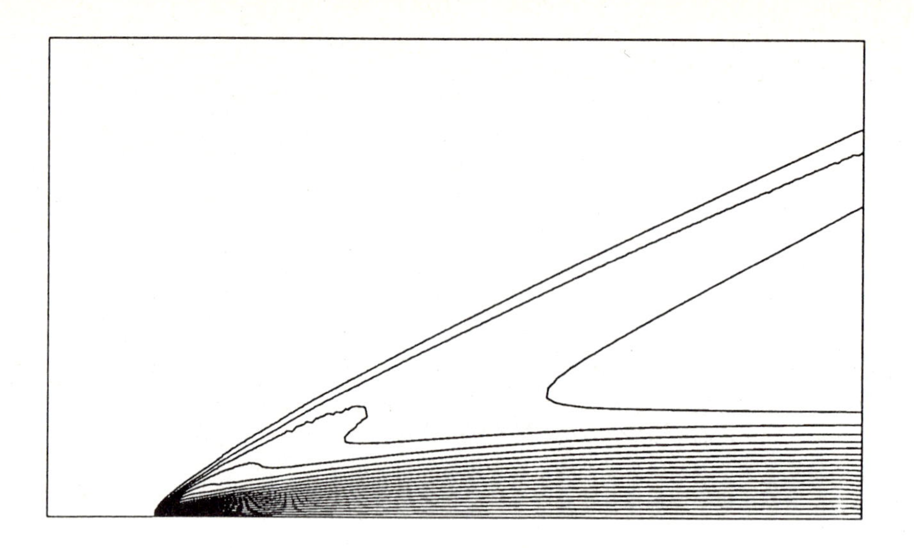

Figure 3: Mach contours with the enriched mesh

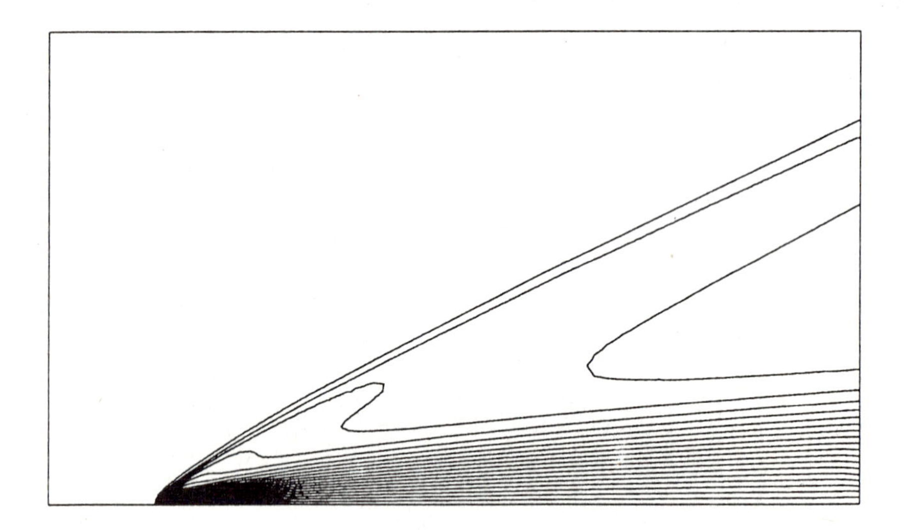

Figure 4: Mach contours for the enriched and deformed mesh

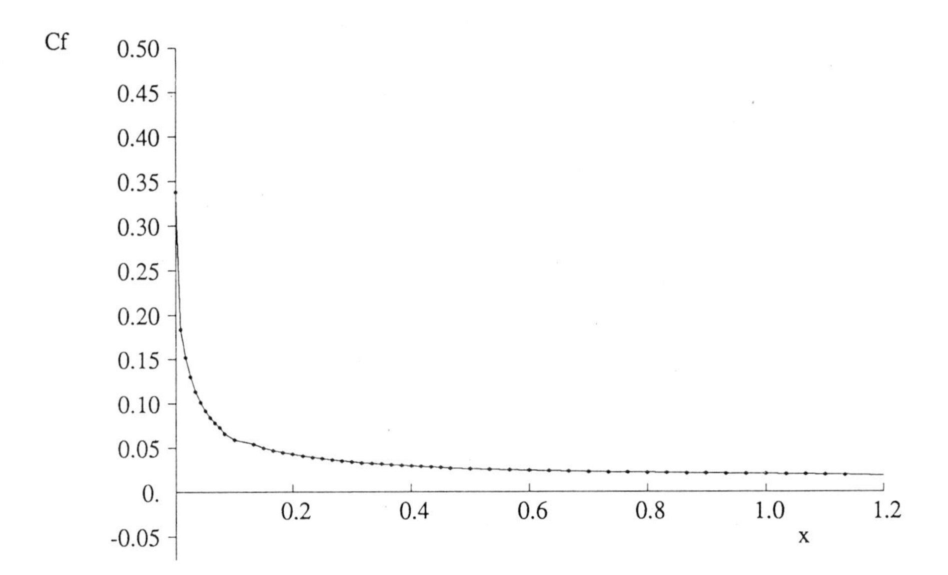

Figure 5: Friction coefficient with the enriched mesh

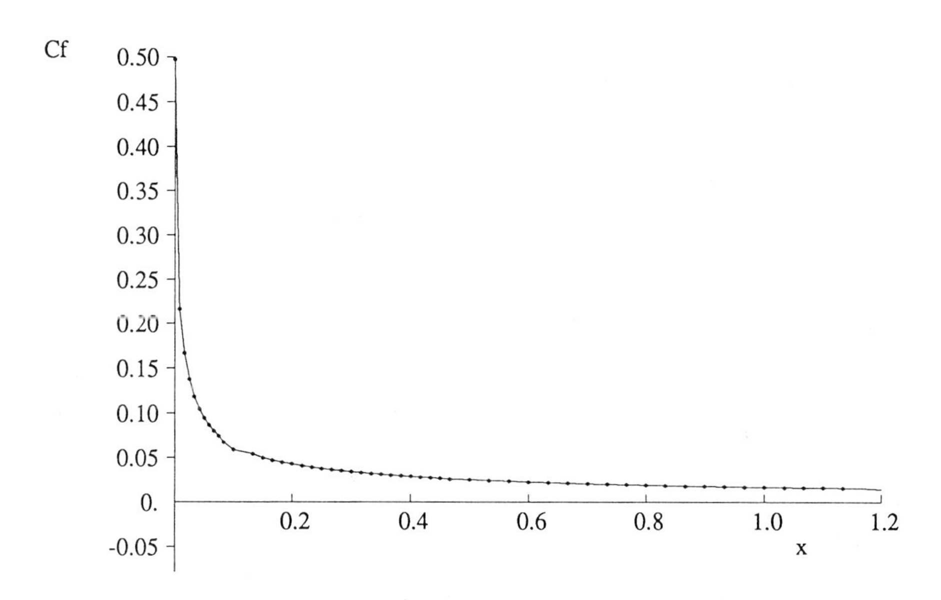

Figure 6: Friction Coefficient for the enriched and deformed mesh

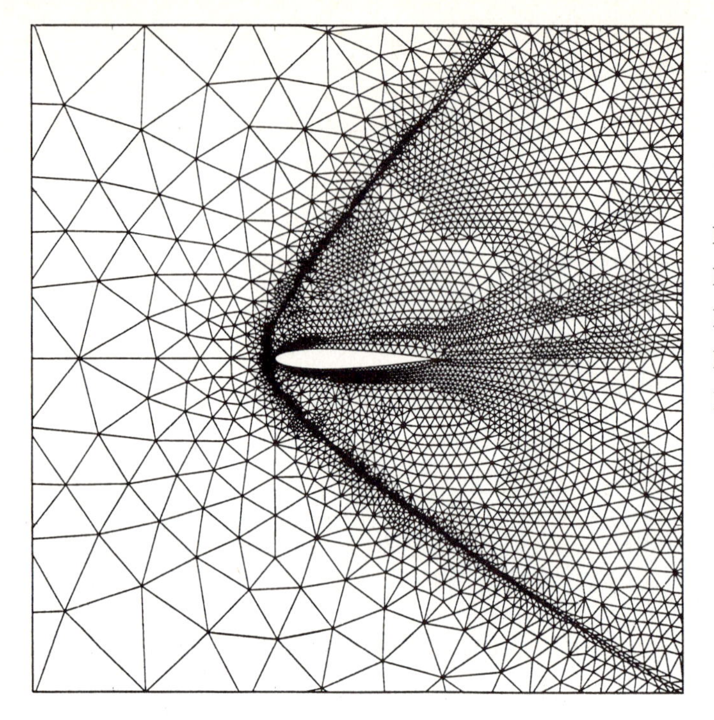

Figure 7:
NACA0012 airfoil
Mach=2
Reynolds=1000
Final mesh

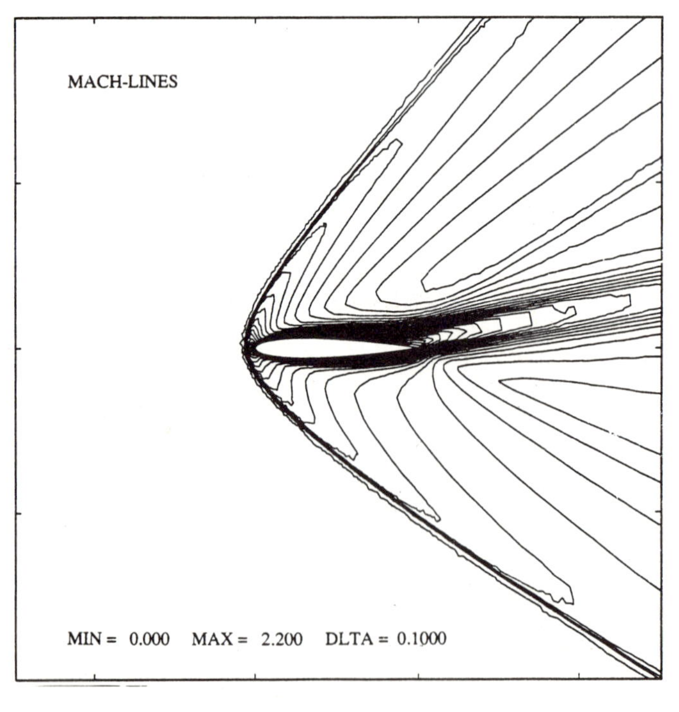

Figure 8:
NACA0012 airfoil
Mach=2
Reynolds=1000
Mach contours

ACKNOWLEDGEMENTS

We thank Loula Fezoui for her contribution in the conception of the global computer code used in this study.

REFERENCES

[1] Ph.M. GRESHO, R.L. LEE, Dont' suppress the wiggles. They are telling you something !, in Finite Element methods for Convection Dominated flows, New York, Dec. 2-7, 1979.

[2] L. FEZOUI, Résolution des équations d'Euler par un schéma de van Leer en éléments finis, INRIA Report 358, 1985.

[3] B. STOUFFLET, J. PERIAUX, L. FEZOUI, A. DERVIEUX, Numérical simulation of 3-D hypersonic Euler flows around space vehicles, 25th AIAA Conf., Reno, 1987, AIAA paper 87-0560.

[4] L. FEZOUI, S. LANTERI, B. LARROUTUROU,C. OLIVIER, Résolution numérique des équations de Navier-Stokes pour un fluide compressible en maillage triangulaire, INRIA Report, 1989.

[5] Ph. ROSTAND, B. STOUFFLET, Finite volume Galerkin methods for viscous gas dynamics, INRIA Report 863, 1988.

[6] B. PALMERIO, A. DERVIEUX, A 3-D unstructured–mesh adaption relying on physical analogy, Comm. to Numerical Grid Generation in Computational Fluid Mechanics, Miami, Dec. 1988.

[7] B. van LEER, W.A. MULDER, Relaxation methods for hyperbolic equations, in Numerical Methods for the Euler Equations of Fluid Dynamics, Angrand et al. Eds., SIAM (1985)

[8] Ph. L. ROE, Approximate Riemann solvers, parameters vectors and difference schemes, J. of Comp. Phys., 43, 357-371 (1981)

[9] V. BILLEY, A. DERVIEUX, L. FEZOUI, J. PERIAUX, V. SELMIN, B. STOUFFLET, Recent improvement in Galerkin and upwind Euler solvers and application to 3-D transonic flow in aircraft design, in Eighth Int. Conf. on Computer Methods in Applied Science and Engineering, Versailles, dec. 1987, North Holland 1988

[10] M. O. BRISTEAU, R. GLOWINSKI, J. PERIAUX, H. VIVIAND, (Eds) Numerical simulation of compressible Navier-Stokes flows, Note on Numerical Fluid Mechanics, 18, Vieweg, Braunschweig (1987)

COMPUTATION OF THE VISCOUS FLOW PAST A PROLATE SPHEROID AT INCIDENCE .

Jean **PIQUET** & Patrick **QUEUTEY**

CFD Group , LHN , URA 1217 CNRS
ENSM , 1 Rue de la Noë , 44072 NANTES Cedex , FRANCE

SUMMARY

The computation of the threedimensional viscous flow past a prolate spheroid at incidence and its wake is investigated. An iterative technique resting on the fully elliptic mode is applied to the Reynolds-Averaged-Navier-Stokes-Equations (RANSE) written down in a non-orthogonal curvilinear body-fitted coordinate system. Results of the computation are compared to available experiments such as the DFVLR experiments and the ONERA experiments .

-I- INTRODUCTION

Among the most well documented data relative to vortical threedimensional flows which can be used as benchmark test cases for the validation of Navier-Stokes solvers, the prolate spheroid at incidence is interesting as archetypal of complex flows involving streamwise separation over the aft portion of the body and crossflow separation over the forward portions of the body [1],[2]. Moreover, the geometry is very simple and a potential flow solution, valid for low angles of attack, is analytically known. Endly, the boundary layer characteristics of the flow have been studied with a lot of details by numerous authors .

The evidence that emerged from these boundary layer studies is that viscous-inviscid interaction is already significant on the aft part of the body for angles of attack as low as ten degrees, although boundary layer computations have the ability to give acceptable predictions on the most important part of the body , both for the skin friction, the velocity profiles and the location of the crossflow separation. The most important prerequisite is to pre-determine, in some empirical way, the location of the transition zone .

Now, if one considers the numerical investigations of the problem which have been performed with Navier-Stokes type solvers, the situation is quite contrasted with thin boundary layer solvers as fewer solutions are available. They involve either parabolized or elliptic Navier-Stokes equations [3],[4], strong coupling techniques or thin boundary layer Navier-Stokes equations [4],[5],[6],[7]. The present paper brings some more information with an incompressible Navier-Stokes solver used in a fully elliptic mode of iteration, but without any other approximations in the momentum equations than those dictated by the turbulence model .

Previous computations suggest that, if the pressure field is rather insensitive to the location of the transition, velocity profiles and skin friction depend on it so that this problem must be either solved or bypassed if one wants to use the experimental data to validate the Navier-Stokes solver. Computation over the entire body is performed here only for laminar flows and is a first step towards fully elliptic and turbulent computations over more complex geometries at incidence.

-II- THE EQUATIONS

II.1 THE BASIC EQUATIONS

We consider the equations of motion in cartesian (x,y,z) coordinates for incompressible flows. The exact RANSE of continuity and momentum of the mean flow in dimensionless form are given by equations (1), (2) and (3):

$$\operatorname{div} \mathbf{U} = 0 \tag{1}$$

$$\frac{\partial \mathbf{U}}{\partial t} + \nabla \mathbf{U}.\mathbf{U} + \nabla p + \mathbf{T} = \frac{1}{Re} \nabla^2 \mathbf{U} \tag{2}$$

$$\mathbf{T} = \begin{pmatrix} \frac{\partial}{\partial x}(\overline{uu}) + \frac{\partial}{\partial y}(\overline{uv}) + \frac{\partial}{\partial z}(\overline{uw}) \\ \frac{\partial}{\partial x}(\overline{uv}) + \frac{\partial}{\partial y}(\overline{vv}) + \frac{\partial}{\partial z}(\overline{wv}) \\ \frac{\partial}{\partial x}(\overline{uw}) + \frac{\partial}{\partial y}(\overline{vw}) + \frac{\partial}{\partial z}(\overline{ww}) \end{pmatrix} . \tag{3}$$

U, **p** and **T** are respectively the Velocity vector, the Pressure field and the Turbulence vector. The barred quantities are the Reynolds stresses modelled by means of the classical k-ε turbulence model in which the Reynolds stress is linearly related to the mean rate of strain tensor by an isotropic eddy viscosity as follows :

$$\overline{uu} = -\nu_T\left(2\frac{\partial U}{\partial x}\right) + \frac{2}{3}k \qquad \overline{uv} = -\nu_T\left(\frac{\partial U}{\partial y} + \frac{\partial V}{\partial x}\right)$$
$$\overline{vv} = -\nu_T\left(2\frac{\partial V}{\partial y}\right) + \frac{2}{3}k \qquad \overline{vw} = -\nu_T\left(\frac{\partial V}{\partial z} + \frac{\partial W}{\partial y}\right)$$
$$\overline{ww} = -\nu_T\left(2\frac{\partial W}{\partial z}\right) + \frac{2}{3}k \qquad \overline{uw} = -\nu_T\left(\frac{\partial W}{\partial x} + \frac{\partial U}{\partial z}\right) . \tag{4}$$

The eddy viscosity ν_T is classically given by :

$$\nu_T = C_\mu . \frac{k^2}{\varepsilon} \tag{5}$$

and the adimensional turbulent kinetic energy k and its dimensionless rate of dissipation ε are governed by the following transport equations :

$$\frac{\partial k}{\partial t} + U\frac{\partial k}{\partial x} + V\frac{\partial k}{\partial y} + W\frac{\partial k}{\partial z} = \frac{\partial}{\partial x}\left(\frac{1}{\sigma_k Reff}\frac{\partial k}{\partial x}\right) + \frac{\partial}{\partial y}\left(\frac{1}{\sigma_k Reff}\frac{\partial k}{\partial y}\right) + \frac{\partial}{\partial z}\left(\frac{1}{\sigma_k Reff}\frac{\partial k}{\partial z}\right) + G - \varepsilon \tag{6}$$

$$\frac{\partial \varepsilon}{\partial t} + U\frac{\partial \varepsilon}{\partial x} + V\frac{\partial \varepsilon}{\partial y} + W\frac{\partial \varepsilon}{\partial z} = \frac{\partial}{\partial x}\left(\frac{1}{\sigma_\varepsilon Reff}\frac{\partial \varepsilon}{\partial x}\right) + \frac{\partial}{\partial y}\left(\frac{1}{\sigma_\varepsilon Reff}\frac{\partial \varepsilon}{\partial y}\right) + \frac{\partial}{\partial z}\left(\frac{1}{\sigma_\varepsilon Reff}\frac{\partial \varepsilon}{\partial z}\right) + C_{\varepsilon 1}\frac{\varepsilon}{k}G - C_{\varepsilon 2}\frac{\varepsilon^2}{k} \tag{7}$$

where G is the turbulence generation term :

$$G = \nu_T\left\{ 2\left[\left(\frac{\partial U}{\partial x}\right)^2 + \left(\frac{\partial V}{\partial y}\right)^2 + \left(\frac{\partial W}{\partial z}\right)^2\right] + \left(\frac{\partial W}{\partial x} + \frac{\partial U}{\partial z}\right)^2 + \left(\frac{\partial W}{\partial y} + \frac{\partial V}{\partial z}\right)^2 \right\} . \tag{8}$$

The effective Reynolds number Reff has been defined by :

$$\frac{1}{Reff} = \nu_T + \frac{1}{Re} \ . \tag{9}$$

Unless specified, the constants in the previous equations are taken to :

$$C_\mu = 0.09 \qquad C_{\varepsilon 1} = 1.44 \qquad C_{\varepsilon 2} = 1.92 \qquad \sigma_k = 1.00 \qquad \sigma_\varepsilon = 1.30 \ . \tag{10}$$

II.2 THE EQUATIONS IN THE TRANSFORMED COORDINATE SYSTEM

For most practical applications, the complexity of the geometry prevents the cartesian or cylindrical coordinates system to be used. Analytic or numerical coordinate transformations are highly desirable in that they greatly facilitate the application of the boundary conditions and transform the physical domain in which the flow is studied into a parallelepipedic computational domain. The basic idea of a body-fitted coordinate system is to find a transformation such that the curvilinear boundary surfaces of the physical domain are transformed into boundaries of a simple rectangular domain in the computational space $\{\xi^i\}=\{\xi,\eta,\zeta\}$. The most commonly used is that of Thompson et Al [8] which consists in solving a set of Poisson equations. The partially transformed RANSE are given by the following relations in which the contravariant components of the velocity are defined by $\{u^i\}=\{u,v,w\}$ and the physical components by $\{U_i\}=\{U,V,W\}$:

$$\frac{1}{J}\frac{\partial}{\partial \xi j}(J.u^j) = 0 \quad ; \quad Ju^i = b^i_j U_j \tag{11}$$

$$J.\frac{\partial \phi}{\partial t} + \frac{\partial}{\partial \xi}(Ju.\phi) + \frac{\partial}{\partial \eta}(Jv.\phi) + \frac{\partial}{\partial \zeta}(Jw.\phi)$$

$$= \qquad\qquad \phi = U \, , V \text{ or } W$$

$$\frac{\partial}{\partial \xi}\left[\frac{Jg^{11}}{Reff}\frac{\partial \phi}{\partial \xi}\right] + \frac{\partial}{\partial \eta}\left[\frac{Jg^{22}}{Reff}\frac{\partial \phi}{\partial \eta}\right] + \frac{\partial}{\partial \zeta}\left[\frac{Jg^{33}}{Reff}\frac{\partial \phi}{\partial \zeta}\right] + S_\phi \tag{12}$$

$$S_\phi = \frac{\partial}{\partial \xi}\left[\frac{Jg^{12}}{Reff}\frac{\partial \phi}{\partial \eta}\right] + \frac{\partial}{\partial \xi}\left[\frac{Jg^{13}}{Reff}\frac{\partial \phi}{\partial \zeta}\right]$$
$$+ \frac{\partial}{\partial \eta}\left[\frac{Jg^{21}}{Reff}\frac{\partial \phi}{\partial \xi}\right] + \frac{\partial}{\partial \eta}\left[\frac{Jg^{23}}{Reff}\frac{\partial \phi}{\partial \zeta}\right]$$
$$+ \frac{\partial}{\partial \zeta}\left[\frac{Jg^{31}}{Reff}\frac{\partial \phi}{\partial \xi}\right] + \frac{\partial}{\partial \zeta}\left[\frac{Jg^{32}}{Reff}\frac{\partial \phi}{\partial \eta}\right] - s_\phi \ . \tag{13}$$

It should be noticed that the generic source term $S_\phi = -s_\phi$ for an orthogonal mesh

$$s_U = \frac{\partial}{\partial \xi}\left[b^1_1(p+\frac{2}{3}k)\right] + \frac{\partial}{\partial \eta}\left[b^2_1(p+\frac{2}{3}k)\right] + \frac{\partial}{\partial \zeta}\left[b^3_1(p+\frac{2}{3}k)\right] - \left\{\frac{\partial}{\partial \xi^i}\left[\frac{\nu_T b^i_m b^j_1}{J}.\frac{\partial U_m}{\partial \xi^j}\right]\right\}$$

$$s_V = \frac{\partial}{\partial\xi}\left[b_2^1(p+\frac{2}{3}k)\right] + \frac{\partial}{\partial\eta}\left[b_2^2(p+\frac{2}{3}k)\right] + \frac{\partial}{\partial\zeta}\left[b_2^3(p+\frac{2}{3}k)\right] - \left\{\frac{\partial}{\partial\xi^i}\left[\frac{\nu_T b_m^i b_2^j}{J}\cdot\frac{\partial U_m}{\partial\xi^j}\right]\right\}$$

$$s_W = \frac{\partial}{\partial\xi}\left[b_3^1(p+\frac{2}{3}k)\right] + \frac{\partial}{\partial\eta}\left[b_3^2(p+\frac{2}{3}k)\right] + \frac{\partial}{\partial\zeta}\left[b_3^3(p+\frac{2}{3}k)\right] - \left\{\frac{\partial}{\partial\xi^i}\left[\frac{\nu_T b_m^i b_3^j}{J}\cdot\frac{\partial U_m}{\partial\xi^j}\right]\right\} \tag{24}$$

The additional source terms s_ϕ, for a turbulence length model, or for laminar studies, are easily deduced from the three previous relations, (14), by setting $k \equiv 0$ and the bracketed terms $\{...\} \equiv 0$. The momentum equations for the turbulence quantities are also written :

$$J.\frac{\partial\Psi}{\partial t} + \frac{\partial}{\partial\xi}(Ju.\Psi) + \frac{\partial}{\partial\eta}(Jv.\Psi) + \frac{\partial}{\partial\zeta}(Jw.\Psi) \qquad \Psi = k \text{ or } \varepsilon$$

$$= \frac{\partial}{\partial\xi}\left[Jg^{11}A_\Psi\frac{\partial\Psi}{\partial\xi}\right] + \frac{\partial}{\partial\eta}\left[Jg^{22}A_\Psi\frac{\partial\Psi}{\partial\eta}\right] + \frac{\partial}{\partial\zeta}\left[Jg^{33}A_\Psi\frac{\partial\Psi}{\partial\zeta}\right] + S_\Psi \tag{15}$$

$$S_\Psi = \frac{\partial}{\partial\xi}\left[Jg^{12}A_\Psi.\frac{\partial\Psi}{\partial\eta}\right] + \frac{\partial}{\partial\xi}\left[Jg^{13}A_\Psi.\frac{\partial\Psi}{\partial\zeta}\right]$$
$$+ \frac{\partial}{\partial\eta}\left[Jg^{21}A_\Psi.\frac{\partial\Psi}{\partial\xi}\right] + \frac{\partial}{\partial\eta}\left[Jg^{23}A_\Psi.\frac{\partial\Psi}{\partial\zeta}\right]$$
$$+ \frac{\partial}{\partial\zeta}\left[Jg^{31}A_\Psi.\frac{\partial\Psi}{\partial\xi}\right] + \frac{\partial}{\partial\zeta}\left[Jg^{32}A_\Psi.\frac{\partial\Psi}{\partial\eta}\right] - J.s_\Psi \tag{16}$$

$$\Psi = k \ : \ s_k = G - \varepsilon \ ; \quad A_k = [\sigma_k \mathrm{Reff}\,]^{-1}$$

$$\Psi = \varepsilon \ : \ s_\varepsilon = C_{\varepsilon 1}\frac{\varepsilon}{k}G - C_{\varepsilon 2}\frac{\varepsilon^2}{k} \ ; \quad A_\varepsilon = [\sigma_\varepsilon \mathrm{Reff}\,]^{-1}. \tag{17}$$

Endly, the metric coefficients involved in the transformation are given. They are the contravariant base, $\{\mathbf{b}^i\}$, normalized by the Jacobian, J, of the transformation and the metric tensor $\mathbf{g}$:

$$\mathbf{b}^1 = \begin{bmatrix} y_\eta z_\zeta - y_\zeta z_\eta \\ x_\zeta z_\eta - x_\eta z_\zeta \\ x_\eta y_\zeta - x_\zeta y_\eta \end{bmatrix} \quad \mathbf{b}^2 = \begin{bmatrix} y_\zeta z_\xi - y_\xi z_\zeta \\ x_\xi z_\zeta - x_\zeta z_\xi \\ x_\zeta y_\xi - x_\xi y_\zeta \end{bmatrix} \quad \mathbf{b}^3 = \begin{bmatrix} y_\xi z_\eta - y_\eta z_\xi \\ x_\eta z_\xi - x_\xi z_\eta \\ x_\xi y_\eta - x_\eta y_\xi \end{bmatrix} \tag{18}$$

$$J = x_\xi b_1^1 + y_\xi b_2^1 + z_\xi b_3^1 = x_\eta b_1^2 + y_\eta b_2^2 + z_\eta b_3^2 = x_\zeta b_1^3 + y_\zeta b_2^3 + z_\zeta b_3^3 \quad : \ b_2^1 \equiv (\mathbf{b}^1)_2 \tag{19}$$

$$g^{ij} = g^{ji} = \frac{\mathbf{b}^i.\mathbf{b}^j}{J^2} \ . \tag{20}$$

III DISCRETISATION

III.1 MOMENTUM EQUATIONS

The momentum equations are re-written in a generic way for the dependant variables ϕ= U,V,W, k or ε :

$$J.\frac{\partial \phi}{\partial t} + \frac{\partial}{\partial \xi^i}(C_i \phi) = \frac{\partial}{\partial \xi^i}\left(D_i \frac{\partial \phi}{\partial \xi^j} \right) + S_\phi \quad . \tag{21}$$

The C_i and D_i are respectively the Convective and Diffusive coefficients :

$$C_i = J.u^i \quad ; \quad D_i = \frac{J.g^{ii}}{Reff} \tag{21.bis}$$

and S contains the source terms and the pressure gradients. A modified hybrid scheme (MHS) is used, based on a possible dicretization for the following convection/diffusion type equation :

$$\frac{\partial}{\partial x}\left(C.\frac{\partial \phi}{\partial x} \right) = \frac{\partial}{\partial x}(D.\phi) + S \quad . \tag{22}$$

In order to avoid the undesirable limitation on the mesh Peclet number $P=C.\Delta x/D$, the discretization of the diffusion rests on a classical three-point centered formulae but the convection should be up- or down-centered. This is done gradually means of a continuous discretization of the first derivative :

$$\frac{\partial}{\partial x}(b\phi) \approx \frac{1}{\Delta x}.[(\alpha_n b_n + (\alpha_s-1)b_s).\phi_C - \alpha_s b_s.\phi_S - (\alpha_s-1)b_n.\phi_N] \tag{23}$$

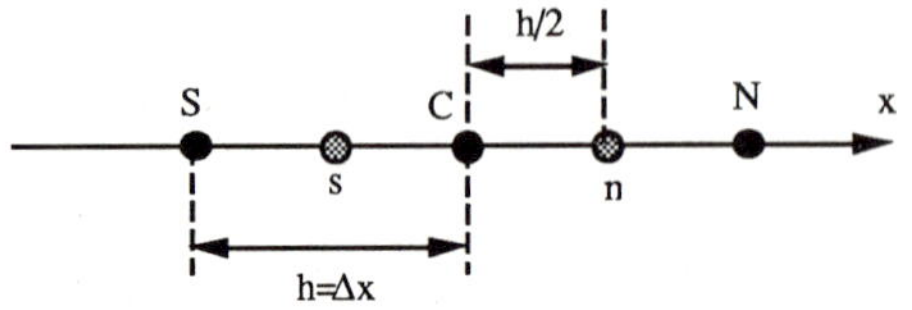

h/2
S s C n N x
h=Δx

Fig.1

$$\alpha = \frac{\omega + 1}{2}$$

$$\begin{cases} \omega(Px) = L(P - Pc) & \text{if } Px > Pc \\ \omega(Px) = 0 & \text{if } -Pc \leq Px \leq +Pc \\ \omega(Px) = -L(P+Pc) & \text{if } Px < -Pc \end{cases}$$

and Pc=2 (Critical Peclet) , $L(X) = 1 - 1/Cosh(X)$, $Px = Ch/D$. (24)

Then the basic momentum equation (21) is discretized and yields a seven points relation for the nodal values around the central point C:

$$A_C\phi_C = A_N\phi_N + A_S\phi_S + A_E\phi_E + A_W\phi_W + A_U\phi_U + A_D\phi_D - \left[J.\frac{\partial \phi}{\partial t} - S_\phi \right]_C \tag{25}$$

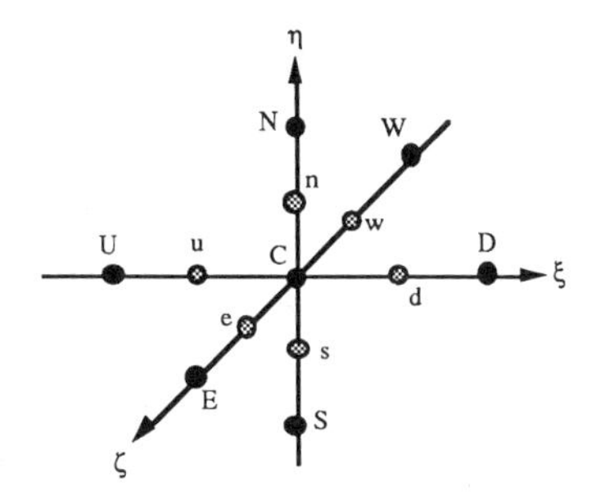

$\Delta\xi=\Delta\eta=\Delta\zeta=1$ Fig.2

As a *collocated* arrangement is used, the coefficents A are independent of the components(k) of the momentum equation and are given by the previous relations (24).The central coefficient is simply :

$$A_C = \sum_{nb=N,S,..} A_{nb}\phi_{nb} + [\, C_{1d} - C_{1u} + C_{2n} - C_{2s} + C_{3e} - C_{3w} \,] . \tag{26}$$

The unsteady term is backward differenced and the following generic discretized momentum equation is obtained :

$$\phi_C \cdot \left[1 + \left(\frac{JC}{\Delta t}\right)_C \right] = \sum_{nb=N,S,..} C_{nb}\phi_{nb} + \phi_C^0 \cdot \left(\frac{JC}{\Delta t}\right)_C + C_C \cdot (S_\phi)_C \tag{27}$$

in which the C-coefficients are the A-coefficients normalised with A_C and $\phi=\phi(t)$, $\phi^0=\phi(t-\Delta t)$. The time step is chosen so that a certain amount of diagonal dominance, E, is enforced. This is why the (local)time step is small in the regions where the continuity equation (whose a centered discretization appears in the A_C-coefficient) is not satisfied and high elsewhere. Most of the results have been obtain with E=0.5 (i.e 50% of diagonal dominance). Of course, such a procedure allows only steady flows to be captured.

III.2 PRESSURE/VELOCITY COUPLING

The coupling between the Pressure field and the Velocity is in the spirit of the PISO and SIMPLER algorithms, [9], in which the "pressure equation" is extracted from the continuity constraint. The starting point consists in isolating *all the pressure gradients* contribution into the k-generic momentum equation :

$$U_k^C = U_k^{*C} - C_C \cdot \left[\frac{\partial}{\partial \xi^j}(J a_k^j p) \right]_C . \tag{28}$$

Where the "homogenous" field $\mathbf{U}^*$ is partially detailed by :

$$U_k^{*C} \equiv \left\{ \sum_{nb=N,S,..} C_{nb} U_k^{nb} - \left[\frac{CJ}{\Delta t}\right]_C . (U_k^C - U_k^{C0}) + C_C \Sigma_k \right\} . \tag{29}$$

Then the continuity equation yields a pressure equation when applying Div to equation (28) :

$$\mathrm{Div}\left(\overrightarrow{U^*} \right) = \frac{1}{J} \cdot \frac{\partial}{\partial \xi^i}\left[J g^{ij} . (CJ) . \frac{\partial p}{\partial \xi^j} \right] . \tag{30}$$

In order to simplify the problem, only the orthogonal terms are retained in the previous pressure operator, the non-orthogonal terms are treated as source terms at the previous time step : (p^0). In accordance with the Poisson type nature of the pressure operator, equation (30) is simply discretized using central differences around the central point C :

$$D^* = C^P_C p_C + \sum_{nb=N,S,..} C^P_{nb} p_{nb}$$

$$D^* = J.Div\left(\overrightarrow{U^*} \right) - \left(\frac{\partial}{\partial \xi^i} \left[Jg^{ij}.(CJ).\frac{\partial p^0}{\partial \xi^j} \right]_{i \neq j} \right)_C \quad (31)$$

with the coefficients for the seven-point basis molecule :

$$C^P_C = - \sum_{nb=N,S,..} C^P_{nb}$$

$$C^P_D = (C_\tau .Jg^{11})_d \ , \quad C^P_U = (C_\tau .Jg^{11})_u \qquad C_\tau = C.J$$

$$C^P_N = (C_\tau .Jg^{22})_n \ , \quad C^P_S = (C_\tau .Jg^{22})_s$$

$$C^P_E = (C_\tau .Jg^{33})_e \ , \quad C^P_W = (C_\tau .Jg^{33})_w \ . \quad (32)$$

Endly, once the pressure field is known, a velocity correction is applied in order to enforce the incompressibility of the velocity field. This is done, in one step, with the help of equation (28) seen as a correction equation for the physical components of the velocity field.
The velocity field equations are solved elliptically using an SLOR iterative scheme and the elliptic pressure operator is inverted by means of an incomplete LU-preconditionned conjugate gradient solver [10].

IV RESULTS

An O-O grid, partially shown in figure I , is used for the computation of the flow around the complete 6:1 prolate spheroid. The sting support is not retained in the present work. An exponential stretching is used in the normal direction and the normalized spacing is about 10^{-5} near the wall for resolving the thin viscous layers present in high Reynolds number flows. A typical laminar run of the code on the Cray-2 needs 400 iterations to reduce the maximum initial residual by three orders of magnitude and takes 3000 seconds with a $60(\xi)x40(\eta)x40(\zeta)$ mesh at the Reynolds number of 1.6 millions and 10° incidence .

The figure II shows the computed skin-friction lines at the wall for the unwrapped coordinate system ($\xi,\eta,\zeta=1$). We can see a good agreement with the nice results of Thomas et Al [6] : Two main convergence lines emerge as revealed by experimental observations and they are the starting point of two longitudinal vortices on either side of the spheroid. The letters P,R and S of figure II stand for respectively the Primary separation (main vortex) , the Reattachment line and the Secondary separation (Secondary vortex) .

The figure IV is the "physical" equivalent of figure II but the secondary separation is not very well detected by the particle traces used to build the friction lines. This picture is colored by the modulus of the friction vector , the blue color denotes low absolute levels of friction (for exemple following the main convergence line) and the red color denotes high levels (for exemple in near the nose region). The location of the main convergence line is slightly different from those found in the experiments, probably because of the effects of the transition to turbulence before the separation and by the lack of resolution in the circumferential direction ζ.

The figure III shows the wall pressure distributions in the leeward plane (0°) and in the windward plane (180°). The agreement is quite good over most of the body and the main differences occur on the leeward side, near the trailing edge, where lower pressures than those expected are predicted. The sting support is not taken into account and is suspected to be responsible for such discrepancies. Finally the figure V, through the iso-U(axial velocity) of a 30° incidence test case, Re=10000, gives an idea of higher order incidence effects on the topology of the flow field where the main vortices are quite much more developped than those generated by a 10° incidence.

Acknowledgments

Partial support of DRET, through contract 86-104, is gratefully acknowledged. Thanks are also due to the Scientific Committee of CCVR and the DS/SPI for attributions of Cpu on the Cray2 and on the VP200 .

REFERENCES

[1] H.P. Kreplin, H.U. Meier & A. Maier "*Wind-tunnel model and measuring techniques for the investigation of threedimensional boundary layers*" Proc. AIAA 10th Aero.Testing Conf., San Diego (1982).

[2] H.P. Kreplin, H. Vollmers & H.U. Meier "*Measurements of the wall-stress on an inclined prolate spheroid*" Zeitchrift für Flugwissenschaften und Welttaumferschung , Vol.6 heft 4 , pp 248-252 (1982) .

[3] T. Cebeci, R.S. Hirsh & K. Kaups "*Calculation of three dimensional boundary layers on bodies of revolution at incidence*" ; Mc. Donnell Douglas Rept. MDCJ7643 (contract N60921-76-C-0089), (1978) .

[4] M. Rosenfeld, M. Israeli & M. Wolfshtein "*Numerical study of the skin friction on a spheroid at incidence*" , AIAA Journal , Vol.26, N°2 , Feb.(1988) .

[5] T.C. Wong, O.A. Kandil & C.H. Liu "*Navier-Stokes computation of separated vortical flows past prolate spheroid at incidence*" AIAA 89-0553 (1989] .

[6] V.N. Vatsa, J.L. Thomas & B. Wedan "*Navier-Stokes computation of prolate spheroids at angle of attack*", AIAA 87-2627-CP, (1987).

[7] D. Pan & T.H. Pulliam "*The computation of steady three dimensional flows over aerodynamic bodies at incidence and yaw*" AIAA P. 86-0109, Jan.(1986).

[8] J.F. Thompson "*Numerical grid generation*" Elsevler Publ. Co. N-Y.

[9] J. Piquet & M. Visonneau "*Steady threee dimensional viscous flow past a shiplike hull*" ,Proc. 7th GAMM Conf. on Num. Methods in Fluid Mechanics; Louvain, Belg.; Notes Num. Fluid Mech., Vol.20 ; Vieweg; M. Deville Eds. (1987).

[10] J. Piquet, P. Queutey & M. Visonneau "*Computation of the three dimensional wake of a shiplike hull*" Proc. 11th Int. Conf. On Num. Methods in Fluid Dynamics; Williamsburg, USA; Dwoyer, D.L.; Hussaini & M.Y. Voigt , R.G. Eds. Springer Verlag Lect. Notes in Physics ; Vol.323 (1988)

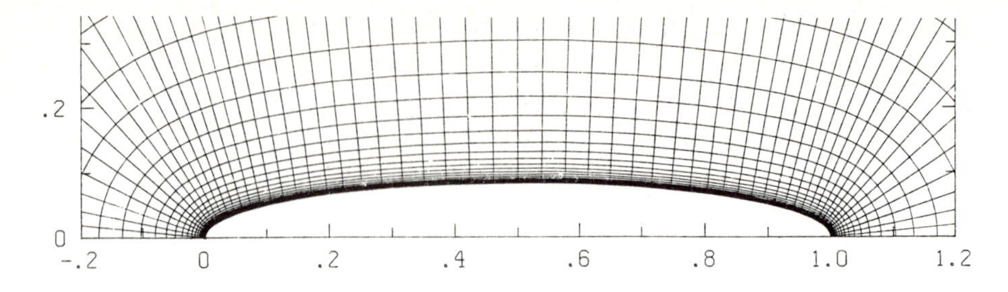

Figure I : Partial view of the computational grid (no sting) .

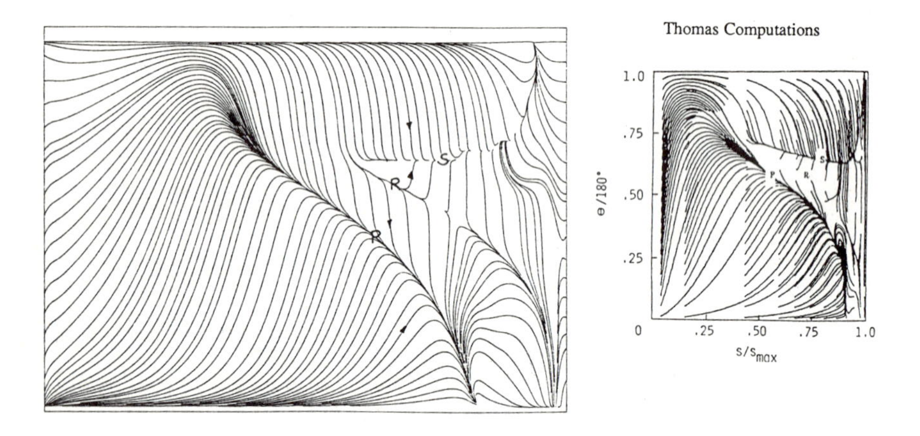

Figure II : Surface streamlines for the unwrapped coordinates and comparisons with Thomas computations . Re=1.6 10^6 , α=10°

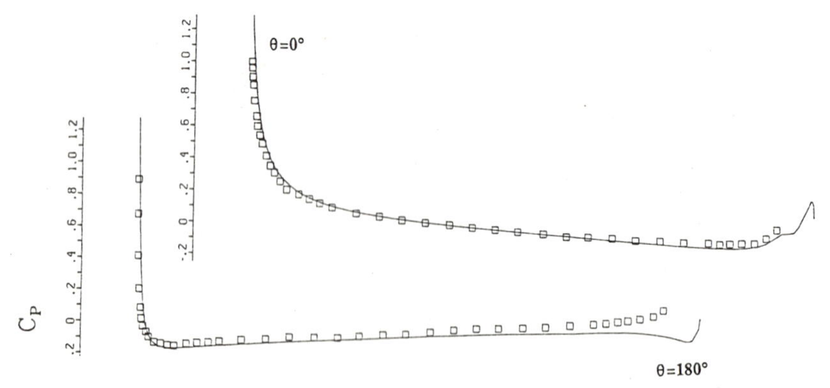

Figure III:Longitudinal pressure distribution comparisons at symmetry planes. Re=1.6 10^6 , α=10°

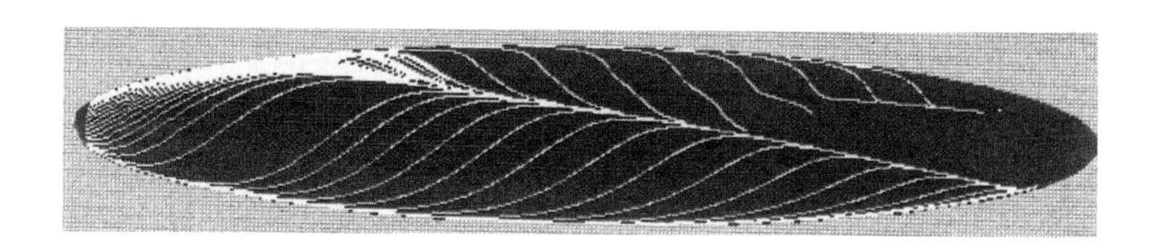

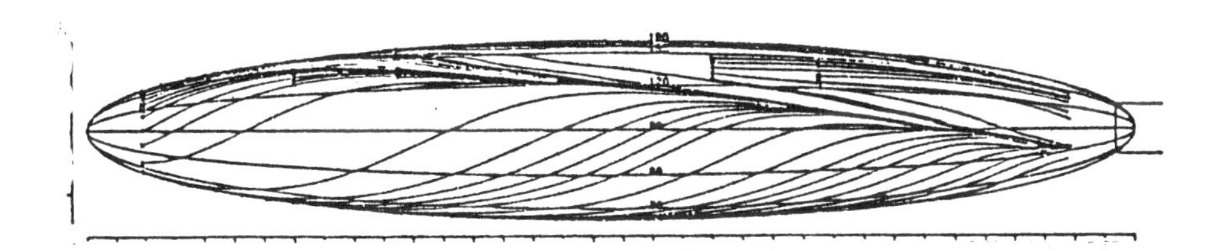

Figure IV : Surface streamlines on the physical surface and comparison with experiment. Re=1.6 10^6 , α=10°

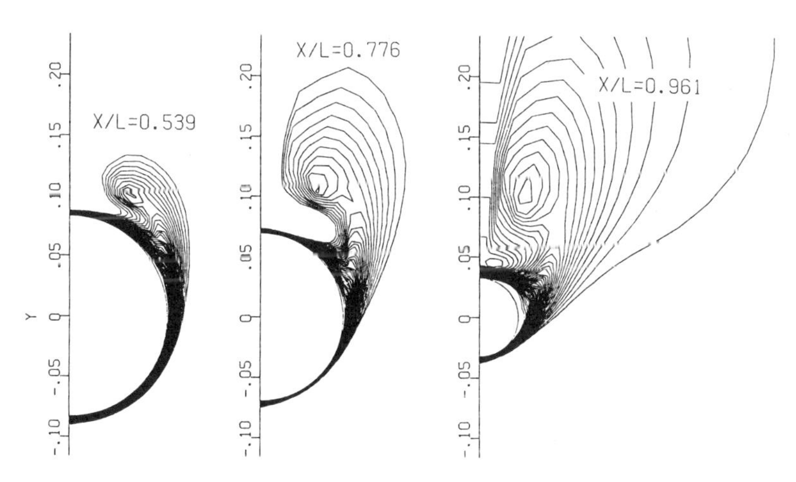

Figure V : Iso-U contours at fixed axial locations . Re=10000 , α=30° .

SPARSE QUASI–NEWTON METHOD FOR NAVIER–STOKES SOLUTION

Ning QIN and Bryan E. RICHARDS

Department of Aerospace Engineering
University of Glasgow, Glasgow G12 8QQ, Scotland, U.K.

SUMMARY

A sparse finite difference Newton method and a sparse quasi–Newton method have been applied to the Navier–Stokes solution. Much faster convergence to the steady state has been achieved compared to the conventional time marching method. For multidimensional applications, a block line Gauss–Seidel iterative method has been used for the solution of the resulting linear system. The methods have been demonstrated for hypersonic flow solution around a sharp cone using Osher's flux difference splitting scheme for spatial discretization.

1. INTRODUCTION

Difficulties have been experienced in predicting aerodynamic heating in hypersonic flows by solving the Navier–Stokes equations due to its strong dependence on the cell Reynolds number. This necessitates the use of a very fine grid to resolve the high gradients in the thin shear layer near the wall. One reasonable way to achieve the high resolution of the shear layer is to stretch the grid towards the wall. It is also necessary to employ a discretization scheme which can capture accurately not only strong shock waves but also thin shear layers in the flowfield.

High stretch of the grid and the use of high order sophisticated upwind schemes, e.g. van Leer's flux vector splitting, Roe's or Osher's flux difference splitting and Harten's or Yee's TVD schemes, have created convergence problems in the Navier–Stokes solutions. The two most currently used approaches to accelerate convergence, i.e., (i) the multigrid technique and (ii) implicit time marching, become unsatisfactory. The efficiency of multigrid techniques has been found to be significantly reduced on a highly stretched grid for Navier–Stokes solutions. Furthermore, implicit procedures involve linearization which, if exact, implies analytical calculation of the Jacobian of the non–linear system and therefore, creates difficulties in application to the Navier–Stokes solution using high resolution schemes. Approximate linearization commonly used in practice, e.g. using only first order upwind inviscid terms in the implicit operator, generally still results in poor convergence for Navier–Stokes solutions because of the unbalanced left and right hand sides.

Avoiding the difficulties in linearization of implicit schemes, the present authors introduced in the last GAMM Conference on Numerical Methods in Fluid Mechanics [1] a new approach for fast steady state solution of CFD problems through the use of a sparse quasi–Newton method [2,3]. Its general formulation and application to the solution of Euler equations using high shock resolution schemes were presented in that paper. The present paper will concern its application to Navier–Stokes solutions, where the convergence problem is more pronounced.

2. SPATIAL DISCRETIZATION OF NAVIER–STOKES EQUATIONS

2.1. *The Navier–Stokes Equations*

For hypersonic flows around conical geometries, consider the following set of Navier–Stokes equations, which are derived from the steady three dimensional compressible

Navier–Stokes equations written in a general coordinate system through the use of a locally conical approximation.

$$\frac{\partial \bar{F}(Q)}{\partial \eta} + \frac{\partial \bar{G}(Q)}{\partial \zeta} + H(Q) = 0, \qquad (1)$$

where

$$\bar{Q} = Q/J, \quad Q = [\rho, \rho u, \rho v, \rho w, e]^T,$$

$$\bar{E} = (\frac{\xi_x}{J})(E_I - E_V) + (\frac{\xi_y}{J})(F_I - F_V) + (\frac{\xi_z}{J})(G_I - G_V), \ \ldots$$

$$E_I = [\rho u, \rho u^2+p, \rho uv, \rho uw, (e+p)u]^T, \ \ldots$$

$$E_V = [0, \tau_{xx}, \tau_{xy}, \tau_{xz}, u\tau_{xx}+v\tau_{xy}+w\tau_{xz}-q_x]^T, \ \ldots$$

$$H = (2\bar{E}_I - \bar{E}_V)/\xi .$$

In the above equations, ξ is the coordinate along the radial direction measured from the tip of the conical geometry. The source term H is introduced from the term $(\partial \bar{E}/\partial \xi)$ in 3D NS equations after the locally conical approximation. J is the Jacobian of the transformation. And the x, y, z, u, v, w, ρ, T, μ, p and e are nondimensionalized to L, V_∞, ρ_∞, T_∞, μ_∞, $\rho_\infty V_\infty^2$ and ρV_∞^2, respectively. Then

$$p = (\gamma-1)[e-\rho(u^2+v^2+w^2)/2], \qquad (2)$$

$$T = \gamma M_\infty^2 p/\rho . \qquad (3)$$

The viscosity μ is calculated from temperature T through the Sutherland formula.

If the angle of attack of the geometry to the flow is zero, a simpler set of Navier–Stokes equations can be derived from the axisymmetric Navier–Stokes equations.

$$\frac{\partial \bar{F}(Q)}{\partial \eta} + H(Q) = 0, \qquad (4)$$

where

$$\bar{Q} = Q/J, \quad Q = [\rho, \rho u, \rho v, e]^T,$$

$$\bar{E} = (\frac{\xi_x}{J})(E_I - E_V) + (\frac{\xi_y}{J})(F_I - F_V), \ \ldots$$

$$E_I = [\rho u, \rho u^2+p, \rho uv, (e+p)u]^T, \ \ldots$$

$$E_V = [0, \tau_{xx}, \tau_{xy}, u\tau_{xx}+v\tau_{xy}-q_x]^T, \ \ldots$$

$$H = H_1 + H_2 .$$

where H_1 is the contribution from the source term of the axisymmetric Navier–Stokes equations and H_2 is the contribution from the source term of the locally conical Navier–Stokes equations. Eq.(4) is the simplest set of Navier–Stokes equations governing realistic flows, which provides a case for extensively testing different discretization schemes and convergence acceleration for the Navier–Stokes equations. Eq.(1) and Eq.(4) are considered in the present paper for testing the new iterative methods proposed.

2.2. *The Osher's Flux Difference Splitting Scheme*

For the Navier–Stokes solution, the viscous or diffusion terms are all discretized using a central differencing scheme. The discretization of the inviscid or convection terms are discussed in detail below.

For accurate heat transfer prediction in hypersonic flows, the accurate capturing of shear layers, i.e. the accurate modelling of diffusion, becomes as important as the accurate modelling of shock waves. Sophisticated upwind schemes can usually capture a shock wave sharply without adding artificial viscosity terms as is necessary for central differencing schemes. But it does not necessarily mean that they can capture shear layers as well [4]. The Osher's flux difference splitting(FDS) scheme is preferred for the following two reasons: (i) the scheme can capture well both shock waves and shear layers, which is important for accurate heat transfer prediction in hypersonic Navier–Stokes solutions [5]; (ii) the finite difference operator is continuously differentiable, which makes the application of Newton–type methods possible for fast steady state solution.

A finite–volume discretization of the equations is employed. The conserved variables $\bar{Q}$ are evaluated at cell centers and represent cell–averaged values. The fluxes $\bar{F}$ and $\bar{G}$ are evaluated at cell interfaces. The spatial derivatives are represented as a flux balance across a cell as, for example, for Eq.(1),

$$\left(\frac{\partial \bar{F}_I}{\partial \eta}\right)_{i,j} = (\tilde{F}_I)_{i+\frac{1}{2},j} - (\tilde{F}_I)_{i-\frac{1}{2},j}. \tag{5}$$

The interface flux is determined from a state–variable interpolation and a local one–dimensional model of wave interactions normal to the cell interfaces. With the flux difference splitting (FDS) model developed by Osher [6,7] , the interface flux can be written as

$$\tilde{F}_I = \frac{1}{2}\left[\bar{F}_I(q^-) + \bar{F}_I(q^+) - \int_{q^-}^{q^+} \left|\frac{\partial \bar{F}_I}{\partial q}\right| dq \right], \tag{6}$$

where the integral is carried out along a path piecewise parallel to the eigenvectors of $\partial\bar{F}_I/\partial q$. Osher [6,7] proposed a reverse ordering of the subcurves while a natural ordering, which is more time saving, has been employed here according to Spekreijse [8] in calculating the integral in the above interface flux.

2.3. *High Order Accuray*

The state–variable interpolations determine the resulting accuracy of the scheme. It was found in the present research that the use of primitive variables $q=[\rho,u,v,w,p]^T$ in the interpolation is more robust than the use of conserved variables Q in the sense that non–physical states from the interpolation, such as, negative pressure, are easier to avoid. First order differencing leads to

$$\begin{aligned} q^-_{i+\frac{1}{2},j} &= q_{i,j} \\ q^+_{i+\frac{1}{2},j} &= q_{i+1,j} \end{aligned} \tag{7}$$

and a k–parameter family of higher–order schemes can be written as

$$\begin{aligned} q^-_{i+\frac{1}{2},j} &= q_{i,j} + \{(s/4)[(1-ks)\Delta_- + (1+ks)\Delta_+]q\}_{i,j} \\ q^+_{i+\frac{1}{2},j} &= q_{i+1,j} - \{(s/4)[(1+ks)\Delta_- + (1-ks)\Delta_+]q\}_{i+1,j} \end{aligned} \tag{8}$$

where Δ_+ and Δ_- denote forward and backward difference operators, respectively, in the η direction. The parameter k determines the spatial accuracy of the difference approximation: $k=-1$ corresponds to a fully-upwind second order scheme, $k=1$ to a central difference scheme, and $k=1/3$ to a third order upwind-biased scheme. In the present research the third order upwind-biased option is used.

The parameter s serves to limit higher-order terms in the interpolation to cell interfaces in order to avoid oscillations at discontinuities in the solutions such as shock waves. The limiting is implemented by locally modifying the difference values in the interpolation to ensure monotone interpolation as

$$s = \frac{2\Delta_+ q \Delta_- q + \epsilon}{(\Delta_+ q)^2 + (\Delta_- q)^2 + \epsilon} \tag{9}$$

where ϵ is a small number preventing division by zero in regions of null gradients.

3. SPARSE QUASI-NEWTON METHOD

3.1. *Explicit, Implicit and Newton methods*

After spatial discretization of the Navier-Stokes equations, a large sparse nonlinear system results, which can be written as

$$R(Q) = 0. \tag{10}$$

Time dependent approaches are generally used for steady state solutions of Euler or Navier-Stokes equations. These appoaches introduce a unsteady term, i.e.

$$\frac{\partial Q}{\partial t} + R(Q) = 0. \tag{11}$$

An obvious advantage of these approaches is its robustness in convergence to the steady state because they try to follow a physical path. On the other hand, a penalty to this is its slow convergence to the steady state.

Explicit methods are usually very slow due to the restriction on the time step by the stability condition. Better convergence can be achieved through various techniques which relax the requirement on time accuracy. Local time stepping accelerates the convergence by using local maximum time steps on nonuniform grids. And multigrid techniques propagate information more efficiently by using multilevel grids.

In order to get rid of the strict stability restriction of explicit methods, implicit methods can be used. For example, if a backward Euler scheme is used for the time discretization of Eq.(11), one can obtain

$$[\,(I/\Delta t) - (\partial R/\partial Q)^k\,]\,(Q^{k+1}-Q^k) = R(Q^k). \tag{12}$$

One of the difficulties with implicit methods is the linearization of R, which, in fact, means the evaluation of the Jacobian of the nonlinear system (10). Therefore approximation is generally made in the linearization in implicit methods. A commonly used approximation is to use only the first order inviscid terms in the implicit operator which makes the implicit procedure simpler but may also degrade the convergence due to the unbalanced left and right hand sides.

It is interesting to look at the limiting case of the implicit scheme Eq.(12) when $\Delta t \to \infty$. Then

$$-(\partial R/\partial Q)^k\,(Q^{k+1}-Q^k) = R(Q^k). \tag{13}$$

This reveals the relation between a Newton method and an implicit method with exact linearization. The quadratic convergence of a Newton method is desirable but the Jacobian is generally not available due to the complexity of R, which is due to (1) the high resolution spatial discretization scheme used and (2) the physics modelled, especially if turbulence or chemistry is to be included.

3.2. *Sparse Finite Difference Newton Method*

As indicated above, the complexity of the nonlinear system prohibits the use of the Newton method. However the difficulty in obtaining the Jacobian analytically can be avoided by numerical evaluation of the Jacobian by a finite difference method.

Supposing Eq.(1) has been discretized by the FDS scheme on a INxJN mesh. Then we have INxJNx5 equations in Eq.(10) for INxJNx5 unknowns. Thus the evaluation of the Jacobian needs INxJNx5 evaluations of R, which is prohibitive. It is necessary to make use of the sparsity structure of the Jacobian. At each point, the discretization scheme gives a 9 point stencil with five points in each coordinate direction. So if we disturb not only one point but every fifth point in both directions in one evaluation of R, we can obtain the Jacobian much more efficiently since only 5x5x5 evaluations of R are needed. If the disturbance in calculating the Jacobian is properly chosen according to the machine zero, the sparse finite difference Newton method can also give a quadratic convergence similar to the Newton method. A more detailed formulation can been found in [1].

3.3. *Sparse Quasi–Newton Method*

The sparse finite difference Newton(SFDN) method still involves many times of function evaluations of R. This can be avoided by using a sparse quasi– Newton(SQN) method. The basic idea of the quasi– Newton method is to approximate the Jacobian of the nonlinear system using only function values already calculated. To make a full use of sparsity in the quasi– Newton updating of the Jacobian, the sparse quasi– Newton method [2,3] can be used, which is described below. Ref.[1] gives a more detailed formulation for three point and 5 point stencils.

Define the matrix projection operator $P_J : R^{n\times n} \to R^{n\times n}$ by

$$(P_J(M))_{i,j} = \begin{cases} 0, & \text{if } J(U)_{i,j} = 0 \quad \text{for all } U \in R^n \\ M_{i,j}, & \text{otherwise} \end{cases} \tag{14}$$

Similarly we define $S_i \in R^n$ by

$$(S_i)_j = \begin{cases} 0, & \text{if } J(U)_{i,j} = 0 \quad \text{for all } U \in R^n \\ S_j, & \text{otherwise} \end{cases} \tag{15}$$

The procedure of the sparse quasi– Newton method may be written as following:

$$\begin{aligned} &\text{Given } R: R^n \to R^n,\ Q^0 \in R^n,\ A^0 \in R^{n\times n} \\ &\text{DO for } k = 0,1,2,\ldots \\ &\qquad \text{Solve} \quad A^k s^k = -R(Q^k) \text{ for } s^k, \\ &\qquad\qquad Q^{k+1} = Q^k + s^k, \\ &\qquad\qquad A^{k+1} = A^k + P_J[D^+ R(Q^{k+1})(s^k)^T]. \end{aligned} \tag{16}$$

Here $D^+ \in R^{n\times n}$ is a diagonal matrix with

$$(D^+)_{ii} = \begin{cases} 1/(S_i)^T(S_i), & \text{if } (S_i)^T(S_i) \neq 0 \\ 0, & \text{otherwise} \cdot \end{cases} \tag{17}$$

Under standard assumptions, this procedure has been proved to be locally q– superlinearly convergent. However it is important to note that, as for the Newton method or the SFDN method, the fast convergence is a local property. The basic idea in forming a successful nonlinear algorithm is to combine a fast local convergence strategy with a global convergence strategy in a way that derives benefit from both.

3.4. *Initialization*

Both the SFDN method and the SQN method need a reasonably good initial value Q^0 to start with. For problems in CFD, a natural and robust way to initialize the procedure is the time marching approach, although as an iterative method for steady state solutions it may be extremely slow.

An explicit 4– stage Runge– Kutta method (RK) has been used with local time steps determined by the stability condition. The time marching is switched to the SFDN or SQN iteration as soon as the solution goes into the convergent region. Because the theory on the convergent region for the SFDN or SQN method is absent, this switching point has to be determined by experimentation. In the SQN method, initial approximation A^0 to the Jacobian is required. For a given Q^0, A^0 is obtained by evaluating $J(Q^0)$ using the SFDN method.

3.5. *Iterative Solution of the Linear System for Multidimensional Problem*

For both the SFDN and the SQN methods, a large sparse regularly structured linear system needs to be solved in each iteration. For one dimensional problems, the matrix is block pentadiagonal, which can be written as

$$D_i S_i + B_i S_{i-1} + B1_i S_{i-2} + C_i S_{i+1} + C1_i S_{i+2} = R_i \tag{18}$$

and the system can be solved directly. However, for multidimentional problems, the matrix cannot be solved directly so far in an efficient way. An iterative approach has been employed in the present research. With the 9 point stencil in the spatial discretization, the linear system can be written as

$$\begin{aligned} D_{i,j}S_{i,j} &+ B_{i,j}S_{i-1,j} + B1_{i,j}S_{i-2,j} + C_{i,j}S_{i+1,j} + C1_{i,j}S_{i+2,j} \\ &+ E_{i,j}S_{i,j-1} + E1_{i,j}S_{i,j-2} + F_{i,j}S_{i,j+1} + F1_{i,j}S_{i,j+2} = R_{i,j} \end{aligned} \tag{19}$$

where D, B, B1, C, C1, E, E1, F and F1 are 5x5 matrices and S, R 5x1 vectors. This linear system is solved by a block line Gauss– Seidel iterative method, which can be written as

$$\begin{aligned} D_{i,j}S^{l+1}_{i,j} &+ B_{i,j}S^{l+1}_{i-1,j} + B1_{i,j}S^{l+1}_{i-2,j} + C_{i,j}S^{l+1}_{i+1,j} + C1_{i,j}S^{l+1}_{i+2,j} \\ &= R_{i,j} - E_{i,j}S^{l}_{i,j-1} - E1_{i,j}S^{l}_{i,j-2} - F_{i,j}S^{l}_{i,j+1} - F1_{i,j}S^{l}_{i,j+2}. \end{aligned} \tag{20}$$

4. RESULTS AND DISCUSSION

The results presented here are for hypersonic flow around Tracy's sharp cone[9]. The half cone angle is 10°. The flow conditions are

$$M_\infty = 7.95, \quad Re = 4.2\text{x}10^6/\text{m}, \quad T_\infty = 55.4 \text{ K}, \quad T_W = 309.8 \text{ K}, \quad r = 0.1049 \text{ m}.$$

Case (1): $\alpha = 0°$. For the zero incidence case, Eq.(4) can be use to solve the problem. The problem reduces to a 1D– like problem because η or θ is the only independent variable and ξ or r acts as a parameter. Figs.1 and 2 illustrate the temperature distributions from the Navier– Stokes solutions on (i) a uniformly distributed 65 point grid and (ii) a stretched 65 point grid. It is clear that the solution on the stretched grid gives much better resolution of the shear layer and the shock wave is well captured on both grids as demonstrated by the presence of only one intermediate point. Figs.3 and 4 plot the convergence histories on the two different grids. The stretch of the grid evidently slows down the convergence of the RK method. The procedures switch to SFDN or SQN iteration after 600 RK iterations and converge suddenly to the steady state. The following tables show the corresponding CPU time for convergence.

Table 1. CPU time for convergence for case (1)

grid	method	iteration number	CPU(second)
65 uniform	RK RK+SFDN RK+SFDN+SQN	2119 602 605	72 20 21
65 stretched	RK RK+SFDN RK+SFDN+SQN	11384 604 632	397 22 23

From this table, one can see that both the SFDN method and the SQN method have saved considerable computing time for the solution to reach the steady state. Compared with the SFDN method, the SQN method takes more iterations to converge but less CPU per iteration. Thus the total CPU is similar for both the methods in this case.

Case (2): $\alpha = 12°$. For this case, Eq.(1) needs to be used. The independent variables are η and ζ or θ and φ. Again ξ or r only serves as a parameter in the Eq.(1) and the problem is 2D– like. Fig.5 shows the temperature contours from the Navier– Stokes solution on a 33x17 mesh, which is stretched in the η direction. The shock wave and the shear layer around the cone are both reasonably well captured by the FDS scheme. Fig.6 plots the convergence histories against the iteration number for the three different methods. The following table compares the CPU time for the different methods.

Table 2. CPU time for convergence for case (2)

grid	method	iteration number	CPU(second)
33x17 stretched	RK RK+SFDN RK+SFDN+SQN	2858 603 606	1816 691 724

Due to the inner iteration for solving the linear system, each SFDN or SQN iteration spends much longer time than the RK iteration. However, the very fast convergence of these two methods makes the total CPU much less than those for the RK method. Better convergence can be achieved by improving the inner iteration.

Again the SFDN method and the SQN method show similar total CPU for convergence. Note that the SFDN method needs 125 evalutions of R per iteration while the SQN method needs only one evaluation of R. This difference has been overshadowed by the CPU of the inner iteration for solving the linear system.

5. CONCLUDING REMARKS

The sparse finite difference Newton(SFDN) method and the sparse quasi−Newton(SQN) method have been applied to the Navier−Stokes solution. Much faster convergence to the steady state can be achieved compared to the conventional time marching method. This arises from the combination of the local quadratic or q−superlinear convergence characteristics of these methods with the robustness of the time dependent approach.

For the cases studied, the SQN method has not demonstrated significant advantage over the SFDN method. It is expected that the SQN method will show its potential over the SFDN method in application to more complicated problems when the function evaluation becomes more expensive. Further improvement for multidimensional problems can be made through the development of a more efficient inner iteration strategy.

REFERENCES

1. N. QIN AND B.E. RICHARDS, *Notes on Numer. Fluid Mech.* **20**(1988), 310.
2. L.K. SCHUBERT, *Math. Comp.* **24**(1970), 27.
3. C.G. BROYDEN, *Math. Comp.* **25**(1971), 285.
4. B. VAN LEER, J.L. THOMAS, P. ROE AND R.W. NEWSOME, AIAA−87−1104.
5. N. QIN AND B.E. RICHARDS, *Proc. Int. Conf. on Hypersonic Aerodynamics*, Sept. 1989, Manchester, paper 26.
6. S. OSHER AND F. SOLOMON,*Math. Comp.* **38**(1982), 339.
7. S. OSHER AND S.R. CHAKRAVARTHY, *J. Comp. Phys.* **50**(1983), 447.
8. S.P. SPEKREIJSE, Ph.D. thesis, CWI, Amsterdam, (1987).
9. R.R. TRACY, Aero. Lab. Memo. No.69, CIT, (1963)

Acknowledgement

This research was supported by the Science and Engineering Research Council grant No.GR/E/89056.

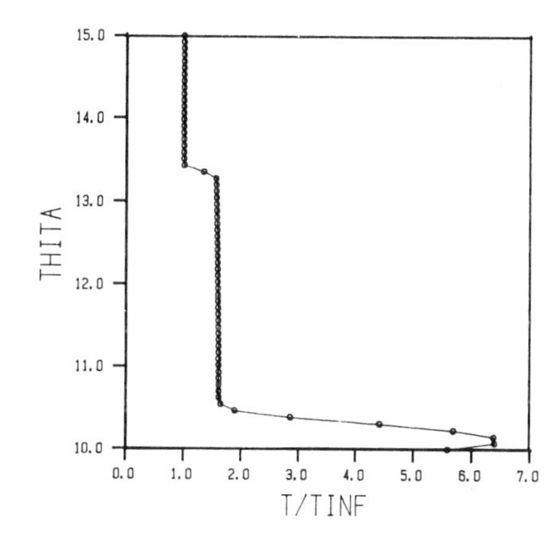

Fig.1. Temperature distribution on uniform grid for Case(1)

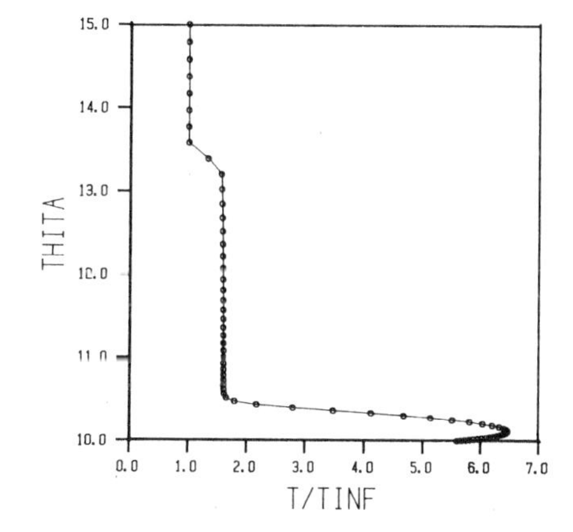

Fig.2. Temperature distribution on streched grid for Case(1)

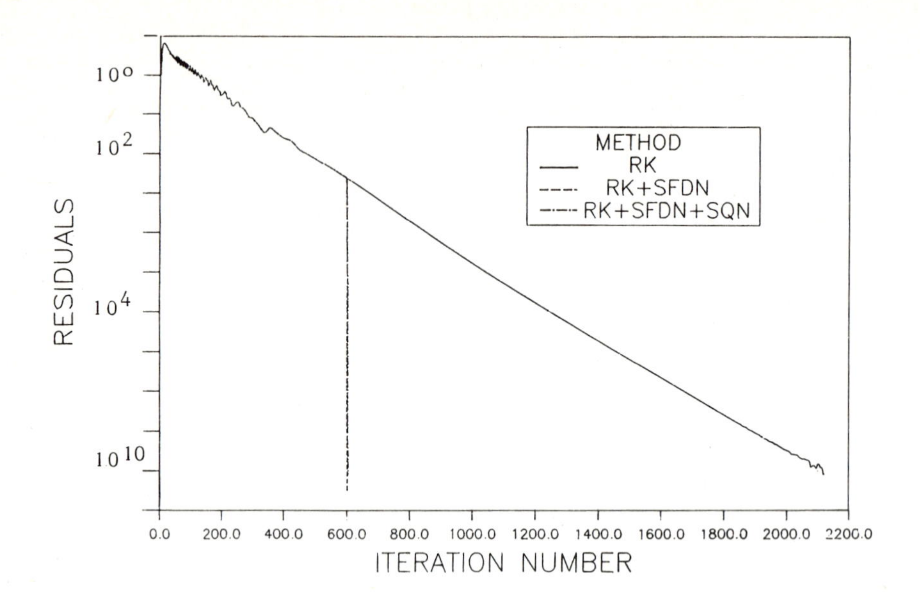

Fig.3. Convergence histories for Case(1) on uniform grid

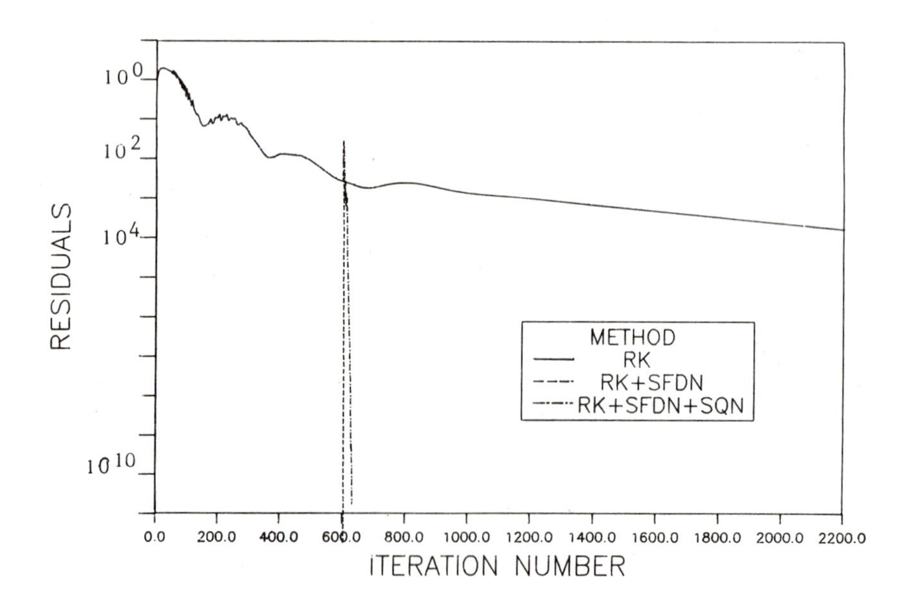

Fig.4. Convergence histories for Case(1) on stretched grid

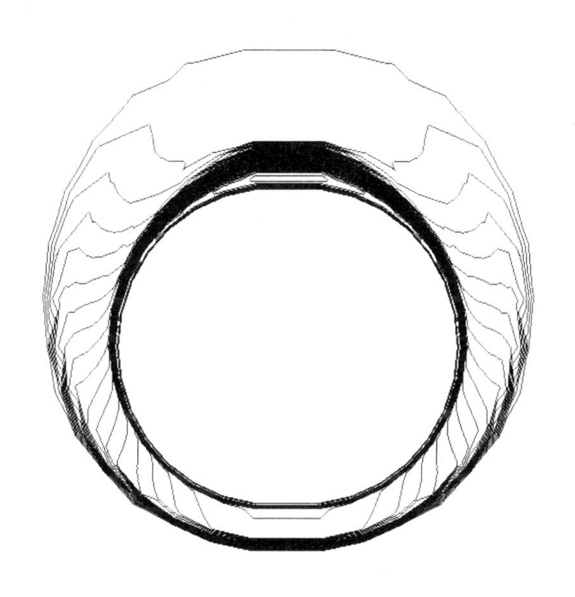

Fig.5. Temperature contours for Case(2)

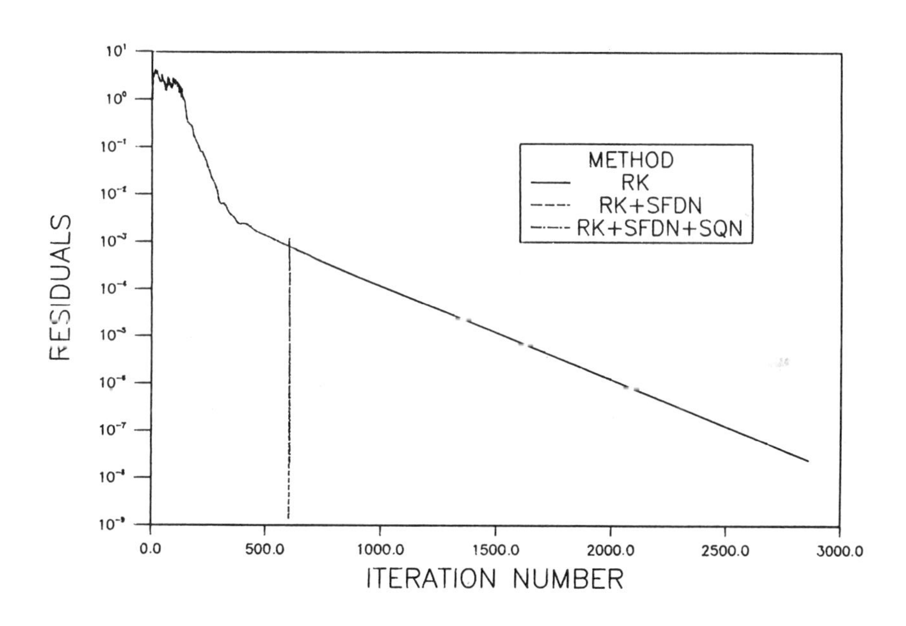

Fig.6. Convergence histories for Case(2)

Numerical Investigation of Three Dimensional Laminar Flows in a Channel with a Built-in Circular Cylinder and Wing-Type Vortex Generators

M. Sanchez, N.K. Mitra, M. Fiebig

Institut für Thermo- und Fluiddynamik
Lehrstuhl für Wärme- und Stoffübertragung
Ruhr-Universität Bochum, Postfach 102148
4630 Bochum, FRG

SUMMARY

Laminar flows around a circular cylinder in a rectangular channel with a pair of built-in vortex generators in form of delta winglets on the bottom plate of the channel have been investigated by solving the complete Navier Stokes equations with a Marker-and-cell technique. Results show that longitudinal vortices from the vortex generators placed in the wake of the cylinder not only control the spread of the wake of the cylinder but also completely damp the periodic vortex street.

INTRODUCTION

In the gas side of a fin plate cross flow heat exchanger, the heat transfer between the gas and the fins deteriorates in the wake of the tubes through which the other fluid (generally a liquid) flows. The spreading of the wake can be controlled by suitably placing longitudinal vortex generators.
As vortex generators, small triangular or rectangular pieces of the fin can be punched out in such a way that while their base remain attached to the fin, these pieces stick out in the flow with an angle of attack to the main flow direction. Depending on their shapes these vortex generators can be termed as delta wing, rectangular wing, or delta winglet pair. Besides controlling the spreading of the wake, the longitudinal vortices additionally increase the heat transfer by inducing in the flow large azimuthal velocities which continuously mix the hot and cold fluids and disturb the boundary layer growth on the fin.
A complete numerical investigation of the flow field around a bank of tubes between two fins even without built-in vortex generators is probably beyond the capacity of even modern computers. Moreover, the flow field will be too complex to analyse. However, the basic structure of the flow, in particular the interaction of longitudinal vortices

with the cylinder wake (vortex street) can be better investigated, if a simplified flow configuration of a channel with a cylinder and one pair of vortex generators is considered. In previous works the structure of the flow in simplified configurations such as a channel with a built-in cylinder and a channel with built-in vortex generators have been investigated/1,2/. The purpose of the present work is a detailed numerical investigation of the three dimensional laminar flows in a channel with a built-in circular cylinder and a pair of vortex generators.

BASIC EQUATIONS AND METHOD OF SOLUTION

Figure 1a shows the computational domain consting of a rectangular channel of height H, width B, with a built-in cylinder of diameter D and a pair of delta winglets of height of 0.2D and span of 0.4D. The flow field in this domain is calculated by solving complete unsteady Navier Stokes equations for incompressible flow with constant viscosity. The basic equations in nondimensional form in cartesian coordinates are

$$\partial u_i/\partial x_i = 0 \ ,$$

$$\partial u_i/\partial t + u_j \ (\partial u_i/\partial x_j) + (\partial p_i/\partial x_i) = \nabla^2 u_i/Re_H \ .$$

Here Re_H is the Reynolds number defined as

$$Re_H = \frac{u_{av} \ H}{\nu} \ ,$$

where u_{av} is the average velocity at the channel inlet.The other symbols have usual meanings.
Fully developed laminar channel flow is used as initial condition at the channel inlet. No-slip condition is used at the top and bottom walls of the channel and the symmetry condition is used at the side boundaries. At the channel exit, the second derivatives of velocity components are taken equal to zero.
The basic equations are solved by a modified version of the marker and cell (MAC) technique /3/.The computational domain has been divided into rectangular parallelopiped cells of unequal size,see fig. 1b.
In the first step of the computation, the cylinder surface has been represented with cartesian grids, in the second step the neighborhood of the cylinder surface (up to a cylinder radius) has been divided in polar grids and the solution of the first stage in the cartesian grids have been interpolated into the polar grid. The solution of the basic equations in the polar coordinates have been continued until a convergent (steady or periodic) solution has been

obtained. The vortex generators are assumed to be of zero thickness. The numerical simulation of the vortex generator surfaces is achieved by using no-slip boundary conditions at the suitable grid points /2/.
MAC uses a staggered grid formulation in which the pressure is defined at the midpoint of the cell and the velocity components on the normal surface of the cell.
The computation with MAC proceeds in two steps. In the first step, the momentum equations are computed explicitly in order to determine the velocity field at a new time step. In the second step the continuity equation is iteratively solved by upgrading the pressure and the velocity components at each cell.Details of MAC can be found in ref. /3/.

RESULTS AND DISCUSSION

Numerical results have been obtained for Re_H = 1000 with a grid of 56 x 42 x 13 cells
Figures 1a and 1b show the allocation of the grids in the channel.
Figure 2 shows velocity vector plots at different channel cross sections. At sections 21 and 25 (see also Fig. 1b) the traces of the horse shoe vortices which have built themselves on the upper and the lower forward stagnation areas and then have bent around the cylinder in form of longitudinal vortices (fig. 2a) are noticed. At the cross section 31 to 38, the fluid passes over the winglet pair on the bottom plate. The formtion of the longitudinal vortices is quite apperent (see fig. 2b and 2c). At the 50th cross section one can see the formation of the induced vortices close to the central plane (fig. 2d and 2e). At the channel exit (55th cross section) one can see two vortices at the bottom and two induced vortices near the top (fig. 2f). The vortices near the top are closer to each other than those near the bottom. All vortices are deformed, hence the overall structure of the wake becomes very complex.
Even on the transverse midplane of the channel, a periodic vortex street has not been observed in the computed flow field. The absence of the vortex street can be atributed to the presence of the vortex generators and the damping from the top and the bottom walls. In order to reduce the damping of the channel walls, computations are carried out in a configuration where the channel height H is equal to the cylinder diameter D. Even for this case, the velocity plots at the transverse midplane fail to show a vortex street. The isobar plots show in front of the winglets (in the stagnation area) high pressure zones which hinder the flow separation on the cylinder and hence the formation of the vortex street. These phenomena are shown in figs. 3 and 4.
In fig. 3 we see the pressure distribution around the cylinder at the lowest horizontal plane of the channel. One notices the presence of a high pressure stagnation zone in the front of the winglets. This high pressure pushes the

boundary layer against the cylinder wall and avoids a separation.
In fig.4a the same effect is shown with velocity vectors. The fluid particles move along the cylinder wall and remain attached. There is no recirculation immediately behind the cylinder, but further downstream a recirculation zone appears. In fig. 4c the velocity vectors for flow in the same geometry as in fig. 4a , but without winglets are presented. Here we notice that in contrast to fig. 4a the recirculation zone reaches up to the rear surface of the cylinder.The boundary layer on the cylinder separates at approximately 100° from the stagnation point. The separation bubble is much larger for the case without winglets than the recirculation zone that appears downstream for the flow with winglets
In fig.4b we show the velocity vectors for the same case as for fig. 4a but here with H/D=1.0.
Here the influence of the longitudinal vortices appears further towards the axis. The recirculation zone is smaller and it is displaced further downstream than in the case of fig. 4a.
The cross section 35 passes trough the winglets, hence the high density of contours in fig. 5a. The cross section 39 (fig. 5b) shows isobars directly behind the winglet. One notices the vortex structure with low pressure zone in the core.

CONCLUSION

The winglets in the rear of the cylinder produce high pressure stagnation zones in the wake of the cylinder and hinder the separation of the boundary layer from the cylinder. However, recirculation zones still appear in the wake a little downstream from the cylinder. The azimuthal velocity generated by the longitudinal vortices disturb the recirculation zones and reduces their size.

REFERENCES

/1/ P. Kiehm, N.K.Mitra, M. Fiebig in Proc. 6 GAMM Conf. on numerical Methods in Fluid Mechanics, edited by D.Rues, W.Kordulle,oct. 1985,Gottingen pp.153-160.
/2/ U.Brockmeier, N.K. Mitra, M. Fiebig, in Proc. 7 GAMM Conf. on Numerical Methods in Fluid Mechanics, edited by M.Deville,Sep.1987 Louvain-La-Neuve, pp.48-55.
/3/ C.W.Hirt, B.D.Nichols,N.C.Romero, SOLA A Numerical Solution Algorithm for Transient Fluid Flow, Los Alamos Scientific Laboratory, Report, LA-5652, New Mexico, 1975.

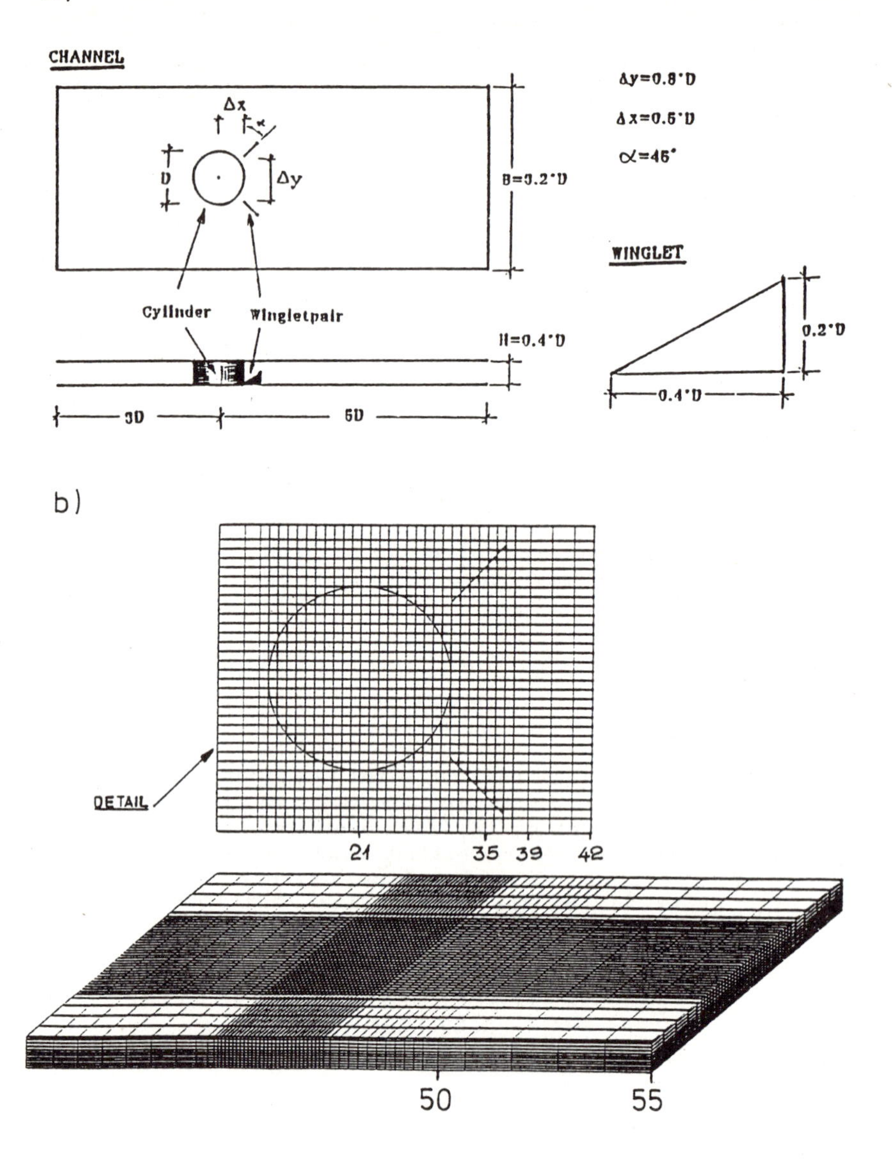

Fig. 1 Computational domain
a. Channel showing the location and the dimension of the cylinder and the pair winglets. All dimensions are normalized with the cylinder diameter D .
b. Non-equal grids in form of rectangular parallelopiped Detail shows the top view of the channel with the cylinder and the trace of the winglet.

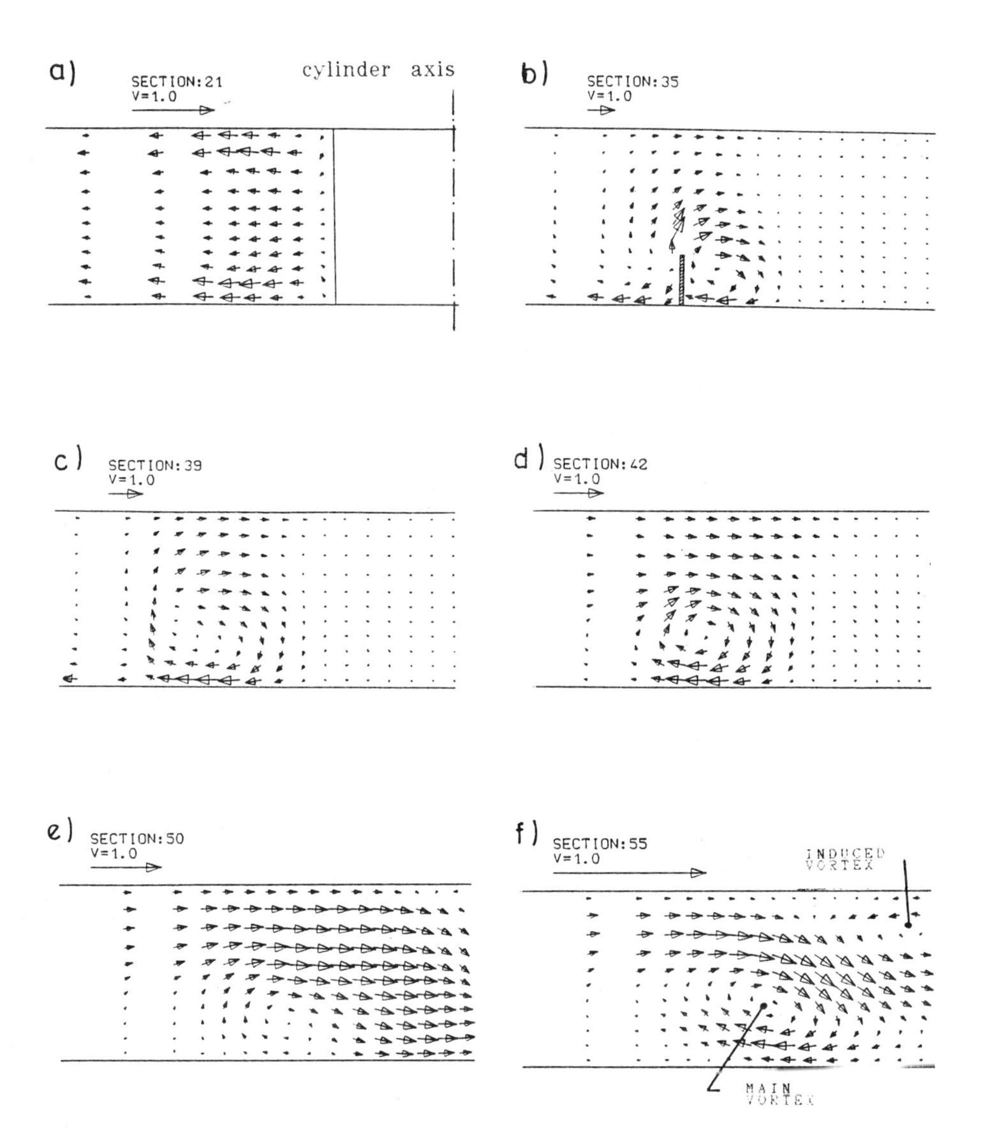

Fig. 2 Velocity vectors at different channel cross section (see also fig. 1) in an area between the cylinder axis and one diameter (D) in the lateral direction.

a. At cylinder axis
b. Across the winglet
c. and d. downstream of the winglet
e. At the 50th cross section
f. At the end of the channel

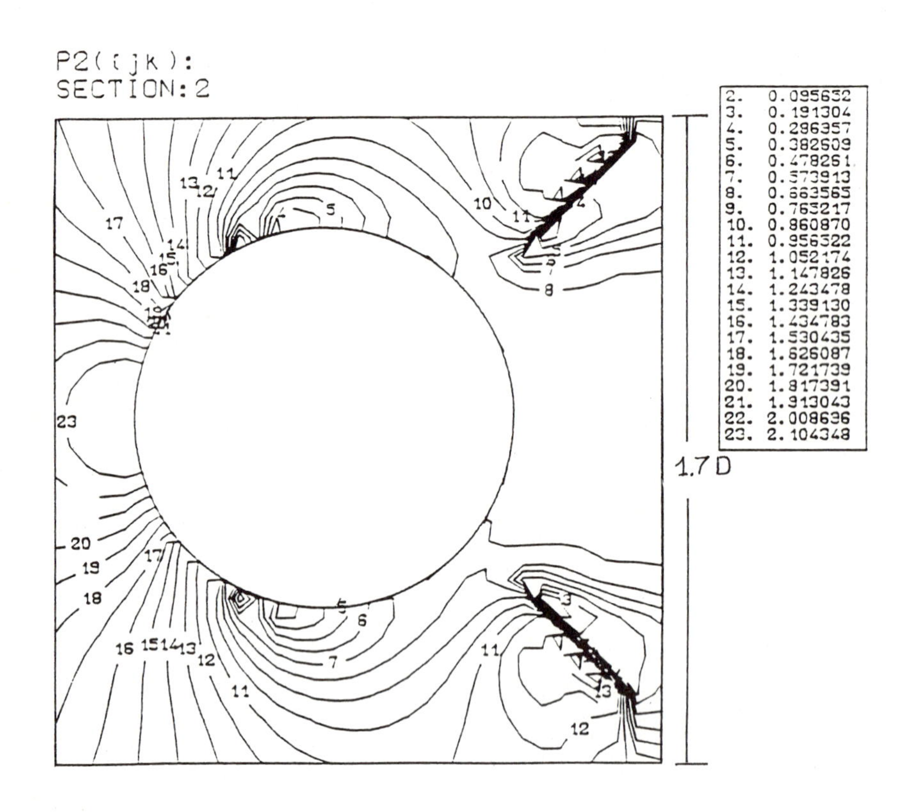

Fig. 3 Isobars on the bottom plate around the cylinder in an area of 1.7D x 1.7D .
The pressure field is symmetric on the both sides of the cylinder. Low pressure zones (isobar 3) appear on the cylinder at 80^{0} from the stagnation point where the high pressure zone (isobar 23) appears. The large density of isobar contours represents the winglet . Here also high pressure zones (isobar 13) appear on the luv side and the low pressure zones (isobar 3) on the lee side.

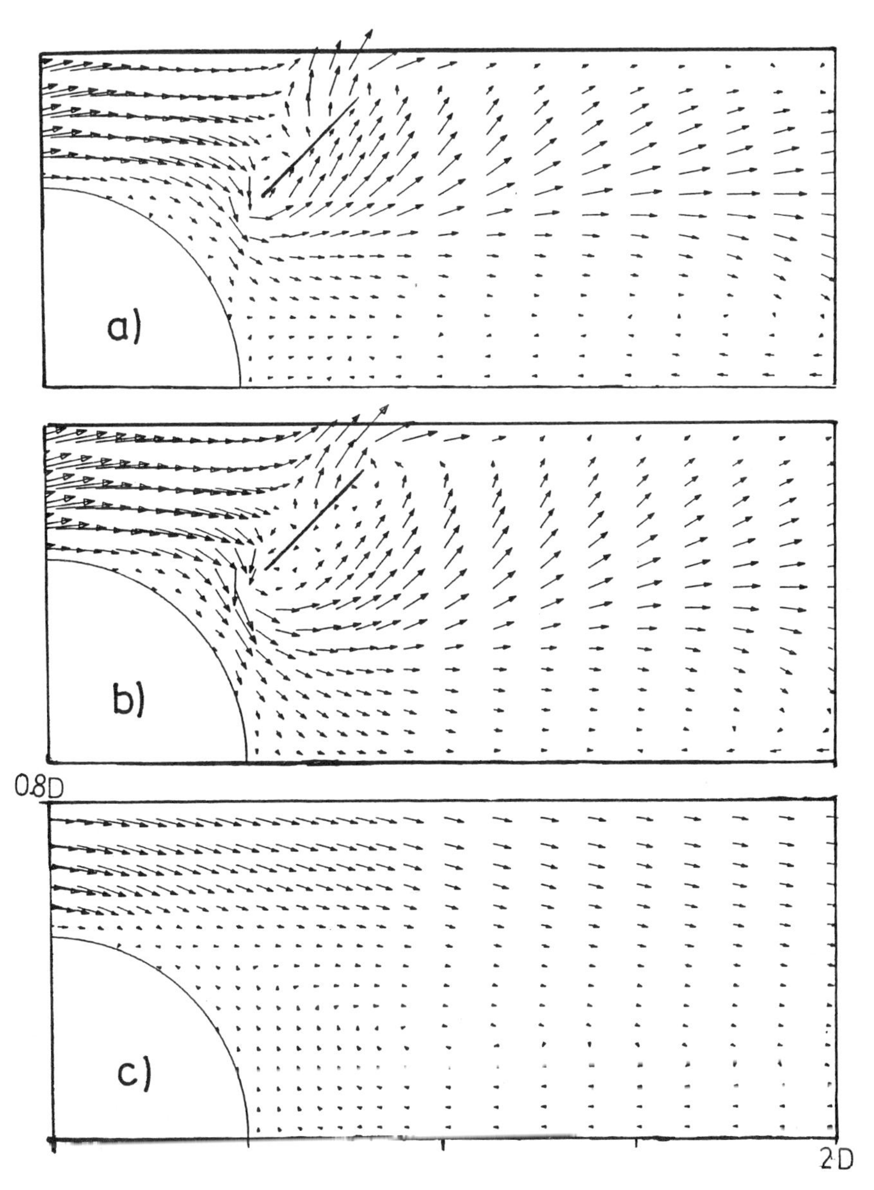

Fig. 4 Velocity vectors at lateral cross section at a distance of D/30 from the bottom plate.
a. Channel with H/D=0.4 , with winglets
b. Channel with H/D=1.0 , with winglets
c. Channel with H/D=0.4 , without winglets
Figures a. and b. show the formation of longitudinal vortices. The recirculation zones appear slightly downstrean of the cylinder, in (a) and (b) but directly behind the cylinder in (c).

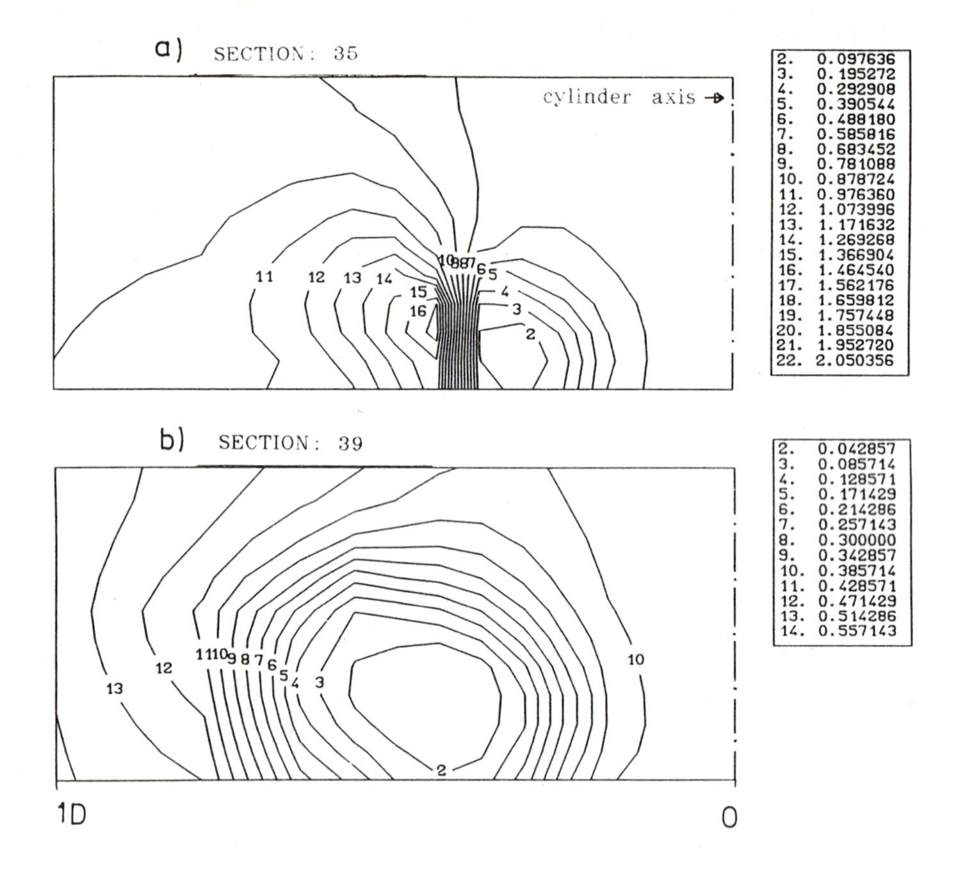

Fig. 5 Isobars at part of the channel cross sections 35 and 39 (see fig. 1).
Fig. 5a shows the large density of isobars near the winglet and the high and low pressure zones in the luv and lee sides respectively. Fig. 5b shows the isobars in the cross section of the longitudinal vortex with the low pressure zone in the vortex core.

A DYNAMIC GRID GENERATOR AND A MULTI-GRID METHOD FOR NUMERICAL FLUID DYNAMICS

G.H. Schmidt
Koninklijke/Shell Exploratie en Produktie Laboratorium
Volmerlaan 6, 2288 GD Rijswijk, The Netherlands

SUMMARY

This paper reviews development of a computational method for flow of fluids through porous media based on flexible gridding, mixed-finite-element discretisation, multi-grid and use of characteristics. The flexible gridding is based on dynamic introduction and deletion of gridpoints (with fixed position). Numerical results show the efficiency of multi-grid and (for linear convection) the very strong synergetic effect of merging this type of flexible gridding with use of characteristics. Extension to non-linear convection is discussed.

1. INTRODUCTION

The two components in convection-diffusion problems have diverging demands for computational methods. The accurate local information (on position and shape of fronts etc.) required for convection needs flexible gridding and integration along streamlines or characteristics, which has resulted in front-tracking and similar methods (see e.g. [2] and references given there). On the other hand, diffusion involves in principle an infinite velocity and therefore needs efficient computational methods for the solution of coupled equations, which has resulted in computations on more or less regular, usually fixed, grids (see e.g. [4] and references given there).

Startingpoint for this approach is the flexible gridding which tends to be a compromise between flexibility and regularity as pointed out in [1]. It consists of a regular, orthogonal, non-equidistant base grid, which is refined and unrefined dynamically by the basic refinement, which divides a gridblock into four (2D) or eight (3D) identical smaller ones.

Merging this gridding with the use of characteristics makes a drastic reduction of numerical diffusion, as compared with direct discretisation on fixed grids, possible at least in principle. Theoretical considerations for the case of a linear convection term predict the numerical diffusion to be proportional to meshwidth**2/timestep, whereas for e.g. one-point upsteam discretisation this diffusion is proportional to the meshwidth. Moreover, there is no Courant-Friedrichs constraint on the timestep, which is almost a must at locally refined grids.

The current program is restricted to the (2D) convection-diffusion (c.d.) equations stated in section 2. To solve these equations, we use two tools: an efficient data-stucture for locally refined grids, and a multi-grid technique to solve (an inhomogeneous, anisotropic generalisation of) Helmholtz equation on locally refined grids. The 2D forms of these tools have been reported in [5,6] and their 3D forms will be reported in [7]. Since these tools can also be applied to other, more general, equations, this should be considered as a model for a more general program.

For easy reference, we give in sections 3 through 5 a description of the tools just mentioned. In section 6 the execution of a timestep for the c.d. equations is given in detail. In each timestep two different updates of the grid are applied, one before the handling of convection and one after. Multi-grid is used separately three times: for the computation of the total flow field, for a conservation-correction and for handling the diffusive effects.

The equations in section 2 have a linear convection term. Extension to non-linear convection is discussed in section 8.

2. CONVECTION-DIFFUSION EQUATIONS

The equations for which we give numerical results are:

$$\nabla . \vec{u} = q; \qquad \vec{u} = - \vec{\vec{k}} \; (\nabla p - \rho \vec{g}) \tag{2.1a,b}$$

$$\partial s / \partial t + \nabla . \vec{v} = q' + \nabla . (\vec{\vec{D}} \; \nabla s) \; ; \quad \vec{v} = s \vec{u} . \tag{2.2a,b}$$

The quantities in these equations can be interpreted as:

- $\vec{u}$: total mass flow, i.e. sum of all flowing material,
- q : source term for the total mass due to wells and/or precipitation,
- $\vec{\vec{k}}$: permeability tensor,
- p : pressure,
- ρ : mass-density of fluid,
- $\vec{g}$: acceleration due to gravity,
- s : mass-fraction of fluid of a specific component; to be called concentration henceforth,
- $\vec{\vec{D}}$: diffusivity tensor,
- $\vec{v}$: specific mass flow,
- q' : source term for the specific component.

In addition to (2.1,2) we assume no-flow boundary conditions and given values of s as initial conditions.

We consider p, $\vec{u}$, s and $\vec{v}$ as <u>primary unknowns</u> and all other quantities in (2.1,2) may depend algebraically on the primary unknowns and on time- and space-coordinates.

3. DATA-STRUCTURE AND FLEXIBLE GRIDDING

In the <u>Euler-approach</u> to flexible gridding (fixed gridpoints, dynamically switched "on and off") an efficient data-structure is essential. We give here a very short description of our FORTRAN code for the data-structure. It is constructed to be favourable for the coding of multi-grid and of mixed-finite-element (m.f.e.) discretisation.

In view of the mixed finite elements, <u>both blocks and interfaces</u> of the grid are numbered (index i <u>for blocks</u> and j for <u>faces</u>) and stored. All blocks and faces in the base grid (see introduction) have <u>level</u> 1 and the basic refinement generically creates blocks and <u>faces</u> one level higher. The numbering of blocks and faces is in order of increasing level. Pointers are used to code the tree-structure, induced by the basic refinement, for both blocks and faces. The geometric structure is coded by pointers from blocks to its neighbouring faces and reverse.

At this point we make distinctive the discretisation grid and the series of coarser grids obtained from that grid by deletion of the blocks of highest level. The first one serves to discretise the equations; the latter ones occur only in the multi-grid process. The discretisation grid usually is irregular and, moreover, dynamically changed by (un-)refinement during simulation time.

Faces between blocks of different level are called green faces. The composition of green faces making up one side of a block is called a green side.

A dynamic update of the grid (introduction and/or deletion of blocks and faces) implies an update of the numbering and of the pointers. The CPU time used for these operations in our (2D and 3D) codes is minute as compared with other computational work in the program.

4. THE MIXED-FINITE-ELEMENT (M.F.E.) DISCRETISATION

All quantities in the c.d. equations are scalar, vector or tensor fields. Here we give the discretisation of the scalar and vector fields; for the tensor fields and also for details we refer to [5,6]. The discretisation essentially is the one of lowest order in the class of Raviart and Thomas, see [3], applied to the grids introduced above.

The approximation of scalar fields is blockwise constant. Hence the value of the field in a block i is determined by one value, denoted by a subscipt i.

For the approximation of a vector field we introduce values, with subscript j associated with face j, representing the normal component of the vector. This normal component is continuous and constant along the face. The value of the approximation in a point in a block is obtained component-wise by linear interpolation in the direction of the component. Consequently, the tangential components at the faces are discontinuous. However, the divergence of the approximation exists in the classical sense.

At the green sides we may impose the green constraint, which states that all faces making up one green side have the same value. Only if the green constraint holds, the approximation of the vector fields is block wise a first order polynomial.

With these approximations, the discretisation of equations involving only the divergence is straight-forward. The discretisation of e.g. (2.1a) is:

$$\sum_j (\pm)_{ij}\, h_j\, u_j = a_i q_i , \qquad (4.1)$$

where the summation is over the faces neighbouring block i, and:

$(\pm)_{ij}$ = +1, if face j is a positive neighbour of block i,
-1, if face j is a negative neighbour of block i,

h_j = meshwidth of face j,

a_i = area of block i.

The discretisation of the gradient is more complicated. It is not given here, but see [5,6].

5. MULTI-GRID

In this section we specify precisely which equations are solved by multi-grid. How this is embedded in the complete solution process is described in the next section.

The multi-grid process solves the scalar field ϕ and the vector field $\vec{u}$ from the m.f.e. discretisation of the equations:

$$c\phi + \nabla.\vec{u} = q; \quad \nabla\phi + \vec{\vec{w}}\vec{u} = \vec{g} \tag{5.1a,b}$$

where c is a non-negative scalar field and $\vec{\vec{w}}$ is positive-definite tensor field. All coefficients in (5.1) are arbitrary functions of the space-coordinates. Boundary conditions are:

$$\phi + a\ \vec{u}.\vec{\nu} = \beta \tag{5.2}$$

on part of the boundary and given values of $\vec{u}.\vec{\nu}$ on the remaining part. Here $\vec{\nu}$ is the unit outward normal and a is non-negative.

The efficiency of the computational process, on locally refined grids and for extreme values of the coefficient-functions, has been discussed in [5,6].

Equivalent to solving the boundary value problem is minimising the functional

$$F = \int\{1/2c\phi^2 + (1/2\vec{\vec{w}}\vec{u}-\vec{g}).\vec{u}\} + \int(1/2a\vec{u}.\vec{\nu} - \beta)(\vec{u}.\vec{\nu}) \tag{5.3}$$

within the constraint of (5.1a) and the given values of $u.\nu$. The boundary integral in (5.3) is over the part where (5.2) holds.

Equations in the form (5.1,2), or in the form (5.1a,3), will be solved at a number of places in the numerical solution of the c.d. equations. All variables in this section are dummy variables which will represent various quantities.

6. EXECUTION OF A TIMESTEP

The evolution of the c.d. equations is computed in discrete timesteps, which are descibed in this section. Between timesteps it is necessary to store the data-structure of the grid and the concentrations. Pressures and total flows may be stored in order to speed up their numerical computation in the next timestep.

The complete timestep consists of the six operations described in subsections 6a through 6f, executed sequentially. The size of a timestep Δt should be based on accuracy, but is considered as given here.

6a. Computation of the total flow field.

The current values of the concentrations are substituted into the sourceterms and permeabilities in (2.1). After multiplication of (2.1b) by $\vec{k}^{-1}$, the equations are of the form (5.1). Hence p and $\vec{u}$ are computed from (2.1) using multi-grid.

6b. Pre-adaptation of the grid.

Each timestep involves two updates of the grid; one as described here and one as described in subsection 6e. Both updates aim at accurate representation of the fields, while economizing on blocks (i.e. memory-space and computational work). This one anticipates on the shift due to convection; the next one mainly deletes blocks at the old position of fronts and other singularities in the solution.

Both updates are done at fixed simulation time. Hence the transfer of fields (e.g. concentration) from the old grid to the new one is trivial for blocks (and faces) that are not changed, is some interpolation for blocks (and faces) created by refinement, and is some averaging for blocks (and faces) created by unrefinement. The interpolations and averages should be such that conservable quantities are conserved (see also 6d).

The computational scheme for the update of this subsection is as follows. Starting at the center of a gridblock of the old grid, we integrate along the streamlines $\vec{u}$:

$$d\vec{x} = \vec{u}\, dt \; ; \qquad t \rightarrow t + \Delta t \; , \tag{6.1}$$

where $\vec{x}$ represents the vector of space coordinates. The endpoints of these integrations form a collection of spacepoints. With each point in the collection we associate a blocksize, equal to the blocksize at the beginning of the integration. The gridupdate applies refinements to the grid till the size of each block is less than all sizes associated with points contained in it.

We make a remark concerning the integration (6.1). As stated in section 4, $\vec{u}$ is blockwise a first-order polynomial in the space coordinates. Hence the integration can be done blockwise analytically. Consequently, the integration is efficient and without loss of accuracy.

In case of large values of Δt the collection of points is possibly scattered too much. Then it might be better to start integration not only at block centers, but at more points per block.

An example of this update is given in Fig. 1: a. is the grid before this update; b. is the grid after.

Note that this update does only refinements. The update of subsection 6e does both refinements and unrefinements.

6c. Handling of the convection term.

We now describe the integration of (2.2) over Δt with disregard of the diffusion term and with time-independent values of $\vec{u}$. The linearity of the convection term allows us to integrate along the characteristics, hence we solve (6.1) extended with:

$$ds = (q' - s\nabla.\vec{u})\, dt \; . \tag{6.2}$$

Note that the integrand at the r.h.s. of (6.2) is blockwise constant, hence integration can also be done blockwise analytically.

The initial values for the integration are computed as follows. Initial space-coordinates are chosen such that the integration (6.1) ends in a block center. This is achieved by integrating (6.1) "backwards" from $t + \Delta t$ to t. The initial space-coordinates do in general not coincide with a block center. The associated (initial) s-value is obtained by interpolation, which is e.g. done by tessellation of the region into triangles and quadrilaterals, the block centers acting as vertices. The s-values for the vertices (equal to the values associated with the blocks) are interpolated over the triangles and quadrilaterals.

6d. Conservation-correction.

Equation (2.2a) is in the form of a conservation equation. Integration over the computational region and over the timestep gives with discretisation as in section 4:

$$\sum_i a_i \left((s_i - s_i^{old})/\Delta t - q_i' \right) = 0 \ , \tag{6.3}$$

where the superscript old refers to the previous timelevel. As a consequence of integrating along the characteristics, equation (6.3) is not satisfied exactly after the operation of subsection 6c. In this subsection we describe a correction to mend this.

A conservative way to solve (2.2a) (with disregard of diffusion) is to solve the m.f.e. discretisation of

$$(s - s^{old})/\Delta t + \nabla.\vec{v} = q' \ . \tag{6.4}$$

The concentration values computed in subsection 6c, henceforth denoted by $\tilde{s}$, serve as a "prediction" for the solution of (6.4). Similarly we compute a "prediction" $\tilde{\vec{v}}$ for $\vec{v}$. We then compute the solution of (6.4) "nearest" to the predicted values. This induces the following computational scheme.

The predicted value for $\vec{v}$ follows from

$$df = s^{old}\, dt/\Delta t \ , \qquad \tilde{\vec{v}} = f\vec{u}. \tag{6.5a,b}$$

At the integration of (6.5a), s^{old} is evaluated at space-points along the streamline (6.1). Since we need a value for v for each face j, we integrate here over the characteristic ending at the center of face j. Initial value for f is zero and the final value of f is substituted in (6.5b). Equation (6.5) holds since $\Delta t v$ is the transport of specific fluid through face j, and this transport equals u times the time-integral over the concentration at the face, which is expressed in the concentration at the previous timelevel by going back along the streamline.

The corrected concentration values are the solution of (6.4) which minimises

$$\int \{ (s - \tilde{s})^2 + \gamma(\vec{v} - \tilde{\vec{v}})^2 \} \ , \tag{6.6}$$

where γ is a weigh-factor. These equations have the same form as (5.1a) and (5.3), so the corrected values are computed by multi-grid.

If we set $\gamma=0$, the predicted values for $\vec{v}$ are disregarded and the correction on the concentrations is a shift over a constant value. With very large values of γ we disregard the predicted values for s. The corrected concentrations are computed fully on basis of $\nabla.\vec{v}$, which results in instability, as appears from numerical results. Optimal values for γ must be determined experimentally.

6e. Post-adaptation of the grid.

The update of the grid in this section refines and unrefines blocks such that finally blocks are small where the fields have strong gradients and are large where the fields are smooth. The following simple strategy to achieve this seems to work well in practice.

For each block an "error-indicator" e_i is computed, which equals the variation of the concentration over block i, as computed by some interpolation with neighbouring blocks. Blocks with $e_i < \epsilon$ are deleted by unrefinement (if the geometry of the grid allows so) and blocks with $e_i > 2\epsilon$ are refined. The value of the tolerance ϵ must result from the balance between desired accuracy and reasonable computational work (number of blocks). Each (un-)refinement induces transfer of fields as described in subsection 6b.

Fig. 1 illustrates also this update; b. is the grid before; c. is the grid after.

6f. Handling of the diffusion term.

Finally we incorporate the effect of diffusion. The effect of diffusion is a correction which smoothes the concentration. The corrected values are computed from the m.f.e. discretisation of:

$$(s - s')/\Delta t + \nabla.\vec{v}' = 0; \qquad \overset{\Rightarrow}{D}(\nabla s) + \vec{v}' = 0, \qquad (6.7a,b)$$

where s' is the concentration after the operation of subsection 6e, and where $\overset{\Rightarrow}{D}$ is evaluated at that value. After multiplication of (6.7b) by $\overset{\Rightarrow}{D}^{-1}$, the equations are of the form of (5.1), and hence the concentrations are computed from (6.7) by multi-grid. This correction is conservative as follows from (6.7a).

7. NUMERICAL PERFORMANCE

Here we give some indications on performance, which follow from the first numerical experiments done with a FORTRAN code executing the program described above on a VAX 8800.

Fig. 1 indicates the typical grids used. The coarsest grid (level 1) is 2 by 2 and the maximal refinement level is 7. On the average over the whole simulation the grid before pre-adaptation consists of about 1600 blocks, mainly of level 7. The grid in between pre- and post-adaptation (Fig. 1b) usually contains twice as many blocks.

Covering the whole computational domain with the mesh of level 7 would result into 16384 blocks. The gain in accuracy would be minute, since the fine mesh in the adaptive grid essentially covers the whole transition region of the concentrations.

The computing times required by the six operations, described in the subsections 6a through 6f, are of the same order. The total CPU time amounts about 1 minute per timestep. Recall that the use of the characteristics allows large values of the timestep, such that a complete simulation consists of e.g. 10 timesteps.

8. EXTENSION TO NON-LINEAR CONVECTION

A non-linear convection term implies $\vec{v} = f_{(s)}\vec{u}$ instead of (2.2b). To compute the prediction for s we now integrate

$$d\vec{x} = f'\vec{u}\,dt, \quad ds = (q' - f\nabla.\vec{u})\,dt, \qquad (8.1a,b)$$

and for the predicted value for v we integrate

$$df = f_{(s^{old})}\,dt/\Delta t\ , \qquad (8.2)$$

in combination with (8.1a) and (6.5b). Here f' denotes the derivative df/ds. Note that the space-time path of the characteristic changes, but not its space path. The non-uniqueness due to crossing space-time pathes inhibits direct application of this method. We discuss here briefly two distinctive ways to incorporate non-linearity into the method.

8a. Approximating linear convection term.

The function f is writen as:

$$f_{(s)} = \lambda s + \tilde{f}_{(s)}\ , \qquad (8.3)$$

where λ is a constant parameter to be chosen optimally. The operation of subsection 6c is executed with disregard of the non-linear term in the r.h.s. of (8.3), which results into concentration values s'. A second operation deals with the non-linear term by solving

$$(s - s')/\Delta t + \nabla.\vec{v}' = q', \quad \vec{v}' = \tilde{f}_{(s)}\vec{u}. \qquad (8.4a,b)$$

These two operations induce a shift of the front and a deformation of the front respectively. The shift is done practically without numerical diffusion. There will be numerical diffusion induced by the operation (8.4), but we expect this to be much less than the diffusion associated with (8.4) applied to the full convection term (f in the r.h.s. of (8.4b)). This expectation stems from the localized action (no shift) of the second operation. Effective computational techniques (discretisation and solution) for (8.4) on the locally refined grids still have to be developed.

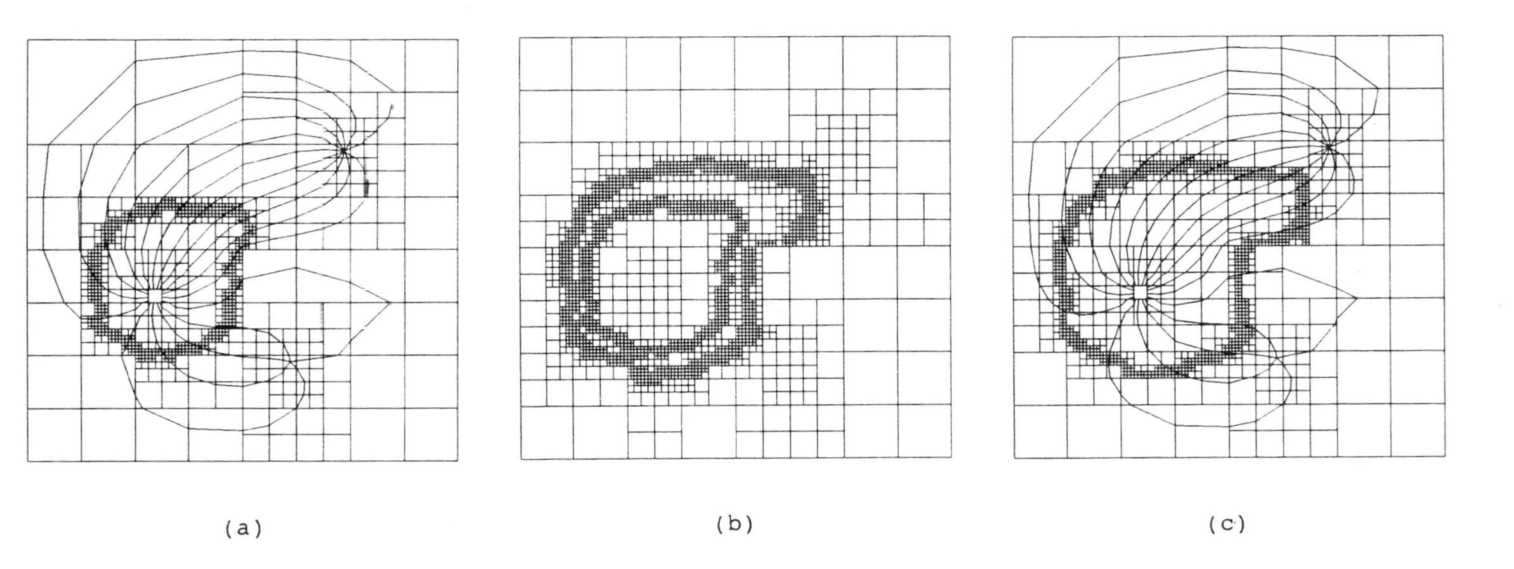

Fig. 1: The discretisation grid before pre-adaptation (a), between pre- and post-adaptation (b), and after post-adaptation (c). Streamlines (from the injection well at the left to the two production wells at the right) are plotted in (a) and (c), but not in (b). In these plots the streamlines are represented by blockwise straight lines. This is in contrast with the computational work, where they are smoother, see subsection 6b.

Computation of the total flow field and handling of the diffusion term is executed on grid (a) (and (c)); handling of the convection term and the associated conservation correction is done on grid (b).

These figures don't give concentration values. Yet, the position of the front can be speculated from the position of the fine mesh. Note that the shift of the front in (c), as compared to (a), varies from about 5 to 20 times the meshwidth of the fine grid.

8b. Iterative execution of the time step.

The idea outlined in this subsection is based on the fact that in analytical methods for convection equations, the non-uniqueness due to crossing characteristics is removed such that global conservation holds. Therefore we expect that iterative application of the (modified) operations of subsections 6c and 6d will converge to the correct solution.

The modification of the operations consists of making the evaluations of s and f in (8.1,2) unique. There are three aspects in which we have to make choices:

- integration along the characteristic forward or backward,
- choice of the value of the weight-factor γ, and
- evaluation at the old or (in the second and subsequent iterations) new timelevel.

Numerical experiments are indispensible for a good choice at these alternatives.

REFERENCES

[1] A. Brandt, "Multi level adaptive solutions to boundary value problems", Math. of Comp., 31, 138, 1977, pp. 333-390.
[2] J. Glimm, B. Lindquist, O. McBryan and L. Padmanabhan: "A front tracking reservoir simulator, five-spot validation studies and the water coning problem", in: "The mathematics of reservoir simulation", R.E. Ewing ed., SIAM, Philadelphia, 1983.
[3] P.A. Raviart and J.M. Thomas: "A mixed finite element method for 2nd order elliptic problems", Lecture Notes in Math. Vol. 606, Springer-Verlag, New York/Berlin, 1977, pp. 292-315.
[4] T.F. Russel and M.F. Wheeler: "Finite element and finite difference methods for continuous flows in porous media", in SIAM, see [2].
[5] G.H. Schmidt and F.J. Jacobs: "Adaptive local grid refinement and multi-grid in numerical reservoir simulation", J. Compt. Phys., Vol. 77, No. 1, 1988, pp. 140-165.
[6] G.H. Schmidt "Improvements in a multi-grid method for locally refined grids", to appear in J. Compt. Phys.
[7] G.H. Schmidt and E. de Sturler: "Multi-grid on locally refined 3D grids", to appear.

SOME ASPECTS OF THE NUMERICAL SIMULATION OF COMPRESSIBLE VISCOUS FLOW AROUND BLUFF BODIES AT LOW MA-NUMBER

B. Schulte-Werning, U. Dallmann, B. Müller
DLR, Institute for Theoretical Fluid Mechanics
Bunsenstr. 10, D-3400 Göttingen, FR Germany

SUMMARY

A numerical code of the Beam-Warming type is used to simulate the two-dimensional symmetric form of the flow around a circular cylinder and a sphere. The calculations are based on the viscous compressible equations of motion. The first aim is an analysis of the solution behaviour of the algorithm in the range of medium Re-numbers and low Ma-numbers to set up the limits for the simulation of steady separation structures. We investigate the influence of initial conditions and small compressibility on the solution. Also we are looking to what extent we can derive stable and steady solutions from the unsteady procedure and how do we have to resolve large gradients in the flow field. The second aim is to study the structure of the separation at higher *Re* in the axisymmetric sphere case. A comprehensive comparison is done with other numerical and experimental results.

1. INTRODUCTION

In order to investigate the development and the physical mechanisms which underlie three-dimensional separation behind bluff bodies, it is necessary to choose a test configuration which allows for successive reduction of imposed temporal and spatial symmetry conditions. According to that it is possible to take into account more and more complex separation phenomena in a step by step manner. Furthermore for the comparison and beyond this for the validation of any two simulations it is necessary that the sequences of topological changes that occur in the fields of physical variables are identical in both numerical and/or experimental simulations (Dallmann & Schulte-Werning [4]). Hence, we simulate flows around the geometric most simple configurations where such sequences of spatial and temporal flow changes occur. We choose the flows around a circular cylinder and around a sphere as basic configurations for our investigations of separated flow phenomena. With these geometrical configurations the bifurcation from two-dimensional, steady forms of flow to three-dimensional, steady or unsteady ones under symmetric boundary conditions can be examined successively.

Here the essential results of a numerical simulation of the circular cylinder flow with forced symmetry and the axisymmetric sphere flow are presented, for further details see Schulte-Werning et al. [16]. One of the aims in our present investigation is to clarify to what extent a numerical code, which had been developed for aerodynamic application of calculating supersonic high Re-number flows around blunt bodies including flow separation is able to resolve those local flow properties required for the physical modelling of complex separated flows and their local and global topological changes. The above mentioned purpose can be expressed in the following questions which will be tackled in the present investigation: **a)** Under what conditions can we derive stable, steady state solutions in the time-asymptotic limit by use of time-de-

pendent algorithms? **b)** How does such a steady or the obtained unsteady solution depend on initial conditions? **c)** What is the effect of small compressibility on the solution as the Mach number drops to smaller and smaller values? **d)** To what extent does one have to resolve large gradients in the separating shear layers in order to describe the physically correct vorticity dynamics?
The second aim, especially in the sphere case, is a study of the development of the steady separation structure with increasing Re-number. This was done as a preliminary work for investigating the bifurcation from two- to three-dimensional separation which is strongly associated with the break-up of axisymmetric closed separation bubbles.

2. MATHEMATICAL TREATMENT

The present calculations are based on the two-dimensional viscous equations of motion for a thermal and caloric perfect gas in non-dimensional conservative form. A transformation is used which maps the physical onto the computational domain (fig. 1). The dimensionless parameters governing the flow field (Re, Ma, Sr, Pr) are definded in the usual way where the cylinder resp. sphere diameter is the characteristic length. In the following we assume that Stokes' hypothesis and the Sutherland law hold, the Pr-number is set to 0.723 and the reference temperature is $293K$.
For the subsonic flow configuration investigated here (fig. 1) the boundary conditions at the in- and outflow are derived from the admissible compatibility equations. Therefore at the inflow density and velocities are prescribed whereas the pressure is extrapolated and at the outflow the pressure is set to its freestream value and density and velocities are extrapolated. At the symmetry line use is made of the fact that density, u-velocity and pressure are even functions and v-velocity is an odd function of the y-coordinate which in the axisymmetric case represents the radius. On the body surface the no-slip condition holds and the pressure is derived from the wall-normal momentum equation. In addition the adiabatic wall condition is used. As initial condition the impulsively started body is employed where the initialisation is done with the freestream quantities, and in addition for the cylinder flow a special form of a step function for $Re = Re(t)$ is tested, too.
To solve the equations of motion numerically the implicit factorized finite-difference scheme of Beam and Warming is employed which is second order accurate both in space and time. Fourth-order damping terms are added explicitly to damp high frequency oscillations, second-order damping terms are added implicitly to extend the stability range of the algorithm. Furthermore a non-linear second-order difference of the conservative variables sensed by the discretized second derivative of the pressure and scaled by the ratio of local and freestream Ma-numbers can be inserted to damp strong gradients. In the following calculations these non-linear damping terms are switched off, unless otherwise mentioned explicitly. This numerical code has been used up to now for the simulation of supersonic flow with good agreement in comparison with experimental and other numerical data (e.g. Müller [12]).
In the cylinder case a grid of C-type is used where the inflow resp. outflow boundary is put 30 diameters upstream and 100 diameters downstream of the body. Thereby the influence of the boundary distance on the solution could be kept small. In the wall-normal resp. wall-tangential direction 81×201 grid points are employed, where 91 points define the cylinder. In the sphere case a grid of O-type with 111×111 points is used where the outer boundary is put 120 diameters away from the body. For a better resolution of the physical situation the grid points are exponentially clustered in the boundary layer. In addition, in the cylinder case grid clustering was

necessary near the rear stagnation point and in the sphere case near the rear symmetry line.

3. RESULTS AND DISCUSSION

3.1 Circular cylinder flow with forced symmetry

It is wellknown that the incompressible flow equations can be obtained by dropping the freestream Ma-number in the compressible flow equations to zero. For the comparison of numerical solutions of the compressible equations of motion with those of the incompressible equations it is necessary to reduce the freestream *Ma*-number in the calculations to values where the compressibility effect is negligible. Unfortunately in the discretized form often the solution procedure will diverge in such a case.
On the other hand the rate of convergence to a steady state solution in most cases is greater and thus computation time is smaller if one could use *Ma*-numbers as high as possible in the subsonic range. Hence, a lower bound for *Ma* is given by convergence and CPU-time requirements and an upper *Ma*-bound is imposed by the condition that compressibility should not affect the solution. Thereby the smallest stable computable *Ma*-number depends on the simulated flow problem and the numerical method. In the present case for the circular cylinder flow, which is a blunt body problem, the solution diverged below $Ma = 0.25$, and the numerical instabilities grew faster (measured in dimensionless time) for decreasing *Ma*-number. In contrast to this *Ma*-limit additional calculations of the flat plate problem led to stable solutions for $Re = 100$ and $Re = 10^7$ at low *Ma*-values such as $Ma = 0.005$ (!). Thus we conclude that for a numerical calculation of the flow around slender as well as blunt bodies the minimal, numerically stable *Ma*-number will depend on the appropriate thickness parameter and will be situated between the above stated limit values.
Fig. 2 shows the dimensionless length of the wake bubble L/D measured from the rear stagnation point as a function of the *Re*-number. The symmetry condition stabilizes the wake which in experiments is physically unstable for $Re > 40$. The straight line connects the results of Fornberg [6], [7]. He solved the steady incompressible equations of motion and his solution gave an $L/D = O(Re)$ relation. In the comparison to the results of other authors it is a striking feature that the deviations are evident in that Re-range where the physical instability of the wake is artificially suppressed by imposing symmetry. It should be noted that no other author was able to support Fornberg's solution, only the data of Ghia et al. [9] show a certain tendency to fit his solution (basic equations time-dependent (TD) or steady (ST)). The present compressible calculations show the maximum for L/D at $Re \approx 300$ and then a decrease to a constant value up to $Re = 2000$. In the legend to fig. 2 a principal sketch of the wake bubble is also shown. Fornberg computed an increase of the wake width W near $Re = 300$ from $W/D = O(\sqrt{Re})$ to $W/D = O(Re)$, and thus a significant thickening for higher *Re*. The present compressible solution likewise shows a wake thickening near $Re = 300$ but with nearly constant values for higher *Re*. Fig. 3 shows the development of the total drag coefficient c_D as a function of *Re*. Except in the case of very small *Re* the drag shows only very small sensitivity to compressibility and the good agreement with the data of this integral quantity of other authors is quite astonishing. For $Re > 600$ no further data were available, the present calculations exhibit an asymptotically decreasing drag the limit of which seems to be near $c_D = 0.2$.
Although the flow behaviour near a point of flow separation is unaffected by local compressibility and temperature variation (Oswatitsch [13]), compressibility effects

may enter the solution in a twofold way. On the one side compressibility influences the pressure field of the wake due to varying local Ma-numbers and on the other side this affected pressure field may have a feedback on the separation point on the body surface itself. Fig. 4 presents the geometrical ratio length to width L/W of the wake bubble as a function of Ma for $Re = 600$. The flow was simulated for $Ma = 0.25/$ $0.40/$ $0.50/$ 0.54 and the strong nonlinear dependency of L/W is obvious. According to Moore & Pullin [11] the axis ratio a/b of the elliptical core of a compressible vortex is strongly dependent on the square of the Ma-number.
Also assuming a quadratic relation for L/W our investigation gives the dependency

$$\frac{L}{W} = 4.60 + 12.82 \cdot (Ma - 0.211)^2 \tag{1}$$

which exhibits its minimum value for L/W at $Ma = 0.211$. This low Ma-number flow was impossible to be calculated with the present compressible code as mentioned before. In fig. 4 there is also shown the value for L/W of Fornberg's incompressible solution [7] which exhibits a distinct deviation from our solution. Anyway it should be kept in mind that L and W alone are essentially higher in that incompressible case. To clarify the solution behaviour for $Ma \rightarrow 0$, the steady state solution $Ma = 0.25$, $Re = 600$ was given as input into an incompressible code (Strykowski [18]) and under elimination of the total energy the solution was advanced as an incompressible one. Unfortunately the bad convergency in this case forced us, with respect to computation economy to stop the calculation without being sure of having reached the new steady state. The (now incompressible) value of L/W is listed in fig. 4, too. A convergence to the value of Fornberg is not established. The cause for the strong variation of the results of different numerical investigations is difficult to term. The differences in the basic equations used, the boundary conditions, the solution algorithms and computational grids are great and their influence on the solution must be investigated elsewhere.
As seen from fig. 5, where the surface pressure c_p, the surface vorticity ω_0 and in addition their wall-tangential derivatives are shown for $Re = 600$ the physical situation on the cylinder is nearly unaffected by compressibility. Especially the development of the derivatives of these quantities in the separation point can be regarded as identical for different Ma, which is in full agreement with Oswatitsch [13]. Thus for $Ma = 0.25/$ $0.40/$ 0.54 the separation angle $\theta_s = 91.8°/$ $92.4°/$ $92.7°$ differs only slightly, Fornberg [7] quoted $\theta_S = 94°$ and Patel [14] $\theta_S = 91°$ for their incompressible flow solutions. It should be noted that the numerical stagnation pressure p_{stag}/p_∞ differs from the analytical value only by 0.2%/ 0.2%/ 0.1%.
The former results were obtained with impulsive start as initial condition and up to $Re = 2000$ a steady state was achieved. In the cylinder case a step function for $Re = Re(t)$ was also tested to investigate the solution dependency on different initial conditions. The impulsively started cylinder flow for $Ma = 0.40$, $Re = 600$ was calculated until a steady state was reached and the obtained solution then was advanced by setting $Re = 1090$ and calculated up to a new steady state, the same was done for $Re = 1500/$ $1750/$ 2000. This form of initial condition is similar to the one in experiments where the freestream velocity is reached by a constant acceleration. Up to $Re = 1750$ the new steady state solution was nearly identical with the one of the impulsively started cylinder. But the jump $Re = 1750 \rightarrow 2000$ gave rise to a time periodic behaviour thus indicating a steady-unsteady bifurcation of the solution. Fig. 6 shows the u-velocity development at a fixed point in the separation zone as a function of time after this jump (a) and the corresponding Fourier spectrum (b), clearly exhibiting the behaviour of a self-exited nonlinear oscillator with dominating

Strouhal number $Sr = 0.4$. The area where considerable unsteadiness appears is the separation zone near the cylinder apex while the wake bubble remains nearly uninfluenced in its geometrical shape. Beyond $Re = 2500$ only periodic separation at the cylinder apex occurred, independently of the initial condition used. From this we conjecture that the separating shear layer on top of the cylinder, where two boundary layers with vorticity of opposite sign merge, is hydrodynamically subcritically unstable in the range $1750 < Re < 2500$ and in this case the disturbances introduced by the step function in Re which are finite in amplitude, may cause a steady-unsteady solution bifurcation due to nonlinear effects. For the complex topological structure of the steady separation for $Re = 2000$ see chapter 3.3.
For $Re = 10^4$ and 10^5 a non-convergence breakdown of the solution for the impulsively started cylinder was observed. This may be connected to an inadequate resolution of the complex flow structure accompanied by large gradients in the thin shear layer along the surface. Fig. 7 shows the instantaneous streamlines and vorticity contours at $Ma = 0.30$, $Re = 10^4$ for different time steps, the area of the later occuring convergence breakdown is marked by an arrow. Fig. 8 shows wall pressure and vorticity as well as their wall-tangential derivatives for $t = 1.48$. Especially the vorticity gradient increases strongly thus overcharging the resolution ability of the algorithm. The breakdown could be avoided by switching on the nonlinear damping of second order. After having advanced the solution beyond the critical transient phase it was even possible to switch off again the nonlinear damping without endangering the solution stability. But in this case no statement is possible about the influence of temporarily used damping terms on the character of the obtained periodic solution.

3.2 Axisymmetric sphere flow

The sphere flow was numerically simulated in the range $50 \leq Re \leq 5000$ with $Ma = 0.40$ fixed. Although in experiments first instabilities of the sphere wake are visible for $Re > 130$, the calculated drag coefficient for the axisymmetric sphere flow agrees very well with other numerical and experimental data up to $Re = 1000$, while for higher Re the deviations grow (not shown here, see Schulte-Werning et al. [16]). The dimensionless length of the wake bubble L/D as a function of Re (resp. log Re) is shown in fig. 9 in comparison with the data of other authors (the line connects the results of Taneda [19]). As in the cylinder case the deviations occur beyond the critical Re-number. Fornberg's [8] solution specifies nearly $L/D = O(\log Re)$ whereas the present compressible flow solution follows this relation up to $Re = 500$ and then L/D decreases to a constant value. A striking feature of Fornberg's solution is the resemblance of the wake bubble with Hill's spherical vortex (see Batchelor [1]), which is an analytical solution of the inviscid theory, both in the streamlines and in the vorticity distribution.
Up to now there is no evidence that a numerical solution of the axisymmetric sphere flow should converge to this exact solution. However, in the present case with $Ma = 0.40$ the length and width of the wake bubble vary such as to lead to a spherical shape of the bubble forming streamlines with increasing Re (fig. 10). But the vorticity distribution in the wake of the compressible flow exhibits a strong deviation from the one of incompressible flow, as seen in fig. 11 for $Re = 1000$. In the incompressible case (Fornberg [8]) the vorticity grows linearly with the radial coordinate up to the wake edge, thus clearly exhibiting a ramp. Contrarily to this the compressible flow shows a significant peak in the vorticity the position of which is nearly identical with the wake bubble axis and the density minimum of the whole flow field (note that different scalings for the vorticity ω are used in fig. 11). This may be a hint that the transport of scalar vorticity is influenced even by low compressibility effects,

which must be further discussed elsewhere. However, the vorticity distribution close to the body in the separation zone of the sphere is nearly the same for compressible and incompressible flow as seen in fig. 12 for $Re = 500$, where the same nonequidistant isolines of the vorticity are displayed.

3.3 Development of the separation structure

Due to the suppressing of physical instabilities by an artificial symmetry condition in the wake it is possible to investigate the development of the steady separation structure for increasing *Re*-number. The sequence of the changes in the separation topology is found to be the same for the cylinder and the sphere flow, only the *Re*-numbers at which the different stages occur are lower in the sphere case. With increasing *Re* the rotating fluid inside the wake bubble induces a secondary flow with increasing strength in the rear of the body. This causes a small closed secondary separation bubble downstream of the primary separation which grows in its geometrical shape and separates a free recirculation zone from the wake eddy. Fig. 13 shows streamlines near the body apex and the topological structure for the cylinder and the sphere flow for $Re = 2000$. Bouard & Coutanceau [2] experimentally discovered the same topological structure in the transient development of the separation of the flow around an impulsively started cylinder for higher *Re*. As mentioned in chapter 3.1 in the cylinder case this separation structure is sensitive to perturbations introduced by different initial conditions, in the sphere case this dependency was not observed. For the sphere Fornberg [8] obtained the same separation structure without indicating it.

4. CONCLUSIONS

The numerical simulations of the two-dimensional compressible cylinder and sphere flows yield good agreement for the integral quantities in comparison with the results of other authors in that *Re*-range, where the physical instabilities do not have to be strongly suppressed. In both cases the geometry of the wake bubble shows significant alteration for higher *Re* in contrast to the incompressible solution, thus indicating a non-vanishing influence of compressibility. However, the physical situation in the shear layer near the body is unaffected by flow compressibility. The impulsively started cylinder shows strong gradients in the transient phase for higher *Re* which led to solution breakdown, unless strong numerical damping is used for a finite time. For higher *Re* the complex topological structure of the steady separation near the body apex agrees well with other data and can be sensitive to different initial conditions.

REFERENCES

[1] Batchelor, G.K.: "An Introduction to Fluid Dynamics", Cambridge University Press, 1967.

[2] Bouard, R. & Coutanceau, M.: "The early stage of development of the wake behind an impulsively started cylinder for $40 < Re < 10^4$", J. Fluid Mech., Vol.101 (1980).

[3] Coutanceau, M. & Bouard, R.: "Experimental determination of the main features of the viscous flow in the wake of a circular cylinder in uniform translation. Part 1: Steady flow. Part 2: Unsteady flow", J. Fluid Mech., Vol.79 (1977).

[4] Dallmann, U. & Schulte-Werning, B.: "Topological Changes of Axisymmetric and Non-Axisymmetric Vortex Flows", IUTAM Symposium on Topological Fluid Mechanics, Cambridge, UK, 1989, to appear.
[5] Dennis, S.C.R. & Chang, G.-Z.: "Numerical solutions for steady flow past a circular cylinder at Reynolds numbers up to 100", J. Fluid Mech., Vol.42 (1970).
[6] Fornberg, B.: "A numerical study of steady viscous flow past a circular cylinder", J. Fluid Mech., Vol.98 (1980).
[7] Fornberg, B.: "Steady viscous flow past a circular cylinder up to Reynolds number 600", J. Comp. Phys., Vol.61 (1985).
[8] Fornberg, B.: "Steady Viscous Flow past a Sphere at high Reynolds numbers", J. Fluid Mech., Vol. 190, 1988.
[9] Ghia, R.N.; Ghia, U.; Osswald, G.A. & Liu, C.A.: "Simulation of separated flow past a bluff body using Navier-Stokes Equations", In: Smith, F.T. & Brown, S.N. (Eds.) Boundary Layer Separation, Springer Verlag Berlin, 1987.
[10] Lin, C.L. & Lee, S.C.: "Transient State Analysis of Separated Flow around a Sphere", Comp. and Fluids Vol.1, 1973.
[11] Moore, D.W. & Pullin, D.I.: "The compressible vortex pair", J. Fluid Mech., Vol.185 (1987).
[12] Müller, B.: "Calculation of Laminar Supersonic Flow over Ramps and Cylinder-Flares", DFVLR IB 221-88 A 14 (1988).
[13] Oswatitsch, K.: "Die Ablösebedingung von Grenzschichten", IUTAM Symposium on Boundary Layer Research, Freiburg, 1957.
[14] Patel, V.A.: "Time dependent solution of the viscous incompressible flow past a circular cylinder by the method of series truncation", Comp. and Fluids, Vol.4 (1976).
[15] Rimon, Y. & Cheng, S.I.: "Numerical Solution of a uniform flow over a sphere at intermediate Reynolds number", Phys. Fluids 12, 1969, 949.
[16] Schulte-Werning, B., Dallmann, U. & Müller, B.: "Zur numerischen Lösung der zweidimensionalen, symmetrischen Zylinder- und Kugelumströmung.", DLR-IB 221-89 A 14, 1989.
[17] Son, J.S. & Hanratty, T.J.: "Numerical solution for the flow around a circular cylinder at Reynolds numbers of 40, 200 and 500", J. Fluid Mech., Vol.35 (1969).
[18] Strykowski, P.: private communication, 1988.
[19] Taneda, S.: "Experimental Investigation of the Wake behind a Sphere at Low Reynolds Numbers", J. Phys. Soc. Japan 11, 10, 1956.

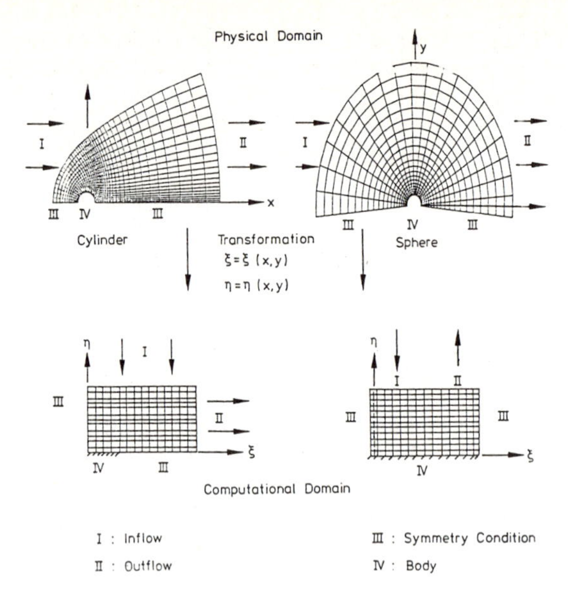

Fig. 1 Transformation of physical onto computational domain.

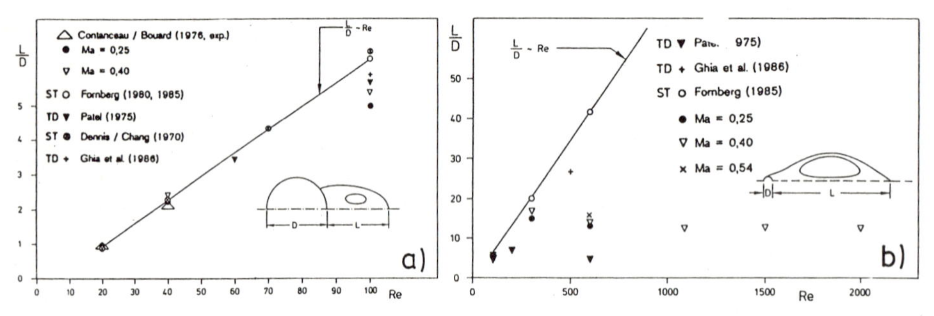

Fig. 2 Cylinder wake length for a) $Re \leq 100$ and b) $Re \leq 2000$.

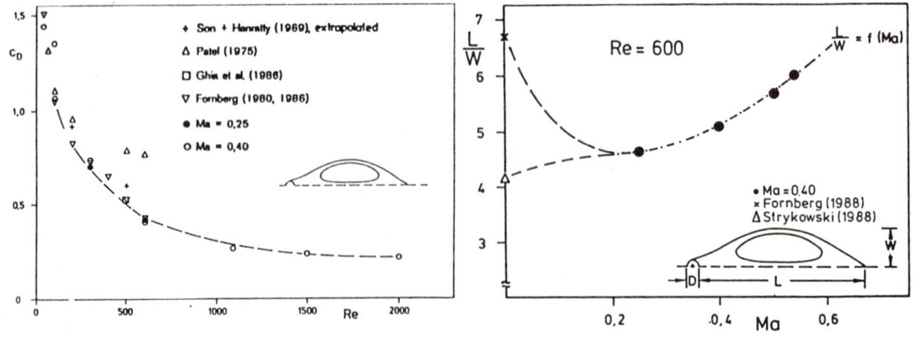

Fig. 3 Cylinder drag for $Re \leq 2000$.

Fig. 4 Cylinder wake geometry.

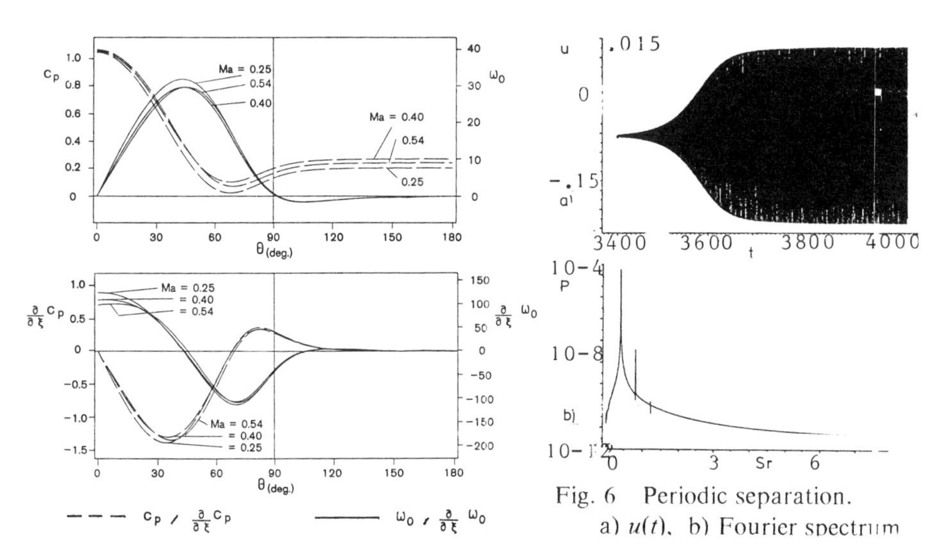

Fig. 6 Periodic separation.
a) $u(t)$, b) Fourier spectrum

Fig. 5 Wall quantities on the cylinder for $Re = 600$.

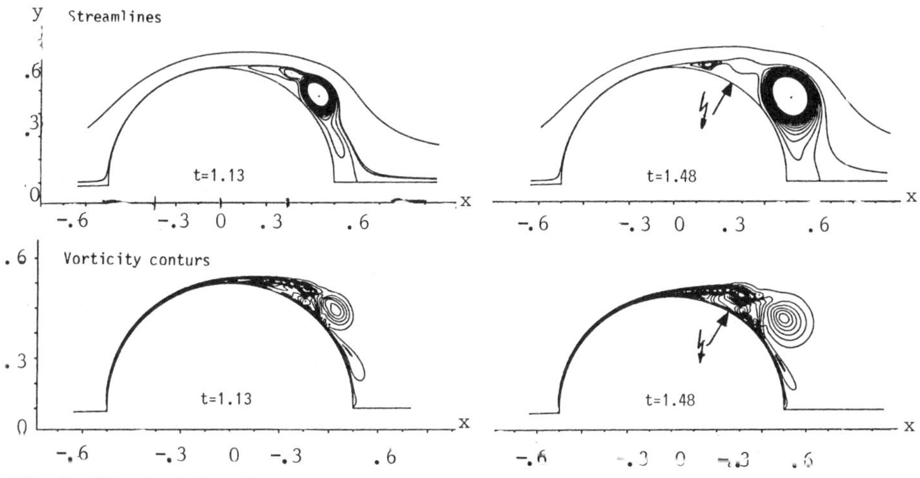

Fig. 7 Physical situation near the cylinder in the transient phase for $Re = 10^4$.

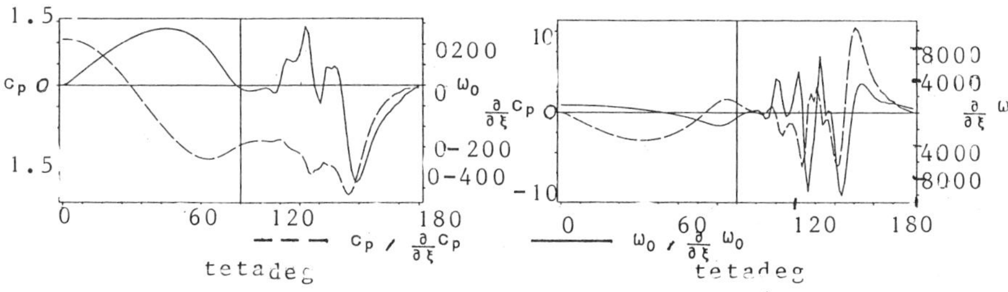

Fig. 8 Wall quantities on the cylinder at $t = 1.48$ for $Re = 10^4$.

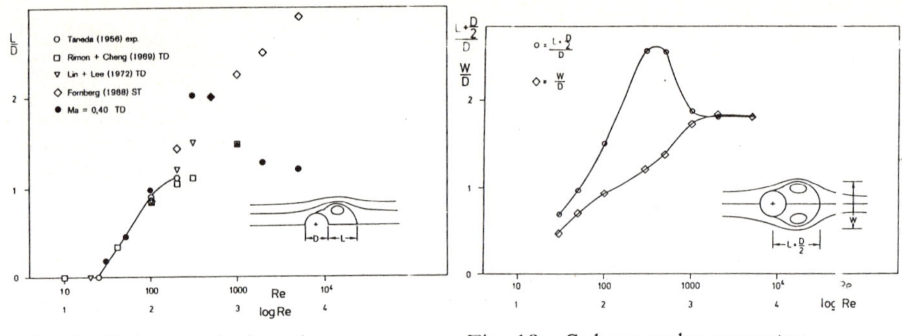

Fig. 9 Sphere wake length.

Fig. 10 Sphere wake geometry.

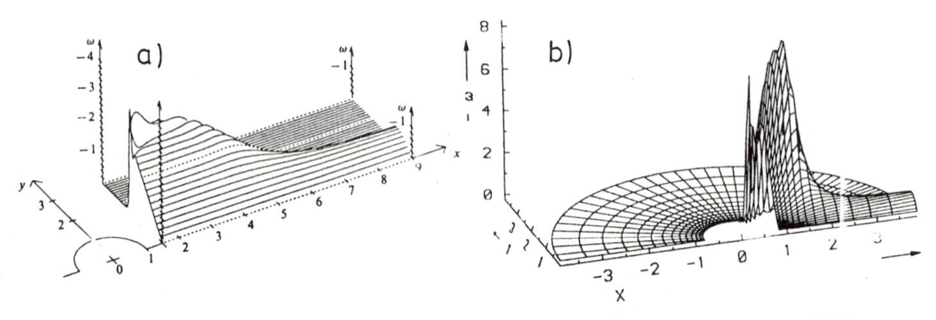

Fig. 11 Vorticity distribution in the wake bubble of the sphere for $Re = 1000$, a) Fornberg [8], b) $Ma = 0.40$.

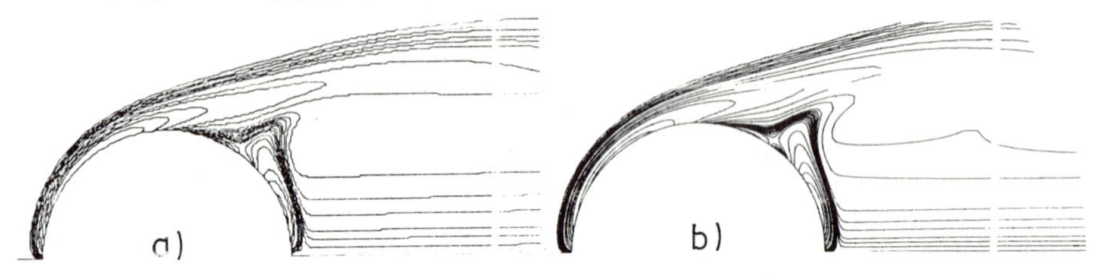

Fig. 12 Vorticity distribution near the sphere for $Re = 500$, a) Fornberg [8], b) $Ma = 0.40$.

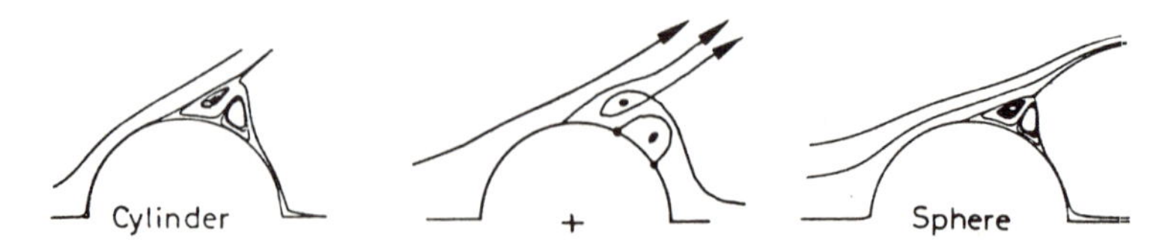

Fig. 13 Streamlines and topological structure of steady separation for $Re = 2000$

NUMERICAL CALCULATION OF STATIONARY SUBSONIC GAS DYNAMICS PROBLEMS

Yu.I.Shokin, G.S.Khakimzyanov

Computing Centre of USSR Academy of Sciences, Siberian Division, Krasnoyarsk

Introduction

The present work considers algorithm of numerical solution of the problems on two-dimensional subsonic ideal gas flows in channels of complicated shape with inlet, outlet and impermeable parts on the boundary. At present the problems on steady ideal gas flows are most often solved by the stabilization method which as a solution of stationary equations selects infinite time solution of the respective nonstationary equations of gas dynamics. The disadvantage of this method lies in its slow convergence for subsonic flows [1] , for in this case the initial perturbations extend with finite velocity both downstream and upstream along the characteristics of the hyperbolic system of equations interacting between themselves and with the boundaries of the region. Such motion of the perturbations within the region can be rather long requiring much computer time to obtain the steady state flow.

Another approach which is different from the stabilization method is constructing numerical algorithms directly for the stationary equations of gas dynamics. The works [2-5] pursuing this trend, suggest to solve the two-dimensional stationary problem by iteration methods. However, the possibilities to use each of these methods are to this or that extent limited. These limitations which are essential in constructing numerical algorithms may be a requirement of small deviations in the flow direction, or the condition of incompressibility of the fluid, or assumption of the simple flow geometry and so on. The work [6] states the algorithm of numerical solution of the problems on two - dimensional steady internal subsonic flows of an ideal compressible fluid in the channels of complicated shape with no limitations and disadvantages mentioned above. The method consists in splitting the system of stationary equations of gas dynamics into elliptic and hyperbolic parts and is based on the iteration process whose every step is successive solution of relatively simple boundary-value problems on rectangular irregular grid and allows to obtain numerical solution in practicable time accurately enough for a broad class of flows. This work develops this method for the case when calculation is performed for an arbitrary curvilinear grid covering the flow region.

Mathematical Statement

Mathematical statement of the problem under consideration is as follows: in a limited simply connected region Ω of $x0y$ plane a solution is to be found for a system of equations

$$\frac{\partial}{\partial x}\rho u y^{\mu} + \frac{\partial}{\partial y}\rho v y^{\mu} = 0, \quad (1)$$

$$\frac{\partial}{\partial x}\rho u^2 y^{\mu} + \frac{\partial}{\partial y}\rho u v y^{\mu} + y^{\mu}\frac{\partial p}{\partial x} = 0, \quad (2)$$

$$\frac{\partial}{\partial x}\rho u v y^{\mu} + \frac{\partial}{\partial y}\rho v^2 y^{\mu} + y^{\mu}\frac{\partial p}{\partial y} = 0, \quad (3)$$

$$\frac{\partial}{\partial x}\rho u H y^{\mu} + \frac{\partial}{\partial y}\rho v H y^{\mu} = 0, \quad (4)$$

$$H = e + p/\rho + V^2/2, \quad (5)$$

satisfying the boundary conditions

$$\rho\vec{v}\Big|_{\Gamma_1} = \vec{\nu}_1(s), \quad \vec{v}\cdot\vec{n}\Big|_{\Gamma_0} = 0, \quad \rho\vec{v}\cdot\vec{n}\Big|_{\Gamma_2} = \nu_2(s), \quad H\Big|_{\Gamma_1} = H_1(s), \quad (6)$$

where x and y are either Cartesian coordinates (for the plane-parallel flow, $\mu = 0$), or axial and radial coordinates (for the axisymmetrical flow, $\mu = 1$), u , v are the components of velocity vector $\vec{v}$ over the axes x and y , respectively, $V = |\vec{v}|$, ρ is density, p is pressure, H is the total enthalpy, e is specific internal energy satisfying the equation of state $e = e(p, \rho)$.

The boundary $\Gamma = \partial\Omega$ is assumed to consist of three parts: Γ_0 is the impermeable part of the boundary, Γ_1 is inlet into Ω , Γ_2 is outlet from Ω . In formulae (6) setting the boundary conditions S denotes the arc length from a certain point of the boundary, $\vec{n} = (n_1, n_2)$ is the unit vector of the external normal to the boundary Γ , $\vec{\nu}_1(S)$, $\nu_2(S)$, $H_1(S)$ are the set functions of S , such that the mass balance relation is satisfied

$$\oint_\Gamma \nu(s)\, y^\mu\, ds = 0\,, \tag{7}$$

where the function $\nu(S)$ is defined by the formula at the right

$$\nu(S) = \begin{cases} \vec{\nu}_1(S)\cdot\vec{n}(S)\,, & S\in\Gamma_1 \\ 0\,, & S\in\Gamma_0 \\ \nu_2(S)\,, & S\in\Gamma_2 \end{cases}$$

In addition to boundary conditions(6) it is assumed that at a certain point $M_0 \in \Gamma_1$ specified is the value of the pressure:

$$p(M_0) = p_0 \tag{8}$$

Let there be nongenerate one-to-one mapping of unit square $D = (0;1)\times(0;1)$ over the region Ω :

$$x = x(x^1, x^2)\,, \quad y = y(x^1, x^2) \tag{9}$$

After the transition to new independent variables x^1 , x^2 the system takes the following form

$$\frac{\partial}{\partial x^i}\sqrt{g}\,\rho v^i = 0\,, \tag{10}$$

$$\frac{\partial}{\partial x^i}\sqrt{g}\,(\rho v^\alpha v^i + p^{\alpha i}) + \sqrt{g}\,G^\alpha_{ij}\,(\rho v^i v^j + p^{ij}) = 0\,, \quad \alpha = 1,2\,, \tag{11}$$

$$\frac{\partial}{\partial x^i}\sqrt{g}\,\rho v^i H = 0\,, \tag{12}$$

where summing is performed over i and j indices $(i,j = 1,2)$, v^i are the contravariant components of the vector $\vec{v}$, $G^\alpha_{i,j}$ are Cristoffel symbols, $p^{ij} = p g^{ij}$, $g = det(g_{ij})(y^\mu)^2$, g_{ij} , g^{ij} are the covariant and contravariant components of the metric tensor, respectively.

Introduce new dependent variables - the stream function ψ and the vortex function ω :

$$\sqrt{g}\,\rho v^1 = \frac{\partial \psi}{\partial x^2}, \quad \sqrt{g}\,\rho v^2 = -\frac{\partial \psi}{\partial x^1}, \tag{13}$$

$$\omega = \frac{1}{\sqrt{g}} \left(-\frac{\partial v_1}{\partial x^2} + \frac{\partial v_2}{\partial x^1} \right), \tag{14}$$

where v_i are the covariant components of the velocity vector $\vec{v}$, $v_i = g_{ij} v^j$.

Then, instead of (11) we have the following equations:

$$(-1)^{\alpha} \rho \sqrt{g}\, v^{3-\alpha} \omega + \rho \frac{\partial}{\partial x^{\alpha}} \left(\frac{V^2}{2} \right) + \frac{\partial p}{\partial x^{\alpha}} = 0, \qquad \alpha = 1,2 \tag{15}$$

from which ensues an equation of hyperbolic type for ω

$$\sum_{i=1}^{2} \frac{\partial}{\partial x^i} \sqrt{g}\, \rho v^i \omega + (-1)^{i+1} \frac{\partial}{\partial x^i} \left(\rho \frac{\partial}{\partial x^{3-i}} \frac{V^2}{2} \right) = 0 \tag{16}$$

and equivalent to (15) integral relation for determination of p :

$$p(M) = p_0 + \int_{\gamma(M_0,M)} \rho \sqrt{g}\, \omega v^2 dx^1 - \rho \sqrt{g}\, \omega v^1 dx^2 - \rho\, d(V^2/2), \tag{17}$$

where $M \in D$, $\gamma(M_0, M)$ is an arbitrary curve connecting the points M_0 and M . The value of $p(M)$ in (17) does not depend on the choice of the curve γ .

Obtain for the stream function an equation of elliptic type with mixed derivatives

$$\frac{\partial}{\partial x^i} \left(k_{ij} \frac{\partial \psi}{\partial x^j} \right) = -\sqrt{g}\, \omega, \tag{18}$$

where summing is performed over i and j indices, $k_{11} = g_{22}/\sqrt{g}\,\rho$, $k_{12} = k_{21} = -g_{21}/\sqrt{g}\,\rho$, $k_{22} = g_{11}/\sqrt{g}\,\rho$.

For the new dependent variables the problem is formulated as follows: found are to be the functions ψ , ω , p , ρ , H , satisfying the equations (5), (12), (16)-(18) in the square D , condition

$$\psi(M) = \psi(M_0) + \int_0^s \sqrt{g}\, \nu(s)\, ds, \qquad M \in \Gamma \tag{19}$$

all over the boundary Γ and over the section Γ_1 the condition for H as well as the following condition

$$g_{i1}\tau^i \frac{\partial \psi}{\partial x^2} - g_{i2}\tau^i \frac{\partial \psi}{\partial x^1}\Big|_{\Gamma_1} = \sqrt{g}\,\vec{\tau}\cdot\vec{\nu}_1 \tag{20}$$

In formulae (19), (20) $\psi(M_0)$ is the arbitrary constant, $\vec{\tau}$ is the unit vector tangent to the inlet, Γ, Γ_1, Γ_2, Γ_0, M_0 are to be understood as the inverse images of the corresponding elements in transformation (9).

Difference Equations

Cover the region $\bar{D}$ with a rectangular regular grid with the steps $h_\alpha = 1/N_\alpha$, $\alpha = 1,2$. Refer the grid functions ψ, p, ρ, $g_{\alpha\beta}$, g to the nodes x_{i_1,i_2} (dark circles in Fig. 1), ω, H - to the mesh centres $x_{i_1+1/2,\, i_2+1/2}$ (light circles), v^1 and v^2 - to the middles of the vertical and horizontal sides of the mesh, respectively (crosses and rectangles in Fig. 1).

The difference equations are written by the integro-interpolation method, and beforehand the differential equations for ψ, ω, H are replaced by the integral relations:

$$\oint_C \left(k_{11}\frac{\partial\psi}{\partial x^1} + k_{12}\frac{\partial\psi}{\partial x^2}\right)dx^2 - \left(k_{21}\frac{\partial\psi}{\partial x^1} + k_{22}\frac{\partial\psi}{\partial x^2}\right)dx^1 = -\iint_{D_C} \sqrt{g}\,\omega\, dx^1 dx^2, \tag{21}$$

$$\oint_C \sqrt{g}\,\rho v^1 \omega\, dx^2 - \sqrt{g}\,\rho v^2 \omega\, dx^1 = -\oint_C \rho\, d(V^2/2), \tag{22}$$

$$\oint_C \sqrt{g}\,\rho v^1 H\, dx^2 - \sqrt{g}\,\rho v^2 H\, dx^1 = 0, \tag{23}$$

where D_C is the arbitrary subregion of the region D with the boundary C homeomorphic to the circumference.

The contours C of integration used in approximating relations (21) and (22) are shown by dash lines in Fig. 1 and Fig. 2, respectively. Replacing the integrals in (21) by the quadrature rectangles formula have nine-point difference equation for ψ, approximating (18) with the second order of accuracy:

$$(\Lambda\psi)_{i_1,i_2} \equiv \Big(\frac{1}{2}\sum_{\alpha=1}^{2}\left[(k_{\alpha\alpha}\psi_{\bar{x}^\alpha})_{x^\alpha} + (k_{\alpha\alpha}\psi_{x^\alpha})_{\bar{x}^\alpha}\right] + \frac{1}{4}\sum_{\alpha\neq\beta}^{1\div 2}\left[(k_{\alpha\beta}\psi_{\bar{x}^\beta})_{x^\alpha} + (k_{\alpha\beta}\psi_{x^\beta})_{\bar{x}^\alpha} + (k_{\alpha\beta}\psi_{x^\beta})_{x^\alpha} + (k_{\alpha\beta}\psi_{\bar{x}^\beta})_{\bar{x}^\alpha}\right]\Big)_{i_1,i_2} = -(\sqrt{g}\,\omega)_{i_1,i_2} \tag{24}$$

where $(\ )_{x^\alpha}$ and $(\ )_{\bar{x}^\alpha}$ denote the right and left differences, respectively.

Integrals in the left-hand side of (22) are approximated allowing for the signs of the contravariant components v^j of the velocity $\vec{v}$. So, for instance, assume

$$\int_{(BC)} \sqrt{g}\,\rho v^1 \omega\, dx^2 \sim (\psi_{i_1+1,\, i_2+1} - \psi_{i_1+1,\, i_2}) \cdot \omega_{BC}\,, \tag{25}$$

where $\omega_{BC} = \omega_3$, if $v^1_{i_1+1,\, i_2+1/2} < 0$ and $\omega_{BC} = \omega_0$, if $v^1_{i_1+1,\, i_2+1/2} \geqslant 0$. Hence, for each grid node $x_{i_1+1/2,\, i_2+1/2}$ have a difference equation for ω type of equations with oriented differences approximating (16) with the first order of accuracy

$$\Big(\sum_{j=0}^{4} \beta_j \omega_j\Big)_{i_1+1/2,\, i_2+1/2} = F_{i_1+1/2,\, i_2+1/2}\,. \tag{26}$$

The difference equations for H are of the same form as (26). They differ only in the right-hand side which is equal to zero. To calculate the pressure we used the method of consistent approximation [7] of ratio (17), therefore the value of p_{i_1, i_2} does not depend on the choice of the broken γ connecting the node x_{i_1, i_2} with the point M_0.

Algorithm

The obtained system of difference equations is solved by the iteration process, its each step consisting of the following stages:

1. solution of the Dirichlet difference problem to determine ψ^{k+1}. The operator corresponding to the system of difference equations (24) is self-adjoint and positive, therefore ψ can be found, say, with iteration method of successive overrelaxation which will converge;
2. calculation of the vortex over the inlet:

$$\omega^{k+1}\Big|_{\Gamma_1} = \delta_\omega \bar{\omega} + (1 - \delta_\omega)\, \omega^k\Big|_{\Gamma_1}$$

where $\omega^k|_{\Gamma_1}$ and $\bar{\omega}$ are the value of the inlet vortex taken from the previous k-th iteration and, respectively, calculated by the formula approximating the right-hand side of (14) with allowance for condition (20) on Γ_1, $\delta_\omega > 0$ is the relaxation parameter;

3. calculation of the vortex within the region D . Assuming to have no closed stream lines and gas rest points, then from the maximum principle it ensues that the system of equations (26) has a unique solution and it can be found by noniteration method analogous to implicit upwind scheme method;

4. calculation of the pressure by consistent approximation method;

5. calculation of the total enthalpy H by the same generalized implicit upwind scheme;

6. determination of the density $\bar{\rho}$ using the difference analogue (5). The final value of the density taken as $\rho^{k+1} = \delta_\rho \bar{\rho} + (1-\delta_\rho)\rho^k$, where $\delta_\rho > 0$ is the relaxation parameter for the density.

Note, that all newly calculated values are immediately used at the following stages of the given step of the iteration process.

Zero approximation to start the iteration process is the numerical solution of the problem on the ideal incompressible fluid which is obtained by the iteration process consisting of the first three stages of the process described above for gas.

The algorithm was applied in solving numerically a number of problems on ideal incompressible and compressible fluid. For example, we solved the problem on plane-parallel vortex flow of an incompressible fluid in a smooth curved wall channel (Fig. 3) as well as in a channel with one wall bent under the direct angle (Fig. 4) and the problem on vortex nonisoenergetic axisymmetrical flow in the subsonic part of the nozzle (Fig. 5). Exact solutions of these problems are known [8] . We solved, also, the problem on subsonic vortex isoenergetic flows in an axisymmetrical channel of a complicated shape, shown in Fig. 6. Calculations performed for a series of grids with finer mesh showed this method to be of the first order of accuracy and very economical.

References

1. J.A.Essers, Intern. J. Computers and Fluids 8 (1980), 351
2. B.D.Moiseenko, B.L.Rozhdestvensky, J. Vychisl. Matem. i Matem. Fiziki 10 (1970), 499
3. B.G.Gurov, N.N.Yanenko, I.K.Yaushev, Chisl. Metody Sploshnoj Sredy 2, 1 (1971), 3

4. M.Feistauer, Aplikace Mathematiky 16 (1971), 265
5. I.L.Osipov, V.P.Pashchenko, A.V.Shipilin, J.Vychisl. Matem. i Matem. Fiziki 18 (1978), 964
6. G.S.Khakimzyanov, I.K.Yaushev, Preprint No 4-87, ITPM SO AN SSSR, Novosibirsk, 1987
7. G.S.Khakimzyanov, I.K.Yaushev, J.Vychisl. Matem. i Matem. Fiziki 24 (1984), 1557
8. O.V.Kaptsov, DAN SSSR 298 (1988), 597

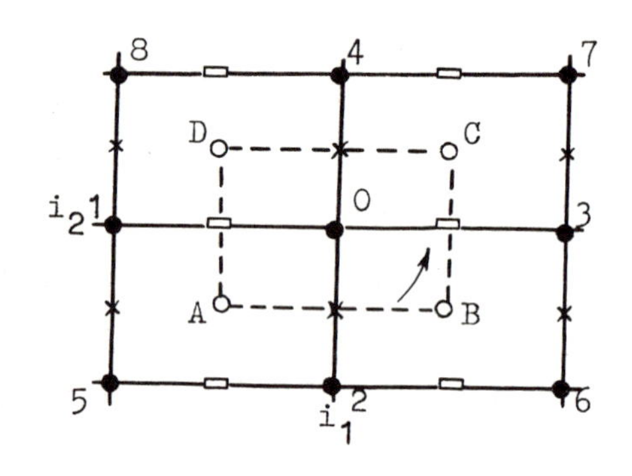

Fig. 1

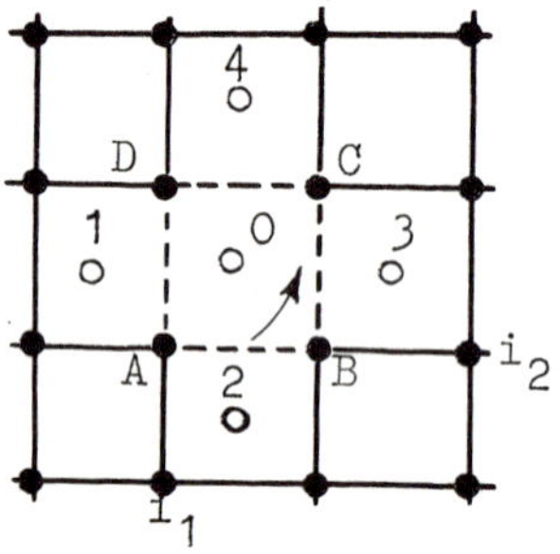

Fig. 2

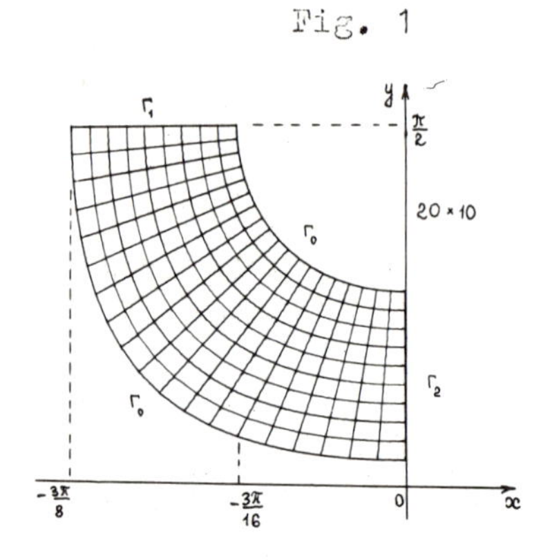

Fig. 3

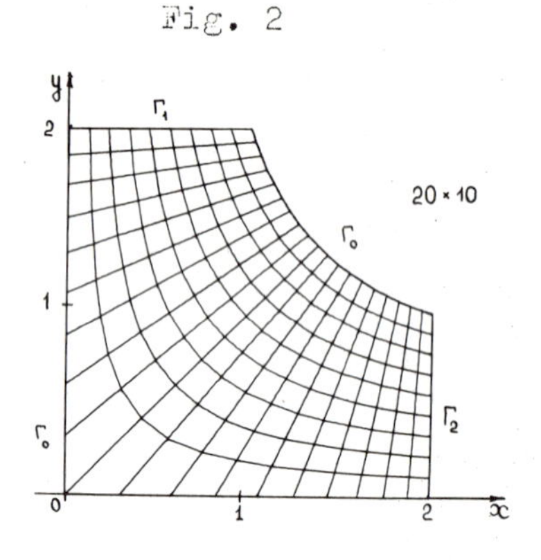

Fig. 4

Exact solution:

$u = a \cdot \cos x \cdot \cos y$

$v = a \cdot \sin x \cdot \sin y$

$\omega = 2a \cdot \cos x \cdot \sin y$

$p = p_0 - a^2 \cdot (\cos 2x - \cos 2y)/4$

$a = \text{const}$

Exact solution:

$$u = \frac{2\,\mathrm{th}(x/2)}{\mathrm{ch}^2(y/2)(1+f^2)}$$

$$v = -\frac{2\,\mathrm{th}(y/2)}{\mathrm{ch}^2(x/2)(1+f^2)}$$

$$\omega = -4\,\frac{f(1-f^2)}{(1+f^2)^2}$$

$f = \mathrm{th}(x/2)\ \mathrm{th}(y/2)$

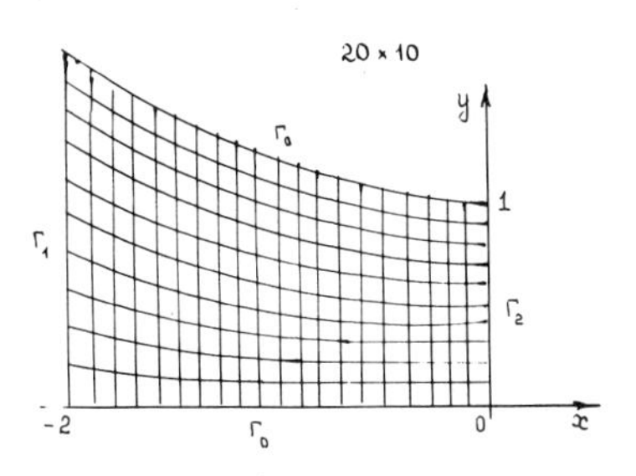

Fig. 5

Exact solution:

$$u=T^{-1}\left[2c_3T^2(1-a\varkappa T^{\varkappa-1}/(\varkappa-1))\right]^{1/2}$$

$$v=y\left[2c_3T^2(1-a\varkappa T^{\varkappa-1}/(\varkappa-1))\right]^{1/4}$$

$p=c_2T^{\varkappa}$, $\rho=c_1T$,

c_1, c_2, $c_3>0$ are the arbitrary constants, $a=c_2/(c_1c_3)$,

$\varkappa$ is the ratio of specific heats,

$T(x)$ is the solution of equation

$$T'=\frac{\left[2T^2(1-a\varkappa T^{\varkappa-1}/(\varkappa-1))\right]^{3/4}}{c_3^{1/4}(a\varkappa(\varkappa+1)T^{\varkappa-1}/(2(\varkappa-1))-1)}, \quad T(0)=T_0$$

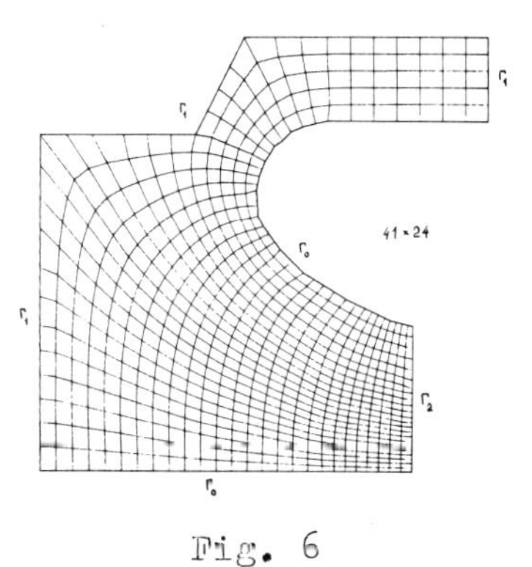

Fig. 6

DERIVATION OF A ROE SCHEME FOR AN N-SPECIES CHEMICALLY REACTING GAS IN THERMAL EQUILIBRIUM

S.P. Spekreijse, R. Hagmeijer
National Aerospace Laboratory NLR
P.O. Box 90502, 1006 BM Amsterdam, The Netherlands

SUMMARY

A Roe-type flux-difference-splitting scheme is developed for an n-species chemically reacting gas in 3D. Thermal equilibrium is assumed. By focusing on the construction of the numerical flux function, it is shown that many details of the construction method are eventually unimportant and this observation leads to much more simple expressions for the numerical flux function than generally encountered in the literature.
As a special type case (n=1), a Roe scheme is also presented for a thermally perfect gas.

1 INTRODUCTION

The recent interest in hypersonic flow simulation has stimulated a need for accurate computations of nonequilibrium and equilibrium flows. As a result many papers have been published where the original flux-difference-splittings of Roe and Osher and the flux-vector-splittings of Van Leer and Steger and Warming are extended and made suitable for both equilibrium and nonequilibrium flows [1, 2, 3, 4, 5, 7, 8].
For nonequilibrium flows, the extensions of the original Roe scheme [6] are more complicated than necessary and in this paper it is shown how the numerical flux functions can be computed in a more efficient and transparent way. No unnecessary assumptions, approximations and auxiliary quantities are introduced. Furthermore, the expression for the speed of sound at the Roe-averaged state is obtained in a very efficient and elegant way.
Section 2 concerns chemically nonequilibrium flow. The gas is made of n species. Thermal equilibrium is assumed (i.e. the internal energy of each species is a function of the temperature only). A comprehensive derivation of the numerical flux function is given. The assumption of thermal equilibrium does not facilitates the computation of the numerical flux function and the reader who wants to construct the numerical flux function in case of nonequilibrium chemistry and thermodynamics can do it along the same lines as presented in this section.
In section 3 the numerical flux function is presented for the special case n = 1. Then we have a thermally perfect gas (i.e. the internal energy is a function of the temperature only). It is also shown that in case of a calorically perfect gas ($p=(\gamma-1)\rho e$), the obtained scheme is exactly the same as the original Roe scheme.
Finally, in section 4 some conclusions are listed. Due to space limitations, no numerical experiments with the presented schemes are given.

2 ROE-TYPE FLUX-DIFFERENCE-SPLITTING FOR NONEQUILIBRIUM FLOW

In this section a Roe-type flux-difference-splitting scheme is developed for a general, inviscid chemically nonequilibrium flow in thermal equilibrium. For an inviscid, chemically reacting flow in thermal equilibrium where the gas is made of n species the governing equations are given by

$$\frac{\partial q}{\partial t} + \frac{\partial}{\partial x} f(q) + \frac{\partial}{\partial y} g(q) + \frac{\partial}{\partial z} h(q) = \Omega(q) \tag{2.1}$$

where

$$\begin{aligned}
q &= (\rho_1 \;,\ldots,\; \rho_n \;,\; \rho u \;,\; \rho v \;,\; \rho w \;,\; E)^T \\
f(q) &= (\rho_1 u \;,\ldots,\; \rho_n u \;,\; \rho u^2+p,\; \rho uv \;,\; \rho uw \;,\; (E+p)u)^T \\
g(q) &= (\rho_1 v \;,\ldots,\; \rho_n v \;,\; \rho uv \;,\; \rho v^2+p,\; \rho vw \;,\; (E+p)v)^T \\
h(q) &= (\rho_1 w \;,\ldots,\; \rho_n w \;,\; \rho uw \;,\; \rho vw \;,\; \rho w^2+p,\; (E+p)w)^T \\
\Omega(q) &= (\Omega_1(q),\ldots,\; \Omega_n(q),\; 0 \;,\; 0 \;,\; 0 \;,\; 0)^T .
\end{aligned} \tag{2.2}$$

Here ρ denotes the density of the mixture, u, v and w are velocity components in the x, y and z directions, p is the pressure and E the total energy. The density of species i is denoted by ρ_i and therefore

$$\rho = \sum_{i=1}^{n} \rho_i . \tag{2.3}$$

The total energy E is related to the internal energy e of the mixture according to

$$E = \rho e + \tfrac{1}{2}\rho(u^2+v^2+w^2) . \tag{2.4}$$

Under the assumption of thermal equilibrium the internal energy of species i, e_i, is solely a function of the temperature T, i.e.

$$e_i = e_i(T) \qquad i = 1,\ldots,n . \tag{2.5}$$

Also, the internal energy of the mixture is computed by summation of the internal energy of the species i in the following way

$$e = \sum_{i=1}^{n} c_i e_i(T) \tag{2.6}$$

where $c_i = \rho_i/\rho$ is the mass fraction of species i.
The pressure p is computed according to Dalton's law (mixture of perfect gases):

$$p = \sum_{i=1}^{n} \rho_i R_i T \tag{2.7}$$

where R_i is the specific gas constant of species i.
The equation of state

$$p = \bar{p}\,(\rho_1,\ldots,\rho_n,e) \tag{2.8}$$

is implicitly given by (2.6) and (2.7).

The source term $\Omega_i(q)$ describes the production of species i due to chemical reactions, therefore

$$\Omega_i = \Omega_i(\rho_1,\dots,\rho_n,T) \qquad i=1,\dots,n \; . \tag{2.9}$$

The construction of the Roe scheme is independent of $\Omega(q)$ and therefore no detailed description of the source $\Omega(q)$ is needed here. For the construction of a Roe-type scheme we have to face the following problem. Given two states q_L, q_R (left and right), compute eigenvectors $\tilde{R}_1,\dots,\tilde{R}_{n+4}$, eigenvalues $\tilde{\lambda}_1,\dots,\tilde{\lambda}_{n+4}$ and coefficients $\tilde{\alpha}_1,\dots,\tilde{\alpha}_{n+4}$ such that

$$q_R - q_L = \sum_{i=1}^{n+4} \tilde{\alpha}_i \tilde{R}_i \tag{P1}$$

$$f(q_R) - f(q_L) = \sum_{i=1}^{n+4} \tilde{\alpha}_i \tilde{\lambda}_i \tilde{R}_i \; . \tag{P2}$$

The third property which should also be fulfilled is

$$q_R \to q_L \to q \quad \Longrightarrow \quad \tilde{\lambda}_i \to \lambda_i \text{ and } \tilde{R}_i \to R_i, \; i=1,\dots,n+4 \tag{P3}$$

where λ_i and R_i are the eigenvalues and eigenvectors of the Jacobian matrix $\frac{df}{dq}(q)$.

Once a method has been developed to compute, for a general pair (q_L,q_R), $\tilde{R}_i(q_L,q_R)$, $\tilde{\lambda}_i(q_L,q_R)$, $\tilde{\alpha}_i(q_L,q_R)$ $i=1,..,n+4$ such that the properties P1, P2, P3 are valid, then the numerical flux function $f(q_L,q_R)$ becomes

$$f(q_L,q_R) = \frac{1}{2}\left\{ f(q_L) + f(q_R) - \sum_{i=1}^{n+4} \tilde{\alpha}_i |\tilde{\lambda}_i| \tilde{R}_i \right\} . \tag{2.10}$$

Roe's approximate Riemann solver is completely determined by the numerical flux function $f(q_L,q_R)$. In the remainder of this section we shall show how to compute the right-hand-side of (2.10).

Due to property P3, a first step is the computation of the Jacobian matrix $\frac{df}{dq}(q)$ and the corresponding eigenvectors and eigenvalues.

It is easily verified that

$$\frac{df}{dq} = \begin{pmatrix}
u(1-c_1) & -c_1 u & \cdots\cdots & -c_1 u & c_1 & 0 & 0 & 0 \\
-c_2 u & u(1-c_2) & & -c_2 u & c_2 & \vdots & \vdots & \vdots \\
-c_3 u & -c_3 u & & \vdots & \vdots & \vdots & \vdots & \vdots \\
\vdots & \vdots & & & & & & \\
-c_n u & -c_n u & \cdots\cdots & u(1-c_n) & c_n & 0 & 0 & 0 \\
\frac{\partial p}{\partial \rho_1} - u^2 & \frac{\partial p}{\partial \rho_2} - u^2 & \cdots & \frac{\partial p}{\partial \rho_n} - u^2 & \frac{\partial p}{\partial \rho u} + 2u & \frac{\partial p}{\partial \rho v} & \frac{\partial p}{\partial \rho w} & \frac{\partial p}{\partial E}
\end{pmatrix}$$

$$\begin{pmatrix} -uv & -uv & \dots & -uv & v & u & 0 & 0 \\ -uw & -uw & \dots & -uw & w & 0 & u & 0 \\ u(\frac{\partial p}{\partial \rho_1} - H) & u(\frac{\partial p}{\partial \rho_2} - H) & \dots & u(\frac{\partial p}{\partial \rho_n} - H) & H + u\frac{\partial p}{\partial \rho u} & u\frac{\partial p}{\partial \rho v} & u\frac{\partial p}{\partial \rho w} & u(1+\frac{\partial p}{\partial E}) \end{pmatrix} \tag{2.11}$$

where H is the total enthalpy

$$H = \frac{E+p}{\rho} = e + \frac{p}{\rho} + \frac{1}{2}(u^2 + v^2 + w^2) \tag{2.12}$$

and

$$\frac{\partial p}{\partial \rho_i} = \frac{\partial \bar{p}}{\partial \rho_i} + \frac{\partial \bar{p}}{\partial e}\left\{\frac{u^2+v^2+w^2}{\rho} - \frac{E}{\rho^2}\right\} \qquad i=1,\dots,n$$

$$\frac{\partial p}{\partial \rho u} = -\frac{u}{\rho}\frac{\partial \bar{p}}{\partial e} \qquad \frac{\partial p}{\partial \rho w} = -\frac{w}{\rho}\frac{\partial \bar{p}}{\partial e}$$

$$\frac{\partial p}{\partial \rho v} = -\frac{v}{\rho}\frac{\partial \bar{p}}{\partial e} \qquad \frac{\partial p}{\partial E} = +\frac{1}{\rho}\frac{\partial \bar{p}}{\partial e} \ . \tag{2.13}$$

Define

$$a_i^2 = \frac{\partial \bar{p}}{\partial \rho_i} + \frac{p}{\rho^2}\frac{\partial \bar{p}}{\partial e} \qquad i=1,\dots,n$$

$$b_i = \frac{E}{\rho} - \rho\,\frac{\partial \bar{p}}{\partial \rho_i}\Big/\frac{\partial \bar{p}}{\partial e} \qquad i=1,\dots,n$$

$$a^2 = \sum_{i=1}^{n} c_i a_i^2 = \sum_{i=1}^{n} c_i \frac{\partial \bar{p}}{\partial \rho_i} + \frac{p}{\rho^2}\frac{\partial \bar{p}}{\partial e} \ . \tag{2.14}$$

Note that a is the so-called "frozen speed of sound". From these definitions and (2.13) it follows that

$$u\frac{\partial p}{\partial E} + \frac{\partial p}{\partial \rho u} = 0 \qquad v\frac{\partial p}{\partial E} + \frac{\partial p}{\partial \rho v} = 0 \qquad w\frac{\partial p}{\partial E} + \frac{\partial p}{\partial \rho w} = 0$$

$$\frac{\partial p}{\partial \rho_i} + u\frac{\partial p}{\partial \rho u} + v\frac{\partial p}{\partial \rho v} + w\frac{\partial p}{\partial \rho w} + H\frac{\partial p}{\partial E} = a_i^2 \qquad i=1,\dots,n$$

$$\frac{\partial p}{\partial \rho_i} + u\frac{\partial p}{\partial \rho u} + v\frac{\partial p}{\partial \rho v} + w\frac{\partial p}{\partial \rho w} + b_i\frac{\partial p}{\partial E} = 0 \qquad i=1,\dots,n \ . \tag{2.15}$$

With these results it is easily verified that the eigenvectors and eigenvalues of $\frac{df}{dq}$ are given by

$$
\begin{array}{llllllllllll}
R_1 & = (1 & 0 & 0 & \dots & 0 & u & v & w & b_1\ )^T & \lambda_1 & = u \\
R_2 & = (0 & 1 & 0 & \dots & 0 & u & v & w & b_2\ )^T & \lambda_2 & = u \\
\cdot & & & & & & & & & & & \cdot \\
\cdot & & & & & & & & & & & \cdot \\
R_n & = (0 & 0 & 0 & \dots & 1 & u & v & w & b_n\ )^T & \lambda_n & = u \\
R_{n+1} & = (c_1 & c_2 & c_3 & \dots & c_n & u-a & v & w & H-ua)^T & \lambda_{n+1} & = u-a \\
R_{n+2} & = (0^1 & 0^2 & 0^3 & \dots & 0^n & 0 & 1 & 0 & v\ )^T & \lambda_{n+2} & = u \\
R_{n+3} & = (0 & 0 & 0 & \dots & 0 & 0 & 0 & 1 & w\ )^T & \lambda_{n+3} & = u \\
R_{n+4} & = (c_1 & c_2 & c_3 & \dots & c_n & u+a & v & w & H+ua)^T & \lambda_{n+4} & = u+a\ .
\end{array}
\qquad (2.16)
$$

These results and the results of the Roe-type linearization for real gases presented in [1] immediately suggest to define $\tilde{R}_i$, $\tilde{\lambda}_i$ and $\tilde{\alpha}_i$, as follows

$$
\begin{array}{llllllllllll}
\tilde{R}_1 & = (1 & 0 & 0 & \dots & 0 & \tilde{u} & \tilde{v} & \tilde{w} & \tilde{b}_1\ )^T & \tilde{\lambda}_1 & = \tilde{u} \\
\tilde{R}_2 & = (0 & 1 & 0 & \dots & 0 & \tilde{u} & \tilde{v} & \tilde{w} & \tilde{b}_2\ )^T & \tilde{\lambda}_2 & = \tilde{u} \\
\cdot & & & & & & & & & & & \cdot \\
\cdot & & & & & & & & & & & \cdot \\
\tilde{R}_n & = (0 & 0 & 0 & \dots & 1 & \tilde{u} & \tilde{v} & \tilde{w} & \tilde{b}_n\ )^T & \tilde{\lambda}_n & = \tilde{u} \\
\tilde{R}_{n+1} & = (\tilde{c}_1 & \tilde{c}_2 & \tilde{c}_3 & \dots & \tilde{c}_n & \tilde{u}-\tilde{a} & \tilde{v} & \tilde{w} & \tilde{H}-\tilde{u}\tilde{a})^T & \tilde{\lambda}_{n+1} & = \tilde{u}-\tilde{a} \\
\tilde{R}_{n+2} & = (0^1 & 0^2 & 0^3 & \dots & 0^n & 0 & 1 & 0 & \tilde{v}\ )^T & \tilde{\lambda}_{n+2} & = \tilde{u} \\
\tilde{R}_{n+3} & = (0 & 0 & 0 & \dots & 0 & 0 & 0 & 1 & \tilde{w}\ )^T & \tilde{\lambda}_{n+3} & = \tilde{u} \\
\tilde{R}_{n+4} & = (\tilde{c}_1 & \tilde{c}_2 & \tilde{c}_3 & \dots & \tilde{c}_n & \tilde{u}+\tilde{a} & \tilde{v} & \tilde{w} & \tilde{H}+\tilde{u}\tilde{a})^T & \tilde{\lambda}_{n+4} & = \tilde{u}+\tilde{a}
\end{array}
\qquad (2.17)
$$

$$
\tilde{\alpha}_i = \Delta\rho_i - \tilde{c}_i \frac{\Delta p}{\tilde{a}^2} \quad i=1,\dots,n \qquad\qquad \tilde{\alpha}_{n+1} = \frac{1}{2\tilde{a}^2}(\Delta p - \tilde{\rho}\tilde{a}\Delta u)
$$

$$
\tilde{\alpha}_{n+2} = \tilde{\rho}\Delta v \qquad \tilde{\alpha}_{n+3} = \tilde{\rho}\Delta w \qquad \tilde{\alpha}_{n+4} = \frac{1}{2\tilde{a}^2}(\Delta p + \tilde{\rho}\tilde{a}\Delta u)\ . \qquad (2.18)
$$

In these formulae the operator Δ is defined by

$$
\Delta(.) = (.)_R - (.)_L \qquad (2.19)
$$

and furthermore

$$
\tilde{\rho} = \sqrt{\rho_L \rho_R} \qquad\qquad \theta = \frac{\sqrt{\rho_L}}{\sqrt{\rho_L} + \sqrt{\rho_R}}
$$

$$
\tilde{u} = \theta u_L + (1-\theta)u_R \qquad \tilde{v} = \theta v_L + (1-\theta)v_R \qquad \tilde{w} = \theta w_L + (1-\theta)w_R
$$

$$
\tilde{H} = \theta H_L + (1-\theta)H_R \qquad \tilde{c}_i = \theta c_i^L + (1-\theta)c_i^R \qquad i=1,\dots,n\ , \qquad (2.20)
$$

Although $\tilde{a}$ and $\tilde{b}_i$ $i=1,\dots,n$ are still undetermined we can investigate to what extent the properties P1 and P2 are satisfied. Writing the properties

P1 and P2 out in full we have

$$\Delta\rho_i = \tilde{\alpha}_i + \tilde{c}_i\tilde{\alpha}_{n+1} + \tilde{c}_i\tilde{\alpha}_{n+4} \qquad i=1,\dots,n \tag{D1}$$

$$\Delta\rho u = \tilde{u}\sum_{i=1}^{n}\tilde{\alpha}_i + (\tilde{u}-\tilde{a})\,\tilde{\alpha}_{n+1} + (\tilde{u}+\tilde{a})\,\tilde{\alpha}_{n+4} \tag{D2}$$

$$\Delta\rho v = \tilde{v}\sum_{i=1}^{n}\tilde{\alpha}_i + \tilde{v}\,\tilde{\alpha}_{n+1} + \tilde{\alpha}_{n+2} + \tilde{v}\,\tilde{\alpha}_{n+4} \tag{D3}$$

$$\Delta\rho w = \tilde{w}\sum_{i=1}^{n}\tilde{\alpha}_i + \tilde{w}\,\tilde{\alpha}_{n+1} + \tilde{\alpha}_{n+3} + \tilde{w}\,\tilde{\alpha}_{n+4} \tag{D4}$$

$$\Delta E = \sum_{i=1}^{n}\tilde{\alpha}_i\tilde{b}_i + (\tilde{H}-\tilde{u}\,\tilde{a})\,\tilde{\alpha}_{n+1} + \tilde{v}\,\tilde{\alpha}_{n+2} + \tilde{w}\,\tilde{\alpha}_{n+3} + (\tilde{H}+\tilde{u}\,\tilde{a})\,\tilde{\alpha}_{n+4} \tag{D5}$$

$$\Delta\rho_i u = \tilde{u}\,\tilde{\alpha}_i + \tilde{c}_i\,(\tilde{u}-\tilde{a})\,\tilde{\alpha}_{n+1} + \tilde{c}_i\,(\tilde{u}+\tilde{a})\,\tilde{\alpha}_{n+4} \qquad i=1,\dots,n \tag{D6}$$

$$\Delta(\rho u^2+p) = \tilde{u}^2\sum_{i=1}^{n}\tilde{\alpha}_i + (\tilde{u}-\tilde{a})^2\,\tilde{\alpha}_{n+1} + (\tilde{u}+\tilde{a})^2\,\tilde{\alpha}_{n+4} \tag{D7}$$

$$\Delta(\rho uv) = \tilde{u}\,\tilde{v}\sum_{i=1}^{n}\tilde{\alpha}_i + (\tilde{u}-\tilde{a})\,\tilde{v}\,\tilde{\alpha}_{n+1} + \tilde{u}\,\tilde{\alpha}_{n+2} + (\tilde{u}+\tilde{a})\,\tilde{v}\,\tilde{\alpha}_{n+4} \tag{D8}$$

$$\Delta(\rho uw) = \tilde{u}\,\tilde{w}\sum_{i=1}^{n}\tilde{\alpha}_i + (\tilde{u}-\tilde{a})\,\tilde{w}\,\tilde{\alpha}_{n+1} + \tilde{u}\,\tilde{\alpha}_{n+3} + (\tilde{u}+\tilde{a})\,\tilde{w}\,\tilde{\alpha}_{n+4} \tag{D9}$$

$$\Delta((E+p)u) = \tilde{u}\sum_{i=1}^{n}\tilde{\alpha}_i\tilde{b}_i + (\tilde{u}-\tilde{a})\,(\tilde{H}-\tilde{u}\,\tilde{a})\,\tilde{\alpha}_{n+1} + \tilde{u}\,\tilde{v}\,\tilde{\alpha}_{n+2} + \tilde{u}\,\tilde{w}\,\tilde{\alpha}_{n+3} + (\tilde{u}+\tilde{a})\,(\tilde{H}+\tilde{u}\,\tilde{a})\,\tilde{\alpha}_{n+4}\,. \tag{D10}$$

The verification of D1,..,D4 and D6,..,D9 becomes straightforward by means of the following two useful relations (φ and Ψ arbitrary functions):

$$\Delta\rho\varphi = \tilde{\rho}\,\Delta\,\varphi + \tilde{\varphi}\,\Delta\,\rho \qquad \Delta\rho\varphi\Psi = \tilde{\rho}\,\tilde{\varphi}\,\Delta\,\Psi + \tilde{\rho}\,\tilde{\Psi}\,\Delta\,\varphi + \tilde{\varphi}\,\tilde{\Psi}\,\Delta\,\rho \tag{2.21}$$

where

$$\tilde{\varphi} = \theta\,\varphi_L + (1-\theta)\,\varphi_R \qquad \tilde{\Psi} = \theta\,\Psi_L + (1-\theta)\,\Psi_R\,. \tag{2.22}$$

Thus independent on the definitions of $\tilde{a}$ and $\tilde{b}_i$ it is found that D1,..,D4 and D6,..,D9 are satisfied. From D5 it follows that $\tilde{b}_i$, i=1,..,n must be defined in such a way that

$$\sum_{i=1}^{n}\tilde{\alpha}_i\,\tilde{b}_i = \Delta E - \frac{\tilde{H}}{\tilde{a}^2}\,\Delta p - \tilde{\rho}\tilde{u}\Delta u - \tilde{\rho}\tilde{v}\Delta v - \tilde{\rho}\tilde{w}\Delta w \tag{2.23}$$

and it is easily seen that if this relation is satisfied then both D5 and D10 are fulfilled.
Suppose that formulae for $\tilde{b}_i$ have been constructed such that (2.23) is fulfilled. Then the properties P1 and P2 are satisfied and the numerical flux function $f(q_L, q_R)$ can be computed according to (2.10). Because $\tilde{b}_i$ enters the righthandside of (2.10) only via the expression $|\tilde{u}| \sum_{i=1}^{n} \tilde{\alpha}_i \tilde{b}_i$, it follows that $f(q_L, q_R)$ is independent on the precise formulae of $\tilde{b}_i$; we can use (2.23) directly for the computation of the numerical flux function.
With the abbreviate notation $f_i = f_i(q_L, q_R)$ we find that

$$f_1 = \tfrac{1}{2}\{(\rho_1 u)_L + (\rho_1 u)_R - (\tilde{\alpha}_1|\tilde{u}| + \tilde{\alpha}_{n+1}|\tilde{u}-\tilde{a}|\tilde{c}_1 + \tilde{\alpha}_{n+4}|\tilde{u}+\tilde{a}|\tilde{c}_1)\}$$

$$\vdots$$

$$f_n = \tfrac{1}{2}\{(\rho_n u)_L + (\rho_n u)_R - (\tilde{\alpha}_n|\tilde{u}| + \tilde{\alpha}_{n+1}|\tilde{u}-\tilde{a}|\tilde{c}_n + \tilde{\alpha}_{n+4}|\tilde{u}+\tilde{a}|\tilde{c}_n)\}$$

$$f_{n+1} = \tfrac{1}{2}\{(\rho u^2+p)_L + (\rho u^2+p)_R - (\tilde{u}|\tilde{u}|(\Delta\rho - \frac{\Delta p}{\tilde{a}^2}) + \tilde{\alpha}_{n+1}\,|\tilde{u}-\tilde{a}|(\tilde{u}-\tilde{a})$$
$$+ \tilde{\alpha}_{n+4}|\tilde{u}+\tilde{a}|(\tilde{u}+\tilde{a}))\}$$

$$f_{n+2} = \tfrac{1}{2}\{(\rho uv)_L + (\rho uv)_R - (\tilde{v}|\tilde{u}|(\Delta\rho - \frac{\Delta p}{\tilde{a}^2}) + \tilde{\alpha}_{n+1}\,|\tilde{u}-\tilde{a}|\tilde{v}$$
$$+ \tilde{\alpha}_{n+2}|\tilde{u}| + \tilde{\alpha}_{n+4}|\tilde{u}+\tilde{a}|\tilde{v})\}$$

$$f_{n+3} = \tfrac{1}{2}\{(\rho uw)_L + (\rho uw)_R - (\tilde{w}|\tilde{u}|(\Delta\rho - \frac{\Delta p}{\tilde{a}^2}) + \tilde{\alpha}_{n+1}\,|\tilde{u}-\tilde{a}|\tilde{w}$$
$$+ \tilde{\alpha}_{n+3}|\tilde{u}| + \tilde{\alpha}_{n+4}|\tilde{u}+\tilde{a}|\tilde{w})\}$$

$$f_{n+4} = \tfrac{1}{2}\{((E+p)u)_L + ((E+p)u)_R - (|\tilde{u}|(\Delta E - \frac{\tilde{H}}{\tilde{a}^2}\,\Delta p - \tilde{\rho}\tilde{u}\Delta u)$$
$$+ \tilde{\alpha}_{n+1}\,|\tilde{u}-\tilde{a}|\,(\tilde{H}-\tilde{u}\tilde{a}) + \tilde{\alpha}_{n+4}\,|\tilde{u}+\tilde{a}|\,(\tilde{H}+\tilde{u}\tilde{a}))\}\,. \tag{2.24}$$

Upto now $\tilde{a}$ has not been defined. The expression for $\tilde{a}$ is obtained as follows. We have assumed that $\tilde{b}_i$, $i=1,\ldots,n$ exist such that (2.23) is fulfilled. Therefore it is necessary that if $\tilde{\alpha}_i=0$, $i=1,\ldots,n$ then the right-hand-side of (2.23) is also equal to zero. This observation leads to a unique expression for $\tilde{a}$ as follows.
Assume $\tilde{\alpha}_i=0$, $i=1,\ldots,n$. It is easily seen that this is equivalent to

$$\Delta c_i = 0 \qquad i=1,\ldots,n \qquad \text{and} \qquad \Delta p = \tilde{a}^2\,\Delta\rho. \tag{2.25}$$

Then it follows that the right-hand-side of (2.23) becomes:

$$\text{RHS (2.23)} = \Delta\rho e + \tfrac{1}{2}(\tilde{u}^2 + \tilde{v}^2 + \tilde{w}^2)\,\Delta\rho - \tilde{H}\,\Delta\rho$$

$$= \tilde{\rho}\Delta e + (\tilde{e} + \tfrac{1}{2}(\tilde{u}^2 + \tilde{v}^2 + \tilde{w}^2) - \tilde{H})\,\Delta\rho$$

$$= \tilde{\rho}\Delta e - \frac{\tilde{p}}{\tilde{\rho}}\,\Delta\rho \tag{2.26}$$

where

$$\tilde{e} = \theta e_L + (1-\theta)\, e_R$$

$$\tilde{p} = \tilde{\rho}(\tilde{H} - \tilde{e} - \tfrac{1}{2}(\tilde{u}^2 + \tilde{v}^2 + \tilde{w}^2)). \tag{2.27}$$

Thus $\tilde{a}$ must be defined in such a way that

$$\Delta e = \frac{\tilde{p}}{\tilde{\rho}^2}\,\Delta\rho\ . \tag{2.28}$$

From $\Delta c_i = 0$ $i = 1,\ldots,n$ it follows that

$$\Delta p = \Delta \sum_{i=1}^{n} \rho\, c_i R_i T = (\tilde{\rho}\Delta T + \tilde{T}\Delta\rho) \sum_{i=1}^{n} \tilde{c}_i R_i \tag{2.29}$$

and

$$\Delta e = \Delta \sum_{i=1}^{n} c_i e_i(T) = (\sum_{i=1}^{n} \tilde{c}_i \frac{\Delta e_i}{\Delta T})\,\Delta T \tag{2.30}$$

where

$$\tilde{T} = \theta T_L + (1-\theta)\, T_R \tag{2.31}$$

$$\frac{\Delta e_i}{\Delta T} = \begin{cases} \dfrac{\Delta e_i}{\Delta T} & \text{if } \Delta T \neq 0 \\[2ex] \dfrac{de_i}{dT} & \text{if } \Delta T = 0\ . \end{cases} \tag{2.32}$$

From (2.29), (2.30) and $\Delta p = \tilde{a}^2\Delta\rho$ it is easily seen that (2.28) is satisfied if

$$\tilde{a}^2 = \frac{\tilde{p}}{\tilde{\rho}} \left\{ \frac{\tilde{\rho}\,\tilde{T} \sum_{i=1}^{n} \tilde{c}_i R_i}{\tilde{p}} + \frac{\sum_{i=1}^{n} \tilde{c}_i R_i}{\sum_{i=1}^{n} \tilde{c}_i \frac{\Delta e_i}{\Delta T}} \right\}. \tag{2.33}$$

From this expression it does not follow immediately that the speed of sound at the Roe-averaged state converge to the speed of sound when the left- and right-state converage to each other. But from (2.6), (2.7), (2.8) and (2.14) we find that

$$a^2 = \frac{p}{\rho}\{1 + \frac{\sum_{i=1}^{n} c_i R_i}{\sum_{i=1}^{n} c_i \frac{de_i}{dT}}\} . \tag{2.34}$$

Thus, indeed $\tilde{a} \to a$ if $q_R \to q_L \to q$.

Remark: Entropy fix.
It is well known that (2.10) is not consistent with an entropy inequality and the scheme might converge to a nonphysical solution. A slight modification of the coefficient of numerical viscosity can remedy the entropy violation problem [9]. This is done by replacing the terms $|\tilde{u}-\tilde{a}|$ and $|\tilde{u}+\tilde{a}|$ in (2.24) by

$$|\tilde{u}-\tilde{a}| \to \tilde{a}\Psi(|\tilde{m}-1|) \qquad\qquad |\tilde{u}+\tilde{a}| \to \tilde{a}\Psi(|\tilde{m}+1|) \tag{2.35}$$

where $\tilde{m} = \tilde{u}/\tilde{a}$,

$$\Psi(x) = \begin{cases} x & \text{if } x \geqq \delta \\ \frac{x^2+\delta^2}{2\delta} & \text{if } 0 \leqq x \leqq \delta \end{cases} \tag{2.36}$$

and $\delta = \varepsilon(|\tilde{m}| + 1)$.
Note that δ depends on $\tilde{m}$. This is especially useful for hypersonic flows [9]. Numerical experiments show that increasing the coefficient ε has the effect of speeding up the convergence rate but smears the discontinuities slightly, since the bigger the ε the larger the numerical dissipations being added. Typical values of ε are $0.1 \leqq \varepsilon \leqq 0.5$.

3 ROE-TYPE FLUX-DIFFERENCE-SPLITTING FOR THERMALLY PERFECT GAS

In case of a thermally perfect gas the construction of a Roe type scheme can be obtained along the same lines as for an n species chemically reacting gas in thermal equilibrium. The results are presented in this section.

For a thermally perfect gas the state vector q is given by

$$q = (\rho,\ \rho u,\ \rho v,\ \rho w,\ E)^T \tag{3.1}$$

and the flux vector f(q) is given by

$$f(q) = (\rho u,\ \rho u^2+p,\ \rho uv,\ \rho uw,\ (E+p)u)^T. \tag{3.2}$$

The pressure p is given by the equation of state

$$p = \rho RT \tag{3.3}$$

and the internal energy is a function of the temperature only:

$$e = e(T) \; . \tag{3.4}$$

The construction of the numerical flux function $f(q_L,q_R)$ can be obtained in exactly the same way as in section 2. With the abbreviate notation $f_i = f_i(q_L,q_R)$ we find that

$$f_1 = \tfrac{1}{2}\{(\rho u)_L + (\rho u)_R - (\tilde{\alpha}_1|\tilde{u}| + \tilde{\alpha}_2|\tilde{u}-\tilde{a}| + \tilde{\alpha}_5|\tilde{u}+\tilde{a}|)\}$$

$$f_2 = \tfrac{1}{2}\{(\rho u^2+p)_L + (\rho u^2+p)_R - (\tilde{\alpha}_1|\tilde{u}|\tilde{u} + \tilde{\alpha}_2|\tilde{u}-\tilde{a}|(\tilde{u}+\tilde{a}) + \tilde{\alpha}_5|\tilde{u}+\tilde{a}|(\tilde{u}+\tilde{a}))\}$$

$$f_3 = \tfrac{1}{2}\{(\rho uv)_L + (\rho uv)_R - (\tilde{\alpha}_1|\tilde{u}|\tilde{v} + \tilde{\alpha}_2|\tilde{u}-\tilde{a}|\tilde{v} + \tilde{\alpha}_3|\tilde{u}| + \tilde{\alpha}_5|\tilde{u}+\tilde{a}|\tilde{v})\}$$

$$f_4 = \tfrac{1}{2}\{(\rho uw)_L + (\rho uw)_R - (\tilde{\alpha}_1|\tilde{u}|\tilde{w} + \tilde{\alpha}_2|\tilde{u}-\tilde{a}|\tilde{w} + \tilde{\alpha}_4|\tilde{u}| + \tilde{\alpha}_5|\tilde{u}+\tilde{a}|\tilde{w})\}$$

$$f_5 = \tfrac{1}{2}\{((E+p)u)_L + ((E+p)u)_R - (|\tilde{u}|(\Delta E - \frac{\tilde{H}}{\tilde{a}^2}\Delta p - \tilde{\rho}\tilde{u}\Delta u)$$

$$+ \tilde{\alpha}_2|\tilde{u}-\tilde{a}|(\tilde{H}-\tilde{u}\tilde{a}) + \tilde{\alpha}_5|\tilde{u}+\tilde{a}|(\tilde{H}+\tilde{u}\tilde{a}))\} \tag{3.5}$$

where $\tilde{\rho}$, θ, $\tilde{u}$, $\tilde{v}$, $\tilde{w}$, $\tilde{H}$ are defined as in (2.20) and

$$\tilde{\alpha}_1 = \Delta\rho - \frac{\Delta p}{\tilde{a}^2} \qquad \tilde{\alpha}_2 = \frac{1}{2\tilde{a}^2}(\Delta p - \tilde{\rho}\tilde{a}\Delta u)$$

$$\tilde{\alpha}_3 = \tilde{\rho}\Delta v \qquad \tilde{\alpha}_4 = \tilde{\rho}\Delta w \qquad \tilde{\alpha}_5 = \frac{1}{2\tilde{a}^2}(\Delta p + \tilde{\rho}\tilde{a}\Delta u) \; . \tag{3.6}$$

The speed of sound $\tilde{a}$ is computed according to

$$\tilde{a}^2 = \frac{\tilde{p}}{\tilde{\rho}} \left\{ \frac{\tilde{\rho} R \tilde{T}}{\tilde{p}} + \frac{R}{\frac{\Delta e}{\Delta T}} \right\} \tag{3.7}$$

where $\tilde{p}$ and $\tilde{T}$ are defined as in (2.27) and (2.31).

Furthermore $\frac{\Delta e}{\Delta T}$ is replaced by $\frac{de}{dT}$ if $\Delta T = 0$.

For the special case of a calorically perfect gas we have

$$e = c_v T \tag{3.8}$$

where c_v is constant. Then it follows that

$$\Delta e = c_v \Delta T, \; \tilde{e} = c_v \tilde{T} \tag{3.9}$$

and therefore

$$\tilde{a}^2 = (\gamma - 1)\,(\tilde{e} + \frac{\tilde{p}}{\tilde{\rho}}) = (\gamma - 1)\,(\tilde{H} - \frac{1}{2}(\tilde{u}^2 + \tilde{v}^2 + \tilde{w}^2)) \tag{3.10}$$

where $\gamma - 1 = R/c_v$. Hence in this case the scheme is exactly the same as the original scheme presented in [6].

4 CONCLUSIONS

A numerical flux function has been derived for an n-species chemically reacting gas in thermal equilibrium. Both the derivation and the obtained expressions are much more simple than generally encountered in the literature. The special case n = 1 corresponds to a thermally perfect gas. For completeness, the numerical flux function is also given for this special case and it is shown that the scheme is exactly the same as the original Roe scheme in case of a calorically perfect gas.

REFERENCES

[1] GLAISTER, P., "An Approximate Linearized Riemann Solver for the Euler Equations for Real Gases", J. Comput. Phys. 74, 382, 1988.

[2] GROSSMAN, B., WALTERS, R.W., "An Analysis of Flux-Split Algorithms for Euler's Equations with Real Gases", AIAA Paper 87-1117-CP, 1987.

[3] GROSSMAN, B., CINELLA, P., "The Computation of Nonequilibrium Chemically Reacting Flows", Computers & Structures, Vol. 30, No. 1/2, pp. 79-93, 1988.

[4] LIOU, M. VAN LEER, B., SHUEN, J., "Splitting of Inviscid Fluxes for Real Gases", NASA TM 100856, 1988.

[5] MONTAGNÉ, J.L., YEE, H.C., VINOKUR, M., "Comparative Study of High-Resolution Shock-Capturing Schemes for a Real Gas", NASA TM 100004, 1987.

[6] ROE, P.L., "Approximate Riemann Solvers, Parameter Vectors, and Difference schemes", J. Comput. Phys. 43, 357, 1981.

[7] SHUEN, J., LIOU, M., "Flux Splitting Algorithms for Two-Dimensional Viscous Flows with Finite-Rate-Chemistry", AIAA-89-0388, 1989.

[8] VINOKUR, M., LIU, J., "Equilibrium Gas Flow Computations II: An Analysis of Numerical Formulations of Conservation Laws", AIAA Paper 88-0127, 1988.

[9] YEE, H.C., "Upwind and Symmetric Shock Capturing Schemes, NASA TM 89464, 1987.

Multidimensional upwind schemes for the Euler equations using fluctuation distribution on a grid consisting of triangles.

R. Struijs and H. Deconinck,
Von Karman Institute for Fluid Dynamics,
Waterloose steenweg 72, 1640 Sint-Genesius-Rode, Belgium

Summary

The basic elements are presented for an Euler solver with genuinely multidimensional upwind space discretization. The fluctuation splitting approach developed by Roe is used on a grid consisting of triangles. The Euler Equations are decomposed in characteristics using the Deconinck-Hirsch optimal decomposition method. Several upwind discretization methods, with different behaviour concerning monotonicity and diffusion have been incorporated and tested, on scalar problems and on the Euler equations. The resulting upwind schemes show little grid dependency. Convergence for the Euler equations is obstructed by noise in the decomposition normals. The effect on the outcome of the calculations is minimal, as can be shown by calculations with the decomposition normal frozen. However, this problem remains one of the principal subjects of investigation.

1. Introduction

In one space dimension, upwind methods have reached a high level of sophistication. The basic ingredient of these methods is an approximate Riemann solver which recognizes the nonlinear phenomena such as contacts, shocks, and expansion waves, and treats them in a proper way. A straightforward extension is constructed in two or three space dimensions based on the one dimensional methods: in a finite volume context, the normals to the cell faces serve as the local space coordinates along which the one dimensional Riemann solver is applied. However, this introduces an unphysical grid dependency of the solution. Therefore, true multidimensional upwind concepts are needed.

During recent years, attempts have been made to develop multidimensional upwind methods for the numerical solution of the Euler equations. Basic ingredients are a pattern recognition phase, in which the local flow gradients are decomposed in relevant simple wave patterns, and a distribution step where the wave information is used to distribute the fluctuations over the nodal points in an upwind manner.

In section 2, a brief discussion is presented on upwind distribution methods for scalar equations. The theoretical framework for these methods is supplied by Roe [1], [2], [3], [4], [5]. Basis of the method is the calculation of the flux balance over a volume, which is proportional to the time derivative of the unknowns in the volume. This fluctuation is to be distributed over the nodes of the volume. Upwind discretizations are obtained by making the weights of the distribution function of the convection speed.

As for the pattern recognition, Roe [2], [3] proposed to use gradients in the data to construct a model flow of four acoustic waves, an entropy wave and a shear wave. Deconinck and Hirsch [6] proposed a decomposition method based on 2D characteristic theory. Characteristic compatibility equations are selected which lead to an optimal decoupling of the Euler equations into a set of scalar 2D convection equations. Four compatibility equations are used, representing an entropy wave, a shear wave and two acoustic waves. As in Roe's decomposition, the propagation velocities depend on the local flow gradients. A conservative cell vertex scheme based on the latter decomposition is reported by Powell and van Leer [3]. A third order accurate evaluation of the flux balance over a quadrilateral cell is distributed over the cell vertices with upwind weights based on the genuinely 2D decomposition. Lacor and Hirsch [8] show results based on a genuinely 2D finite volume method for subsonic and transonic flow in channels

In the present work, the Deconinck-Hirsch decomposition is used on a grid which consists of triangles with cell vertex unknowns (section 3). The flux balance over a triangle is distributed to the vertices with upwind weights related to the convection speeds obtained from the decomposition. Results are shown with two different first order schemes, one of which shows little diffusion, but is not monotone. The simple wave decomposition allows for an easy application of characteristic boundary conditions, where the information going into the domain is replaced by numerical characteristic boundary conditions (see section 4). This however poses some problems concerning conservation.

The solver has been tested on a simple supersonic testcase. It is clearly demonstrated in the results that the use of the optimally decoupling upwinding directions leads to better shock resolution than e.g. streamline upwind directions.

2. The fluctuation splitting method for a scalar convection equation

In a number of articles [1],[2],[3], Roe has described the fluctuation splitting method for the scalar linear hyperbolic equation, in two space dimensions :

$$\frac{\partial u}{\partial t} + a\frac{\partial u}{\partial x} + b\frac{\partial u}{\partial y} = 0 \qquad \text{or} \qquad \frac{\partial u}{\partial t} + \vec{a}\cdot\vec{\nabla}u = 0\ . \tag{2.1}$$

For readability of the following, the main points of his approach are briefly resumed in this section.

The fluctuation ϕ over a triangle, evaluated with Gauss' Theorem, is given by

$$\phi = \iint u_t\, dS = -\iint \vec{a}\cdot\vec{\nabla}u\, dS = \oint u(\vec{a}\cdot d\vec{n})\ , \tag{2.2}$$

where $\vec{n}$ is the inward normal at a point on the circumference of the triangle and $\vec{a}$ is the convection speed $\binom{a}{b}$. Approximating the integral by assuming the data to be piecewise linear over the triangle gives

$$\phi = \tfrac{1}{2}[\ (u_1+u_2)(\vec{n}_3\cdot\vec{a}) + (u_2+u_3)(\vec{n}_1\cdot\vec{a}) + (u_3+u_1)(\vec{n}_2\cdot\vec{a})\ ] \tag{2.3}$$

where the *scaled* inward normal $\vec{n}_i$ is opposite of node i, see fig. 2. The sum of the normals adds up to zero. Defining

$$k_i = \tfrac{1}{2}(\vec{n}_i\cdot\vec{a})\ , \qquad \phi \text{ can be rewritten as} \qquad \phi = -\textstyle\sum_{i=1}^{3} k_i u_i\ . \tag{2.4}$$

An explicit scheme is obtained as follows : Associate to each node of a triangle T a part of the area S_i^T and a positive coefficient α_i^T such that

$$\alpha_1^T + \alpha_2^T + \alpha_3^T = 1\ , \qquad S_1^T + S_2^T + S_3^T = S^T \tag{2.5}$$

with S^T the total area of the triangle.
The discrete form of conservation law (2.1) written for triangle T is then given by

$$\frac{\sum_{i=1}^{3}\Delta u_i^T S_i^T}{\Delta t} = \phi^T, \qquad \Delta u = u^{n+1} - u^n \tag{2.6a}$$

and with each node an increment

$$S_i^T\Delta u_i^T = \Delta t\alpha_i^T\phi^T \tag{2.6b}$$

can be associated such that (2.6a) is satisfied.

The coefficient α_i determines which part of the fluctuation is sent to node i. The global update for point i is then obtained by summing up all the contributions of its surrounding triangles :

$$S_i\Delta u_i = \sum_T S_i^T\Delta u_i^T = \sum_T \Delta t\alpha_i^T\phi^T \tag{2.7}$$

or :

$$\Delta u_i = \frac{1}{S_i}\sum_T \Delta t\alpha_i^T\phi^T \tag{2.8}$$

where $S_i = \sum_T S_i^T$
To see that the scheme is conservative, consider the sum of (2.8) over all nodes

$$\frac{\sum_i S_i\Delta u_i}{\Delta t} = \sum_i\sum_T \alpha_i^T\phi_T \tag{2.9}$$

and since the α_i^T sum up to 1 for each triangle,

$$\frac{\sum_i S_i\Delta u_i}{\Delta t} = \sum_T \phi^T\ . \tag{2.10}$$

Clearly, all the contributions of internal sides in the right-hand-side cancel, and only boundary terms are left in the flux balance, and the left-hand-side is a consistent discretization of the time variation of u over the domain.

The area's S_i^T in the present work have been chosen as the subdivisions of the triangles obtained by drawing its medians $\vec{r}_i$. They have the property $S_i^T = \frac{1}{3}S^T, i = 1,2,3$. As a result of this, the area $S_i = \sum_T S_i^T$ is the Voronoi region around point i, see fig. 1.

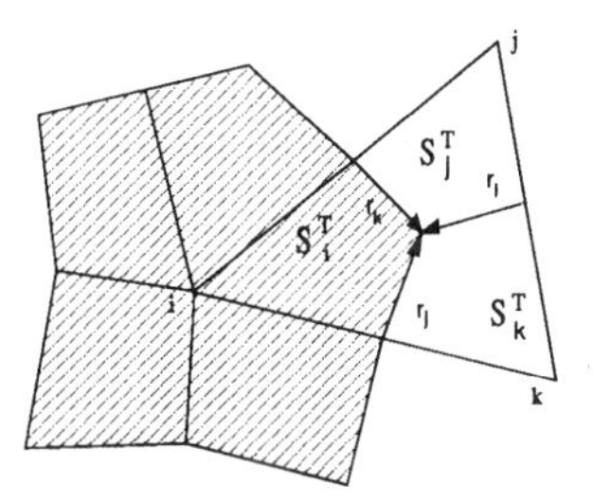

Fig. 1. Voronoi region obtained by subdivision of the triangles.

A wide variety of schemes can be constructed by changing the definition of the α_i, some of which are described in [1], [2], [3].

Upwind schemes are obtained if the α_i depend on the orientation of the convection speed $\vec{a}$. For the distribution, two cases may be considered. In case of one inflow side (fig. 2a) the fluctuation can be sent to the downwind vertex. In case of two inflow sides (fig. 2b), the fluctuation can be split over the two downwind vertices. The details of the coefficients are given in the following paragraphs. For non-upwind schemes like Lax-Wendroff, all vertices get a contribution, irrespective of the relative orientation of the convection speed with respect to the triangle.

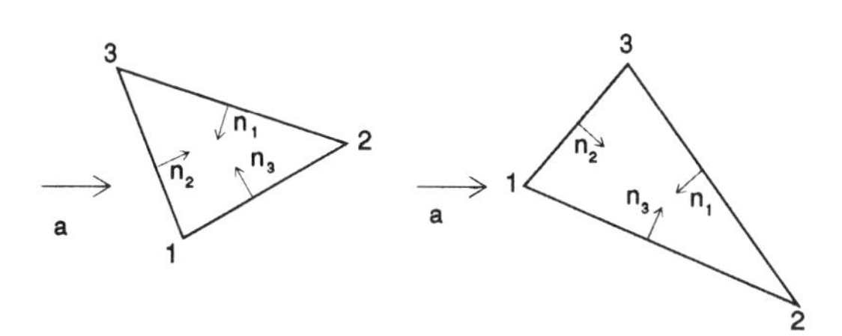

Fig. 2a. One inflow side. *Fig. 2b. Two inflow sides.*

In the following, two particular schemes are discussed. One of them satisfies a local monotonicity property, as defined by Roe [2]. Writing

$$u_p^{n+1} = u_p^n + \sum_T U_T(u_{T_1}^n, u_{T_2}^n, u_{T_3}^n) \tag{2.11}$$

with U_T the increment of u_p coming from triangle T with nodes T_1, T_2 and T_3, local monotonicity requires

$$\frac{\partial U_T}{\partial u_{T_i}} \geq 0 \qquad T_i \neq p\,, \tag{2.12a}$$

$$1 + \sum_T \frac{\partial U_T}{\partial u_p} \geq 0 \qquad T_i = p\,. \tag{2.12b}$$

A monotone upwind scheme

A monotone scheme satisfying equation (2.12) is obtained by defining the coefficients α_i as follows : For the case of two inflow sides (fig. 2b)

$$\alpha_1 = 0 \; , \qquad \alpha_2 = \frac{k_2(u_2 - u_1)}{\sum k_i u_i} \; , \qquad \alpha_3 = \frac{k_3(u_3 - u_1)}{\sum k_i u_i} \; . \tag{2.13}$$

The two coefficients add up to unity, ensuring conservation. In the case of one inflow side, the entire fluctuation is sent to the downstream point (fig. 2a), which means

$$\alpha_2 = 1 \; , \qquad \alpha_1 = \alpha_3 = 0 \; . \tag{2.14}$$

The scheme is monotone and therefore stable under the timestep restriction for point i :

$$\Delta t_i \leq \frac{S_i}{\sum_T max(0, k_i^T)} \; . \tag{2.15}$$

This introduces a local stability condition $CFL \leq 1$ with CFL defined as

$$CFL = \frac{\Delta t_i \sum_T max(0, k_i^T)}{S_i} \; . \tag{2.16}$$

A non-monotone low diffusion upwind scheme

A simple geometric interpretation for the case with two inflow sides is sketched in fig. 3. The amount sent to a vertex depends on the projection of the corresponding edge on the normal to the convection speed vector.

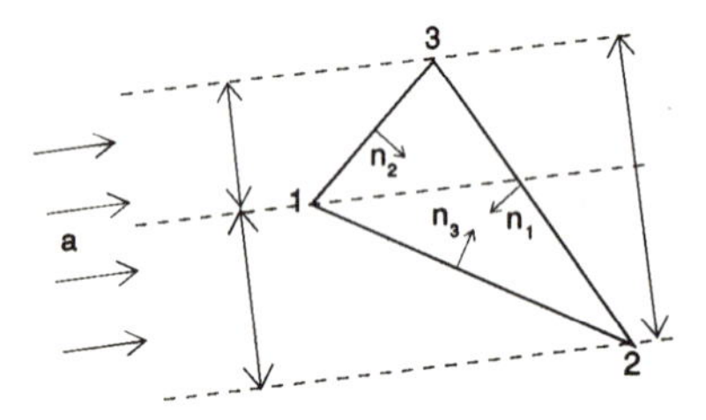

Fig. 3. Geometrical interpretation for the non-monotonic upwind scheme.

$$\alpha_1 = 0 \; , \qquad \alpha_2 = -\frac{k_2}{k_1} \; , \qquad \alpha_3 = -\frac{k_3}{k_1} \; . \tag{2.17}$$

The two coefficients again add up to unity ensuring conservation, but the partial derivatives don't satisfy equation (2.12). The resulting scheme is therefore not monotone.
The same scheme is referred to in finite element literature as the σ-weighting scheme [14].

Other schemes

A number of other fluctuation splitting schemes have been formulated [2] and tested, and will be discussed in a forthcoming paper. These include a monotone upwind scheme identical to the classical finite volume scheme on unstructured grids as well as a Lax-Wendroff fluctuation splitting scheme, and some schemes very close to shock capturing finite element Streamwise Upwind Petrof Galerkin schemes (SUPG).

Application to non-linear hyperbolic equations

Up till now, the theory has been written down for the linear convection equation (2.1), where the convection speed $\vec{a}$ was independent of the solution. For the non-linear case however, the equation in conservation form is given by

$$\frac{\partial u}{\partial t}+\frac{\partial f(u)}{\partial x}+\frac{\partial g(u)}{\partial y}=0 , \quad \text{or} \quad \frac{\partial u}{\partial t}+\vec{\nabla}\cdot\vec{f}=0 , \tag{2.18a}$$

and the convection speed $\vec{a}=\binom{a}{b}$ is obtained from

$$a=\frac{\partial f}{\partial u} \quad \text{and} \quad b=\frac{\partial g}{\partial u} . \tag{2.18b}$$

The distribution functions used for the various schemes all depend on the convection speeds. In order to obtain these speeds from the flux functions f and g, even in points where the derivatives with respect to u are not defined, as in a shock region, a two-dimensional Roe type linearization is applied. This means that $\vec{a}(u_i,u_j,u_k)$ is chosen to fulfill

$$\oint \vec{f}\cdot\vec{n}ds=\vec{a}\oint u\vec{n}ds , \quad \text{or} \quad \sum_i (f,g)_i^T\cdot\vec{n}_i=\vec{a}\cdot[\sum_i u_i\vec{n}_i] . \tag{2.19}$$

For consistency with equation (2.18b), this definition of $\vec{a}$ can be further restricted by taking the equations for the x and y components separately

$$\sum_i f_i n_i^x = a\sum_i u_i n_i^x , \qquad \sum_i g_i n_i^y = b\sum_i u_i n_i^y . \tag{2.20}$$

Using this definition of a and b ensures conservation. A very attractive consequence of this linearization is the reduction to the Roe averaging [13] in case the flow becomes one-dimensional in an arbitrary direction. Take e.g. $u_i=u_j\neq u_k$ in equation (2.20). Substitution gives

$$a=\frac{f_k-f_i}{u_k-u_i} \quad \text{and} \quad b=\frac{g_k-g_i}{u_k-u_i} , \tag{2.21}$$

which is just the Roe linearization for a one-dimensional scheme in a direction normal to $\vec{ij}$. The linearization consequently obtains the correct shock speed.

An extended discussion on scalar schemes, and results on both linear and non-linear hyperbolic equations will be given in a forthcoming paper.

3. Application to the 2D Euler equations for inviscid flow.

The solution of a hyperbolic system of equations by means of simple wave patterns implies the use of what is called a pattern recognition method. This means that at a given timelevel the data have to be interpreted in order to find out which are the significant wave phenomena present in the flow. Once this is achieved, the problem is reduced to scalar transport equations, and can be solved with the results of the previous section.

Deconinck and Hirsch [2] proposed a decomposition method where characteristic compatibility equations are selected which lead to an optimal decoupling of the Euler equations into a set of scalar 2D convection equations. Starting from the Euler equations in divergence form,

$$\frac{\partial U}{\partial t}+\frac{\partial F}{\partial x}+\frac{\partial G}{\partial y}=0 , \quad \text{or} \quad \frac{\partial U}{\partial t}+\vec{\nabla}\cdot\vec{F}=0 , \quad U=(\rho,\rho u,\rho v,E)^T , \tag{3.1}$$

the system of characteristic variables is obtained by a transformation using a 4×4 matrix L given in [6], [7] or [9] :

$$\partial U=L^{-1}\partial W \qquad \partial W=L\partial U \tag{3.2a}$$

$$L^{-1} = (r^1, r^2, r^3, r^4)\,, \qquad L = (l^1, l^2, l^3, l^4)^T. \tag{3.2b}$$

The column vectors r^α and row vectors l^α are closely related to the right respectively left eigenvectors of the Jacobians. In vector notation, the transformed system becomes

$$\frac{\partial W}{\partial t} + D_x \frac{\partial W}{\partial x} + D_y \frac{\partial W}{\partial y} + Q = 0\,, \qquad Q = Q_x + Q_y \tag{3.3}$$

where D_x and D_y are diagonal matrices, which have as diagonal elements the convection speeds in x respectively y direction. The terms Q_x and Q_y represent off diagonal terms, which appear since in general the Jacobians of the fluxes F and G do not commute. The different terms in equation (3.3) are given by :

$$\partial W = \begin{pmatrix} \partial\rho - \frac{1}{c^2}\partial p \\ \vec{s}^{(1)} \cdot \partial\vec{u} \\ \vec{\kappa}^{(2)}\partial\vec{u} + \frac{1}{\rho c}\partial p \\ -\vec{\kappa}^{(2)}\partial\vec{u} + \frac{1}{\rho c}\partial p \end{pmatrix}, \quad D_x = \begin{pmatrix} u & & & \\ & u & & \\ & & u + c\kappa_x^{(2)} & \\ & & & u - c\kappa_x^{(2)} \end{pmatrix}, \quad D_y = \begin{pmatrix} v & & & \\ & v & & \\ & & v + c\kappa_y^{(2)} & \\ & & & v - c\kappa_y^{(2)} \end{pmatrix}, \tag{3.4a}$$

$$Q_x = \begin{pmatrix} 0 \\ \frac{1}{\rho} s_x^{(1)} \partial_x p \\ c s_x^{(2)} (s_x^{(2)} \partial_x u + s_y^{(2)} \partial_x v) \\ c s_x^{(2)} (s_x^{(2)} \partial_x u + s_y^{(2)} \partial_x v) \end{pmatrix}, \quad Q_y = \begin{pmatrix} 0 \\ \frac{1}{\rho} s_y^{(1)} \partial_y p \\ c s_y^{(2)} (s_x^{(2)} \partial_y u + s_y^{(2)} \partial_y v) \\ c s_y^{(2)} (s_x^{(2)} \partial_y u + s_y^{(2)} \partial_y v) \end{pmatrix}. \tag{3.4b}$$

The convection speeds for the four characteristic components are therefore

$$\vec{\lambda}^{(1)} = \vec{u}, \quad \vec{\lambda}^{(2)} = \vec{u}, \quad \vec{\lambda}^{(3)} = \vec{u} + c\vec{\kappa}^{(2)}, \quad \vec{\lambda}^{(4)} = \vec{u} - c\vec{\kappa}^{(2)}. \tag{3.5}$$

The vectors $\vec{s}^{(1)}$ and $\vec{s}^{(2)}$ are normal to the vectors $\vec{\kappa}^{(1)}$ respectively $\vec{\kappa}^{(2)}$. The selection of the optimal characteristics is done by choosing two normals $\vec{\kappa}^{(1)}$ and $\vec{\kappa}^{(2)}$ which render the transformation to give the minimal coupling. The first decomposition normal $\vec{\kappa}^{(1)}$ is proportional to the pressure gradient, while the second decomposition normal $\vec{\kappa}^{(2)}$ is a function of the strain rate tensor. More details on the decomposition can be found in the original paper [2] and [6], [7], [9].

Note that both normals are a function of derivatives of the flow variables. This implies that the decomposition process may be sensitive to perturbations in the flow field with a short wavelength. Although this may deteriorate the convergence, the accuracy of the solution may suffer less from this effect. The noise in the direction of the normals will appear only for small gradients, which goes along with small residuals, thus small changes in the flow field.

The parts of the residual which are convected in x and y - direction are given by respectively

$$D_x \frac{\partial W}{\partial x} = L \frac{\partial F}{\partial x} - Q_x \quad \text{and} \quad D_y \frac{\partial W}{\partial y} = L \frac{\partial G}{\partial y} - Q_y. \tag{3.6}$$

These convective residuals have to be distributed over the nodes of a triangle in an upwind manner. The non monotone low diffusion scheme of equation (2.17) applied to characteristic equation β uses

$$\alpha_1^\beta = 0, \quad \alpha_2^\beta = -\frac{k_2^\beta}{k_1^\beta}, \quad \alpha_3^\beta = -\frac{k_3^\beta}{k_1^\beta} \tag{3.7}$$

where $k_i^\beta = \frac{1}{2}(\vec{n}_i \cdot \vec{\lambda}^\beta)$. The contribution to node 2 is then

$$\alpha_2^\beta [l^\beta \vec{\nabla} \cdot \vec{F} - Q^\beta] \tag{3.8}$$

or in discretized form

$$\alpha_2^\beta [l^\beta \sum_i \frac{1}{2} \vec{F}_i \cdot \vec{n}_i - Q^\beta S]\,. \tag{3.9}$$

At present, the source term is split equally over the nodes. Other possibilities exist, the simplest one being the use of the same coefficients as for the convective residuals. Thus, the convective parts add up to $l^\beta \vec{\nabla} \cdot \vec{F} - Q^\beta$ while the source terms add up to Q^β . Conservation is therefore guaranteed.

For the monotone scheme of equation (2.13) applied to characteristic equation β , the distribution coefficients are :

$$\alpha_1^\beta = 0, \quad \alpha_2^\beta = \frac{[\frac{1}{2}\vec{\lambda}^\beta \cdot \vec{n}_2] l^\beta (U_2 - U_1)}{l^\beta \sum \frac{1}{2}\vec{F}_i \cdot \vec{n}_i - Q^\beta S} , \quad \alpha_3^\beta = \frac{[\frac{1}{2}\vec{\lambda}^\beta \cdot \vec{n}_3] l^\beta (U_3 - U_1)}{l^\beta \sum \frac{1}{2}\vec{F}_i \cdot \vec{n}_i - Q^\beta S} . \tag{3.10}$$

In order to assure conservation, the choice of $\vec{\lambda}^\beta$ has to be taken carefully. Adding up the contributions sent to the nodes, it follows that the following definition ensures conservation, for the x component :

$$\lambda_x^\beta l^\beta \frac{\partial U}{\partial x} = l^\beta \frac{\partial F}{\partial x} - Q_x^\beta , \tag{3.11}$$

which is the system equivalent of equation (2.20). In discretized form, after integrating over a triangle,

$$\lambda_x^\beta = \frac{l^\beta \frac{1}{2} \sum F_i n_{ix} - Q_x^\beta S}{l^\beta \frac{1}{2} \sum U_i n_{ix}} . \tag{3.12}$$

When the fluctuation has been distributed over the nodes of the triangle, the characteristic information can be transformed back again to changes in conservative variables, using the discretized version of equation (3.2) :

$$\Delta U = \sum r^\beta \Delta W^\beta . \tag{3.13}$$

The transformation matrices are at present evaluated on basis of the triangle averaged conservative variables.

Boundary conditions

The presence of all characteristic information invites to apply characteristic boundary conditions. For a solid wall the following procedure was used, based on work of Chakravarthy [10], Pandolfi [11] and Borsboom [12] : The outgoing characteristics, which are not to be affected by the boundary condition imposed, give a certain contribution to the conservative variables at the wall. In general, these updated values at the wall will not satisfy the solid wall boundary condition, which says that the flow should be tangential to the wall. The strength of the ingoing characteristic is adjusted such that the resulting conservative variables satisfy the desired boundary conditions.

The disadvantage of this method is the effect it has on the fluctuation calculated in the boundary volume. Since the fluctuation after changing the ingoing characteristic is not any more proportional to the flux through the boundary, conservation is lost at the boundary.

4. Results for a supersonic flow testcase

The results in the present report are obtained without any smoothing of the gradients used for the decomposition normals. The only step taken is storage of the four convection speeds and the two decomposition normals for use if the gradients dropped below 10^{-6} times a reference gradient calculated from the flow variables.

As a testcase the oblique shock reflection on a flat plate has been taken. Fig. 4 shows the grid used. It is constructed to have the hypotenuse of the triangles in four directions, to prevent grid bias due to asymmetry. Fig. 5 shows the isomachlines with an interval of 0.05 for the solution obtained with the non monotone scheme. At the reflection point of the shock, the effect of lack of conservation can be observed. It is also apparent from the cuts at Y=0. and Y=0.5 (fig. 6). The shock is captured in typically two grid cells, both for the impinging and for the reflected shock. This is exceptional for a first order accurate scheme.

The solution is converged to a level of about 10^{-2} for the density residual after 600 iterations (fig. 7). The convergence then slows down and stagnates. When the normals are frozen and the calculation is continued, convergence to machine accuracy is obtained without changing the solution within plotting accuracy. The decomposition normal $\vec{\kappa}^{(1)}$ given in fig. 8, which is proportional to the pressure gradient, shows that in the shock the gradient is well captured. The scatter in both $\vec{\kappa}^{(1)}$ and $\vec{\kappa}^{(2)}$ (fig. 9) is due to the uniformity of the flow field and the non monotone character of the scheme. The velocity field for the first (or second) characteristic, and for the third characteristic are given in fig. 10 respectively fig. 11.

When for the non monotone scheme the decomposition normals are fixed to be parallel with the velocity field of fig. 10, the shocks tend to be more diffused (fig. 12) compared with the correct 2D decomposition (fig. 5). . For this choice of $\vec{\kappa}^{(1)}$, the source term is maximized instead of minimized. This illustrates clearly the influence of the decomposition.

The results for the monotone scheme (fig. 13 and 14) show that the solution is monotone indeed, but the same conservation problems at the solid wall boundary appear. This scheme is more diffusive than the non monotone scheme as was known from the scalar test results (not shown). The convergence behaviour is similar to the non monotone scheme.

5. Conclusion

Two fluctuation splitting schemes proposed by Roe have been implemented and tested on the Euler equations. The newly developed method to obtain the convection speed for non-linear hyperbolic equations has the correct limiting behaviour for one-dimensional flow cases. The extension to systems of equations is possible, while retaining the conservation property.

The effect of the proper choice of the decomposition normals $\vec{\kappa}^{(1,2)}$ is apparent when the results are compared with a streamwise upwind discretization for the same discretization scheme.

Further investigation is under way for the implementation of conservative boundary conditions.

Acknowledgements

We thank Mr. Roe for the stimulating discussions on the scalar convection schemes.

References

1 P.L. Roe ; Numerical Algorithms for the Linear Wave Equation ; Royal Aircraft Establishment Technical Report 81047, 1981

2 P.L. Roe ; Linear Advection Schemes on Triangular Meshes ; Cranfield Institute of Technology CoA Report No 8720, Cranfield, Bedford, U.K., November 1987

3 P.L. Roe ; A Basis for Upwind Differencing of the Two-Dimensional Unsteady Euler Equations Numerical Methods for Fluid Dynamics II, pp 55-80, eds. K.W. Morton, M.J. Baines, Oxford University Press, 1986

4 P.L. Roe ; The use of the Riemann problem in Finite Difference Schemes ; Proc. of the 7th International Conference on Numerical Methods for Fluid Dynamics, Springer Verlag, 1981

5 P.L. Roe Discrete Models for the Numerical analysis of Time-Dependent Multidimensional Gas Dynamics Journal of Computational Physics 63, 458-476 (1986)

6 H. Deconinck, C Hirsch, J. Peuteman ; Characteristic Decomposition Methods for the Multidimensional Euler Equations ; Lecture Notes in Physics 264, pp 216-221, Springer, 1986

7 K.G. Powell, B. van Leer ; A Genuinely Multi-Dimensional Upwind Cell-vertex Scheme for the Euler Equations ; AIAA 89-0095, january 9-12, 1989, Reno, Nevada

8 Ch Hirsch, C. Lacor ; Upwind Algorithms Based on a Diagonalization of the Multidimensional Euler Equations ; AIAA 89-1958, June 13-15, 1989, Buffalo, New York.

9 H. Deconinck ; A Survey of Upwind Principles for the Multidimensional Euler Equations ; Von Karman Institute Lecture Series, VKI-LS 87-04

10 S. Chakravarthy ; Euler Equations - Implicit Scheme and Boundary Condition ; AIAA-J, 21, pp 699-706

11 M. Pandolfi ; Upwind Formulation for the Euler Equations ; Von Karman Institute Lecture Series, VKI-LS 87-04

12 M. Borsboom ; A Numerical Solution Method for the Steady-State Compressible Navier-Stokes Equations, with Applications to Subsonic Channel and Blade-to-Blade Flow ; Thesis, University of Liège, 1987

13 P.L. Roe ; Approximate Riemann Solvers, Parameter Vectors and Difference Schemes ; JCP 43, 357-372, 1981

14 T.J.R. Hughes, M. Mallet, A. Mizukami ; A new finite-element formulation for computational fluid dynamics, II Beyond SUPG. Comp. Meth. Appl. Mech & Eng 84, No. 3, March 1986.

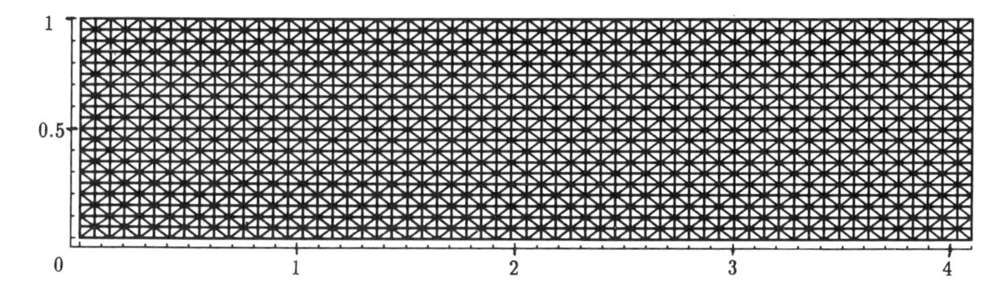

Fig. 4. Diamond shaped grid of 61 × 21 nodes for the oblique shock testcase.

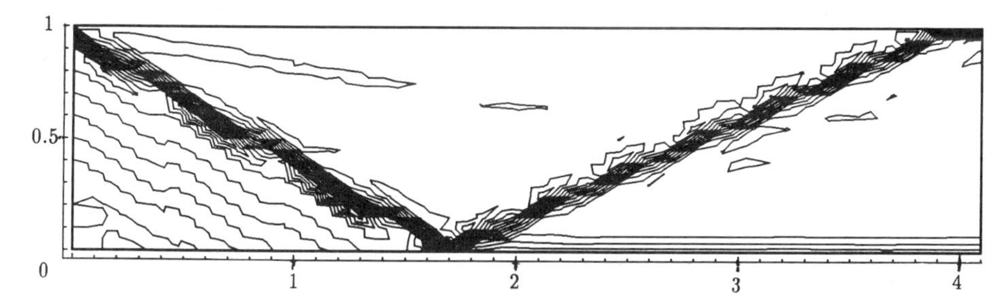

Fig. 5. Iso Machlines obtained with the non monotone scheme for the oblique shock testcase. Increment = 0.05.

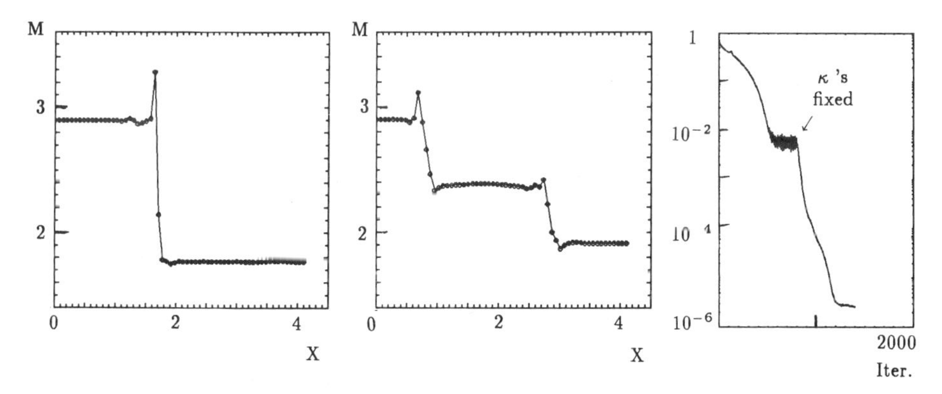

Fig. 6. Mach number plotted for Y=0. and Y = 0.5.

Fig. 7. Convergence history. 800 iterations with κ 's free 600 iterations with κ 's frozen

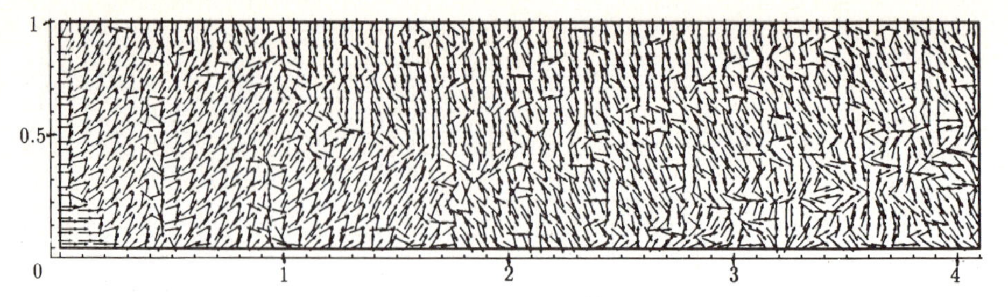

Fig. 8. The pressure dependent decomposition normal $\vec{\kappa}^{(1)}$.

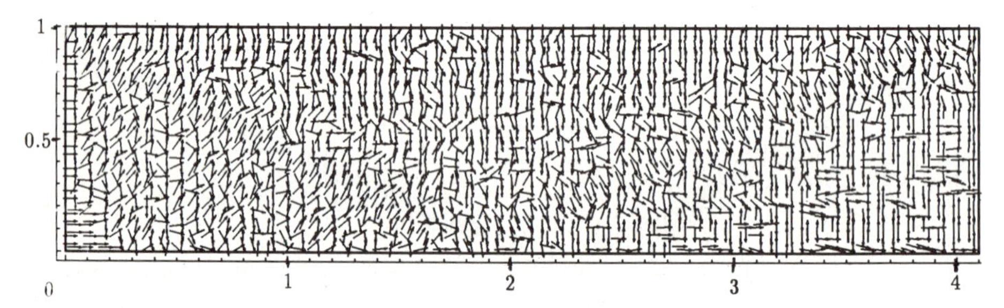

Fig. 9. Decomposition normal $\vec{\kappa}^{(2)}$.

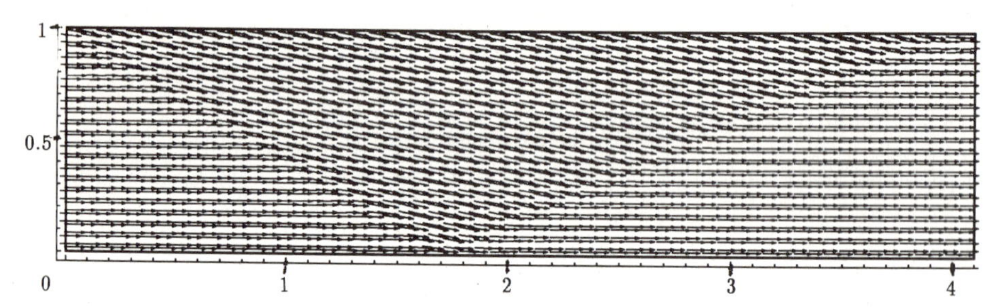

Fig. 10. Velocity field for the first characteristic variable $\vec{u}$.

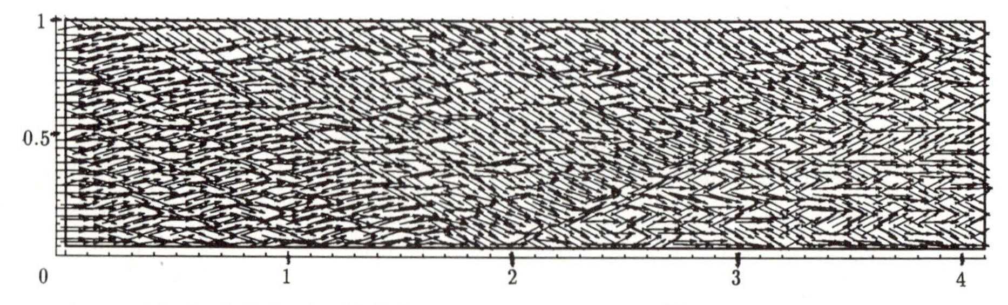

Fig. 11. Velocity field for the third characteristic variable $\vec{u} + c\vec{\kappa}^{(2)}$.

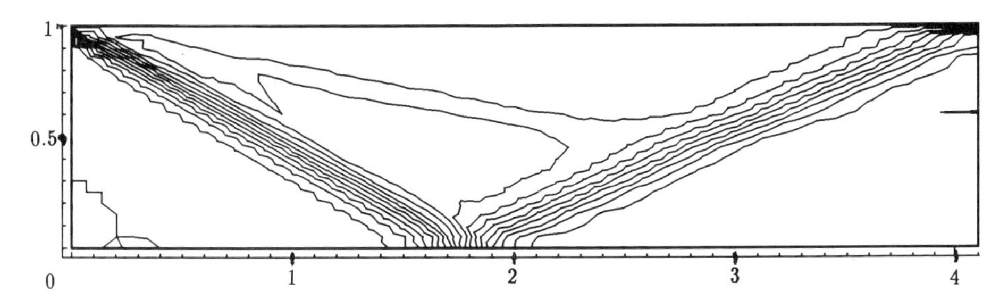

Fig 12. Iso Machlines obtained with the non monotone scheme with the decomposition normals fixed parallel to the velocity field (streamwise upwinding).

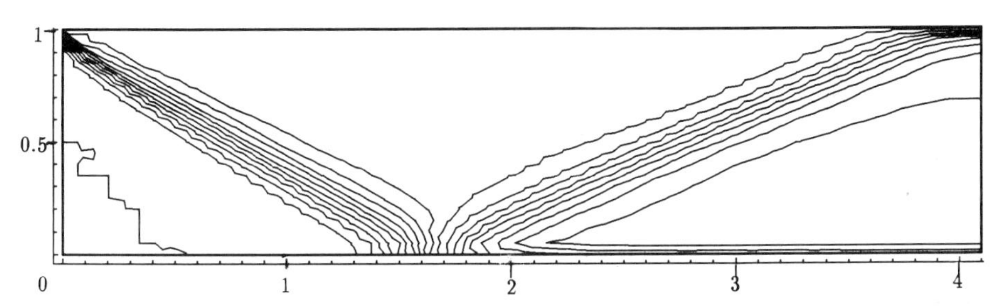

Fig. 13. Iso Machlines obtained with the monotone scheme.

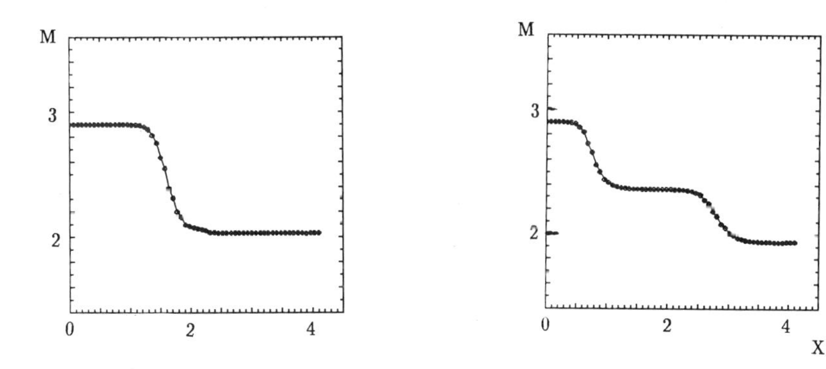

Fig. 14. Mach number plotted for Y=0. and Y = 0.5 for the monotone scheme.

Calculation of a buoyancy-driven 3D cavity flow using a multi-processor system

F.F. van der Vlugt
D.A. van Delft
A.F. Bakker
A.M. Lankhorst

Department of Applied Physics
Delft University of Technology
P.O. Box 5046, 2600 GA Delft, Netherlands.

SUMMARY

The multi-processor system ATOMS has been applied to three-dimensional Navier-Stokes problems. ATOMS was originally designed at AT&T Bell Laboratories for Molecular Dynamics calculations. To use ATOMS for solving Navier-Stokes equations certain hardware features were included in the design to permit data transfer between processor boards as a closely coupled linear array of processors. In this mode we refer to the multi-processor system as DNSP (AT&T's/Delft Navier-Stokes Processor).

An algorithm for calculating the buoyancy driven laminar/turbulent flow in a three-dimensional cavity has been implemented on the DNSP. Turbulence is accounted for by using the k-ε model. For this algorithm an efficiency of 35% (which amounts to 14 Mflops) is obtained. This high efficiency can be reached thanks to the strong coupling between the algorithm and the architecture of the multi-processor system.

Solutions have been obtained for several cavity configurations.

INTRODUCTION

At AT&T Bell Laboratories the multi-processor system ATOMS was designed for Molecular Dynamics calculations, described by Bakker *et al.* [1]. State of the art devices were used to build the multi-processor system which has minisupercomputer speed for a certain class of algorithms at a considerably lower price than the price of a minisupercomputer. We have applied this multi-processor system ATOMS to 2D and 3D Navier-Stokes problems. The 2D implementation and several results of the calculation of a turbulent flow of air in a 2D square cavity are described by Van der Vlugt *et al.* [6]. To use ATOMS for solving Navier-Stokes equations certain hardware features were included in the design to permit data transfer between processor boards as a closely coupled linear array of processors. In this mode we refer to the multi-processor system as DNSP (AT&T's/Delft Navier-Stokes Processor). Because the DNSP (ATOMS) is designed for a class of algorithms, we call it an algorithm oriented processor. The main feature of the DNSP (ATOMS) is that it is very suitable for algorithms with strong local interaction. The algorithms which are used to solve the Navier-Stokes equations have often a local character, although the influence of the equations is not pure local. The equations are discretized with a discretization method (finite differences, finite elements) and the resulting matrices are solved with some point or line iteration algorithm, or more sophisticated algorithms such as the multigrid algorithm. During the research period one fixed algorithm is used but algorithm improvements can be incorporated into the processor software after successful implementation of the fixed algorithm.

A large calculation speed is obtained for the DNSP by using vector operations, parallel execution and pipelining. The cost/performance ratio is very small due to the strong coupling between the architecture of the multi-processor system and the algorithm. The DNSP has been kept simple, the result is low maintenance costs and a large MTBF (mean time between failure).

Because the DNSP is built for the problem involved, very large problems can be solved. Typically problems which can only be solved at extreme costs by using general purpose computers. For example grid refinement can be performed to a rather fine level to test the grid dependence of the solution. Because the DNSP is rather cheap it can be used as a private machine and time sharing with other users is not needed. Furthermore, it is expandable from a one processor board to a sixteen processor board configuration.

A C-like assembler language is used to program the DNSP, because it is very difficult to build a compiler which takes full advantage of the possibilities of the multi-processor system. Of course it is a disadvantage not being able to use a high level programming language, but this is outweighted by the advantage of the much higher performance that is reached with the assembler language. Furthermore it is the aim of the DNSP to solve very time consuming problems with a fixed algorithm. So when the software for the problem has been implemented it will be used heavily. In order to reduce the software effort a number of macros are developed for data transport and general program structures (e.g. do-loops).

THE THREE-DIMENSIONAL FLOW ALGORITHM OF THE DNSP

An algorithm which is frequently used in Computational Fluid Dynamics is implemented on the DNSP. This algorithm is used to calculate the steady buoyancy-driven laminar/turbulent flow of a fluid (e.g. air or water) in a three-dimensional cavity. In most situations the flow is turbulent, therefore the turbulent flow algorithm is discussed. The laminar flow can be calculated with the same algorithm by setting some variables to zero.

The turbulent flow of a Newton fluid in a 3D cavity can be modeled with the time averaged Navier-Stokes equations (Reynolds decomposition) and energy equation.

$$\frac{\partial \rho u_j}{\partial x_j} = 0 \tag{1}$$

$$\frac{\partial(\rho u_i u_j)}{\partial x_j} = \frac{\partial}{\partial x_j}\left\{(\mu + \mu_t)\left[\frac{\partial u_i}{\partial x_j} + \frac{\partial u_j}{\partial x_i}\right]\right\} - \frac{\partial p}{\partial x_i} - \rho g \delta_{i,2} \tag{2}$$

$$\frac{\partial(\rho u_j T)}{\partial x_j} = \frac{\partial}{\partial x_j}\left\{\left[\frac{\mu}{Pr} + \frac{\mu_t}{\sigma_T}\right]\frac{\partial T}{\partial x_j}\right\} \tag{3}$$

where x_j, ρ, u_j, p, T, μ, μ_t, g, Pr, σ_T, are the coordinate in the jth direction, density, velocity in the jth direction, pressure, temperature, molecular dynamic viscosity, turbulent dynamic viscosity, gravitational acceleration, Prandtl number, turbulent Prandtl number for T, respectively. Further $\delta_{i,j}$ is the Kronecker delta function, $\delta_{i,j} = 1$ if $i = j$ else 0. The Einstein summation convention is used; when a subscript is repeated in a term a summation of three terms is implied. Turbulence is modeled by introducing a turbulent dynamic viscosity (the eddy viscosity concept), the effective dynamic viscosity is the sum of the laminar and turbulent dynamic viscosity. For modeling of the turbulent dynamic viscosity the k-ε model described by Rodi [5] is used, as given in (4).

$$\frac{\partial(\rho u_j \phi)}{\partial x_j} = \frac{\partial}{\partial x_j}\left\{\left[\mu + \frac{\mu_t}{\sigma_\phi}\right]\frac{\partial \phi}{\partial x_j}\right\} + S_\phi \qquad , \phi = k,\ \varepsilon \tag{4}$$

$$\mu_t = c_\mu \rho \frac{k^2}{\varepsilon}$$

where k is the turbulent kinetic energy, ε the rate of turbulent energy dissipation, σ_ϕ is the turbulent Prandtl number for ϕ, and c_μ is a constant. More details of the used k-ε model are given by Van der Vlugt *et al.* [6].

To complete the system of equations a number of boundary conditions is required. For the temperature a number of boundary conditions can be applied, which is discussed later. The boundary conditions for k and ε at the wall are derived from the logarithmic wall function for forced convection. The three velocity components on the wall are set to zero.

The partial differential equations (1)-(4) are discretized on a non-equidistant staggered grid by using the finite volume method. For the discretization of convection and diffusion the hybrid difference scheme (or a central difference scheme) is used, as described by Patankar [4]. The SIMPLE algorithm, as described by Patankar [4], is applied for the iterative adjustment of the pressure and velocity components to fulfill the continuity equation. The result of the discretization is a set of difference equations which have the general form as given in (5).

$$\begin{aligned} ap^\phi_{i,j,k}\ \phi_{i,j,k} = {} & ae^\phi_{i,j,k}\ \phi_{i+1,j,k} + aw^\phi_{i,j,k}\ \phi_{i-1,j,k} + \\ & an^\phi_{i,j,k}\ \phi_{i,j+1,k} + as^\phi_{i,j,k}\ \phi_{i,j-1,k} + \\ & af^\phi_{i,j,k}\ \phi_{i,j,k+1} + ab^\phi_{i,j,k}\ \phi_{i,j,k-1} + s_{i,j,k} \end{aligned} \tag{5}$$

$$i = 1, ..., N_1;\ j = 1, ..., N_2;\ k = 1, ..., N_3$$

where ϕ is one of the variables. The coefficients a_p, a_e, etc. are functions of the variables ϕ, thus an iterative solution method has to be used to solve the set of equations. The coefficients are evaluated at the previous iteration level. The difference equations for each variable can be solved on the whole calculation domain by using some direct or iterative solution method or with a method which is restricted to solving the equations on a 2D-plane or on a vertical line in the 3D calculation domain. We use the restriction to a vertical line because for this method only a small number of coefficients have to be stored, so the memory of the DNSP is used efficiently. Furthermore the line method is more local than the whole field method. An iteration on a line is performed by solving the difference equations, as given by (6), one after the other.

$$l_j \ \phi_{j-1} + m_j \ \phi_j + r_j \ \phi_{j+1} = ll_j \qquad j = 1, \ldots, N_2 \tag{6}$$
$$l_1 = 0, \ r_{N_2} = 0.$$

The difference equations (6) can be solved by applying Gauss elimination. An iteration on a 2D plane (x-y direction) is made by stepping through the 2D plane and performing a line iteration on each vertical line. A sweep on the whole 3D calculation domain is made by stepping through the 3D calculation domain and performing an iteration on each of the 2D planes. The solution of the difference equations is obtained by performing a sufficient number of sweeps.

THE ARCHITECTURE OF THE DNSP

The DNSP consists of a control board, a large memory and several processor boards which are connected with each other in a one-dimensional way, as depicted in figure 1. All boards are connected with a host computer via a common bus. The host computer is connected with several devices such as terminals, disks, grafic workstations, printers, etc..

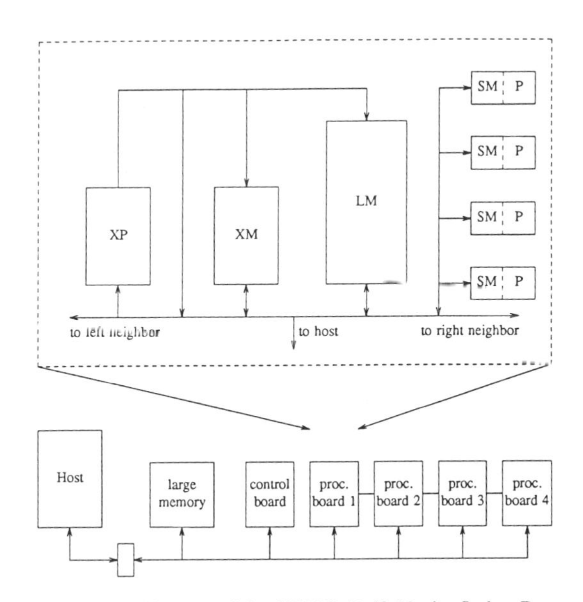

Fig. 1. The architecture of the AT&T's/Delft Navier-Stokes Processor.

The host computer can write (or read) numerical data only to (from) the large memory, which is used to store initialization data, intermediate data or solution data. The large memory can be expanded to fit the problem size. Furthermore the host computer can send instructions to the control board and receive some status signals in return. A large number of instructions can be performed such as transport of data between the large memory and one or more processor boards, transport of data between a processor board and it's neighbors via the special interconnection, or starting the execution on one ore more processor boards.

The control board supervises the processor boards and the large memory. The processor boards are divided into two sides; the data management side and the floating-point side. At the data management side communication with the control board takes place and data storage and transport are arranged. Data can be stored on the data management side in two large memories, the LM memory (short: LM) which is a 1M words (32 bit) dynamic RAM memory and the XM memory (short: XM) which is a 64k words static RAM memory. At the floating-point side the floating-point operations are performed by four processors with a small memory (called SM, 2×32 words) attached to each of them. The SM memories are divided into two parts, at the same moment one can be accessed by the data management side and the other by the floating-point side. The operations of one processor are performed parallel by two floating-point chips, one for multiplication and the other for addition, subtraction and division. The four processors work SIMD (Single Instruction Multiple Data), they all perform the same instructions on different data. The four processors are used to perform do-loops, or scalar operations if needed. The asymptotic performance of one processor is about 2.5 Mflops (floating-point operations per second) double precision (64 bit), which results in an asymptotic performance of about 10 Mflops double precision for one processor board. When data transport is fast enough the floating-point calculations are performed with only small interrupts and the performance of the processors is close to the asymptotic performance.

To run an application on the DNSP a number of programs have to be written. For the control board a control program is required which controls data transport and the execution process on the processor boards. It can be a standard program which is used by all users. For the processor boards a program is required for the data management (transport) side and for the four processors. The programs for the DNSP can be downloaded from the host computer into the processor boards. After downloading the programs and loading numerical data into the large memory the execution can be started.

THE IMPLEMENTATION OF THE 3D FLOW ALGORITHM ON THE DNSP

Before starting the implementation a data-base structure has to be defined. This is of great importance since the data-base structure effects the storage requirements and the efficiency of the calculation process. The problem is split into several subproblems to be able to let all processor boards work independent and parallel. We have chosen for a domain decomposition technique in which the 3D calculation domain is split into a number of slices. In figure 2 a schematic diagram is given of this domain decomposition technique for the DNSP. For the DNSP with four processor boards the calculation domain is divided into four equal slices (or almost equal slices). The data of one slice is stored on one of the processor boards. A sweep starts in the middle of the slice, as depicted in figure 2. First two half sweeps are performed on the right side of every slice. At this stage the new data on the right side of the slice of the processor boards can be transported to the right neighbor. Next two half sweeps are performed on the left side of every slice and new data on the left side of the slice of the processor boards can be transported to the left neighbor.

On a processor board the variables are stored in LM. For each variable a number of 2D planes are stored one after the other. The maximum grid size can be obtained from the size of LM which is fixed to 1 Mwords and the number of processor boards which is four. Several examples of the largest possible grids are: $66 \times 66 \times 32$ (139,400 grid points), $52 \times 52 \times 52$ (140,600 grid points). Data which is used more than once for the calculation of a vertical line is stored in XM, because transport between XM and SM is faster than transport between LM and SM. Therefore the data of the vertical line which is calculated and the data of it's neighbors is stored in XM. After solving the variables on the vertical line the results are transported from XM to LM and the variables needed for the calculation of a next line are transported from LM to XM.

Given the data-base structure in XM a calculation procedure is found which is the most efficient. The solution procedure for the variables is made up of a number of do-loops. Several of the do-loops for the solution of the difference equations (6) contain recurrences. These do-loops can only be calculated with one processor, or the do-loop should be changed to a non-recursive do-loop. The four processors can be used to calculate simple do-loops consisting of for example adding two vectors or of more difficult do-loops like calculating the coefficients of one of the variables. A do-loop which does not contain recurrences can be calculated in a general way. Every do-loop is split into several groups of four loops. Then a number of times four loops are calculated by the four processors. To start-up floating-point execution the four SM memories have to be loaded (maximal sixteen 64 bit words). Next the floating-point calculation can be started

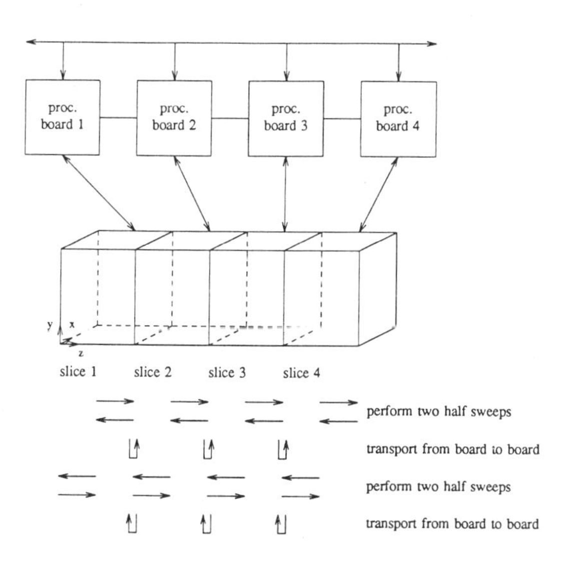

Fig. 2. Schematic diagram of the iteration procedure.

by the four processors and new data can be loaded into the second part of the SM memories. It should be noted that the data transport and floating-point calculations are performed parallel. After both the calculation and data transport are finished the new floating-point calculations can be started. At the same moment the results of the previous calculations can be transported from the SM memories to the XM memory and new data can be loaded into the SM memories. When data transport is fast enough the floating-point calculations are performed with only small interrupts and the performance of the processors is close to the asymptotic performance.

To solve one of the variables on a vertical line the coefficients and source terms are calculated and next the resulting difference equations (6) are solved. The computer code for calculating the coefficients and source terms of the difference equation of a variable is made up of a number of non-recursive do-loops which have to be calculated after each other. These do-loops are calculated in the general way.

The Gauss algorithm is implemented to solve the difference equations. However, only one processor can be used for the calculation, because the algorithm contains several recurrences. The algorithm has to be modified to be able to use four processors efficiently. Several algorithms for solving a tridiagonal matrix on more than one processor have been introduced in literature. An algorithm for Gauss elimination from the bottom and the top of the tridiagonal matrix is described by Van der Vorst [7]. One of the processors starts the Gauss elimination on the top and the other on the bottom of the matrix. During the eliminations the first and second processor arrive on two neighboring points in the middle of the matrix, then a 2×2 matrix is solved and the elimination can be finished. This algorithm is about 1.5 times faster than the Gauss elimination with one processor. An algorithm for solving tridiagonal matrices parallel with more than two processors is described by Wang [8]. The tridiagonal matrix is split into four parts which are calculated parallel on the four processors. For the implementation a modification of the Wang algorithm described by Michielse & Van der Vorst [3] is used. This modification ensures a better distribution of the calculation process over the four processors. This algorithm is slightly better than the elimination from two sides.

To find the processing speed several experiments have been done with the 3D program on the DNSP. In table 1 the results for the processing speed for several grid sizes are given. The processing speed depends on the vector length and for vector lengths between 20 - 60 the speed is 2.7 - 3.4 Mflops for one processor board. For the four board system we find 10.8 - 13.6 Mflops. From the table values can be found that the asymptotic speed on one processor board is almost 3.8 Mflops and $n_{½} \approx 9$ ($n_{½}$ is the half performance vector length, defined by Hockney & Jesshope [2]). The calculation

Table 1. A comparison of the speed of the DNSP and the Convex C220.

grid	speed (Mflops) DNSP 1 proc. board	speed (Mflops) Convex C220 1 processor
20 × 10 × 10	1.93	2.83
20 × 20 × 10	2.66	5.07
20 × 40 × 10	3.16	6.56
20 × 60 × 10	3.39	7.36
20 × 80 × 10	3.49	7.80
20 × 100 × 10	3.57	8.06

speed is about 30% - 35% of the maximum performance of the processor system. To compare the DNSP with other computers an estimation has to be made of the performance/cost. The performance/cost of the DNSP is about 190 flops/dollar (host: 20k dollar, boards: 40k dollar, etc: 20k dollar, asymptotic performance: 15.2 10^6 flops).

Several timing experiments have been done on the Convex C220 (operating system V7.0). Only one processor was used. The timing results are given in table 1, from these values the asymptotic speed can be found which is 9.0 Mflops and $n_{1/2} \approx 15$. All the do-loops were vectorized except for the tridiagonal matrix calculation. We have not used a special algorithm for the tridiagonal matrices on the C220. If we had done so probably an improvement of 10% - 20% was obtained. Running with two instead of one processor gives a reduction of the efficiency of about 20% - 30% depending on the vector length. So from the two effects described above it can be concluded that the C220 with two processors has an asymptotic speed of about 18 Mflops. The resulting performance/cost is about 12 flops/dollar (the price of the system is based on literature, 1988).

THE BUOYANCY-DRIVEN TURBULENT FLOW IN A 3D CAVITY

A large variety of cavity flows can be calculated with the 3D program on the DNSP. For the temperature several boundary conditions have been implemented. Several standard boundary conditions are: a constant temperature on the wall, a temperature which is changing linearly with one of the three coordinates or the heat transfer through the wall can be set to zero (adiabatic wall). For these boundary conditions it is assumed that the thickness of the cavity walls is zero. For the comparison of the numerical results with the experimental results the wall can be given a certain thickness. The heat transfer in the walls and heat losses through the walls to the surrounding fluid (Robbins boundary condition) can be calculated.

Solutions have been obtained for several cavity configurations, here one specific configuration is discussed. For this configuration also several experiments have been done. The cavity is filled with air and has the dimensions: 1m × 1m × 0.1m. The left wall has a temperature of 45.8^oC and the right wall 5.4^oC. For the thermal boundary conditions of the bottom and top walls experimentally obtained temperature distributions have been used. The front and back wall have a thickness 6 10^{-3}m and a conductivity of 0.9 W/mK, thus heat is conducted through the walls. The heat transfer to the ambient is set to zero. The Rayleigh number for the above described situation is $Ra = 3.88\ 10^9$. The flow in the experiments and calculations is turbulent.

Table 2. Several characteristic quantities of the flow for various grids.

	Nu_l	Nu_f	v_{max}	u_{max}
60 × 60 × 30	10.299 10^1	9.029 10^1	0.269	0.163
40 × 40 × 40	10.253 10^1	8.995 10^1	0.271	0.161
40 × 40 × 20	10.125 10^1	8.915 10^1	0.275	0.163
20 × 20 × 20	9.464 10^1	8.349 10^1	0.286	0.153
20 × 20 × 10	8.916 10^1	7.655 10^1	0.265	0.146

The solution is obtained on the following grids: 20 × 20 × 10, 20 × 20 × 20, 40 × 40 × 20, 40 × 40 × 40, 60 × 60 × 30. Several characteristics of the flow for the grids are given in table 2. Nu_l is the Nusselt number (dimensionless heat transfer) averaged over a vertical line in the center of the left wall, Nu_f is the mean Nusselt number on the left wall, v_{max} is the maximum v velocity on a line from the center of the left wall to the center of the cavity, and u_{max} is the maximum u velocity on a line from the center of the bottom wall to the center of the cavity. For the grid in the x-direction there are more grid points close to the walls than for the y- and z-direction. This has been done because there are thin boundary layers along the left and right wall. As can be seen in the table the solutions on the three finest grids deviate just a little for the characteristic numbers, maximum deviation about 2%. Of course this does not prove that the solution on the finest grid (60 × 60 × 30) is close to the solution of the difference equations for $h \rightarrow 0$, where h is some measure for the grid spacing. We try to give more evidence by

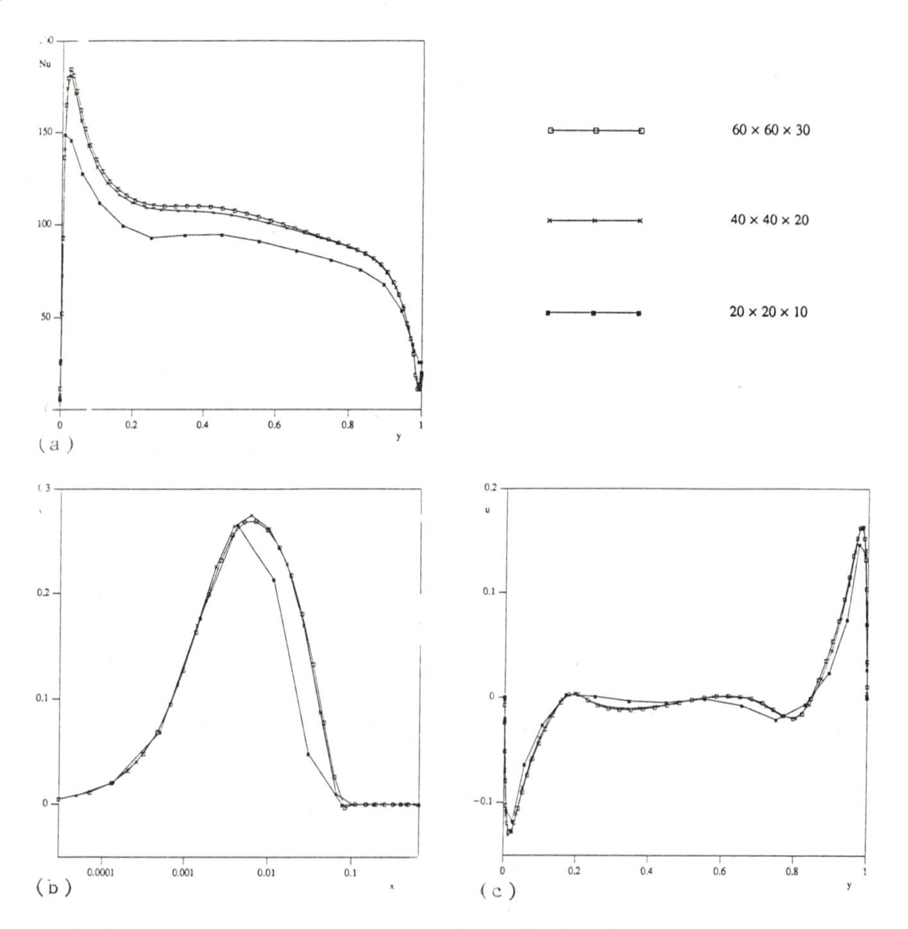

Fig. 3. The Nusselt number along the vertical line in the middle of the left wall (a), the v velocity on the line from the middle of the left wall to the center of the cavity (b), the u velocity on the line from the middle of the bottom wall to the middle of the top wall (c).

examining three profiles. In figure 3a the Nusselt number along the vertical line in the center of the left wall is given for the three grids. In figure 3b the v velocity on the line from the middle of the left wall to the center of the cavity is given, and in figure 3c the u velocity on the line from the middle of the bottom wall to the middle of the top wall is given. These three profiles represent the characteristics of the flow and we use them to check the accuracy of the solution. We realize that in fact the seven (variable) fields should be examined on the entire domain, but we think that the three profiles are a good representation. We compare the $20 \times 20 \times 10$, $40 \times 40 \times 20$ and $60 \times 60 \times 30$ grids. To find the accuracy of the solution (profiles) we should do some error estimation. We expect that the error in the solution in every point behaves as $O(h^p)$, where p is constant. We tried to find p but found that the solution on the $20 \times 20 \times 10$ grid is not in the asymptotic region. In fact we need a solution on a $80 \times 80 \times 40$ grid, assuming that the solution of the $40 \times 40 \times 20$ lies already in the asymptotic region. However, a solution on a $80 \times 80 \times 40$ grid can not be obtained with the DNSP at this moment. So the grid can not be made fine enough to find the accuracy of the solution, on the other hand we think that the three profiles give enough evidence to conclude that the solution on the $60 \times 60 \times 30$ grid is close to the solution of the difference equations for $h \rightarrow 0$.

ACKNOWLEDGEMENTS

These investigations in the program of the Foundation for Fundamental Research on Matter (FOM) have been supported (in part) by the Netherlands Technology Foundation (STW).

REFERENCES

[1] BAKKER, A.F., GILMER, G.H., GRABOW, M.H. & THOMPSON, K.: "A Special Purpose Computer for Molecular Dynamics Calculations", Journal of Computational Physics (in press).

[2] HOCKNEY, R.W. & JESSHOPE, C.R.: "Parallel Computers (Architecture Programming and Algorithms)", Adam Hilger Ltd, 1981.

[3] MICHIELSE, P.H. & VAN DER VORST, H.A.: "Data transport in Wang's partition method", Parallel Computing, 7 (1988) pp. 87-95.

[4] PATANKAR, S.V.: "Numerical Heat Transfer and Fluid Flow", McGraw-Hill, New York 1980.

[5] RODI, W.: "Turbulence models and their application in hydraulics", I.A.H.R., Delft 1980.

[6] VAN DER VLUGT, F.F., VAN DELFT, D.A., BAKKER, A.F. & VAN DER MEER, Th.H.: "The Buoyancy-driven Turbulent Flow in a Square Cavity calculated with the DNSP" , Proc. Sixth Int. Conf. on Num. Meth. in Laminar and Turbulent Flow, 6 (1989) pp. 973-983.

[7] VAN DER VORST, H.A.: "Large tridiagonal and block tridiagonal linear systems on vector and parallel computers", Parallel Computing, 5 (1987) pp. 45-54.

[8] WANG, H.H.: "A Parallel Method for Tridiagonal Equations", ACM Transactions on Mathematical Software, 7 (1981) pp. 170-183.

A 2D Euler Solver for Real Gas Flows using an Adaptive Grid

J.B. Vos
Hydraulic Machines and Fluid Mechanics Institute (IMHEF)
Swiss Federal Institute of Technology - Lausanne
1015 Lausanne, Switzerland

E. Bonomi
GASOV/Project d'Ecole ASTRID
Swiss Federal Institute of Technology - Lausanne
1015 Lausanne, Switzerland

C.M. Bergman
European Centre for Research and Advanced Training
in Scientific Computing (CERFACS)
42 Avenue Gustave Coriolis
F-31057 Toulouse, France

Summary.

Within the framework of the collaboration between the Swiss Federal Institute of Technology-Lausanne and CERFACS in Toulouse, a 2D Euler solver for Real gas flows has been developed. Three different methods to account for the effects of Equilibrium air chemistry in the flowfield were incorporated in the solver. These methods are discussed and compared in this paper. To improve the shock capturing capabilities of the numerical scheme, an adaptive grid generator based on molecular dynamics has been connected to the Euler solver. At present, the adaptive grid is used to capture both internal and external shock waves. Several strategies to adapt the grid are discussed and compared. It is concluded that the use of an adaptive grid improves the solution near shock waves, and can lead to a faster convergence.

Introduction.

Stimulated by the ambitious plan to build an European Space Shuttle, the HERMES, many research groups in Europe have (re-) initiated research on Hypersonic flows. These flows are characterized by high Mach numbers. Strong shock waves exist in the flowfield, and temperatures behind these may be so high that the air molecules start to dissociate. This influences strongly the flow field, and hence the aerodynamic coefficients as for instance C_L, C_D and C_M.
Experimental simulation of hypersonic flows over the whole range of possible flight conditions is impossible due to the extreme flow conditions encountered which are impossible to simulate experimentally. In the development program for the HERMES numerical simulation of hypersonic flows therefore is used to determine the aerodynamic coefficients for those flight situations.
When making a numerical simulation of hypersonic flows, it is vital to use the correct thermodynamic model which connects the two independent thermodynamic variables, the density ρ and the internal energy e, to the dependent variables as pressure and temperature. In the first part of this paper, several thermodynamic models describing dissociating air at chemical equilibrium are discussed. These models are incorporated into a 2D Euler solver and results obtained with the different models are compared. Special attention is given to the computational efficiency of the different models since

the computational resources today still are limited, especially if one wants to make 3D calculations.
The second part of this paper deals with the use of an adaptive grid generator to improve the shock capturing capability of the numerical scheme. The Euler equations are solved using a cell centered scheme offering a limited resolution of (strong) shock waves. The adaptive grid is therefore used to cluster grid points in regions with strong gradients, hereby improving the resolution of shock waves. At this stage both internal and external shock waves are captured.

Numerical Method.

In two dimensions and in cartesian coordinates, the unsteady Euler equations can be written in conservative form as

$$\frac{\partial}{\partial t}\mathbf{w} + \frac{\partial}{\partial x}\mathbf{F}(\mathbf{w}) + \frac{\partial}{\partial y}\mathbf{G}(\mathbf{w}) = 0 \tag{1}$$

where

$$\mathbf{w} = \begin{pmatrix} \rho \\ \rho u \\ \rho v \\ \rho E \end{pmatrix}, \quad \mathbf{F}(\mathbf{w}) = \begin{pmatrix} \rho u \\ \rho u^2 + p \\ \rho u v \\ u(\rho E + p) \end{pmatrix}, \quad \mathbf{G}(\mathbf{w}) = \begin{pmatrix} \rho v \\ \rho u v \\ \rho v^2 + p \\ v(\rho E + p) \end{pmatrix}. \tag{2}$$

ρ is the density, u and v the cartesian velocity components, p the pressure and E the total energy. This system of equations is completed by the thermodynamic relation between pressure and internal energy, $p = p(\rho,e)$, and the relation between internal and total energy

$$e = E - \frac{u^2 + v^2}{2} . \tag{3}$$

Equation (1) is discretized in space using the finite volume method, resulting in a system of ordinary differential equations in time,

$$\frac{d}{dt}(S_{i,j}\mathbf{w}_{i,j}) + Q_{i,j} = 0 \tag{4}$$

where $S_{i,j}$ is the cell area of cell (i,j), $\mathbf{w}_{i,j}$ an approximation of $\mathbf{w}$ in this cell and $Q_{i,j}$ is the net flux leaving or entering the cell. The fluxes at the cell sides are calculated from Eq.(2) using the average of the state vectors $\mathbf{w}$ having the cell side in common. On a cartesian grid this results in a centered scheme which is second order accurate in space.
Since second order schemes produce spurious oscillations near discontinuities, an artificial dissipation term based on a second order difference is added near discontinuities. This makes the scheme locally first order in these regions. Moreover a fourth order dissipation term is added to prevent odd/even oscillations. The method proposed in [1] is used to calculate these dissipation terms. Equation (4) can now be written as

$$\frac{d}{dt}(S_{i,j}\mathbf{w}_{i,j}) + Q_{i,j} - D_{i,j} = 0 \tag{5}$$

where Di,j is the net dissipative flux in cell (i,j). Equation (5) is integrated in time using the four stage explicit Runge Kutta scheme [2].
Boundary conditions have to be prescribed at the four sides of the calculation domain. At solid walls, the tangency condition is imposed, and the pressure at the body is calculated from the normal momentum equation. Outflow boundaries are assumed to be supersonic, and the state vector **w** is extrapolated using linear extrapolation For the calculations discussed in the first part of this paper, it is assumed that the external bow shock is fitted, and the Rankine Hugoniot relations for a moving shock wave together with the compatibility relation for a backward running characteristic are used to calculate the position of the shock wave and the state vector **w** behind the shock wave [3].

Modelling of Equilibrium Air chemistry.

As mentioned in the Introduction, using the correct thermodynamic model is vital when simulating hypersonic flows. This can be easily understood if one calculates for instance the stagnation temperature behind a shock wave for a flow with a free stream Mach number of 25. If the gas is taken as calorically perfect (which is a normal assumption for supersonic flows) and the free stream temperature is assumed to be 230 K, one finds that the stagnation temperature equals 28980 K. However if it is assumed that the air is in chemical equilibrium, the stagnation temperature will be between 6000 and 9000 K, depending on the value of the free stream pressure. Because temperature, pressure and density are related to each other by the equation of state, this strong reduction in temperature also affects the pressure and density.

For most of the flight conditions of HERMES, it can be assumed that the flow is in chemical equilibrium. For these type of flows, the Euler equations (Eqs.(1) and (2)) are the same as for caloric perfect gas flows. Only the relation between pressure, density and internal energy is changed.
There exist two classes of methods to relate pressure to the density and internal energy for air in chemical equilibrium. The first class is using experimental data which provide the pressure as function of density and internal energy. These data are available in tables, or are fitted in polynomials. This method is direct since no information about the equilibrium air composition nor the equilibrium air temperature is required. The second class of methods is based on the calculation of the equilibrium composition and the equilibrium temperature, and the equation of state then is used to calculate the pressure from the composition, density and temperature. This class of methods requires the knowledge of the variation of internal energy and specific heat as function of temperature and composition, and it requires the knowledge of the equilibrium constants for the equilibrium reactions as function of temperature.
In this paper, one method of the first class, and two methods of the second class are compared.

The routine TGAS [4] is a widely used routine which belongs to the first class of methods. In TGAS tabulated thermodynamic properties of high temperature air are given in the form of polynomials. Besides the pressure, also the temperature and speed of sound can be calculated using TGAS. Furthermore, TGAS is based on the use of an "effective γ", $\tilde{\gamma}$. For a caloric perfect gas, the equation of state can be written as:

$$p = (\gamma - 1)\, \rho\, e \qquad (6)$$

where γ is the ratio of specific heats at constant pressure and constant volume, equal to 1.4 for air. When using TGAS, Eq.(6) is used to relate pressure to density and internal energy, but with γ replaced by $\tilde{\gamma}$, which is calculated from a polynomial. For explicit schemes

using a low CFL number it is not necessary to calculate $\tilde{\gamma}$ each time step. In general once per 100 time steps is sufficient.
The two methods belonging to the second class of methods both assume that high temperature air can be considered to consist of five species, N_2, O_2, NO, O and N. This is a valid assumption for temperatures below 9000 K, where ionization of air has not become important yet. The equilibrium composition is calculated by solving two elemental mass fraction equations, together with three reaction equations or "laws of mass action":

$$O_2: \quad Y_{O_2} + Y_O + \frac{1}{2}\frac{M_{O_2}}{M_{NO}} Y_{NO} = Z_{O_2} \quad ,$$

$$N_2: \quad Y_{N_2} + Y_N + \frac{1}{2}\frac{M_{N_2}}{M_{NO}} Y_{NO} = Z_{N_2} \quad ,$$

$$\tfrac{1}{2}O_2 \leftrightarrow O: \quad \rho^{1/2}\left(\frac{Y_O}{M_O}\right)\left(\frac{Y_{O_2}}{M_{O_2}}\right)^{-\frac{1}{2}} = K_{C,1} \quad , \tag{7}$$

$$\tfrac{1}{2}N_2 \leftrightarrow N: \quad \rho^{1/2}\left(\frac{Y_N}{M_N}\right)\left(\frac{Y_{N_2}}{M_{N_2}}\right)^{-\frac{1}{2}} = K_{C,2} \quad ,$$

$$\tfrac{1}{2}O_2 + \tfrac{1}{2}N_2 \leftrightarrow NO: \quad \left(\frac{Y_{NO}}{M_{NO}}\right)\left(\frac{Y_{O_2}}{M_{O_2}}\right)^{-\frac{1}{2}}\left(\frac{Y_{N_2}}{M_{N_2}}\right)^{-\frac{1}{2}} = K_{C,3} \quad .$$

In these equations, Z denotes the elemental mass fraction (which is constant), Y the mass fractions, M the molar mass and K_C the equilibrium constant based on concentrations.
The first model of this class, called GIBBS, uses the Gibbs free energy to calculate the equilibrium constant K_C [3]. In this model, the enthalpy, specific heat at constant pressure and the entropy are for each species given as fifth degree polynomials in temperature [5]. These polynomials are valid for temperatures up to 25000 K. The mixture enthalpy and internal energy are related by:

$$e = \sum_s Y_s h_s - \frac{p}{\rho} \tag{8}$$

and the pressure is calculated from

$$p = \rho R^\circ T \sum_s \frac{Y_s}{M_s} \tag{9}$$

where R^o is the universal gas constant and h the enthalpy. The temperature T can be solved iteratively from Eq.(8).
In the second model, called PARK, the equilibrium constants are given as a polynomials in temperature. The enthalpies of the different species are calculated as the sum of the translational, rotational and vibrational energies (the latter two energies for molecules only) [7]. For the species O, N and NO, the heat of formation must be added to the enthalpy. Equation (8) is used to relate the mixture enthalpy to the internal energy, and the pressure is calculated from Eq.(9).
To save computational costs when using the two models above, an effective $\tilde{\gamma}$ was defined as

$$\tilde{\gamma} = 1 + \frac{p}{\rho e} = 1 + \frac{R^\circ T}{e} \sum_s \frac{Y_s}{M_s} \quad . \tag{10}$$

Equation (6) with γ replaced by $\tilde{\gamma}$ is used to calculate the pressure, and $\tilde{\gamma}$ is recalculated each 100 time steps.

Calculation Results for the different Equilibrium Chemistry Models.

Calculations with the three different models were carried out for the flow around a cylindre. In all calculations, a 40x30 grid was used, and the free stream temperature was equal to 230 K. Calculations were carried out for free stream Mach numbers of 5, 15 and 25, and for each Mach number, free stream pressures of 10, 10^3 and 10^5 Pa were used. All other input parameters were kept constant, except for the CFL number, which had to be decreased from 2.0 to 1.75 for the calculations with $M_\infty = 25$, and $p_\infty = 10$ and 10^3 Pa. To compare the computational costs of an equilibrium calculation, the same calculations were also carried out for a caloric perfect gas flow. All calculations were carried out on the Cray 2 of the Computing Centre of the Swiss Federal Institute of Technology - Lausanne.
The calculated pressures, temperatures and Mach numbers using the three different thermodynamic models were for each of the nine calculations compared, and the maximum differences between one model and the two others were determined. Moreover, the calculated shock stand off distances, and for the models GIBBS and PARK, the body values of the calculated mass fractions were compared. The results of these calculations are summarized in Table 1.

Table 1. Comparing Equilibrium Air calculations

TGAS - GIBBS	average ΔT = 1.20%	max ΔT = 2.16%	for $M_\infty = 15$, $p_\infty = 10$
	average Δp = 0.86%	max Δp = 1.57%	for $M_\infty = 15$, $p_\infty = 10$
	average ΔM = 0.87%	max ΔM = 1.34%	for $M_\infty = 25$, $p_\infty = 10$
TGAS - PARK	average ΔT = 1.03%	max ΔT = 2.09%	for $M_\infty = 25$, $p_\infty = 10^5$
	average Δp = 1.05%	max Δp = 2.08%	for $M_\infty = 15$, $p_\infty = 10$
	average ΔM = 0.77%	max ΔM = 1.51%	for $M_\infty = 25$, $p_\infty = 10$
GIBBS - PARK	average ΔT = 0.91%	max ΔT = 1.32%	for $M_\infty = 15$, $p_\infty = 10^5$
	average Δp = 0.88%	max Δp = 1.96%	for $M_\infty = 15$, $p_\infty = 10^5$
	average ΔM = 0.60%	max ΔM = 1.00%	for $M_\infty = 25$, $p_\infty = 10^5$

Comparing the mass fractions calculated with GIBBS and PARK showed differences up to 26%.These differences were for the species with only low concentrations and for this reason did not affect the temperature. Comparing the shock stand off distances showed differences in the order of 1%. From Table 1, it can be concluded that the maximum difference between the calculations carried out for the different thermodynamic models is below 2.2%. The average difference between the different thermodynamic models is around 1%, which is within the accuracy of the numerical scheme itself.
Comparing the computational costs showed that for eight of the nine calculations, the number of time steps to obtain a converged solutions was about 10 - 15% higher when using TGAS. For the ninth calculation, GIBBS and PARK needed both 15% more steps than TGAS. Comparing GIBBS and PARK with the perfect gas calculations showed that at high Mach numbers, GIBBS and PARK required between 25 and 80% more time steps to reach convergence. The reason for this is the shock fitting procedure. For caloric perfect gas flows, the shock stand off distance becomes constant for increasing Mach, while for equilibrium flow calculations, the shock stand off distance is decreasing with increasing Mach number [3], and with decreasing free stream pressure. This implies that at high Mach and low pressures, the grid cells are much smaller than for low Mach and high

pressure, decreasing the global time step, hence more time steps are required to reach convergence.
Comparing the computational costs per time step showed that GIBBS is 14%, TGAS 218% and PARK 9% more expensive than a caloric perfect gas calculation. The reason for the large difference between TGAS and the other two models is that TGAS is not vectorizable, while both GIBBS and PARK are fully vectorized.

Adaptive Grid Generator.

To improve the shock capturing capabilities of the numerical scheme an adaptive grid generator has been incorporated in the solver. At the present stage, the adaptive grid is used to capture both internal and external shock waves.
To limit truncation errors, a grid used in a numerical simulation must be smooth, and for the numerical method used here, it is desirable to have the grid as orthogonal as possible. Both these requirements imply a strong correlation between a grid point and its neighbours. Similar to molecular dynamics, this can be represented by attraction or repulsion forces between grid points. The grid is created as follows [8]. First of all, grid points are distributed over the calculation domain. Then grid points are moved in order to minimize a function which is a measure of the smoothness and orthogonality of the grid. Although this is comparable to a variational approach, there is a difference since in our approach it is possible to change locally forces between grid points. Hence it is easy to adapt the grid locally. Once an initial solution is known, regions with strong gradients of an observable (e.g. density or pressure) are determined, and the coupling constants between grid points in these regions are adjusted to cluster grid points locally.

Adaptation Strategy.

When adapting a grid to a solution one has to make several choices. First of all, one has to select the observable to which the grid will be adapted. Secondly one has to decide if the grid is adapted only in one direction (e.g. only in the direction normal to the body) or in two directions. In the latter case one has to decide if grid points on boundary curves are allowed to move or not. If the grid is adapted during the Euler calculation one has to decide when, and how often the grid is to be adapted. Finally, it is possible to control the clustering of grid points by specifying a parameter by which the coupling constants between grid points are multiplied.
Initial calculations showed that the moment and the number of times the grid is adapted during an Euler calculation is very important. This can be easily understood if one keeps in mind that the Euler calculation is started using free stream values every where. In the initial phase of the calculation, the external shock wave must be found as part of the solution, and has to move to the correct position. If the grid is adapted in this initial phase, then grid points will be clustered in the wrong region, and another adaptation of the grid is required to obtain the clustering at the correct position. It was observed that if the number of time steps between two adaptations was to large, the calculation stopped to converge, and the residual became constant. Only after a new adaptation the residual started to decrease further. Adapting the grid too often also poses problems since each grid adaptation results into a perturbation of the time stepping procedure, and a certain number of time steps without adaptation are required to damp this perturbation.
A measure of the quality of the (adapted) grid has been defined as follows. First of all, remark that the coupling constants between grid points are a measure of the gradient in an observable. Hence, the difference between the coupling constant at time level n and the coupling constant after the last adaptation, indicated by the superscript k, is a measure of how good the present grid is aligned with the gradient in the observable. Define $cd_{i,j}$ as the coupling constant for cell (i,j), and a^n as

$$a^n = \frac{\sqrt{\sum_{i,j} \left(cd^n_{i,j} - cd^k_{i,j}\right)^2}}{\text{number of points}} \quad . \tag{11}$$

a^n is the average value of the difference between the coupling constants at time level n and the coupling constants after adaptation k. It is obvious that, just after an adaptation, a^n will be small, and then will start to increase. The grid is now adapted (again) if

$$\delta = \left| \frac{a^n - a^{n-1}}{a^{n-1}} \right| \leq \varepsilon \quad . \tag{12}$$

This criteria ensures that the grid is adapted only if the solution is only slowly varying. In practice a^n is not calculated each time step, but only each 50 time steps. Moreover the value of ε is reduced after each adaptation.
Figure 1 shows an example of the values of a^n and δ as function of the time step. As can be seen from this figure, a^n increases rapidly the first 800 iterations due to the fact that in this phase the external shock wave is appearing as part of the solution, and is moving towards its position. If the shock wave has found its position, a^n is becoming constant, and the grid is adapted for the first time.

Calculation Results.

Calculations using the adaptive grid were carried out for a caloric perfect gas flow around a double ellipse at Mach 25 with zero angle of attack. Two grid systems were used in the calculations, a grid of 177x20 points, and a grid with 177x25 points. Calculations were carried out using the pressure and density as observable to adapt the grid to, and using 1D and 2D adaptation. In the latter case, grid points on the body were allowed to move. Values of the parameter used to multiply the coupling constants with were 2 and 3.
Initial calculations were carried out varying the initial value of ε, and varying the factor used to decrease ε after each adaptation. These calculations showed that the initial value of ε and the factor used to decrease ε had only a small influence. In all the calculation, the initial value of ε was set to 10^{-2}, and after each adaptation, ε was divided by 5.
Comparing the calculated results using the pressure and density as observable showed that adaptation on the pressure gradient led to a faster convergence, and yielded a better result. For this reason, only results obtained with adaptation on the pressure are discussed in the following.
Figure 2a shows the calculated pressure contours on the upper body of the double ellipse obtained with the 177x25 grid without adaptation, and Figs. 2b and 2c show these contours using 1D and 2D adaptation respectively. In these calculations, the coupling constants were multiplied by 2. It can be seen that the with the adaptive grid the internal shock wave is less spread out, and the external shock wave is captured on fewer grid points. Comparing 1D and 2D adaptation shows that 2D adaptation gives a better results on capturing the external shock wave, while 1D adaptation yields a better results on capturing the internal shock wave.
Figure 3a and 3b show the final grid systems for the upperbody of the calculations with adaptation. The shock waves are clearly visible in the grid system. Moreover, it can be seen that when using 2D adaptation grid points on the boundary are allowed to move, resulting in a clustering of grid points near the canopy.
Figure 4 shows the residual histories for the calculations shown in Fig. 2. It can be seen that adaptation of the grid results in a perturbation of the residual. However, directly after the adaptation the program resumes convergence at a much faster rate, and for both 1D

and 2D adaptation, convergence was obtained within less time steps than for the non adapted grid.
Multiplying the coupling constants with 3 improved the resolution of the shock waves even more, but at the expense of a slower convergence. This is due to the smaller grid cells, resulting in a smaller global time step.
The calculations with the 177x20 grid showed a similar behaviour in so far it concerns the resolution of the internal and external shock waves. However, all calculations with adaptation of the grid during the calculation required more time steps to reach convergence than the calculation without adapting the grid.

Conclusions

A comparison of three different methods to account for equilibrium air chemistry has been made. The differences in calculated results using these methods is at maximum 2.2%, while the average difference is around 1%. These differences are within the accuracy of the numerical scheme itself and it is concluded that all the three models produce reliable results.
Comparing the computational costs on a vector computer showed that the model PARK is the most efficient, followed by the model GIBBS. TGAS is more than twice times as expensive as PARK. The computational cost per timestep is for the model PARK only 9% higher compared to caloric perfect gas flows.
Calculations were carried out using an adaptive grid to capture both internal and external shock waves. An adaptation criteria was defined. It is concluded that adaptation on the pressure gives better results than adaptation on the density. Adapting the grid during the calculation improves the resolution of shock waves. Moreover, it is concluded that using the adaptive grid can in some cases result in a faster convergence.

Acknowledgements.

The research at the Swiss Federal Institute of Technology was performed under contract from Avions Marcel Dassault - Breguet Aviation, with financial support by the Commission Suisse d'Encouragement des Recherches Scientifiques.

References.

[1] Jameson, A., "Steady-state Solution of the Euler Equations for Transonic Flow by a Multigrid Method", Advances in Scientific Computing, p 37, Academic Press 1982.
[2] Jameson, A., "Numerical Solution of the Euler Equations for Compressible Inviscid Fluids", In: Numerical Methods for the Euler Equations of Fluid Dynamics, Ed.: Angrand,F., e.a., SIAM, Philidelphia 1985
[3] Vos, J.B., Bergman, C.M. and Petersson, N.A., "An efficient Euler solver for Real Gas flows", Report IMHEF T-89-1, CERFACS RF/89/1, Lausanne, May 1989.
[4] Srinivasan, S., Tannehill, J.C. and Weilmuenster, K.J.,"Simplified Curve Fits for the Thermodynamic Properties of Equilibrium Air",NASA Report 1988.
[5] Sawley, M.L., Wuthrich, S. and Vos, J.B.,"Modellization and Calculation of Laminar, Hypersonic Boundary Layer Flows",EPFL - IMHEF Report (in preparation).
[6] Park, C.,"On Convergence of Chemically Reacting Flows", AIAA 85-0247.
[7] Anderson Jr., J.D. ,"Modern Compressible Flow", McGraw-Hill, New York, 1982
[8] Bonomi, E., "Generation of Structured Adaptive Grids based upon Molecular Dynamics",1987

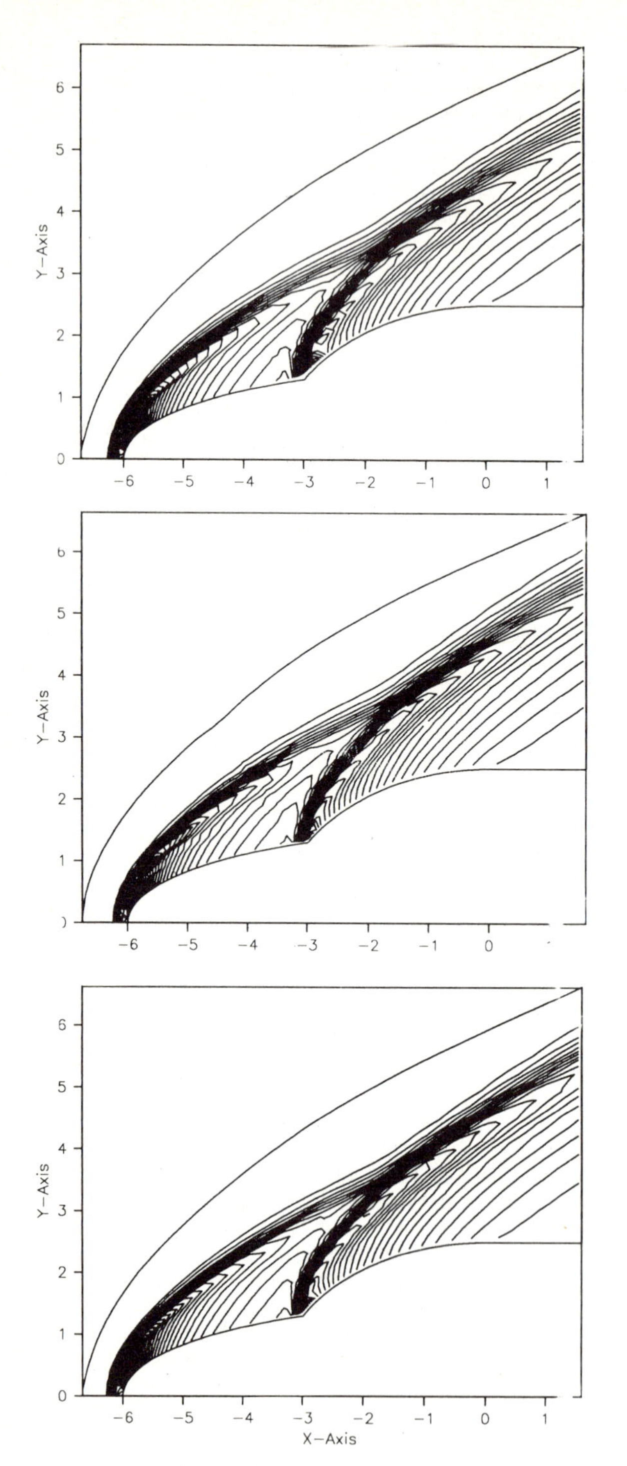

Fig. 2a. Iso pressure lines not adapted grid

Fig. 2b. Iso pressure lines 1D Adapted grid

Fig. 2c. Iso pressure lines 2D Adapted grid

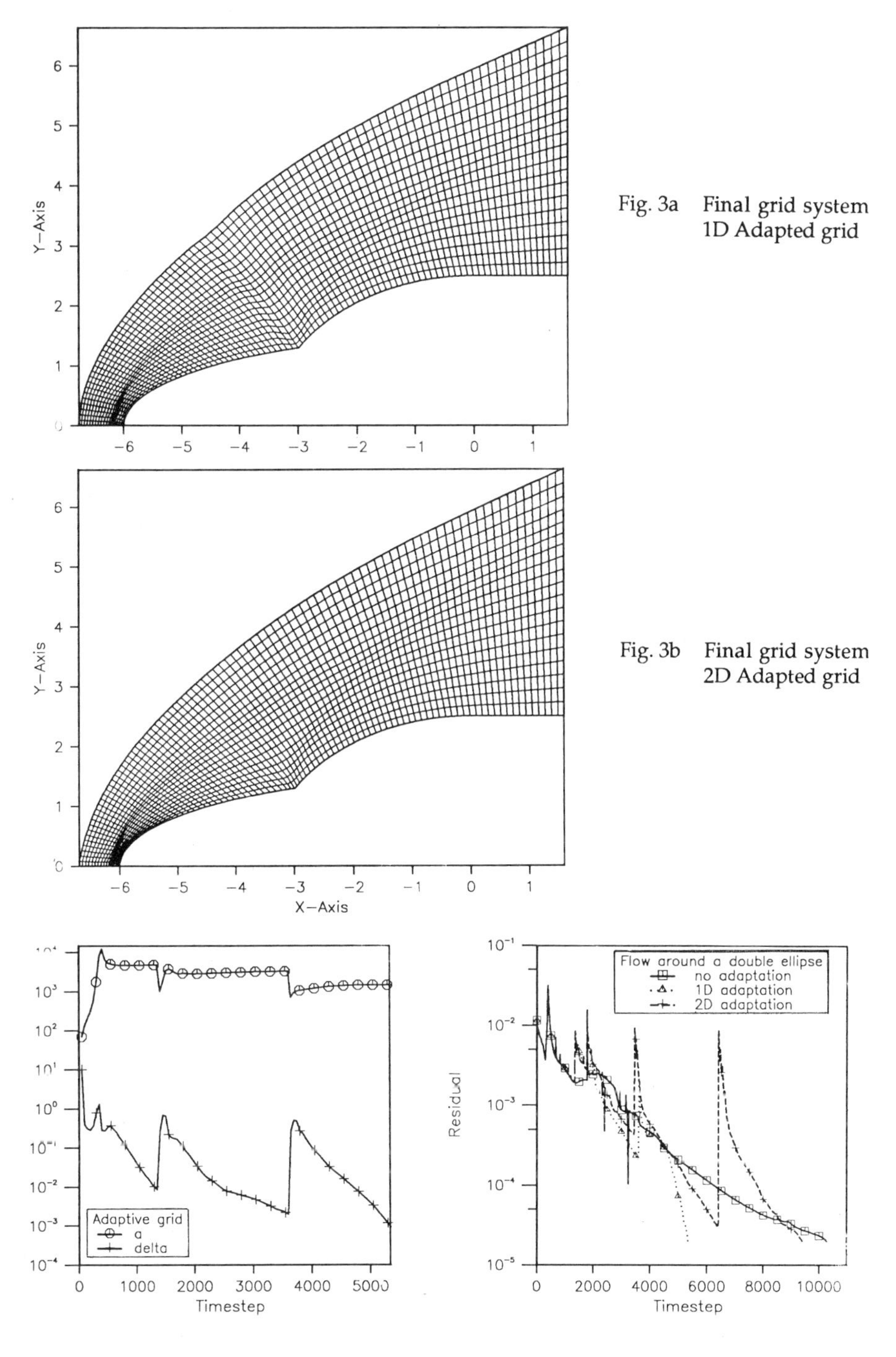

Fig. 3a Final grid system 1D Adapted grid

Fig. 3b Final grid system 2D Adapted grid

Fig. 1. Histories of a and δ.

Fig. 4. Residual histories.

AN EIGENVALUE ANALYSIS OF FINITE-DIFFERENCE APPROXIMATIONS FOR HYPERBOLIC IBVPs

ROBERT F. WARMING AND RICHARD M. BEAM

NASA AMES RESEARCH CENTER, MOFFETT FIELD, CA 94035, USA

SUMMARY

The eigenvalue spectrum associated with a linear finite-difference approximation plays a crucial role in the stability analysis and in the actual computational performance of the discrete approximation. We investigate the eigenvalue spectrum associated with the Lax-Wendroff scheme applied to a model hyperbolic equation. For an initial-boundary-value problem (IBVP) on a finite domain, the eigenvalue or normal mode analysis is analytically intractable. A study of auxiliary problems (Dirichlet and quarter-plane) leads to asymptotic estimates of the eigenvalue spectrum and to an identification of individual modes as either *benign* or *unstable*. The asymptotic analysis establishes an intuitive as well as quantitative connection between the algebraic tests in the theory of Gustafsson, Kreiss, and Sundström and Lax-Richtmyer L_2 stability on a finite domain.

1. INTRODUCTION

A classical method for carrying out a stability analysis of a discrete hyperbolic IBVP is the normal mode analysis of Gustafsson, Kreiss, and Sundström (GKS) [1]. The GKS theory avoids the analytical intractability of the finite-domain normal mode analysis by analyzing *related* quarter-plane problems. On the other hand, when one performs numerical experiments to verify stability and/or accuracy predictions, the computations are on a finite domain and one typically uses the discrete L_2 norm and not the GKS norm used to prove stability. Thus in practice, we have the dichotomy of *analyzing* quarter-plane problems with GKS norms and *computing* on finite domains with L_2 norms.

The goal of this paper is twofold. First we present asymptotic limits for the normal modes of the discrete (Lax-Wendroff) IBVP on a finite domain. These limits lead to a delineation of the normal modes of the finite-domain problem into three classes. Next we use the asymptotic estimates to make a direct algebraic connection between the normal modes of the GKS quarter-plane analysis and the classes of normal modes of the finite-domain problem. This leads to an interpretation of (unstable) GKS modes which is readily understandable in terms of the Lax-Richtmyer stability in the L_2 norm. In this paper we give only a brief outline of our analysis. A detailed exposition is given in [2].

2. IBVP FOR A MODEL HYPERBOLIC EQUATION

We consider the scalar hyperbolic equation

$$\frac{\partial u}{\partial t} = c\frac{\partial u}{\partial x}, \quad 0 \leq x \leq L, \quad t > 0 \tag{2.1}$$

where $u = u(x,t)$ and c is a real constant. For a well-posed IBVP on a finite domain one must specify initial data, $u(x,0) = f(x)$, $0 \leq x \leq L$, and an *analytical* boundary condition at $x = L$,

$$u(L,t) = g(t) \quad \text{for} \quad c > 0. \tag{2.2}$$

3. A PROTOTYPE FINITE-DIFFERENCE APPROXIMATION

To obtain a difference approximation of the model equation (2.1) a mesh is introduced in (x, t) space with increments Δx and Δt and indexing defined by $x = j\Delta x$ and $t = n\Delta t$. The spatial domain $0 \leq x \leq L$ is divided into J equally spaced increments, i.e., $J\Delta x = L$. As a prototype (explicit) finite-difference approximation for the model equation (2.1), we consider the Lax-Wendroff scheme

$$u_j^{n+1} = u_j^n + \frac{\nu}{2}(u_{j+1}^n - u_{j-1}^n) + \frac{\nu^2}{2}(u_{j+1}^n - 2u_j^n + u_{j-1}^n) \tag{3.1}$$

where $\nu = c\Delta t/\Delta x$ is defined to be the Courant number. In our analysis the analytical boundary condition (2.2) for the difference approximation is assumed to be homogeneous, i.e.,

$$u_J^n = 0. \tag{3.2}$$

If we apply the Lax-Wendroff scheme at the outflow boundary ($j = 0$), the computational stencil protrudes one point to the left of the boundary. It is clear that an additional *numerical boundary scheme* (NBS) is required to calculate u_0^{n+1}, i.e., the solution on the outflow boundary at time level $n+1$. As a prototype NBS we choose the spatially one-sided scheme:

$$u_0^{n+1} = u_0^n + \nu[-\alpha u_2^n + (1 + 2\alpha)u_1^n - (1 + \alpha)u_0^n] \tag{3.3}$$

where α is a (real) parameter. If $\alpha = 0$, then (3.3) is simply

$$u_0^{n+1} = u_0^n + \nu(u_1^n - u_0^n). \tag{3.4}$$

The Lax-Wendroff scheme (3.1) together with the analytical boundary condition (3.2) and the NBS (3.3) is called a *discrete* IBVP. For our purposes it is convenient to rewrite the NBS (3.3) as an equivalent space extrapolation formula [2]:

$$h(E)u_{-1}^n = 0 \quad \text{where} \quad h(E) = (E - 1)^2[2\alpha E - (1 - \nu)]. \tag{3.5}$$

The shift operator E is defined by $Eu_j = u_{j+1}$ and $h(E)$ is a polynomial in E.

4. LAX-RICHTMYER STABILITY OF A DISCRETE IVP OR IBVP

For the stability analysis of a discrete initial value problem (IVP) or IBVP with requisite homogeneous boundary conditions, it is appropriate to write the discrete approximation in vector-matrix form:

$$\mathbf{u}^{n+1} = \mathbf{C}\mathbf{u}^n. \tag{4.1}$$

The Lax-Wendroff scheme (3.1) with the analytical boundary condition (3.2) and NBS (3.3) can be written in vector-matrix form (4.1) where

$$\mathbf{u}^n = \begin{bmatrix} u_0^n \\ u_1^n \\ \cdot \\ \cdot \\ \cdot \\ u_{J-2}^n \\ u_{J-1}^n \end{bmatrix}, \quad \mathbf{C} = \begin{bmatrix} s & v & w & & & & \\ p & q & r & & & O & \\ & \cdot & \cdot & \cdot & & & \\ & & \cdot & \cdot & \cdot & & \\ & & & \cdot & \cdot & \cdot & \\ & O & & & p & q & r \\ & & & & & p & q \end{bmatrix} \tag{4.2a,b}$$

and

$$p = -\nu(1-\nu)/2, \quad q = 1-\nu^2, \quad r = \nu(1+\nu)/2 \tag{4.3a}$$

$$s = 1-\nu(1+\alpha), \quad v = \nu(1+2\alpha), \quad w = -\nu\alpha. \tag{4.3b}$$

Here the matrix size is $J \times J$.

The determination of the *eigensolutions* of the first-order system (4.1) is sometimes called the *normal mode* analysis. If $\mathbf{C}$ has a complete set of eigenvectors, then the general solution of (4.1) can be written as

$$\mathbf{u}^n = \sum_{\ell=0}^{J-1} \alpha_\ell z_\ell^n \boldsymbol{\phi}_\ell \tag{4.4}$$

where z_ℓ and $\boldsymbol{\phi}_\ell$ denote the ℓth eigenvalue and eigenvector of the matrix $\mathbf{C}$ and the α_ℓ's are complex constants determined from a specified initial vector $\mathbf{u}^0$. Thus the eigensolutions are the *normal modes* and it is obvious from (4.4) that they act independently.

A discrete IVP or IBVP represented by (4.1) is Lax-Richtmyer stable if there exists a constant $K \geq 1$ such that for any initial condition $\mathbf{u}^0$

$$\|\mathbf{u}^n\| \leq K\|\mathbf{u}^0\| \tag{4.5}$$

for all $n \geq 0$, $0 \leq n\Delta t \leq T$ with T fixed and $\Delta t/\Delta x$ fixed. A necessary condition for Lax-Richtmyer stability is that there exists a nonnegative constant w such that

$$\rho(C) \leq 1 + \frac{w}{J} \tag{4.6}$$

for all $n \geq 0$, $0 \leq n\Delta t \leq T$ with T fixed and $\Delta t/\Delta x$ fixed. Here $\rho(\mathbf{C})$ denotes the spectral radius of $\mathbf{C}$. Inequality (4.6) is referred to as the *spectral radius condition.*

5. EIGENVALUE SPECTRUM OF A DISCRETE IVP

A necessary condition for the stability of a discrete IBVP is the stability of the corresponding pure IVP or Cauchy problem. In this section we review the eigenvalue spectrum of the IVP for the Lax-Wendroff scheme. The solution of the discrete IVP is assumed to be spatially periodic with period $L = J\Delta x$, and hence

$$u_j^n = u_{j+J}^n. \tag{5.1}$$

Consequently, the Lax-Wendroff scheme can be written in vector-matrix form (4.1) where $\mathbf{C}$ is a $J \times J$ circulant matrix and the eigenvalues z_ℓ are given analytically by

$$z_\ell = 1 - 2\nu^2 \sin^2(\theta_\ell/2) + i\nu \sin\theta_\ell, \quad \ell = 0, 1, \cdots, J-1 \tag{5.2}$$

where $\theta_\ell = 2\ell\pi/J$ and $\nu = c\Delta t/\Delta x$ is the Courant number. The eigenvalue locus given by (5.2) is an ellipse in the complex z-plane. (The eigenvalue *locus* is defined to be a curve through the eigenvalues in the complex z-plane.) The eigenvalue loci for $\nu = 0.5$ and $\nu = 1.1$ are shown in Fig. 5.1.

The ellipse is contained within the unit circle for $|\nu| < 1$ (Fig. 5.1a) and if $|\nu| = 1$, the locus is the unit circle. If $|\nu| > 1$, the ellipse contains the unit circle (Fig. 5.1b). Since a circulant matrix is normal, the spectral radius condition (4.6) is necessary and

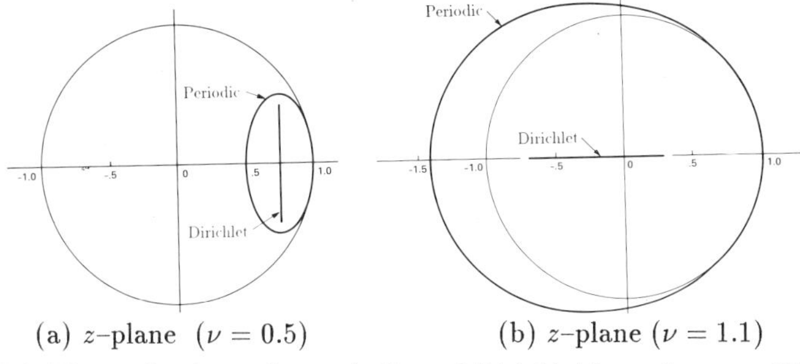

(a) z–plane $(\nu = 0.5)$ (b) z–plane $(\nu = 1.1)$

Fig. 5.1. Eigenvalue locus for periodic and Dirichlet boundary conditions.

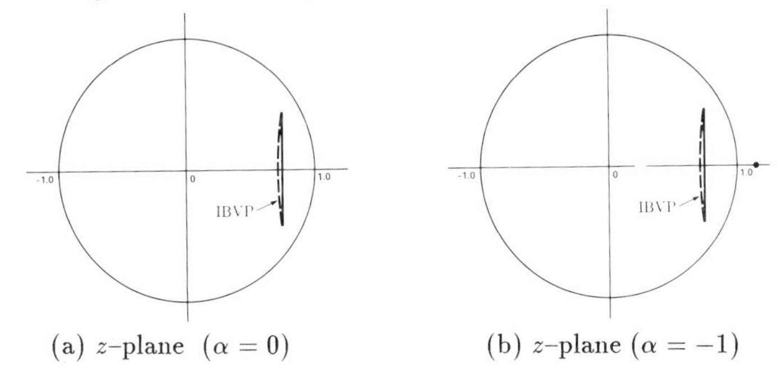

(a) z–plane $(\alpha = 0)$ (b) z–plane $(\alpha = -1)$

Fig. 6.1. Dashed curve is eigenvalue locus of the matrix (4.2b) for ν=0.5.

sufficient for stability. Consequently, as is well-known, the Lax-Wendroff scheme is stable for the IVP (i.e., Cauchy stable) in the L_2 norm for $|\nu| \leq 1$.

6. EIGENVALUE SPECTRUM OF A DISCRETE IBVP

Before we present asymptotic limits for the eigenvalues of the IBVP matrix (4.2b) it is advantageous to examine the eigenvalue spectrum by computing the eigenvalues numerically. As an example if we choose $\alpha = 0$ in (3.3) we obtain the NBS (3.4). A sketch of the eigenvalue locus is shown by the dashed curve in Fig. 6.1a for $\nu = 0.5$. If we increase the number of spatial increments J one finds that the eigenvalue locus approaches the solid vertical line for increasing J.

As a second numerical example, we consider $\alpha = -1$ in (3.3). This NBS leads to an unstable discrete IBVP. The eigenvalue locus is shown in Fig. 6.1b. The eigenvalue locus is qualitatively similar to the locus shown in Fig. 6.1a except for a single eigenvalue outside the unit circle shown by the solid symbol in the figure. To plotting accuracy the eigenvalue indicated by the solid symbol appears fixed for increasing values of J. The eigenvalue locus indicated by the dashed curve approaches the solid vertical line for increasing values of J. If one does a GKS quarter-plane analysis, one finds that the lone eigenvalue outside the unit circle is a GKS eigenvalue.

7. AN AUXILIARY DIRICHLET PROBLEM

The GKS stability analysis involves three auxiliary problems: the Cauchy problem and the left- and right-quarter plane problems. In this section we consider a fourth auxiliary problem which we call the Dirichlet problem on a finite domain. The importance of the Dirichlet problem accrues from the fact that, with the exception of isolated eigenvalues detected by the GKS quarter-plane analysis, the eigenvalue spectrum of a discrete IBVP is a perturbation of the eigenvalue spectrum of the auxiliary Dirichlet problem.

The auxiliary Dirichlet problem is constructed by equating to zero any grid function value u_j^n which is required by the interior difference approximation but falls outside the computational domain $0 \leq x \leq L$. Hence the auxiliary Dirichlet problem for the Lax-Wendroff scheme can be written in the vector-matrix form (4.1) where the matrix operator $\mathbf{C}$ is a $J \times J$ tridiagonal matrix with elements p, q, r. The eigenvalues of $\mathbf{C}$ can be determined analytically:

$$z_\ell = (1-\nu^2) + i\nu\sqrt{1-\nu^2}\,\cos\theta_\ell, \qquad \ell = 1, 2, \cdots, J \tag{7.1}$$

where $\theta_\ell = \ell\pi/(J+1)$. If $|\nu| < 1$, the eigenvalues are complex and the eigenvalue locus is a vertical line centered at the point $(1-\nu^2, 0)$ in the complex z-plane. If $|\nu| = 1$ all the eigenvalues degenerate to the single point $z_\ell = 0$. For $|\nu| > 1$ the eigenvalues are real. The eigenvalue spectra of the pure IVP (periodic boundary conditions) and the auxiliary Dirichlet problem are compared in Fig. 5.1. By some elementary calculations [2] one finds the rather remarkable result that the eigenvalue locus of the auxiliary Dirichlet problem is simply a straight line segment joining the foci of the ellipse which is the eigenvalue locus of the IVP.

8. NORMAL MODE ANALYSIS

In this section the relevant formulas for the normal mode analysis of the finite-domain problem and the quarter-plane problem are summarized. A detailed analysis is given in [2].

8.1 Finite-Domain Normal-Mode Analysis – Summary. An eigensolution or normal mode of the finite-domain IBVP is determined by looking for a solution of the form

$$u_j^n = z^n \phi_j \tag{8.1}$$

which satisfies the Lax-Wendroff scheme (3.1) with the analytical boundary condition (3.2) and the NBS (3.3) written as an extrapolation formula (3.5).

The eigenvalue z is given by

$$z = 1 + \frac{\nu}{2}(\kappa - \frac{1}{\kappa}) + \frac{\nu^2}{2}(\kappa - 2 + \frac{1}{\kappa}) \tag{8.2}$$

and the components ϕ_j of the eigenvector $\boldsymbol{\phi}$ are

$$\phi_j = a[\kappa^j - (-\kappa^2/\zeta)^J(-\zeta/\kappa)^j] \tag{8.3}$$

where $\kappa = \sqrt{\zeta}\hat{\kappa}$ and $\hat{\kappa}$ is a root of the characteristic equation

$$h(\sqrt{\zeta}\hat{\kappa}) - (-\hat{\kappa}^2)^{J+1} h(-\sqrt{\zeta}/\hat{\kappa}) = 0. \tag{8.4}$$

The parameter ζ, which is positive for $-1 < \nu < 1$, is defined by

$$\zeta = \frac{1-\nu}{1+\nu}. \tag{8.5}$$

The coefficient a on the right-hand side of (8.3) is an arbitrary constant. The polynomial $h(\sqrt{\zeta}\hat{\kappa})$ depends solely on the NBS (3.5), i.e., $h(\sqrt{\zeta}\hat{\kappa})$ is the polynomial associated with the NBS written as an extrapolation formula. If one could solve for the roots of the characteristic equation (8.4), then the eigenvalues z and the eigenvectors $\boldsymbol{\phi}$ would follow directly from (8.2) and (8.3). The normal mode analysis on a finite domain is, in general, analytically intractable because one cannot solve for the roots of the characteristic equation (8.4).

For the auxiliary Dirichlet problem, the polynomial $h(\sqrt{\zeta}\hat{\kappa})$ is unity and (8.4) reduces to

$$(-\hat{\kappa}^2)^{J+1} = 1. \tag{8.6}$$

One can solve (8.6) by using the roots of unity formula and the normal modes can be found analytically.

8.2 Quarter-Plane Normal-Mode Analysis – Summary. For the right quarter-plane problem one also looks for a solution of the form (8.1) which satisfies the Lax-Wendroff scheme (3.1) and the NBS (3.5). The details of the GKS normal mode analysis are given in [2, Appendix A]. The eigenvalue z is given by (8.2) and the components ϕ_j of the eigenvector $\boldsymbol{\phi}$ are

$$\phi_j = a\kappa^j \tag{8.7}$$

where $\kappa = \sqrt{\zeta}\hat{\kappa}$ and $\hat{\kappa}$ is a root of the quarter-plane characteristic equation

$$h(\sqrt{\zeta}\hat{\kappa}) = 0. \tag{8.8}$$

The (α, ν) parameter space for which the discrete IBVP is GKS stable is shown by the cross-hatched region of Fig. 8.1.

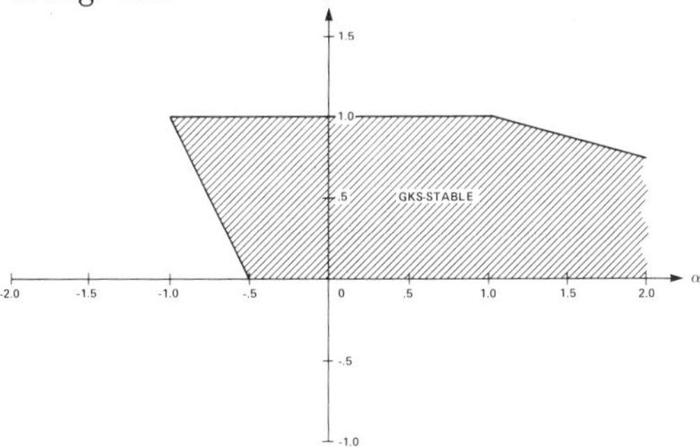

Fig. 8.1. GKS-stability region for Lax-Wendroff (3.1) with NBS (3.3).

9. ASYMPTOTIC ROOTS OF CHARACTERISTIC EQUATION

There exists a small class of discrete IBVP's which are sometimes called *borderline cases*. Borderline cases are unstable according to the GKS theory but they may be Lax-Richtmyer stable or unstable in the L_2 norm on a finite domain. Borderline approximations can be characterized by the presence of a *stationary* mode for the finite-domain problem. A stationary mode is defined to have the property that a $\hat{\kappa}$ root of the characteristic equation (8.4) is independent of J, i.e., $\hat{\kappa}$ remains fixed in the complex plane as J increases. It is important to note that the detection of a stationary mode requires no asymptotic (large J) analysis because there is no J dependence.

Although there can be stationary modes which are independent of J, almost all roots of the characteristic equation (8.4) depend upon J and we write $\hat{\kappa} = \hat{\kappa}(J)$. One can show that there is no loss in generality in assuming $|\hat{\kappa}| \leq 1$ and we write

$$|\hat{\kappa}| = |\hat{\kappa}(J)| = 1 - \epsilon, \quad 0 < \epsilon(J) < 1 \tag{9.1}$$

where

$$\text{either} \quad \epsilon(J) \geq \delta > 0 \quad \text{or} \quad \epsilon(J) \to 0 \qquad \text{as} \quad J \to \infty. \tag{9.2a,b}$$

In particular we are interested in the conditions under which the characteristic equation (8.4) reduces to the quarter-plane characteristic equation (8.8) in the limit $J \to \infty$. Obviously this depends on the asymptotic behavior of $|\hat{\kappa}(J)|^J$. There are only two possibilities, either

$$I: \quad \lim_{J \to \infty} |\hat{\kappa}(J)|^J = \text{constant} > 0, \qquad \text{or} \quad II: \quad \lim_{J \to \infty} |\hat{\kappa}(J)|^J = 0. \tag{9.3a,b}$$

In case II it is obvious that (8.4) reduces to (8.8) as $J \to \infty$. In case I one can show that the roots of (8.4) asymptotically approach the roots of the auxiliary Dirichlet problem (8.6) as $J \to \infty$. The details of the asymptotic estimates are in [2].

10. CLASSIFICATION OF NORMAL MODES

The normal modes of a discrete IBVP can be divided into three classes according to their asymptotic behavior as $J \to \infty$. We associate the following nomenclature with the three classes:

$$\begin{aligned} &I. \quad \text{Dirichlet-}\textit{like} \text{ modes} \\ &II. \quad \text{Quarter-plane-}\textit{like} \text{ modes} \\ &III. \ \text{Stationary modes.} \end{aligned} \tag{10.1}$$

The adjective *like* is used to imply that one can identify for finite J a particular mode that becomes either a Dirichlet mode or quarter-plane mode in the limit $J \to \infty$.

For a given difference approximation almost all the normal modes are in class I. These modes have a generic eigenvalue spectrum which is easy to describe; to wit, for a given Courant number ν and finite J the spectrum is simply a perturbation of the Dirichlet spectrum. A typical generic case is shown in Fig. 6.1a. The solid vertical line is the Dirichlet locus. The dotted line slightly to the left is the eigenvalue locus of the class I modes of the difference approximation to the IBVP. One can show [2] that the dotted eigenvalue locus approaches the Dirichlet locus at least as fast as $O(1/J)$. Hence as the mesh is refined, i.e., $J \to \infty$, the dotted locus collapses onto the Dirichlet locus. Hence the terminology Dirichlet-*like* modes.

The class I modes are always *benign* in the sense that they do not introduce unstable modes into a difference approximation which is Cauchy stable. Only modes of class II and III can introduce *unstable* modes into a difference approximation which is Cauchy stable. If modes of class II and III exist, they are created by the NBS.

The modes in class II are related to the GKS stability theory in the sense that they become quarter-plane modes as $J \to \infty$, i.e., as the the mesh is refined. Finally, the stationary modes which constitute class III are common to both the finite-domain problem and the quarter-plane problems.

11. SKETCHES OF ROOT AND EIGENVALUE DISTRIBUTIONS

In this section we give a pictorial description of the roots of the characteristic equation (8.4) and the corresponding eigenvalue spectra associated with each of the three classes (10.1). The $\hat{\kappa}$ roots plotted in the examples were computed numerically from the characteristic equation (8.4) and the corresponding eigenvalues were computed using (8.2).

The roots for the characteristic equation (8.6) of the auxiliary Dirichlet problem are plotted in Fig. 9.1a for $J = 19$. The corresponding eigenvalue locus is the vertical line

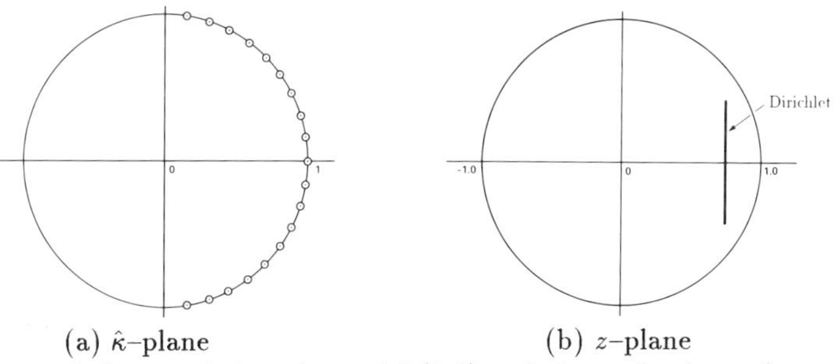

(a) $\hat{\kappa}$–plane (b) z–plane

Fig. 9.1. Roots of characteristic polynomial (8.6) and eigenvalue locus for $\nu = 0.5$.

shown in Fig. 9.1b. In this figure and the figures to follow we plot the eigenvalue locus rather than individual eigenvalues because of the small size of the figures.

11.1 Dirichlet-*like* modes. The roots of the characteristic equation (8.4) which correspond to modes in class I have a generic root locus in the complex $\hat{\kappa}$-plane which is simply a perturbation (inside the unit circle) of the Dirichlet root locus which is on the unit circle. As an example we consider a stable case from the shaded region of Fig. 8.1 by choosing parameter values $\alpha = 0$ and $\nu = 0.5$. The roots of (8.4) for $J = 19$ are plotted in Fig. 9.2a.

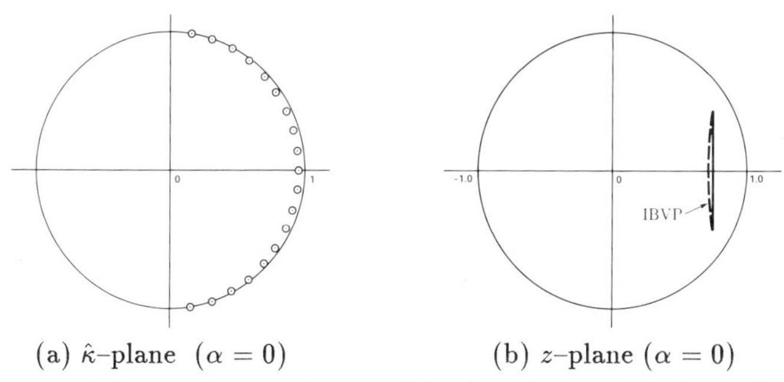

(a) $\hat{\kappa}$–plane ($\alpha = 0$) (b) z–plane ($\alpha = 0$)

Fig. 9.2. Roots of characteristic polynomial (8.4) and eigenvalue locus for $\nu = 0.5$.

The eigenvalue locus is indicated by the dashed curve of Fig. 9.2b. The solid vertical line is the eigenvalue locus of the auxiliary Dirichlet problem. As J increases the root locus approaches the unit circle and the eigenvalue locus moves toward the Dirichlet locus at least as fast as $O(1/J)$.

11.2 Quarter-plane-*like* modes. The previous example had only class I modes, i.e., Dirichlet-*like* modes, while the examples of this section have both class I and class II modes. Modes in class II are related to the GKS theory in the sense that they become quarter-plane modes as $J \to \infty$. In addition, there are only a few modes in this class and the maximum number is known exactly.

The following two examples are unstable discrete IBVP's. Even though a discrete IBVP is unstable, there is no *dramatic* change in the eigenvalue locus in the sense that it remains a perturbation of the Dirichlet eigenvalue locus but with the addition of one or two eigenvalues near or strictly outside the unit circle. These additional eigenvalues correspond to GKS eigenvalues or generalized eigenvalues in the limit $J \to \infty$.

11.2a Unstable quarter-plane-*like* mode – GKS eigenvalue. For the first example we choose $\alpha = -1$ and $\nu = 0.5$ from the unshaded region of Fig. 8.1. The roots of (8.4) are shown in Fig. 9.3a for $J = 19$.

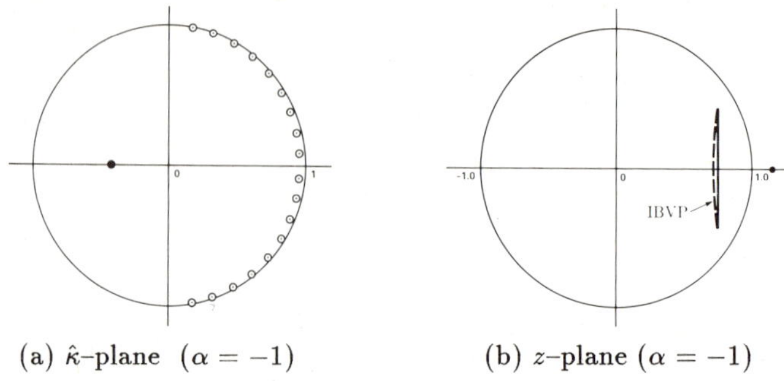

(a) $\hat{\kappa}$–plane $(\alpha = -1)$ (b) z–plane $(\alpha = -1)$

Fig. 9.3. Roots of characteristic polynomial (8.4) and eigenvalue locus for $\nu = 0.5$.

From the figure it is apparent that there is a single isolated root indicated by the solid symbol in the figure. The corresponding eigenvalue is indicated by the solid symbol in Fig. 9.3b. This single eigenvalue remains strictly outside the unit circle as $J \to \infty$ and consequently the approximation is unstable (GKS eigenvalue).

11.2b Unstable quarter-plane-*like* mode – GKS generalized eigenvalue. One should expect the discrete IBVP to be unstable for $\nu < 0$ for otherwise one would have a stable approximation for an ill-posed IBVP. As an example we choose $\alpha = 0$ and $\nu = -0.5$ from the unshaded region of of Fig. 8.1. The roots of (8.4) are depicted in Fig. 9.4a for $J = 19$.

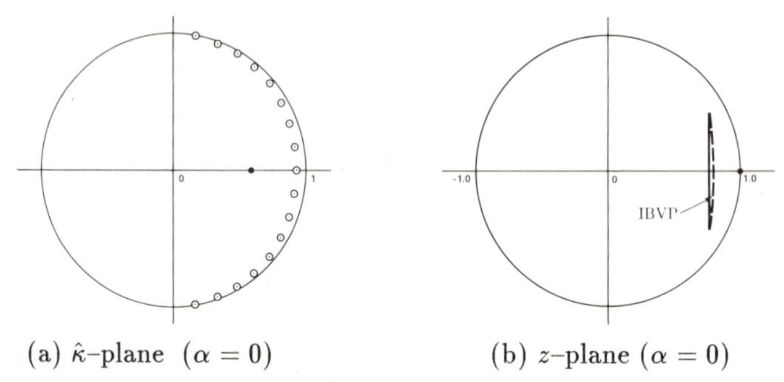

(a) $\hat{\kappa}$–plane $(\alpha = 0)$ (b) z–plane $(\alpha = 0)$

Fig. 9.4. Roots of characteristic polynomial (8.4) and eigenvalue locus for $\nu = -0.5$.

There are two (nearly coincident) isolated roots as shown by the solid symbol of the figure. The corresponding eigenvalues are $z \approx 1$ in Fig. 9.4b. For finite J, one of these eigenvalues is slightly inside the unit circle and the other is slightly outside. In this example the origin of the instability is rather subtle. The instability is not due to a violation of the spectral radius condition (4.6) but rather is due to the introduction of a solution (proportional to the eigenvector) whose norm cannot be uniformly bounded by the norm of the initial data as the mesh is refined, i.e., there is algebraic growth. In the nomenclature of the GKS theory, there is a generalized eigenvalue.

11.3 Stationary mode. As a final example we consider a *borderline* case. For parameter values we pick $\alpha = -0.75$ and $\nu = 0.5$ which is a point on the *left* boundary between stability and instability in Fig. 8.1. The roots of (8.4) are shown in Fig. 9.5a for $J = 19$.

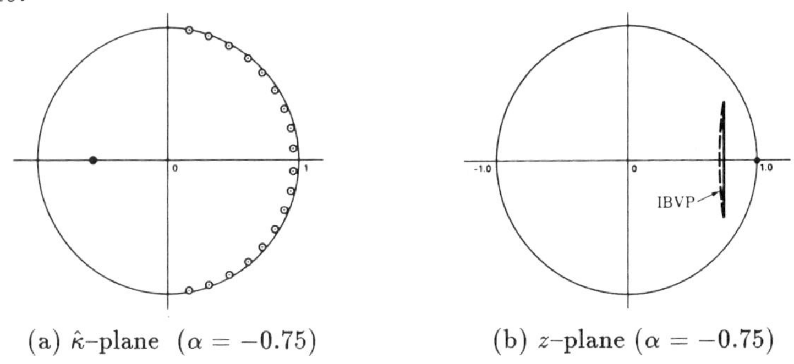

(a) $\hat{\kappa}$–plane ($\alpha = -0.75$) (b) z–plane ($\alpha = -0.75$)

Fig. 9.5. Roots of characteristic polynomial (8.4) and eigenvalue locus for $\nu = 0.5$.

There is one isolated root shown by the solid symbol of the figure. The corresponding eigenvalue is $z = 1$. This eigenvalue of unity is independent of J. Since $z^n = 1$, the stability or instability of the difference approximation devolves to the behavior of the corresponding eigenvector as the mesh is refined. This approximation happens to be Lax-Richtmyer stable but GKS unstable. The details of this example are worked out in [3].

12. CONCLUSIONS

We have investigated the eigenvalue spectrum for the Lax-Wendroff scheme applied to a model hyperbolic IBVP. For the discrete IBVP on a finite domain, the normal mode analysis is analytically intractable even for the simple prototype difference approximation. On the basis of an asymptotic normal mode analysis (large J), we have classified the normal modes of the finite-domain problem into three classes. The resulting classification leads to a simple description of the asymptotic eigenvalue distribution. For a given Courant number, the spectrum is simply a perturbation of the spectrum of the auxiliary Dirichlet problem plus whatever eigenvalues are present in the GKS normal mode analysis for the related quarter-plane problems.

Almost all the modes are Dirichlet-*like* modes and are *benign* in the sense that they do not introduce unstable modes into a difference approximation which is Cauchy stable. Only quarter-plane-*like* modes and stationary modes can introduce *unstable* modes into an approximation which is Cauchy stable. Consequently, if an instability exists it is caused by the NBS and is detected in the GKS analysis.

REFERENCES

[1] B. Gustafsson, H. O. Kreiss and A. Sundström, *Stability theory of difference approximations for mixed initial boundary value problems. II*, Mathematics of Computation 26, (1972), pp. 649-686.

[2] R. F. Warming, R. M. Beam: *Stability of finite-difference approximations for hyperbolic initial-boundary-value problems I: an eigenvalue analysis*, to appear.

[3] R. F. Warming, R. M. Beam: *Stability of finite-difference approximations for hyperbolic initial-boundary-value problems II: stationary modes*, to appear.

The Use of Fast Solvers in Computational Fluid Dynamics

Gabriel Wittum

Sonderforschungsbereich 123

Universität Heidelberg

Im Neuenheimer Feld 294

D-6900 Heidelberg

Summary

The recent development in Computational Fluid Dynamics leads to larger and larger numbers of unknowns. Thus the use of fast solvers becomes more and more important. In the present paper, we focus on multi-grid methods and Navier-Stokes equations. Recent results demonstrate the effectivity of such solvers over standard approaches applied to the compressible as well as to the incompressible, the stationary as well as to the non-stationary case. All the approaches mentioned above use special ingredients to obtain full multi-grid efficiency.

After an overview of multi-grid in CFD, we incorporate the smoothers proposed by Vanka, [11], into the r-transforming framework from [14] and report on some numerical comparisons of these ones.

1. Introduction

The recent development in CFD, e.g. the treatment of three-dimensional problems, complicated geometries, resolution of boundary layers, modelling turbulent flows etc. leads to larger and larger numbers of unknowns. Using classical solvers, the work necessary to resolve the governing equations increases non-linearly with growing number of unknowns. Bringing a computing power on the desks, numerical analysts never dreamt of some years ago, the fast hardware development only improves the constant, but does not change the substantial problem. Thus the use of fast solvers is inevitable in CFD. As the problems to be treated are discretizations of strongly nonsymmetric and non-linear partial differential equations, the main interest focuses on multi-grid methods, which are asymptotically optimal.

The first multi-grid method for the Navier-Stokes equations was presented already in 1977 by P. Wesseling, [12], based on a stream-funciton vorticity formulation. In the following there are two main approaches. The first one is based on constructing multi-grid methods for the Euler equations and then incorporating the elliptic part. The main problem here is to introduce sufficient discrete ellipticity without loosing sharpness of the approximation. Substantial suggestions in this area are due to Jamesson, [5], Hemker and Koren, [4], Spekreijse, [10], Dick, [2], and Hänel et al. [3]. Multi-grid is mainly used to accelerate steady-state computations, as there the gains are most drastic, however it can be used too in time dependent calculations e.g with implicit time-discretizations but it may also help to reduce the influence of stability restricitions on the stepsize in semi-implicit calculations.

The second approach starts with the work by Brandt-Dinar, [1], which is based on a method for the Stokes equations in primal variables. Incorporating the nonlinear terms it covers the case of incompressible flow. Besides the discretization of the convective terms, the main problem is to construct a suitable smoother, handling the indefiniteness on the one hand and being robust, i.e. not sensitive to the singular perturbation, on the other one. There are several suggestions for smoothers, see [1], [7], [6], [8], [11], [14], [16], which all are comprised in the concept of transforming smoothing, see [14], as explained below.

In the present paper we consider several details of constructing a r-transforming smoother for the incompressible, steady-state Navier-Stokes equations

$$\begin{aligned} -\Delta \boldsymbol{u} + Re \cdot \boldsymbol{u} \cdot \nabla \boldsymbol{u} + \nabla p &= f \\ \nabla \cdot \boldsymbol{u} &= 0 \end{aligned} \qquad \text{in } \Omega \subset \mathbb{R}^2 \tag{1.1}$$

$$\boldsymbol{u} = \boldsymbol{g} \qquad \text{on } \partial\Omega.$$

So we discuss local transforming smoothers which now also include the methods as suggested by Vanka, [11], and finally compare the r-transforming smoothers of DGS and SIMPLE type using the standard driven cavity test problem for Reynolds-numbers $0 \leq Re \leq 1000$.

2. R-Transforming Smoothers

We discretize (1.1) on staggered grids as described in [14]. The resulting discrete operator

$$K_\ell(u_\ell^0) = \begin{pmatrix} Q_\ell(u_\ell^0) & \nabla_\ell \\ \nabla_\ell^T & 0 \end{pmatrix} \tag{2.1a}$$

with

$$Q_\ell(u_\ell^0) = -\Delta_\ell + N_\ell(u_\ell^0),\ N_\ell(u_\ell^0) = u_\ell^0 \cdot \nabla_\ell + \nabla u_\ell^0 \cdot I_\ell \tag{2.1b}$$

is indefinite. Thus a classical splitting with

$$K_\ell(u_\ell^0) = M - N,\ M \text{ regular, "easily invertible"} \tag{2.2}$$

is not possible. A simple remedy is to use a r-transforming iteration for smoothing as introduced in [14]. To that end we construct an auxiliary matrix $\overline{K_\ell}$ such that a splitting

$$K_\ell \overline{K_\ell} \quad = \quad M - N \tag{2.3}$$

makes sense. The corresponding r–transforming iteration to solve

$$K_\ell \, x \;=\; b \tag{2.4}$$

reads

$$x^{(i+1)} \;=\; x^{(i)} - \overline{K_\ell} \, M^{-1} (K_\ell x^{(i)} - b). \tag{2.5}$$

Two r-transformations yield common methods for (1.1). First the DGS and TILU-method as introduced in [1] and [14] using

$$\overline{K_\ell} \;=\; \begin{pmatrix} I_\ell & \nabla_\ell \\ 0 & -Q'(u_\ell^0) \end{pmatrix} \tag{2.6}$$

with

$$K_\ell \overline{K_\ell} \;=\; \begin{pmatrix} Q_\ell(u_\ell^0) & W \\ -\nabla_\ell^T & -\Delta_\ell^N \end{pmatrix}, \tag{2.7}$$

second the SIMPLE methods, introduced in [7] using

$$\overline{K_\ell} \;=\; \begin{pmatrix} I_\ell & -D_\ell^{-1} \nabla_\ell \\ 0 & I_\ell \end{pmatrix} \tag{2.8}$$

where D_ℓ^{-1} is some approximate inverse for $Q_\ell(u_\ell^0)$ and

$$K_\ell \overline{K_\ell} \;=\; \begin{pmatrix} Q_\ell(u_\ell^0) & (I_\ell - Q_\ell(u_\ell^0) D_\ell^{-1}) \nabla_\ell \\ -\nabla_\ell^T & \nabla_\ell^T D_\ell^{-1} \nabla_\ell \end{pmatrix}. \tag{2.9}$$

The DGS/TILU methods have been originally designed as multi-grid smoothers whereas SIMPLE methods are widely used as iterative solvers in technical production software. In multi-grid context it was first used by Lonsdale, [6], and even together with some turbulence model by Scheuerer et al. in [8].

A crucial point of the realization of SIMPLE is the choice of D_ℓ^{-1}. As discussed in [14] it is reasonable to use a **perturbed r-transforming smoother**. Here we leave the upper right block of $K_\ell \overline{K_\ell}$ for splitting (2.4) but still use the original $\overline{K_\ell}$ for the transformation in (2.5). Additionally, we are free to prescribe boundary conditions for $\overline{K_\ell}$ and $K_\ell \overline{K_\ell}$ provided both operators still are stable and yield a convergent method. The choice of boundary conditions may have considerable influence on the effectivity of the resulting method as discussed below.

One of the main requirements for a multi-grid method which is to be used for fluid-mechanical problems is **robustness**, meaning that the algebraic solver is insensitive to singular perturbations (see [13], [15]). Robust methods for scalar problems can be constructed via incomplete LU-smoothers (ILU). The r-transforming approach allows the extension of these scalar methods to indefinite systems like the Navier-Stokes equations (in [14]).

3. Local R-Transformations

In context of single-step methods like Gauß-Seidel we have the possiblity to apply such transformations **locally**. I.e. marching through the grid we collect the equations in a grid-point or cell resp., solve the corresponding transformed equations and »distribute« the resulting corrections by $\overline{K}_\ell$. That is the way Vanka, [11], performs his **coupled** method. Taking the momentum equations in the $\boldsymbol{u}$ and $\boldsymbol{v}$ points of the box shown in fig. 3.1.

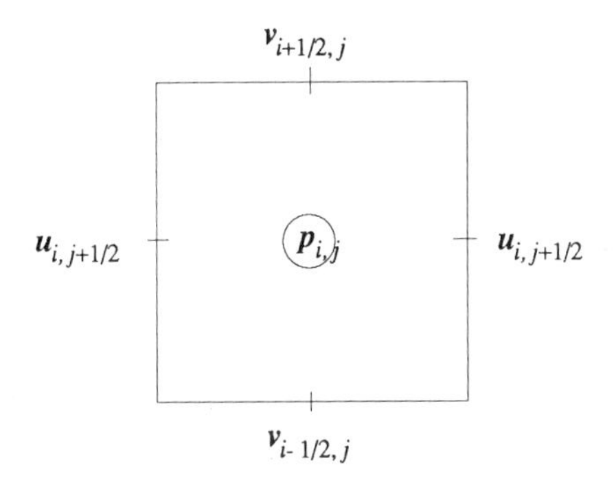

Fig. 3.1: One cell of the MAC-grid as used in [14].

We obtain the local equations

$$\begin{pmatrix} \Omega_u & 0 & \begin{matrix} h^{-1} \\ -h^{-1} \end{matrix} \\ 0 & \Omega_v & \begin{matrix} h^{-1} \\ -h^{-1} \end{matrix} \\ h^{-1} \quad -h^{-1} & h^{-1} \quad -h^{-1} & 0 \end{pmatrix} \cdot \begin{pmatrix} u_{i,j-\frac{1}{2}} \\ u_{i,j+\frac{1}{2}} \\ v_{i-\frac{1}{2},j} \\ v_{i+\frac{1}{2},j} \\ p_{i,j} \end{pmatrix} = \begin{pmatrix} d_{i,j-\frac{1}{2}} \\ d_{i,j+\frac{1}{2}} \\ d_{i-\frac{1}{2},j} \\ d_{i+\frac{1}{2},j} \\ d_{i,j} \end{pmatrix} \tag{3.1a}$$

with

$$\mathfrak{Q}_u = \begin{pmatrix} q^u_{m,i,j-\frac{1}{2}} & q^u_{r,i,j-\frac{1}{2}} \\ q^u_{\ell,i,j+\frac{1}{2}} & q^u_{m,i,j+\frac{1}{2}} \end{pmatrix} \tag{3.1b}$$

and

$$\mathfrak{Q}_v = \begin{pmatrix} q^v_{m,i-\frac{1}{2},j} & q^v_{r,i-\frac{1}{2},j} \\ q^v_{\ell,i+\frac{1}{2},j} & q^v_{m,i+\frac{1}{2},j} \end{pmatrix} \tag{3.1c}$$

where the $q_{...}$ are the coefficients of Q_ℓ from (2.1b) corresponding to the points of the box shown above and the right-hand side d consists of the defect of the old iterate in the resp. points.

Computing the inverse in (3.1) directly is a rather time consuming job, but instead we can use the transformation $\overline{K_\ell}$ from (2.6). This yields the product system

$$\begin{pmatrix} \mathfrak{Q}_u & & 0 & & 0 \\ & 0 & & \mathfrak{Q}_v & 0 \\ h^{-1} & -h^{-1} & h^{-1} & -h^{-1} & s \end{pmatrix} \tag{3.2}$$

with

$$s := \begin{pmatrix} h^{-1} & -h^{-1} & h^{-1} & -h^{-1} \end{pmatrix} \cdot \begin{pmatrix} \mathfrak{Q}_u & 0 \\ 0 & \mathfrak{Q}_v \end{pmatrix} \cdot \begin{pmatrix} h^{-1} \\ -h^{-1} \\ h^{-1} \\ -h^{-1} \end{pmatrix} = s_u + s_v ,$$

$$s_{u/v} := h^{-2} \, \frac{\left(q^{u/v}_{m,i,j-\frac{1}{2}} + q^{u/v}_{r,i,j-\frac{1}{2}} + q^{u/v}_{\ell,i,j+\frac{1}{2}} + q^{u/v}_{m,i,j+\frac{1}{2}}\right)}{\det\left(\mathfrak{Q}_{u/v}\right)} .$$

which is easily inverted. Thereafter we »distribute« the corrections by

$$\overline{K_\ell} := \begin{pmatrix} I & 0 & -\mathfrak{Q}_u^{-1} \cdot \begin{pmatrix} h^{-1} \\ -h^{-1} \end{pmatrix} \\ 0 & I & -\mathfrak{Q}_v^{-1} \begin{pmatrix} h^{-1} \\ -h^{-1} \end{pmatrix} \\ & & 1 \end{pmatrix} . \tag{3.3}$$

The original version by Vanka uses the simplifications

$$\mathfrak{Q}_u = \begin{pmatrix} q^u_{m,i,j-\frac{1}{2}} & 0 \\ 0 & q^u_{m,i,j+\frac{1}{2}} \end{pmatrix} \tag{3.4a}$$

and

$$\mathfrak{Q}_{\mathrm{v}} = \begin{pmatrix} \overset{\mathrm{v}}{q}_{m,i-\frac{1}{2},j} & 0 \\ 0 & \overset{\mathrm{v}}{q}_{m,i+\frac{1}{2},j} \end{pmatrix} \tag{3.4b}$$

and consequently

$$s = h^{-2} \left(\frac{1}{\overset{u}{q}_{m,i,j-\frac{1}{2}}} + \frac{1}{\overset{u}{q}_{m,i,j+\frac{1}{2}}} + \frac{1}{\overset{\mathrm{v}}{q}_{m,i-\frac{1}{2},j}} + \frac{1}{\overset{\mathrm{v}}{q}_{m,i+\frac{1}{2},j}} \right) \tag{3.5}$$

for s from (3.2).

As shown in Section 4 the simplification gives almost the same results. Usually, the corrections obtained by this process are scaled by relaxation parameters, which is necessary for optimal performance. Nevertheless the method may serve as smoother without applying any parameters.

On the same basis we can write the linewise coupled smoother introduced by Shah, Mayers and Rollett in [9], just extending K from (3.1) over a whole line (or column resp.) of our grid. Then the scalar entry s in $K\overline{K}$ from (3.2a) changes into a tridiagonal matrix, corresponding to a one-dimensional Laplacian, while applying the same transformation to the whole grid at once causes a two-dimensional Laplacian in this place. This clearly shows how all these often very complicated looking schemes can be easily incorporated into the framework of r-transforming smoothers. In the very same manner the »distributive relaxation« by Brandt and Dinar, [1], may be applied too.

4. Numerical Results

In [16] we gave a comparison between several implementations of r-transforming ILU_β with SIMPLE transformation (2.6) and Brandt-Dinar transformation (2.8). We now compare the two variants of Vankas method described above in the same framework and finally compare this to the TILU methods.

As benchmark problem we took the standard lid-driven cavity in two dimensions. To obtain reasonable solutions we used a defect correction process with central differences for the convective terms. This process is known to converge up to $Re \leq 5000$. The nonlinearity is handled by a simplified Newton iteration, see [14]. We measured the residual convergence of the outer defect correction iteration. The comparison of the exact and the approximate inversion in Vankas method is shown in fig 4.1.

The comparison clearly shows that an exact inversion does not pay, since the operation count of the exact method is larger than the approximate one by a factor of 13/11 which is

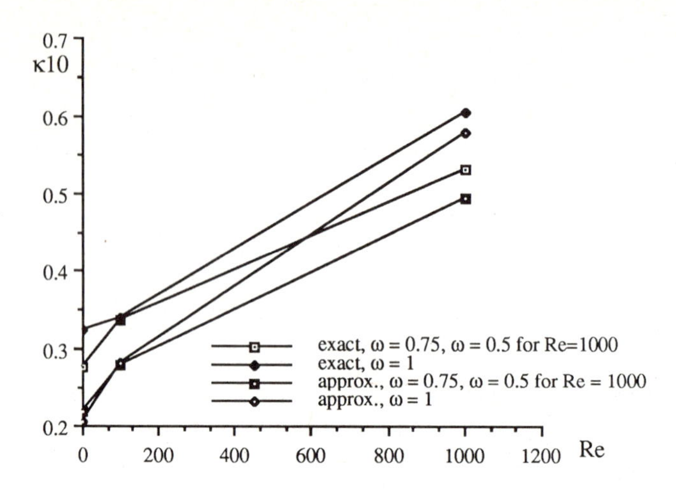

Fig. 4.1: Comparison of the two variants of Vanka's method, e relaxation parameter $\omega = \omega_u = \omega_v$, $\omega_p = 1$. Two smoothing step, V-cycle for $Re = 0$, W-cycle else, $h = 1/64$.

already incorporated in the figure. Second, the figure tells that depending on the Reynolds-number the appropriate relaxation factor improves the performance, however, the method works without relaxation too. The relaxation factors were taken from [9] where they are given as the optimal ones. Figure 4.2 shows the comparison of the efficiency of the optimal variant of Vanka's algorithm and the two $TILU_\beta$ methods from [16].

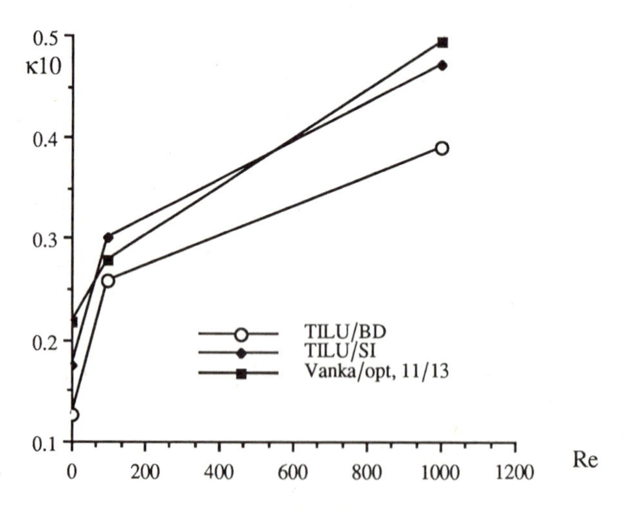

Figure 4.2: Comparison between the smoothers from [16] and the optimal variant of Vanka's smoother. κ_{10} (residual convergence averaged over 10 iterations) vs. *Re*.

From fig. 4.2 we see that the TILU/BD method is slightly superior, in particular as it is (almost) free of parameters. However, it should be mentioned that Vanka's method needs less memory and may be easier to program. Thus all the methods are of practical interest.

References

[1] Brandt,A, Dinar,N: Multigrid solutions to elliptic flow problems. *ICASE Report* Nr. 79-15 (1979).

[2] Dick,E.: A multigrid method for steady incompressible Navier-Stokes equations in primitve variables form. in: *Proceedings of the 7th GAMM-Conference on numerical methods in fluid mechanics*. Louvain-la-neuve, 1987

[3] Hänel,D., Meinke,M., Schröder,W.: Application of the multigrid methods in solutions of the compressible Navier-Stokes equations. *Preprint, Aerodynamisches Institut, RWTH Aachen*, April 1989.

[4] Hemker,P.W., Koren,B. :Multigrid, defect correction and upwind schemes for the steady Navier-Stokes equations. in: W. Hackbusch (ed.): Robust multi-grid methods. *Proceedings of the 4th GAMM-Seminar Kiel 1988.*

[5] Siclari,M.J., Del Guidice,P., Jameson,A.: A multigrid finite volume method for solving the Euler and Naiver-Stokes equations ....*AIAA paper* 89-0283 (1989)

[6] Lonsdale,G. : Solution of a rotating Navier-Stokes problem by a nonlinear multigrid algorithm. *Report* Nr. 105, Manchester University, (1985).

[7] Patankar, S.V. , D.B. Spalding : A calculation procedure for heat and mass transfer in three-dimensional parabolic flows. *Int. J. Heat Mass Transfer*, 15 (1972), 1787-1808

[8] Periç,M., Rüger,M., Scheuerer,G.: A finite volume multigrid method for calculating turbulent flows. *TSF 7 Stanford University*, Aug. 1989

[9] Shah,T.M.,Mayers,D.F.,Rollett,J.S.: Analysis and application of a line solver for re circulationg flows. in *Proceedings of the 5th GAMM Seminar Kiel,* 1989, to appear

[10] Spekreijse,S.:Multigrid solution of monotone second order discretizations of hyperbolic conservation laws. *Math Comp* 49 (179), 135-155 (1987).

[11] Vanka,S.P.: Block implicit multigrid solution of the Navier-Stokes equations in primitve vartiables. *J. Comput. Phys.*, (65) 138-158, 1986

[12] Wesseling,P.: Numerical solution of the stationery Navier-Stokes equation... *Report* NA-18, TU Delft (1977)

[13] Wesseling,P.: Theoretical and practical aspects of a multigrid method. *SISSC* 3 (1982), 387-407.

[14] Wittum,G.: Multi grid methods for Stokes and Navier-Stokes equations Transforming smoothers - algorithms and numerical results. *Numer. Math.*, 54 , 543-563 (1989).

[15] Wittum,G. : On the robustness of ILU-smoothing. *SISSC*, 10, 699-717 (1989)

[16] Wittum,G. : R-transforming smoothers for the incompressible Navier-Stokes equations.. *Proceedings of the 5th GAMM Seminar Kiel*, 1989, to appear

MmB — A New Class of Accurate High Resolution Schemes for Conservation Laws in Two Dimensions

Wu Huamo Yang Shuli

Computing Center Academia Sinica

Beijing 100080 P.R. of China

Abstract

In this paper we present two classes of 2-D second-order accurate MmB schemes, which preserve the local maximum and minimum bounds of the initial data in the smallest union of mesh elements of previous time step containing the domain of dependence of the solution on the mesh element with center at point P under consideration. In 1-D, the MmB schemes are almost identical with TVD schemes. It is proved that 1-D discrete MmB (or TVD) and 1-D semi-discrete TVD schemes may have second-order accuracy at (nonsonic) critical points, but cannot be of uniformly second-order accurate in the whole neighborhood of the critical points. Numerical results for 1-D and 2-D test problems are given.

Introduction

Considerable effort has been expended seeking numerical methods which enable us to solve the fluid dynamics problems accurately and resolve the discontinuities sharply, especially in two and three space dimensions.

Encouraged by the success of the TVD schemes in 1-D, one wants to extend the TVD schemes for two dimensions. Unfortunately, any conservative TVD scheme for solving scalar conservation laws in two space dimensions is at most first-order accurate [1], although the splitting methods using second-order 1-D TVD schemes seem to work quite well for practical problems. Hence, it may be worthy of creating new conception beyond TVD in two dimensions.

Here we present a new class of second- order accurate high resolution schemes — local Maxima and minima Bounds preserving (MmB) schemes.

MmB Schemes in One Dimension

First we give the concept of MmB schemes in general case. Consider the following initial value problem

$$\frac{\partial u}{\partial t} + \sum_{i=1}^{m} \frac{\partial f_i(u)}{\partial x_i} = 0 \quad , \quad |x_i| < +\infty,\ t \leq T$$

$$u(0,x) = g \qquad , \quad x \in (l)$$

for scalar function $u(t,x)$, where (l) is a m–dimensional initial hyperplane.

Let $w_h^{n+1}(P)$ be a mesh element with center at P in the plane $t = t_{n+1}$, Ω_h^n be the mesh consists of mesh elements $w_h^n(P_j)$ with centers at P_j in the plane $t = t_n$. Let $I\left(w_h^{n+1}(P)\right)$ be the domain of dependence on the plane $t = t_n$ of u at points on the domain $w_h^{n+1}(P), I_h\left(w_h^{n+1}(P)\right)$ be the smallest union of mesh elements $w_h^n(P_j)$ of Ω_h^n, such that

$$I\left(w_h^{n+1}(P)\right) \subseteq I_h\left(w_h^{n+1}(P)\right) = \bigcup_{j=1}^{S(P)} w_h^n(P_j)\,.$$

A scheme

$$u_h(P) = L_h u_h(\Omega_h^n)\,, \quad P \in \Omega_h^{n+1}\,,$$

is called as MmB scheme, if

$$\min_{1\leq j\leq S(P)} u_h(P_j) \leq u_h(P) \leq \max_{1\leq j\leq S(P)} u_h(P_j)\,. \tag{1}$$

which is equivalent to the following equation

$$u_h(P) = \sum_{j=1}^{S(P)} \alpha_j^{(P)} \cdot u_h(P_j)\,, \quad \alpha_j^{(P)} \geq 0\,, \quad \sum_{j=1}^{S(P)} \alpha_j^{(P)} = 1\,, \tag{2}$$

where the coefficients $\alpha_j^{(P)}$ may depend on the values of u_h at points beyond the union $I_h\left(w_h^{n+1}(P)\right)$.

For the sake of simplicity in writting, we use the notation $u_h^n = u_h$ in the case with no risk of any confusion.

Consider the 1-D scalar equation

$$u_t + u_x = 0$$

Let $\lambda = a\Delta t/\Delta x$ be the Courant number $|\lambda| \leq 1$.
According to the definition the two time-level explicit MmB schemes have the forms

if $a > 0$,

$$u_j^{n+1} = u_j - c_{j-\frac{1}{2}}(u_j - u_{j-1}), \qquad 0 \leq c_{j-\frac{1}{2}} \leq 1; \tag{3}$$

if $a < 0$,

$$u_j^{n+1} = u_j + d_{j+\frac{1}{2}}(u_{j+1} - u_j), \qquad 0 \leq d_{j+\frac{1}{2}} \leq 1; \tag{4}$$

if a changes the sign,

$$u_j^{n+1} = u_j - c_{j-\frac{1}{2}}(u_j - u_{j-1}) + d_{j+\frac{1}{2}}(u_{j+1} - u_j), \tag{5}$$
$$c_{j-\frac{1}{2}} \geq 0, \quad d_{j+\frac{1}{2}} \geq 0, \quad c_{j-\frac{1}{2}} + d_{j+\frac{1}{2}} \leq 1.$$

Recall that TVD conditions for schemes (5) are [2]

$$c_{j-\frac{1}{2}} \geq 0, \quad d_{j+\frac{1}{2}} \geq 0, \quad 0 \leq c_{j+\frac{1}{2}} + d_{j+\frac{1}{2}} \leq 1.$$

Therefore, for schemes of the form (3) or (4), especially for the upwind schemes [3] of the form

$$\begin{aligned} u_j^{n+1} \quad = \quad & u_j - \frac{|\lambda| + \lambda}{2}(u_j - u_{j-1}) - \frac{\lambda + |\lambda|}{4}(1 - \lambda)\Delta_-[\varphi_j^+ \Delta_{j+\frac{1}{2}} u] \\ & - \frac{\lambda - |\lambda|}{2}(u_{j+1} - u_j) + \frac{\lambda - |\lambda|}{4}(1 + \lambda)\Delta_-[\varphi_{j+1}^- \Delta_{j+\frac{1}{2}} u] \end{aligned} \tag{6}$$

with $\Delta_- f_i = f_j - f_{j-1} = \Delta_{j-\frac{1}{2}} f$, MmB and TVD schemes are identical. But in general, for schemes of the form (5), they may be different.

MmB Schemes in Two Dimensions

Consider the difference schemes for 2-D scalar equation

$$\frac{\partial u}{\partial t} + a\frac{\partial u}{\partial x} + b\frac{\partial u}{\partial y} = 0, \quad a \geq 0, \quad b \geq 0.$$

Let $\lambda = a\Delta t/\Delta x$, $\mu = b\Delta t/\Delta y \geq 0$, be the Courant numbers.

We present here two classes of second-order accurate MmB schemes in two space dimension (see also [4]).

The first class of second-order accurate MmB schemes in 2-D consists of splitting schemes using 1-D second-order accurate MmB schemes.

Theorem 1. *Let L_x, L_y be 1-D MmB operators, then the products*

$$L_x L_y, \quad L_y L_x, \quad L_x L_y L_y L_x, \quad L_y L_x L_x L_y$$

are MmB in 2-D.

Applying Theorem 1 for symmetric product $L_x L_y L_y L_x$ or $L_y L_x L_x L_y$ using second- order accurate 1-D MmB schemes L_x, L_y, we have second-order accurate MmB schemes in 2-D.

Second class consists of the nonsplitting modifications of 2-D Lax-Wendroff scheme.

Consider the scalar equation with constant coefficients $a > 0$, $b > 0$.

The following two flux limited upwind second-order accurate schemes

$$\begin{aligned} u_{j,k}^{n+1} &= u_{j,k} - \lambda\left(u_{j,k} - u_{j-1,k}\right) - \frac{\lambda(1-\lambda)}{2}\Delta_-^x\left[\varphi_{j,k}\left(u_{j+1,k} - u_{j,k}\right)\right] \\ &\quad -\mu\left(u_{j,k} - u_{j,k-1}\right) - \frac{\mu(1-\mu)}{2}\Delta_-^y\left[\psi_{j,k}\left(u_{j,k+1} - u_{j,k}\right)\right] \\ &\quad +\lambda\mu\left(u_{j,k} - u_{j-1,k} + u_{j-1,k-1} - u_{j,k-1}\right), \end{aligned} \tag{7}$$

$$\begin{aligned} u_{j,k}^{n+1} &= u_{j,k} - \lambda\Delta_{j-\frac{1}{2},k}u - \frac{\lambda(1-\lambda)}{2}\left(\varphi_{j,k}\Delta_{j+\frac{1}{2},k}u - \varphi_{j-1,k}\Delta_{j-\frac{1}{2},k}u\right) \\ &\quad +\frac{\beta\lambda\mu}{2}\left[\Theta_{j,k-\frac{1}{2}}\Delta_{j,k-\frac{1}{2}}u - \Theta_{j-1,k-\frac{1}{2}}\Delta_{j-1,k-\frac{1}{2}}u\right] \\ &\quad -\mu\Delta_{j,k-\frac{1}{2}}u - \frac{\mu(1-\mu)}{2}\left(\psi_{j,k}\Delta_{j,k+\frac{1}{2}}u - \psi_{j,k-1}\Delta_{j,k-\frac{1}{2}}u\right) \\ &\quad +\frac{\alpha\lambda\mu}{2}\left[\chi_{j-\frac{1}{2},k}\Delta_{j-\frac{1}{2},k}u - \chi_{j-\frac{1}{2},k-1}\Delta_{j-\frac{1}{2},k-1}u\right] \end{aligned} \tag{8}$$

are unsplitting MmB schemes, where $\varphi(r)$, $\psi(s)$, $\Theta(p)$, $\chi(q)$ are flux limiters satisfying some conditions.

Extension to Nonlinear Systems

Consider the 2-D system

$$\frac{\partial u}{\partial t} + \frac{\partial F(u)}{\partial x} + \frac{\partial G(u)}{\partial y} = 0, \quad u = (u^{(1)}, \ldots, u^{(m)}). \tag{9}$$

We assume that system (9) is hyperbolic, $A = \partial F/\partial u$ and $B = \partial G/\partial u$ have eigenvalues $\lambda_1, \ldots, \lambda_m$, and $\mu_1, \ldots, \mu_m$ and corresponding complete sets of eigenvectors. Define the following diagonal matrices:

$$\begin{aligned} \Lambda &= \operatorname{diag}(\lambda_1, \ldots, \lambda_m), & M &= \operatorname{diag}(\mu_1, \ldots, \mu_m), \\ |\Lambda| &= \operatorname{diag}(|\lambda_1|, \ldots, |\lambda_m|), & |M| &= \operatorname{diag}(|\mu_1|, \ldots, |\mu_m|), \\ \Lambda^\pm &= (\Lambda \pm |\Lambda|)/2, & M^\pm &= (M \pm |M|)/2. \end{aligned}$$

There are nonsingular matrices R, S such that

$$A = R\Lambda R^{-1}\,, \quad B = SMS^{-1}\,.$$

Define

$$\begin{aligned} A^{\pm} &= R\Lambda^{\pm}R^{-1}\,, \quad B^{\pm} = SM^{\pm}S^{-1}\,, \\ \widetilde{\Lambda} &= \frac{\Delta t}{\Delta x}|\Lambda|\left(1 - \frac{\Delta t}{\Delta x}|\Lambda|\right)\,, \\ \widetilde{M} &= \frac{\Delta t}{\Delta y}|M|\left(1 - \frac{\Delta t}{\Delta y}|M|\right)\,. \end{aligned}$$

The usual 1-D Mac Cormack scheme is denoted by $\big(MC_x(u)\big)$.

The scheme (7) is extended for system (9) in the following form:

$$\begin{aligned} u_{jk}^{n+1} &= \big(MC_x(u)\big)_{j,k} + \big(MC_y(u)\big)_{j,k} - u_{j,k} \\ &+\frac{1}{2}\Delta_-^x\left\{R_{j+\frac{1}{2},k}\widetilde{\Lambda}_{j+\frac{1}{2},k}R^{-1}_{j+\frac{1}{2},k}\left(1-\varphi_{j+\frac{1}{2},k}\right)(u_{j+1,k}-u_{j,k})\right\} \\ &+\frac{1}{2}\Delta_-^y\left\{S_{j,k+\frac{1}{2}}\widetilde{M}_{j,k+\frac{1}{2}}S^{-1}_{j,k+\frac{1}{2}}\left(1-\psi_{j,k+\frac{1}{2}}\right)(u_{j,k+1}-u_{j,k})\right\} \\ &+\frac{1}{2}\Delta_-^x\Big\{A^+_{j,k-\frac{1}{2}}B^+_{j,k-\frac{1}{2}}(u_{j,k}-u_{j,k-1}) \\ &+A^+_{j,k+\frac{1}{2}}B^-_{j,k+\frac{1}{2}}(u_{j,k+1}-u_{j,k}) \\ &+A^-_{j+1,k-\frac{1}{2}}B^+_{j+1,k-\frac{1}{2}}(u_{j+1,k}-u_{j+1,k-1}) \\ &+A^-_{j+1,k+\frac{1}{2}}B^-_{j+1,k+\frac{1}{2}}(u_{j+1,k+1}-u_{j+1,k})\Big\} \\ &+\frac{1}{2}\Delta_-^y\Big\{B^+_{j-\frac{1}{2},k}A^+_{j-\frac{1}{2},k}(u_{j,k}-u_{j-1,k}) \\ &+B^-_{j-\frac{1}{2},k+1}A^+_{j-\frac{1}{2},k+1}(u_{j,k+1}-u_{j-1,k+1}) \\ &+B^+_{j+\frac{1}{2},k}A^-_{j+\frac{1}{2},k}(u_{j+1,k}-u_{j,k}) \\ &+B^-_{j+\frac{1}{2},k+1}A^-_{j+\frac{1}{2},k+1}(u_{j+1,k+1}-u_{j,k+1})\Big\}\,. \end{aligned} \tag{10}$$

Scheme (8) can be extended for the nonlinear systems (9) in the similar form(**10**)but with flux limiters for mixed derivatives.

The generalization of splitting schemes to (9) is straightforward using the 1-D MmB schemes for nonlinear conservation laws.

Consider the following 1-D MmB (or TVD) scheme [3]

$$u_j^{n+1} = u_j - \lambda(u_j - u_{j-1}) - \frac{\lambda(1-\lambda)}{2}\left[\varphi_j(u_{j+1} - u_j) - \varphi_{j-1}(u_j - u_{j-1})\right],$$

and semi-discrete TVD scheme

$$\frac{\partial u_j}{\partial t} = -\frac{a}{\Delta x}(u_j - u_{j-1}) - \frac{a}{2\Delta x}\left[\varphi_j(u_{j+1} - u_j) - \varphi_{j-1}(u_j - u_{j-1})\right] \quad (11)$$

with Lipschitz continuous φ.

Theorem 2. *The necessary and sufficient condition for MmB or TVD scheme using flux limiters to be of second-order accurate at critical points is* [5], [6]

$$\varphi(3) + \varphi(-1) = 2\,, \quad (12)$$

and second-order accurate at the point far from critical points is

$$\varphi(1) = 1\,. \quad (13)$$

Now consider the question of global accuracy of TVD (MmB) schemes in the whole neighborhood of critical points.
Second-order accuracy in whole region of smooth solution, requires that the flux limiter $\varphi(r)$ must be designed such that

$$(1-2\alpha)\varphi\left(\frac{2\alpha+1}{2\alpha-1}\right) + (1+2\alpha)\varphi\left(\frac{2\alpha+3}{2\alpha+1}\right) = 2 \quad (14)$$

is satisfied for any finite number α with condition

$$\varphi(0) = 0\,.$$

It has unique solution

$$\varphi(r) = r\,,$$

which behaves outside the MmB region of the scheme for r large.

We summarize above analysis into the following theorem.

Theorem 3. *There is no TVD or MmB scheme of the form using flux limiters, which has uniform second-order accuracy in the whole neighborhood of the critical points.*

Now, consider the analysis of accuracy problem for 2-D MmB schemes. It is evident that symmetric splitting methods $L_x L_y L_y L_x$ and $L_y L_x L_x L_y$ are second-order accurate in the regions where L_x, L_y have second-order accuracy, the non-splitting methods (7) and (8) have second-order accuracy in the regions of smooth u far away from the critical point where $u_x = 0$ or $u_y = 0$ if (13) is fulfilled. The design of non-splitting schemes with second- order accuracy at critical points requires further careful considerations.

In order to improve the accuracy of the numerical solution, we suggest to use high-order accurate limiters

$$\varphi_{W2N}(r) = \max\Big(0, \min\big(2/(1-\lambda),\ 2r/\lambda,\ (1+r)/2\big)\Big),$$

$$\varphi_{W3N}(r) = \max\Big(0, \min\big(2/(1-\lambda),\ 2r/\lambda,\ (2-\lambda+(1+\lambda)r)/3\big)\Big),$$

$$\varphi_{COM}(r) = \begin{cases} \min\big(2/(1-\lambda), r\big) & \text{if } r \geq 0 \\ \max\Big(1-2/\lambda,\ \min\big(0, 1+(1+\lambda)r/(1-\lambda)\big)\Big), & \text{if } r \leq 0 . \end{cases}$$

The second one will yield a third-order accurate MmB (TVD) scheme.

Numerical Experiments

Some numerical experiments are performed to test the resolution abilities and the accuracy of the MmB schemes of this paper. Scheme UNO2 is taken from [7], schemes COM, W2N, W3N are MmB schemes using the flux limiters φ_{COM}, φ_{W2N}, φ_{W3N}.

1 The Convection Problems

$$u_t + u_x = 0 , \quad u(0,x) = \sin \pi x , \quad 0 < t \leq 2 .$$

$$\Delta x = 2/N,\ N = 20,\ 40,\ 80.\ \lambda = \Delta t/\Delta x = 0.8,\ 0.5.$$

$$\varepsilon_{\Delta x} = \max |u - u_{\Delta x}| = 0(\Delta x^P).\quad P = \log\big(\varepsilon_{2\Delta x}/\varepsilon_{\Delta x}\big)(\log 2)^{-1}.$$

Table 1. Errors and orders

Schemes	λ	$\varepsilon_{2/20}$	$\varepsilon_{2/40}$	$\varepsilon_{2/80}$	P(20–40)	P(40–80)
UNO 2		.71-2	.16-2	.40-3	2.14	2.01
COM	0.8	.313-1	.699-2	.24-2	2.161	1.541
W2N		.249-1	.779-2	.291-2	1.680	1.415
W3N		.215-1	.676-2	.215-2	1.667	1.654
UNO 2		.902-2	.114-2	.149-3	2.98	2.94
COM	0.5	.866-1	.207-1	.723-2	2.068	1.514
W2N		.394-1	.104-1	.288-2	1.928	1.844
W3N		.394-1	.104-1	.288-2	1.928	1.844

1-D convection test problem [7] (Fig. 1).

$$u_t + u_x = 0 \ ,$$

$$u(0,x) = \begin{cases} -x\sin(3\pi x^2/2)\ , & -1 < x < -1/3\ , \\ |\sin(2\pi x)|\ , & |x| < 1/3\ , \\ 2x - 1 - 1/6\sin(3\pi x), & 1/3 < x < 1\ . \end{cases}$$

$$u(0, x+2) = u(0,x)\ , \quad \Delta x = 0.05,\ \lambda = \Delta t/\Delta x = 0.8,\ t = 2.$$

2 1-D Sod shock tube problem [8] (Fig. 2).

3 2-D Riemann problems for the 2-D Burgers' equation

$$\frac{\partial u}{\partial t} + \frac{\partial}{\partial x}(u^2/2) + \frac{\partial}{\partial y}(u^2/2) = 0$$

with initial condition of four constant states in four quadrants. Fig.3 shows the contours of the numerical solutions using splitting and nonsplitting second- order accurate MmB schemes.

4 Shock Reflection. Standard oblique shock reflection from a flat plate, $M_\infty = 2.9$, angle of incidence = 29° (Fig. 4). Blast wave reflection from a 45° ramp, $M_\infty = 5$ (Fig. 5).

5 Blund Body in Hypersonic Flow (21 × 17 mesh,Fig.6).

References

[1] J. B. Goodman and R. J. Le Veque: *On the accuracy of stable schemes for 2D conservation Laws.* Math. Comp. **45** No. 171, pp. 15–21 (1985).

[2] A. Harten: *High resolution schemes for conservation Laws.* J. Comput. Phys. **49**, pp. 357–393 (1983).

[3] P. K. Sweby: *High resolution schemes using flux limiters for hyperbolic conservation laws.* SIAM J. Num. Anal. **21**, pp. 995–1011 (1984).

[4] S. L. Yang and H. M. Wu: *A class of second-order accuracy MmB schemes for solving initial value problems of 2D linear hyperbolic equations.* Preprint, Computing Center Academia Sinica (1989).

[5] H. L. Ren: *Private communication on the accuracy of TVD schemes at critical points* (1988).

[6] H. M. Wu: *On the possible accuracy of TVD schemes.* Preprint SC-89-3, Konrad-Zuse-Zentrum für Informationstechnik Berlin (1989).

[7] A. Harten and S. Osher: *Uniformly high-order accurate non-oscillatory difference schemes I.* SIAM J. Num. Anal. **24**, pp. 279– 309 (1987).

[8] G. A. Sod: *A survey of several finite difference methods for systems of non-linear hyperbolic conservation laws.* J. Comput. Phys. **27**, pp. 1–31 (1978).

Figures

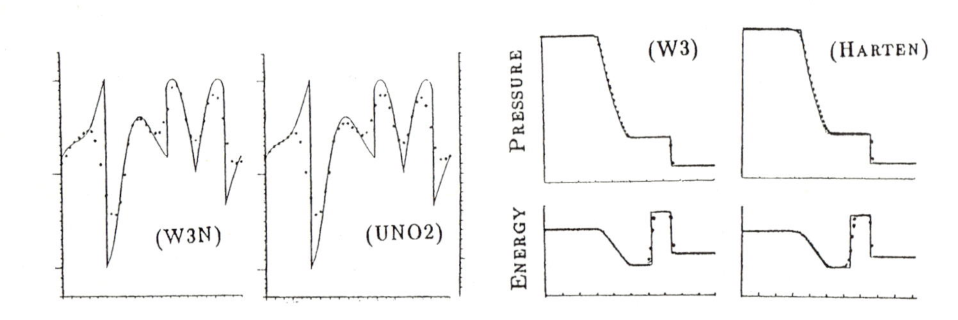

Figure 1 $u_t + u_x = 0$, $t = 2$, $\lambda = 0.8$
— exact ... numerical

Figure 2 Sod's shock tube problem
$\Delta x = 0.01$, $\Delta t = 0.0035385$, $t = 0.14154$

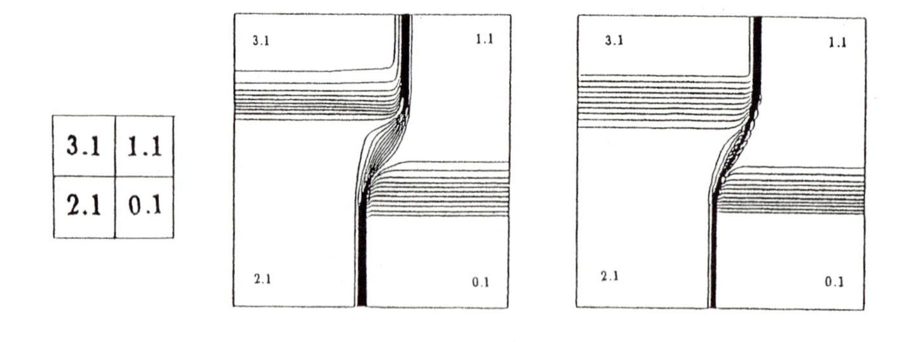

nonsplitting MmB scheme splitting MmB scheme

Figure 3 Two-dimensional Riemann problem of Burgers' equation

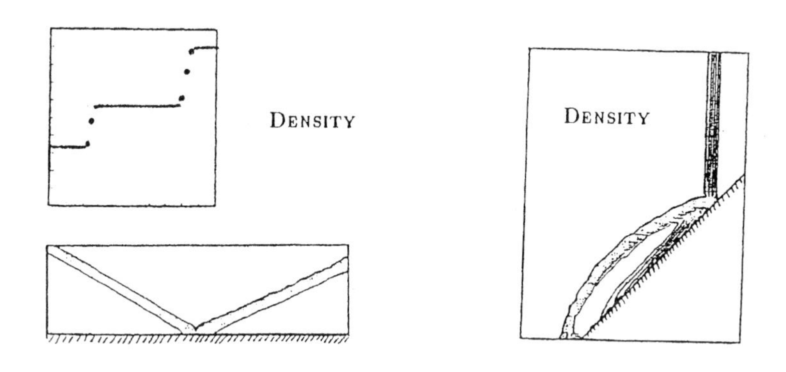

Figure 4 Oblique shock reflection from a flat plate

Figure 5 Regular blast wave reflection from a ramp

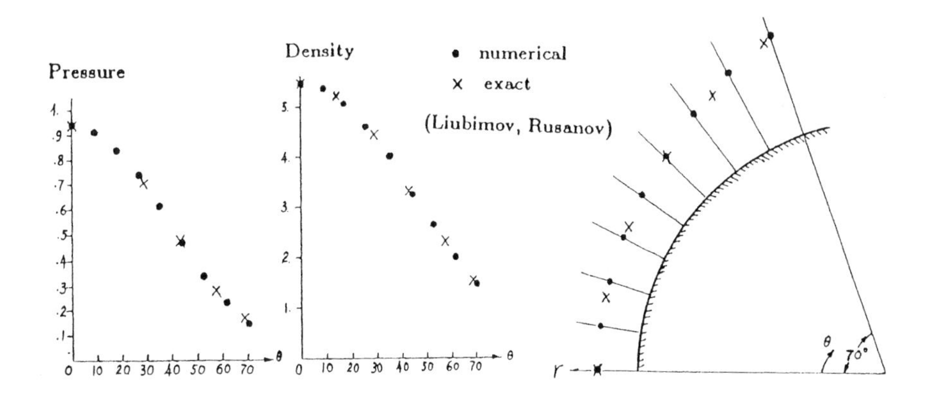

Figure 6 Bow shock and distribution of pressure and density in Θ

AN INVERSE BOUNDARY LAYER PROCEDURE WITH APPLICATION TO 3-D WING FLOW

L. Xue and F. Thiele
Technische Universität Berlin,
Hermann-Föttinger-Institut für Thermo- und Fluiddynamik
Straße des 17. Juni 135, D-1000 Berlin 12

SUMMARY

For three-dimensional boundary layers an inverse solution procedure is described to be used in connection with the interactive approach. Due to the finite-difference approximation applied the pressure gradients as well as the displacement thicknesses can be directly incorporated into the numerical scheme. For this reason the method developed differs only slightly from the direct mode with prescribed external velocity components. Calculations are reported for the turbulent flow over a flat plate and a wing flow. The results confirm that the inverse procedure provides accurate solutions within few iterations even in regions where the flow tends to separate.

INTRODUCTION

In modern aircraft design it is necessary to predict accurately the flow around finite swept and tapered wings. Since the numerical solution of the full Navier-Stokes equations requires high computing time and large memory, interactive methods based on the coupling of the viscous and inviscid flow are an alternative approach. In two-dimensional flows the interactive method has proved to be an accurate and reliable tool to numerically calculate the flow around an airfoil [1]. Even though the interactive procedure has been popular for airfoil calculations the main difficulty arises in connection with flow separation. For a prescribed pressure distribution the boundary-layer equations become singular at the separation point and the solution breaks down. However, prescribing for example a displacement thickness the external velocity is calculated as part of the solution. This inverse procedure allows the boundary-layer calculation to be extended into regions of flow separation.

When the interactive method is extended to three-dimensional flow problems the main obstacle of the inverse procedure is associated with the fact that two external velocity components have to be determined. This requires two prescribed boundary-layer parameters, for example the displacement thicknesses δ_1 and δ_2. For the infinite swept wing problem Stock [2] proposed an integral procedure which needs only one prescribed boundary-layer property. Délery and Formery [3] considered two displacement thicknesses for their formulation which results, however, in additional differential equations. Therefore it is desirable that an inverse 3-d method has similar features as the direct ones.

In this paper a new inverse method for 3-d boundary-layer flows is developed which differs only slightly from the direct mode with prescribed external velocity components. Using the definition of a two-component vector potential and introducing a similarity transformation the boundary-layer equations are reduced to a system of two third-order differential equations. The integration along the surface is performed such that only ordinary differential equations have to be solved numerically. Special attention is drawn to the relationship between the displacement thicknesses and the external velocity components. The features of the inverse procedure presented are demonstrated by calculating the flow over a flat plate with either induced pressure gradients [4] or attached cylinder [5] as well as to the wing flow [6].

BASIC EQUATIONS AND BOUNDARY CONDITIONS

The governing boundary-layer equations for three-dimensional wing flow in a curvilinear coordinate system (Fig. 1) are given by [7]:

continuity equation:

$$\frac{\partial}{\partial x}\left[\varrho\, u\, h_2 \sin\theta\right] + \frac{\partial}{\partial z}\left[\varrho\, w\, h_1 \sin\theta\right] + \frac{\partial}{\partial y}\left[\overline{\varrho v}\, h_1\, h_2 \sin\theta\right] = 0 \quad (1)$$

x-momentum equation

$$\frac{\varrho\, u}{h_1}\frac{\partial u}{\partial x} + \frac{\varrho\, w}{h_2}\frac{\partial u}{\partial z} + \overline{\varrho v}\frac{\partial u}{\partial y} - \varrho \cot\theta\, K_1\, u^2 + \varrho \csc\theta\, K_2\, w^2 + \varrho\, K_{12}\, u\, w$$

$$= -\frac{\csc^2\theta}{h_1}\frac{\partial p}{\partial x} + \frac{\cot\theta \csc\theta}{h_2}\frac{\partial p}{\partial z} + \frac{\partial}{\partial y}\left[\mu \frac{\partial u}{\partial y} - \varrho\, \overline{u'v'}\right] \quad (2)$$

z-momentum equation

$$\frac{\varrho\, u}{h_1}\frac{\partial w}{\partial x} + \frac{\varrho\, w}{h_2}\frac{\partial w}{\partial z} + \overline{\varrho v}\frac{\partial w}{\partial y} - \varrho \cot\theta\, K_2\, w^2 + \varrho \csc\theta\, K_1\, u^2 + \varrho\, K_{21}\, u\, w$$

$$= \frac{\csc\theta \cot\theta}{h_1}\frac{\partial p}{\partial x} - \frac{\csc^2\theta}{h_2}\frac{\partial p}{\partial z} + \frac{\partial}{\partial y}\left[\mu \frac{\partial w}{\partial y} - \varrho\, \overline{w'v'}\right]. \quad (3)$$

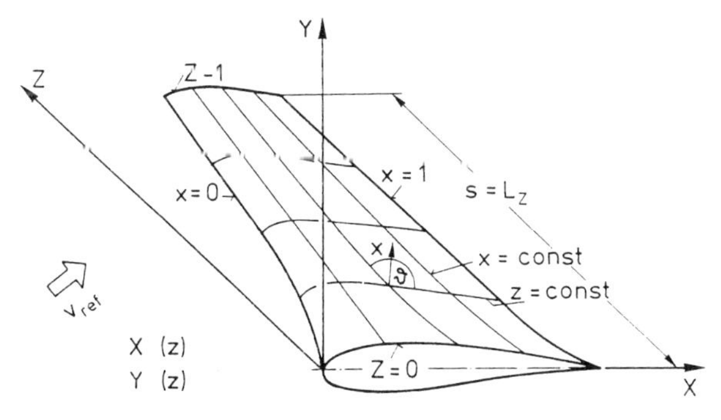

Fig. 1 The surface orientated coordinate system.

The boundary conditions are at the solid surface

$$y = 0 : \quad u = 0 \, , \quad v = v_w \, , \quad w = 0 \tag{4}$$

and at the outer edge of the boundary layer

$$y = \delta: \begin{cases} u = u_e \, , \; w = w_e & \text{direct mode} \\ \left.\begin{aligned} \int_0^\delta \left(1 - \frac{\rho}{\rho_e} \frac{u}{u_e} \right) dy = \delta_{1x} \\ \int_0^\delta \left(1 - \frac{\rho}{\rho_e} \frac{w}{w_e} \right) dy = \delta_{1z} \end{aligned} \right\} & \text{inverse mode .} \end{cases} \tag{5}$$

The turbulence model applied is based on the eddy-viscosity concept and relates the Reynolds stresses to the mean velocity profiles by

$$- \overline{u'v'} = \varepsilon_t \frac{\partial u}{\partial y} \qquad - \overline{w'v'} = \varepsilon_t \frac{\partial w}{\partial y} \, . \tag{6}$$

Here, we use the formulation of Michel [8]

$$\varepsilon_t = L^2 F^2 \left[\left(\frac{\partial u}{\partial y}\right)^2 + \left(\frac{\partial w}{\partial y}\right)^2 + 2 \cos\theta \frac{\partial u}{\partial y} \frac{\partial w}{\partial y} \right]^{1/2} \tag{7}$$

where the mixing length L and the damping factor F are defined by

$$\begin{aligned} \frac{L}{\delta} &= \frac{L_e}{\delta} \tanh \left(0.41 \frac{y/\delta}{L_e/\delta} \right) & \frac{L_e}{\delta} &= 0.085 \\ F &= 1 - \exp \left(- \frac{y}{A} \right) & A &= 26 \frac{\nu}{u_\tau} \left(\frac{\rho}{\rho_w} \right)^{1/2} . \end{aligned} \tag{8}$$

This model has been preferred because it assures a continuous distribution of the eddy-viscosity across the boundary layer.

TRANSFORMATION OF THE BASIC EQUATIONS

The boundary-layer equations can be solved either in physical coordinates or in transformed coordinates. Here, the latter ones are favoured because the transformed coordinates allow larger steps to be taken in the longitudinal and transverse directions. The introduction of a two-component vector potential

$$\frac{\partial \Psi}{\partial y} = \rho u h_2 \sin\theta \qquad \frac{\partial \Phi}{\partial y} = \rho w h_1 \sin\theta \qquad -\left(\frac{\partial \Psi}{\partial x} + \frac{\partial \Phi}{\partial z} \right) = \overline{\rho v} h_1 h_2 \sin\theta \tag{9}$$

which satisfies the continuity equation reduces the number of equations to be solved. Introducing the transformation

$$d\eta = \frac{\rho \, u_0}{\alpha \, (\rho_e \, u_0 \, \mu_e \, s_1)^{1/2}} \, dy \tag{10}$$

and the functions f and g

$$\Psi = ß\ h_2 \sin\theta\ f\ , \qquad \Phi = ß\ h_1 \sin\theta\ g. \tag{11}$$

we can write the momentum equations as x-momentum equation

$$(bf'')' + m_1 ff'' - m_2 f'^2 - m_5 f'g' + m_6 f''g - m_8 g'^2$$

$$\begin{cases} + m_{11} c & \text{direct mode} \\ + c\left[m_2 f_e'^2 + m_5 f_e' g_e' + m_8 g_e'^2 + m_{10} f_e' \dfrac{\partial f_e'}{\partial x} + m_7 g_e' \dfrac{\partial f_e'}{\partial z}\right] & \text{inverse mode} \end{cases}$$

$$= m_{10}\left(f' \frac{\partial f'}{\partial x} - f'' \frac{\partial f}{\partial x}\right) + m_7 \left(g' \frac{\partial f'}{\partial z} - f'' \frac{\partial g}{\partial z}\right) \tag{12}$$

z-momentum equation

$$(bg'')' + m_1 fg'' - m_3 g'^2 - m_4 f'g' + m_6 gg'' - m_9 f'^2$$

$$\begin{cases} + m_{12} c & \text{direct mode} \\ + c\left[m_3 g_e'^2 + m_4 f_e' g_e' + m_9 f_e'^2 + m_{10} f_e' \dfrac{\partial g_e'}{\partial x} + m_7 g_e' \dfrac{\partial g_e'}{\partial z}\right] & \text{inverse mode} \end{cases}$$

$$= m_{10}\left(f' \frac{\partial g'}{\partial x} - g'' \frac{\partial f}{\partial x}\right) + m_7 \left(g' \frac{\partial g'}{\partial z} - g'' \frac{\partial g}{\partial z}\right). \tag{13}$$

The boundary conditions of the physical plane result in

$$\eta = 0 : \quad f_w' = 0\ , \quad g_w' = 0\ , \tag{14}$$

$$m_1 f_w + m_6 g_w + m_{10}\left(\frac{\partial f}{\partial x}\right)_w + m_7 \left(\frac{\partial g}{\partial z}\right)_w = -\left(\frac{s_1}{\varrho_e\ \mu_e\ u_0}\right)^{1/2} \varrho_w v_w \tag{15}$$

$$\eta = \eta_e: \quad f_e' = \frac{u_e}{u_0}\ , \quad g_e' = \frac{w_e}{u_0} \qquad \text{direct mode}$$

$$\left.\begin{aligned} f_e + f_e' \left[\delta_{1x} \frac{u_r}{ß} - \int_0^{\eta_e} c\ d\eta\right] &= f_w \\ g_e + g_e' \left[\delta_{1z} \frac{u_r}{ß} - \int_0^{\eta_e} c\ d\eta\right] &= g_w \end{aligned}\right\} \quad \text{inverse mode}\ . \tag{16}$$

Here we can define a semi-inverse procedure by

$$\eta = \eta_e: \quad f_e + f_e' \left[\delta_{1x} \frac{u_r}{ß} - \int_0^{\eta_e} c\ d\eta\right] = f_w\ , \qquad g_e' = \frac{w_e}{u_0}\ . \tag{17}$$

It can be shown that the boundary conditions f_w and g_w have to be chosen arbitrarily such that the condition (15) is satisfied.

NUMERICAL SOLUTION PROCEDURE

The general principle to obtain a stable and accurate solution of the 3-d boundary-layer equation requires to identify zones of influence and dependence for any grid point. Here, we use a combination of the zig-zag scheme [10] and

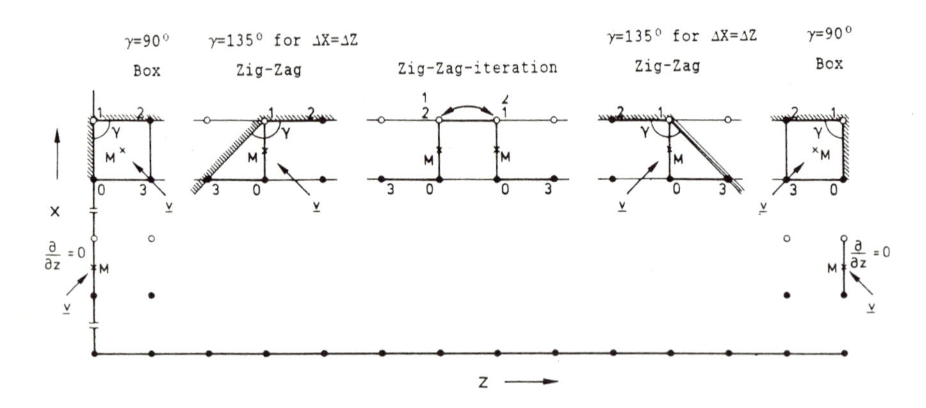

Fig. 2 Discretization scheme and finite-difference molecule.

the box scheme [7] as shown in Fig. 2. Depending on the scheme the finite-difference approximation can be written as

$$\text{zig-zag:}\quad \frac{\partial F}{\partial x} = \frac{F_1 - F_0}{Dx}, \qquad \frac{\partial F}{\partial z} = \frac{F_1 + F_3 - F_2 - F_0}{2Dz}, \quad F_m = \frac{F_0 + F_1}{2} \tag{18}$$

$$\text{box:}\quad \frac{\partial F}{\partial x} = \frac{F_1 + F_2 - F_0 - F_3}{2Dx}, \quad \frac{\partial F}{\partial z} = \frac{F_1 + F_0 - F_2 - F_3}{2Dz}, \quad F_m = \frac{F_0 + F_1 + F_2 + F_3}{4}. \tag{19}$$

This marching procedure reduces the boundary-layer equations (12) and (13) after Newton-Raphson linearization to ordinary equations at any grid point in the surface plane.

$$\begin{aligned} B^*f''' + C^*f'' + Df^*f' + Ef^*f + Hf^*g' + Sf^*g + Ff^*f_e' + Gf^*g_e' &= Rf^* \\ B^*g''' + C^*g'' + Dg^*f' + Eg^*f + Hg^*g' + Sg^*g + Fg^*f_e' + Gg^*g_e' &= Rg^*. \end{aligned} \tag{20}$$

The approximation of equation (20) is performed by the accurate finite-difference method of Hermitian type [9]. The advantage of this method is that the first derivatives at the grid points are unknowns of the finite-difference approximation. Therefore, the pressure gradients as well as complex boundary conditions, such as for the inverse mode, can be directly incorporated into the numerical scheme. The system of finite-difference equations is given by equation (21). It is obvious that the direct and inverse mode differ only slightly by the column for the external velocity components.

$$
\begin{bmatrix}
\mathbf{TA}_1 & 0 & & & & & 0 \\
\mathbf{TB}_2 & \mathbf{TA}_2 & \mathbf{TC}_2 & & 0 & & \mathbf{TR}_2 \\
 & \mathbf{TB}_3 & \mathbf{TA}_3 & \mathbf{TC}_3 & & & \mathbf{TR}_3 \\
 & \cdot & \vdots & \vdots & \vdots & \cdot & \vdots \\
 & & & \mathbf{TB}_{M2} & \mathbf{TA}_{M2} & \mathbf{TC}_{M2} & \mathbf{TR}_{M2} \\
 & 0 & & & \mathbf{TB}_{M1} & \mathbf{TA}_{M1} & \mathbf{TC}_{M1} \\
 & & & & \mathbf{TU}_{M2} & \mathbf{TB}_{M} & \mathbf{TA}_{M}
\end{bmatrix}
*
\begin{bmatrix}
\delta_1 \\ \delta_2 \\ \delta_3 \\ \vdots \\ \delta_{M2} \\ \delta_{M1} \\ \delta_M
\end{bmatrix}
=
\begin{bmatrix}
\mathbf{r}_1 \\ \mathbf{r}_2 \\ \mathbf{r}_3 \\ \vdots \\ \mathbf{r}_{M2} \\ \mathbf{r}_{M1} \\ \mathbf{r}_M
\end{bmatrix}
\qquad (21)
$$

$$
\delta_j = \begin{bmatrix} f_j \\ f'_j \\ g_j \\ g'_j \end{bmatrix}
$$

$\mathbf{TA}_j$, $\mathbf{TB}_j$, $\mathbf{TC}_j$, $\mathbf{TR}_j$, $\mathbf{TU}_j$ are 4×4-matrices

δ_j, $\mathbf{r}_j$ vectors .

The quasi 4×4 block tridiagonal system is solved directly by using a LU-decomposition technique. In this way the two third-order differential equations (20) are solved simultaneously together with the boundary condition for the inverse mode. Due to this procedure the iteration process is stable and converges within few iterations.

RESULTS AND DISCUSSION

The inverse 3-d boundary-layer procedure was first applied to the flow (MUKR 79) on a flat plate generated by guide walls [4]. To verify the solution procedure the calculation was performed in the direct mode with specified external velocity components. The displacement thicknesses obtained from the numerical results have been used as input for the inverse calculation. The results of the skin friction and the wall flow angle are presented in Fig. 3 for the x,z-plane.

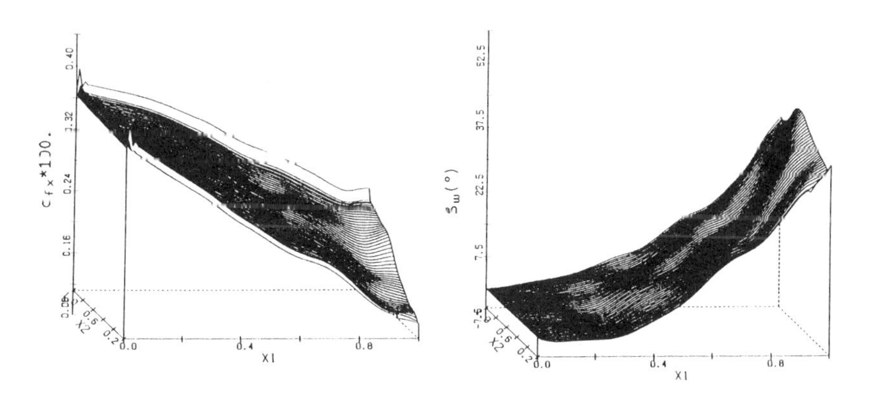

Fig. 3 Distribution of skin friction and wall flow angle for MUKR 79.

In Fig. 4 the velocity profiles at station 6,5 are shown (streamwise u, crosswise w). The agreement between the calculation and the experiment as well as the numerical results of previous investigations may be regarded as satisfactory.

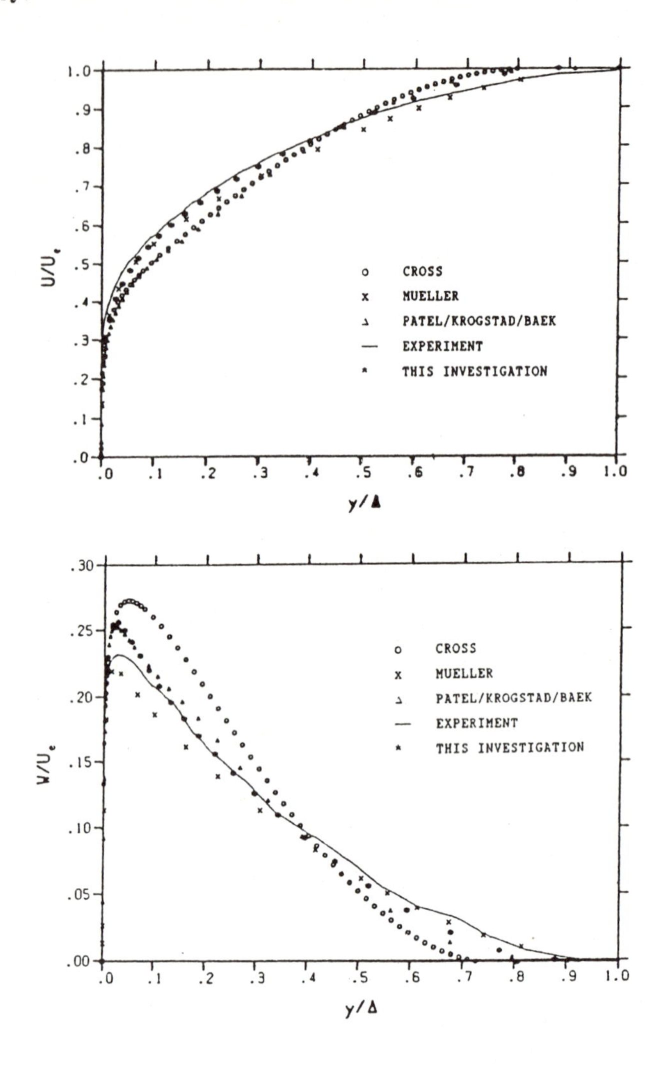

Fig. 4 Velocity profiles at station 6,5 for MUKR 79.

Fig. 5 compares the results of the direct and inverse procedure, respectively. For Re_θ the error is less than 1°/oo. The results of the inverse calculation are presented at $x/L = 0.6$ for the variation of the skin friction with z/L.

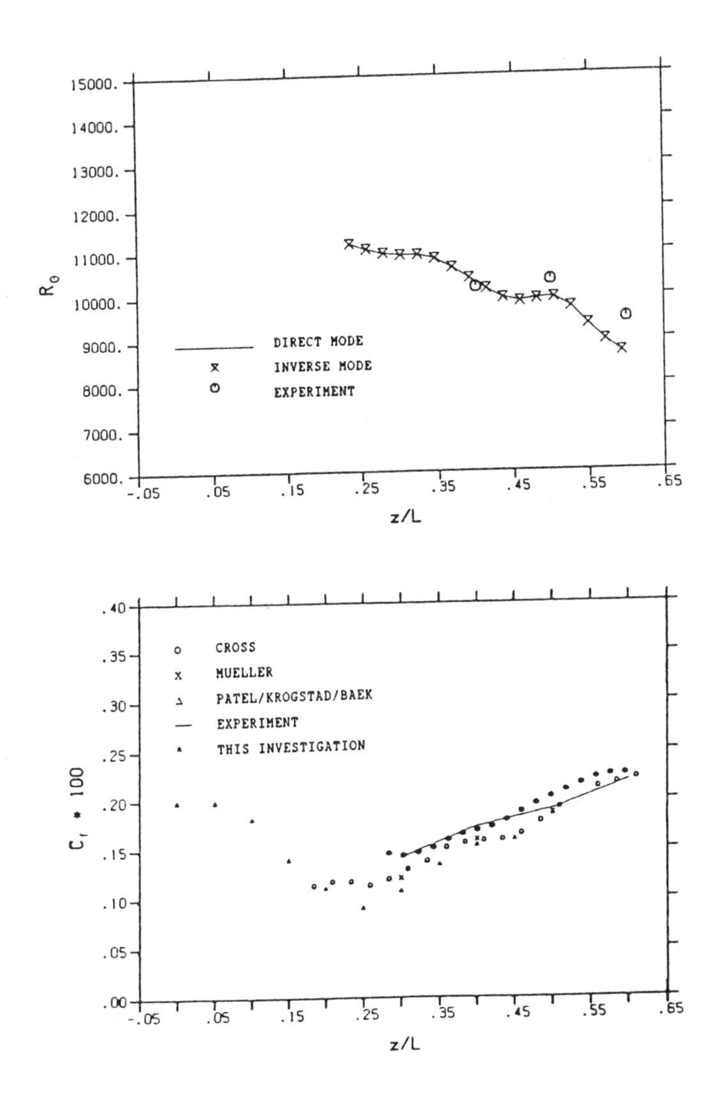

Fig. 5 Results of the inverse procedure for momentum thickness, Reynolds number, and skin friction.

The second test case considered (DEFE 77) concerns the boundary-layer flow generated on a flat plate by a cylindrical body on the plate [5]. Here, the semi-inverse method has been applied since the W-velocity vanishes at the symmetry axis. The momentum thicknesses were obtained from the direct calculation and extrapolated into the region of separation. Fig. 6 shows a comparision between a shear-stress vector plot generated from the numerical results and the experimental oil flow visualisation.

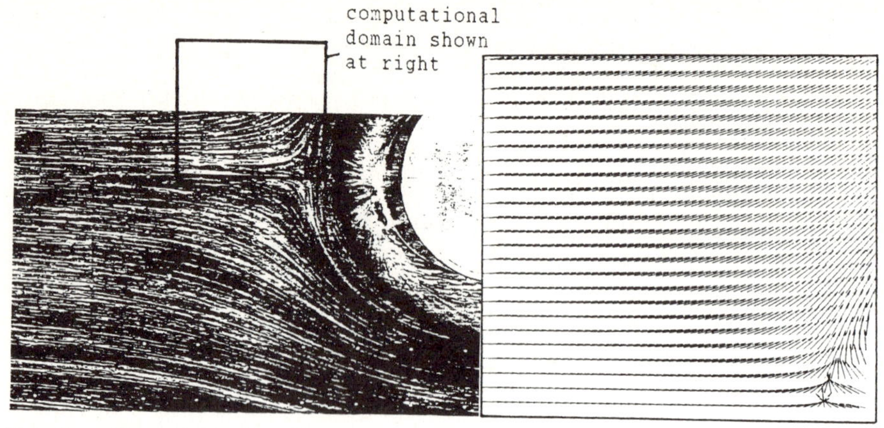

Fig. 6 Flow field visualization and vector plot for DEFE 77.

The variation of skin friction and wall flow angle are presented in Fig. 7. The agreement between the experiment and the numerical result is good except in the region where the flow approaches separation.

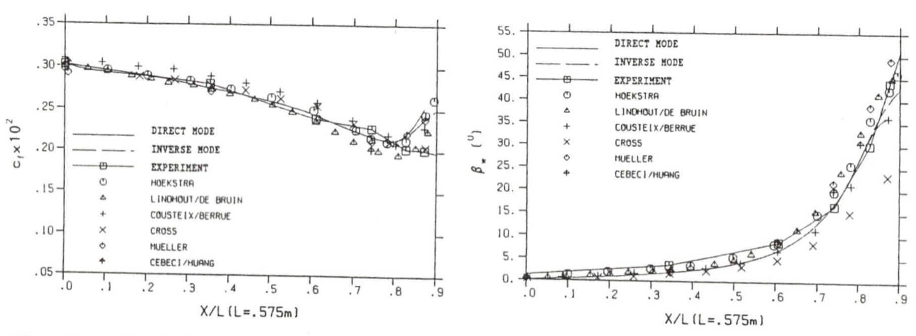

Fig. 7 Variation of skin friction and wall flow angle for DEFE 77.

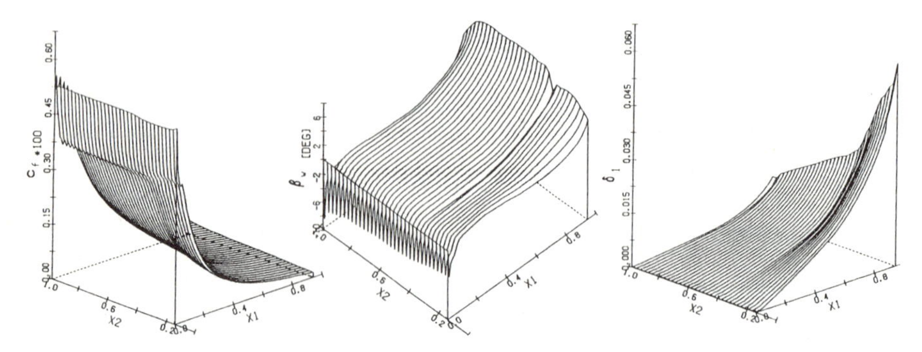

Fig. 8 Variation of skin friction, wall flow angle, momentum thickness Reynolds number, and momentum thickness for TA11-1V wing.

As an application to wing flow the upper side of a TA11-1V wing [6] is calculated. Some results are presented in Fig. 8, where the variation of skin friction, wall flow angle, momentum thickness Reynolds numbers, and momentum thickness are shown. The numerical results obtained by the inverse mode are in good agreement with the direct calculation. To solve the ordinary differential equation 3-4 iterations were required for both, direct and inverse mode.

CONCLUSIONS

An implicit finite-diffence procedure to solve the 3-d boundary-layer equations is described. The inverse boundary-layer method developed is no more complicated than the corresponding direct more with prescribed external velocity. This is due to the fact that all terms of the differential equation including the complex boundary conditions are directly incorporated into the finite-difference scheme of Hermitian type. The calculation procedure converged within few iterations for the test cases considered. From the investigations performed it can be concluded that there is essentially no difference in the accuracy of the numerical results between the direct procedure and the inverse or semi-inverse method, respectively. However, the inverse method is able to proceed the calculation into regions, where the flow approaches separation.

ACK NOWLEDGEMENT

This research war partially supported by the MBB GmbH, Bremen and the BMFT, Bonn.

REFERENCES

[1] CEBECI, T. (Editor).: Numerical and Physical Aspects of Aerodynamic Flows II, Springer-Verlag New York (1984).

[2] STOCK, H.W.: An Inverse Boundary Layer Method for Turbulent Flows on Infinite Swept Wings. ZFW 12 (1988) pp. 51-62.

[3] DÉLERY, J.M.; FORMERY, M.J.: A Finite Differnce Method for the Inverse Solution of 3-D Turbulent Boundary Layer Flow. ONERA TP no. 1983-4.

[4] MÜLLER, U.R.: Messung von Reynoldschen Spannungen und zeitlich gemittelten Geschwindigkeiten in einer drei-dimensionalen Grenzschicht mit nichtverschwindenden Druckgradienten. Ph.D. Thesis, University of Aachen (1979).

[5] DECHOW, R.; FELSCH, K.O.: Measurements of the Mean Velocity and of the Reynolds Stress Tensor in a Three-dimensional Turbulent Boundary Layer Induced by a Cylinder Standing on a Flat Wall. First Symp. "Turb. Shear Flows"; Univ. Park, Pennsylvania (1977).

[6] HANSEN, H.; GREFF, E.: Entwurf des TA11-Basisflügels. MBB-Report, TE2-1358 (1984).

[7] CEBECI, T.; KAUPS, K.; RAMSEY, J.A.: A General Method for Calculating Three-dimensional Compressible Laminar and Turbulent Boundary Layers on Arbitrary Wings. NASA-CR-2777 (1977).

[8] MICHEL, R.; QUÉMARD, C.; DURANT, R.: Hypotheses on the Mixing Length and Application to the Calculation of the Turbulent Boundary Layer. AFOSR-LFP Stanford Conf., ed.: S.J. Kline, M.V. Markovin, G. Sovran, D.J. Cockrell (1968).

[9] THIELE, F.: Accurate Numerical Solutions of Boundary Layer Flows by the Finite-difference Method of Hermitian Type. J. Comp. Physics 27 (1978), pp. 138-159.

[10] KRAUSE, E.; HIRSCHEL, E.H.; BOTHMANN, Th.: Die numerische Integration der Bewegungsgleichungen dreidimensionaler laminarer kompressibler Grenzschichten. DGLR-Fachbuchreihe Band 3, Braunschweig (1969).

DEVELOPMENT OF A GENERAL MULTIBLOCK FLOW SOLVER FOR COMPLEX CONFIGURATIONS

N. J. Yu, H. C. Chen, T. Y. Su, and T. J. Kao
Boeing Commercial Airplanes
P. O. Box 3707
Seattle, Washington 98124, USA

SUMMARY

A configuration- and grid topology-independent flow solver has been developed for the analysis of complex flow fields over geometrically complex configurations. The approach used in the present study is based on a multiblock method where the complete flow field is divided into a number of topologically simple blocks, such that within each block, surface-fitted grids and efficient flow solution algorithms can easily be constructed. In the present study, the field grid is generated using a combination of algebraic and elliptic methods. The flow solver utilizes an explicit time-stepping scheme to solve the Euler equations. A multiblock version of the multigrid method is implemented to improve the convergence of the calculations. The generality of the present method is demonstrated through the analysis of a number of complex configurations at various flow conditions. Results are checked against available data for code validation.

INTRODUCTION

In the development of a flow code for the analysis of complex 2D or 3D configurations using a surface-fitted grid, one encounters two major problems: the generation of field grids and the development of a configuration-independent flow solver. In the past several years, a number of flow codes using a specific grid topology, such as O-type, C-type, and H-type grids have been produced [1-5]. These codes are not only grid topology-dependent, but also configuration-dependent due to the use of a particular grid type. A completely general flow solver based on an unstructured, finite-element grid has been explored extensively in the past few years by a number of researchers [6-8]. However, due to the requirement of large central memory, and the relatively low execution efficiency compared to the structured grid methods, the unstructured grid flow solver requires further development for practical applications. Another approach is to use Cartesian grids throughout the complete flow field [9-11]. This non-surface fitted approach has been proven to be very effective for inviscid analysis [12], although its applicability for the Navier-Stokes solver remains to be explored.

This paper explores an alternative method of solving complex flow fields around complex configurations using a surface-fitted structured grid. The flow solver algorithm is completely independent of the grid topology and the geometry of the configuration, and is capable of solving the flow problem using any type of structured grids. The basic method employs a multiblock concept [13-17], where the complete flow field is divided into an arbitrary number of topologically simple blocks. Within each block, smooth and well structured grids can be generated. Two new features are introduced in the present multiblock flow solver, i.e., 1) generalized block-to-block connectivity - each block can have arbitrary number of neighbors, and 2) general boundary conditions - the block face can have any combination of boundary conditions. These new features relieve the blocking restrictions encountered in most existing multiblock methods, where each face of the block can only have one neighbor, and each face can only have one type of boundary condition.

The basic flow solution algorithm utilizes Jameson et al's explicit time-stepping scheme [3] to solve the unsteady Euler equations. To accelerate the convergence rate of the iterative process, a multigrid scheme is used in a block-by-block manner. The advantages and the shortcomings of this simplified multigrid method are discussed in a later section.

Computed results for a variety of complex configurations are included. To validate the accuracy of the present code, the results are checked against other numerical solutions, as well as available test data.

MULTIBLOCK GRID GENERATION

An essential element in the present general flow solver development is the generation of surface-fitted grid over a geometrically complex configuration. It is clear that neither analytical mapping nor numerical grid generation using simple grid topology can produce usable grids for complex configuration analysis. One needs to divide the flow field into a number of simple regions, as illustrated in Figure 1, such that within each region, smooth and well distributed grids with a desirable grid topology can be generated. The grid size of each block is limited by the central memory of the machine to be used. For a typical Euler analysis, each block contains less than 50,000 points, in order to fit the flow-field data and the coordinates into the 2,000,000-word central memory of a Cray XMP. Along the common face of two adjacent blocks, identical grid points are used. Each block contains six boundary faces in the computational space. Some boundary faces may have no neighbor, such as the far-field boundary or the configuration surface. Others may have multiple neighbors, such as the interface boundary. The configuration surface or surfaces must lie on the block-boundary face or faces. Along the boundary face of each block, a combination of boundary conditions may be used.

The complete grid generation process can be divided into four steps:

1. Blocking of the complete flow field

The very first step in the grid generation procedure is to divide the complete flow field into a number of desirable blocks. Simple geometry, such as an airfoil or a wing, needs one or two blocks in order to generate smooth grids for the complete flow field. More complex geometry, such as a complete airplane, may require a large number of blocks in order to divide the flow field into a number of topologically simple regions. The blocking process is done effectively using interactive graphics, where the vertices of each block can be placed at the proper location. The block-boundary edges are then created and discretized according to a desired curvature and space distributions.

2. Configuration and block-boundary surface discretization

Once the blocking of the complete flow field is completed, the grids on each face of the block boundary are then generated. If the boundary surface is part of a configuration surface, the grid is interpolated in the parametric space of the original geometry definition to insure that the grid lies on the configuration surface. If the boundary surface is either an interblock boundary or the far-field boundary surface, the grid can be generated either by an algebraic method or by an elliptic grid-generation method. This part of the grid-generation process, together with the global blocking set up, is done interactively using the Boeing-developed geometry package, Aerodynamics Grid and Panel System (AGPS) [18].

3. Field-grid generation

After the surface grids of each block have been properly prepared, the field grid within each block can easily be produced using an algebraic/transfinite interpolation, or an elliptic grid-generation method. Several grid-generation packages are available for field-grid generation. In the case of a complex 2D configuration, the AGPS geometry package [18] can also be used for field-grid generation. For simple 3D configuration, the grid-generation program of reference 4 is used. The EAGLE grid-generation package acquired from the U.S. Air Force [19] is used for complex 3D configurations.

4. Block-to-block relation and block-boundary condition setup

The block-to-block relation file establishes the essential flow-field communications between two adjacent blocks. The block-boundary condition file provides the proper information for the type of boundary conditions to be applied along every boundary surface of each individual block. This information is most conveniently set up during the grid-generation process. The user has detailed knowledge about the complete block setup and the corresponding boundary conditions to be applied along each block surface. This information is then processed to produce the block-to-block relation and the block-boundary condition files needed for the flow solver.

The block-to-block relation is set up in a way that along every common block boundary, a unique record number is assigned to it, together with the 2D grid index associated with that boundary. The record number of its direct neighbor is also identified. The flow-field data along the two adjacent planes parallel to the block boundary are saved on this preassigned record such that they can easily be retrieved and used for the convection and dissipation flux calculations for the adjacent block.

The block-boundary condition is assigned in patches. Each patch represents a unique boundary condition with a given 2D grid index associated with that patch. Each face can have any combination of the following types of boundary conditions: 1) interface, 2) solid surface, 3) inlet, 4) exhaust, 5) far field, 6) transpiration, 7) centerline, 8) mirror image, and 9) actuator disk. More details on the block-to-block relation and the implementation of different type of boundary conditions will be discussed later.

GENERAL FLOW SOLVER ALGORITHM

The main objective of the present paper is to develop a flow solver that can accept any combination of grid topologies for the analyses of complex flows over complex configurations. To achieve this objective, the flow solver must be grid-topology independent and configuration-geometry independent. The multiblock approach described in the previous sections is designed specifically for this purpose. Any surface that requires the specification of boundary conditions is defined as a block boundary or part of a block boundary in the grid-generation process. The interior of every block contains regular flow-field grid points only, such that the flow field of every block can be solved using an identical flow solver. The boundary conditions are updated at every iteration. The complete flow field is solved in a block-by-block manner until the solution converges to a prescribed level of accuracy.

The present solution strategy employs an efficient use of the solid state disk (SSD) of a Cray XMP machine. The entire flow field, including both the grid coordinates and the flow variables, is stored on the SSD. During the iteration process, the data are brought into the central memory one block at a time. The block-boundary data needed for the calculation are also brought into the central memory. After all the data of one block are brought into memory, the solution is then advanced one time step for the finest level of mesh. One then performs the multigrid procedure within that block up to the allowable level of grids. At the end of each iteration cycle, the complete flow-field data, together with two layers of flow variables adjacent to all block boundaries, are written out onto the SSD in separate records with proper arrangements of the data, such that they can be readily read in for the neighboring block as appropriate boundary conditions. Each set of boundary data are written onto the record identified by the block-to-block relation file, and each record has the same prescribed record length in order to simplify the I/O process.

The multigrid strategy employed in the present study is a simple V-cycle multigrid applied within each individual block of the flow field. The main reason for using V-cycle multigrid in a block-by-block manner is to minimize data transfer from main memory to the secondary memory. It also reduces the programming complexity significantly for the present multiblock solver. However, this approach restricts the multigrid calculation to one block at a time. The residual collection and the correction interpolation at the grid cell near the block boundary could not be done precisely in the same way as the single block approach. This results in a deterioration of the overall convergence rate. It may also cause numerical oscillations for the case of supercritical wings with rooftop-type pressure distributions. A simple way to alleviate the problem is to turn off the multigrid calculation after the solution converges to a certain level of accuracy, and continue the single grid calculation for an additional 100 to 200 iterations. Numerical experiments indicate that this simple fix is very effective in eliminating undesirable numerical oscillations due to the use of simplified multiblock, multigrid method. A better way to correct the problem is to go through the fine mesh calculation for all blocks, and collect the residual for the coarser mesh calculations in the same way as that of the single block approach. This will involve additional data transfer for saving and retrieving the residuals for every block, an additional overhead to be paid in order to achieve the same convergence rate of a single block method. The benefit of this approach is under evaluation.

The cell-centered scheme based on Jameson et al's finite volume method [3] for solving Euler equations has been implemented in the present flow solver. The method is accurate and reliable provided that the grids in the flow field are smooth and well distributed. However, the accuracy and the reliability of the method deteriorates if the flow field contains irregular or singular grids, such as collapsed grids on the upper surface of a wing/nacelle/strut configuration (Fig. 2a), and the fictitious corner along the fan cowl surface (Fig. 2b). Special treatments along these grids are needed in order to obtain accurate and reliable solution for a general 3D configuration. First, the convective flux and the dissipative flux along these grid lines must be formulated in a fully conservative form. For the cell-centered scheme used in the present study, this is relatively easy to do. One computes flux terms along each face of the cell and uses them consistently for the flux balance of the two cells adjacent to the face. Another improvement which has been implemented to the flow solver is the new dissipation formulation based on a spectral radius scaling [20,21]; that is, the dissipation term $D(w)_i$ added to the governing equations is reformulated

$$D(w)_i = d_{i+1/2,j,k} - d_{i-1/2,j,k}$$

$$d_{i+1/2,j,k} = \lambda_{i+1/2,j,k} \left(\varepsilon^{(2)} - \varepsilon^{(4)} \delta_\xi{}^2 \right) \left(w_{i+1/2,j,k} - w_{i-1/2,j,k} \right)$$

with $\lambda_{i+1/2,j,k} = \left(q_{n\,i+1/2,j,k} + a_{i+1/2,j,k} \right) S_{i+1/2,j,k}$.

Here, $\lambda_{i+1/2,j,k}$, $q_{ni+1/2,j,k}$, $a_{i+1/2,j,k}$, and $S_{i+1/2,j,k}$ represent the spectral radius, normal velocity, speed of sound and face area at cell face i+1/2, j, k respectively. $\varepsilon^{(2)}$ and $\varepsilon^{(4)}$ are the coefficients for the second-order and fourth-order dissipation terms. It has been shown numerically that the new dissipation formulations give better numerical accuracy than the original one [3], especially for skewed grid and high aspect-ratio grid [22].

The present multiblock Euler solver can be used directly for 2D analysis. In such case, the convective and the dissipative fluxes in the third dimension are set to zero together with minor changes in the far-field boundary conditions to account for circulation contributions.

The boundary condition implemented along the configuration surface is the conventional solid-surface boundary condition, where the normal momentum equation is used to evaluate the pressure on the surface. At the far field, characteristic boundary conditions based on inflow and outflow conditions are implemented. The nacelle inlet and exhaust face, as well as the propeller disk surface are simulated in the same way as in reference 4. The nacelle center line, as well as any grid line in the flow field representing a collapsed grid plane, is treated as solid boundary with zero face area. Also included in the program are the transpiration boundary condition for simulating boundary-layer-displacement thickness effects, and the mirror-image boundary condition for producing exact symmetry solutions.

RESULTS AND DISCUSSIONS

To demonstrate the generality of the present method, both 2D and 3D configurations were analyzed. Figure 3 shows the surface grid of a wing-wingtip device combination. The field grid used for this configuration is a combination of an H-type in the streamwise direction and an O-type at the wingtip region, as shown in Figure 4. Computed isobars near the wingtip region are shown in Figure 5. Notice that a detailed flow field for this multiple lifting configuration can easily be simulated. To check the accuracy of the present code, a two-element airfoil representing a section cut of the wingtip device is analyzed, and the results are compared with panel method solution in Figure 6 for a subsonic flow condition. The agreement between the two solutions is quite good.

The next test case for the multiblock flow solver is a 3D turbofan nacelle with dual exhausts. The field grid and the flow-field Mach contours are shown in Figure 7. The fan exhaust and the core exhaust are set at different total pressures and temperatures to simulate the engine at cruise condition. The expansion and contraction of the fan exhaust plume and the detailed flow field of the core exhaust are well captured in the present analysis.

Figure 8 shows the surface grids of a wing/nacelle/strut configuration. The complete flow field consists of 20 blocks with a total number of 600,000 grid points. An H-type grid is used for all external flow field, and a cylindrical grid is used for nacelle inlet and exhaust flows. Figure 9 shows the isobars on the configuration surface. Notice that the collapsed grids on the upper surface of the wing have no effect on the solution. The results also show that detailed flow resolutions in the nacelle-strut region have been obtained with the present multiblock solver. This capability is very useful for the study of engine installation effects.

Finally, to validate the computational results with test data, a wing-mounted propfan configuration in a wind tunnel, shown in Figure 10, is analyzed. The propeller power input in terms of total pressure, total temperature, and swirl is prescribed along the propeller disk surface. The computed results are compared with test data in Figure 11. Excellent agreement has been obtained.

In conclusion, a general multiblock Euler solver has been developed. The code has been tested for a number of complex configurations at different flow conditions. Good agreement between the present analysis results, the test data, and other numerical results have been obtained. Further validation and improvement in the areas of grid generation and flow solver technology are needed in order to make the multiblock flow solver program a useful engineering tool.

ACKNOWLEDGEMENTS

The authors would like to thank Dr. P. Buning of NASA Ames research center in helping the wing-mounted propfan analysis. They would also like to acknowledge the Advanced Design Group of the Boeing de Havilland in providing flow conditions and test data of the wing-mounted propfan configuration for code validation. Special thanks are due to Dr. A. Chen of Boeing Commercial Airplanes in providing grids and the panel method solution for the two-element airfoil analysis.

REFERENCES

1. Jameson, A., Schmidt, W., and Turkel, E., "Numerical Solution of the Euler Equations by Finite Volume Methods using Runge-Kutta Time-Stepping Schemes," AIAA paper 81-1259, 1981

2. Caughey, D. A., "Multigrid Programs for Calculation of Transonic Flows Past Wing and Wing Fuselage Combinations," Computational Fluid Dynamics, A Workshop Held at the University of Tennessee Space Institute, Tullahoma, Tennessee, 1984

3. Jameson, A., and Baker, T. J., "Solution of the Euler Equations for Complex Configurations," AIAA 83-1929, 1983

4. Yu, N. J., Kusunose, K., Chen, H. C., and Sommerfield, D. M., "Flow Simulations for a Complex Airplane Configuration Using Euler Equations," AIAA 87-0454, 1987

5. Wigton, L. B., "Application of MACSYMA and Sparse Matrix Technology to Multielement Airfoil Calculation," AIAA 87-1142, 1987

6. Mavriplis, D. J., "Accurate Multigrid Solution of The Euler Equations on Unstructured and Adaptive Meshes," ICASE Report No. 88-40, 1988

7. Jameson, A., and Baker, T. J., "Improvements to The Aircraft Euler Method," AIAA 87-0452, 1987

8. Peraire, J., Peiro, J., Formaggia, L., and Morgan, K., "Adaptive Numerical Solutions of The Euler Equations in 3D Using Finite Elements," 11th International Conference on Numerical Methods in Fluid Dynamics, Williamsburg, Virginia, 1988

9. Wedan, B., and South, J. C., "A Method for Solving the Transonic Full Potential Equation for General Configurations," AIAA Paper 83-1889,1983

10. Reyhner, T. A., "Three-Dimensional Transonic Potential Flow About Complex Three-Dimensional Configurations," NASA CR-3814, 1984

11. Samant, S. S., Bussoletti, J. E., Johnson, F. T., Burkhart, R. H., Everson, B. L., Melvin, R. G., Young, D. P., Erickson, M. D., and Woo, A. C., "TRANAIR: A Computer Code for Transonic Analyses of Arbitrary Configurations," AIAA Paper 87-0034, 1987

12. Bieterman, M. B., Bussoletti, J. E., Johnson, F. T., Melvin, R. G., Samant, S. S., and Young, D. P., "Solution Adaptive Local Rectangular Grid Refinement for Transonic Aerodynamic Flow Problems," Eighth GAMM Conference on Numerical Methods in Fluid Mechanics, September, 1989, Delft University of Technology, the Netherlands

13. Fritz, W., Haase, W., and Seibert, W., "Mesh Generation for Industrial Application of Euler and Navier Stokes Solvers," Conference on Automated Mesh Generation and Adaption, Grenoble, France, October, 1987

14. Amendola,A., Tognaccini, R., Boerstoel, J. W., and Kassies, A., "Validation of A Multi-Block Euler Flow Solver with Propeller-Slipstream Flows," AGARD Fluid Dynamics Panel Symposium, Lisbon, 1988

15. Raj, P., et al., "Three-Dimensional Euler Aerodynamic Method (TEAM) Volumes 1-3," AFWAL-TR-87-3074, 1987

16. Sawada, K., and Takanashi, S., "A Numerical Investigation on Wing/Nacelle Interferences of USB Configurations," AIAA 87-0455, 1987

17. Flores, J., Holst, T. L., Kaynak, U., Gundy, K., and Thomas, S. D., "Transonic Navier Stokes Wing Solution Using a Zonal Approach, Part 1, Solution Methodology and Code Validation," AGARD 58th Fluid Dynamics Panel Symposium, Aix-en-Provence, France, April, 1986

18. Snepp, K. D., and Pomeroy, R. C., "A Geometry System for Aerodynamic Design," AIAA 87-2902, 1987

19. Thompson, J. F., "Program EAGLE User's Manual," USAF Armament Laboratory Technical Report, AFATL-TR-88-117, Eglin AFB, 1988

20. Jameson, A., "Successes and Challenges in Computational Aerodynamics," AIAA Paper 87-1184, 1987

21. Swanson, R. C., and Turkel, E., "Artificial Dissipation and Central Difference Schemes for the Euler and Navier-Stokes Equations," AIAA Paper 87-1107, 1987

22. Chen, H. C., and Yu, N. J., "Development of a Highly Efficient and Accurate 3D Euler Flow Solver," Lecture Notes in Physics, D. L. Dwoyer, M. Y. Hussaini and R. G. Voigt (Eds.), 11th International Conference on Numerical Methods in Fluid Dynamics, 1988, pp 187-191.

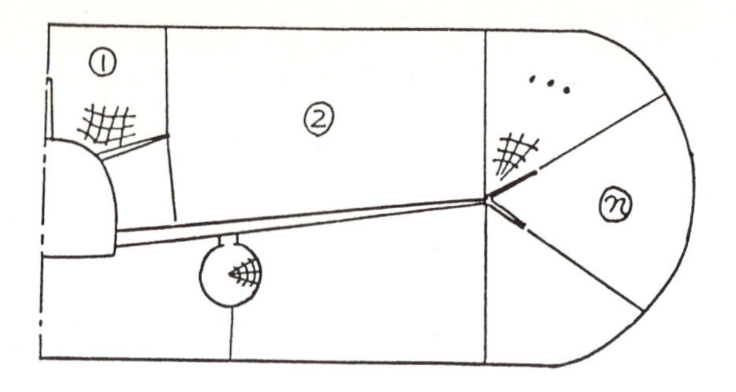

Fig. 1 Illustration of block layout for an airplane configuration

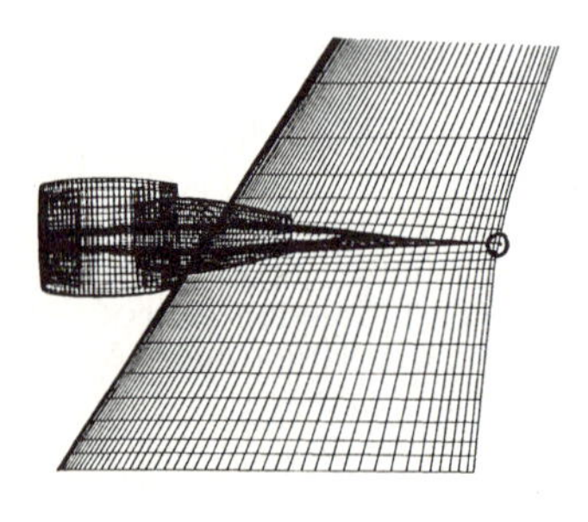

a. Collapsed grid

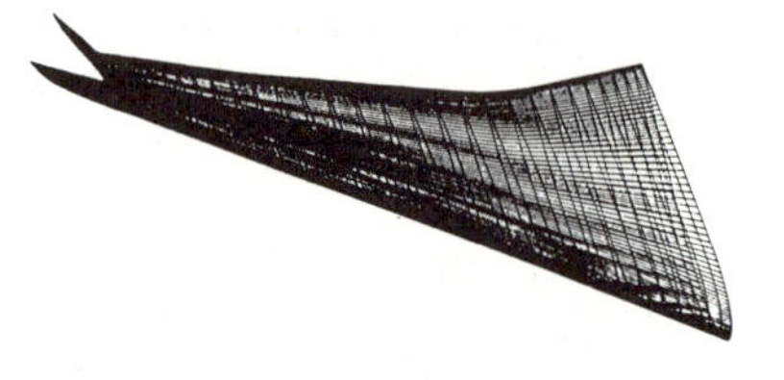

Fig. 3 Surface grids for a wing-wingtip device

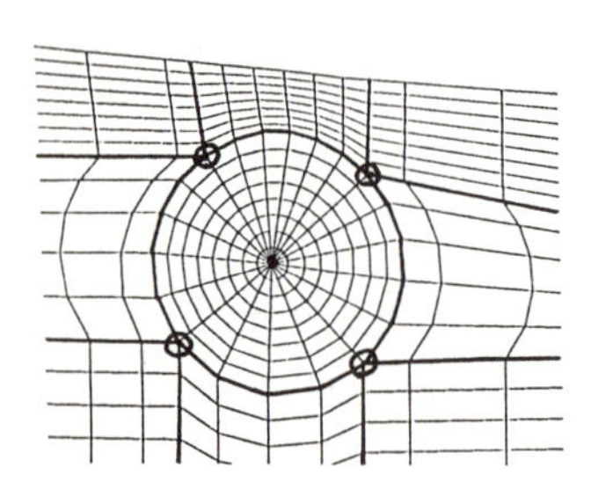

b. Fictitious corners

Fig. 2 Irregular of singular grids for a general multiblock solver

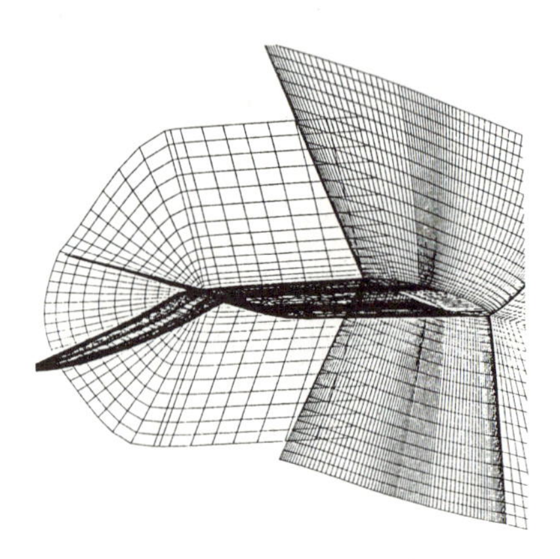

Fig. 4 Field grids for a wing-wingtip device

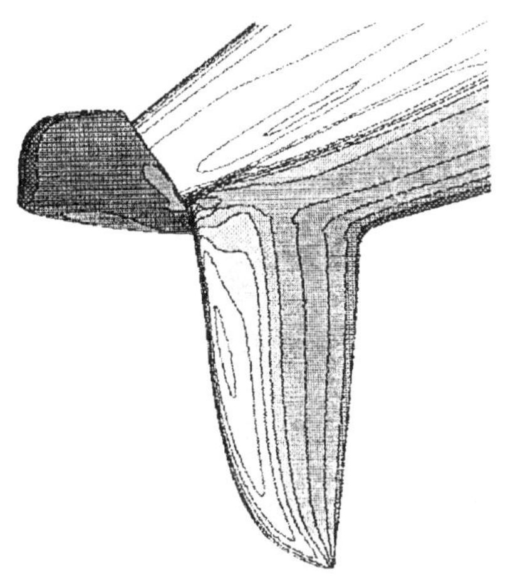

Fig. 5 Isobars at wingtip region, $M_\infty = .8$, $\alpha = 1.89°$

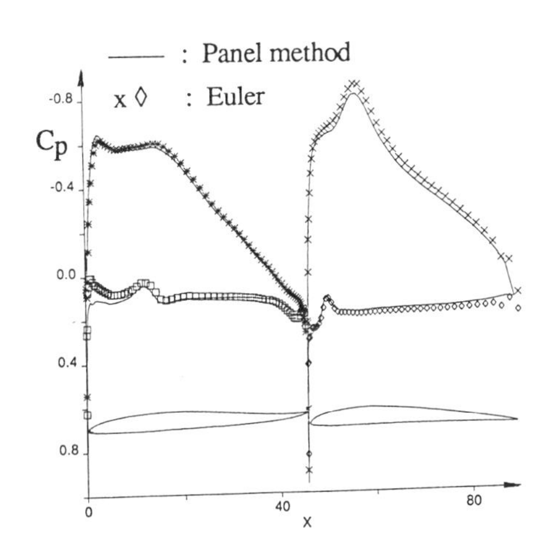

Fig. 6 Comparisons of C_p for a two element airfoil at $M_\infty = .3$, $\alpha = 0°$

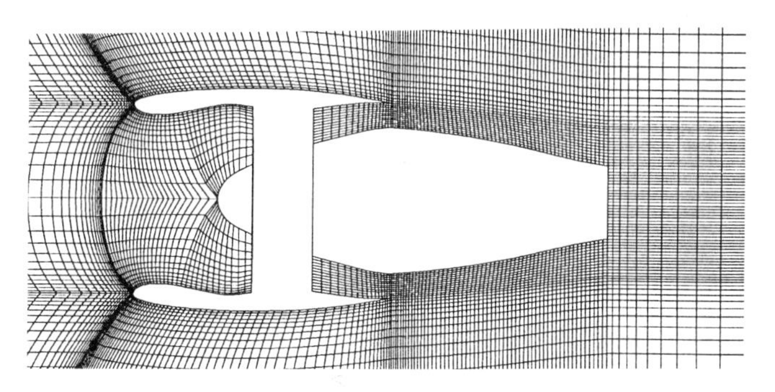

a. Field grids

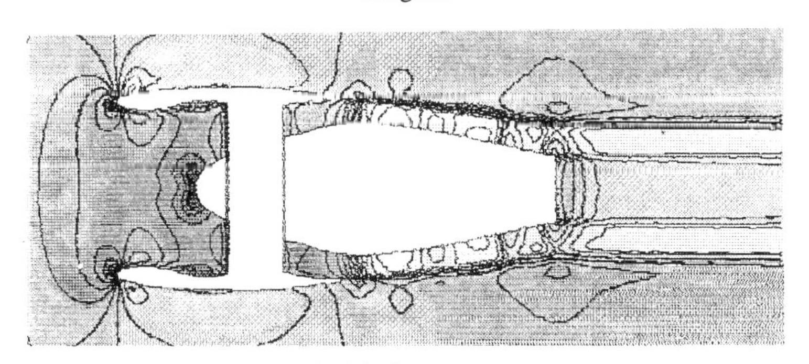

b. Mach contours

Fig. 7 Turbofan nacelle at $M_\infty = .8$, $\alpha = 0°$

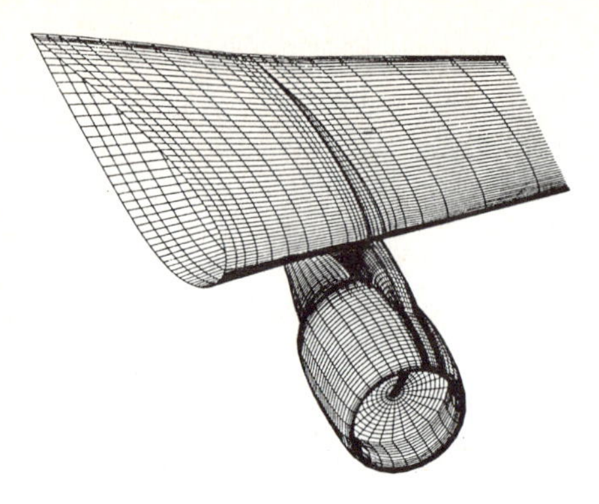

Fig. 8 Surface grids for a wing/nacelle/strut

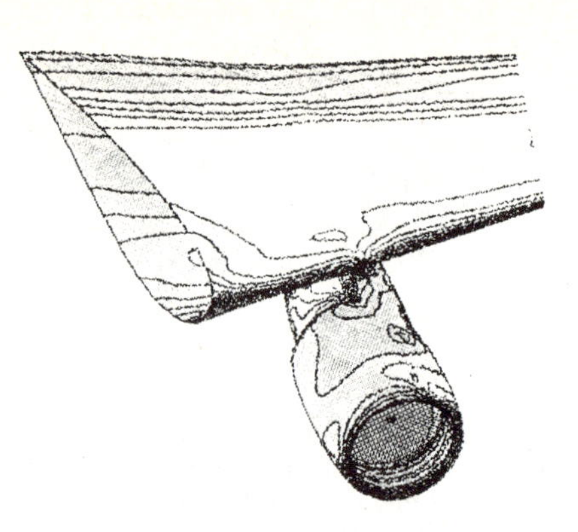

Fig. 9 Isobars for a wing/nacelle/strut at $M_\infty = .8$, $\alpha = 2°$

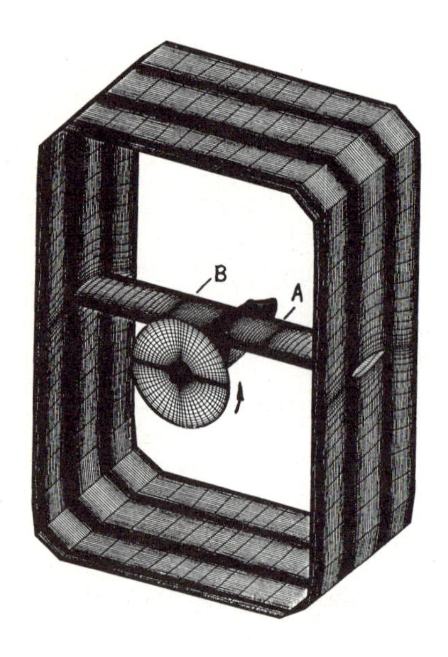

Fig. 10 Wing-mounted propfan configuration in a wind tunnel

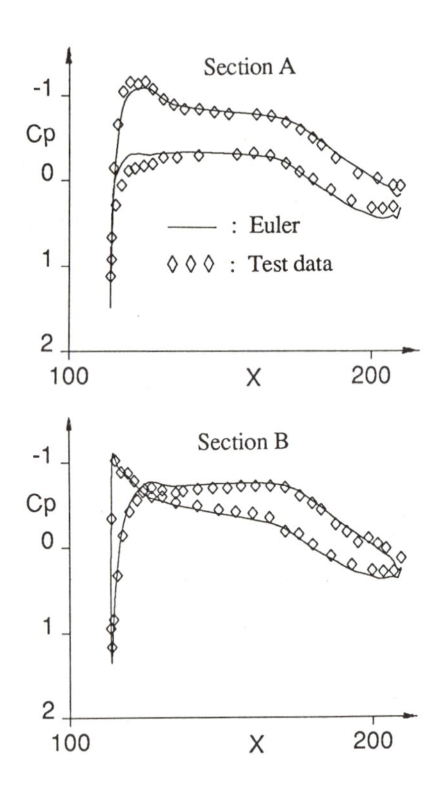

Fig. 11 Test/theory comparisons for a wing-mounted propfan at $M_\infty = .167$, $\alpha = 0°$

REPORT ON WORKSHOP :"NUMERICAL SIMULATION OF OSCILLATORY CONVECTION IN LOW-Pr FLUIDS".

B. Roux
Institut de Mécanique des Fluides, 1 rue Honnorat 13003 Marseille, France

SUMMARY

The workshop concerns 2D time-dependent Navier-Stokes and Boussinesq equations for low-Prandtl-number fluids in rectangular cavities, differentially heated. Various methods belonging in the four main classes (finite differences, finite volumes, finite elements and spectral) were assessed. Very accurate benchmark solutions computed by some of the contributors are now available and can be used for future comparison purposes.

1. INTRODUCTION

The primary goals of the Workshop [1] were to assess numerical methods for solving the time-dependent Navier-Stokes and Boussinesq equations and to provide benchmark solutions. The numerical simulation of such equations being extremely time-consuming, the benchmark problems were limited to 2D flows in a rectangular geometry; whether this corresponds to reality or not was not questioned.

Twenty-eight groups contributed to the benchmark, confirming that the problem of laminar, time-dependent flow is important not only for the problem of crystal growth from melts which motivated this workshop but also for many other applications.The contributions can be divided into four main categories of methods: Finite Differences (12 papers), Finite Volumes (5 papers), Finite Elements (6 papers) and Spectral (5 papers).The performance of these methods are compared for the mandatory cases concerning rectangular cavities (R-R case) or open boats (R-F case), for a few given values of the Grashof number, Gr. In addition to these mandatory cases, several groups considered various related problems, such as the accurate determination of the threshold for the onset of oscillations, $Gr_{c,osc}$, the thermocapillary effect in the R-F case, and 3D simulations. For Gr >> $Gr_{c,osc}$, interesting behaviours have been emphasized in the R-R case, with a succession of various regimes (period doubling, quasi-periodic behaviour, reverse transition) and hysteresis loops.

A detailed comparison of the mandatory results have been made for each category of methods by Behnia [2], Henkes & Segal [3], Sani [4] and Deville [5]. The workshop was also open to complementary contributions from stability theory and experimental observations in order to assess the domain of validity of the 2D (and 3D) numerical simulations (see Chap.6 and 7 of Ref.1). The experiments were performed for a parallelepiped cavity (height:width:length) of dimensions 1:1:4 and 1:2:4. But, only the computational contributions are reported herein.

2. BENCHMARK DEFINITION

The benchmark is devoted to convection of low-Pr fluids subject to buoyancy forces in a long rectangular cavity whose vertical walls are maintained at constant temperatures: T_1 and T_2. The cavity has a rigid bottom wall, and two types of conditions are considered for its upper surface: (i) rigid (denoted R-R) and (ii) shear-stress free (denoted R-F). Experiments [6] and [7] establish that the buoyancy-driven flow which ensues as soon as $T_1 \neq T_2$ becomes oscillatory beyond a certain critical value, Gr_{osc}, of the Grashof number which is defined as $Gr = g\beta\gamma H^4/\nu^2$, where $\gamma = \Delta T/L$ and $\Delta T = T_2 - T_1$. Here H and L denote the height and the length of the cavity, and A (=L/H) denotes the aspect ratio (see Fig. 1).

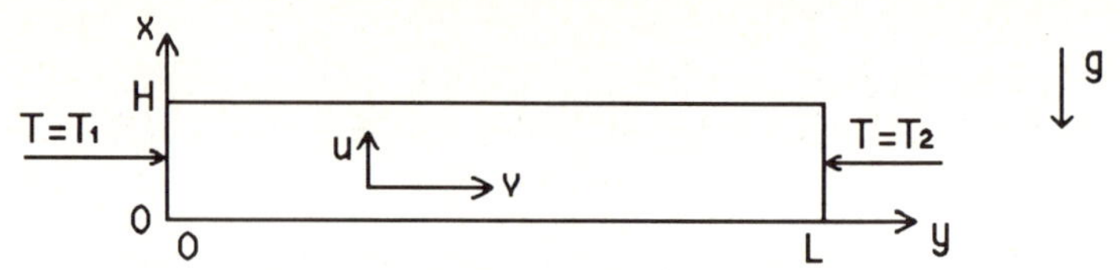

Fig. 1 Geometry of the problem

The aim of the workshop was to compare numerical techniques for situations corresponding to slightly subcritical (steady) state and slightly supercritical (oscillatory) state, according to the bifurcation theory results obtained by Winters [8-9]. The velocity is small enough such that the flow can be considered laminar. In addition, the fluid is assumed to be quasi- incompressible (Boussinesq approximation).The mandatory cases are for Pr=0 where energy equation become uncoupled. Recommanded cases are restricted to Pr=0.015 which corresponds to Hurle et al. experiments [6], and to conducting horizontal walls (temperature varying linearly in y).

2.1. Governing equations

After Ostrach [10] a suitable nondimensionalization for values of Gr where inertia balances buoyancy (typically here for $Gr \geq 10^4$), is obtained by taking H, $v_{ref} = \nu Gr^{0.5}/H$ and H^2/ν for length, velocity and time scales, respectively. So, the Navier-Stokes equations can be written in the streamfunction and vorticity formulation, ψ and Ω, as :

$$\Omega_t + Gr^{0.5}[u\,\Omega_x + v\,\Omega_y] = [\Omega_{xx} + \Omega_{yy}] - Gr^{0.5}\,\theta_y \qquad (1)$$

$$\psi_{xx} + \psi_{yy} + \Omega = 0 \qquad (2)$$

with : $u = \psi_y \quad ; \quad v = -\psi_x \quad \text{and} \quad \Omega = v_x - u_y\,.$ (3)

The transport equation for energy is

$$\theta_t + Gr^{0.5}[u\,\theta_x + v\,\theta_y] = Pr^{-1}[\theta_{xx} + \theta_{yy}] \qquad (4)$$

where : $\theta = (T - T_1)/T_{ref}$, with $T_{ref} = \Delta T / A$.

2.2. Boundary conditions

$u = v = 0$, on the rigid walls (x=0 , y=0 , y=A) (5a)
$u = 0$ and $\partial v/\partial x = 0$, on the upper free boundary (x=1), for the **R-F** case (5b)
$u = 0$ and $v=0$, on the upper rigid boundary (x=1), for the **R-R** case (5c)
$\theta(x,0) = \theta_1 = 0$; $\theta(x, A) = \theta_2 = A$, for the two isothermal vertical walls (6)
$\theta(y) = y$, for conducting horizontal walls. (7)

2.3. Requested test cases (A=4)

All the requested cases are for one given aspect ratio, A = 4 ; they defined as follows:

- as a mandatory case , at least one of the two following cases , at Pr = 0 ,

Case A : **R-R** , for $Gr = 2\times10^4$; 2.5×10^4 ; 3×10^4 and 4×10^4
Case B : **R-F** , for $Gr = 10^4$; 1.5×10^4 and 2×10^4

- 2 "recommended" cases , corresponding to the condition (7), at Pr = 0.015

Case C : **R-R** , for $Gr = 2\times10^4$; 2.5×10^4 ; 3×10^4 and 4×10^4
Case D : **R-F** , for $Gr = 10^4$; 1.5×10^4 and 2×10^4.

3. MANDATORY CASE A (R-R ; Pr=0)

For $Gr=2x10^4$ and $2.5\ x10^4$, the flow is found to be steady by almost all the contributors. The solutions which are centro-symmetric correspond to a main central cell and two adjacent small cells, this structure is denoted S12. The best results were obtained by Le Quéré [11] with two kinds of spectral methods using Chebyshev polynomials (pseudo-spectral method on staggered grid, and tau method). Also the "zero-grid" solution of Behnia & de Vahl Davis [12], extrapolated from 21x81, 41x161 and 81x321 grids, can be proposed as benchmark solution (see Table 1). Note that three contributors using F.E.M. with a too coarse resolution found oscillatory (P1) solutions.

For $Gr=3x10^4$, most of the results exhibited a time dependent character, but with varying signatures. Generally the solution was found to be monoperiodic (denoted P1) for several tens of periods, with a frequency ranging from $17.4<f<17.9$. A time history of this P1-solution is given at four instants over one period in Fig.2., where the beginning of the period (denoted t=0) corresponds to the maximum of v_{max}. This P1-solution which is still centro-symmetric suddenly changes to a quasi-periodic one (noted QP) regime, after a long enough integration time and sufficiently accurate scheme, as shown in Fig.3 (after Pulicani & Peyret [13]). Such a behaviour is also found by Le Quéré [11], Randriamampianina et al. [14], Haldenwang et al.[15], Ben Hadid & Roux [16] and Henry & Buffat [17]; it implies that the P1-solution is not stable.The QP solution is shown to be no longer centro-symmetric. However it is also not proved to be definitely stable!

A steady-state two-cell solution (denoted S2), also exists. It was found by several authors utilizing different strategies. One strategy, used by Le Quéré [11], Pulicani & Peyret [13], Ben Hadid & Roux [16], Le Garrec & Magnaud [18] and Winters [9] is to start from the S2-solution obtained at higher Gr (e.g. $4x10^4$). Another way adopted by Henry & Buffat [17] is to give a strong perturbation to the time-dependent solution. A third strategy is to use a steady algorithm (see Schneidesch et al. [19]). This S2-solution is shown to be stable by Winters [9]. Characteristics of all these solutions are given in Table 1.The mesh resolution influence has been discussed by most of the authors (see for example Segal et al. [20]).

For $Gr= 4x10^4$, the solution also exhibits very stable oscillations which persist for several tens of periods, followed by a rapid transient to a steady-state solution; this is observed by most of the contributors, during or after the workshop (e.g. Grötzbach [21]). This evolution of P1- to steady S2-solution is shown in Fig.4, after Henkes & Hoogendoorn [22]. Again, the interpretation is that the initial monoperiodic solution is not stable. In fact such behaviour was originally mentioned by Gresho & Sani [23], and observed by Roux et al. [24], for $Gr= 5x10^4$. Clearly this transient state between the P1- and the steady-state solution is associated with the effect of an asymmetric perturbation to which the P1-solution is unstable.

Thus imposing an asymmetry enhances the transition of the P1-solution to the transient regime and finally to the steady-state solution. If no asymmetry is imposed, not even through the initial conditions, only the accumulation of round-off errors can lead to a strong enough asymmetric perturbation, leading ultimately to the steady-state solution after a transient regime. In that way the higher is the grid resolution, the longer is the duration of the (unstable) P1-regime! The final (stable) solution is again centro-symmetric and is a two-cell structure (S2).

4. MANDATORY CASE B (R-F ; Pr=0)

For $Gr=10^4$, all the contributors found a steady-state solution (which is no longer centro-symmetric). The solution for $Gr=1.5x10^4$ is oscillatory (P1) with a frequency close to 13.2. Very accurate computations have been also performed by Le Quéré [11] using two different methods. For $Gr=2x10^4$, most of contributors also found a monoperiodic (P1) solution. The most accurate results by Behnia & de Vahl Davis [12] and Pulicani & Peyret [13] give a frequency of approximately 16.2. The main characteristics of all the mandatory R-F solutions are presented in Table 2.

5. RECOMMENDED CASE C (R_c-R_c; Pr=0.015)

The flow pattern is steady and centro-symmetric (S12-solution) for Gr=$2x10^4$ and $2.5x10^4$. A good agreement is observed between the results of various authors using an appropriate resolution. As in Pr=0 case, different solution forms exist at Gr=$3x10^4$ and Gr=$4x10^4$, including time-dependent and steady-state solutions. A P1-solution is found at Gr=$3.x10^4$ using the initial condition from a smaller Gr; the frequency is quite similar to that of Case A (18.05 instead of f=17.87, after Behnia & de Vahl Davis [12]), but here there is no evidence of destabilization of the P1-solution (e.g. to reach a QP solution as for (Pr=0). At Gr=$4x10^4$, unlike the Pr=0 case, a P1-solution is observed. However, after a very long integration time a period doubling (P2) can be reached, where the frequency f is dominant with respect to f/2. This P2-solution has been carefully studied by Pulicani & Peyret [13], with various initial conditions and grid resolutions; it is also found by Haldenwang & Elkeslassy [15], Ben Hadid & Roux [16], Daube & Rida [25] and Shimizu [26]. Like case A, a steady-state S2-solution is possible for Gr=$3x10^4$ and Gr=$4x10^4$, as shown by Ben Hadid & Roux [16].

6. RECOMMENDED CASE D (R_c-F_c ; Pr=0.015)

For Gr=10^4 the flow is steady (S11-solution). While for Gr=$1.5x10^4$ and Gr=$2x10^4$ a P1-solution is found (like the Pr=0 case) with frequency increasing with Gr from nearly 13 to 16. Again, a good agreement is obtained between results of authors using appropriate resolution.

CONCLUSIONS

The workshop did attain its goals and not only were various solution methods assessed but also very accurate benchmark solutions computed by some of the contributors are now available. Various flow structures and regimes have been exhibited which can be drastically influenced by the grid resolution. This grid effect which can strongly affect the frequency (or even suppress oscillations), should be useful information for those interested in the accurate solution of either 2D or 3D time-dependent problems.

REFERENCES

[1] Numerical Simulation of oscillatory convection in Low-Pr Fluids. A GAMM-Workshop, B. Roux ed., Notes on Numerical Fluid Mechanics, Vieweg, Vol. 27.

[2] BEHNIA M. (1989), Notes Num. Fluid Mech.,Vieweg, 27, 265-272.

[3] HENKES R.A.W. & SEGAL A. (1989), Notes Num. Fluid Mech.,Vieweg, 27, 273-276.

[2] SANI R.L. (1989), Notes Num. Fluid Mech.,Vieweg, 27, 277-280.

[2] DEVILLE M. (1989), Notes Num. Fluid Mech.,Vieweg, 27, 280-284.

[6] HURLE D.T.J., JAKEMAN E. & JOHNSON C.P. (1974), J. Fluid Mech. 64 , 565-576.

[7] HART J.E. (1983), *Int. J. Heat Mass Transfer*, 26 , 1069-1076.

[8] WINTERS K.H. (1988), Int. J. Num. Methods Engng., 25, 401-414.

[9] WINTERS K.H. (1989), Notes Num. Fluid Mech.,Vieweg, 27, 319-326.

[10] OSTRACH S. (1976), *Proc. Secd. European Symp. on Material Sciences in Space* , ESA-SP 114.

[11] LE QUERE P. (1989), Notes Num. Fluid Mech.,Vieweg, 27, 227-236.

[12] BEHNIA M. & de VAHL DAVIS G. (1989), Notes Num. Fluid Mech.,Vieweg, 27, 11-18.

[13] PULICANI J.P. & PEYRET R. (1989), Notes Num. Fluid Mech., Vieweg, 27, 237-244.

[14] RANDRIAMAMPIANINA A., CRESPO E., FONTAINE J.P. & BONTOUX P. (1989), Notes Num. Fluid Mech., Vieweg, 27, 245-255.

[15] HALDENWANG P. & ELKESLASSY S. (1989), Notes Num. Fluid Mech.,Vieweg, 27, 217-226.

[16] BEN HADID H. & ROUX B. (1989), Notes Num. Fluid Mech., Vieweg, 27, 25-34 .

[17] HENRY D. & BUFFAT M. (1989), Notes Num. Fluid Mech., Vieweg, 27, 182-187.

[18] LE GARREC S. & MAGNAUD J.P. (1989), Notes Num. Fluid Mech., (1989), Notes Num. Fluid Mech., Vieweg, 27, 189-198

[19] SCHNEIDESCH C., DEVILLE M. & DEMARET P. (1989), Notes Num. Fluid Mech.,Vieweg, 27, 256-262.

[20] SEGAL A., CUVELIER C. & KASSELS C. (1989), Notes Num. Fluid Mech.,Vieweg, 27,199-206.
[21] GRÖTZBACH G. (1989), Notes Num. Fluid Mech.,Vieweg, 27,157-164.
[22] HENKES R.A.W & HOOGENDOORN C.J. (1989), Notes Num. Fluid Mech.,Vieweg, 27,144-152.
[23] GRESHO Ph. & SANI R.L. (1984), private communication
[24] ROUX B., BONTOUX P. & HENRY D. (1985), Lect. Notes Physics, Springer, 230, 202-213.
[25] DAUBE O. & RIDA S. (1989), Notes Num. Fluid Mech., Vieweg, 27, 43-48.
[26] SHIMIZU T. (1989), Notes Num. Fluid Mech., Vieweg, 27, 207-214.

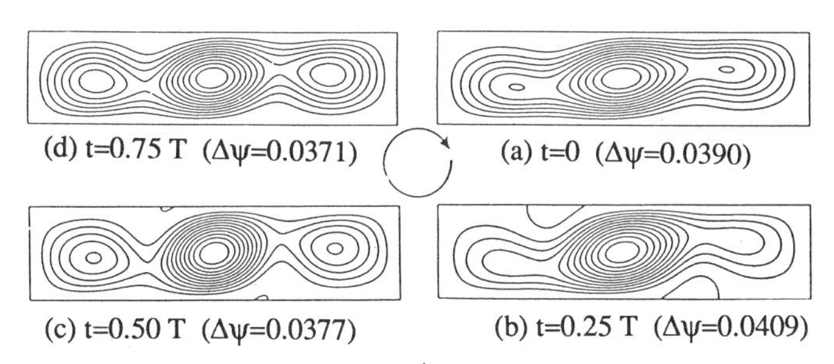

Fig.2 Oscillatory (P1) solution at Gr= $3x10^4$, after Behnia & de Vahl Davis [12], in R-R case at Pr=0; time evolution of the iso-streamfunction patterns.

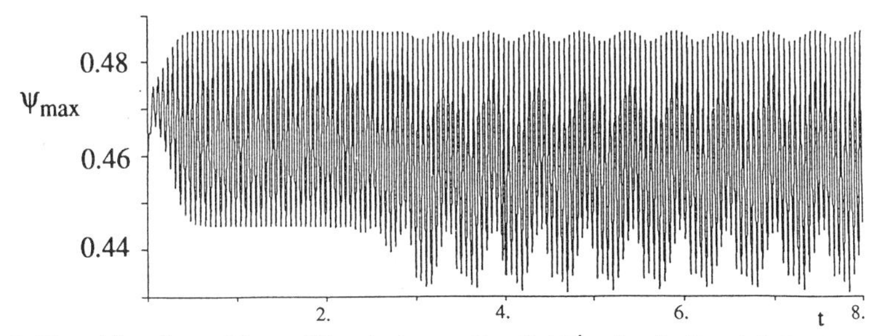

Fig.3 Transition from P1- to QP-solution at Gr= $3x10^4$, after Pulicani & Peyret[13], in R-R case at Pr=0; time evolution of the streamfunction maximum.

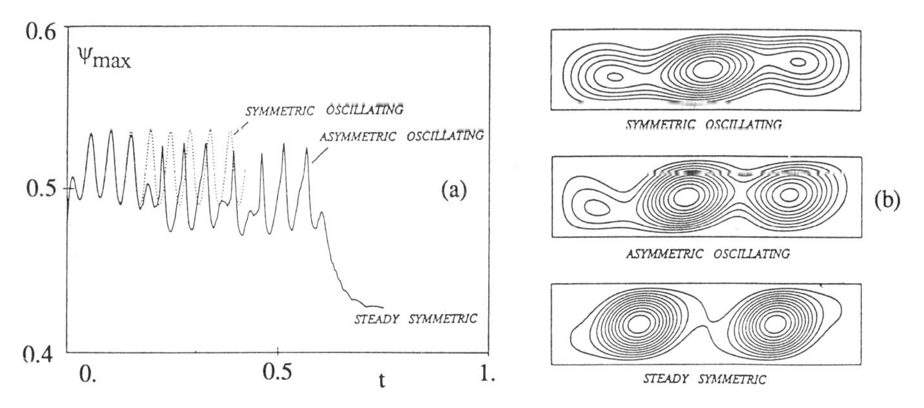

Fig.4 Transition from P1- to S2-solution at Gr= $4x10^4$, after Henkes & Hoogendoorn [22], in R-R case at Pr=0; (a) time evolution of the streamfunction maximum ; (b) iso-streamfunction patterns.

Table 1. Results obtained for the R-R case at Pr=0.

Gr	Authors	$U^*_{max} : U^*_{min}$ at x=1/2	$V^*_{max} : V^*_{min}$ at y=1	$\Psi^*_{max} : \Psi^*_{min}$	f
2.10^4	[1]	0.5226	0.6750	0.4155	—
(S12)	[2]	0.5225	0.6753	0.4156	—
$2.5\ 10^4$	[1]	0.6286	0.7022	0.4457	—
(S12)	[2]	0.6284	0.7035	0.4456	—
3.10^4	[1]	.9230 : .5079	.9436 : .4877	.4877 : .4429	17.87
(P1)	[2]	.9197 : .5106	.9408 : .4967	.4873 : .4434	17.88
	[2]	.935 : .509	1.051 : .450	.4874 : .4406	17.6
(QP)	[3]	—	—	.4885 : .4415	17.55
	[4]	.9388 : 5126	1.0584 : 0.4356	—	17.63
	[5]	—	—	.4847 : .4732	17.46
	[3]	—	—	0.3883	—
(S2)	[4]	0.7217	1.1874		—
	[5]	—	—	0.3841	—
4.10^4	[1]	0.8500	1.203	0.4167	—
(S2)	[2]	0.8499	1.2036	0.4168	—

[1] Behnia & de Vahl Davis (F.D.M); extrapolated solution from uniform 21x81,41x161 and 81x321meshes
[2] Le Quéré (S.M.); pseudo-spectral Chebyshev (staggered grid); 20x50 to 32x64 polynomials
[3] Ben Hadid & Roux (F.D.M.); 35x101 non-uniform grid
[4] Pulicani & Peyret (S.M.); Tau-Chebyshev ; 80x100 polynomials
[5] Randriamampianina et al. (S.M.); Tau-Chebyshev+LSODA; 16x30 polynomials

Table 2. Results obtained for the R-F case at Pr=0.

Gr	Authors	$U^*_{max} : U^*_{min}$ at x=1/2	$V^*_{max} : V^*_{min}$ at y=1	$\Psi^*_{max} : \Psi^*_{min}$	f
1.10^4	[1]	1.058	1.945	0.5573	
	[2]	1.0581	1.9465	0.55736	—
(S11)	[3]	1.0577	1.9447	—	—
$1.5\ 10^4$	[1]	1.3329 : 1.1149	2.3069 : 1.9616	.6440 : .5953	13.20
	[2]	1.3528 : 1.0910	2.3241 : 1.9162	.6540 : .5877	13.24
(P1)	[3]	1.3583 : 1.0946	2.3340 : 1.9106	—	13.04
2.10^4	[1]	1.5833 : 1.0138	2.5587 : 1.7311	.6926 : .5781	16.18
(P1)	[3]	1.5872 : 1.0280	2.5576 : 1.7313	—	16.20

[1] Behnia-de Vahl Davis (F.D.M); uniform 81x321 mesh solution
[2] Le Quéré (S.M.); pseudo-spectral Chebyshev (staggered grid), with 24x64 polynomials, and Tau-Chebyshev, with 24x50 polynomials.
[3] Pulicani & Peyret (S.M.); spectral Chebyshev method, with 24x30 polynomials

Addresses of the editors of the series "Notes on Numerical Fluid Mechanics":

Prof. Dr. Ernst Heinrich Hirschel (General Editor)
Herzog-Heinrich-Weg 6
D-8011 Zorneding
Federal Republic of Germany

Prof. Dr. Kozo Fujii
High-Speed Aerodynamics Div.
The ISAS
Yoshinodai 3-1-1, Sagamihara
Kanagawa 229
Japan

Prof. Dr. Keith William Morton
Oxford University Computing Laboratory
Numerical Analysis Group
8-11 Keble Road
Oxford OX1 3QD
Great Britain

Prof. Dr. Earll M. Murman
Department of Aeronautics and Astronautics
Massachusetts Institute of Technology (MIT)
Cambridge, MA 02139
USA

Prof. Dr. Maurizio Pandolfi
Dipartimento di Ingegneria Aeronautica e Spaziale
Politecnico di Torino
Corso Duca Degli Abruzzi, 24
I-10129 Torino
Italy

Prof. Dr. Arthur Rizzi
FFA Stockholm
Box 11021
S-16111 Bromma 11
Sweden

Dr. Bernard Roux
Institut de Mécanique des Fluides
Laboratoire Associè au C.R.N.S. LA 03
1, Rue Honnorat
F-13003 Marseille
France